INTRODUCTION TO DIGITAL CIRCUITS

THEODORE F. BOGART, JR.

University of Southern Mississippi

GLENCOE

Macmillan/McGraw-Hill

Lake Forest, Illinois
Columbus, Ohio
Mission Hills, California
Peoria, Illinois

Introduction to Digital Circuits
International Edition 1992

Exclusive rights by McGraw-Hill Book Co-Singapore. This book cannot be re-exported from the country to which it is consigned by McGraw-Hill.

2 3 4 5 6 7 8 9 KKP UPE 9 6 5 4 3 2

ISBN 0-02-819941-3

Library of Congress Cataloging-in-Publication Data

Bogart, Theodore F., Jr.
     Introduction to digital circuits /Theodore F. Bogart, Jr.
       p.   cm.
     Includes index.
     ISBN 0-02-819941-3
     1.Digital electronics. 2. Logic circuits. I. Title.
TK7868.D5B64 1991
621.381 '5—dc20   90-26441
                                                CIP

When ordering this title use ISBN No:0-07-112555-8.

Cover photograph: Computer chip, C. O'Rear/Westlight.

Printed in Singapore

PREFACE

ORGANIZATION AND SCOPE

Introduction to Digital Circuits contains sufficient material for a comprehensive two-semester course in digital logic and circuits but is readily adaptable to a one-semester course in either of those areas. The first nine chapters are used at the author's institution in a TAC/ABET-accredited program in electronics engineering technology to support a sophomore-level course in digital fundamentals. Chapters 10 through 17 are used in the succeeding junior-level course (the second of a three-semester sequence culminating in a course on microprocessors).

Part I, *Number Systems and Logical Analysis*, requires no background in circuit theory or electronics and could be used for an introductory course. Part II, *Pulse and Digital Circuits*, requires a knowledge of circuit theory and should probably be reinforced by a prerequisite or concurrent course in semiconductor devices. However, the initial thrust of this material is to encourage students to view diodes and transistors as *voltage-controlled switches*, so a rigorous understanding of the subtleties of semiconductor behavior is not essential. The text contains notes designed to help instructors select topics that are appropriate to students having different levels of preparation. In most cases, coverage of more advanced topics requiring a greater theoretical background can be omitted without loss of continuity. For example, Chapter 7, "Waveshaping and Waveform Analysis," may be beyond the scope of some digital courses but can be omitted completely, since it contains no topics that are prerequisite to any subsequent material. The material in Chapter 7 can be considered optional for a course whose principal focus is digital devices. Waveshaping and waveform analysis are topics traditionally treated in pulse and digital circuits courses, but they are not prerequisites for any subsequent material in this book. Possible exceptions are Schmitt triggers and analog voltage comparators, which are mentioned peripherally in some later discussions. (Also, operational amplifiers are used in the discussion of some A/D and D/A converters in Chapter 17.) Part III, *Storage Devices and Sequential Logic*, covers traditional topics in flip-flops, shift registers, coding circuits, and counters. Little circuit theory is required, so most of this section could be covered immediately after Part I to complete a one-semester course. Part IV, *Data Processing, Manipulation, and Storage* is word-oriented, in that it concerns digital devices and systems used for processing and storing data words.

The selection of topics for this book was guided by two principal concerns: that it reflect the most modern industrial practice and that it be comprehensive in scope. Toward that end, the latest developments in current and emerging technologies are included, as are traditional topics in logic design and switching circuits. The scope and depth of the book are intended to make it suitable for courses that emphasize either digital logic or digital circuitry, or both.

ANSI/IEEE STANDARD SYMBOLS

ANSI/IEEE notation is appearing in much of the new product literature published by manufacturers, and documentation prepared under government contract is now required to employ that notation. Despite a historical reluctance on the part of many educators and practitioners to accept the standards, it seems likely that universal adoption will eventually prevail. For these compelling reasons, the majority of the chapters in *Digital Electronics* contain sections that explain how dependency notation is used in the new logic symbols. Many also contain end-of-chapter exercises that require students to interpret or construct logic diagrams using the notation. However, most logic diagrams in the book are drawn using traditional symbols, since at this writing a universal changeover appears to be still several years distant.

SPICE

The powerful circuit-modeling program SPICE—Simulation Program with Integrated Circuit Emphasis—is the subject of Appendix A to *Introduction to Digital Circuits*. Emphasizing the application of SPICE to pulse and digital circuitry, this material can be used to instruct students having no previous background in computer modeling or for those who have already used SPICE in linear circuit simulation. Many chapters contain examples of logic circuit simulation using SPICE. Each such chapter also has end-of-chapter SPICE exercises.

EXAMPLES AND EXERCISES

The book contains a very large number of end-of-chapter exercises, with answers provided for odd-numbered exercises in Appendix B. One reason for the large quantity of exercises is that every effort was made to include at least one odd-numbered exercise and one even-numbered exercise for every important topic or concept discussed in the book. Thus, instructors who prefer to assign exercises with answers and those who prefer exercises without answers will both find an ample selection.

To promote immediate reinforcement of problem-solving skills, each worked-out example is followed by a drill exercise. The drill exercises are typically slight variations of the examples that precede them, so students are forced to work through a new set of computations. This kind of activity is a valuable way for students to learn of any gaps that may exist in their understanding or in their computational skills—learning that does not usually follow from a cursory reading of an example.

ACKNOWLEDGMENTS

I am particularly indebted to Pam Miller and Patricia Webb for their extensive typing services during the preparation of the manuscript. Their skills as well as their cooperative attitudes, often under pressure imposed by deadlines, contributed greatly to the timely release of the book.

I also wish to thank my students and colleagues for much constructive feedback based on classroom experience with the manuscript.

CONTENTS

Part I

NUMBER SYSTEMS AND LOGICAL ANALYSIS

Chapter 1

INTRODUCTION

1-1   TERMINOLOGY AND DEFINITIONS

Electronic circuits and systems are classified as being either *analog* or *digital*. The distinction is not so much in the kinds of semiconductor devices used to construct these circuits as it is in the way they are *operated*—i.e., in the voltage and current variations that occur when each type of circuit is performing the useful function for which it was designed. Analog circuits are those in which voltages and currents vary *continuously* throughout some range. Thus, for example, the output voltage from an audio amplifier might be any one of (an infinite number of) values between -10 V and $+10$ V at any particular instant of time. Other examples of analog devices include sinewave signal generators, radio-frequency transmitters and receivers, power supplies, electric motors and speed controllers, and many analog-type instruments—those having "pointers" that move in a continuous arc across a calibrated scale. By contrast, a digital circuit is one in which voltage levels alternate among a finite number of distinct values. In virtually all modern digital circuits, there are just *two* such voltage levels.*

Of course, the voltages in a digital circuit repeatedly *change* from one level to another, and during the time such a change occurs, every voltage between the two levels is produced by the circuit. However, these transition times are extremely brief and we do not regard the in-between voltage levels as being useful output. We say that the voltage levels in a digital circuit are *switched* from one value to another, and digital circuits are often called *switching circuits*.

Because digital computers are now so widely available and so important to modern industry, there is a tendency to associate digital technology exclusively with applications involving numerical computing ("number crunching"). Indeed, as we shall presently learn, *number systems* are important tools used to study, understand, and design digital circuitry. However, digital circuits are also used in many non-computational applications because of certain advantages they enjoy. In fact, many traditionally analog-type products

---

*As we discover in a later discussion, each of the two voltage levels in a practical digital system can actually be a narrow *band,* or range, of voltages.

are now being produced in digital form, including the analog circuit examples cited previously.

The word *linear* is often used loosely to mean analog, or non-digital. We hear, for example, of linear integrated circuits, meaning non-digital integrated circuits. The terminology can be confusing because the word *non-linear* is frequently used to mean analog circuits having certain properties, rather than "non-analog," or digital.

Discrete is another term that is widely used in reference to electronic devices and circuits. Discrete means *distinct*—i.e., having individually identifiable values or elements. As we have already indicated, the useful voltages in digital circuitry usually have just two values, so we say that digital circuitry produces discrete outputs. In this context, the opposite of discrete is *continuous*. In another context, *discrete circuits* are those whose components are individual transistors, diodes, resistors, etc., as opposed to integrated circuits, all of whose components are fabricated on a single crystalline chip. Most modern digital and analog circuitry is constructed in integrated circuits rather than in discrete form. However, discrete components are still used in applications where heavy power dissipations occur.

Hybrid circuits contain both integrated and discrete components. For example, the output of a digital integrated circuit is often connected to a discrete transistor that drives a long cable or a set of indicator lamps. In another context, an integrated circuit is itself hybrid if it contains both analog and digital circuitry. Examples include analog-to-digital converters and digitally controlled analog switches.

Digital circuitry is often called *logic* circuitry because the level of each output voltage depends on the levels of several input voltages, and the inputs may appear in many different combinations. The circuitry effectively makes logical decisions, producing one output level if certain input combinations are present and a different output level if other input combinations are present. *Logic gates*, which we study in detail, are the building blocks of digital systems, and the art of interconnecting such gates to achieve prescribed outcomes is called *logic design*.

1–2   THE DIGITAL REVOLUTION—A HISTORICAL PERSPECTIVE

Despite the fact that digital technology is now widely used in applications unrelated to computing, the technology evolved from a desire to perform numerical computations quickly and accurately. Thus, up until recent years, the history of digital technology has been the history of digital computation. The Latin word *digitus*, meaning finger or toe, is the origin of the words digital and digit, and the association of these words with computational activity is linked to the way early people relied on their fingers for counting. The decimal number system, containing 10 distinct symbols, is similarly linked to early finger counting as a computational method. Unlike modern computers, the design of early computing machines was based on the decimal number system.

One of the earliest calculating machines based on the decimal number system was invented by French scientist Blaise Pascal in 1642. It contained geared wheels and could be used to add and subtract numbers directly but not to multiply or divide. In 1670, Baron von Liebnitz, a German mathematician, improved on Pascal's design by adding wheels that made the machine capable of performing multiplication and division as well as extraction of roots. His design was the forerunner of the mechanical calculators widely used as business machines before the advent of electronic calculators.

The first calculating machine to employ rudimentary principles found in modern computers was conceived by the British mathematician Charles Babbage in the early 1800s. Although Babbage's "analytical engine" was never completely operational, it represented the first attempt to incorporate memory into a computing machine and had other components whose functional counterparts are found in modern computers.

In 1854, British mathematician George Boole published a now-classical treatise that laid the foundation for computations based on two-valued (binary) numbers instead of decimal numbers. Entitled *An Investigation of the Laws of Thought,* Boole's work showed how logical propositions can be expressed mathematically and analyzed to determine whether they are true or false. Many years later, his system of mathematics became the *Boolean algebra* that is used today to analyze and design digital logic circuits. C. E. Shannon is credited with developing *switching theory* based on Boolean principles ("A Symbolic Analysis of Relay and Switching Circuits," 1938).

In the late nineteenth century, an American statistician employed by the United States Census Bureau developed a punched-card system for coding and sorting information. As part of that work, Herman Hollerith devised the code that bears his name and is still in use today. The Hollerith code was used to punch data into stiff cards, later to be known as IBM cards, which became the most popular means for storing and processing data in the 1960s and early 1970s. Hollerith founded the Computing-Tabulating-Recording Company, which later became International Business Machines.

The first large-scale electronic computer was the Electronic Numerical Integrator and Computer (ENIAC), built by Eckert and Mauchly in 1946 at the University of Pennsylvania. The ENIAC was the first computer to use active electronic devices (vacuum tubes) as data-storage elements. Although it laid the foundation for electronic computation, the ENIAC was a monster by today's standards: It contained 18,000 vacuum tubes and weighed nearly thirty tons! Greater computing power is now available in hand-held calculators. In addition to frequent breakdowns caused by tube failures, a major shortcoming of the ENIAC was that it had to be programmed using "hard-wired" plug-in boards. The concept of storing a program in computer memory was a major breakthrough, pioneered by IBM in its Selective Sequence Electronic Calculator (1948) and by Remington Rand in its UNIVAC I (1951), the first electronic computer produced commercially.

Advances in electronic computer technology since the 1950s have closely paralleled—and in many cases have been the direct result of—advances in the development of electronic devices. Among the most important of these advances was the invention of the transistor in 1947 by Shockley, Brattain, and Bardeen of Bell Telephone Laboratories. Logic circuits constructed from discrete transistors soon replaced the early vacuum-tube machines, which are now known as first-generation computers. The transistorized second-generation computers were far more reliable, compact, and efficient. Third-generation computers developed in the 1960s were characterized by the development of sophisticated programming languages and operating systems (internal programs designed to control the operation of various components within and external to the computer).

The invention of the integrated circuit (at Texas Instruments, Inc.) eventually led to the fourth-generation computers of today. Early integrated circuits contained only a few logic gates per chip and are now known as examples of small-scale integration (SSI). The continuing quest for packing ever more circuitry into ever smaller space has led to medium, large, and very large scale integration (MSI, LSI, VLSI), with concurrent improvements in reliability, speed, and compactness. RCA pioneered the development of metal-oxide-semiconductor (MOS) devices, which enabled greater packing densities and reduced power consumption dramatically. The faster CMOS (complementary MOS) design followed, as did high-speed CMOS, developed by National Semiconductor and Motorola in 1981. Most fourth-generation computers have solid-state memories instead of the magnetic-core memories that were popular in the 1960s and early 1970s.

The invention of the *microprocessor* by the Intel Corporation in 1971 launched another computer revolution, albeit of a different kind. A microprocessor is essentially a computer on a chip, although a relatively slow one in comparison to mainframe fourth-generation computers. It quickly became a popular ingredient of a wide variety of consumer products where compactness was more important than computing speed, including personal computers, video games, and appliance controllers. In many such applications, the device is used principally as a controller rather than for numerical computation. Many large computer systems use microprocessors simply to control internal operations and peripheral devices. The microprocessor is responsible in large part for the increasingly dominant role that digital technology has assumed in circuit applications that have been traditionally analog in nature. Notable examples include laboratory instruments, motor controllers, sound recording and playback systems, and radio-frequency and data communications equipment.

Chapter 2

NUMBER SYSTEMS

2-1   NUMBERS

Although we use numbers on a daily basis to keep track of all sorts of quantities as well as to perform many useful computations, it is an interesting fact that numbers themselves cannot be *defined* in a strictly rigorous manner. Numbers are simply a symbolic representation of certain intuitive ideas that we are so accustomed to using that we accept their validity without question. Numbers defy definition because their most fundamental properties can be described only in terms of other numbers. Consider, for example, our *decimal number system,* which is based on the symbols 0, 1, 2, . . . , 9. If we are willing to accept, without formal definition, our intuitive notion of the meaning of the number 10, then we can say that the decimal number system has 10 distinct symbols. As such, we say that the decimal number system has *base* 10, or *radix* 10, and we can describe its properties in terms of that base. The reason we want to examine such fundamental concepts in detail is so that we can apply the same concepts to describe and understand number systems having other bases. In particular, we will see that the binary (base 2), octal (base 8), and hexadecimal (base 16) number systems are extremely useful in digital technology.

2-2   THE DECIMAL NUMBER SYSTEM

As we know, decimal numbers are represented by arranging the 10 symbols 0 through 9, called the decimal *digits,* in various sequences. The position of each digit in a sequence has a certain numerical *weight,* and each digit is a multiplier of the weight of its position. The decimal number system is therefore an example of a *weighted, positional* number system. The weight of each position is a power of the base number 10. The *value* of a number is the *sum* of the products obtained by multiplying each digit by the weight of its respective position. Consider, for example, the number 825. The following diagram shows that 8 is in the hundreds position (10^2), 2 is in the tens position (10^1), and 5 is in the units position (10^0).

$$8 \times 10^2 \quad 2 \times 10^1 \quad 5 \times 10^0$$

$$8 \; 2 \; 5$$

The value of the number is, therefore, $(8 \times 10^2) + (2 \times 10^1) + (5 \times 10^0) = 800 + 20 + 5 = 825$. Notice that the weight of any position is one power of 10 greater than the weight of the position to its right. Thus, the weights are decreasing powers of 10 as we proceed from left to right. When we reach the units position (10^0), we insert a decimal point, and, continuing to the right, we assign *negative* powers of 10 to succeeding positions. For example, the value of the number 368.49 is computed as follows:

$$3 \times 10^2 \quad 6 \times 10^1 \quad 8 \times 10^0 \quad 4 \times 10^{-1} \quad 9 \times 10^{-2}$$

$$3 \; 6 \; 8 \; . \; 4 \; 9$$

$$368.49 = (3 \times 10^2) + (6 \times 10^1) + (8 \times 10^0) + (4 \times 10^{-1}) + (9 \times 10^{-2})$$
$$= 300 + 60 + 8 + 0.4 + 0.09$$
$$= 368.49$$

The leftmost digit in any number representation, which has the greatest weight, is called the *most significant digit*, and the rightmost digit, which has the least weight, is called the *least significant digit*. The digits on the left side of the decimal point form the *integer part* of a decimal number, and those on the right side form the *fractional part*.

Inserting a zero (0) in any number position simply implies that no multiple of the weight corresponding to that position is added to the sum. For example,

$$60.02 = 6 \times 10^1 + 0 \times 10^0 + 0 \times 10^{-1} + 2 \times 10^{-2}$$
$$= 60 + 0 + 0 + 0.02 = 60.02$$

Of course, any number of zeros can be inserted in leading positions (to the left of the most significant digit) or in trailing positions (to the right of the least significant digit) without affecting the value of a number. Such zeros are called leading and trailing zeros, respectively. In many computer operations, all numbers must be expressed using the same number of positions, so leading or trailing zeros must be inserted as necessary to fill each position with a digit.

2–3   THE BINARY NUMBER SYSTEM

As discussed in Section 1–1, the useful voltage levels in digital circuitry have just two values. These levels are sometimes called high and low, or on and off, or true and false. Whatever terminology is used, the two-valued nature of the circuitry means that digital logic lends itself readily to analysis using a number system that has only two symbols. The binary number system, having the two symbols 0 and 1, is just such a system. The base, or radix, of the binary number system is 2. As in the decimal number system, the base equals the number of symbols, and the base itself is not one of those symbols.

A binary digit is called a *bit*. A binary number consists of a sequence of bits, each of which can be either 0 or 1 and each of which multiplies a different power of 2. The weight of each bit position is one power of 2 greater than the weight of the position to its right. Note the similarity of these properties to those of the decimal number system. The value of a binary number is the sum of all its bits multiplied by the weights of their respective positions. Finding the decimal value of a binary number is the same as *converting* the binary number to its decimal equivalent. Following is an example:

$$
\begin{array}{cccc}
2^3 & 2^2 & 2^1 & 2^0 \\
\downarrow & \downarrow & \downarrow & \downarrow
\end{array}
$$
$$
1 \quad 1 \quad 0 \quad 1 \; = \; (1 \times 2^3) + (1 \times 2^2) + (0 \times 2^1) + (1 \times 2^0)
$$
$$
= \; 8 + 4 + 0 + 1 \doteq 13
$$

Note that the binary number 1101 in this example could be misinterpreted to be the decimal number one thousand, one hundred one. To avoid this kind of confusion, we use a right parenthesis followed by a subscript equal to the base in order to indicate the base of the system in which a number is expressed. For example:

$$
1101)_2 = 13)_{10}
$$

In this book, we use the subscripted-base notation wherever there is a possibility of confusion but omit it if the value of the base is clear from the context.

As in the decimal number system, bit positions to the right of the units (2^0) position have weights that are negative powers of the base. A *binary point*, rather than a decimal point, separates the integer and fractional parts of a binary number. The next example illustrates how a binary number containing a fractional part is converted to its decimal equivalent.

Example 2–1

Find the decimal equivalent of the binary number $10010.011)_2$.

Solution

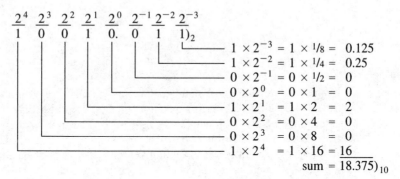

$$
\begin{array}{ccccccccc}
2^4 & 2^3 & 2^2 & 2^1 & 2^0 & 2^{-1} & 2^{-2} & 2^{-3} \\
1 & 0 & 0 & 1 & 0. & 0 & 1 & 1)_2
\end{array}
$$

$$
\begin{aligned}
1 \times 2^{-3} &= 1 \times \tfrac{1}{8} = 0.125 \\
1 \times 2^{-2} &= 1 \times \tfrac{1}{4} = 0.25 \\
0 \times 2^{-1} &= 0 \times \tfrac{1}{2} = 0 \\
0 \times 2^{0} &= 0 \times 1 = 0 \\
1 \times 2^{1} &= 1 \times 2 = 2 \\
0 \times 2^{2} &= 0 \times 4 = 0 \\
0 \times 2^{3} &= 0 \times 8 = 0 \\
1 \times 2^{4} &= 1 \times 16 = \underline{16} \\
\text{sum} &= 18.375)_{10}
\end{aligned}
$$

Drill 2–1

Find the decimal equivalent of $111011.101)_2$.
Answer: $59.625)_{10}$.

2–4 BINARY ARITHMETIC

Binary Addition

Let us begin our study of binary addition by finding all possible sums that result from adding two single-bit binary numbers. In other words, letting X represent a single-bit number (so X can be either 0 or 1) and letting Y be another single-bit number (Y likewise can be either 0 or 1), we are interested in finding the sum $X + Y$ for every possible combination of values of X and Y. Adding 0 to any number does not change the number, so the following sums are obvious:

X	Y	$X + Y$
0	0	$0 + 0 = 0$
0	1	$0 + 1 = 1$
1	0	$1 + 0 = 1$

Now consider the case $X = 1$ and $Y = 1$. For binary addition to be consistent with its decimal counterpart, the binary sum of two numbers should be equivalent to the decimal sum obtained by adding the same two numbers in decimal. In decimal, $1)_{10} + 1)_{10} = 2)_{10}$. Therefore, $1)_2 + 1)_2$ should equal the binary equivalent of $2)_{10}$, namely, $10)_2$. Thus,

X	Y	$X + Y$
1	1	$1 + 1 = 10$

In this case, adding two single-bit numbers produces the 2-bit sum 10. We say that the sum is 0 with a *carry* of 1. To extend this idea to the previous combinations of X and Y—i.e., to allow the sum in every case to be a 2-bit number—we can say that the carry is 0 in each of those previous cases. Summarizing these ideas, we have

X	Y	$X + Y$	*Sum Bit*	*Carry Bit*	*Two-bit Sum*
0	0	$0 + 0$	0	0	00
0	1	$0 + 1$	1	0	01
1	0	$1 + 0$	1	0	01
1	1	$1 + 1$	0	1	10

When adding two binary numbers containing an arbitrary number of bits, we simply add the bits in each column (bit position) one column at a time, so we are, in effect, repeatedly adding two 1-bit numbers. Thus, the conclusions we derived from adding 1-bit numbers apply to the addition of bits in any two binary numbers. Following the usual rules of addition, a carry bit generated in a given bit position must be added to the bits in the next most significant position. Therefore, in general, adding the bits in a column actually requires that we form a 2-bit sum and then add a third (carry) bit to it. Adding carries from one column into the next most significant column is the same procedure we follow when adding decimal numbers. To illustrate these ideas, we add the two 4-bit binary numbers $X = 0101$ and $Y = 0110$:

$$
\begin{array}{cccccc}
 & 2^3 & 2^2 & 2^1 & 2^0 & \\
 & \downarrow & \text{(Carry)} & & & \\
 & 1 & & & & \\
Y = & 0 & 1 & 1 & 0 & \\
X = +0 & & 1 & 0 & 1 & \\
\hline
\text{sum} = & 1 \;(1)\; 0 & 1 & 1 & & = 1011 \\
 & \underbrace{\quad}_{1\,+\,1\,=\,10} & & & &
\end{array}
$$

Notice that the carry bit is 0 in the two least significant columns and 1 in the third (2^2) column. This carry bit is added to the bits in the most significant column to obtain a most significant sum bit of $1 + 0 + 0 = 1$. The binary sum can be checked by performing the same addition using decimal equivalents and verifying that the binary sum corresponds to the decimal sum. Decimal conversions are facilitated by writing the decimal weights 1, 2, 4, 8, . . . above the binary positions corresponding to $2^0, 2^1, 2^2, 2^3, ...$. Verifying the preceding example, we find:

$$
\begin{array}{ccccll}
8 & 4 & 2 & 1 & & \\
\hline
0 & 1 & 1 & 0 & = & 6)_{10} \\
+0 & 1 & 0 & 1 & = & + \ 5)_{10} \\
\hline
1 & 0 & 1 & 1 & = & 11)_{10}
\end{array}
$$

Since it is possible that a carry bit of 1 may have to be added to a column that already contains two 1s, we must also consider the sum $1 + 1 + 1$. Since $1)_{10} + 1)_{10} + 1)_{10} = 3)_{10} = 11)_2$, we see that the addition $1)_2 + 1)_2 + 1)_2$ produces a sum bit of 1 and a carry bit of 1. The next example illustrates such a case.

$$
\begin{array}{rccccc}
 & 2^4 & 2^3 & 2^2 & 2^1 & 2^0 \\
\hline
\text{carry bits:} & \mathbf{1} & \mathbf{1} & \mathbf{1} & & \\
X = & 0 & 1 & 1 & 1 & 1 \\
Y = & 0 & 0 & 1 & 1 & 0 \\
\hline
\text{sum} = & 1 & 0 & 1 & 0 & 1
\end{array}
$$

Notice that the carry generated in the 2^1 column is added to the bits in the 2^2 column to produce a sum bit of 1 and another carry bit of 1. The latter carry is then added to the bits in the 2^3 column to produce yet another carry of 1. This is an example of *carry propagation*, a process that can reduce the speed at which a computer performs arithmetic computations.

Binary numbers containing fractional parts are added in exactly the same way as those containing no fractional parts. Carries produced in the 2^{-1} position on the fractional side propagate into the 2^0 position on the integer side. If a number containing no fractional part is to be added to a number containing a fractional part, *insert 0s as necessary to make both numbers have the same number of bits*. In fact, any two binary numbers should be adjusted by inserting leading or trailing 0s, so that both have the same number of bits, before any arithmetic operations are performed. This adjustment is good practice because binary numbers in a computer all have the same length—i.e., the same number of bits—called the *word size*, and all arithmetic operations are performed on equal-length words. Furthermore, certain binary computations we discuss later can produce erroneous results unless the numbers are adjusted to have the same length before the computations are performed.

Example 2-2

Find the binary sum $X + Y$ for each of the following pairs of binary numbers.

a. $X = 1010, Y = 0011.1$

b. $X = 111001.1, Y = 0.0101$

c. $X = 10011.010, Y = 11111.110$

In each case, verify the computation by finding the decimal equivalents of X and Y and the decimal equivalent of the sum.

Solution

a. We adjust the length of X by inserting a trailing 0 to make it have the same length as Y:

$$
\begin{array}{cccccccc}
\underline{8} & \underline{4} & \underline{2} & \underline{1} & \underline{\tfrac{1}{2}} & & \\
 & & & 1 & & & \\
Y = & 0 & 0 & 1 & 1 & .1 & = & 3.5)_{10} \\
X = & +1 & 0 & 1 & 0 & .0 & = & +10.0)_{10} \\
\hline
X + Y = & 1 & 1 & 0 & 1 & .1 & = & 13.5)_{10}
\end{array}
$$

b. We must add trailing 0s to X and leading 0s to Y to make both have the same length:

$$
\begin{array}{ccccccccccccc}
\underline{32} & \underline{16} & \underline{8} & \underline{4} & \underline{2} & \underline{1} & & \underline{\tfrac{1}{2}} & \underline{\tfrac{1}{4}} & \underline{\tfrac{1}{8}} & \underline{\tfrac{1}{16}} & & \\
Y = & 0 & 0 & 0 & 0 & 0 & 0 & . & 0 & 1 & 0 & 1 & = & 0.3125)_{10} \\
X = & +1 & 1 & 1 & 0 & 0 & 1 & . & 1 & 0 & 0 & 0 & = & +57.5000)_{10} \\
\hline
X + Y = & 1 & 1 & 1 & 0 & 0 & 1 & . & 1 & 1 & 0 & 1 & = & 57.8125)_{10}
\end{array}
$$

c. In this example, X and Y are both 8-bit numbers, but a preliminary computation will show that the sum is a 9-bit number. We should therefore insert leading 0s to make X and Y both 9-bit numbers. If the word size in a computer performing this computation were 8 bits, then the carry generated in the most significant position would be called an *overflow*.

$$
\begin{array}{cccccccccccc}
\underline{32} & \underline{16} & \underline{8} & \underline{4} & \underline{2} & \underline{1} & & \underline{\tfrac{1}{2}} & \underline{\tfrac{1}{4}} & \underline{\tfrac{1}{8}} & \\
1 & 1 & 1 & 1 & 1 & 1 & & 1 & & & \\
Y = & 0 & 1 & 1 & 1 & 1 & 1 & . & 1 & 1 & 0 & & 31.75)_{10} \\
X = & +0 & 1 & 0 & 0 & 1 & 1 & . & 0 & 1 & 0 & = & +19.25)_{10} \\
\hline
& 1 & 1 & 0 & 0 & 1 & 1 & . & 0 & 0 & 0 & = & 51.00)_{10}
\end{array}
$$

**Drill
2–2**

Find the sum $X + Y + Z$ when $X = 101.01$, $Y = 110$, and $Z = 10.1$.
Answer: $1101.11 = 13.75)_{10}$.

Binary Subtraction

Subtracting 0 from a number does not change its value, and subtracting 1 reduces its value by 1. Therefore, the following results are obvious:

$$0 - 0 = 0$$
$$1 - 0 = 1$$
$$1 - 1 = 0$$

In order to subtract 1 from 0, we must *borrow* a 1 (from the next most significant bit in a multi-bit number). The borrowed 1 is placed to the left of the 0, converting it to 10. Thus, the result of the subtraction is $10 - 1 = 1$:

$$0 - 1 = 1 \qquad \text{with a borrow of 1}$$

This procedure is the same as that followed in decimal subtraction when a borrow is necessary. Consider, for example, the subtraction $73)_{10} - 48)_{10}$:

$$\begin{array}{r} 73 \\ -48 \end{array} = \begin{array}{r} \overset{6}{\cancel{7}}\ \overset{13}{\cancel{3}} \\ -4\ \ 8 \\ \hline 2\ \ 5 \end{array} \quad\overset{\neararrow\ \text{borrowed 1}}{}$$

In this example, notice that the 1 is borrowed from the 7 in the next most significant column, reducing the 7 to 6, and that 3 becomes 13. Similarly, when a 1 is borrowed from the next most significant bit in a binary number, that bit is reduced by 1. Following is an example:

$$\begin{array}{r} 1\ 0\ 1\ 1 \\ -0\ 1\ 0\ 1 \end{array} = \begin{array}{r} \overset{0\ \ 10}{\cancel{1}\ \cancel{0}\ 1\ 1} \\ -0\ 1\ 0\ 1 \\ \hline 0\ 1\ 1\ 0 \end{array} \quad\overset{\text{borrowed 1}}{}$$

Notice that borrowing 1 from the 1 in the most significant position reduced the value of that 1 to 0 ($1 - 1 = 0$). As in binary addition, binary subtraction can be checked by performing the same computation in decimal. It can also be checked by adding the *subtrahend* to the *difference* and verifying that the sum equals the *minuend*. Checking the previous example,

$$\begin{array}{c} \begin{array}{cccc} 8 & 4 & 2 & 1 \\ \hline 1 & 0 & 1 & 1 \end{array} = 11)_{10} \\ \begin{array}{cccc} -0 & 1 & 0 & 1 \end{array} = -6)_{10} \\ \hline \begin{array}{cccc} 0 & 1 & 1 & 0 \end{array} = 5)_{10} \end{array} \qquad \begin{array}{rcccc} \text{minuend} = & 1 & 0 & 1 & 1 \\ \text{subtrahend} = & -0 & 1 & 0 & 1 \\ \text{difference} = & +0 & 1 & 1 & 0 \\ \hline & 1 & 0 & 1 & 1 = \text{minuend} \end{array} \Big\}\text{sum} =$$

If the digit from which a borrow is attempted is 0, then recall in decimal subtraction that it is necessary to borrow from the next non-zero digit to the left. In such a case, that 0 and all intervening 0s are changed to 9s. Following are examples:

$$\begin{array}{r} 602 \\ -\ 56 \end{array} = \begin{array}{r} \overset{5}{\cancel{6}}\ \overset{9}{\cancel{0}}\ \overset{12}{\cancel{2}} \\ -\ \ 5\ \ 6 \\ \hline 5\ \ 4\ \ 6 \end{array}$$

$$\begin{array}{r} 8003 \\ -\ 125 \end{array} = \begin{array}{r} \overset{7}{\cancel{8}}\ \overset{9}{\cancel{0}}\ \overset{9}{\cancel{0}}\ \overset{13}{\cancel{3}} \\ -\ \ \ 1\ \ 2\ \ 5 \\ \hline 7\ \ 8\ \ 7\ \ 8 \end{array}$$

Similarly, in binary subtraction, it is necessary to borrow from a more significant bit if the bit from which a borrow is attempted is 0. In this case, each intervening zero is changed to a 1, as in the following example:

$$\begin{array}{r} 1001 \\ -0110 \end{array} = \begin{array}{r} \overset{0}{\cancel{1}}\ \overset{1}{\cancel{0}}\ \overset{10}{\cancel{0}}\ 1 \\ -0\ 1\ 1\ 0 \\ \hline 0\ 0\ 1\ 1 \end{array}$$

Example
2–3

Perform the subtraction $X - Y$ for each of the following pairs of binary numbers.

   a. $X = 111.01$, $Y = 100.10$

   b. $X = 10001$, $Y = 110$

   c. $X = 1001.01$, $Y = 0011.10$

Solution

a. Bits in the fractional parts of binary numbers are subtracted in the same way as bits in the integer parts:

$$
\begin{array}{r}
111.01 \\
-100.10 \\
\end{array}
\quad = \quad
\begin{array}{r}
1 \; 1 \; 1. \; 0 \; 1 \\
-1 \; 0 \; 0. \; 1 \; 0 \\
\hline
0 \; 1 \; 0. \; 1 \; 1 \\
\end{array}
$$

with borrow notation "0 10" above the $1.\,0$ columns, where the 1 and 0 are crossed out.

b. We insert leading 0s in Y to give it the same length as X. Notice that it is necessary to change two 0s to 1s in order to borrow in this example:

$$
\begin{array}{r}
10001 \\
-00110 \\
\end{array}
\quad = \quad
\begin{array}{r}
1 \; 0 \; 0 \; 0 \; 1 \\
-0 \; 0 \; 1 \; 1 \; 0 \\
\hline
0 \; 1 \; 0 \; 1 \; 1 \\
\end{array}
$$

with borrow notation "0 1 1 10" above and crossed-out digits.

c. In this example, it is necessary to borrow twice:

first borrow / second borrow 10

$$
\begin{array}{r}
1001.01 \\
-0011.10 \\
\end{array}
=
\begin{array}{r}
1 \; 0 \; 0 \; 1. \; 0 \; 1 \\
-0 \; 0 \; 1 \; 1. \; 1 \; 0 \\
\hline
. \; 1 \; 1 \\
\end{array}
=
\begin{array}{r}
1 \; 0 \; 0 \; 1. \; 0 \; 1 \\
-0 \; 0 \; 1 \; 1. \; 1 \; 0 \\
\hline
0 \; 1 \; 0 \; 1. \; 1 \; 1 \\
\end{array}
$$

Notice that the 1 (in the 2^0 column) that was changed to 0 because of the first borrow was changed to 10 when the second borrow was made.

Drill
2–3

Perform the binary subtraction $10101.001 - 01110.110$.
Answer: 00110.011.

Binary Multiplication

The rules for multiplying 1-bit binary numbers are quite straightforward. Since any number multiplied by 0 equals 0 and any number multiplied by 1 equals the number, we have

$$0 \times 0 = 0$$
$$0 \times 1 = 0$$
$$1 \times 0 = 0$$
$$1 \times 1 = 1$$

The procedure for multiplying binary numbers containing any number of bits is exactly the same as that used for multiplying decimal numbers. Beginning with the least significant bit of the multiplier, we multiply each bit of the multiplicand, one bit at a time, and write the results in a line. The line of results is called a *partial product*. The process is repeated for the next most significant bit in the multiplier, remembering to shift the next partial product one position to the left. After the process has been repeated for every bit in the multiplier, the results are added. Following is an example:

$$
\begin{array}{r}
110 \quad \text{multiplicand} \\
\times\ 101 \quad \text{multiplier} \\
\hline
110 \\
000 \\
110 \\
\hline
11110 \quad \text{product}
\end{array}
$$

110   multiplicand
× 101   multiplier
110 ⎫
000 ⎬ partial products
110 ⎭
11110   product

Binary multiplication is a simple process because, as demonstrated in the preceding example, each partial product is either an exact duplicate of the multiplicand or a line of 0s. Of course, if the multiplier or multiplicand are relatively large, it will be necessary to add a substantial number of partial products, an addition task that could become rather formidable. However, with some practice, the student will find that adding several binary numbers simultaneously is not as difficult as it seems. This fact is illustrated in Example 2–4.

When multiplying binary numbers containing fractional parts, we set the binary point in the product using the same procedure followed for decimal multiplication. Recall that we make the number of places (bit positions) in the fractional part of the product equal to the sum of the number of places in the fractional parts of the multiplicand and multiplier. The next example illustrates the procedure.

Example 2–4

Perform the binary multiplication 10110.01×1101.1.

Solution

10110.01 ←——— 2-bit fractional part
×   1101.1 ←——— 1-bit fractional part
1011001   ┌— 3-bit fractional part
1011001
0000000
1011001
1011001
100101100.011 ←

Notice that the partial products are written without binary points and that the fractional part of the product has $2 + 1 = 3$ positions. Let us examine how the sum and carry bits are found in the 2^0, 2^1, and 2^2 columns of the product, repeated here:

$$
\begin{array}{r}
\underline{2^2}\ \ \underline{2^1}\ \ \underline{2^0} \\
\text{carry:}\quad \mathbf{1}\ \ \mathbf{10}\ \ \mathbf{1} \\
0\ \ 1\ \ 1 \\
1\ \ 1\ \ 0 \\
0\ \ 0\ \ 0 \\
0\ \ 0\ \ 1 \\
0\ \ 1 \\
\hline
1\ \ 0\ \ 0
\end{array}
$$

The carry of 1 generated in the 2^0 column is added to the 2^1 column, so we must add four 1s in that column. Now, $1)_{10} + 1)_{10} + 1)_{10} + 1)_{10} = 4)_{10} = 100)_2$. Therefore, the sum bit in the 2^1 column is 0 and *we have a 2-bit carry of* 10 to add to the 2^2 column. Adding 10 to the 1 in the 2^2 column gives 11—i.e., a sum bit of 1 and a carry bit of 1.

Drill
2–4

Perform the binary multiplication $111011.11 \times 10011.01$.
Answer: 10001111110.0011.

Binary Division

As in long division with decimals, binary division is basically a trial-and-error procedure to determine how many times a number (the divisor) divides another number (the dividend). However, binary division is simpler than its decimal counterpart because at each step in the process, a divisor either divides into a group of bits (giving a quotient bit of 1) or it doesn't (giving a quotient bit of 0). Whether or not the divisor divides a group of bits is determined simply by whether the divisor has a value less than or equal to the value of those bits. The following example illustrates the procedure and shows its similarity to decimal long division:

$$
\begin{array}{r}
101 \quad \text{quotient} \\
\text{divisor} \quad 110\overline{)11110} \quad \text{dividend} \\
-110 \\
\hline
011 \\
-000 \\
\hline
110 \\
-110 \\
\hline
000
\end{array}
$$

Since 110 is smaller than the value of the first 3 bits of the dividend (111), the first quotient bit is 1. Subtracting 110 from 111 and bringing down the next 1 from the dividend gives 011, which is smaller than 110. Consequently, the next quotient bit is 0. Bringing down the 0 from the dividend gives 110, which equals the divisor, so the last quotient bit is 1. There is no remainder in this example. Division can be extended beyond the least significant bit of the dividend by inserting trailing 0s, just as in decimal division. The next example illustrates such a case.

Example
2–5

Divide 11001 by 100.

Solution. Note that we insert trailing 0s in the dividend so that the quotient can be expressed with two additional places:

$$
\begin{array}{r}
110.01 \\
100\overline{)11001.00} \\
-100 \\
\hline
100 \\
-100 \\
\hline
001 \\
-000 \\
\hline
010 \\
-000 \\
\hline
100 \\
-100 \\
\hline
000
\end{array}
$$

Drill   Divide 1110011 by 1000.
2–5   *Answer:* 1110.011.

2–5   OCTAL NUMBERS AND ARITHMETIC

Octal Numbers

One disadvantage of the binary number system is that it is an awkward and inefficient way for humans to communicate with a computer. It is, for example, considerably more difficult to write down and to gain a sense of the magnitude of the binary number 1010001010 than it is to do the same with its decimal equivalent, 650. On the other hand, digital systems must operate in binary, and it is a cumbersome task (as we shall see) to convert from decimal to binary. What we need is a number system that is "close" to decimal, yet easy to convert to binary. The *octal* number system, with base 8, is such a system. It contains the octal digits 0, 1, 2, 3, 4, 5, 6, and 7; and because its base is a power of 2 (2^3), it is easy to convert from octal to binary (as we shall see). Note that the digits 8 and 9 do not belong to the octal number system.

Octal numbers have exactly the same structure as binary and decimal numbers. Each digit of an octal number is multiplied by a power of the base 8. For example, the decimal equivalent of the octal number $326.4)_8$ is found as follows:

$$
\begin{array}{cccc}
8^2 & 8^1 & 8^0 & 8^{-1} \\
\hline
3 & 2 & 6 & . & 4
\end{array}
$$

$$
\begin{aligned}
4 \times 8^{-1} &= 4 \times {}^1\!/_8 = & 0.5 \\
6 \times 8^0 &= 6 \times 1 = & 6 \\
2 \times 8^1 &= 2 \times 8 = & 16 \\
3 \times 8^2 &= 3 \times 64 = & \underline{192} \\
& \text{sum} = & 214.5)_{10}
\end{aligned}
$$

Octal Arithmetic

The sum of two octal digits is the same as their decimal sum, provided the decimal sum is less than 8. If the decimal sum is 8 or greater, subtract 8 to obtain the octal digit. A carry of 1 is produced when the decimal sum is corrected this way. Following are some examples:

$$
\begin{aligned}
4)_8 + 2)_8 &= 6)_8 \\
6)_8 + 7)_8 &\rightarrow 13 - 8 = 5)_8, \quad \text{carry 1} \\
5)_8 + 3)_8 &\rightarrow 8 - 8 = 0)_8, \quad \text{carry 1}
\end{aligned}
$$

To obtain the sum of multi-digit octal numbers, the procedure just described is applied to each column of digits. Following is an example:

$$
\begin{array}{r}
1\,1 \\
126)_8 \\
+\,255)_8 \\
\hline
403)_8
\end{array}
$$

Note that the sum of the least significant digits is 3 $(11 - 8)$, with a carry of 1, which is added to the next column. In that column, the sum $(1 + 2 + 5)$ is 0 $(8 - 8)$, with a carry propagated to the most significant column. Since the sum of the digits in that column $(1 + 1 + 2)$ is 4, no correction is necessary.

If the decimal sum of several octal digits is 16 or greater, subtract 16 and carry 2. In general, we can express any decimal sum in octal by repeatedly subtracting 8 until the result is one of the octal digits 0 through 7. Each time 8 is subtracted, the amount of the carry is increased by 1. The next example illustrates the procedure.

Example 2–6

Add the octal numbers $377)_8$, $175)_8$, $450)_8$, and $563)_8$.

Solution

$$
\begin{array}{r}
\text{carries:} \quad \mathbf{231} \\
377)_8 \\
175)_8 \\
450)_8 \\
+ \; 563)_8 \\
\hline
2027)_8
\end{array}
$$

- $(15 - 8)$, carry 1
- $(26 - 3 \times 8)$, carry 3
- $(16 - 2 \times 8)$, carry 2

Drill 2–6

Add the octal numbers $655)_8$, $272)_8$, $177)_8$, and $566)_8$.
Answer: $2134)_8$.

Octal subtraction is accomplished most easily using the complement method, which we describe in a later section. Octal multiplication can be performed by converting decimal partial products to octal, using the procedure described for addition. Following is an example:

$$
\begin{array}{r}
\text{carries:} \quad \begin{cases} \mathbf{1} \\ \mathbf{2} \end{cases} \\
15)_8 \\
\times \; 24)_8 \\
\hline
64 \longleftarrow (20 - 16 = 4, \;\; \text{carry 2}) \\
32 \longleftarrow (10 - 8 = 2, \;\; \text{carry 1}) \\
\hline
404)_8
\end{array}
$$

2–6   HEXADECIMAL NUMBERS AND ARITHMETIC

Hexadecimal Numbers

The hexadecimal number system, with base 16, is another number system that is particularly useful for human communications with a computer. Although it is somewhat more difficult to interpret than the octal number system, it has become the most popular means for direct data entry and retrieval in digital systems. Since its base is a power of

2 (2^4), it is easy to convert hexadecimal numbers to binary and vice versa. Furthermore, as we shall see, these conversions are very conveniently performed in terms of 4-bit binary numbers, called *nibbles,* that are fundamental parts of larger binary words.

The hexadecimal number system contains 16 symbols. We use the 10 decimal symbols 0 through 9 but must supplement those with 6 new symbols to provide a full set. For that purpose, the first six letters of the alphabet are incorporated into the set, each letter having the decimal value shown:

Hexadecimal Digit	Decimal Value
A	10
B	11
C	12
D	13
E	14
F	15

The decimal value of a hexadecimal number is computed in the same way decimal values are found for numbers with other bases. The position of each digit has a weight equal to a power of 16. Following are some examples:

$$4 \quad 7)_{16} = (4 \times 16^1) + (7 \times 16^0) = 64 + 7 = 71)_{10}$$

$$3 \quad B)_{16} = (3 \times 16^1) + (11 \times 16^0) = 48 + 11 = 59)_{10}$$

$$A \quad F \quad 5)_{16} = (10 \times 16^2) + (15 \times 16^1) + (5 \times 16^0)$$
$$= 2560 + 240 + 5 = 2805)_{10}$$

$$9. \quad 4 \quad C = (9 \times 16^0) + \left(4 \times \frac{1}{16}\right) + \left(12 \times \frac{1}{256}\right)$$
$$= 9 + 0.25 + 0.046875 = 9.296875)_{10}$$

Hexadecimal Arithmetic

The sum of two hexadecimal digits is the same as their equivalent decimal sum (with any sum from 10 through 15 expressed as a letter from A through F), provided the decimal equivalent is less than 16. If the decimal sum is 16 or greater, subtract 16 to obtain the hexadecimal sum. A carry of 1 is produced when this correction is necessary. Following are some examples:

$$7)_{16} + 2)_{16} = 9)_{16}$$
$$C)_{16} + 2)_{16} = 12)_{10} + 2)_{10} = 14)_{10} = E)_{16}$$
$$8)_{16} + D)_{16} = 8)_{10} + 13)_{10} = 21)_{10} \rightarrow 21 - 16 = 5)_{16}, \quad \text{carry 1}$$
$$E)_{16} + F)_{16} = 14)_{10} + 15)_{10} = 29)_{10} \rightarrow 29 - 16 = 13)_{10}$$
$$= D)_{16}, \quad \text{carry 1}$$

Hexadecimal addition is somewhat awkward because of the unfamiliar symbols and the computational labor involved in converting sums between hexadecimal and decimal. The

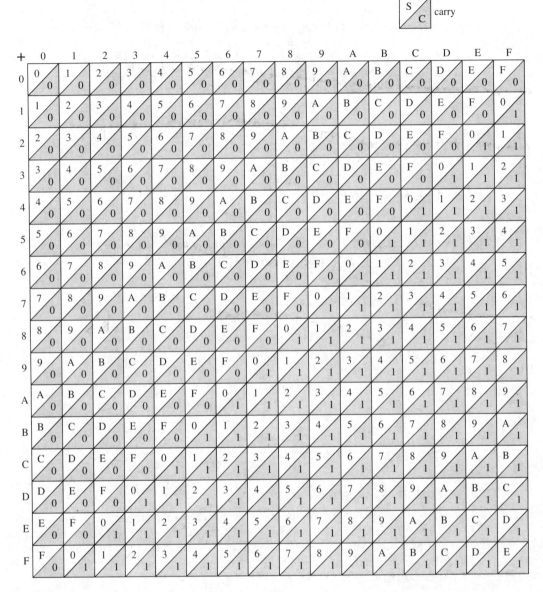

Table 2.1
Hexadecimal addition table.

process becomes routine for those who must compute such sums on a regular basis, but others may prefer to use a hexadecimal *addition table*, such as the one shown in Table 2.1. The table shows the sum and carry digits resulting from every possible sum of two hexadecimal digits.

Example
2-7

Perform the addition $94.B)_{16} + 2E.2)_{16}$.

Solution

$$
\begin{array}{r}
\text{carry:} \quad \mathbf{10} \\
94.B)_{16} \\
2E.2)_{16} \\
\hline
C2.D)_{16}
\end{array}
$$

From Table 2.1:

$B + 2 = D, \quad$ carry 0

$4 + E = 2, \quad$ carry 1

$1 + 9 + 2 = A + 2 = C, \quad$ carry 0

Drill
2-7

Perform the addition $7F0.8A)_{16} + F19.B1)_{16}$.
Answer: $170A.3B)_{16}$.

Hexadecimal subtraction is best accomplished using the complement method, which is described later. Hexadecimal multiplication can be performed by converting the decimal equivalents of partial products to hexadecimal, as shown in the following example:

$$
\begin{array}{lll}
2)_{16} \times 6)_{16} + 1)_{16} & \quad \overset{\mathbf{1}}{6A} & 2)_{16} \times A)_{16} \\
= 2)_{10} \times 6)_{10} + 1)_{10} & \quad \underline{\times 12} & = 2)_{10} \times 10)_{10} \\
= 13)_{10} = D)_{16} \longrightarrow & \longrightarrow D4 \longleftarrow & = 20)_{10} = 4)_{16}, \quad \text{carry 1} \\
& \quad \underline{6A} & \\
& \quad 774 &
\end{array}
$$

2-7 COUNTING

In many digital systems it is necessary to keep track of a variety of events, represented by electrical pulses, that may occur over a period of time. Electronic *counters,* whose structure we study in a later chapter, are designed to perform this important counting function. To understand how counters operate, it is necessary to be able to count in the various number systems we have discussed. Furthermore, the ability to count, particularly in binary, is a fundamental skill required in the design and analysis of logic circuitry.

To gain some insights that can be applied to counting in other number systems, let us review the process we use to count in decimal. In keeping with the requirement described earlier that all number representations in a given system should have the same length, we count up to 9999 using 4-digit decimal numbers, as follows:

10^3	10^2	10^1	10^0									
0	0	0	0		0	0	0	6		0	0	1	2
0	0	0	1		0	0	0	7		0	0	1	3
0	0	0	2		0	0	0	8				.	
0	0	0	3		0	0	0	9				.	
0	0	0	4		0	0	1	0		0	0	2	9
0	0	0	5		0	0	1	1		0	0	3	0

(continued on page 22)

$$\begin{array}{cccc} \underline{10^3} & \underline{10^2} & \underline{10^1} & \underline{10^0} \\ 0 & 0 & 3 & 1 \end{array}$$

10^3	10^2	10^1	10^0					
0	0	3	1		0	9	9	9
					1	0	0	0
		.					.	
		.					.	
		.					.	
0	0	9	9				.	
0	1	0	0				.	
		.			9	9	9	9

Notice that the count begins by listing the 10 symbols 0 through 9 in the least significant (10^0) column. When the list of symbols is exhausted, we add 1 to the next more significant (10^1) column and repeat the list of symbols in the least significant column. This process continues, adding 1 to the 10^1 column each time the set of symbols 0 through 9 is repeated in the 10^0 column, until we reach 99. A 1 is then added to the 10^2 column and the entire process just described is repeated in the 10^0 and 10^1 columns. The pattern thereafter is clear: add 1 to any column after all columns to its right contain 9, and repeat the counting process in those columns.

To count with 4-bit binary numbers, we follow the same process. In this case, the list of symbols consists of just 0 and 1, so when we exhaust that list in the least significant (2^0) column, we add 1 to the 2^1 column:

2^3	2^2	2^1	2^0		2^3	2^2	2^1	2^0
0	0	0	0		1	0	0	0
0	0	0	1		1	0	0	1
0	0	1	0		1	0	1	0
0	0	1	1		1	0	1	1
0	1	0	0		1	1	0	0
0	1	0	1		1	1	0	1
0	1	1	0		1	1	1	0
0	1	1	1		1	1	1	1

As in decimal counting, a new count adds 1 to any column after all columns to its right contain the largest digit in the symbol list. Those rightmost columns are all *reset* to 0. For example, the next count after 0111 is 1000. With a little practice, counting in binary becomes as simple as counting in decimal. A good way to obtain practice is to write down an arbitrary binary number and then write the numbers that immediately precede and immediately follow it. Of course, increasing a count by 1, called *incrementing* the count, is the same as mathematically adding 1 to it, so practice in incrementing arbitrary numbers can always be checked. Similarly, decreasing a count by 1, called *decrementing* the count, can be checked by subtracting 1.

Example 2–8   Write the binary count that results when each of the following is incremented and when each is decremented.

a. 10101

b. 11011

c. 01000

d. 11111

Solution

	Incremented Count	Decremented Count
a. 10101	10110	10100
b. 11011	11100	11010
c. 01000	01001	00111
d. 11111	100000	11110

Drill 2–8

Write the binary count that results (a) when 0011101111 is incremented and (b) when 1110010000 is decremented.

Answer: (a) 0011110000; (b) 1110001111.

Since the octal number system contains the eight symbols 0 through 7, a new count will add 1 to any column when all columns to its right contain 7 and will reset those columns to 0. Following are some examples:

.	.	.
.	.	.
.	.	.
05	075	3575
06	076	3576
07	077	3577
10	100	3600
11	101	3601
.	.	.
.	.	.
.	.	.

The largest digit in the hexadecimal system is F, so a new count will add 1 to any column when all columns to its right contain F and will reset those columns to 0:

.	.	.	.
.	.	.	.
.	.	.	.
00D	0AD	09FD	3FF9
00E	0AE	09FE	3FFA
00F	0AF	09FF	3FFB
010	0B0	0A00	3FFC
011	0B1	0A01	3FFD
.	.	.	3FFE
.	.	.	3FFF
.	.	.	4000
			.

The following are two important facts in connection with counting:

1. The largest decimal number that can be represented with n digits in a base-b number system (the greatest count) is $b^n - 1$.
2. The total number of counts (different numbers) that can be obtained with n digits in a base-b number system is b^n.

For example, a 5-bit binary counter can count up to $11111)_2 = 31)_{10} = 2^5 - 1$, and the total number of counts is $2^5 = 32$. Notice that the total number of counts is 1 greater than the largest count because 00000 is the first count.

A 4-bit binary number is called a *nibble,* and an 8-bit number is called a *byte.* Many modern microcomputers operate with a 16-bit word size, so each word constitutes two bytes. One byte can represent $2^8 = 256$ different numbers, whereas a 16-bit word can represent $2^{16} = 65,536$ different numbers.

Example 2–9

A hexadecimal counter capable of counting up to at least $10,000)_{10}$ is to be constructed. What is the minimum number of hexadecimal digits that the counter must have?

Solution. Since $16^3 = 4,096)_{10}$ and $16^4 = 65,536)_{10}$, we see that at least four hexadecimal digits are required.

In situations where we must find the number of digits necessary to meet a certain requirement, it is often possible to use logarithms to solve the equation specifying that requirement. In the present example, since the largest decimal number that can be represented by n hexadecimal digits is $16^n - 1$, we have

$$16^n - 1 = 10,000$$
$$16^n = 10,001$$

Taking the logarithm of both sides and recalling that $\log x^n = n \log x$, we find

$$\log 16^n = \log 10,001$$
$$n \log 16 = \log 10,001$$
$$n = \frac{\log 10,001}{\log 16} = 3.32$$

Since n must be an integer (whole number), we see that four digits are required.

Drill 2–9

What is the minimum number of bits that a binary counter must have to be capable of counting to at least $10,000)_{10}$?
Answer: 14.

2–8   CONVERSIONS BETWEEN NUMBER SYSTEMS

We have already seen how binary, octal, and hexadecimal numbers are converted to their decimal equivalents. We now wish to investigate conversions from decimal to each of those other number bases and conversions between them.

Decimal to Binary

The most direct way to convert an *integer* decimal number to binary is to use a kind of trial-and-error method that decomposes the number into powers of 2. We begin by writing the powers of 2 from right to left, with $2^0 = 1$ on the right, $2^1 = 2$ next, then 4, 8, 16, ... If X is the decimal number to be converted, we continue writing the powers of 2 until the largest value less than X is reached. For example, if $X = 24$, we write

$$16 \quad 8 \quad 4 \quad 2 \quad 1$$

Then, beginning at the left, write a 1 under any power of 2 that when added to other such designated powers on its left produces a cumulative sum that is less than or equal to X. Otherwise, write a 0 under the power of 2. The following example shows how to convert $37)_{10}$ to binary:

Step 1: $\underline{32}\ \underline{16}\ \underline{8}\ \underline{4}\ \underline{2}\ \underline{1}$
 $\qquad\quad 1$ $\qquad\qquad\qquad\quad 32 < 37$

Step 2: $\underline{32}\ \underline{16}\ \underline{8}\ \underline{4}\ \underline{2}\ \underline{1}$
 $\qquad\quad 1\ \ \ 0$ $\qquad\qquad\qquad 32 + 16 = 48 > 37$

Step 3: $\underline{32}\ \underline{16}\ \underline{8}\ \underline{4}\ \underline{2}\ \underline{1}$
 $\qquad\quad 1\ \ \ 0\ 0$ $\qquad\qquad\qquad 32 + 8 = 40 > 37$

Step 4: $\underline{32}\ \underline{16}\ \underline{8}\ \underline{4}\ \underline{2}\ \underline{1}$
 $\qquad\quad 1\ \ \ 0\ 0\ 1$ $\qquad\qquad\quad 32 + 4 = 36 < 37$

Step 5: $\underline{32}\ \underline{16}\ \underline{8}\ \underline{4}\ \underline{2}\ \underline{1}$
 $\qquad\quad 1\ \ \ 0\ 0\ 1\ 0$ $\qquad\qquad 32 + 4 + 2 = 38 > 37$

Step 6: $\underline{32}\ \underline{16}\ \underline{8}\ \underline{4}\ \underline{2}\ \underline{1}$
 $\qquad\quad 1\ \ \ 0\ 0\ 1\ 0\ 1$ $\qquad\quad 32 + 4 + 1 = 37$

Thus, $37)_{10} = 100101)_2$. Of course, it is not necessary to write out all six steps in the manner shown, since all the calculations can be performed as we move from left to right under a single line of powers.

A more systematic method for converting an integer decimal number to binary is to divide the number by 2, divide the resulting quotient by 2, divide that quotient by 2, and so forth. The sequence of *remainders* obtained from these divisions is the binary equivalent of the decimal number, where the first remainder obtained is the least significant bit and the last remainder is the most significant bit. The divide-by-2 process continues until we finally divide 2 into 1, giving 0 with a remainder of 1. The following example shows how the method is used to convert $26)_{10}$ to binary:

$$
\begin{array}{rl}
& 1 \\
& \underline{13} \\
2\overline{)}\,& 26 \\
& \underline{26} \\
& 0 \leftarrow \text{least significant bit} \\
\\
& 6 \\
2\overline{)}\,& 13 \\
& \underline{12} \\
& 1 \\
\\
& 3 \\
2\overline{)}\,& 6 \\
& \underline{6} \\
& 0
\end{array}
\qquad
\begin{array}{rl}
& 1 \\
2\overline{)}\,& 3 \\
& \underline{2} \\
& 1 \\
\\
& 0 \\
2\overline{)}\,& 1 \\
& \underline{0} \\
& 1 \leftarrow \text{most significant bit}
\end{array}
$$

Writing the remainders in sequence, we obtain $26)_{10} = 11010)_2$.

To convert a decimal *fraction* to binary, we first multiply the fraction by 2. The resulting product is a decimal number whose integer part is either 0 or 1. That integer is the first (most significant) bit of the binary equivalent. The fractional part of the product is then multiplied by 2, producing another decimal number whose integer part is either 0 or 1. This integer is the second bit of the binary equivalent. The process continues indefinitely, unless we reach a step where we are multiplying a zero fractional part by 2. In that case, all subsequent multiplications will produce 0s, corresponding to non-significant trailing 0s, so we can terminate the conversion. The following example shows how the method is used to convert 0.6875 to binary:

$$
\begin{aligned}
2 \times 0.6875 &= 1.375 \\
2 \times 0.375 &= 0.750 \\
2 \times 0.750 &= 1.500 \\
2 \times 0.500 &= 1.000
\end{aligned}
$$

most significant bit — points to the 1 in 1.375

least significant bit — points to the 1 in 1.000

Thus, $0.6875 = 0.1011$. Notice that the process terminated when we multiplied $2 \times 0.500 = 1.000$, because all subsequent multiplications of 0.000 will produce 0.000.

To convert a decimal number having both an integer and a fractional part to its binary equivalent, we simply convert each part separately, using the methods already described. The next example demonstrates such a case.

Example 2–10

Convert $43.6)_{10}$ to binary. Express the fractional part in 5 bits.

Solution.   Using the division process to convert the integer part, 43, to binary, we find

$$
\begin{array}{cccccc}
21 & 10 & 5 & 2 & 1 & 0 \\
2\overline{)43} & 2\overline{)21} & 2\overline{)10} & 2\overline{)5} & 2\overline{)2} & 2\overline{)1} \\
\underline{42} & \underline{20} & \underline{10} & \underline{4} & \underline{2} & \underline{0} \\
1 & 1 & 0 & 1 & 0 & 1
\end{array}
$$

Thus, $43)_{10} = 101011)_2$.

Using the multiplication process to convert the fractional part, 0.6, we find

$$
\begin{aligned}
2 \times 0.6 &= 1.2 \\
2 \times 0.2 &= 0.4 \\
2 \times 0.4 &= 0.8 \\
2 \times 0.8 &= 1.6 \\
2 \times 0.6 &= 1.2
\end{aligned}
$$

The first 5 bits of the fractional equivalent are therefore 0.10011. Notice that this conversion could continue indefinitely, producing, in fact, a repeating, non-terminating binary fraction. Therefore, 0.10011 is actually a 5-bit approximation of $0.6)_{10}$. Verify that

$0.10011)_2 = 0.59375)_{10}$. We see that terminating decimal fractions do not necessarily have terminating binary equivalents.

Combining the integer and fractional binary parts, we express $43.6)_{10}$ as $101011.10011)_2$.

Drill
2–10

Convert $16.7)_{10}$ to a binary number whose value is within 0.1% of $16.7)_{10}$.
Answer: 10000.1011.

Decimal to Octal and Hexadecimal

The process for converting decimal numbers to octal or hexadecimal parallels that for converting from decimal to binary. In converting integers, only the division method is recommended. We repeatedly divide by 8 (for octal) or by 16 (for hexadecimal) and keep track of the remainders. In hexadecimal conversions, remainders greater than 9 should be expressed as equivalent hexadecimal digits (A through F). To illustrate, we convert $379)_{10}$ to both octal and hexadecimal:

$$
\begin{array}{ll}
\begin{array}{r}
47 \\
\overline{8\,)\,379} \\
376 \\
\hline
3 \\
\end{array}
\leftarrow \text{least significant} \atop \text{octal digit}
&
\begin{array}{r}
23 \\
\overline{16\,)\,379} \\
368 \\
\hline
11 = B \\
\end{array}
\leftarrow \text{least significant} \atop \text{hexadecimal digit}
\end{array}
$$

$$
\begin{array}{ll}
\begin{array}{r}
5 \\
\overline{8\,)\,47} \\
40 \\
\hline
7 \\
\end{array}
&
\begin{array}{r}
1 \\
\overline{16\,)\,23} \\
16 \\
\hline
7 \\
\end{array}
\end{array}
$$

$$
\begin{array}{ll}
\begin{array}{r}
0 \\
\overline{8\,)\,5} \\
0 \\
\hline
5 \\
\end{array}
\leftarrow \text{most significant} \atop \text{octal digit}
&
\begin{array}{r}
0 \\
\overline{16\,)\,1} \\
0 \\
\hline
1 \\
\end{array}
\leftarrow \text{most significant} \atop \text{hexadecimal digit}
\end{array}
$$

We see that $379)_{10} = 573)_8 = 17B)_{16}$.

To convert a decimal fraction to octal or hexadecimal, we repeatedly multiply decimal fractions by 8 or 16, respectively, keeping track of the integer parts of the results. For example, to convert $0.546875)_{10}$ to octal and hexadecimal, we compute as follows:

$$
\begin{array}{ll}
8 \times 0.546875 = 4.375 & 16 \times 0.546875 = 8.75 \\
8 \times 0.375 = 3.000 & 16 \times 0.75 \quad = 12.00 = C.00
\end{array}
$$

Thus, $0.546875)_{10} = 0.43)_8 = 0.8C)_{16}$.

Conversions Between Binary, Octal, and Hexadecimal

Recall that a principal advantage of octal and hexadecimal numbers is the ease by which they are converted to binary numbers. To find the binary equivalent of a number whose base is of the form 2^n, *each digit* of the number is replaced by its *n*-bit binary equivalent. Since octal numbers have base $8 = 2^3$, each octal digit is replaced by a 3-bit binary equivalent. For example, to convert $271)_8$ to binary, we write

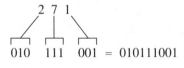

$$010 \quad 111 \quad 001 \ = \ 010111001$$

Similarly, hexadecimal numbers, with base $16 = 2^4$, are converted to binary by replacing each hexadecimal digit with its 4-bit binary equivalent. For example, $A94.0D)_{16}$ is converted to binary as follows:

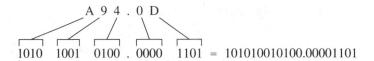

$$1010 \quad 1001 \quad 0100 \ . \ 0000 \quad 1101 \ = \ 101010010100.00001101$$

In each of these examples, note that the *full* 3- or 4-bit binary equivalent must be used to replace each octal or hexadecimal digit. For example, the hexadecimal digit 0 must be replaced by 0000. Table 2.2 lists the binary equivalents of all octal and hexadecimal digits for quick reference.

Of course, the relationships just described can also be used to convert binary numbers to octal and hexadecimal equivalents. In these cases, the binary number must be partitioned into groups of 3 or 4 bits, so that each group of 3 can be replaced by its octal equivalent or each group of 4 by its hexadecimal equivalent. For example, to convert $101100110101)_2$ to octal and hexadecimal, we write

Table 2.2
Binary equivalents of octal and hexadecimal digits.

Octal or hexadecimal digit	Binary equivalent for octal numbers	Binary equivalent for hexadecimal numbers
0	000	0000
1	001	0001
2	010	0010
3	011	0011
4	100	0100
5	101	0101
6	110	0110
7	111	0111
8		1000
9		1001
A		1010
B		1011
C		1100
D		1101
E		1110
F		1111

$$\underbrace{101100110101}_{\text{B35})_{16}}$$

$$\underbrace{101100110101}_{\text{5465})_{8}}$$

The following are two important points to remember in connection with the grouping of bits:

1. Grouping always begins at the binary point. Thus, groups are formed by moving left through the integer part and moving right through the fractional part of a binary number.
2. Insert as many leading and trailing 0s as necessary to make the binary number begin and end with a group of 3 (or 4) bits.

The next example illustrates the importance of these points.

Example 2-11

Convert $1111100101.00111)_2$ to octal and hexadecimal.

Solution. To find the octal equivalent, we begin at the binary point and form groups of 3 bits, moving to the left and to the right (as shown by the arrows):

Insert two leading 0s. Insert one trailing 0.

$$\underbrace{001111100101.001110}_{1745.16)_8}$$

Notice how leading and trailing zeros are inserted to form groups of 3 bits at each end of the binary number. If the trailing 0 were not inserted, the 11 at the right end might be misinterpreted as $3)_8$ instead of $6)_8$.

To find the hexadecimal equivalent, we begin at the binary point and form groups of 4 bits:

Insert two leading 0s. Insert three trailing 0s.

$$\underbrace{001111100101.00111000}_{3E5.38)_{16}}$$

Drill 2-11

Use the methods just described to find the hexadecimal equivalent of $117.446)_8$.
Answer: 4F.93.

2-9   COMPLEMENTS AND COMPLEMENT SUBTRACTION

In most computers, subtraction is accomplished by performing certain addition operations. When the difference $X - Y$ is required, the computer forms the *complement* of the subtrahend Y and adds that complement to X. Thus, the complement of Y is the number that added to another number X produces the difference $X - Y$. There is more than one kind of complement, and—depending on the kind used—other operations may be necessary to obtain the true difference $X - Y$ from the complement method.

When using the complement method, it is important that both numbers, X and Y, be expressed using the same number of digits.

Decimal Complements

Although arithmetic operations are performed in computers using binary numbers, a kind of hybrid decimal arithmetic involving *binary-coded decimals* is sometimes encountered. For that reason and because the study of decimal complements introduces a new concept in terms of a familiar number system, we begin our study of the complement method using decimal numbers. The 9's *complement* of a decimal number is obtained by subtracting each digit of the number from 9. For example, the 9's complement of 147 is $999 - 147 = 852$, and the 9's complement of 53.28 is 46.71.

The difference $X - Y$ is found by adding the 9's complement of Y to X. *The carry digit produced in the most significant column of this addition is added to the least significant digit of the sum.*

Example 2–12

Use 9's-complement subtraction to find each of the following.

a.   762
    $- 143$

b.   501
    $- 85$

c.   385.16
    $- 72.9$

Solution

a. The 9's complement of 143 is $999 - 143 = 856$.

$$
\begin{array}{r}
762 \\
-143 \\
\hline
619
\end{array}
\quad\Longrightarrow\quad
\begin{array}{r}
762 \\
+\ 856 \\
\hline
\text{carry}\ \ \textcircled{1}\,618 \\
+\ \ \longrightarrow 1 \\
\hline
619
\end{array}
$$

Note that the carry produced in the most significant column is added to the least significant digit to obtain the true difference, 619.

b. Since both numbers must be expressed using the same number of digits, we insert a leading 0 in the subtrahend: 085. The 9's complement is then $999 - 085 = 914$.

$$
\begin{array}{r}
501 \\
-\ 85 \\
\hline
416
\end{array}
\quad\Longrightarrow\quad
\begin{array}{r}
501 \\
+\ 914 \\
\hline
\textcircled{1}\,415 \\
+\ \longrightarrow 1 \\
\hline
416
\end{array}
$$

c. Leading and trailing 0s are added to the subtrahend, so the subtraction is expressed as $385.16 - 072.90$. The 9's complement of 072.90 is $999.99 - 072.90 = 927.09$.

$$
\begin{array}{r}
385.16 \\
-\ 072.90 \\
\hline
312.26
\end{array}
\quad\Longrightarrow\quad
\begin{array}{r}
385.16 \\
+\ 927.09 \\
\hline
\textcircled{1}\,312.25 \\
+\ \longrightarrow 1 \\
\hline
312.26
\end{array}
$$

Notice that the carry digit is added to the least significant digit of the sum, which, in this example, is in the 10^{-2} position. Also note that this operation is quite different from adding 1 to the sum (a common error).

Drill
2–12

Use 9's-complement subtraction to find $1097.6)_{10} - 64.38)_{10}$.
Answer: $1033.22)_{10}$.

The 10'*s complement* of a decimal number is found by adding the digit 1 to the least significant digit of the 9's complement. When the subtraction $X - Y$ is performed by adding the 10's complement of Y to X, *the carry in the most significant position is discarded (ignored)*. Following are two examples:

1. $73 - 34$

$$
\begin{array}{rr}
\text{9's complement of 34} = & 65 \\
+ & 1 \\
\hline
\text{10's complement of 34} = & 66 \\
\end{array}
$$

$$
\begin{array}{rr}
73 \\
- 34 \\
\hline
39 \\
\end{array}
\Longrightarrow
\begin{array}{rr}
73 \\
+ 66 \\
\hline
\cancel{1}\,39 \\
\end{array}
$$
└────── Discard carry.

2. $105.2 - 28.96 = 105.20 - 028.96$

$$
\begin{array}{rr}
\text{9's complement of 028.96} = & 971.03 \\
+ & 1 \\
\hline
\text{10's complement of 028.96} = & 971.04 \\
\end{array}
$$

$$
\begin{array}{rr}
105.20 \\
- 028.96 \\
\hline
076.24 \\
\end{array}
\Longrightarrow
\begin{array}{rr}
105.20 \\
+ 971.04 \\
\hline
\cancel{1}\,076.24 \\
\end{array}
$$
└────── Discard carry.

Binary Complements

The 1'*s complement* of a binary number is found by subtracting each bit from 1. For example, the 1's complement of 101101.10 is $111111.11 - 101101.10 = 010010.01$. Notice that the 1's complement can be formed very easily by simply changing every 0 to a 1 and every 1 to a 0. Thus, the 1's complement of 101.01 is seen immediately to be 010.10. By now, the reader may have wondered how the complement method avoids the need for subtraction, since the complement of a number must itself be found by subtraction. We see now that the 1's complement of a binary number can be found by changing each bit to its opposite, a quick and easy operation in a computer, so no formal subtraction operation is required. Changing a bit to its opposite is called *complementing* the bit.

As with 9's-complement decimal subtraction, a carry produced in the most significant position is added to the least significant bit of the sum. This modification of the sum to produce the true difference between two binary numbers is often called *end-around carry*. Following is an example:

$$
\begin{array}{r}
11010 \\
- \;01101 \\
\hline
01101
\end{array}
\qquad \Longrightarrow \qquad
\begin{array}{r}
11010 \\
+ \;10010 \longleftarrow \text{1's complement of 01101} \\
\hline
①\;01100 \\
+ \quad\longrightarrow 1 \longleftarrow \text{end-around carry} \\
\hline
01101
\end{array}
$$

The *2's complement* of a binary number is found by adding a 1 to the least significant bit of the 1's complement. For example, to find the 2's complement of 10011.10 we write

$$
\begin{array}{rr}
\text{1's complement of } 10011.10 = & 01100.01 \\
+ & 1 \\
\hline
\text{2's complement of } 10011.10 = & 01100.10
\end{array}
$$

Another way to find the 2's complement of a binary number is to use the following procedure: Beginning at the least significant bit and moving left, write down each bit exactly as it is, up to and including the first 1 encountered. Complement every bit thereafter.

Using this method in the previous example, we find

$$
\begin{array}{l}
10011.10 = \text{original number} \\
01100.10 = \text{2's complement}
\end{array}
$$

These bits are —————— complements of original bits.
———— These bits are identical to original bits.

When subtraction is performed using the 2's complement method, the carry generated in the most significant position is discarded.

Example
2–13

Use the 2's complement method of subtraction to compute $1100110.11 - 11101.1$.

Solution. Expressing each number with the same number of bits, we must find $1100110.11 - 0011101.10$. The 2's complement of 0011101.10 is found as follows:

$$
\begin{array}{rr}
\text{1's complement of } 0011101.10 = & 1100010.01 \\
+ & 1 \\
\hline
\text{2's complement of } 0011101.01 = & 1100010.10
\end{array}
$$

(As an exercise, verify this result using the alternative method for finding a 2's complement.)

$$
\begin{array}{r}
1100110.11 \\
- \;0011101.10 \\
\hline
1001001.01
\end{array}
\qquad \Longrightarrow \qquad
\begin{array}{r}
1100110.11 \\
+ \;1100010.10 \\
\hline
⊗\;1001001.01
\end{array}
$$

———— Discard carry.

Drill
2–13

Use the 2's complement method of subtraction to compute $10000 - 0.0001$.
Answer: 01111.1111.

In all the subtraction examples presented so far, a carry has been generated in the most significant position of the sum. The reason for this consistent result is that every subtraction has produced a positive difference—i.e., the minuend has been greater than the subtrahend. To investigate cases in which the difference is negative, we must be able to represent *signed numbers*, which we discuss in the next section.

Octal and Hexadecimal Subtraction Using the Complement Method

One reasonably quick way to perform octal and hexadecimal subtraction is to convert the numbers to binary, perform the subtraction, and convert the result back to octal or hexadecimal. However, the complement methods already described for binary and decimal subtraction can also be used. The 7's and 8's complements for octal numbers and 15's and 16's complements for hexadecimal numbers are found and used in ways that should be obvious now. For example, the 7's complement of an octal number is found by subtracting each digit from 7, and the carry produced in the addition operation is added to the least significant digit of the sum. The 8's complement is found by adding a 1 to the least significant digit of the 7's complement, and the carry in the sum is discarded. The 15's and 16's complements of hexadecimal numbers are found and used similarly. The following example shows the computation of $62.5)_8 - 13.4)_8$ using the 7's complement method:

$$
\begin{array}{r}
62.5)_8 \\
-\ 13.4)_8 \\
\hline
47.1)_8
\end{array}
\implies
\begin{array}{r}
62.5)_8 \\
+\ \ \ 64.3)_8 \longleftarrow \text{7's complement} \\
\hline
①\ \ 47.0)_8 \quad \text{of } 13.4)_8 \\
+\ \ \ \longrightarrow 1 \\
\hline
47.1)_8
\end{array}
$$

2–10   SIGNED NUMBERS

The bit in the most significant position of a binary number can be interpreted as a *sign bit*, indicating whether the number represented by the remaining bits is positive or negative:

$$\text{sign bit} = 0 \implies \text{positive}$$
$$\text{sign bit} = 1 \implies \text{negative}$$

The choice of 0 for positive and 1 for negative is not arbitrary. In fact, this choice makes it possible to add sign bits in binary addition, just as other bits are added, and thereby to obtain the correct sign bit for the sum. Thus, we emphasize that the most significant bit *can* (optionally) be interpreted as a sign bit or can serve its usual role as the most significant magnitude bit of a binary number. Addition operations in a computer are performed the same way on all binary numbers, and it is left to the user to interpret the results as representing signed or unsigned quantities. These ideas will become clear after some additional discussion and illustrative examples.

When the most significant bit is interpreted as a sign bit, a 0 in that position means that the remaining bits represent the true magnitude of a number. For example,

$$
\text{sign bit} \longrightarrow \underset{+\ \ \ \ 75)_{10}}{0\underbrace{1001011}} = +75)_{10}
$$

On the other hand, when the sign bit is 1, the remaining bits represent the (1's or 2' *complement* of the number. For example, in 1's-complement arithmetic,

sign bit ⟶

$$1\underbrace{1001011}_{} = -52)_{10}$$

1's complement
of $0110100 = 52)_{10}$

Notice that, as an *unsigned* number, 11001011 could also be interpreted to represent $203)_{10}$.

In 2's complement arithmetic, $-52)_{10}$ is expressed as

sign bit ⟶

$$1\underbrace{1001100}_{} = -52)_{10}$$

2's complement
of $0110100 = 52)_{10}$

Whether negative numbers are represented in 1's- or 2's-complement form depends on the kind of arithmetic operations performed by a given computer—that is, on whether it is designed for 1's- or 2's-complement arithmetic. Most modern computers are designed for 2's-complement operation. Notice that positive numbers are represented identically in both kinds of arithmetic: a 0 sign bit followed by the true magnitude bits.

As previously indicated, sign bits are added in the same way as other bits. For example, the addition $(+87) + (-26) = +61$ is performed in 8-bit 2's-complement arithmetic as follows:

sign
bits
↓

$$\begin{array}{r} 01010111 = +87 \\ 11100110 = -26 \\ \hline \cancel{①}\,00111101 = +61 \end{array}$$

As usual in 2's-complement operations, the carry bit generated in the most significant position is discarded. Notice that the sign bit of the sum (0, for positive) is then automatically correct. Also notice that the addition in this example *could* be interpreted as addition of the unsigned numbers $01010111 = 87$ and $11100110 = 230$. With that interpretation, the carry out of the most significant position indicates an *overflow*. Nine bits are required to represent $87 + 230 = 317 = 100111101$.

It is clear that a subtraction problem, $X - Y$, can be treated as the addition of two signed numbers: $(+X) + (-Y)$. Furthermore, the use of signed numbers allows us to obtain negative differences, which occur when $Y > X$, and to obtain a negative sum when two negative numbers are added. When a sum resulting from the addition of 1's-complement or 2's-complement numbers is negative, the sign bit of the sum is 1 and *the magnitude of the sum automatically appears in complement form.* The next example illustrates such cases in both 1's- and 2's-complement arithmetic.

Example 2-14

Perform each of the following computations using signed, 8-bit words (7 magnitude bits plus a sign bit) in 1's-complement and 2's-complement binary arithmetic:

a. $(+95)_{10} + (-63)_{10}$
b. $(+42)_{10} + (-87)_{10}$
c. $(-13)_{10} + (-59)_{10}$
d. $(+38)_{10} + (-38)_{10}$
e. $(-105)_{10} + (-120)_{10}$

Solution

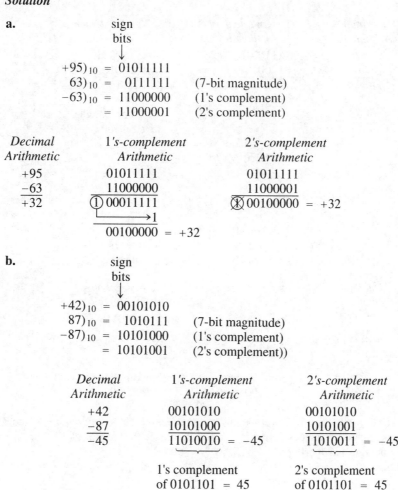

a.

sign bits

$$+95)_{10} = 01011111$$
$$63)_{10} = 0111111 \quad \text{(7-bit magnitude)}$$
$$-63)_{10} = 11000000 \quad \text{(1's complement)}$$
$$= 11000001 \quad \text{(2's complement)}$$

Decimal Arithmetic	*1's-complement Arithmetic*	*2's-complement Arithmetic*
+95	01011111	01011111
−63	11000000	11000001
+32	① 00011111	⊗ 00100000 = +32
	⟶1	
	00100000 = +32	

b.

sign bits

$$+42)_{10} = 00101010$$
$$87)_{10} = 1010111 \quad \text{(7-bit magnitude)}$$
$$-87)_{10} = 10101000 \quad \text{(1's complement)}$$
$$= 10101001 \quad \text{(2's complement))}$$

Decimal Arithmetic	*1's-complement Arithmetic*	*2's-complement Arithmetic*
+42	00101010	00101010
−87	10101000	10101001
−45	11010010 = −45	11010011 = −45
	1's complement of 0101101 = 45	2's complement of 0101101 = 45

Notice that the sign bit is 1 in each sum, confirming a negative result, and that the magnitude of each sum therefore appears in complement form. Also note that the true magnitude of a number appearing in complement form can be obtained by finding the complement of that complement. In other words, the (1's or 2's) complement of the complement of a number equals the original number.

c.

sign
bits
↓

$$13)_{10} = 0001101 \quad \text{(7-bit magnitude)}$$
$$-13)_{10} = 11110010 \quad \text{(1's complement)}$$
$$= 11110011 \quad \text{(2's complement)}$$
$$59)_{10} = 0111011 \quad \text{(7-bit magnitude)}$$
$$-59)_{10} = 11000100 \quad \text{(1's complement)}$$
$$= 11000101 \quad \text{(2's complement)}$$

Decimal Arithmetic	*1's-complement Arithmetic*	*2's-complement Arithmetic*
-13	11110010	11110011
-59	11000100	11000101
-72	① 10110110	⊗ 10111000 = -72
	└──→ +1	
	10110111 = -72	2's complement of 1001000 = 72

1's complement
of 1001000 = 72

d.

sign
bits
↓

$$+38)_{10} = 00100110$$
$$+38)_{10} = 11011001 \quad \text{(1's complement)}$$
$$= 11011010 \quad \text{(2's complement)}$$

Decimal Arithmetic	*1's-complement Arithmetic*	*2's-complement Arithmetic*
$+38)_{10}$	00100110	00100110
$-38)_{10}$	11011001	11011010
$0)_{10}$	11111111 = -0	⊗ 00000000 = $+0$

1's complement
of 0000000 = 0

2's complement
of 0000000 = 0

In this example, 1's complement arithmetic produces a *negative zero,* and 2's complement arithmetic produces a *positive zero.* Of course, the magnitude is zero in either case, and the sign is superfluous.

e.

sign
bits
↓

$$105)_{10} = 1101001 \quad \text{(7-bit magnitude)}$$
$$-105)_{10} = 10010110 \quad \text{(1's complement)}$$
$$= 10010111 \quad \text{(2's complement)}$$
$$120)_{10} = 1111000 \quad \text{(7-bit magnitude)}$$
$$-120)_{10} = 10000111 \quad \text{(1's complement)}$$
$$= 10001000 \quad \text{(2's complement)}$$

Decimal Arithmetic	1's-complement Arithmetic	2's-complement Arithmetic
−105	10010110	10010111
−120	10000111	10001000
−225	① 00011101	⊗ 00011111 = +31
	└──→ +1	
	00011110 = +30	

Notice that the sums resulting from both the 1's- and 2's-complement arithmetic are *incorrect*. The reason for these results is that the largest number that can be represented by the 7 magnitude bits is $1111111 = 2^7 - 1 = 127)_{10}$. Since the correct magnitude, $225)_{10}$, requires a minimum of 8 bits, an overflow has occurred.

Drill 2–14 Perform the computation $(-16,830)_{10} + (-977)_{10}$ using signed 16-bit words (15 magnitude bits plus a sign bit) in 1's-complement and 2's-complement binary arithmetic. *Answer:* 1's complement: 1011101001110000; 2's complement: 1011101001110001.

EXERCISES

Section 2–3

2.1 Find the decimal value of each of the following binary numbers.
(a) 1001
(b) 101011
(c) 1111111
(d) 110.01
(e) 11011101.110

2.2 Find the decimal value of each of the following binary numbers.
(a) 1110111
(b) 10000101
(c) 11011.011
(d) 0.0111
(e) 1111110.1010

Section 2–4

2.3 Perform each of the following additions in binary. Verify each sum by performing the addition using decimal equivalents.

(a)   101101
    +010000

(b)   1001101
    +1010101

(c)   111111
    +100111

(d)   101101.111
    +   111.101

(e) $(110.11) + (101.1) + (1.01)$

2.4 Perform each of the following additions in binary. Verify each sum by performing the addition using decimal equivalents.

(a)   1001011
    +0001001

(b)   111111
    +101110

(c) $(101.01) + (11100.1)$
(d) $(1011110.1) + (111.001)$
(e) $(0.1011) + (0.11) + (0.01)$

2.5 Perform each of the following subtractions in binary. Check each result in binary by verifying that the subtrahend plus the difference equals the minuend.

(a)   1101
    −0100

(b)   1011010
    −0010101

(c) $(10001) - (110)$
(d) $(1100.01) - (11.11)$
(e) $(101010.11) - (1111.1)$

2.6 Perform each of the following subtractions in binary. Check each result in binary by verifying that the subtrahend plus the difference equals the minuend.

(a)   110111
      − 100011

(b)   1001.10
      − 0101.01

(c) $(10101.1) - (1011.11)$

(d) $(0.1001) - (0.011)$

(e) $(10000.01) - (1111.1)$

2.7 Perform each of the following multiplications in binary. Verify each product by performing the multiplication using decimal equivalents.

(a)   101
      × 101

(b)   1101
      × 1001

(c)   101.10
      ×   1.11

(d)   1.0011
      × 0.0011

2.8 Perform each of the following multiplications in binary. Verify each product by performing the multiplication using decimal equivalents.

(a)   101
      × 111

(b)   1101
      × 1110

(c) $(1011.1) \times (11.01)$

(d) $(11110) \times (0.111)$

2.9 Perform each of the following divisions in binary. Carry each division as far out as necessary to obtain a terminating fraction (if any) in the quotient. Check each result by verifying that the binary product of the quotient and the divisor equals the dividend.

(a) 101⟌1111   (c) 100⟌11010

(b) 111⟌1011011   (d) 10⟌11110.1

2.10 Perform each of the following divisions in binary. Carry each division as far out as necessary to obtain a terminating fraction (if any) in the quotient. Check each result by verifying that the binary product of the quotient and the divisor equals the dividend.

(a) 110⟌111100   (c) 100⟌100111

(b) 100⟌1001000   (d) 101⟌100000110.1

Section 2–5

2.11 Find the decimal value of each of the following octal numbers.

(a) $35)_8$

(b) $507)_8$

(c) $64.24)_8$

(d) $1100.4)_8$

(e) $4)_8$

2.12 Find the decimal value of each of the following octal numbers.

(a) $42)_8$

(b) $3165)_8$

(c) $77.77)_8$

(d) $0.064)_8$

(e) $3)_8$

2.13 Perform each of the following additions in octal. Check each sum by performing the addition using decimal equivalents.

(a) $125)_8 + 351)_8$

(b) $53)_8 + 72)_8$

(c) $14.5)_8 + 65.7)_8$

(d) $777)_8 + 77)_8$

2.14 Perform each of the following additions in octal. Check each sum by performing the addition using decimal equivalents.

(a) $102)_8 + 55)_8$

(b) $265)_8 + 345)_8$

(c) $67.7)_8 + 52.1)_8$

(d) $1001)_8 + 77)_8$

2.15 Perform each of the following multiplications in octal. Check each product by performing the multiplication using decimal equivalents.

(a) $3)_8 \times 15)_8$

(b) $21)_8 \times 35)_8$

(c) $63)_8 \times 57)_8$

(d) $2.4)_8 \times 1.5)_8$

2.16 Perform each of the following multiplications in octal. Check each product by performing the same multiplication using decimal equivalents.

(a) $4)_8 \times 21)_8$

(b) $25)_8 \times 104)_8$

(c) $45)_8 \times 76)_8$

(d) $0.47)_8 \times 3.4)_8$

Section 2–6

2.17 Find the decimal value of each of the following hexadecimal numbers.

(a) $25A)_{16}$
(b) $9)_{16}$
(c) $3BC.0F)_{16}$
(d) $FD5.E)_{16}$

2.18 Find the decimal value of each of the following hexadecimal numbers.
(a) $1B6)_{16}$ (c) $7)_{16}$
(b) $ABC.DE)_{16}$ (d) $F.F)_{16}$

2.19 Perform each of the following additions in hexadecimal. Check each sum by performing the addition using decimal equivalents.
(a) $4)_{16} + 9)_{16}$
(b) $B)_{16} + D)_{16}$
(c) $A53)_{16} + 2E9)_{16}$
(d) $80F.C)_{16} + 13)_{16}$
(e) $13AF6)_{16} + D19)_{16}$

2.20 Perform each of the following additions in hexadecimal. Check each sum by performing the addition using decimal equivalents.
(a) $5)_{16} + 7)_{16}$
(b) $A4)_{16} + 37)_{16}$
(c) $2B5C)_{16} + 97F2)_{16}$
(d) $785)_{16} + A.E)_{16}$
(e) $9D4B)_{16} + FA)_{16}$

Section 2–7

2.21 Write the binary number that results when each of the following is incremented and when each is decremented.
(a) 10011
(b) 100000
(c) 1101111
(d) 0010100

2.22 Write the binary number that results when each of the following is incremented and when each is decremented.
(a) 110111
(b) 100
(c) 001001
(d) 111000

2.23 Write the octal or hexadecimal number that results when each of the following is incremented and when each is decremented.
(a) $777)_8$
(b) $A00)_{16}$
(c) $20FF)_{16}$
(d) $1000)_8$
(e) $1000)_{16}$

2.24 Write the octal or hexadecimal number that results when each of the following is incremented and when each is decremented.
(a) $1077)_8$
(b) $30C0)_{16}$
(c) $3DFF)_{16}$
(d) $40200)_8$
(e) $40200)_{16}$

2.25 The word size in a computer is 3 bytes.
(a) How many different numbers can a word represent?
(b) What is the largest decimal number that can be represented by one word?

2.26 What minimum number of bits should a binary counter have if it must be capable of counting at least $30,000)_{10}$ events?

Section 2–8

2.27 Find the binary equivalent of each of the following decimal numbers.
(a) 43
(b) 100
(c) 365
(d) 82.25
(e) 241.125
(f) 1.625

2.28 Convert each of the following decimal numbers to an equivalent binary number having an 8-bit integer part and a 5-bit fractional part. (Terminate the fractional conversion after 5 bits.)
(a) 182.75
(b) 255.20
(c) 76.34
(d) 15.1

2.29 Convert each of the following decimal numbers to equivalent octal and hexadecimal numbers.
(a) 125
(b) 6046
(c) 25.5
(d) 677.5625

2.30 Convert each of the following decimal numbers to equivalent octal and hexadecimal numbers.
(a) 39
(b) 1208
(c) 100.3125
(d) 1791.25

2.31 Convert each of the following numbers to equivalent binary numbers.
(a) $653)_8$
(b) $40B.6)_{16}$
(c) $A193.E)_{16}$
(d) $0.567)_8$

2.32 Convert each of the following numbers to equivalent binary numbers.
(a) $375)_8$
(b) $9A2.D)_{16}$
(c) $4F3.8C)_{16}$
(d) $1.07)_8$

Section 2–9

2.33 Use the 9's-complement method of subtraction to perform each of the following decimal computations.
(a) $38 - 17$
(b) $642 - 75$
(c) $1056 - 385.2$
(d) $812.45 - 97.1$

2.34 Use the 9's-complement method of subtraction to perform each of the following decimal computations.
(a) $65 - 32$
(b) $574 - 51$
(c) $9085 - 637.4$
(d) $581.075 - 95.2$

2.35 Repeat Exercise 2.34 using the 10's-complement method.

2.36 Repeat Exercise 2.33 using the 10's-complement method.

2.37 Perform each of the following binary subtractions using the 1's-complement method.
(a) $10110 - 10011$
(b) $1001100 - 11001$
(c) $11100.101 - 101.01$
(d) $100000 - 0.11$

2.38 Perform each of the following binary subtractions using the 1's-complement method.
(a) $110011 - 101100$
(b) $1001001 - 11001$
(c) $10111101.1 - 110$
(d) $111111 - 0.01$

2.39 Repeat Exercise 2.38 using the 2's-complement method.

2.40 Repeat Exercise 2.37 using the 2's-complement method.

2.41 Perform each of the following computations using the method indicated.
(a) $A47)_{16} - 6B)_{16}$ (convert numbers to binary and the difference back to hexadecimal)
(b) $462.1)_8 - 357)_8$ (8's complement)
(c) $D08.7)_{16} - 1F.C)_{16}$ (15's complement)
(d) $35D)_{16} - A.4)_{16}$ (16's complement)

2.42 Repeat Exercise 2.41 using the following methods for each part.
(a) 16's complement
(b) 7's complement
(c) Convert numbers to binary and the difference back to hexadecimal.
(d) 15's complement

Section 2–10

2.43 Express each of the following decimal numbers as a signed 8-bit binary number (sign bit plus 7 magnitude bits) as necessary for both 1's- and 2's-complement arithmetic.
(a) -56
(b) $+124$
(c) -68
(d) $+12$
(e) -100

2.44 Express each of the following decimal numbers as a signed 16-bit binary number (sign bit plus 15 magnitude bits) as necessary for both 1's- and 2's-complement arithmetic:
(a) -6438
(b) $+400$
(c) -12
(d) $+15,621$
(e) $-25,000$

2.45 Find the decimal value of each of the following bytes assuming each represents (i) an 8-bit binary number; (ii) a signed number in 1's-complement form; (iii) a signed number in 2's-complement form.
(a) 01011101
(b) 11011111
(c) 11100100
(d) 10001110
(e) 00011010

2.46 Find the decimal value of each of the following hexadecimal numbers by converting each to binary and assuming it represents (i) a

12-bit binary number; (ii) a signed number in 1's-complement form; (iii) a signed number in 2's-complement form.

(a) A3D

(b) 7F4

(c) FFE

(d) 40C

(e) 9B6

2.47 Perform each of the following computations in binary using 8-bit, signed binary numbers in 1's-complement arithmetic and in 2's-complement arithmetic.

(a) $(+75)_{10} + (-16)_{10}$

(b) $(+42)_{10} + (+63)_{10}$

(c) $(-91)_{10} + (+24)_{10}$

(d) $(-83)_{10} + (-25)_{10}$

(e) $(+14)_{10} + (-14)_{10}$

2.48 Each of the following additions is performed by a computer using 8-bit, signed numbers and 2's-complement arithmetic. Find the signed binary and decimal numbers corresponding to each hexadecimal number and the signed 2's-complement, binary sum. Convert the binary sum to hexadecimal and to a signed decimal number.

(a) 3A + 2B

(b) C3 + 10

(c) 2D + D3

(d) AA + AA

CHALLENGING EXERCISES

2.49 The sum of the most and least significant digits of a three-digit decimal integer is 10, and the sum of all three digits is 18. When the integer is divided by its least significant digit, the quotient is 171. What is the number?

2.50 If each bit in a binary number is moved one position to the left and a 0 is inserted in place of the least significant bit of the original number, a new binary number is formed. What is the mathematical relation of this new number to the original number? Explain.

2.51 Perform the following addition in binary.

$$
\begin{array}{r}
1101 \\
1011 \\
1100 \\
0111 \\
+ 1010 \\
\end{array}
$$

2.52 The sum of two binary numbers is 100000 and their difference is 1100. What are the binary numbers?

2.53 Perform the following addition using base-5 arithmetic: $2414)_5 + 3423)_5$.

2.54 How many octal numbers have the same decimal value as hexadecimal numbers having the same symbolic representation as the octal numbers?

2.55 Perform the following addition in hexadecimal.

$$
\begin{array}{r}
AB3 \\
2CD \\
F5E \\
+ C96 \\
\end{array}
$$

2.56 The maximum decimal value to which a 4-digit, base-b counter can count is $255)_{10}$. What is the value of b?

2.57 A base-32 number system that utilizes all the letters of the alphabet is used as a code to send messages in binary form. In this code, the digit A has decimal value 0, B has value 1, C has value 2, etc., up to Z, which has decimal value 25. The remaining six digits in the code are represented by special symbols and punctuation marks. Decode the following message by converting the binary numbers to alphabetic characters. (*Hint:* $32 = 2^5$.)

0011001110   001010111010001   0100010011

2.58 Find two 8-bit binary numbers whose 2's-complement representations are the same as their own binary representations.

2.59 Determine a test based on sign bits alone that will reveal whether the sum of two signed 8-bit numbers represents an overflow.

Chapter 3

LOGIC GATES

3-1   INTRODUCTION

In Chapter 1, we described *logic gates* as the fundamental building blocks of digital systems. Recall that the name *logic gate* is derived from the ability of such a device to make decisions, in the sense that it produces one output level when some combinations of input levels are present and a different output when other combinations are present. In this chapter, we learn that there are just three basic types of gates, each of which is quite elementary. The fact that large-scale digital systems, such as computers, are able to perform very complex logic operations stems from the way these elementary gates are interconnected. The interconnection of gates to achieve prescribed outcomes is called *logic design,* a topic that we study in detail in a later chapter.

Logic gates are electronic devices constructed in a wide variety of forms. Many are embedded in integrated circuits with a vast number of other devices and, as such, are not easily accessible or even identifiable. Gates are also constructed in small scale integrated circuits (SSI), where they appear with a few others of the same type. In these devices, the inputs and outputs of all gates are accessible, in the sense that external connections can be made to them. Thus, gates can be interconnected to perform a variety of logical operations. Because of their accessibility, gates constructed in this manner are often called *discrete* logic gates, despite the fact they appear in integrated circuits. Logic gates embedded in large and very large scale integrated circuits (LSI and VLSI), which are not, therefore, accessible for making external connections, are said to be *dedicated* to specific logic operations. Discrete logic gates are often used to *interface* (interconnect) LSI and VLSI circuits.

We study the various kinds of electronic devices used to construct logic gates in a later chapter. For our purposes now, we concentrate on the *functional* behavior of gates—i.e., on the nature of the logical decisions they perform—without reference to their electronic structure. In that context, we can represent logic gates by certain standard symbols that reveal the logic functions performed rather than the specific electronic devices used to perform them. Schematic diagrams show electronic devices and circuit connections, whereas *logic diagrams* employ these standard gate symbols to reveal decision-making properties only.

3-2   THE AND GATE

The two levels produced by digital logic circuitry are referred to variously as HIGH and LOW, TRUE and FALSE, ON and OFF, or simply 1 and 0. The last-mentioned terminology is preferred when discussing logic functions, as opposed to logic circuitry, particularly because of the association of 1 and 0 with the binary number system. In practice, a 1 may be represented by a high voltage and a 0 by a low voltage (or vice versa), but, as mentioned earlier, those considerations are irrelevant to our investigation of logic *functions*. Therefore, in this and the next two chapters, we use 1 and 0 exclusively to refer to the two possible logic levels, keeping in mind that practical electronic gates operate with two corresponding different voltage levels.

An AND *gate is a device whose output is* 1 *if and only if all its inputs are* 1. In discussing logic functions, it is assumed that any output (or input) that is not 1 must necessarily be 0, and conversely, since there are only two possible levels. Thus, the output of an AND gate is 0 if any one or more of its inputs are 0. AND gates have a single output and can have two or more inputs.

To describe the behavior of a gate more fully, it is convenient to regard inputs and outputs as digital *variables* and to assign them symbols such as A, B, X, and Y. Each variable can have only the values 0 or 1. For example, if A and B are the inputs to an AND gate, we say that the output is 1 if and only if both A *and* B are 1. Using this notation, we can make a list of all possible combinations of values that A and B can have and show the output of the gate corresponding to each input combination. Such a list is called a *truth table*. Following is the truth table for a 2-input AND gate:

Inputs		
A	**B**	**Output**
0	0	0
0	1	0
1	0	0
1	1	1

Truth table for a 2-input AND *gate.*

Note that the truth table confirms our definition of an AND gate: the output is 1 only when A and B are both 1; the output is 0 when any one or more inputs are 0.

The logical AND function of two or more variables is expressed symbolically in the same way that *multiplication* is expressed in algebra—i.e., by using parentheses or a dot or by simply writing the two variables adjacent to each other. Thus, the logical AND function of the variables A and B can be expressed as

$$(A)(B) \quad \text{or} \quad A \cdot B \quad \text{or} \quad AB$$

Each of these is read as "A *and* B" (not "A times B"). The output of a 3-input AND gate can be designated by ABC, read as "A and B and C."

Figure 3.1(*a*) shows the traditional symbol* used to represent an AND gate in a logic

---

*Section 3-10 shows the new (ANSI/IEEE) standard symbol for the AND gate and the standard symbols for other logic gates discussed in this chapter.

Figure 3.1
Logic symbol and corresponding truth table for a 2-input AND gate.

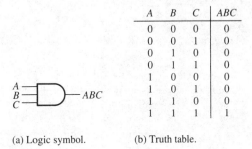

A	B	AB
0	0	0
0	1	0
1	0	0
1	1	1

(a) Logic symbol.   (b) Truth table.

Figure 3.2
Logic symbol and truth table for a 3-input AND gate.

A	B	C	ABC
0	0	0	0
0	0	1	0
0	1	0	0
0	1	1	0
1	0	0	0
1	0	1	0
1	1	0	0
1	1	1	1

(a) Logic symbol.   (b) Truth table.

diagram. Note that the output is labeled *AB*, as is the column of outputs in the truth table shown with it.

The reader may have noticed that the list of input combinations in the truth table is exactly the same as counting in binary from 00 to 11. In this case, counting with two bits up to the maximum possible count produces each and every one of the $2^2 = 4$ possible combinations of values that the input variables can have. In the same way, we can generate all possible input combinations for an *n*-input gate by counting in binary through all 2^n numbers. Figure 3.2 shows the logic symbol and truth table for a 3-input AND gate. Notice that all $2^3 = 8$ input combinations are found by counting from 000 through 111 (i.e., from $0)_{10}$ through $7)_{10}$). Also notice once again that the output is 1 only when all (three) inputs are 1.

3–3 THE OR GATE

An OR gate is a device whose output is 1 if at least one of its inputs is 1. Thus, the output of an OR gate is 0 only when all its inputs are 0. The OR function is symbolized using a + sign. For example, the output of a two-input OR gate is written $A + B$ and read "*A or B*" (not "*A plus B*". The name of the gate is derived from the fact that its output is 1 if either *A or B* (or both) is 1. Figure 3.3 shows the logic symbol and the truth table for a 2-input OR gate.

Of course, it is not necessary to restrict variable names to the letters *A*, *B*, *C*, and so on. In fact, variables are often given names that identify their source or their purpose in a digital system. The next example shows a practical illustration of this fact.

Figure 3.3
Logic symbol and truth table for a 2-input OR gate.

A	B	A + B
0	0	0
0	1	1
1	0	1
1	1	1

(a) Logic symbol.   (b) Truth table.

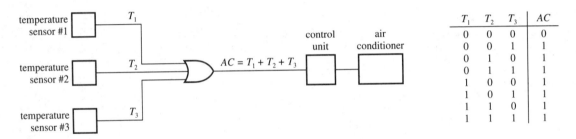

T_1	T_2	T_3	AC
0	0	0	0
0	0	1	1
0	1	0	1
0	1	1	1
1	0	0	1
1	0	1	1
1	1	0	1
1	1	1	1

Figure 3.4
Using an OR *gate to control an air-conditioning system (Example 3–1).*

*Example
3–1*

The air-conditioning unit in a building is to be turned on (by a logical 1 received at its control unit) if the temperature at any one of three locations exceeds 78°F. Assuming that the temperature-sensing devices at the locations each produce a logical 1 when the temperature exceeds 78°F and a logical 0 otherwise, design the logic required to control the air conditioning.

Solution. Let T_1, T_2, and T_3 be the variables representing the outputs of the three temperature sensors. Thus, each of T_1, T_2, and T_3 is 1 if the temperature at the sensor it represents exceeds 78°F and is 0 otherwise.

The problem statement requires that the control unit on the air conditioning receive a 1 if at least one of T_1 or T_2 or T_3 is 1. Thus, a 3-input OR gate is required. If we let AC be the input variable to the control unit, then $AC = T_1 + T_2 + T_3$. Figure 3.4 shows the logic diagram and the truth table for the 3-input OR gate.

*Drill
3–1*

In the interest of economy, it was decided that the air conditioner in Example 3–1 should be turned on only if the temperature at all three locations exceeds 78°F. Draw a logic diagram and truth table similar to Figure 3.4 to show how this control logic can be achieved.
Answer: $AC = T_1T_2T_3$.

3–4   THE INVERTER

An inverter has one input and one output. *It is a device whose output is 1 when its input is 0 and whose output is 0 when its input is 1.* Thus, an inverter *complements* a digital variable. An inverter is sometimes given the (unfortunate) name NOT gate because its output is not the same as its input. We avoid this usage. The complement of a variable is shown by drawing a line, or bar, over the variable name. Thus, the complement of A is designated $\overline{A}$, usually read as "A-bar," "not A," or "A-complement." The variable $\overline{A}$ is

A	$\overline{A}$
0	1
1	0

Figure 3.5
Logic symbol and truth table for an inverter.   (a) Logic symbol.   (b) Truth table.

0 when A is 1 and is 1 when A is 0. Figure 3.5 shows the logic symbol and truth table for an inverter.

3–5   NAND AND NOR GATES

All logic operations can be performed using combinations of the three basic functions we have discussed: AND, OR, and INVERT. The remaining gates we study are merely combinations of those functions. They are so frequently encountered that they are fabricated as if they were themselves fundamental entities—i.e., with outputs that represent certain combinations of logic operations performed on their inputs. NAND and NOR gates are examples.

The output of a NAND gate is 1 if at least one of its inputs is 0. Thus, the output is 0 only when both inputs are 1. The name NAND is derived from NOT AND because the gate performs the same logic as an AND gate followed by an inverter. This fact is demonstrated for a 2-input gate in Figure 3.6. Here, we show that the truth table for an AND gate followed by an inverter is identical to that of the NAND gate. (Two logic operations are equivalent if they produce identical truth tables.) Notice that the NAND operation is symbolized by $\overline{AB}$ and that the logic symbol for a NAND gate is like that of an AND gate with a circle, or "bubble," at the output. The bubble symbol is widely used in logic diagrams to imply inversion.

The output of a NOR gate is 1 if and only if all its inputs are 0. Thus, the output is 0 if at least one of its inputs is 1. A NOR (NOT OR) gate performs the same logic function as an OR gate followed by an inverter. This fact is demonstrated for a 2-input gate in Figure 3.7. Note that the NOR operation is symbolized by $\overline{A + B}$ and that the logic symbol is the same as that of an OR gate with the inversion bubble drawn at the output.

3–6   CONSTRUCTING TRUTH TABLES AND LOGIC EXPRESSIONS FROM LOGIC DIAGRAMS

Figures 3.6(*a*) and 3.7(*a*) show examples of how truth tables can be constructed from logic diagrams containing more than one gate. In these simple cases, we constructed a separate column in the truth table for the output of each gate. We follow the same general procedure to construct truth tables from more complex logic diagrams. Beginning at the

Figure 3.6

A NAND *gate performs the same logic function as an* AND *gate followed by an inverter.*

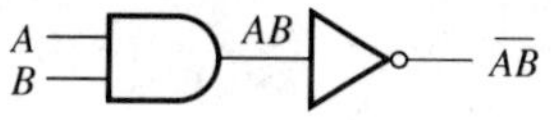

		AND	NAND
A	B	AB	$\overline{AB}$
0	0	0	1
0	1	0	1
1	0	0	1
1	1	1	0

(a) AND gate followed by inverter.

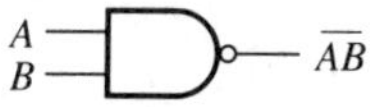

A	B	$\overline{AB}$
0	0	1
0	1	1
1	0	1
1	1	0

(b) Logic symbol and truth table for a NAND gate.

Figure 3.7
A NOR gate performs the same logic functions as an OR gate followed by an inverter.

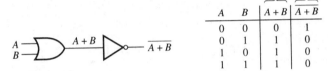

A	B	OR $A+B$	NOR $\overline{A+B}$
0	0	0	1
0	1	1	0
1	0	1	0
1	1	1	0

(a) OR gate followed by inverter.

A	B	$\overline{A+B}$
0	0	1
0	1	0
1	0	0
1	1	0

(b) Logic symbol and truth table for a NOR gate.

gate(s) where the inputs to the system are connected, we construct column(s) showing the first logic operations performed on the inputs. These output(s) are the inputs to other gates, and we construct new columns showing the results of the logic operations performed on these inputs. The process continues until we reach the overall output of the logic diagram. While the analysis is being performed, it is useful to develop an overall logic expression that shows how the overall output is related to the inputs. This expression is developed by gradually expanding—i.e., compounding—the logic operations as they are performed from input to output. For example, if the two inputs to an AND gate are $(A + B)$ and $\overline{C}$, then the output of the AND gate is $(A + B)\overline{C}$. Note that the parentheses are used to show that the *entire* term $(A + B)$ undergoes the AND operation with $\overline{C}$. We say that $(A + B)$ is ANDed with $\overline{C}$.

Example 3–2

Construct a truth table and develop a logic expression for output Z in Figure 3.8(a).

Solution. Since there are three input variables $(A, B, \text{and } C)$, we must construct a truth table showing the $2^3 = 8$ possible combinations of the inputs. As described earlier, these combinations are found by counting in binary from 000 to 111.

The output of the AND gate in Figure 3.8(a) is AB, as shown in Figure 3.8(b). Figure 3.8(c) shows the truth table with a column headed AB. The entries in this column are the results of applying the AND operation to the entries in the input columns headed by A and

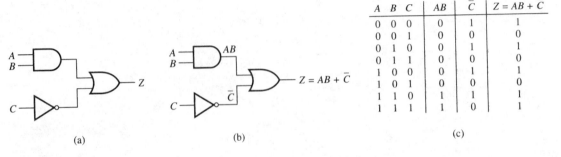

A	B	C	AB	$\overline{C}$	$Z = AB + \overline{C}$
0	0	0	0	1	1
0	0	1	0	0	0
0	1	0	0	1	1
0	1	1	0	0	0
1	0	0	0	1	1
1	0	1	0	0	0
1	1	0	1	1	1
1	1	1	1	0	1

(a)   (b)   (c)

Figure 3.8
Example 3–2.

B. The column headed by $\overline{C}$ shows the results of complementing the entries in the input column headed by *C*. $\overline{C}$ is the output of the inverter, as shown in Figure 3.8*(b)*.

Notice that the output of the AND gate *(AB)* and the output of the inverter *($\overline{C}$)* are the inputs to the OR gate in Figure 3.8*(b)*. We say that *AB* and $\overline{C}$ are ORed together at the output. Thus, $Z = AB + \overline{C}$. The column labeled $Z = AB + \overline{C}$ in the truth table shows the results of applying the OR operation to the columns headed by *AB* and $\overline{C}$. This last column is the overall truth table corresponding to the output: $Z = AB + \overline{C}$.

Drill
3–2

Construct a truth table and develop a logic expression for *Z* when the AND gate and the OR gate in Figure 3.8*(a)* are interchanged.
Answer: $Z = (A + B)\overline{C}$.

Example
3–3

An aircraft engine is equipped with a safety system that turns on a warning light when certain combinations of engine speed, pressure, and temperature occur. The devices that sense these quantities produce a 1 or a 0 according to the following table:

		Output
Speed *(S)*	$S < 5000$ rpm	0
	$S \geq 5000$ rpm	1
Pressure *(P)*	$P \leq 200$ psi	1
	$P > 200$ psi	0
Temperature *(T)*	$T > 180°F$	0
	$T \leq 180°F$	1

Figure 3.9*(a)* shows the logic diagram that controls the warning light in response to input variables *S, P,* and *T*. Assuming that a 1 turns on the light, develop the overall logic expression and construct a truth table to determine if the warning light is on when

a. the speed is 6250 rpm, the pressure is 280 psi, and the temperature is 150°F;
b. the speed is 7400 rpm, the pressure is 180 psi, and the temperature is 200°F.

Solution.   Figure 3.9*(b)* shows the development of the logic expression. Letting *W* represent the output that turns on the warning light, we see that $W = (P + \overline{S})\overline{T}$.

The truth table is shown in Figure 3.9*(c)*. The column headed $P + \overline{S}$ shows the results when input *P* is ORed with $\overline{S}$. Notice that $(P + \overline{S})$ is then ANDed with $\overline{T}$ to produce output *W*.

Referring to the table showing the conditions under which each variable is 0 or 1, we see that

$$\textbf{(a)} \quad \left. \begin{array}{lll} S = 6250 \text{ rpm} & \rightarrow S = 1 \\ P = 280 \text{ psi} & \rightarrow P = 0 \\ T = 150°F & \rightarrow T = 1 \end{array} \right\} \quad SPT = 101$$

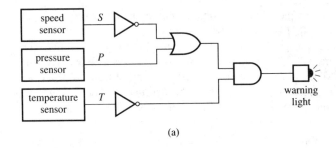

(a)

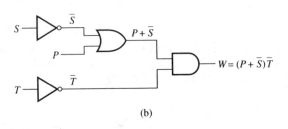

(b)

S	P	T	$\bar{S}$	$P+\bar{S}$	$\bar{T}$	$W=(P+\bar{S})\bar{T}$
0	0	0	1	1	1	1
0	0	1	1	1	0	0
0	1	0	1	1	1	1
0	1	1	1	1	0	0
1	0	0	0	0	1	0
1	0	1	0	0	0	0
1	1	0	0	1	1	1
1	1	1	0	1	0	0

(c)

Figure 3.9
Example 3–3.

(b) $\left.\begin{array}{l} S = 7400 \text{ rpm} \rightarrow S = 1 \\ P = 180 \text{ psi} \quad \rightarrow P = 1 \\ T = 200°F \quad \rightarrow T = 0 \end{array}\right\}$ $SPT = 110$

Referring to the truth table for W, we find that $W = 0$ when $SPT = 101$, so the warning light is off under the conditions specified in (a). On the other hand, $W = 1$ when $SPT = 110$, so the warning light is on under the conditions specified in (b).

Drill
3–3

Determine the logic expression for W and construct its truth table when the AND gate and the OR gate in Figure 3.9(a) are interchanged.
Answer: $W = \bar{S}P + \bar{T}$.

3–7 CONSTRUCTING LOGIC DIAGRAMS AND TRUTH TABLES FROM LOGIC EXPRESSIONS

In a typical design problem, a logic expression is derived first, and a logic diagram that *implements* (performs) the specified logic operations must be constructed. This process is straightforward, in that we simply draw a gate for each logic operation contained in the expression and make certain that the correct variables or combinations of variables are operated on by those gates. Truth tables can be constructed directly from the logic expression or from the logic diagram. The procedure is best explained by way of an example.

Example
3–4

Construct a truth table and draw a logic diagram that implements the expression $W = \overline{X + Y} + X\bar{Z}$.

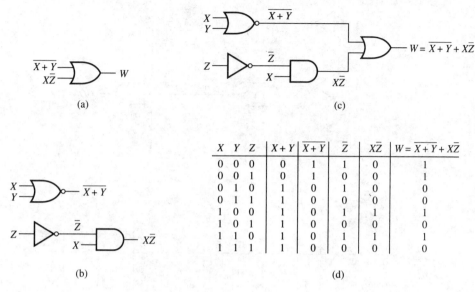

Figure 3.10
Example 3-4.

Solution. It is apparent that W will be the output of an OR gate having the two inputs $\overline{X + Y}$ and $X\overline{Z}$, as shown in Figure 3.10(a). Thus, we must construct logic gates that produce those two inputs. The term $\overline{X + Y}$ is produced by a NOR gate whose inputs are X and Y. The term $X\overline{Z}$ is produced by an AND gate whose inputs are X and $\overline{Z}$, and $\overline{Z}$ is obtained by inverting Z. The implementation of these terms is shown in Figure 3.10(b). In Figure 3.10(c), we show how the gates are finally interconnected to implement the overall logic expression for W.

Figure 3.10(d) shows the truth table for W, which is constructed following the procedures described earlier.

**Drill
3-4**

Draw a logic diagram that implements the expression $Q = (\overline{AB})(A + \overline{C})$.

**Example
3-5**

A computer circuit performs a logic operation on the 3-bit binary number $A_2A_1A_0$, where A_2 is the most significant bit and A_0 is the least significant bit. If the number has certain decimal values, the circuit produces a GO signal that launches a rocket. The circuit produces a 1 in accordance with the expression

$$GO = \overline{(\overline{A_2} + A_1 + A_0)(A_2 + \overline{A_1} + \overline{A_0})(A_2 + A_1 + \overline{A_0})}$$

Draw the logic diagram and construct the truth table. Determine which decimal values the number must have in order for the rocket launch to occur ($GO = 1$).

Solution. The logic expression shows that GO is the output of a 3-input NAND gate. The terms that are NANDed together are $(\overline{A_2} + A_1 + A_0)$, $(A_2 + \overline{A_1} + \overline{A_0})$, and $(A_2 + A_1 + \overline{A_0})$. Each of these is produced by a 3-input OR gate. To simplify the notation, we let $X = (\overline{A_2} + A_1 + A_0)$, $Y = (A_2 + \overline{A_1} + \overline{A_0})$, and $Z = (A_2 + A_1 + \overline{A_0})$. The logic diagram is shown in Figure 3.11(a). Note that $GO = \overline{XYZ}$.

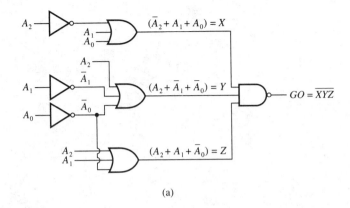

(a)

$A_2\ A_1\ A_0$	$\overline{A}_0$	$\overline{A}_1$	$\overline{A}_2$	$X =$ $(\overline{A}_2 + A_1 + A_0)$	$Y =$ $(A_2 + \overline{A}_1 + \overline{A}_0)$	$Z =$ $(A_2 + A_1 + \overline{A}_0)$	$GO =$ $\overline{XYZ}$
0 0 0	1	1	1	1	1	1	0
0 0 1	0	1	1	1	1	0	1
0 1 0	1	0	1	1	1	1	0
0 1 1	0	0	1	1	0	1	1
1 0 0	1	1	0	0	1	1	1
1 0 1	0	1	0	1	1	1	0
1 1 0	1	0	0	1	1	1	0
1 1 1	0	0	0	1	1	1	0

(b)

Figure 3.11
Example 3–5.

Figure 3.11(*b*) shows the truth table. We see that GO equals 1 when $A_2 A_1 A_0 = 001$, 011, or 100. Thus, the decimal values that launch the rocket are $1)_{10}$, $3)_{10}$, and $4)_{10}$.

Drill 3–5

What decimal values of $A_2 A_1 A_0)_2$ will launch the rocket in Example 3–5 if $GO = (A_2 + \overline{A}_1 + A_0)(\overline{A}_2 + A_1 + \overline{A}_0)(\overline{A}_2 + \overline{A}_1 + \overline{A}_0)$?
Answer: 2, 5, and 7.

3–8 EXCLUSIVE-OR AND COINCIDENCE GATES

The exclusive-OR gate is another example of a gate that implements a widely used combination of the three fundamental logic functions: AND, OR, and INVERT. Unlike previously defined gates, the exclusive-OR gate always has *two* inputs. The output of an exclusive-OR gate is 1 if *exactly one* of its two inputs is 1. Thus, the output is 0 if both inputs are 0 or if both inputs are 1. The truth table is shown in Figure 3.12(*a*). The name exclusive-OR is derived from the fact that the output is 1 when *exclusively* one input is 1 (it excludes the combination where both inputs are 1). By way of contrast, the OR gate we defined previously is sometimes called an *inclusive*-OR gate because it includes that combination. The symbol for the exclusive-OR function of inputs A and B is $A \oplus B$, and the symbol used to represent it on logic diagrams is shown in Figure 3.12(*b*).

Figure 3.12
Truth table and logic symbol for an exclusive-OR gate.

A	B	A⊕B
0	0	0
0	1	1
1	0	1
1	1	0

(a) Truth table.

$A \oplus B$

(b) Logic symbol.

The next example confirms that the exclusive-OR function is a combination of the fundamental logic functions AND, OR, and INVERT.

Example 3–6 Draw a logic diagram of the function $A\bar{B} + \bar{A}B$. By constructing its truth table, show that it is equivalent to $A \oplus B$.

Solution. $A\bar{B} + \bar{A}B$ is implemented by an OR gate whose inputs are the terms $A\bar{B}$ and $\bar{A}B$, each of which is obtained from an AND gate. Figure 3.13(a) shows the logic diagram.

Figure 3.13(b) shows the truth table. Since the column $A\bar{B} + \bar{A}B$ is identical to the column $A \oplus B$ in the exclusive OR truth table, the two functions are equivalent.

Drill 3–6 Draw the logic diagram for $(A + B)(\overline{AB})$ and show that it also is equivalent to $A \oplus B$.

Like the exclusive OR gate, the *coincidence* gate has only two inputs. The output of a coincidence gate is 1 if and only if both its inputs are 1 or both its inputs are 0. In other words, the output is 1 if the inputs *coincide*. It should be apparent that this logic is exactly complementary to that of the exclusive-OR gate. Thus, a coincidence gate can be implemented by an exclusive-OR gate followed by an inverter. In fact, the coincidence gate is also called an *exclusive-NOR* gate. Its truth table and logic symbol are shown in Figure 3.14, p. 54. The symbolic representation of the coincidence function is $A \odot B$.

Example 3–7 Use truth tables to show that the coincidence function is the complement of the exclusive-OR function. Is the exclusive-OR function the complement of the coincidence function?

Figure 3.13
Demonstrating that $A \oplus B = A\bar{B} + \bar{A}B$ (Example 3–6)

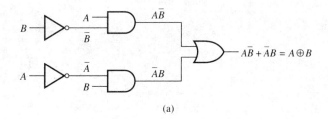

(a)

A	B	$\bar{A}$	$\bar{B}$	$A\bar{B}$	$\bar{A}B$	$A\bar{B}+\bar{A}B$	A	B	A⊕B
0	0	1	1	0	0	0	0	0	0
0	1	1	0	0	1	1	0	1	1
1	0	0	1	1	0	1	1	0	1
1	1	0	0	0	0	0	1	1	0

(b)

Figure 3.14
Truth table and logic symbol for a coincidence (exclusive-NOR) gate.

A	B	$A \odot B$
0	0	1
0	1	0
1	0	0
1	1	1

(a) Truth table.

$$A \odot B$$

(b) Logic symbol.

Figure 3.15
The truth tables show that $\overline{A \oplus B} = A \odot B$ (Example 3–7).

A	B	$A \oplus B$	$\overline{A \oplus B}$
0	0	0	1
0	1	1	0
1	0	1	0
1	1	0	1

A	B	$A \odot B$
0	0	1
0	1	0
1	0	0
1	1	1

Solution.   Figure 3.15 shows the truth table for an exclusive-OR gate followed by an inverter. It is apparent that the result is identical to the truth table of the coincidence gate. If we were to complement the column $A \odot B$ in the truth table for the coincidence gate, we would once again obtain the truth table for the exclusive-OR gate. Thus, $\overline{A \oplus B} = A \odot B$ and $\overline{A \odot B} = A \oplus B$.

Drill 3–7

Draw a logic diagram for the expression $\overline{A}\,\overline{B} + AB$ and show, using truth tables, that it is equivalent to $A \odot B$.

Example 3–8

Derive a logic expession that equals 1 only when the two binary numbers $A_1 A_0$ and $B_1 B_0$ have the same value. Draw the logic diagram and construct the truth table to verify the logic.

Solution.   The numbers $A_1 A_0$ and $B_1 B_0$ have the same value only when their most significant bits (A_1 and B_1) coincide *and* their least significant bits (A_0 and B_0) coincide. Thus, we must AND the coincidence $A_1 \odot B_1$ with the coincidence $A_0 \odot B_0$. Figure 3.16(a) shows the logic diagram for the required expression: $(A_1 \odot B_1)(A_0 \odot B_0)$. Figure 3.16(b) shows the truth table. Note that we have four variables in this example, so we must count in binary from 0000 to 1111 to obtain all 16 possible input combinations. The truth table confirms that the output is 1 only in the four cases where $B_1 B_0 = A_1 A_0$—namely, 00, 01, 10, and 11.

Figure 3.16
Example 3–8.

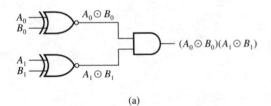

(a)

B_1	B_0	A_1	A_0	$A_0 \odot B_0$	$A_1 \odot B_1$	$(A_0 \odot B_0)(A_1 \odot B_1)$
0	0	0	0	1	1	1
0	0	0	1	0	1	0
0	0	1	0	1	0	0
0	0	1	1	0	0	0
0	1	0	0	0	1	0
0	1	0	1	1	1	1
0	1	1	0	0	0	0
0	1	1	1	1	0	0
1	0	0	0	1	0	0
1	0	0	1	0	0	0
1	0	1	0	1	1	1
1	0	1	1	0	1	0
1	1	0	0	0	0	0
1	1	0	1	1	0	0
1	1	1	0	0	1	0
1	1	1	1	1	1	1

(b)

3-9   INTRODUCTION TO 7400-SERIES LOGIC

In a later chapter, we discuss the circuitry and characteristics of a *family* of integrated circuits called transistor-transistor logic (TTL). This popular series of chips contains a variety of gates whose inputs and outputs are accessible through connections to external *pins* (integrated circuit terminals). The chips are designated by 7400-series numbers, each of which refers to a particular set or combination of logic gates implemented by the chip. For example, the 7402 contains four NOR gates, each of which has two inputs, whereas the 7404 contains six inverters.

The reason for introducing 7400 series logic now is so that you can perform laboratory experiments, if desired, using TTL logic to verify basic concepts. The material that follows is very rudimentary and may be skipped by readers who already have experience using integrated circuit chips.

The voltage levels, called *logic levels,* used to represent 1 and 0 in 7400-series logic are $1 = +5$ V and $0 = 0$ V. It is important to realize that 0 V means *ground,* or *common,* the low side of the $+5$ V power source that supplies dc power to a chip. Zero volts does *not* mean open-circuit or no connection. In fact, if an input to a TTL gate is left open (unconnected), the gate behaves as if a logic 1 were connected to the input. Such inputs are said to "float high."

For making external connections, integrated-circuit chips have pins, or terminals, that are numbered in a standard sequence. Figure 3.17 shows examples. Note that a groove or dot identifies the top of a chip and that the numbers begin at the upper left-hand pin, run down and across (counterclockwise), and end at the upper right-hand pin.

Each pin of an integrated-circuit chip is the terminal for an input or output of a gate or for a power-supply connection. Diagrams in manufacturers' data sheets must be consulted to determine which pin corresponds to which terminal. (Sometimes pins are not used and are designated NC for *no connection.*) All 7400-series chips require a $+5$ Vdc power connection, designated V_{CC}, and a ground connection. Figure 3.18, pp. 56–57,

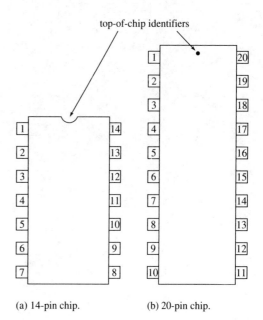

Figure 3.17
Pin numbering system for integrated-circuit chips. Top views shown.

(a) 14-pin chip.   (b) 20-pin chip.

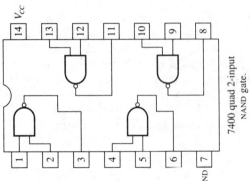

7402 quad 2-input
NOR gate.

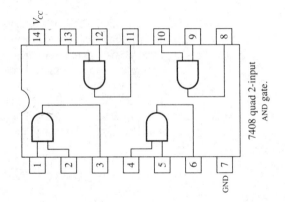

7410 triple 3-input
NAND gate.

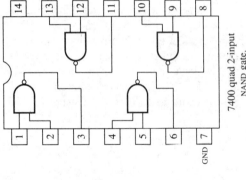

7400 quad 2-input
NAND gate.

7408 quad 2-input
AND gate.

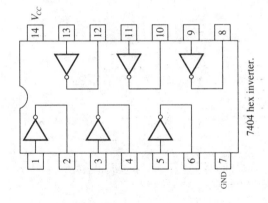

7404 hex inverter.

Figure 3.18
TTL integrated-circuit logic gates

56

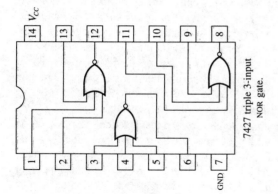

7427 triple 3-input
NOR gate.

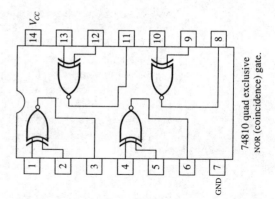

74810 quad exclusive
NOR (coincidence) gate.

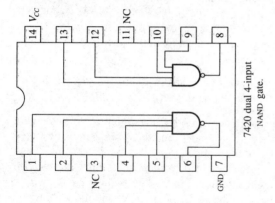

7420 dual 4-input
NAND gate.

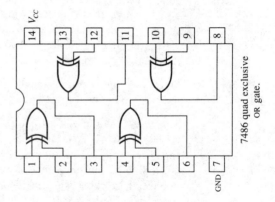

7486 quad exclusive
OR gate.

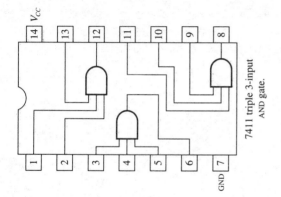

7411 triple 3-input
AND gate.

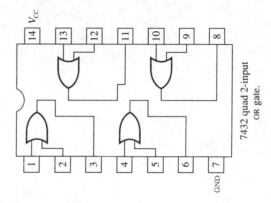

7432 quad 2-input
OR gate.

shows some of the most commonly used 7400-series chips and their pin connections. It should be noted that integrated circuits are available in different *package* styles and that pin assignments vary with the style selected. (Figure 3.18 shows pin assignments used by Texas Instruments in their J- and N-style packages.) Also, the 7400-series chips are available with *open-collector outputs,* which are discussed in a later chapter. These versions require the connection of external resistors and, in general, have pin assignments different from those shown in Figure 3.18. When connecting circuits, consult manufacturers' data sheets for specific information on pin assignments.

In practice, the 7400-series numbers have letters, or groups of letters, that appear as prefixes, suffixes, or even in the middle of the numbers. Prefixes identify the manufacturer. For example, Texas Instruments uses the prefix SN to identify its products, so SN7403 designates a 7403 chip manufactured by that company. Letters appearing immediately after the 74 in the 7400-series designations indicate special characteristics resulting from variations in chip construction. For example, the SN74H01 is a high-power, high-speed version of the 7401. We discuss that variation and others in a later chapter. Suffix letters usually indicate a chip whose performance specifications are slightly different from those of the unsuffixed version. We should note that 7400-series chips are also available in a corresponding 5400 series. These are comparable to the 7400 chips but can operate over a wider temperature range, to meet military specifications.

Although the 7400-series logic levels are *nominally* 0 V and 5 V, experimenters should be aware that measured values of gate output voltages may vary slightly from those values. Depending on the *load* connected to the output, a logic 0 output may in fact be a few tenths of a volt positive, and a logic 1 output may be as low as +2.4 V.

Example 3–9

Draw a wiring diagram using the 7400-series chips with the pin assignments shown in Figure 3.18 to create a logic circuit that implements the expression $X = \overline{AB + CD + EF}$.

Solution. We first construct a logic diagram showing the gates required to implement X. The terms AB, CD, and EF are produced by AND gates, and those terms must be NORed together, as shown in Figure 3.19(a). The three 2-input AND gates can be obtained from a 7408 chip, and the 3-input NOR gate can be obtained from a 7427 chip. One way to connect these chips using the pin assignment data from Figure 3.18 is shown in the wiring diagram in Figure 3.19(b). The wiring diagram shows the *physical* layout of the chips and the actual conducting paths between them. When constructing a circuit in a laboratory setting, a less cluttered connection diagram, such as the one shown in Figure 3.19(c), is preferred. Note that pin numbers are shown in parentheses at each input and output. Since only three of the four AND gates in the 7408 are used, we identify them as ¾ 7408. Similarly, only one of three NOR gates is used, so it is identified as ⅓ 7427.

Drill 3–9

Draw a wiring diagram using the 7400-series chips with the pin assignments shown in Figure 3.18 to create a logic circuit that implements $Z = \overline{AB + CD}$.

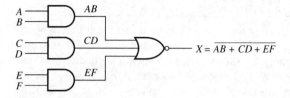

(a) Logic diagram.

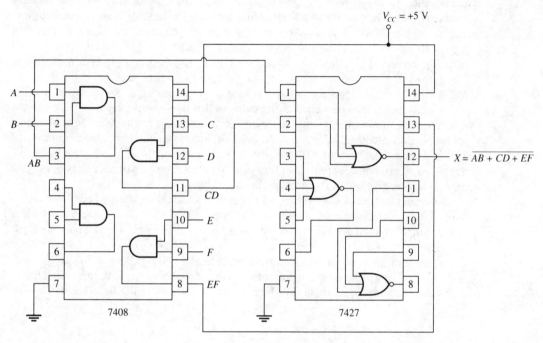

(b) Wiring diagram.

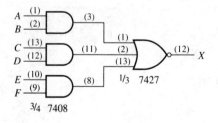

(c) Connection diagram.

Figure 3.19
Example 3-9.

3-10   ANSI/IEEE STANDARD LOGIC SYMBOLS

The symbols used to represent logic gates in this chapter have been in use for many years and will no doubt continue to be used by many practicing engineers and technologists. However, a new standard issued jointly by the American National Standards Institute (ANSI) and the Institute for Electrical and Electronics Engineers (IEEE) is now being used by many industries and manufacturers of electronic devices. Called ANSI/IEEE Standard 91-1984, it was introduced many years ago but has been largely ignored until recent years. Its use is now required in all documentation prepared under U.S. military contracts, and it has begun to appear in the product literature of many manufacturers. Instead of using distinctive symbols for the various types of gates, the new standard depicts all gates in rectangular outlines. Characters called *qualifying symbols* are placed inside the rectangular outlines to indicate the type of logic operations performed. Figure 3.20 shows the representation of an AND gate, an OR gate, and an inverter in the ANSI/IEEE standard. Note that the *ampersand* (&) represents the AND operation, the $\geqslant 1$ notation represents the OR operation (the output is 1 if *one or more* inputs are 1), and a 1 with a triangle drawn at the output represents inversion. Gates with more than two inputs are simply drawn with additional input lines, as in traditional notation. Direction of signal flow is assumed to be from left to right—i.e., inputs on the left and outputs on the right—so no arrowheads are required. In cases where it is desired or necessary to indicate right-to-left signal flow, a left-pointing arrowhead may be used.

NAND and NOR gates are represented in the new standard by obvious combinations of AND, OR, and INVERT symbols, as shown in Figure 3.21(a) and (b). The exclusive-OR gate is a rectangle with the symbol $=1$ drawn inside, since the output is 1 if and only if exactly one input is 1. See Figure 3.21(c). Since the coincidence, or exclusive-NOR, function is the complement of the exclusive-OR, it can be depicted as a $=1$ gate with an inverted output, as shown in 3.21(d). It is also represented by the $=$ symbol, meaning that the output is 1 if both inputs are equal (both 0 or both 1), as also shown in the figure.

Figure 3.20
The ANSI/IEEE standard symbols for AND *gates,* OR *gates, and inverters*.

Traditional symbol.   ANSI/IEEE symbol.

(a) AND gate.

Traditional symbol.   ANSI/IEEE symbol.

(b) OR gate.

Traditional symbol.   ANSI/IEEE symbol.

(c) Inverter.

Figure 3.21
The ANSI/IEEE standard symbols for NAND, NOR, *exclusive-*OR, *and coincidence gates*.

Traditional symbol.   ANSI/IEEE symbol.

(a) NAND gate.

Traditional symbol.   ANSI/IEEE symbol.

(b) NOR gate.

Traditional symbol.   ANSI/IEEE symbol.

(c) Exclusive- OR gate.

Traditional symbol.   Alternative
                             ANSI/IEEE symbols.

(d) Coincidence, or exclusive-NOR, gate.

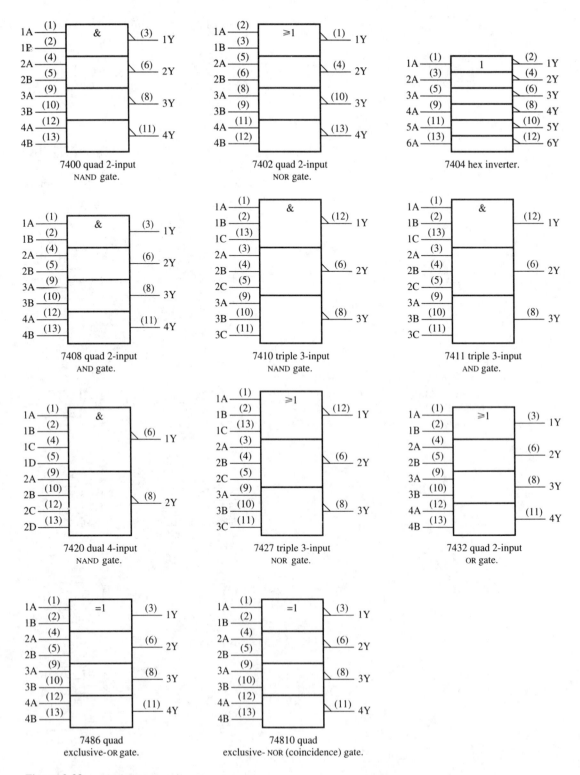

Figure 3.22
ANSI/IEEE standard symbols for 7400-series integrated-circuit gates.

62

The ANSI/IEEE logic symbols for integrated circuits containing multiple logic gates are drawn by stacking rectangles on top of each other, one rectangle for each gate. Figure 3.22 shows the ANSI/IEEE symbols for each of the 7400-series circuits whose pin diagrams are shown in Figure 3.18. Note that a single qualifying symbol is used for each circuit. The logic operation represented by the symbol is assumed to apply to all rectangles (gates) lying below the one in which it appears. Pin numbers for each input and output are shown in parentheses.

Example 3–10

Draw a logic diagram using ANSI/IEEE standard symbols to implement the function $X = \overline{(A + B)}\ \overline{(C + D)}\overline{(E + F)}$. Also draw a connection diagram using ANSI/IEEE symbols for 7400-series integrated circuits.

Solution.   See Figure 3.23.

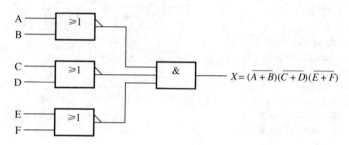

(a) Logic diagram.

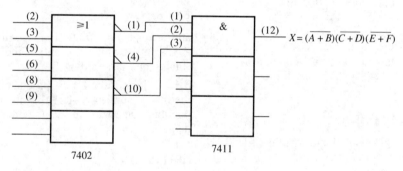

(b) Connection diagram.

Figure 3.23
Example 3–10.

EXERCISES

Sections 3–1 to 3–3

3.1 Draw the logic symbol and construct the truth table for an AND gate having inputs A, B, C, and D.

3.2 An AND gate has inputs V, W, X, Y, and Z. How many different combinations of input levels are possible?

3.3 An OR gate has inputs A_0, A_1, A_2, A_3, A_4, and A_5. How many different combinations of input levels are possible?

3.4 Draw the logic symbol and construct the truth table for an OR gate having inputs B, N, R, and V.

3.5 Each of two temperature sensors produces a logical 0 when the temperature at its location, T_1 or T_2, is above 80°F. A heater is to be turned off by a logical 0 received at its control unit when the temperature at either or both locations is above 80°F. Draw a logic diagram showing the kind of gate that should be used to control the heater.

3.6 Each of three pressure sensors produces a logical 0 if the pressure at its location, P_1, P_2, or P_3, falls below 25 psi. A pump is to be turned on by a logical 0 received at its control unit if the pressure at all three locations falls below 25 psi. Draw a logic diagram showing the kind of gate that should be used to control the pump.

Sections 3–4 to 3–6

3.7 Draw the logic symbol and write the truth table for the operation $\overline{X_1 X_2 Y_1 Y_2}$.

3.8 Draw the logic symbol and write the truth table for the variable UP, given that $UP = \overline{A + X + D + M}$.

3.9 Write the logic expression for OUT in each of the logic diagrams shown in Figure 3.24. Construct a truth table for each logic expression.

3.10 Write the logic expression for ON in each of the logic diagrams shown in Figure 3.25. Construct the truth table for each logic expression.

3.11 Let SUNSHINE be 0 if the sun is shining and 1 if it is not. Let RAIN be 0 if it is raining

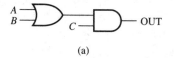

(a)

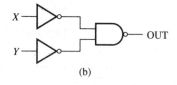

(b)

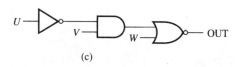

(c)

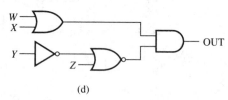

(d)

Figure 3.24
Exercise 3.9.

and 1 if it is not. Draw the logic diagram for RAINBOW, which is 1 only if the sun is shining and it is raining.

3.12 A logic circuit must produce 0 only when the binary number $A_2 A_1 A_0$ has its maximum possible numerical value. Draw the logic diagram.

Section 3–7

3.13 Draw the logic diagram and construct the truth table for each of the following logic expressions.
(a) $W = A + \overline{BC}$
(b) $Z = \overline{X}_1 X_2 (\overline{X_1 + X_3})$
(c) OUT $= \overline{(A + \overline{B})} + \overline{\overline{X}Y}$
(d) IN $= \overline{A}B + \overline{B}C + A\overline{C} + A\overline{B}C$

3.14 Draw the logic diagram and construct the truth table for each of the following logic expressions.
(a) $R = S(\overline{T + P})$
(b) $X = A\overline{\overline{B}}C$

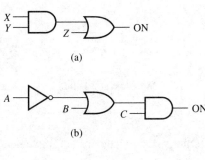

(a)

(b)

(c)

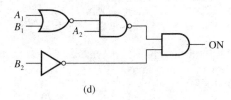

(d)

Figure 3.25
Exercise 3.10.

(c) $Q = (\overline{W_1 + W_2})(W_3 + \overline{W_4})$

(d) $FAN = \overline{\overline{WX}} + \overline{\overline{YZ}} + \overline{\overline{WY}}$

Section 3–8

3.15 Use truth tables to show that $(\overline{A} + \overline{B})(A + B) = A \oplus B$. Draw the logic diagram for this equivalent expression.

3.16 Use truth tables to show that $(\overline{A} + B)(A + \overline{B}) = A \odot B$. Draw the logic diagram for this equivalent expression.

3.17 Using the symbolic representations for the exclusive-OR and coincidence functions, write the logic expression for W in Figure 3.26. Construct the truth table for W.

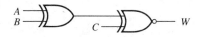

Figure 3.26
Exercise 3.17.

3.18 Draw a logic diagram that implements $A = (X_1 \odot X_2) + (X_2 \oplus X_3)$. Construct the truth table for A.

3.19 A logic circuit is required to produce a 1 if and only if 1 and only 1 bit of the binary number A_1A_0 is 0. Draw the logic diagram required, and write the logic expression.

3.20 Draw the logic diagram of a circuit that will produce the sum bit when two 1-bit binary numbers, A and B, are added.

Section 3–9

3.21 Draw a wiring diagram using the 7400-series chips with the pin assignments shown in Figure 3.18 to create a logic circuit that implements the expression OUT $= \overline{AB\overline{C}D}$. Show all connections between pins.

3.22 Draw a wiring diagram using the 7400-series chips with the pin assignments shown in Figure 3.18 to create a logic circuit that implements the expression $UP = (\overline{X_1 + X_2}) \oplus (\overline{X_3X_4})$.

Section 3–10

3.23 Draw a logic diagram using ANSI/IEEE standard symbols to implement the function $Z = \overline{AB} \oplus \overline{CD}$. Also draw the connection diagram using ANSI/IEEE standard symbols for 7400-series integrated circuits.

3.24 Write the logic expression for the function F implemented by the circuit shown in Figure 3.27.

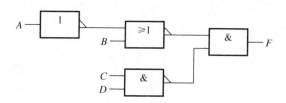

Figure 3.27
Exercise 3.24.

CHALLENGING EXERCISES

3.25 Write the logic expression for OUT in Figure 3.28 and construct its truth table.

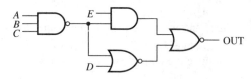

Figure 3.28
Exercise 3.25.

3.26 Write the logic expression for W in Figure 3.29 and construct its truth table.

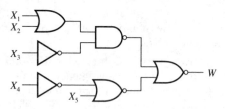

Figure 3.29
Exercise 3.26.

3.27 A logic circuit is required to produce a 1 if and only if the binary number $A_4A_3A_2A_1A_0$ has a value greater than $7)_{10}$ but less than $16)_{10}$. Write the logic expression for the circuit and draw its logic diagram. (*Hint:* Construct the truth table and examine the pattern of the bits.)

3.28 A logic circuit is required to produce a 1 if and only if the binary number $A_4A_3A_2A_1A_0$ has a value equal to an *even* decimal number greater than $15)_{10}$. Write the logic expression

and draw its logic diagram. (*Hint:* Construct the truth table and examine the pattern of the bits.)

3.29 Draw a logic diagram that implements the expression

$$V = \overline{(\overline{AB} + \overline{\overline{CD}})(\overline{\overline{E} + F})\,\overline{AB}}$$

3.30 Draw a logic diagram that implements the expression

$$A = \overline{\overline{\overline{(\overline{X_1} + \overline{X_2} + X_3)}\,\overline{(X_4 + \overline{X_3X_5})}(X_1X_3X_4 + \overline{X_3X_5})}}$$

3.31 Determine which of the following expressions are equivalent to $A \oplus B$ and which are equivalent to $A \odot B$.
(a) $\overline{A} \oplus B$ **(b)** $A \odot \overline{B}$ **(c)** $\overline{A} \oplus \overline{B}$
(d) $\overline{A} \odot B$ **(e)** $A \oplus \overline{B}$ **(f)** $\overline{A} \odot \overline{B}$

3.32 Find the logical equivalent of each of the following expressions.
(a) $A \oplus 0$ **(b)** $A \oplus 1$ **(c)** $A \odot 0$
(d) $A \odot 1$ **(e)** $1 \oplus \overline{B}$ **(f)** $0 \odot \overline{B}$

3.33 A streetlight is controlled by a logical combination of the digital variables A, B, C, and D. The light is turned on when the following expression equals 1:

$$\text{ON} = (\overline{A} + B)(\overline{C} + \overline{D}) + AB + CD$$

Under what conditions is the light off?

3.34 An elevator is controlled by the digital variables W, X, Y, and Z. The elevator travels to the fourth floor when the following expression equals 1:

$$\text{FOUR} = (\overline{W} + X)(Y + \overline{Z})(W\overline{X} + \overline{Y}Z)$$

Under what conditions does the elevator travel to the fourth floor?

Chapter 4

BOOLEAN ALGEBRA

4–1   INTRODUCTION

Boolean algebra is a set of rules used with digital variables and logical operations to develop, manipulate, and simplify logical expressions. Many of the logical expressions we encountered in Chapter 3 can be expressed in equivalent forms (often in *simpler* forms), and Boolean algebra gives us a systematic means for discovering those alternate expressions. It also helps us discover redundancies or contradictions that may arise when we attempt to translate a set of English-language statements into a logical expression. The principal value of Boolean algebra is that it strips away all superfluous verbiage in our formulation of logical propositions and allows us to examine, test, and reformulate those propositions using only symbols and a rigorous set of rules.

The axioms, or postulates, of Boolean algebra are a set of logical expressions that we accept without proof and upon which we can build a set of useful theorems. Actually, the axioms are nothing more than the definitions of the three basic logic operations we studied in Chapter 3: AND, OR, and INVERT. In the list that follows, recall that the AND operation is expressed like multiplication, the OR operation is expressed like addition, and the INVERT operation is shown by a bar. Each axiom can be read and interpreted as the outcome of an operation performed by a logic gate. For example, axiom 1 states that $0 \cdot 0 = 0$, "0 AND 0 equals 0", i.e., the output of an AND gate is 0 when both its inputs are 0.

Axiom 1	$0 \cdot 0 = 0$	**Axiom 6**	$0 + 1 = 1$
Axiom 2	$0 \cdot 1 = 0$	**Axiom 7**	$1 + 0 = 1$
Axiom 3	$1 \cdot 0 = 0$	**Axiom 8**	$1 + 1 = 1$
Axiom 4	$1 \cdot 1 = 1$	**Axiom 9**	$\overline{1} = 0$
Axiom 5	$0 + 0 = 0$	**Axiom 10**	$\overline{0} = 1$

The axioms are easy to remember because they are similar to ordinary arithmetic. Note, however, the *idempotent* relation expressed by Axiom 8: $1 + 1 = 1$ (the output of an OR gate is 1 when both its inputs are 1).

4–2 THEOREMS OF BOOLEAN ALGEBRA

Each of the theorems of Boolean algebra that follow is proved using axioms, truth tables, logic diagrams, or previously proved theorems. Note that the use of truth tables and logic diagrams is simply a convenient way of using the axioms listed in Section 4–1.

The Commutative Properties

Theorem 1a AB = BA

A AND B is the same as B AND A; in effect, it makes no difference which input of an AND gate is connected to A and which is connected to B. The truth tables are identical.

A	B	AB		B	A	BA
0	0	0		0	0	0
0	1	0	=	0	1	0
1	0	0		1	0	0
1	1	1		1	1	1

Theorem 1b A + B = B + A

A OR B is the same as B OR A; in effect, it makes no difference which input of an OR gate is connected to A and which is connected to B. The truth tables are identical:

A	B	A + B		B	A	B + A
0	0	0		0	0	0
0	1	1	=	0	1	1
1	0	1		1	0	1
1	1	1		1	1	1

Note that the commutative properties can be extended to any number of variables. For example, since $A + B = B + A$, it follows that $A + B + C = B + A + C$, and since $A + C = C + A$, it is true that $B + A + C = B + C + A$. Similarly, $ABCD = BACD = BADC = ABDC$, and so on.

The Associative Properties

Theorem 2a (AB)C = A(BC)

A AND B ANDed with C is the same as A ANDed with B AND C.

A	B	C	AB	(AB)C
0	0	0	0	0
0	0	1	0	0
0	1	0	0	0
0	1	1	0	0
1	0	0	0	0
1	0	1	0	0
1	1	0	1	0
1	1	1	1	1

=

A	B	C	BC	A(BC)
0	0	0	0	0
0	0	1	0	0
0	1	0	0	0
0	1	1	1	0
1	0	0	0	0
1	0	1	0	0
1	1	0	0	0
1	1	1	1	1

Note that $(AB)C = A(BC) = ABC$.

Theorem 2b $(A + B) + C = A + (B + C)$

A OR B ORed with C is the same as A ORed with B OR C.

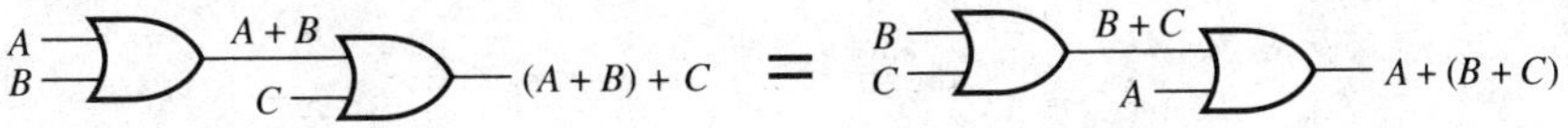

A	B	C	A + B	(A + B) + C
0	0	0	0	0
0	0	1	0	1
0	1	0	1	1
0	1	1	1	1
1	0	0	1	1
1	0	1	1	1
1	1	0	1	1
1	1	1	1	1

=

A	B	C	B + C	A + (B + C)
0	0	0	0	0
0	0	1	1	1
0	1	0	1	1
0	1	1	1	1
1	0	0	0	1
1	0	1	1	1
1	1	0	1	1
1	1	1	1	1

Note that $(A + B) + C = A + (B + C) = A + B + C$.

The associative properties can be extended to any number of variables:

$$A(BCD) = (AB)(CD) = (ABC)D = ABCD$$
$$A + (B + C + D) = (A + B) + (C + D) = (A + B + C) + D$$
$$= A + B + C + D$$

The Idempotent Properties

Theorem 3a $AA = A$

If $A = 0$, then $AA = 0 \cdot 0 = 0 = A$.
If $A = 1$, then $AA = 1 \cdot 1 = 1 = A$.

Theorem 3b $A + A = A$

If $A = 0$, then $A + A = 0 + 0 = 0 = A$.
If $A = 1$, then $A + A = 1 + 1 = 1 = A$.

It is easy to verify that $AAA = A$, $A + A + A = A$, $AAAA = A$, and so on.

The Identity Properties

Theorem 4a $A \cdot 1 = A$

If $A = 0$, then $A \cdot 1 = 0 \cdot 1 = 0 = A$.
If $A = 1$, then $A \cdot 1 = 1 \cdot 1 = 1 = A$.

By the commutative property, it is also true that $1 \cdot A = A$.

Theorem 4b $A + 1 = 1$

If $A = 0$, then $A + 1 = 0 + 1 = 1$.
If $A = 1$, then $A + 1 = 1 + 1 = 1$.

By the commutative property, it is also true that $1 + A = 1$.
    The identity properties apply to any number or any combination of variables. For example,

$$(ABC + D) \cdot 1 = ABC + D$$

and

$$AB + CD + 1 = 1$$

(From the definition of the OR operation, it is clear that *anything* ORed with 1 produces 1.)

The Null Properties

Theorem 5a $A \cdot 0 = 0$

If $A = 0$, then $A \cdot 0 = 0 \cdot 0 = 0$.
If $A = 1$, then $A \cdot 0 = 1 \cdot 0 = 0$.

By the definition of the AND operation, *anything* ANDed with 0 produces 0.

Theorem 5b $A + 0 = A$

If $A = 0$, then $A + 0 = 0 + 0 = 0 = A$.

If $A = 1$, then $A + 0 = 1 + 0 = 1 = A$.

The null properties apply to any number or any combination of variables. For example,

$$(AB + CD) \cdot 0 = 0$$
$$ABC + D + 0 = ABC + D$$

The Distributive Property

Theorem 6 $A(B + C) = AB + AC$

A B C	(B + C)	A(B + C)
0 0 0	0	0
0 0 1	1	0
0 1 0	1	0
0 1 1	1	0
1 0 0	0	0
1 0 1	1	1
1 1 0	1	1
1 1 1	1	1

$=$

A B C	AB	AC	AB + AC
0 0 0	0	0	0
0 0 1	0	0	0
0 1 0	0	0	0
0 1 1	0	0	0
1 0 0	0	0	0
1 0 1	0	1	1
1 1 0	1	0	1
1 1 1	1	1	1

By the commutative property, it is also true that

$$A(B + C) = (B + C)A = BA + CA = AB + AC$$

Note once again that the distributive property applies to combinations of variables as well as to single variables. For example,

$$AB(C + DE) = ABC + ABDE$$

and

$$(A + B)(C + D) = (A + B)C + (A + B)D = AC + BC + AD + BD$$

The distributive property is often used in reverse; i.e., given $AB + AC$, we replace it by its equal, $A(B + C)$. As in ordinary algebra, this process is called *factoring*. Note that we factored A out of the expression $AB + AC$. As another example,

$$XYW + XYZ = XY(W + Z)$$

Negation Properties

Theorem 7a $A\overline{A} = 0$

If $A = 0$, then $\overline{A} = 1$, and $A\overline{A} = 0 \cdot 1 = 0$.
If $A = 1$, then $\overline{A} = 0$, and $A\overline{A} = 1 \cdot 0 = 0$.

Theorem 7b $A + \overline{A} = 1$

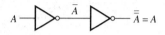

If $A = 0$, then $\overline{A} = 1$, and $A + \overline{A} = 0 + 1 = 1$.
If $A = 1$, then $\overline{A} = 0$, and $A + \overline{A} = 1 + 0 = 1$.

Of course, by commutativity, it is also true that $\overline{A}A = 0$ and $\overline{A} + A = 1$. The negation properties apply to combinations of variables. For example, $ABC + \overline{ABC} = 1$.

The Double Negation Property

Theorem 8 $\overline{\overline{A}} = A$

$$A \;—\!\!\!\triangleright\!\!o\;—\; \overline{A} \;—\!\!\!\triangleright\!\!o\;—\; \overline{\overline{A}} = A$$

If $A = 0$, then $\overline{A} = 1$, and $\overline{\overline{A}} = \overline{1} = 0 = A$.
If $A = 1$, then $\overline{A} = 0$, and $\overline{\overline{A}} = \overline{0} = 1 = A$.

It is easy to show that any odd number of inversions is equivalent to a single inversion and any even number of inversions is equivalent to no inversions. As usual, A can be replaced by a combination of variables. For example,

$$\overline{\overline{A + BC}} = A + BC$$

4–3   SIMPLIFYING LOGIC EXPRESSIONS

The theorems of Boolean algebra are used to simplify logic expressions for two main reasons:

1. A simpler expression can be implemented using fewer gates.
2. Simplification sometimes reveals redundant variables or logic contradictions.

As examples of these situations, consider the two logic expressions $W = ABC + CAB + BAC$ and $X = ABC + AB\overline{C}$. W can be simplified as follows:

$$
\begin{aligned}
W &= ABC + CAB + BAC \\
&= ABC + ABC + ABC \quad \text{(Theorem 1a)} \\
&= ABC + ABC \qquad\qquad \text{(Theorem 3b)} \\
&= ABC \qquad\qquad\qquad\;\; \text{(Theorem 3b)}
\end{aligned}
$$

The simplified expression can be implemented by one AND gate, whereas the original expression required three AND gates and an OR gate.

The logic expression for X is simplified as follows:

$$
\begin{aligned}
X &= ABC + AB\overline{C} \\
&= AB(C + \overline{C}) \quad &\text{(Theorem 6)} \\
&= AB \cdot 1 \quad &\text{(Theorem 7b)} \\
&= AB \quad &\text{(Theorem 4a)}
\end{aligned}
$$

Not only is the simplified expression more economical to implement, the variable C is not even needed. We say that C is *redundant*.

Now consider the logic expression $\text{ON} = X(\overline{X}YZ + \overline{X}Y\overline{Z})$, which might represent the conditions under which a lamp is turned on. Presumably, the logic designer has arrived at this expression by translating a set of English-language statements involving three variables into a Boolean expression containing the three variables X, Y, and Z. Simplifying, we find

$$
\begin{aligned}
\text{ON} &= X(\overline{X}YZ + \overline{X}Y\overline{Z}) \\
&= X\overline{X}YZ + X\overline{X}Y\overline{Z} \quad &\text{(Theorem 6)} \\
&= 0 \cdot YZ + 0 \cdot Y\overline{Z} \quad &\text{(Theorem 7a)} \\
&= 0 + 0 \quad &\text{(Theorem 5a)} \\
&= 0 \quad &\text{(Axiom 5)}
\end{aligned}
$$

If the lamp is turned on when ON equals 1, it is clear that the lamp will *never* be turned on, since ON equals 0 under all circumstances. In this instance, Boolean algebra has helped us discover what is apparently an error in the formulation of the logic expression or has showed us that no lamp is required.

The next example shows how the theorems can be used to simplify a variety of logic expressions.

Example 4–1   Use Boolean algebra to simplify each of the following logic expressions as much as possible.

a. $W = AB(\overline{A}C + B)$
b. $V = \overline{X}Y + \overline{X}$
c. $Z = A\overline{B}\overline{C} + A\overline{B}C + \overline{A}BC + \overline{A}\overline{B}C$
d. $N = (X_1 + X_2)(\overline{X}_1 + X_1X_2) + (\overline{X}_2 + X_1\overline{X}_2)$
e. $A = WXY(W\overline{X} + W\overline{Y}) + W\overline{X}Y(\overline{W}X + X\overline{Y})$

Solution

a.
$$
\begin{aligned}
W &= AB(\overline{A}C + B) \\
&= AB\overline{A}C + ABB \quad &\text{(Theorem 6)} \\
&= A\overline{A}BC + ABB \quad &\text{(Theorem 1a)} \\
&= 0 \cdot BC + AB \quad &\text{(Theorem 7a)} \\
&= 0 + AB \quad &\text{(Theorem 5a)} \\
&= AB \quad &\text{(Theorem 5b)}
\end{aligned}
$$

b. In this example, note that a variable ($\overline{X}$ in this case) can be factored from itself, leaving 1 in its place:

$$V = \overline{X}Y + \overline{X}$$
$$= \overline{X}(Y + 1) \qquad \text{(Theorem 6)}$$
$$= \overline{X} \cdot 1 \qquad \text{(Theorem 4b)}$$
$$= \overline{X} \qquad \text{(Theorem 4a)}$$

c. $Z = A\overline{B}\overline{C} + A\overline{B}C + \overline{A}BC + \overline{A}B\overline{C}$

$$= A\overline{B}(\overline{C} + C) + \overline{A}CB + \overline{A}C\overline{B} \qquad \text{(Theorems 6 and 1a)}$$

$$= A\overline{B} \cdot 1 + \overline{A}C(B + \overline{B}) \qquad \text{(Theorems 7b and 6)}$$

$$= A\overline{B} + \overline{A}C \cdot 1 \qquad \text{(Theorems 4a and 7b)}$$
$$= A\overline{B} + \overline{A}C \qquad \text{(Theorem 4a)}$$

d. $N = (X_1 + X_2)(\overline{X}_1 + X_1X_2) + (\overline{X}_2 + X_1\overline{X}_2)$

$$= X_1(\overline{X}_1 + X_1X_2) + X_2(\overline{X}_1 + X_1X_2) + \overline{X}_2 + X_1\overline{X}_2 \qquad \text{(Theorems 6 and 2b)}$$
$$= X_1\overline{X}_1 + X_1X_1X_2 + X_2\overline{X}_1 + X_2X_1X_2 + \overline{X}_2 + X_1\overline{X}_2 \qquad \text{(Theorem 6)}$$

$$= 0 + X_1X_2 + X_2\overline{X}_1 + X_1X_2X_2 + \overline{X}_2 + X_1\overline{X}_2 \qquad \text{(Theorems 7a, 3a, and 1a)}$$

$$= X_2X_1 + X_2\overline{X}_1 + X_1X_2 + X_1\overline{X}_2 + \overline{X}_2 \qquad \text{(Theorems 1a and 1b)}$$
$$= X_2(X_1 + \overline{X}_1) + X_1X_2 + \overline{X}_2(X_1 + 1) \qquad \text{(Theorem 6)}$$

$$= X_2 \cdot 1 + X_1X_2 + \overline{X}_2 \cdot 1 \qquad \text{(Theorems 7b and 4b)}$$
$$= X_2 + X_1X_2 + \overline{X}_2 \qquad \text{(Theorem 4a)}$$
$$= X_2 + \overline{X}_2 + X_1X_2 \qquad \text{(Theorem 1b)}$$

$$= 1 + X_1X_2 \qquad \text{(Theorem 7b)}$$
$$= 1 \qquad \text{(Theorem 4b)}$$

This result shows that N equals 1 under all circumstances—i.e., no matter what the combination of values of the input variables.

e. $A = WXY(W\overline{X} + W\overline{Y}) + W\overline{X}Y(\overline{W}X + X\overline{Y})$

$$= WXYW\overline{X} + WXYW\overline{Y} + W\overline{X}Y\overline{W}X + W\overline{X}YX\overline{Y} \qquad \text{(Theorem 6)}$$
$$= WX\overline{X}YW + WXY\overline{Y}W + W\overline{W}X\overline{X}Y + WY\overline{X}X\overline{Y} \qquad \text{(Theorem 1a)}$$

$$= W \cdot 0 \cdot YW + WX \cdot 0 \cdot W + 0 \cdot \overline{X}Y\overline{X} + WY \cdot 0 \cdot \overline{Y} \qquad \text{(Theorem 7a)}$$
$$= 0 + 0 + 0 + 0 \qquad \text{(Theorem 5a)}$$
$$= 0 \qquad \text{(Theorem 3b)}$$

This result shows that A equals 0 under all circumstances—i.e., no matter what the combination of values of the input variables.

***Drill
4-1***

Simplify the expression $Y = ABC + AB + A$ as much as possible.
Answer: $Y = A$.

***Example
4-2***

An air-conditioning unit is controlled by the four variables temperature, T; humidity, H; time of day, D; and day of week, W, which are defined as follows:

$$T = \begin{cases} 1 & \text{if temperature is greater than 78°F} \\ 0 & \text{otherwise} \end{cases}$$

$$H = \begin{cases} 1 & \text{if humidity is greater than 85\%} \\ 0 & \text{otherwise} \end{cases}$$

$$D = \begin{cases} 1 & \text{if time of day is between 8 AM and 5 PM} \\ 0 & \text{otherwise} \end{cases}$$

$$W = \begin{cases} 1 & \text{if Monday through Friday} \\ 0 & \text{otherwise} \end{cases}$$

The unit is to be turned on (by a logic 1) under any of the following circumstances:

1. The temperature exceeds 78°F and the time of day is between 8 AM and 5 PM; OR
2. The humidity exceeds 85%, the temperature exceeds 78°F, and it is a weekend; OR
3. The humidity exceeds 85%, the temperature exceeds 78°F, and the time of day is between 8 AM and 5 PM; OR
4. It is Saturday or Sunday and the humidity exceeds 85%.

Write a logic expression for controlling the air-conditioning unit. Simplify the expression as far as possible.

Solution. Each of the four situations listed above is a set of conditions that are ANDed together, and the four sets are themselves ORed together. By referring to the definitions of the variables T, H, D, and W, we see that the sets of conditions are equivalent to the following Boolean terms:

1. TD
2. $HT\overline{W}$
3. HTD
4. $H\overline{W}$

Thus, the control output, ON, is given by

$$\text{ON} = TD + HT\overline{W} + HTD + H\overline{W}$$

Using the commutative properties, we can rearrange this expression as

$$\text{ON} = TD + TDH + \overline{W}H + \overline{W}HT$$

Factoring, we find

$$\begin{aligned} \text{ON} &= TD(1 + H) + \overline{W}H(1 + T) \\ &= TD \cdot 1 + \overline{W}H \cdot 1 \\ &= TD + \overline{W}H \end{aligned}$$

Thus, we can implement the logic using two 2-input AND gates and one 2-input OR gate. If the expression had not been simplified, it would have required four AND gates and a 4-input OR gate.

Drill
4–2

Simplify the expression for ON in Example 4–2 when the following additional condition is included: It is a weekday and the humidity exceeds 85%.
Answer: ON $= TD + H$.

Simplifying a logic expression so it can be implemented with the fewest number of gates is not *always* the most economical approach. For example, if an expression involves one NOR operation and the 7402 TTL chip containing four NOR gates is used to implement that one operation, it would be desirable to find some way to implement the rest of the expression using the other three NOR gates in the chip. Clearly this solution would be more economical than using several different chips containing different kinds of gates to implement the expression. The total number of chips used to implement a logic expression is called the *chip count,* a number that should be minimized in practical designs. In Chapter 5, we investigate methods that can be used to implement logic expressions using only NAND gates or only NOR gates.

4–4 MORE THEOREMS OF BOOLEAN ALGEBRA

The Absorption Properties

Theorem 9a $A + AB = A$

We have encountered the combination $A + AB$ in previous examples and, by factoring A from itself, we have seen that the combination equals A. We wish now to adopt this result as a formal theorem, with proof, so we can apply it without proof in future examples:

$$A + AB = A(1 + B) = A \cdot 1 = A$$

Note that B can be replaced by any number or any combination of variables. For example,

$$A + A(B\overline{C} + D) = A$$

Theorem 9b $A(A + B) = A$

$$A(A + B) = AA + AB = A + AB = A$$

Once again, each variable can be replaced by a combination of variables. For example,

$$X\overline{Y}(X\overline{Y} + WYZ) = X\overline{Y}$$

Theorem 9c $A + \overline{A}B = A + B$

We must use an innovative approach to prove this theorem. First, we note that A can be replaced by its equal, $A + AB$ (from Theorem 9a):

$$A + \overline{A}B = \underbrace{A + AB}_{A} + \overline{A}B$$

Then,

$$A + \overline{A}B = A + AB + \overline{A}B = A + (A + \overline{A})B = A + (1 \cdot B) = A + B$$

Following are examples in which A and B are replaced by other terms. In particular, if we replace A by $\overline{A}$ (so $\overline{A}$ is replaced by $\overline{\overline{A}} = A$), we have

$$\overline{A} + AB = \overline{A} + B$$

Also,

$$X + Y + (\overline{X + Y})Z = X + Y + Z$$

DeMorgan's Theorems

Theorem 10a $\overline{AB} = \overline{A} + \overline{B}$

This very important theorem states that a NAND gate performs the same operation as an OR gate whose inputs are inverted:

A B	AB	$\overline{AB}$		A B	$\overline{A}$	$\overline{B}$	$\overline{A} + \overline{B}$
0 0	0	1		0 0	1	1	1
0 1	0	1	=	0 1	1	0	1
1 0	0	1		1 0	0	1	1
1 1	1	0		1 1	0	0	0

Theorem 10b $\overline{A + B} = \overline{A}\,\overline{B}$

A NOR gate performs the same operation as an AND gate whose inputs are inverted.

A B	A + B	$\overline{A + B}$		A B	$\overline{A}$	$\overline{B}$	$\overline{A}\overline{B}$
0 0	0	1		0 0	1	1	1
0 1	1	0	=	0 1	1	0	0
1 0	1	0		1 0	0	1	0
1 1	1	0		1 1	0	0	0

Note that $\overline{AB}$ does not equal $\overline{A}\,\overline{B}$ and $\overline{A + B}$ does not equal $\overline{A} + \overline{B}$.

DeMorgan's two theorems can be regarded as a *single* theorem by observing the following rule:

Change the logic operation covered by the inversion bar, remove the inversion bar, and complement each variable that was originally covered by the bar.

For example, in the expression $\overline{AB}$, the AND operation between A and B is changed to an OR operation, the bar is removed, and the variables A and B are complemented:

$$\overline{AB} = \overline{A} + \overline{B}$$

Similarly, in the expression $\overline{A + B}$, the OR operation between A and B is changed to an AND operation, the bar is removed, and the variables A and B are complemented:

$$\overline{A + B} = \overline{A}\,\overline{B}$$

As with the other theorems of Boolean algebra, DeMorgan's theorems can be extended to include additional variables or combinations of variables. For example,

$$\overline{ABC} = \overline{A} + \overline{B} + \overline{C} \quad \text{and} \quad \overline{A + B + C} = \overline{A}\,\overline{B}\,\overline{C}$$

Notice how the single-theorem rule is applied in these cases to change every logic operation covered by the bar.

When DeMorgan's theorem is applied to an expression containing the complement of a variable, that complement becomes complemented, and we recover the uncomplemented variable. Following are examples:

$$\overline{\overline{A}B} = \overline{\overline{A}} + \overline{B} = A + \overline{B}$$
$$\overline{A + \overline{B} + C} = \overline{A}\,\overline{\overline{B}}\,\overline{C} = \overline{A}BC$$

Many logic expressions can be simplified by applying DeMorgan's theorem more than one time. Such is the case when an expression contains combinations of variables, which may themselves contain other combinations of variables. When removing complement bars in these cases, care must be exercised to change only those logic operations that join combinations covered by the bar. For example,

$$\overline{A\overline{B} + C} = (\overline{A\overline{B}})\overline{C} = (\overline{A} + \overline{\overline{B}})\overline{C} = (\overline{A} + B)\,\overline{C}$$

Note that only the OR operation between $A\overline{B}$ and C was changed when the first complement bar was removed. DeMorgan's theorem is then applied a second time to remove the bar covering $A\overline{B}$. Here is a more complex example:

$$\overline{(A + \overline{B})\overline{C}D + E} = [\overline{(A + \overline{B})\overline{C}D}]\overline{E}$$
$$= [\overline{(A + \overline{B})} + \overline{\overline{C}D}]\overline{E} = [(A + \overline{B}) + \overline{\overline{C}} + \overline{D}]\overline{E}$$
$$= (A + \overline{B} + C + \overline{D})\overline{E}$$

Note the sequence in which the complement bars are removed and the logic operations that are changed with each removal: The first complement bar covers the two terms $(A + \overline{B})\overline{C}D$ and E, so the OR operation between those terms is changed to AND when the bar is removed. The second complement bar covers the terms $(A + \overline{B})$ and $\overline{C}D$, so the AND operation between those terms is changed to OR when that bar is removed. Since $\overline{\overline{(A + \overline{B})}} = A + \overline{B}$, it is not necessary to apply DeMorgan's theorem to remove those bars. The third application of DeMorgan's theorem removes the bar across $\overline{C}$ and D, changing the AND to an OR and producing $\overline{\overline{C}} + \overline{D} = C + \overline{D}$.

When applying DeMorgan's theorem to remove complement bars from complex expressions, two of the most commonly committed errors are (1) failure to retain parentheses where they belong and (2) attempting to remove too many bars simultaneously. Consider, for example, the expression $F = \overline{ABC + \overline{BC}}$. Observe how an error occurs when one set of parentheses is prematurely dropped:

$$F = \overline{ABC + \overline{BC}}$$

Correct	Incorrect
$F = (\overline{ABC})(\overline{\overline{BC}})$	$F = (\overline{ABC})(\overline{\overline{BC}})$
parentheses retained $= (\overline{A} + \overline{B} + \overline{C})(BC)$	$\neq \overline{A} + \overline{B} + \overline{C}(BC)$ parentheses dropped
$= \overline{A}BC + \overline{B}BC + BC\overline{C}$	$= \overline{A} + \overline{B} + \overline{C}BC$
$\underbrace{}_{0} \underbrace{}_{0}$	$\underbrace{}_{0}$
$= \overline{A}BC$ (**correct** answer)	$= \overline{A} + \overline{B}$ (**incorrect** answer)

Now observe how an error can occur when we attempt to remove the two bars in the original expression simultaneously:

two bars removed simultaneously

$$F = \overline{ABC + \overline{BC}} \neq \overline{ABC}(\overline{B} + \overline{C})$$
$$= (\overline{A} + \overline{B} + \overline{C})(\overline{B} + \overline{C})$$
$$= \overline{B} + \overline{C}   \quad (\textbf{incorrect answer})$$

Example 4–3   Simplify the following logic expressions as much as possible:

a. $T = X\overline{Y} + \overline{Y}ZX$

b. $W = A_1\overline{A}_2A_3(A_1A_3 + \overline{A}_2A_1A_3 + A_1A_2)$

c. $M = WXYZ + \overline{XY}$

d. $R = \overline{\overline{A} + \overline{BC}}$

e. $S = \overline{(\overline{AB})(\overline{C} + \overline{D})}$

f. $V = \overline{(X + \overline{YZ})(\overline{X}YZ)}$

g. $X = \overline{A}(C + \overline{D}) + \overline{C}(A + \overline{B})$

Solution

a. $T = X\overline{Y} + \overline{Y}ZX$
$= X\overline{Y} + X\overline{Y}Z$ (Theorem 1a)
$= X\overline{Y}$ (Theorem 9a, with $A = X\overline{Y}$)

b. $W = A_1\overline{A}_2A_3(A_1A_3 + \overline{A}_2A_1A_3 + A_1A_2)$
$= A_1\overline{A}_2A_3(A_1\overline{A}_2A_3 + A_1A_3 + A_1A_2)$ (Theorems 1a and 1b)
$= A_1\overline{A}_2A_3$ (Theorem 9b, with $A = A_1\overline{A}_2A_3$)

c. $M = WXYZ + \overline{XY}$
$= \overline{XY} + XYWZ$ (Theorems 1a and 1b)
$= \overline{XY} + WZ$ (Theorem 9c, with $A = \overline{XY}$)

d. $R = \overline{\overline{A} + \overline{BC}}$

$= \overline{\overline{A}} \, \overline{\overline{BC}}$ (DeMorgan's theorem)

$= ABC$ (Theorem 8)

e. $S = \overline{\overline{A}\,\overline{B}(\overline{C} + \overline{D})}$

$= \overline{\overline{A}\,\overline{B}} + \overline{(\overline{C} + \overline{D})}$ (DeMorgan's theorem)

$= \overline{\overline{A}} + \overline{\overline{B}} + \overline{\overline{C}\,\overline{D}}$ (DeMorgan's theorem)

$= A + B + CD$ (Theorem 8)

f. $V = \overline{\overline{(X + \overline{YZ})(\overline{X}YZ)}}$

$= \overline{(X + \overline{YZ})} + \overline{\overline{X}YZ}$ (DeMorgan's theorem)

$= (X + \overline{YZ}) + \overline{\overline{X}} + \overline{Y} + \overline{Z}$ (Theorem 8 and DeMorgan's theorem)

$= X + \overline{Y} + \overline{Z} + X + \overline{Y} + \overline{Z}$ (DeMorgan's theorem and theorem 8)

$= X + \overline{Y} + \overline{Z}$ (Theorem 3b)

g. $X = \overline{\overline{A}(C + \overline{D})} + \overline{\overline{C}(A + \overline{B})}$

$= \overline{\overline{A}} + \overline{(C + \overline{D})} + \overline{\overline{C}} + \overline{(A + \overline{B})}$ (DeMorgan's theorem)

$= A + \overline{C}D + C + \overline{A}\overline{\overline{B}}$ (Theorem 8 and DeMorgan's theorem)

$= A + \overline{C}D + C + \overline{A}B$ (Theorem 8)

$= A + \overline{A}B + C + \overline{C}D$ (Theorem 1b)

$= A + B + C + D$ (Theorem 9c)

Drill
4–3

Simplify $W = \overline{\overline{(\overline{A} + \overline{B})(\overline{C} + \overline{AB})}}$ as much as possible.

Answer: $W = \overline{A}B + B\overline{C} + \overline{A}C$.

Example
4–4

Figure 4.1(*a*) shows the logic diagram of a computer circuit used to generate a 1 whenever the decimal value of the binary number $A_3A_2A_1A_0$ is greater than 9. ($A_3A_2A_1A_0$ is called a *binary-coded decimal,* about which we have more to say later.) Find a simpler way to implement the circuit.

Solution.   As shown in Figure 4.1(*a*), the output of the circuit is

$$\overline{\overline{A}_3 + A_2 + \overline{A}_1} + \overline{\overline{A}_3 + \overline{A}_2 + A_1} + \overline{\overline{A}_3 + \overline{A}_2 + \overline{A}_1}$$

Applying DeMorgan's theorem to each term, we have

$$\overline{\overline{A}}_3\overline{A}_2\overline{\overline{A}}_1 + \overline{\overline{A}}_3\overline{\overline{A}}_2\overline{A}_1 + \overline{\overline{A}}_3\overline{\overline{A}}_2\overline{\overline{A}}_1 = A_3\overline{A}_2A_1 + A_3A_2\overline{A}_1 + A_3A_2A_1$$

Simplifying this expression gives

$$A_3\overline{A}_2A_1 + A_3A_2\underbrace{(\overline{A}_1 + A_1)}_{1} = A_3\overline{A}_2A_1 + A_3A_2$$

$$= A_3(A_2 + \overline{A}_2A_1)$$

$$= A_3(A_2 + A_1)$$

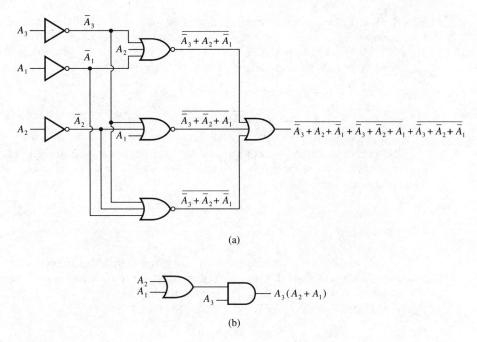

Figure 4.1
Example 4–4.

The simplified expression can be implemented with one AND gate and one OR gate, as shown in Figure 4.1(*b*).

Drill 4–4   Use truth tables to verify that both logic circuits in Figure 4.1 produce 1 if and only if the decimal value of $A_3 A_2 A_1 A_0$ is greater than 9.

4–5   STANDARD FORMS

We have seen that a logic expression can be written in a number of equivalent forms. In some applications, it is useful to be able to write expressions in one (or both) of two standard forms: the *sum-of-products* form and the *product-of-sums* form. The words *sum* and *product* are derived from the symbolic representations of the OR and AND functions by + and · (addition and multiplication), respectively, but we realize that these are not arithmetic operators in the usual sense.

A *product* term is any group of variables that are ANDed together. Examples are ABC, $\overline{X}Y$, and $T_1\overline{T}_2\overline{T}_3 T_4$. A *sum* term is any group of variables that are ORed together, such as $A + B + C$. A *sum of products* is a group of product terms ORed together, as, for example,

$$A\overline{B}C + \overline{A}B + \overline{A}BC$$

A *product of sums* is any group of sum terms ANDed together, as, for example,

$$(A + B + C)(A + \overline{B} + \overline{C})(\overline{A} + \overline{B})$$

We say that a sum of products is a *standard* (or canonical) sum of products if *every* product term involves *every* variable or its complement. For example, $F = A\overline{B}C + ABC + \overline{A}B\overline{C}$ is a standard sum of products if A, B, and C are the only variables pertaining to the logic. Note that this expression is not in simplest form because we can write

$$F = A\overline{B}C + ABC + \overline{A}B\overline{C}$$
$$= AC(\overline{B} + B) + \overline{A}B\overline{C}$$
$$= AC + \overline{A}B\overline{C}$$

The simplified expression is a sum of products but not a *standard* sum of products. A product of sums is in standard form if every sum term involves every variable or its complement. For example,

$$W = (X_1 + \overline{X}_2 + X_3)(\overline{X}_1 + X_2 + \overline{X}_3)$$

is a standard product of sums if X_1, X_2, and X_3 are the only variables pertaining to the logic.

The logic expression corresponding to a given truth table can be written in a standard sum-of-products form by writing one product term for each input combination that produces an output of 1. These product terms are ORed together to create the standard sum of products. Consider, for example, the following truth table:

A	B	C	F	
0	0	0	0	
0	0	1	1	$\longleftarrow \overline{A}\overline{B}C$
0	1	0	0	
0	1	1	0	
1	0	0	1	$\longleftarrow A\overline{B}\overline{C}$
1	0	1	1	$\longleftarrow A\overline{B}C$
1	1	0	0	
1	1	1	0	

We note that F is 1 when $A = 0$, $B = 0$, and $C = 1$, so this particular combination makes the output of an AND gate equal to 1 when $\overline{A}$ and $\overline{B}$ and C are all equal to 1. Thus $\overline{A}\overline{B}C$ is one product term. Similarly, the other two product terms are $A\overline{B}\,\overline{C}$ and $A\overline{B}C$. Thus, the standard sum-of-products form is

$$F = \overline{A}\overline{B}C + A\overline{B}\,\overline{C} + A\overline{B}C$$

The logic expression corresponding to a truth table can also be written in a standard product-of-sums form by writing one sum term for each output 0. The sum terms are expressed by writing the *complement* of a variable when it appears as an input 1, and the variable itself when it appears as an input 0. Consider, for example, the following truth table:

X_1	X_2	W	
0	0	0	$\longleftarrow X_1 + X_2$
0	1	1	
1	0	0	$\longleftarrow \overline{X}_1 + X_2$
1	1	1	

The sum corresponding to input combination 00 is $X_1 + X_2$, and the sum corresponding to input 10 is $\overline{X}_1 + X_2$. Thus, the standard product of sums is

$$W = (X_1 + X_2)(\overline{X}_1 + X_2)$$

A product-of-sums form derived from a truth table is logically equivalent to a sum-of-products form derived from the same truth table. In the previous example, we see that the standard sum-of-products form is $W = \overline{X}_1 X_2 + X_1 X_2$. As an exercise, verify that this is equivalent to the standard product-of-sums form $(X_1 + X_2)(\overline{X}_1 + X_2)$. Although the sum-of-products form is used more frequently in practical applications, a product-of-sums form will yield a simpler expression if the truth table has significantly fewer output 0s than output 1s.

It should be apparent from the foregoing discussion that one way to obtain the standard form of *any* logic expression is to construct its truth table and use the methods described to derive the desired standard form.

Example 4-5   A water-well pump is to be turned on automatically if the water level is low in any two or more of the three reservoirs it supplies. Water-level detectors in the reservoirs each generate a logical 0 if the level in a reservoir is low. Design a logic system that turns on the pump by generating a logical 1 in response to inputs received from the level detectors.

Solution.   Let the reservoir levels be designated L_1, L_2, and L_3. The truth table for turning on the pump is shown in Figure 4.2(a). We note that ON = 1 for four different input combinations. The standard sum-of-products form is

$$\text{ON} = \overline{L}_1 \overline{L}_2 \overline{L}_3 + \overline{L}_1 \overline{L}_2 L_3 + \overline{L}_1 L_2 \overline{L}_3 + L_1 \overline{L}_2 \overline{L}_3$$

Simplifying, we find

$$
\begin{aligned}
\text{ON} &= \overline{L}_1 \overline{L}_2 (\overline{L}_3 + L_3) + \overline{L}_1 L_2 \overline{L}_3 + L_1 \overline{L}_2 \overline{L}_3 \\
&= \overline{L}_1 \overline{L}_2 (1) + \overline{L}_1 L_2 \overline{L}_3 + L_1 \overline{L}_2 \overline{L}_3 \\
&= \overline{L}_1 \overline{L}_2 + \overline{L}_1 L_2 \overline{L}_3 + L_1 \overline{L}_2 \overline{L}_3 \\
&= \overline{L}_1 (\overline{L}_2 + L_2 \overline{L}_3) + L_1 \overline{L}_2 \overline{L}_3 \\
&= \overline{L}_1 (\overline{L}_2 + \overline{L}_3) + L_1 \overline{L}_2 \overline{L}_3 \\
&= \overline{L}_1 \overline{L}_2 + \overline{L}_1 \overline{L}_3 + L_1 \overline{L}_2 \overline{L}_3 \\
&= \overline{L}_2 (\overline{L}_1 + L_1 \overline{L}_3) + \overline{L}_1 \overline{L}_3 \\
&= \overline{L}_2 (\overline{L}_1 + \overline{L}_3) + \overline{L}_1 \overline{L}_3 \\
&= \overline{L}_1 \overline{L}_2 + \overline{L}_2 \overline{L}_3 + \overline{L}_1 \overline{L}_3
\end{aligned}
$$

This expression can be implemented using four gates by recognizing that each term in the sum can be produced by a NOR gate. Thus,

$$\text{ON} = \overline{(L_1 + L_2)} + \overline{(L_2 + L_3)} + \overline{(L_1 + L_3)}$$

The logic diagram is shown in Figure 4.2(b).

Drill 4-5   Write the standard product-of-sums form for ON in Example 4-5.
Answer: ON = $(L_1 + \overline{L}_2 + \overline{L}_3)(\overline{L}_1 + L_2 + \overline{L}_3)(\overline{L}_1 + \overline{L}_2 + L_3)(\overline{L}_1 + \overline{L}_2 + \overline{L}_3)$.

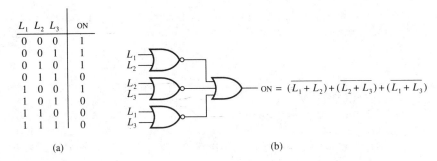

L_1	L_2	L_3	ON
0	0	0	1
0	0	1	1
0	1	0	1
0	1	1	0
1	0	0	1
1	0	1	0
1	1	0	0
1	1	1	0

(a)

$$\text{ON} = \overline{(L_1 + L_2)} + \overline{(L_2 + L_3)} + \overline{(L_1 + L_3)}$$

(b)

Figure 4.2
Example 4–5.

4–6 KARNAUGH MAPS

Although Boolean algebra is a useful tool for simplifying logic expressions, we can never be absolutely certain that an expression simplified by Boolean algebra alone is the simplest possible expression. On the other hand, the *map method* gives us a systematic way for discovering the simplest possible expression, provided the rules for using the method are strictly observed. The basis of this graphical technique is the *Karnaugh map*, also called a *Veitch diagram,* on which 1s are plotted and grouped together in a way that reveals the simplest product terms of a logic expression.

A Karnaugh map is a chart, or grid, containing boxes, called *cells*, each of which represents one of the 2^n possible products that can be formed from n variables. Thus, a 2-variable map contains $2^2 = 4$ cells, a 3-variable map contains $2^3 = 8$ cells, and so forth. Figure 4.3 shows outlines of 2-, 3-, and 4-variable maps.

A convenient way to assign product terms to the cells of a Karnaugh map is to label each row and each column of the map with a variable, with its complement, or with a combination of variables and complements. The product term corresponding to a given cell is then the product of all variables in the row and column where the cell is located. Figure 4.4 shows two ways to label the rows and columns of a 2-variable map and the product terms corresponding to each cell.

Notice this important fact about the cell assignments in Figure 4.4: When moving from one cell to the next along any row or from one cell to the next along any column, *one and only one variable in the product term changes* (to a complemented or to an uncomplemented form). For example, in Figure 4.4(a) the only change that occurs in moving along the top row from $\overline{A}\,\overline{B}$ to $\overline{A}B$ is the change from $\overline{B}$ to B. Similarly, the only change that occurs in moving down the left column from $\overline{A}\,\overline{B}$ to $A\overline{B}$ is the change from

Figure 4.3
Outlines of 2-, 3-, and 4-variable Karnaugh maps.

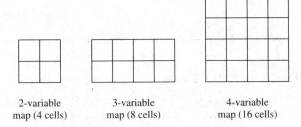

2-variable
map (4 cells)

3-variable
map (8 cells)

4-variable
map (16 cells)

Figure 4.4
Two ways to assign product terms to a 2-variable Karnaugh map.

	$\bar{B}$	B
$\bar{A}$	$\bar{A}\bar{B}$	$\bar{A}B$
A	$A\bar{B}$	AB

	B	$\bar{B}$
A	AB	$A\bar{B}$
$\bar{A}$	$\bar{A}B$	$\bar{A}\bar{B}$

(a)   (b)

$\bar{A}$ to A. *Every* Karnaugh map, no matter how many variables, must be laid out so that a single change occurs when moving between any two horizontally adjacent cells or between any two vertically adjacent cells. Figure 4.5 shows how the rows and columns of 3- and 4-variable maps can be labeled so that this property is realized. These are not the only schemes that satisfy the requirement. Note that the labels along each row and column must conform to the single-change rule.

To plot a logic expression on a Karnaugh map, we write a 1 in each cell that corresponds to a product term in the expression. Thus, a logic expression that is to be simplified using a Karnaugh map should be written (or rewritten) in a sum-of-products form. This form can be obtained by Boolean manipulation if necessary or by constructing a truth table and using the methods of Section 4–5. It is not necessary to obtain a *standard* sum-of-products form, as we demonstrate presently. Figure 4.6 shows examples of plots of 2- and 3-variable logic expressions.

To simplify a logic expression plotted on a map, we first encircle groups of *adjacent* cells containing 1s. Adjacent cells are those that adjoin each other along a row or along a column but *not* on a diagonal. Each group of adjacent cells that we encircle must contain a number of 1s equal to a power of 2—i.e., 1, 2, 4, 8, ... (A group can consist of a single cell containing a 1.) *The groups must be as large as possible,* subject to the requirement that the number of 1s they contain is 1, 2, 4, ... Figure 4.7 illustrates this point. Although two groups of two 1s can be formed as shown (*a*), these must be combined into a single large group containing four 1s, as shown in (*b*).

When grouping cells containing 1s, we regard the two outer (rightmost and leftmost) columns as being adjacent and the upper and lower rows as being adjacent. For example,

Figure 4.5
Row and column designations for 3- and 4-variable Karnaugh maps. Note that only one variable changes between adjacent row designations and between adjacent column designations.

	$\bar{B}\bar{C}$	$\bar{B}C$	BC	$B\bar{C}$
$\bar{A}$	$\bar{A}\bar{B}\bar{C}$	$\bar{A}\bar{B}C$	$\bar{A}BC$	$\bar{A}B\bar{C}$
A	$A\bar{B}\bar{C}$	$A\bar{B}C$	ABC	$AB\bar{C}$

(a)

	$\bar{C}\bar{D}$	$\bar{C}D$	CD	$C\bar{D}$
$\bar{A}\bar{B}$	$\bar{A}\bar{B}\bar{C}\bar{D}$	$\bar{A}\bar{B}\bar{C}D$	$\bar{A}\bar{B}CD$	$\bar{A}\bar{B}C\bar{D}$
$\bar{A}B$	$\bar{A}B\bar{C}\bar{D}$	$\bar{A}B\bar{C}D$	$\bar{A}BCD$	$\bar{A}BC\bar{D}$
AB	$AB\bar{C}\bar{D}$	$AB\bar{C}D$	$ABCD$	$ABC\bar{D}$
$A\bar{B}$	$A\bar{B}\bar{C}\bar{D}$	$A\bar{B}\bar{C}D$	$A\bar{B}CD$	$A\bar{B}C\bar{D}$

(b)

Figure 4.6
Examples of plots of 2- and 3-variable logic expressions. A 1 is written in each cell corresponding to a product term in the expression.

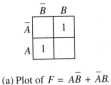

(a) Plot of $F = A\overline{B} + \overline{A}B$.

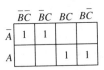

(b) Plot of $F = \overline{A}\,\overline{B}\,\overline{C} + \overline{A}\,\overline{B}C + AB\overline{C} + AB\overline{C}$.

Figure 4.7
Adjacent cells containing 1s must be encircled to form the largest *possible groups (provided the number of 1s is a power of 2).*

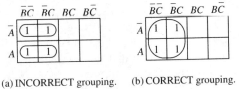

(a) INCORRECT grouping.   (b) CORRECT grouping.

$$F = \overline{A}\,\overline{B}\,\overline{C} + \overline{A}\,\overline{B}C + A\overline{B}\,\overline{C} + A\overline{B}C$$

the columns labeled $\overline{B}\overline{C}$ and $B\overline{C}$ in Figure 4.7 are adjacent. Notice that there is only one variable change (B to $\overline{B}$) when going from the rightmost column to the leftmost column. The rows and columns of a map must be labeled so that these border columns and border rows also obey the single-change rule. We can visualize a map as a cylinder because the two vertical borders are touching and simultaneously as another cylinder with the two horizontal borders touching. Thus, cells containing 1s can be grouped when they occupy border rows or border columns. Figure 4.8 shows examples of this kind of grouping in a 4-variable map.

A second rule concerning the grouping of adjacent cells is that we must form as *few* groups as necessary to include every 1 in at least one group. In some cases, this rule may require that a cell containing a 1 be the one and only group containing that 1. However,

Figure 4.8
Grouping adjacent cells in border columns and border rows.

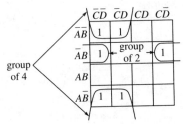

$$F = \overline{A}\,\overline{B}\,\overline{C}\overline{D} + \overline{A}\,\overline{B}\,\overline{C}D + A\overline{B}\,\overline{C}\overline{D} + A\overline{B}\,\overline{C}D + AB\overline{C}\overline{D} + \overline{A}B\overline{C}\overline{D}$$

Figure 4.9
Three adjacent 1s are grouped by overlapping two groups of two 1s.

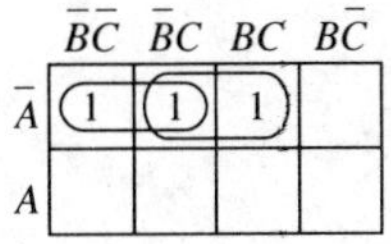

$$F = \overline{A}\overline{B}\overline{C} + \overline{A}\overline{B}C + \overline{A}B\overline{C}$$

groups may overlap, so it is often possible to enclose a 1 in a larger group by including it with 1s in another group. Figure 4.9 shows an example. Notice that

1. All 1s are in at least one group.
2. By overlapping the groups, we are able to form the *largest* possible groups (two groups of 2 instead of one group of 2 and one group of 1).
3. The fewest possible groups (2) are used to enclose all 1s.

When all the 1s on a Karnaugh map have been grouped in accordance with the rules we have described, the logic expression plotted on the map can be simplified. Each group represents a product term whose variables are those—and only those—that are *common* to every cell in the group. For example, only the variables $\overline{A}$ and $\overline{B}$ are common to the leftmost group in Figure 4.9, so that group represents the product term $\overline{A}\overline{B}$. Similarly, only the variables $\overline{A}$ and C are common to the rightmost group, so it represents the product term $\overline{A}C$. The simplified expression is the sum of all the product terms represented by the groups. Thus, in Figure 4.9, the simplified expression is $F = \overline{A}\overline{B} + \overline{A}C$.

Example 4–6

Use a Karnaugh map to simplify each of the following logic expressions.

a. $W = A\overline{B} + AB$

b. $F = \overline{A}\overline{B}\overline{C} + A\overline{B}\overline{C} + \overline{A}\overline{B}C + \overline{A}BC$

c. $V = \overline{X}\overline{Y}\overline{Z} + X\overline{Y}\overline{Z} + XYZ + \overline{X}Y\overline{Z}$

d. $M = \overline{A}\overline{B}\overline{C}\overline{D} + \overline{A}\overline{B}C\overline{D} + \overline{A}B\overline{C}D + \overline{A}BCD + \overline{A}B\overline{C}\overline{D} + \overline{A}BC\overline{D} + A\overline{B}C\overline{D}$

e. $R = (\overline{X}_1 + \overline{X}_2 + X_3 + X_4)(\overline{X}_1 + \overline{X}_2 + X_3 + \overline{X}_4)(\overline{X}_1 + \overline{X}_2 + \overline{X}_3 + \overline{X}_4)$

Solution

a. $W = A\overline{B} + AB$ is a 2-variable expression, and the 2-variable map is shown in Figure 4.10(*a*). We see that the two 1s can be enclosed by a single group whose common term is A. Thus, the simplified expression is $W = A$. In this case, the simplification is easy to verify using Boolean algebra:

$$\begin{aligned} W &= A\overline{B} + AB \\ &= A(\overline{B} + B) \\ &= A(1) = A \end{aligned}$$

b. The 3-variable map for F is shown in Figure 4.10(*b*). Note that we do *not* group the 1 in $\overline{A}\overline{B}\overline{C}$ with the 1 in $\overline{A}BC$ because that would create a third, unnecessary group.

c. The Karnaugh map for this 3-variable expression is shown in Figure 4.10(*c*). Notice how the adjacent 1s in the upper left and upper right corners are combined into the group $\overline{X}\overline{Z}$. Also note that the group $\overline{Y}\overline{Z}$ is drawn to overlap the group $\overline{X}\overline{Z}$, so the 1 in $X\overline{Y}\overline{Z}$ is included in as large a group as possible. The term XYZ has no adjacent 1s, so it must be treated as a single-1 group.

(a) $W = A\bar{B} + AB = A$

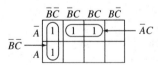

(b) $F = \bar{A}\bar{B}\bar{C} + A\bar{B}\bar{C} + \bar{A}\bar{B}C + \bar{A}BC = \bar{B}\bar{C} + \bar{A}C$

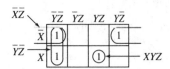

(c) $V = \bar{X}\bar{Y}\bar{Z} + X\bar{Y}\bar{Z} + XYZ + \bar{X}Y\bar{Z}$
$= \bar{X}\bar{Z} + \bar{Y}\bar{Z} + XYZ$

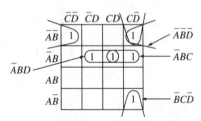

(d) $M = \bar{A}\bar{B}\bar{C}\bar{D} + \bar{A}\bar{B}C\bar{D} + \bar{A}\bar{B}CD + \bar{A}BCD$
$+ \bar{A}BC\bar{D} + \bar{A}B\bar{C}\bar{D} + A\bar{B}C\bar{D}$
$= \bar{A}\bar{B}\bar{D} + \bar{A}BD + \bar{B}C\bar{D} + \bar{A}BC$

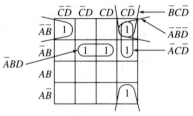

(e) $M = \bar{A}\bar{B}\bar{C}\bar{D} + \bar{A}\bar{B}C\bar{D} + \bar{A}B\bar{C}\bar{D} + \bar{A}BCD + \bar{A}BC\bar{D} + \bar{A}\bar{B}CD + A\bar{B}C\bar{D}$
$= \bar{A}\bar{B}\bar{D} + \bar{A}BD + \bar{B}C\bar{D} + \bar{A}C\bar{D}$

X_1	X_2	X_3	X_4	R
0	0	0	0	1
0	0	0	1	1
0	0	1	0	1
0	0	1	1	1
0	1	0	0	1
0	1	0	1	1
0	1	1	0	1
0	1	1	1	1
1	0	0	0	1
1	0	0	1	1
1	0	1	0	1
1	0	1	1	1
1	1	0	0	0 $\leftarrow X_1 X_2 \bar{X}_3 \bar{X}_4$
1	1	0	1	0 $\leftarrow X_1 X_2 \bar{X}_3 X_4$
1	1	1	0	1
1	1	1	1	0 $\leftarrow X_1 X_2 X_3 X_4$

(f) $R = (\bar{X}_1 + \bar{X}_2 + X_3 + X_4)(\bar{X}_1 + \bar{X}_2 + X_3 + \bar{X}_4)(\bar{X}_1 + \bar{X}_2 + \bar{X}_3 + \bar{X}_4)$

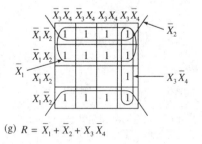

(g) $R = \bar{X}_1 + \bar{X}_2 + X_3 \bar{X}_4$

Figure 4.10
Example 4–6.

d. This 4-variable expression can be simplified in *two* different ways, as shown in Figure 4.10(*d*) and (*e*). The 1 in cell $\bar{A}BC\bar{D}$ can be grouped with either the 1 in cell $\bar{A}BCD$ [Figure 4.10(*d*)] or with the 1 in cell $\bar{A}\bar{B}C\bar{D}$ [Figure 4.10(*e*)]. Thus, the simplified expression for M can include either the term $\bar{A}BC$ or the term $\bar{A}C\bar{D}$. As an exercise, verify that the truth tables for the two simplified expressions are identical.

e. Since the expression for R is not in a sum-of-products form, we must find a way to rewrite it in that form. The easiest way is to construct a truth table, realizing that the three terms given in a product-of-sums form correspond to input combinations for which $R = 0$. Using the methods of Section 4–5, we know that $R = 0$ for the combinations $X_1X_2\overline{X}_3\overline{X}_4$, $X_1X_2\overline{X}_3X_4$, and $X_1X_2X_3X_4$. As shown in the truth table in Figure 4.10(f), every other input combination makes $R = 1$. The 1s are plotted in the Karnaugh map shown in Figure 4.10(g). Note that we can form two groups of eight 1s and one group of four 1s, giving $R = \overline{X}_1 + \overline{X}_2 + X_3\overline{X}_4$.

Drill 4–6

Use a Karnaugh map to simplify $F = (\overline{X}_1 + \overline{X}_2 + X_3 + \overline{X}_4)(\overline{X}_1 + \overline{X}_2 + \overline{X}_3 + \overline{X}_4)$.
Answer: $F = \overline{X}_1 + \overline{X}_2 + \overline{X}_4$.

Example 4–6 illustrates the fact that the larger a group of 1s, the simpler the product term it represents. The number of variables in the product term represented by a group also depends on the total number of variables in the map:

Number of variables in map	Size of group	Number of variables in product term represented by the group
2	1	2
	2	1
3	1	3
	2	2
	4	1
4	1	4
	2	3
	4	2
	8	1

If a logic expression includes a product that contains fewer than the maximum number of variables, we insert a 1 in *every* cell that includes that product. For example, in a 3-variable map, the term $A\overline{B}$ in a logic expression produces a 1 in cells $A\overline{B}\overline{C}$ and $A\overline{B}C$. In the same map, the term C produces 1s in cells $\overline{A}\overline{B}C$, $\overline{A}BC$, $A\overline{B}C$, and ABC. The next example illustrates such a case.

Example 4–7

The specifications for a submarine state that a SURFACE command shall be initiated automatically under certain conditions related to oxygen level (X), battery charge (B), and fresh water supply (W). Each of these variables becomes 0 if the quantity it represents is dangerously low. The SURFACE command is to be initiated if

1. Battery charge and oxygen level are both low; OR
2. Fresh water is low, battery charge is high, and oxygen is low; OR
3. Fresh water is high and oxygen level is low.

Using a Karnaugh map, find the simplest possible logic expression for SURFACE.

Solution.   The product terms corresponding to the combinations requiring a SURFACE command are:

1. $\overline{B}\overline{X}$
2. $\overline{W}B\overline{X}$
3. $W\overline{X}$

Figure 4.11
Example 4–7.

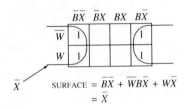

Thus, SURFACE $= \overline{B}\,\overline{X} + \overline{W}B\overline{X} + W\overline{X}$. Figure 4.11 shows the 3-variable Karnaugh map. Note that the term $\overline{B}\,\overline{X}$ produces a 1 in the *two* cells $\overline{W}\,\overline{B}\,\overline{X}$ and $W\overline{B}\,\overline{X}$. Similarly, the term $W\overline{X}$ produces a 1 in the two cells $W\overline{B}\,\overline{X}$ and $WB\overline{X}$. (It does not matter that $\overline{B}\,\overline{X}$ and $W\overline{X}$ both produce a 1 in cell $W\overline{B}\,\overline{X}$; we insert a single 1 in that cell.) Grouping the 1s shows that the logic expression reduces to

$$\text{SURFACE} = \overline{X}$$

In other words, only the oxygen level must be monitored to control the SURFACE command.

Drill
4–7

Use a Karnaugh map to simplify $F = \overline{A}\,\overline{B}\,\overline{C} + \overline{A}B + \overline{A}C + \overline{A} + \overline{C}\,\overline{D} + AB\overline{C}D$.
Answer: $F = \overline{A} + \overline{C}\,\overline{D}$.

Five-Variable Karnaugh Maps

Karnaugh maps are most conveniently used to simplify 2-, 3-, and 4-variable expressions. They can also be used to simplify 5-variable expressions, although the procedure is somewhat more involved. A 5-variable map requires $2^5 = 32$ cells, but adjacent cells are difficult to identify on a single 32-cell map. Therefore, two 16-cell maps are generally used. If the variables are A, B, C, D, and E, two identical 16-cell maps containing A, B, C, and D can be constructed. One map is then used for E and the other for $\overline{E}$. See Figure 4.12. Every cell in one map is adjacent to the corresponding cell in the other map, because only one variable (E) changes between such corresponding cells. Thus, every row on one map is adjacent to the corresponding row (the one occupying the same position) on the other map, as are corresponding columns. Also, the right and left border columns *within* each 16-cell map are adjacent, just as they are in any 16-cell map, as are the top and

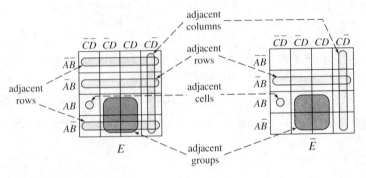

Figure 4.12
A 5-variable Karnaugh map with examples of adjacencies.

Figure 4.13
Example 4–8.

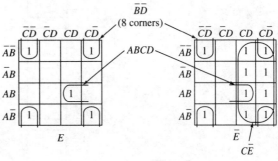

$$F = ABCD + \overline{B}\overline{D} + C\overline{E}$$

bottom rows. However, the right border column of one map is not adjacent to the left border column of the other map, since those are not corresponding columns. Nor is the top row of one map adjacent to the bottom row of the other.

Example 4–8

Use the Karnaugh map method to simplify

$$F = ABCDE + C\overline{D}\overline{E} + \overline{A}\overline{B}CD + \overline{B}C\overline{D} + ABC\overline{D}\overline{E} + CD\overline{E}$$

Solution.   Figure 4.13 shows the plot of F on two 16-cell maps. Note that a 3-variable term in the logic expression produces a 1 in 4 cells, a 4-variable term produces a 1 in 2 cells, and a 5-variable term produces a 1 in 1 cell of a 5-variable Karnaugh map.

The 1s in Figure 4.13 are grouped to produce two 8-cell groups and one 2-cell group. Notice that the eight corners of the two 16-cell maps form one group, which reduces to $\overline{B}\overline{D}$. Also note that cells $ABCDE$ and $ABCD\overline{E}$ are adjacent and that they form the one 2-cell group, $ABCD$. The simplified expression for F is then $F = C\overline{E} + \overline{B}\overline{D} + ABCD$.

Drill 4–8

Use a Karnaugh map to simplify $W = \overline{C}DE + A\overline{B}CD + \overline{A}B\overline{C} + \overline{A}\overline{B}C\overline{D}E + \overline{A}BCD + \overline{C}D\overline{E}$.
Answer: $W = \overline{C}D + \overline{A}BD + A\overline{B}CD$.

4–7   AN EXAMPLE USING ANSI/IEEE SYMBOLS

Example 4–9

Use Boolean algebra to simplify the logic expression implemented by the circuit in Figure 4.14. Then draw the simplified circuit using ANSI/IEEE standard symbols.

Solution.   As shown in Figure 4.15(a), the logic expression implemented by the circuit is

$$F = \overline{\overline{A} + B + C} + \overline{\overline{\overline{A}} + \overline{B} + \overline{D}} + \overline{(\overline{A} + \overline{B} + \overline{C})(\overline{A} + C + D)} + \overline{A}B\overline{C}D + \overline{A}\overline{B}C$$

Removing inversion bars, we obtain

$$F = \overline{A}\overline{B}\overline{C} + \overline{\overline{\overline{A}}\,\overline{\overline{B}}\,\overline{\overline{D}}} + (\overline{A} + \overline{B} + \overline{C}) + (\overline{A} + C + D) + \overline{A}B\overline{C}D + \overline{A}\overline{B}C$$
$$= \overline{A}\overline{B}\overline{C} + ABD + ABC + A\overline{C}\overline{D} + \overline{A}B\overline{C}D + \overline{A}\overline{B}C$$

This expression is plotted on a 4-variable Karnaugh map, as shown in Figure 4.15(b). We see that it simplifies to $F = \overline{A}\overline{B} + AB + \overline{C}\overline{D} = A \odot B + \overline{C + D}$. Thus, the expression

Figure 4.14
Example 4–9.

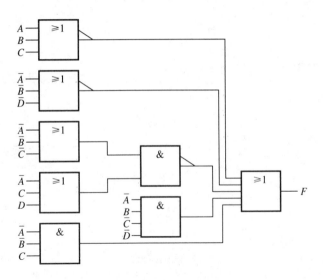

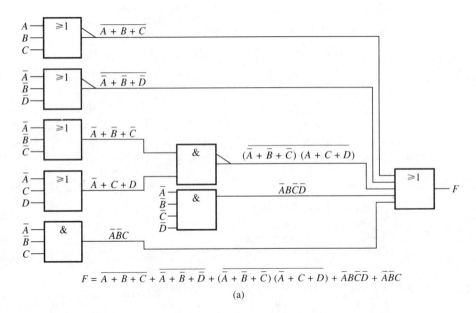

$$F = \overline{A + B + C} + \overline{\overline{A} + \overline{B} + \overline{D}} + \overline{(\overline{A} + \overline{B} + \overline{C})(\overline{A} + C + D)} + \overline{A}B\overline{C}\overline{D} + \overline{A}\overline{B}C$$

(a)

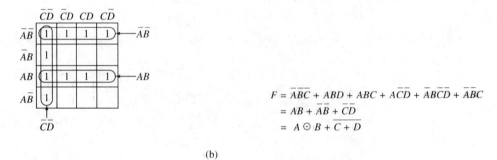

$$F = \overline{A}\,\overline{B}\,\overline{C} + ABD + ABC + A\overline{C}\overline{D} + \overline{A}B\overline{C}\overline{D} + \overline{A}\,\overline{B}C$$
$$= AB + \overline{A}\overline{B} + \overline{C}\,\overline{D}$$
$$= A \odot B + \overline{C + D}$$

(b)

Figure 4.15
Example 4–9.

92

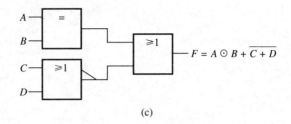

(c)

Figure 4.15
(Continued)

can be implemented with one coincidence gate, one NOR gate, and one OR gate, as shown in Figure 4.15(c). Besides reducing the number of gates required, the simplification has eliminated the need for the complements of the variables.

EXERCISES

Section 4–3

4.1 Using Boolean algebra, simplify the following logic expressions as much as possible.
(a) $X = A\bar{B} + BA$
(b) $Z = \bar{B}C(ABC + A\bar{B}C)$
(c) $M = V\bar{W}(\bar{V} + W)(V\bar{W} + V)$
(d) $T = T_1\bar{T}_2T_3 + T_1\bar{T}_2$
(e) $V = (A + B + C)(\bar{A} + \bar{B} + \bar{C})A$
(f) $B = X_1\bar{X}_2 + X_1X_2 + \bar{X}_1X_2 + \bar{X}_1\bar{X}_2$

4.2 Using Boolean algebra, simplify the following logic expressions as much as possible:
(a) $R = \bar{X}Y + YX$
(b) $Z = AB\bar{C}(AB + \bar{A}C)$
(c) $A = Z_1Z_2 + Z_2\bar{Z}_1 + Z_1Z_2Z_3$
(d) $W = A \oplus B + A \odot B$
(e) $X = AB(\bar{A} + \bar{B})(A + AB)$
(f) $N = (WX + W\bar{Y})(X + W) + WX(\bar{X} + \bar{Y})$

4.3 A computer circuit is to generate a 1 if the decimal value of the binary number $A_2A_1A_0$ is 4 or 5 or 6 or 7. Write a logic expression that performs the required logic and simplify it as much as possible.

4.4 Write the logic expression for OUT in Figure 4.16, simplify it as much as possible, and draw a logic diagram that implements the simplified expression.

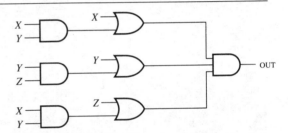

Figure 4.16
Exercise 4–4.

Section 4–4

4.5 Using Boolean algebra, simplify each of the following logic expressions as much as possible.
(a) $Y = A + AB + ABC + ABCD$
(b) $Z = A(A + AB)(A + ABC)(A + ABCD)$
(c) $A = X + Y + \bar{X}\bar{Y}Z$
(d) $B = \overline{\overline{XY}\,\overline{YZ}}$
(e) $W = \overline{\overline{X_1\bar{X}_2} + \overline{\bar{X}_2X_3}}$
(f) $C = (A_1\bar{A}_2A_3)(\overline{A_1\bar{A}_2}) + A_2A_3$

4.6 Using Boolean algebra, simplify each of the following logic expressions as much as possible.
(a) $E = XYZ + X + WZYX + ZW$
(b) $W = (X_3 + X_3X_4)X_3(X_3 + X_1X_2X_3)(X_3 + X_2)$

(c) $Z = \bar{A} + \bar{B} + ABC$

(d) $M = \overline{(\bar{X} + \bar{Y})Y\bar{Z}}$

(e) $X = \overline{A(\overline{BC})} + \bar{A} + \overline{B(\overline{BC})}$

(f) $V = [\bar{Z}_1 + \overline{(Z_2 + Z_1)}][\bar{Z}_2\overline{(Z_2Z_3)}][Z_3\overline{(Z_1Z_2\bar{Z}_3)}]$

4.7 Use DeMorgan's theorem to prove $\overline{A \oplus B} = A \odot B$ and $\overline{A \odot B} = A \oplus B$.

4.8 Prove the following theorem:
$(A + B)(A + C) = A + BC$.

4.9 Draw the simplest possible logic diagram that implements the output of the logic diagram shown in Figure 4.17.

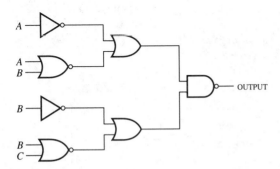

Figure 4.17
Exercise 4–9.

4.10 Draw the simplest possible logic diagram that implements the output of the logic diagram shown in Figure 4.18.

Section 4–5

4.11 Write the standard sum-of-products form and the standard product-of-sums form for the

logic expression whose truth table is each of the following.

(a)

A	B	X
0	0	0
0	1	0
1	0	1
1	1	1

(b)

X_1	X_2	X_3	Z
0	0	0	1
0	0	1	0
0	1	0	0
0	1	1	1
1	0	0	1
1	0	1	1
1	1	0	0
1	1	1	1

4.12 Write the standard sum-of-products form and the standard product-of-sums form for the logic expression whose truth table is each of the following. (Continued on p. 95.)

(a)

A	B	C	W
0	0	0	1
0	0	1	1
0	1	0	1
0	1	1	0
1	0	0	1
1	0	1	0
1	1	0	0
1	1	1	1

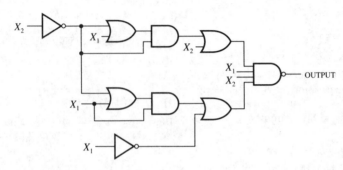

Figure 4.18
Exercise 4–10.

(b)

W	X	Y	Z	F
0	0	0	0	1
0	0	0	1	0
0	0	1	0	1
0	0	1	1	0
0	1	0	0	1
0	1	0	1	0
0	1	1	0	1
0	1	1	1	1
1	0	0	0	1
1	0	0	1	1
1	0	1	0	1
1	0	1	1	0
1	1	0	0	0
1	1	0	1	1
1	1	1	0	1
1	1	1	1	1

Section 4–6

4.13 Use a Karnaugh map to simplify each of the following logic expressions as much as possible.
(a) $F = A\bar{B} + \bar{A}B + AB$
(b) $G = \bar{X}\bar{Y}\bar{Z} + \bar{X}Y\bar{Z} + X\bar{Y}\bar{Z} + \bar{X}YZ + XYZ$
(c) $H = \bar{A}\bar{B}CD + A\bar{B}\bar{C}D + \bar{A}B\bar{C}D + AB\bar{C}D + \bar{A}\bar{B}\bar{C}D + A\bar{B}\bar{C}D + ABCD$

4.14 Use a Karnaugh map to simplify each of the following logic expressions as much as possible.
(a) $X = \bar{A}B + A\bar{B} + \bar{A}\bar{B}$
(b) $W = \bar{X}\bar{Y}Z + \bar{X}YZ + XYZ + X\bar{Y}Z + \bar{X}Y\bar{Z}$
(c) $A = \bar{X}_1\bar{X}_2X_3X_4 + X_1\bar{X}_2X_3\bar{X}_4 + X_1\bar{X}_2X_3\bar{X}_4 + \bar{X}_1\bar{X}_2\bar{X}_3\bar{X}_4 + \bar{X}_1\bar{X}_2X_3\bar{X}_4 + \bar{X}_1\bar{X}_2X_3\bar{X}_4$

4.15 Use a Karnaugh map to simplify each of the following logic expressions as much as possible.
(a) $F = \bar{A}C + \bar{B}C + A\bar{B}\bar{C} + \bar{A}B$
(b) $G = \bar{B}CD + C\bar{D} + \bar{A}\bar{B}C\bar{D} + \bar{A}BC$
(c) $T = \bar{W}X + Y + W\bar{Y}Z + \bar{W}X\bar{Y}Z + \bar{Y}Z$
(d) $W = (\bar{A} + B + \bar{C})(A + B + C)(A + B + \bar{C})$
(e) $B = \bar{A}_1A_2 + \bar{A}_2A_3 + A_1 + \bar{A}_1\bar{A}_2$

4.16 Use a Karnaugh map to simplify each of the following logic expressions as much as possible.
(a) $F = AB + \bar{A}\bar{B} + ABC + \bar{A}B\bar{C}$
(b) $N = AB\bar{C} + A\bar{D} + \bar{A}\bar{B}\bar{C}\bar{D} + \bar{C}D + \bar{A}C\bar{D} + \bar{A}\bar{B}C\bar{D}$
(c) $K = \bar{W}Z + \bar{Z} + W X\bar{Y} + W\bar{X}\bar{Y}Z + W\bar{X}Y$

(d) $M = X_2X_3 + \bar{X}_1\bar{X}_2X_3 + \bar{X}_3 + X_1\bar{X}_2X_3$

(e) $W = (\overline{A\bar{B}\bar{C}})(\overline{A\bar{B}C})(\overline{ABC})$

4.17 Use a Karnaugh map to simplify
$$F = \bar{A}\bar{B}CD + BCDE + A\bar{B}\bar{C}\bar{D}E + CD\bar{E} + \bar{A}B\bar{C}\bar{D}E + A\bar{B}CD + A\bar{B}\bar{C}\bar{E} + \bar{A}BCDE$$

4.18 Use a Karnaugh map to simplify
$G = \overline{UVWXYZ}$ where $U = (A + B + \bar{C})$, $V = (\bar{A} + B + E)$, $W = (\bar{A} + \bar{B} + \bar{C} + D)$, $X = (A + B + C + E)$, $Y = (\bar{A} + B + \bar{C} + D)$, and $Z = (A + B + C + \bar{E})$

4.19 A lawn-sprinkling system at a large estate is to be automatically controlled by certain combinations of the following variables:

Season ($S = 1$ if summer, 0 otherwise)
Moisture content of soil ($M = 1$ if high, 0 if low)
Outside temperature ($T = 1$ if high, 0 if low)
Outside humidity ($H = 1$ if high, 0 if low)

The landscape architect designing the system wants the sprinkler to be turned ON under any of the following circumstances:

1. Moisture content is low in the winter; or
2. Temperature is high and moisture content is low in the summer; or
3. Temperature is high and humidity is high in the summer; or
4. Temperature is low and moisture content is low in the summer; or
5. Temperature is high and humidity is low.

Use a Karnaugh map to find the simplest possible logic expression involving the variables S, M, T, and H for turning on the sprinkler system (ON = 1).

4.20 The inputs to a certain computer circuit are the 4 bits A_3, A_2, A_1, and A_0 of the binary number $A_3A_2A_1A_0$. The circuit is required to produce a 1 if and only if all the following conditions hold:

1. The most significant bit is 1 or any of the other bits are 0; AND
2. A_2 is 1 or any of the other bits are 0; AND
3. Any of the four bits are zero.

Use a Karnaugh map to find the simplest possible logic expression for the circuit.

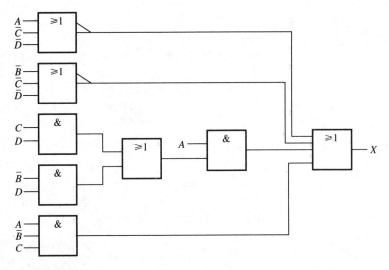

Figure 4.19
Exercise 4–21.

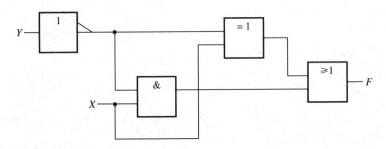

Figure 4.20
Exercise 4–22.

Section 4–7

4.21 Use Boolean algebra to simplify the logic expression implemented by the circuit in Figure 4.19. Then draw the simplified circuit using ANSI/IEEE logic symbols.

4.22 Use Boolean algebra to simplify the logic expression implemented by the circuit in Figure 4.20. Then draw the simplified circuit using ANSI/IEEE standard symbols.

CHALLENGING EXERCISES

4.23 Jones, Smith, and Brown are prosecution witnesses at Wilson's trial. To corroborate their testimony, each witness is given a lie detector test. The jury is willing to convict Wilson if Jones and Smith both pass the lie detector test. The jury is also willing to convict if either Jones or Smith (or both) fail the test but Brown passes it. Use Boolean algebra to find a simpler way to express the terms under which Wilson will be convicted.

4.24 A certain proposition is true when it is not true that conditions A and B both hold. It is also true when conditions A and B both hold but condition C does not. Is the proposition true when it is not true that conditions B and C both hold? Use Boolean algebra to justify your answer.

4.25 Without using truth tables, prove that

$$A[B(\overline{\overline{CD}})](\overline{\overline{ABC}})B\overline{D} =$$

$$\overline{[(\overline{A} + B) + C]} + \overline{(\overline{B} + D)} +$$

$$\overline{A + [\overline{B} + (\overline{\overline{C + D}})]}$$

4.26

(a) Find a way to implement the AND function AB using *only* NOR gates.

(b) Find a way to implement the OR function $A + B$ using *only* NAND gates. (*Hint:* $\overline{A + A} = \overline{A}$ and $\overline{AA} = \overline{A}$.)

4.27 Find the *standard* sum-of-products form for the logic expression

$$W = ABC + B\overline{C}D$$

4.28 Find the *standard* sum-of-products form for the logic expression

$$F = VW\overline{X}Y + \overline{W}XYZ$$

4.29 Use a Karnaugh map to simplify the logic expression

$$F = \overline{A\overline{B}C + A\overline{B}\overline{D} + BCD + B\overline{\overline{C}}}$$

(*Hint:* Find the 1s by plotting the 0s.)

4.30 Use a Karnaugh map to simplify

$$F = \overline{W(\overline{X} + Y)\overline{Z} + X(\overline{Y} + Z)}$$

4.31 Three 1-bit binary numbers, A, B, and C, are added together. Use a Karnaugh map to find the simplest possible expression for the sum bit.

4.32 Three 1-bit binary numbers, X, Y, and Z, are added together. Use a Karnaugh map to find the simplest possible expression for the carry bit in the sum.

Chapter 5

PRACTICAL CONSIDERATIONS IN LOGIC DESIGN

5–1   NAND AND NOR LOGIC

In Chapter 3, we discussed the fact that several logic gates of the same type are often fabricated in one integrated circuit chip. It would obviously be uneconomical to implement a logic expression requiring, say, one NOR gate, using a chip that contains four NOR gates. For that reason, it is useful to be able to implement logic expressions using only gates of a single type. *Any* logic expression can be implemented using only NAND gates or only NOR gates.

Figure 5.1 shows how the three basic logic functions, AND, OR, and INVERT, can be implemented using only NAND gates *(a)* and using only NOR gates *(b)*. Note that the INVERT operation is achieved by connecting the input signal to both (or all) the gate inputs.

To implement a particular logic expression using NAND or NOR logic, we first draw a logic diagram using conventional AND, OR, and INVERT symbols. We can then replace each such symbol with its respective NAND- or NOR-gate equivalent, as shown in Figure 5.1. If the resulting diagram has two inverters in series, they can both be eliminated. The next example demonstrates the procedure.

Example 5–1

Implement the exclusive-OR function using (a) NAND logic and (b) NOR logic.

Solution.   Figure 5.2*(a)* shows the implementation of the exclusive-OR function using conventional gates. Part *(b)* shows how each gate is replaced by its NAND-gate equivalent. Notice in two places there are two NAND gates serving as inverters and connected in series. Each pair can be eliminated, giving the NAND-logic implementation shown in Figure 5.2*(c)*.

Figure 5.3 shows the NOR-logic implementation. Notice again that two pairs of NOR inverters can be eliminated. Also notice that the NOR-logic implementation requires one more gate than its NAND-logic counterpart.

As an exercise, use Boolean algebra to verify that the logic diagrams in Figures 5.2*(c)* and 5.3*(c)* both implement the exclusive-OR.

Drill 5–1

Implement the coincidence function using only NAND gates and again using only NOR gates.

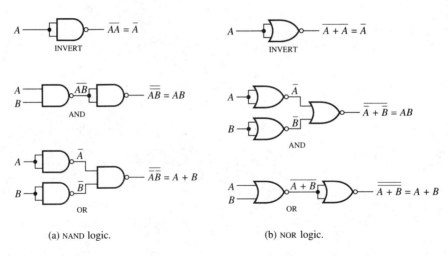

(a) NAND logic.   (b) NOR logic.

Figure 5.1
Implementation of the three basic logic functions using only NAND gates and only NOR gates.

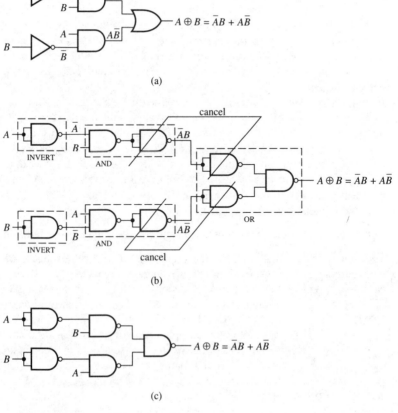

(a)

(b)

(c)

Figure 5.2
Example 5–1.

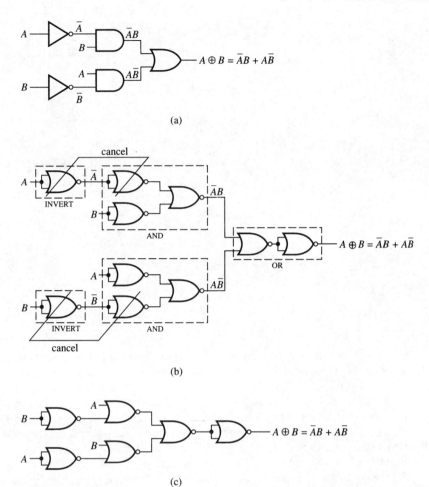

Figure 5.3
Example 5–1.

5–2   ACTIVE-LOW NOTATION

As we have seen in previous examples, logic variables are often given names, such as ON, GO, or START, that reflect the type of action represented or initiated by the variable when it equals 1. In some digital circuits, the action occurs when one (or several) inputs are 0. Such inputs are said to be *active low*. On logic diagrams, an active-low input is shown by placing the inversion bubble at the point where the input signal is connected. We may regard the bubble itself as an inverter. Thus, we can think of a 0 occurring on the external signal line as being inverted and producing a 1 internal to the device and vice versa. This viewpoint, which is simply another way of visualizing logic operations, is often given the unfortunate name ''negative'' logic, as if the circuitry were constructed in a different manner. Such is not the case, and we will avoid that terminology.

Active-low bubbles can be very useful when analyzing logic diagrams, because they can be placed in such a way that they effectively cancel one another. Such cancellations

Figure 5.4
Logically equivalent ways of drawing gates using active-low inputs.

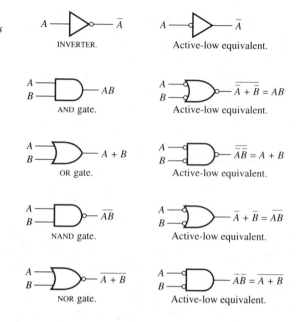

INVERTER.

Active-low equivalent.

AND gate.

Active-low equivalent.

OR gate.

Active-low equivalent.

NAND gate.

Active-low equivalent.

NOR gate.

Active-low equivalent.

eliminate the necessity for writing numerous inversion bars over compound logic expressions. Figure 5.4 shows how some of the gates we have already studied can be redrawn in equivalent ways using active-low inputs. Notice that a NAND gate is equivalent to an OR gate with inverted (active-low) inputs and a NOR gate is equivalent to an AND gate with inverted inputs. Of course, these facts are a direct consequence of DeMorgan's theorem. In practice, logic diagrams are often drawn using one of the logically equivalent forms of a gate, even though the physical circuitry may be different. This practice helps simplify the interpretation of logic expressions, as previously discussed. Figure 5.13 in Section 5–6 shows logically equivalent ways of representing gates using ANSI/IEEE standard symbols and active-low notation.

To illustrate how the active-low notation can simplify writing logic expressions, consider the logic diagram shown in Figure 5.5(*a*). We see that this circuit implements the expression

$$\overline{\overline{AB\overline{C}} + D}$$

With some effort, we can use Boolean algebra to show that this expression is equivalent to

$$AB + C + D$$

Figure 5.5(*b*) shows the same implementation as (*a*) but with active-low equivalents replacing one NAND gate and one inverter. Since consecutive inversion bubbles cancel, as shown, it is readily apparent in this diagram that the output is $AB + C + D$.

5–3 AND/OR AND AND/OR/INVERT LOGIC

Because so many logic functions are expressed in practice as sums of products, manufacturers have developed integrated circuits that implement a variety of sum-of-

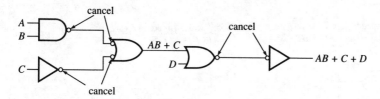

(a) Conventional logic diagram.

(b) Logic diagram using active-low notation.

Figure 5.5
Using active-low notation to simplify interpretation of an expression implemented by a logic diagram.

products forms and their complements. A sum of products is implemented by a group of AND gates whose outputs are ORed together, so these circuits are called AND/OR (AO) gates. Circuits implementing the *complements* of AND/OR functions are available in a greater variety of styles and are called AND/OR/INVERT (AOI) gates. Figure 5.6 shows some of the styles that are available in TTL integrated circuits. Note that a circuit having n AND gates is said to be *n-wide*. The 7452 AO gate shown in the figure is an example of an *expandable gate*, one in which an additional variable — or combination of variables — can be included in the logic operation. For example, if we used one-half of a 7451 to implement $X = \overline{PQ + RS}$ and connected that function to the X-input on the 7452, we would obtain

$$Y = AB + CDE + FG + HI + \overline{PQ + RS}$$

To implement a logic expression using an AOI gate, we must express it as a complemented sum of products. For example, to implement $F = (\overline{A} + B)(C + \overline{D})$ using AOI logic, we can write

$$F = (\overline{A} + B)(C + \overline{D})$$
$$= \overline{\overline{(\overline{A} + B)(C + \overline{D})}}$$
$$= \overline{\overline{(\overline{A} + B)} + \overline{(C + \overline{D})}}$$
$$= \overline{A\overline{B} + \overline{C}D}$$

Example 5–2

Using any of the TTL gates shown in Figure 5.6, draw a logic diagram that implements $F = A\overline{B} + C$ in simplified AOI logic. (Assume each variable and its complement are available as input signals.)

Solution. Figure 5.7(a) shows the truth table for F. From the truth table, we obtain the product-of-sums form

$$F = (A + B + C)(A + \overline{B} + C)(\overline{A} + \overline{B} + C)$$

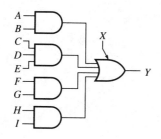

7452 expandable 4-wide AO gate.
$Y = AB + CDE + FG + HI + X$

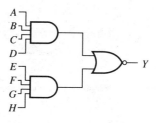

7455 2-wide, 4-input AOI gate.
$Y = \overline{ABCD + EFGH}$

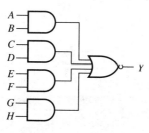

7454 4-wide, 2-input AOI gate.
$Y = \overline{AB + CD + EF + GH}$

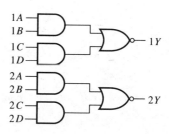

7451 dual 2-wide, 2-input AOI gate.
$Y = \overline{AB + CD}$
(Some versions of this circuit have
two 2-input AND gates and two
3-input AND gates.)

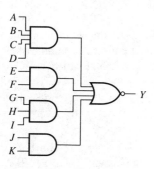

7464 4-2-3-2 AOI gate.
$Y = \overline{ABCD + EF + GHI + JK}$

Figure 5.6
TTL AND/OR (AO) and AND/OR/INVERT (AOI) gates.

Then,

$$F = \overline{\overline{(A + B + C)(A + \overline{B} + C)(\overline{A} + \overline{B} + C)}}$$
$$= \overline{\overline{(A + B + C)} + \overline{(A + \overline{B} + C)} + \overline{(\overline{A} + \overline{B} + C)}}$$
$$= \overline{\overline{A}\,\overline{B}\,\overline{C} + \overline{A}B\overline{C} + AB\overline{C}}$$

The sum-of-products expression under the inversion bar can be simplified using a Karnaugh map, as shown in Figure 5.7(b). Since the entire expression is inverted, we are

Figure 5.7
Example 5–2.

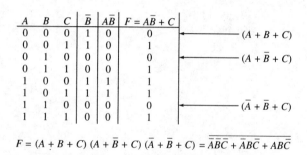

A	B	C	$\bar{B}$	$A\bar{B}$	$F = A\bar{B} + C$	
0	0	0	1	0	0	$(A + B + C)$
0	0	1	1	0	1	
0	1	0	0	0	0	$(A + \bar{B} + C)$
0	1	1	0	0	1	
1	0	0	1	1	1	
1	0	1	1	1	1	
1	1	0	0	0	0	$(\bar{A} + \bar{B} + C)$
1	1	1	0	0	1	

$$F = (A + B + C)(A + \bar{B} + C)(\bar{A} + \bar{B} + C) = \overline{\overline{ABC} + \overline{A}B\overline{C} + AB\overline{C}}$$

(a)

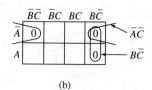

(b)

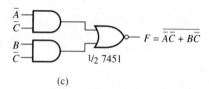

(c)

actually plotting the 0s of F. We see that $F = \overline{\overline{AC} + \overline{BC}}$. Figure 5.7(c) shows how F is implemented using one-half of a 7451 dual 2-wide, 2-input AOI gate.

**Drill
5–2**

Express $A = WX\bar{Y} + Z$ in a simplified form suitable for implementation by AOI logic. *Answer:* $A = \overline{\overline{WZ} + Y\overline{Z} + \overline{X}\overline{Z}}$.

Example 5–2 demonstrates that a sum of products can be expressed as a complemented sum of products by double complementing the product-of-sums form. Note, however, that a simplified AOI expression can be obtained directly by plotting the 0s of the original sum-of-products expression. That is, we can skip the intermediate steps of constructing a truth table and finding the product-of-sums form. It is necessary only to plot the 0s (by first plotting the 1s—all unoccupied cells then contain the 0s), simplify the expression obtained from that 0-plot, and express F as the complement of the simplified expression.

**Example
5–3**

Using any of the TTL gates shown in Figure 5.6, draw a logic diagram that implements $F = B\overline{C}D + \overline{A}B\overline{C} + C\overline{D} + BCD$ in simplified AOI logic. Assume each variable and its complement are available as input signals.

Solution. Figure 5.8(a) shows a plot of the 1s and 0s of F. Grouping the 0s, we find $F = \overline{\overline{A}\overline{B}C + CD}$. Figure 5.8(b) shows the implementation of F using a 7464 AOI gate.

Figure 5.8
Example 5–3.

$$F = B\overline{C}\overline{D} + \overline{A}\overline{B}C + C\overline{D} + B\overline{C}D$$

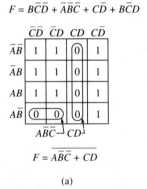

$$F = \overline{A\overline{B}\overline{C} + C\overline{D}}$$

(a)

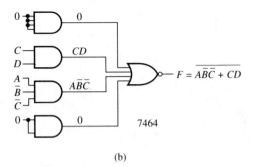

(b)

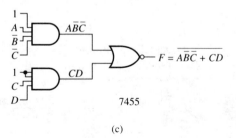

(c)

Note that unused inputs are connected to 0 (ground), since $\overline{A\overline{B}\overline{C} + CD + 0 + 0} = \overline{A\overline{B}\overline{C} + CD}$. Figure 5.8(c) shows the implementation using a 7455 gate. In this case, unused inputs are connected to 1 (+5 V), since $\overline{A\overline{B}\overline{C} \cdot 1 + CD \cdot 1 \cdot 1} = \overline{A\overline{B}\overline{C} + CD}$.

Drill
5–3

Find a simplified expression for $Z = \overline{X}_1 X_2 + X_2\overline{X}_3 + X_1 X_3 X_4$ suitable for implementation by AOI logic.
Answer: $Z = \overline{\overline{X_1\overline{X}_2} + \overline{X_2\overline{X}_3} + X_1 X_3\overline{X}_4}$.

5–4   *DON'T CARE* CONDITIONS

In some practical logic designs, there are combinations of input variables that we *know* will never occur. This is particularly true in logic circuits designed to perform various *coding* operations. We study binary codes in a later chapter and learn that certain bit patterns do not occur because they are not included in the definition of a particular code. These bit patterns correspond to variable combinations that can never appear at the input to a logic circuit. When simplifying a logic expression using a Karnaugh map, we are free to plot any such combination as a 1, if doing so contributes to the simplification, or to disregard it entirely (treat it as a 0). Since it does not matter whether it is treated as a 1 or a 0, such a combination is called a *don't care*. The best way to use *don't cares* on a map is to write *d* in each *don't care* cell. We can then see which *don't cares* should be grouped with other 1s to form larger groups and which *don't cares* should be ignored.

Example 5–4

An 8-4-2-1 *binary-coded decimal* (BCD) is a 4-bit code used to define the decimal digits 0 through 9, as follows:

A_8	A_4	A_2	A_1	Decimal digit
0	0	0	0	0
0	0	0	1	1
0	0	1	0	2
0	0	1	1	3
0	1	0	0	4
0	1	0	1	5
0	1	1	0	6
0	1	1	1	7
1	0	0	0	8
1	0	0	1	9

(Header: **4-bit code** spans A_8 A_4 A_2 A_1; **Decimal digit**)

Notice that the bit combinations 1010 through 1111 are not part of the code.

Design a logic circuit that produces a 1 when its 8-4-2-1 BCD input represents any of the decimal numbers 2, 3, 6, or 7.

Solution. From the truth table, the sum-of-products form for the required expression (OUT) is

$$\text{OUT} = \overline{A_8}\,\overline{A_4}\,A_2\,\overline{A_1} + \overline{A_8}\,\overline{A_4}\,A_2\,A_1 + \overline{A_8}\,A_4\,A_2\,\overline{A_1} + \overline{A_8}\,A_4\,A_2\,A_1$$
$$\quad\;\;(2)\qquad\qquad (3)\qquad\qquad (6)\qquad\qquad (7)$$

Since bit combinations 1010 through 1111 are not part of the code and never appear as inputs, the *don't care* products are

$$A_8\,\overline{A_4}\,A_2\,\overline{A_1} \qquad (1010)$$
$$A_8\,\overline{A_4}\,A_2\,A_1 \qquad (1011)$$
$$A_8\,A_4\,\overline{A_2}\,\overline{A_1} \qquad (1100)$$
$$A_8\,A_4\,\overline{A_2}\,A_1 \qquad (1101)$$
$$A_8\,A_4\,A_2\,\overline{A_1} \qquad (1110)$$
$$A_8\,A_4\,A_2\,A_1 \qquad (1111)$$

The logic expression for OUT and the *don't cares* are plotted on the Karnaugh map shown in Figure 5.9. Notice that four of the *don't cares* are treated as 1s and grouped with the four 1s of OUT to produce a large group of eight. The other two *don't cares* are ignored. *Don't cares* are useful only when grouped with 1s, so the four *don't cares* along row $A_8 A_4$ should not be treated as one group. We see that OUT = A_2, and no logic gates are required.

Figure 5.9
Example 5–4.

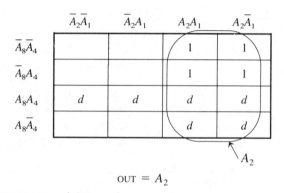

OUT $= A_2$

Find a simplified logic expression for a circuit that produces a 1 when its BCD input represents a decimal number greater than 3.

Answer: OUT $= A_4 + A_8$

5–5   USING CELL NUMBERS IN KARNAUGH MAPS

A simplified notation is often used to facilitate the plotting of logic expressions on Karnaugh maps. In this notation, each cell is assigned a decimal number corresponding to the value of the product term it represents. The assignment of numbers is particularly easy when the bit designations have subscripts equal to the decimal weights they carry. An example is the 8-4-2-1 BCD input $A_8A_4A_2A_1$ of Example 5–4. As shown in Figure 5.10, each cell number is simply the sum of the subscripts of the uncomplemented variables that identify the cell. The notation is also useful when plotting a Karnaugh map directly from a truth table, since it is easy to determine the decimal equivalent of each input combination that produces a 1.

Cell numbers can be used to abbreviate a logic expression in a sum-of-products form as well as *don't care* conditions. The mathematical symbol Σ (sigma), meaning *sum,* is used for this purpose. For example, with reference to Figure 5.10, the expression

$$F = \overline{A_8}\,\overline{A_4}A_2A_1 + \overline{A_8}A_4A_2\overline{A_1} + A_8\overline{A_4}\,\overline{A_2}\,\overline{A_1} + A_8\overline{A_4}\,\overline{A_2}A_1$$

with *don't cares*

$$d = A_8\overline{A_4}A_2\overline{A_1} + A_8\overline{A_4}A_2A_1 + A_8A_4\overline{A_2}\,\overline{A_1} + A_8A_4\overline{A_2}A_1 +$$
$$A_8A_4A_2\overline{A_1} + A_8A_4A_2A_1$$

Figure 5.10
The number of each cell is the sum of the subscripts of the uncomplemented variables that identify the cell.

	$\overline{A_2}\,\overline{A_1}$	$\overline{A_2}A_1$	A_2A_1	$A_2\overline{A_1}$
$\overline{A_8}\,\overline{A_4}$	0	1	3	2
$\overline{A_8}A_4$	4	5	7	6
A_8A_4	12	13	15	14
$A_8\overline{A_4}$	8	9	11	10

can be abbreviated as

$$F = \Sigma\ 3,\ 6,\ 8,\ 9$$
$$d = \Sigma\ 10,\ 11,\ 12,\ 13,\ 14,\ 15$$

Example 5-5   Use the cell-numbering notation to express and simplify a logic expression for a circuit that produces a 1 when its 8-4-2-1 BCD input represents an even number (including 0) less than $10)_{10}$.

Solution.   The even numbers less than $10)_{10}$ are represented by

$$F = \Sigma\ 0,\ 2,\ 4,\ 6,\ 8$$

Since the input is an 8-4-2-1 BCD number, the *don't cares* are

$$d = \Sigma\ 10,\ 11,\ 12,\ 13,\ 14,\ 15$$

The logic expression and the *don't cares* are plotted in Figure 5.11. We see that $F = \overline{A_1}$.

Figure 5.11
Example 5-5.

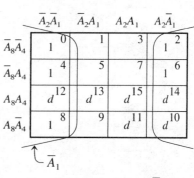

$$F = \Sigma\ 0,\ 2,\ 4,\ 6,\ 8 = \overline{A_1}$$
$$d = \Sigma\ 10,\ 11,\ 12,\ 13,\ 14,\ 15$$

Drill 5-5   Repeat Example 5-5 if the circuit is to produce a 1 when its 8-4-2-1 BCD input represents an odd number less than 9.
Answer: $F = \overline{A_8}A_1$.

5-6   ANSI/IEEE STANDARD SYMBOLS

Active-Low Notation

The triangle that symbolizes inversion in the ANSI/IEEE standards can be used at the input of a device to represent an active-low input. The triangle points in the direction of signal flow, normally from left to right. Figure 5.12 shows the different ways that the triangle can be used to represent active-low inputs and outputs.

Figure 5.13 shows logically equivalent ways of representing gates using ANSI/IEEE standard symbols and active-low notation. This figure is the counterpart of Figure 5.4.

Figure 5.12
Use of the triangular inversion symbol to represent active-low inputs and outputs.

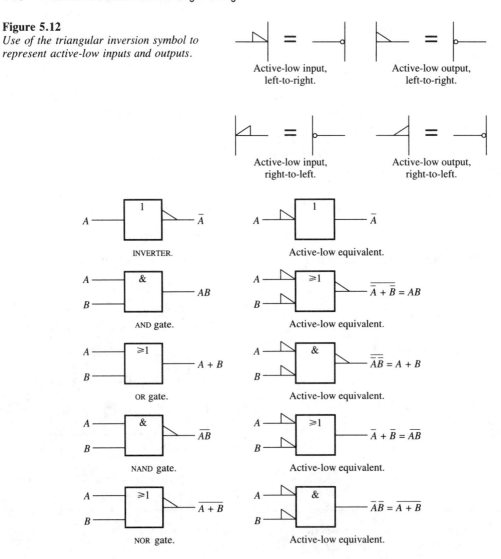

Active-low input,
left-to-right.

Active-low output,
left-to-right.

Active-low input,
right-to-left.

Active-low output,
right-to-left.

INVERTER.

Active-low equivalent.

AND gate.

Active-low equivalent.

OR gate.

Active-low equivalent.

NAND gate.

Active-low equivalent.

NOR gate.

Active-low equivalent.

Figure 5.13
Logically equivalent ways of representing gates using ANSI/IEEE standard symbols and active-low notation.

AOI Gates

The ANSI/IEEE standard symbol for an AND-OR-INVERT (AOI) gate uses AND, OR, and INVERT symbols in the left-to-right sequence that would be expected. Figure 5.14 shows the logic symbol for the 7451 dual 2-wide, 2-input AOI gate. Note that the AND-followed-by-OR sequence is symbolized by placing two AND blocks to the left of and adjacent to the OR block. The omission of connection lines between logic blocks within one device is a convention that is observed in the ANSI/IEEE representation of all logic devices.

Example 5–6   Write the logic expression implemented by the circuit shown in Figure 5.15(a). Remove complement bars from the expression.

Figure 5.14
ANSI/IEEE standard logic symbol for the 7451 dual 2-wide, 2-input AOI gate.

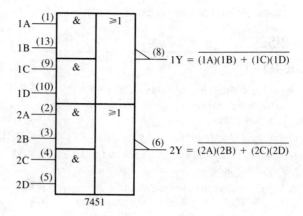

Figure 5.15
Example 5-6.

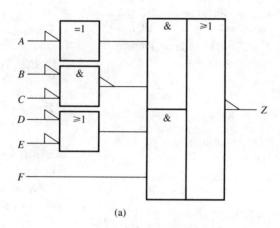

(a)

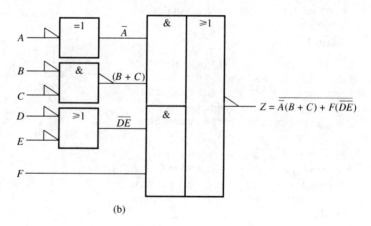

(b)

Solution.   See Figure 5.15*(b)*.

$$Z = \overline{\overline{\overline{A}}(B + C) + F(\overline{\overline{DE}})}$$
$$= \overline{\overline{A}(B + C)} \cdot \overline{F(\overline{\overline{DE}})}$$
$$= [A + \overline{(\overline{B} + \overline{C})}][\overline{F} + \overline{\overline{\overline{DE}}}]$$
$$= (A + \overline{B}\,\overline{C})(\overline{F} + DE)$$

EXERCISES

Section 5–1

5.1 Draw logic diagrams to implement $F = ABC + \overline{B}\overline{C}$ using only NAND gates and again using only NOR gates.

5.2 Draw logic diagrams to implement $P = (\overline{A}_1 + A_2A_3)A_4$ using only NAND gates and again using only NOR gates.

5.3 A logic designer has two NOR gates available in a 7402 quad 2-input NOR gate chip and two NAND gates available in a 7400 quad 2-input NAND gate chip. Show how these gates can be used to implement $F = AB + BC$.

5.4 Show how $F = ABCD$ can be implemented using only *two-input* NAND gates.

Section 5–2

5.5 Using active-low notation, redraw the logic diagram shown in Figure 5.16 so that an expression for W can be written directly with no complement bars.

5.6 Using active-low notation, draw a logic diagram to implement the expression $Z = \overline{(\overline{ABC} + \overline{D})\overline{E}}$ in such a way that an equivalent expression for Z having no complement bars can be written directly from the diagram.

Section 5–3

5.7 Using any of the TTL gates shown in Figure 5.6, draw a logic diagram that implements $F = \overline{A}C + AB$ in simplified AOI logic. Assume each variable and its complement are available as input signals.

5.8 Using any of the TTL gates shown in Figure 5.6, draw a logic diagram that implements $F = AB\overline{C} + \overline{B}CD + B\overline{C} + \overline{B}CD$ in simplified AOI logic. Assume each variable and its complement are available as input signals.

Figure 5.16
Exercise 5.5.

Section 5–4

5.9 Using a Karnaugh map, simplify $F = \overline{A}B + B\overline{C}$ when the *don't care* conditions are ABC and $\overline{B}\overline{C}$.

5.10 Using a Karnaugh map, simplify $W = \overline{X}_2X_3\overline{X}_4 + X_1\overline{X}_2\overline{X}_4 + \overline{X}_1X_2X_3X_4$ when the *don't care* conditions are $\overline{X}_1\overline{X}_2X_3\overline{X}_4$, $X_2\overline{X}_3\overline{X}_4$, and $X_1X_2X_3\overline{X}_4$.

5.11 $A_8A_4A_2A_1$ is an 8-4-2-1 BCD input to a logic circuit whose output is 1 when $A_8 = 0$, $A_4 = 0$, and $A_2 = 1$ or when $A_8 = 0$ and $A_4 = 1$. Find the simplest possible logic expression for the circuit.

5.12 $B_8B_4B_2B_1$ is an 8-4-2-1 BCD input to a logic circuit whose output is 1 when the input represents any of the decimal numbers 1, 2, or 3. Find the simplest possible logic expression for the circuit.

Section 5–5

5.13 Find the simplest possible logic expression for a circuit whose input is $A_8A_4A_2A_1$, given that it implements

$$F = \Sigma\ 7, 11, 12, 13, 15$$

with *don't cares*

$$d = \Sigma\ 0, 6, 8, 10, 14$$

5.14 $A_8A_4A_2A_1$ is an 8-4-2-1 BCD input to a logic circuit designed to produce a 1 if the input represents any of the decimal numbers 2, 4, 5, or 6. Use the cell-numbering notation to express and simplify a logic expression for the circuit.

Section 5–6

5.15 Write the logic expression implemented by the circuit shown in Figure 5.17. Remove complement bars from the expression.

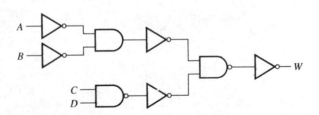

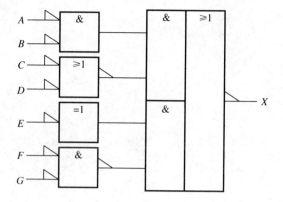

Figure 5.17
Exercise 5.15.

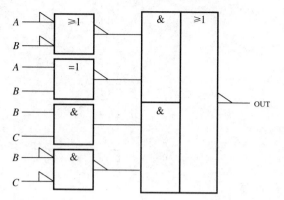

Figure 5.18
Exercise 5.16.

5.16 Simplify the logic expression implemented by the circuit shown in Figure

5.18. Then redraw the simplified circuit using an AOI gate and ANSI/IEEE standard symbols.

CHALLENGING EXERCISES

5.17 Implement $F = \overline{A}\,\overline{B}C + A\overline{B}C + \overline{A}BD + \overline{BCD}$ using no more than five 2-input NAND gates.

5.18 Show how $F = A + B + \overline{C}$ can be implemented with one 2-input NAND gate and one 2-input NOR gate.

5.19 Show how an exclusive-OR gate can be used as an inverter and how an exclusive-NOR gate can be used as an inverter.

5.20 A manufacturer is proposing to develop a line of OR/AND/INVERT gates like that shown in Figure 5.19. Show how this gate could be used to implement $B = \overline{W}X + \overline{Y}Z$. Assume each variable and its complement are available as input signals.

5.21 Show how the OR/AND/INVERT gate in Figure 5.19 could be used to implement

$$W = (\overline{X}_1 + X_2 + X_3)(\overline{X}_1 + X_2 + \overline{X}_3) \cdot (X_1 + \overline{X}_2 + \overline{X}_3)(X_1 + \overline{X}_2 + X_3)$$

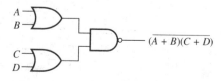

Figure 5.19
Exercise 5.20.

Assume each variable and its complement are available as an input signal.

5.22 The excess-three (XS-3) BCD code defines decimal digits as follows:

X_8	XS-3 code X_4	X_2	X_1	Decimal digit
0	0	0	0	}
0	0	0	1	not defined
0	0	1	0	}
0	0	1	1	0
0	1	0	0	1
0	1	0	1	2
0	1	1	0	3
0	1	1	1	4
1	0	0	0	5
1	0	0	1	6
1	0	1	0	7
1	0	1	1	8
1	1	0	0	9
1	1	0	1	}
1	1	1	0	not defined
1	1	1	1	}

Find the simplest possible expression for a logic circuit that produces a 1 when its input is an XS-3 number representing any odd decimal number greater than $1)_{10}$.

Part II

PULSE AND DIGITAL CIRCUITS

Chapter 6

PULSE AND DIGITAL SIGNALS

6–1 IDEAL AND REAL SWITCHES

Ideal Switches

As discussed in earlier chapters, logic circuits are also called *switching* circuits because the electronic devices from which they are constructed effectively switch their outputs between two distinct voltage levels. In connection with logic circuitry, an *ideal* switch is one that (1) has zero resistance when closed (short-circuit), (2) has infinite resistance when open (open-circuit), and (3) can switch from one such state to the other in *zero time*. Figure 6.1(*a*) and (*b*) shows how the output a gate can be interpreted as a switch. (Although the symbols for conventional mechanical switches are used in the figure, there are, of course, no such devices in a logic gate.) In each case, resistance R is the internal (or Thevenin equivalent) resistance of the gate. Part (*c*) of the figure shows the waveform that the ideal switch would produce if it were alternately closed and opened.

Real Switches

The switching action in most logic gates is performed by an active output device, such as a transistor, which can only approximate the characteristics of an ideal switch. A real switch has small but non-zero resistance when closed, has large but finite resistance when open, and requires short but non-zero time to change from one state to the other. Figure 6.2 shows equivalent circuits of a real switch, which has resistance R_{ON} when it is closed and resistance R_{OFF} when it is open. (The terms ON and OFF convey the idea that the switch is conducting when closed and non-conducting when open.) We see that the output voltage is no longer zero when the switch is closed because of the voltage division that takes place between R and R_{ON}:

$$V_o(\text{low}) = \left(\frac{R_{ON}}{R + R_{ON}}\right)V_{CC} \qquad \textbf{(6.1)}$$

Figure 6.1
Interpretation of the output of a logic gate as the output of an ideal switch.

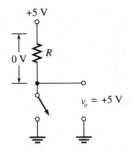

(a) When the switch is open, no current flows through R, and output v_o is +5 V (high).

(b) When the switch is closed, the output is shorted to ground, so $v_o = 0$ V (low). The entire 5 V is dropped across R.

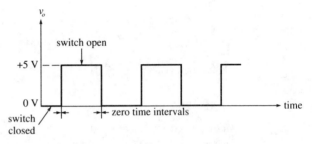

(c) Waveform produced by the <u>ideal</u> switch when alternately closed and opened. Notice that v_o changes from 0 V to +5 V, and from +5 V to 0 V, in <u>zero</u> time.

The output voltage when the switch is open is similarly affected by voltage division:

$$V_o(\text{high}) = \left(\frac{R_{\text{OFF}}}{R + R_{\text{OFF}}}\right)V_{CC} \tag{6.2}$$

Note that equations (6.1) and (6.2) confirm that $V_o(\text{low})$ becomes 0 V in the ideal case, where $R_{\text{ON}} = 0$ Ω, and $V_o(\text{high})$ becomes V_{CC} in the ideal case, where $R_{\text{OFF}} = \infty$.

Resistive Loading of Switches

The output of every practical switching circuit is connected to some sort of *load,* be it an indicating device, a recording instrument, or (most often) the input to another switching circuit. The characteristics of the load, particularly its resistance and capacitance, further affect the behavior of the real switch and determine the extent to which its non-ideal character is detrimental to circuit performance. Figure 6.3 shows equivalent circuits of the real switch with load resistance R_L connected to its output. For the moment, we neglect any capacitance that may be present in the circuit or the load. The equivalent resistance of the parallel combination of R_L and the resistance of the real switch is:

$$\text{switch closed:} \quad R_P(\text{ON}) = R_{\text{ON}} \parallel R_L = \frac{R_{\text{ON}}R_L}{R_{\text{ON}} + R_L} \tag{6.3}$$

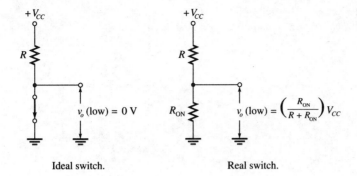

Ideal switch.   **Real switch.**

(a) Switch closed.

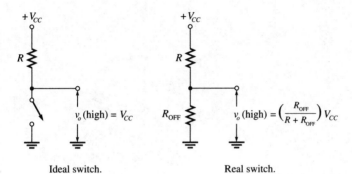

Ideal switch.   **Real switch.**

(b) Switch open.

Figure 6.2
Voltage division affects the output of a real switch, which has non-zero resistance R_{ON} when closed and non-infinite resistance R_{OFF} when open.

$$\text{switch open:}\quad R_P(\text{OFF}) = R_{OFF} \parallel R_L = \frac{R_{OFF}R_L}{R_{OFF} + R_L} \tag{6.4}$$

When the switch is loaded in this way, the output voltages are found once again by voltage division:

$$v_o(\text{low}) = \left[\frac{R_P(\text{ON})}{R + R_P(\text{ON})}\right]V_{CC} \tag{6.5}$$

$$v_o(\text{high}) = \left[\frac{R_P(\text{OFF})}{R + R_P(\text{OFF})}\right]V_{CC} \tag{6.6}$$

Since $R_P(\text{ON})$ is smaller than R_{ON}, $v_o(\text{low})$ is closer to the ideal value of 0 V when the switch is loaded. However, $R_P(\text{OFF})$ is smaller than R_{OFF}, so $v_o(\text{high})$ deviates further from the ideal value of V_{CC} when the switch is loaded.

Manufacturers of commercially available logic gates specify the *maximum* voltage that a particular gate will interpret as a low input and the *minimum* voltage that it will interpret as a high input. Loading of a gate output is therefore particularly relevant to

Figure 6.3
Effects of switch loading on output voltage division.

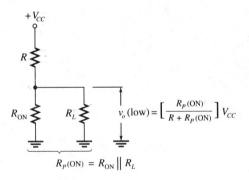

$$v_o(\text{low}) = \left[\frac{R_P(\text{ON})}{R + R_P(\text{ON})}\right] V_{CC}$$

$$R_P(\text{ON}) = R_{\text{ON}} \| R_L$$

(a) Switch closed.

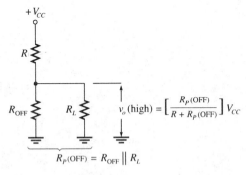

$$v_o(\text{high}) = \left[\frac{R_P(\text{OFF})}{R + R_P(\text{OFF})}\right] V_{CC}$$

$$R_P(\text{OFF}) = R_{\text{OFF}} \| R_L$$

(b) Switch open.

applications where that output is the input to other gates. We will see examples of such specifications in a later chapter. We will also see that many practical switching circuits are *non-linear*, as a result of which the values of parameters such as R and R_L depend on whether the switched output is ON or OFF.

Example 6–1

A manufacturer specifies that the low input to a certain gate can be no greater than 0.5 V and the high input can be no less than 4 V, to ensure satisfactory performance.

The equivalent source resistance of the gate is 250 Ω, its ON-resistance is 10 Ω, and its OFF-resistance is 1 MΩ. If it is operated with a supply voltage of $V_{CC} = +5$ V and a 900-Ω load, will the output levels meet specifications as inputs to another gate of the same type?

Solution

$$R_P(\text{ON}) = R_{\text{ON}} \| R_L = 10\ \Omega \| 900\ \Omega \approx 10\ \Omega$$

$$v_o(\text{low}) = \left(\frac{10\ \Omega}{250\ \Omega + 10\ \Omega}\right) 5\ \text{V} = 0.192\ \text{V}$$

$$R_P(\text{OFF}) = R_{\text{OFF}} \| R_L = 1\ \text{M}\Omega \| 900\ \Omega \approx 900\ \Omega$$

$$v_o(\text{high}) = \left(\frac{900\ \Omega}{250\ \Omega + 900\ \Omega}\right) 5\ \text{V} = 3.91\ \text{V}$$

We see that v_o(low) is less than 0.5 V and therefore meets the specification, but v_o(high) is less than 4.0 V and therefore fails.

Drill
6–1

What minimum value of R_L in Example 4–1 is necessary to meet the specification for v_o(high)?
Answer: 1 kΩ.

Capacitance Effects

Capacitance exists between any two conducting paths, objects, surfaces, or terminals that are separated by an insulator. Consequently, capacitance is always present at the output of a real switching circuit. It is found between the output terminals of a gate, between the input terminals of a load (such as another gate), and between the conducting paths that join the gate to the load. These are examples of what is called *stray capacitance*. Capacitance is also present between semiconductor surfaces *inside* a gate and/or its load. This *interelectrode* capacitance is an inherent property of every semiconductor device because the PN junctions they contain are themselves conducting surfaces separated by depletion (non-conducting) regions.

To switch the output of a gate from low to high, the voltage across the output capacitance must be increased by *charging* the capacitance. To switch the output from high to low, the capacitance must be discharged. Since it is impossible to change the voltage across a capacitance instantaneously, there is always a time delay associated with switching the output of a real gate from one level to another. Large-scale digital systems contain many logic gates and switching circuits, none of which can initiate a change in output level until *after* the time delay(s) of the circuit(s) driving them have elapsed. Thus, delays accumulate in such a system, and we can say that capacitance is the primary culprit in limiting the speed at which digital computers and similar systems can operate.

To study the effect of capacitance on a switching circuit, we can *lump* all the output capacitance into a single equivalent capacitor that *shunts* the output—that is, diverts it into an equivalent capacitor connected in parallel with the output terminals. We will also see that we can lump the effective resistance in the circuit into a single equivalent resistance. The result is an RC circuit such as shown in Figure 6.4. To charge the capacitor, we close the switch in 6.4(*a*). Recall that the capacitor voltage as a function of time after the switch is closed at $t = 0$ is given by

$$V_C(t) = E(1 - e^{-t/\tau}) \text{ volts} \qquad (6.7)$$

where $\tau = RC$ seconds is the *time-constant* of the circuit. Figure 6.4(*b*) shows a plot of $v_C(t)$ versus time, from which it is apparent that the time delay required to fully charge the capacitor to E volts is approximately 5τ.

When the switch in Figure 6.4(*c*) is closed at $t = 0$, the capacitor discharges, and its voltage is given by

$$V_C(t) = Ee^{-t/\tau} \text{ volts} \qquad (6.8)$$

where $\tau = RC$ seconds is the discharge time-constant. Figure 6.4(*d*) shows a plot of $v_C(t)$ as it discharges. It is apparent that a delay of approximately 5τ is required to fully discharge the capacitance.

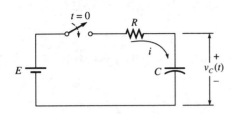

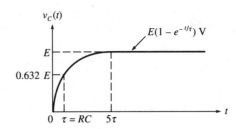

(a) The capacitor charges when the switch is closed at $t = 0$.

(b) Plot of $v_C(t)$ versus t when the capacitor charges. It is, essentially, fully charged to E volts after 5 time-constants ($t = 5\tau = 5RC$).

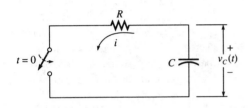

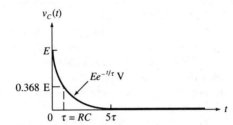

(c) The capacitor discharges when the switch is closed at $t = 0$.

(d) Plot of $v_C(t)$ versus t when the capacitor discharges. It is, essentially, fully discharged after 5 time-constants ($t = 5\tau = 5RC$).

Figure 6.4
Charging and discharging a capacitor.

Example 6–2 For the RC circuit in Figure 6.4, $R = 130\ \Omega$, $C = 50$ pF, and $E = 5$ V.

    a. Write the equations for $v_C(t)$ during charge and discharge.
    b. How long does it take $v_C(t)$ to charge to 90% of E?
    c. How long does it take $v_C(t)$ to discharge to 10% of E?

Solution. The time-constant of the circuit is $\tau = RC = (130)(50 \times 10^{-12}) = 6.5 \times 10^{-9}$ s, or 6.5 ns.

    a. From equation (6.7), the charging voltage is

$$v_C(t) = 5(1 - e^{-t/6.5 \times 10^{-9}})\ \text{V}$$

From equation (6.8), the discharge voltage is

$$v_C(t) = 5e^{-t/6.5 \times 10^{-9}}\ \text{V}$$

    b. We must find the time required for $v_C(t)$ to charge to 90% of 5 V—that is, to $(0.9)(5\ \text{V}) = 4.5$ V. Setting the charging equation equal to 4.5, we have

$$4.5 = 5(1 - e^{-t/6.5 \times 10^{-9}})\text{V}$$

or

$$\frac{4.5}{5} = 0.9 = 1 - e^{-t/6.5 \times 10^{-9}}$$

$$0.1 = e^{-t/6.5 \times 10^{-9}}$$

To solve for t, we must take the natural logarithm (base e) of both sides:

$$\ln(0.1) = \ln(e^{-t/6.5 \times 10^{-9}})$$

$$-2.303 = \frac{-t}{6.5 \times 10^{-9}}$$

Then

$$t = (2.303)(6.5 \times 10^{-9})$$
$$= 15 \times 10^{-9}\ \text{s,}\quad \text{or}\quad 15\ \text{ns}$$

c. 10% of $E = (0.1)(5\ \text{V}) = 0.5\ \text{V}$. Setting the discharge equation equal to 0.5, we have

$$0.5 = 5e^{-t/6.5 \times 10^{-9}}$$

or

$$0.1 = e^{-t/6.5 \times 10^{-9}}$$

Taking the natural log of both sides,

$$-2.303 = \frac{-t}{6.5 \times 10^{-9}}$$

or

$$t = 15 \times 10^{-9}\ \text{s} = 15\ \text{ns}$$

Drill 6–2

What is the total time required for the capacitor in Example 6.2 to charge from 10% of E to 90% of E?

Answer: 14.32 ns.

Consider now an RC circuit in which the capacitor is charged to an initial voltage, V_0, as shown in Figure 6.5(a). If E is greater than V_0, then when the switch is closed at $t = 0$, the capacitor will charge from V_0 to E. This case is shown in Figure 6.5(b). If V_0 is greater than E, the capacitor will discharge from V_0 to E, as shown in Figure 6.5(c). In either case, the equation for the voltage across the capacitor as a function of time is

$$v_C(t) = E - (E - V_0)e^{-t/RC} \tag{6.9}$$

Figure 6.6(a) shows the equivalent circuit of the output of a gate loaded by resistance R_L and shunt capacitance C when the output switches from low to high, corresponding to the gate being switched to its off resistance, R_{OFF}. Do not confuse the switch shown in the equivalent circuit with the switch representation of a gate; opening the gate (shutting it off) corresponds to inserting R_{OFF} by closing the switch shown in the figure. The initial voltage on the capacitance is the logic level V_{LO}. When the switch is closed, the capacitor voltage will rise from V_{LO} to V_{HI}.

The circuit is redrawn in Figure 6.6(b). The (Thevenin) equivalent circuit is shown in Figure 6.6(c). The Thevenin equivalent voltage is the logic level V_{HI} to which the capacitance must charge:

$$E_{TH} = V_{HI} = \frac{R_{P(OFF)}}{R + R_{P(OFF)}} V_{CC} \tag{6.10}$$

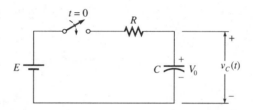

(a) An RC circuit in which C has an initial voltage V_0 (at $t = 0$).

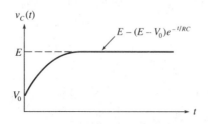

(b) When $E > V_0$, the capacitor charges from V_0 to E.

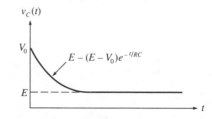

(c) When $V_0 > E$, the capacitor discharges from V_0 to E.

Figure 6.5
Charging and discharging of a capacitor having initial voltage V_0. Note that the charge and discharge equations are identical.

The total equivalent resistance (R_{TH}) is $R \parallel R_{OFF} \parallel R_L$, so the time-constant is

$$\tau = (R \parallel R_{OFF} \parallel R_L)C \text{ s} \qquad (6.11)$$

Notice that this circuit has the same form as the RC circuit in Figure 6.5(b). With reference to equation (6.9), the equation for the output voltage of the gate is, therefore,

$$v_o(t) = V_{HI} - (V_{HI} - V_{LO})e^{-t/\tau} \qquad (6.12)$$

where V_{HI} is given by (6.10) and τ by (6.11). In most practical switching circuits, R_{OFF} is much greater than either R or R_L, so $\tau \approx (R \parallel R_L)C$. Since the time-constant is a measure of the time required to charge the capacitance, we see that this time can be minimized by making R, C, or both as small as possible. Manufacturers strive to develop switching circuits that have a small value of R (the effective source resistance of the gate) as well as small stray and interelectrode capacitance.

Example 6–3
A logic gate has an effective source resistance of 150 Ω and the interelectrode capacitance at its output is 12 pF. Its off-resistance is 2.4 MΩ and its output is connected to a 1-k Ω load. The stray capacitance shunting the output is 68 pF. The low logic level is 0.5 V.

When the gate switches from low to high, the output must rise to 3 V before the load recognizes the output as a high. If $V_{CC} = 5$ V, how long does it take the output to rise to that level?

Solution. The total shunt capacitance is the sum of the interelectrode and stray capacitance:

$$C = 12 \text{ pF} + 68 \text{ pF} = 80 \text{ pF}$$

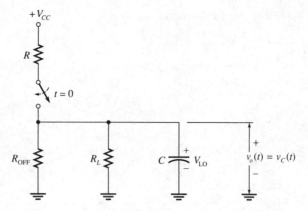

(a) Closing the switch at $t = 0$ connects the off-resistance, R_{OFF} into the circuit.

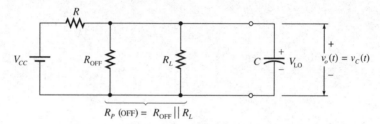

$$R_P \text{ (OFF)} = R_{OFF} \| R_L$$

(b) The circuit equivalent to (a) after the switch is closed.

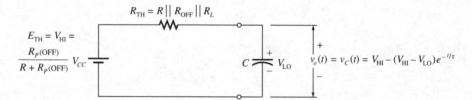

(c) The Thevenin equivalent circuit with respect to the capacitor terminals.

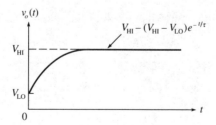

(d) The capacitance charges from V_{LO} to V_{HI} with time constant $\tau = (R \| R_L \| R_{OFF})C$ seconds.

Figure 6.6
Equivalent circuits of a gate output with capacitive loading when it is switched from low to high.

The equivalent resistance at the output is

$$R_{TH} = R \parallel R_{OFF} \parallel R_L = 150 \ \Omega \parallel 2.4 \ M\Omega \parallel 1 \ k\Omega$$

Since R_{OFF} is so much larger than either R or R_L, the parallel combination is essentially determined by R and R_L:

$$R_{TH} \approx 150 \ \Omega \parallel 1 \ k\Omega = \frac{(150)(10^3)}{150 + 10^3} = 130 \ \Omega$$

The time-constant is, therefore,

$$\tau = (130 \ \Omega)(80 \times 10^{-12} \ F) = 1.04 \times 10^{-8} \ s$$

Since $R_P(OFF) = 2.4 \ M\Omega \parallel 1 \ k\Omega \approx 1 \ k\Omega$, we have

$$E_{TH} = V_{HI} = \frac{R_P(OFF)}{R + R_P(OFF)} V_{CC}$$

$$= \left(\frac{1 \ k\Omega}{150 \ \Omega + 1 \ k\Omega}\right) 5 \ V = 4.35 \ V$$

Therefore, from equation (6.12),

$$v_o(t) = 4.35 - (4.35 - 0.5)e^{-t/1.04 \times 10^{-8}} \ V$$
$$= 4.35 - 3.85e^{-t/1.04 \times 10^{-8}} \ V$$

Setting $v_o(t)$ equal to 3 V and solving for t, we find

$$3 \ V = 4.35 - 3.85e^{-t/1.04 \times 10^{-8}}$$
$$1.35 = 3.85e^{-t/1.04 \times 10^{-8}}$$
$$0.3506 = e^{-t/1.04 \times 10^{-8}}$$

$$\ln(0.3506) = -1.048 = \frac{-t}{1.04 \times 10^{-8}}$$
$$t = (1.048)(1.04 \times 10^{-8})$$
$$= 10.9 \times 10^{-9} \ s$$
$$= 10.9 \ ns$$

Drill 6-3

How many *time-constants* must elapse before the output of the gate in Example 6-3 reaches 4.0 V?
Answer: 2.4.

Figure 6.7(a) shows the equivalent circuit of the output of a gate when it is switched from high to low. This action is equivalent to switching R_{ON} into the circuit at $t = 0$. We assume the capacitor is fully charged to V_{HI} volts. Note that V_{HI} is the Thevenin equivalent voltage in Figure 6.6(c)—that is, the maximum output voltage to which the capacitance charges when the output is high. The circuit is redrawn in Figure 6.7(b) and the Thevenin equivalent circuit is shown in Figure 6.7(c). Note that the Thevenin equivalent voltage source is now as shown on the top of page 128.

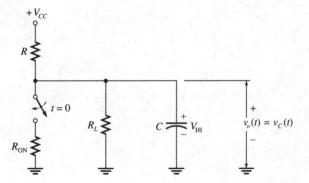

(a) Closing the switch at $t = 0$ connects R_{ON} into the circuit. The capacitor is fully charged.

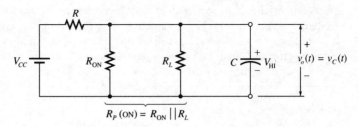

(b) The circuit equivalent to (a) after the switch is closed.

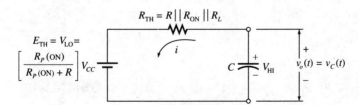

(c) The Thevenin equivalent circuit with respect to the capacitor terminals.

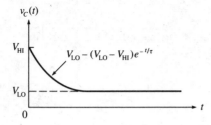

(d) The capacitor voltage decays from V_{HI} to V_{LO}.

Figure 6.7
Equivalent circuits of a gate output with capacitive loading when it is switched from high to low.

$$V_{LO} = E_{TH} = \left(\frac{R_{ON} \| R_L}{R + R_{ON} \| R_L} \right) V_{CC}$$

$$= \frac{R_{P(ON)}}{R + R_{P(ON)}} V_{CC}$$

(6.13)

In this case, the capacitor voltage does not decay all the way to zero but to V_{LO}, as shown in Figure 6.7(d). From equation (6.9), with $V_0 = V_{LO}$ and $E = V_{HI}$,

$$v_C(t) = v_o(t) = V_{LO} - (V_{LO} - V_{HI})e^{-t/\tau} \text{ volts}$$

(6.14)

where

$$\tau = (R \| R_{ON} \| R_L)C$$

Since R_{ON} is usually quite small, the time-constant governing the decay of the output voltage is often much smaller than that governing the rise. In other words, the delay in switching an output from low to high may be quite different from that in switching it from high to low. As we shall see in a later discussion of practical logic gates, manufacturers have developed innovative circuitry that allows capacitance to both charge and discharge through small values of resistance.

Example 6–4

The output of a logic gate driving a 1.5-kΩ load is switched from high to low. The gate has source resistance 120 Ω and ON resistance 10 Ω. If $V_{CC} = 5$ V, $V_{HI} = 4.5$ V, and the total shunt capacitance is 100 pF,

a. Write the equation for the output voltage of the gate;
b. Find the time required for the output voltage to fall to 2.0 V.

Solution

a. The time-constant is

$$\tau = (R \| R_{ON} \| R_L)C = [(120 \ \Omega) \| (10 \ \Omega) \| (1.5 \ k\Omega)](100 \times 10^{-12} \text{ F})$$
$$= 9.17 \times 10^{-10} \text{ s}$$

From equation (6.13),

$$V_{LO} = \left[\frac{(10 \ \Omega \| 1.5 \ k\Omega)}{120 \ \Omega + (10 \ \Omega \| 1.5 \ k\Omega)} \right] 5 \text{ V} = 0.382 \text{ V}$$

From equation (6.14),

$$v_o(t) = 0.382 - (0.382 - 4.5)e^{-t/9.17 \times 10^{-10}} \text{ V}$$
$$= 0.382 + 4.118e^{-t/9.17 \times 10^{-10}} \text{ V}$$

b. Setting $v_o(t)$ equal to 2 V and solving for t, we find

$$2 = 0.382 + 4.118e^{-t/9.17 \times 10^{-10}}$$
$$1.618 = 4.118e^{-t/9.17 \times 10^{-10}}$$
$$0.3929 = e^{-t/9.17 \times 10^{-10}}$$

$$\ln(0.3929) = -0.934 = \frac{-t}{9.17 \times 10^{-10}}$$

$$t = (0.934)(9.17 \times 10^{-10}) = 8.57 \times 10^{-10} \text{ s} = 0.857 \text{ ns}$$

Drill
6–4

What is the total time required for the output voltage in Example 6–4 to fall from 90% of its maximum value to 10% of its maximum value?

Answer: 3.66 ns.

6–2   PULSE FUNDAMENTALS

A positive voltage pulse is a change in voltage from low to high, followed later by a change from high to low, as illustrated in Figure 6.8. A positive current pulse is defined similarly, in terms of changes in current. The *leading edge* of a pulse is the first level transition that occurs and the *trailing edge* is the second. The *amplitude* of a pulse is the total change in value between its low and high levels: $(V_{HI} - V_{LO})$. An *ideal* pulse changes from one level to another in zero time. The ideal pulse shown in Figure 6.8(*a*) is called a *rectangular* pulse. Since pulses are produced by switching circuits, we know that real pulses cannot change levels instantaneously. The *rise-time, t_r,* of a pulse is the total time required for its leading edge to change from 10% of its amplitude to 90% of its amplitude. The *fall-time, t_f,* is the total time required for the trailing edge to fall from 90% of its amplitude to 10% of its amplitude. These time intervals are shown in Figure 6.8(*b*). As the name implies, *pulsewidth, PW,* is the total time between the leading edge and the trailing edge, measured between corresponding levels on each edge. A commonly used level for this definition is the 50% value on each edge, as illustrated in Figure 6.8(*b*).

Figure 6.8
The positive voltage pulse.

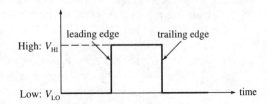

(a) An ideal pulse.

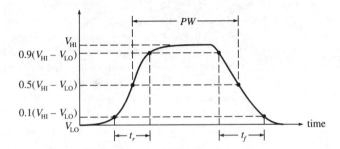

(b) A real pulse, showing definitions of rise-time (t_r), fall-time (t_f), and pulsewidth (*PW*).

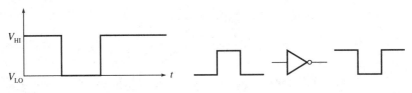

(a) Negative-going pulses.

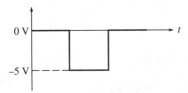

(b) Example of a negative pulse.

Figure 6.9
Negative-going and negative pulses.

In most digital systems, the voltage on a signal line is "normally" low; that is, a low voltage constitutes the *inactive* state, and a transition from low to high initiates some response by another device. A positive pulse is thus a means for activating a signal line for a period of time equal to the pulsewidth. (However, some devices respond only to a leading or trailing *edge*.) In some systems, the inactive state can be considered to be high, whereas the active state is low. Examples are the active-low inputs discussed in Chapter 5. In such systems, a pulse is a high-to-low transition followed by a low-to-high transition. See Figure 6.9(*a*). This kind of pulse is sometimes called a *negative-going* pulse. Note that a negative-going pulse can be produced by supplying a positive pulse to the input of an inverter. Although a negative-going pulse may never reach a negative level, a *negative* pulse by definition undergoes transitions between a positive (or zero) level and a negative level. An example is shown in Figure 6.9(*b*). Note that V_{HI} in this example is 0 V, and V_{LO} is -5 V. It is easy to see how definitions of the pulse parameters we have given can be extended to apply to negative-going and negative pulses.

Example 6–5

A pulse, which for all practical purposes can be considered ideal, is applied to the input of two identical series-connected inverters, as shown in Figure 6.10(*a*). Due to shunt capacitance, each inverter produces a delayed output whose rise- and fall-times are both 40 ns. The inverters do not respond to (invert) a high input until it reaches 4.5 V and do not respond to a low input until it falls to 0.5 V. If the ideal input pulse has a pulsewidth of 0.5 μs, find the total time between the leading edge of the input and the point at which the output of the second inverter falls to 0.5 V. Assume the output of each inverter requires 5 ns to go from 5 V to 4.5 V or from 0 V to 0.5 V.

Solution. The outputs of inverters 1 and 2 are shown in Figure 6.10(*b*). Letting the leading edge of the input be at $t = 0$, the trailing edge will occur at $t = 0.5$ μs, since the pulsewidth is 0.5 μs. The output of inverter 1 begins to rise at the trailing edge of the input. This output requires 5 ns to reach 0.5 V and another 40 ns to reach 4.5 V, since 40 ns is the total time between the 10% and 90% levels. Thus, a total of 45 ns beyond the

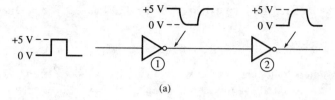

(a)

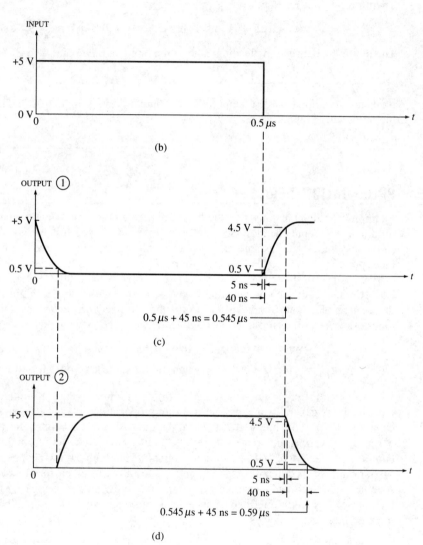

Figure 6.10
Example 6-5.

0.5 μs point is required for output 1 to reach the 4.5 V level. The time at that level is therefore

$$t = 0.5 \text{ μs} + 45 \text{ ns} = 0.5 \text{ μs} + 0.045 \text{ μs} = 0.545 \text{ μs}$$

At the instant output 1 reaches 4.5 V ($t = 0.545$ μs), output 2 begins to respond. We see that the time required for this output to fall to 0.5 V is 5 ns + 40 ns = 45 ns. Therefore, output 2 reaches 0.5 V at

$$t = 0.545 \text{ μs} + 45 \text{ ns} = 0.545 \text{ μs} + 0.045 \text{ μs} = 0.590 \text{ μs}$$

The total time between the leading edge of the input and the point where output 2 falls to 0.5 V is thus 0.590 μs.

Drill
6–5

Find the total time between the point where output 1 in Example 6–5 falls to 0.5 V and the point where output 2 falls to 4.5 V.
Answer: 0.5 μs.

6–3   PROPAGATION DELAY

Example 6–5 demonstrates that the output of a gate does not change at the same instant of time that its input changes. Because of internal capacitance and irrespective of external stray capacitance, the response of every practical logic gate is delayed with respect to an input pulse. This delay is called *propagation delay, t_p.* We have also seen that, in general, different time-constants may govern the rise and fall of the output voltage of a practical gate. Consequently, the propagation delay associated with a low-to-high-level transition in a particular gate may differ from the propagation delay associated with a high-to-low-level transition. To distinguish between these two delays, we define

t_{pLH} = propagation delay when *output* switches from low to high

t_{pHL} = propagation delay when *output* switches from high to low

Although different gates may require different input voltages to initiate a response, manufacturers generally specify and measure propagation delays between voltage points equal to 50% of input and output pulse amplitudes. Figure 6.11 illustrates the definitions of t_{pLH} and t_{pHL} for an inverter. Note that t_{pLH} is measured at the low-to-high change of the *output* of the inverter, and t_{pHL} similarly refers to the high-to-low change at the output. Manufacturers' specifications for t_{pLH} and t_{pHL} are usually given for the case in which a specific amount of load capacitance is connected to the output of a gate, such as 10 pF. Of course, any additional capacitance in a particular application can be expected to increase propagation delays.

Example
6–6

The total equivalent capacitance shunting the output of a certain inverter is 24 pF. Its total equivalent resistance when the output is switching from low to high is 400 Ω, and it is 60 Ω when switching from high to low. The output switches between $V_{LO} = 0.1$ V and $V_{HI} = 4.7$ V. Find t_{pLH} and t_{pHL}.

Solution.   We may assume the input is an ideal pulse and calculate the time required for the output to rise to one-half its maximum value and to fall to one-half its maximum value.

Figure 6.11
Illustration of the definitions of propagation delays t_{pLH} and t_{pHL} for an inverter.

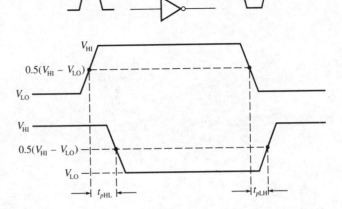

Figure 6.12 shows the equivalent circuits and the waveforms. The time-constant for the low-to-high transition is

$$\tau = (400 \ \Omega)(24 \times 10^{-12} \ \text{F}) = 9.6 \times 10^{-9} \ \text{s}$$

From equation (6.12), the output voltage rises according to

$$v_o(t) = 4.7 - (4.7 - 0.1)e^{-t/9.6 \times 10^{-9}} \ \text{V}$$
$$= 4.7 - 4.6e^{-t/9.6 \times 10^{-9}} \ \text{V}$$

Fifty percent of the amplitude of the pulse is $0.5(4.7 \ \text{V} - 0.1 \ \text{V}) = 2.3 \ \text{V}$. Setting $v_o(t)$ equal to 2.3 V and solving for t, we find

$$2.3 = 4.7 - 4.6e^{-t/9.6 \times 10^{-9}}$$
$$2.4 = 4.6e^{-t/9.6 \times 10^{-9}}$$
$$0.5217 = e^{-t/9.6 \times 10^{-9}}$$
$$\ln(0.5217) = -0.651 = \frac{-t}{9.6 \times 10^{-9}}$$
$$t = (0.651)(9.6 \times 10^{-9}) = 6.25 \times 10^{-9} \ \text{s}$$

Thus, $t_{pLH} = 6.25$ ns.

The time-constant for the high-to-low transition is

$$\tau = (60 \ \Omega)(24 \times 10^{-12} \ \text{F}) = 1.44 \times 10^{-9} \ \text{s}$$

From equation (6.14), the output voltage falls in accordance with

$$v_o(t) = 0.1 - (0.1 - 4.7)e^{-t/1.44 \times 10^{-9}} \ \text{V}$$
$$= 0.1 + 4.6e^{-t/1.44 \times 10^{-9}} \ \text{V}$$

Setting $v_o(t)$ equal to 2.3 V and solving for t gives $t_{pHL} = 1.06$ ns.

**Drill
6–6**

What *additional* load capacitance in Example 6–6 would make $t_{pLH} = 10$ ns?
Answer: 14.4 pF.

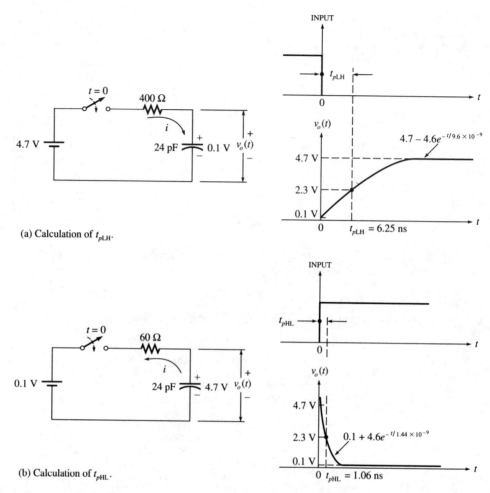

(a) Calculation of t_{pLH}.

(b) Calculation of t_{pHL}.

Figure 6.12
Example 6–6.

6–4   PULSE DISTORTION

We have already seen that shunt capacitance distorts a pulse (i.e., changes its shape) by rounding the leading edge and drawing out the trailing edge. Since the rise- and fall-times of most pulses are small compared to their pulsewidths, a typical pulse viewed on an oscilloscope will appear perfectly sharp and ideal. To view the distortion, it is usually necessary to expand the horizontal sweep so that only the leading or trailing edge is displayed. Other forms of pulse distortion include droop, ringing, dc and ac noise contamination, and bounce.

Droop

Pulse droop, also called tilt, is characterized by a falling-off of the high level. Two commonly encountered forms of droop are shown in Figure 6.13. For a pulse to serve as

Figure 6.13
Two common forms of pulse droop.

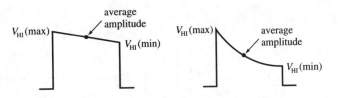

Figure 6.14
Finding the rise- and fall-times of a pulse having droop.

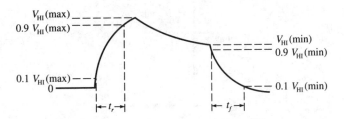

a digital signal, the droop must not be so severe that the high level falls below the minimum value that a gate will recognize as a high. Droop is usually caused by series capacitance and is most often found in very wide pulses. Percent droop is defined by

$$\% \text{ droop} = \frac{\text{change in pulse amplitude}}{\text{average pulse amplitude}} \times 100\% \qquad \textbf{(6.15)}$$

where (see Figure 6.13)

$$\text{change in pulse amplitude} = V_{HI}(\text{max}) - V_{HI}(\text{min})$$

$$\text{average pulse amplitude} = \frac{V_{HI}(\text{max}) + V_{HI}(\text{min})}{2}$$

Rise- and fall-times of a pulse having droop are calculated using $V_{HI}(\text{max})$ as the maximum value for the rise-time and $V_{HI}(\text{min})$ as the maximum value for the fall-time. These points are illustrated in Figure 6.14.

Ringing

When inductance as well as capacitance is present at the output of a switching circuit, an oscillatory *ringing* may occur when the output switches level. A pulse having this kind of ringing is shown in Figure 6.15. The inductance and the capacitance form a *resonant circuit* that, like a filter, responds vigorously to a particular frequency (the resonant frequency) and suppresses others. An abrupt change in level, such as the leading or trailing edge of a pulse, contains a broad band of frequencies, and the LC filter effectively

Figure 6.15
A pulse distorted by ringing.

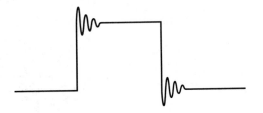

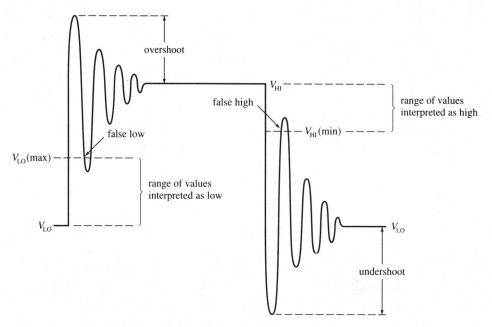

Figure 6.16
Generation of false low and high due to severe ringing.

amplifies the frequency at which it is resonant. The amplification of one frequency and the suppression of others are responsible for the pulse distortion we call ringing. Ringing is an example of a *damped oscillation*—i.e., an oscillation whose amplitude gradually dies out.

Two potential problems associated with ringing are

1. If the leading edge *overshoots* the nominal high level by too great a value or the trailing edge *undershoots* the nominal low level too far, voltage breakdown may occur in devices where the pulse is applied;
2. If the magnitude of the ringing is too great, the high level may drop to a value that is interpreted as a low, or the low level may rise to a value that is interpreted as a high. This situation is illustrated in Figure 6.16.

Overshoot and undershoot are defined to be the maximum voltage excursions beyond the nominal high and low levels of a pulse, as illustrated in Figure 6.16. Percent overshoot and undershoot are referenced to the nominal amplitude, $V_{HI} - V_{LO}$, of a pulse:

$$\% \text{ overshoot} = \frac{\text{overshoot}}{V_{HI} - V_{LO}} \times 100\% \qquad (6.16)$$

$$\% \text{ undershoot} = \frac{\text{undershoot}}{V_{HI} - V_{LO}} \times 100\% \qquad (6.17)$$

Example 6–7

The specifications for a certain pulse generator state that the maximum overshoot of the pulse it produces is 5% and the maximum undershoot is 1%. What are the minimum and maximum voltage excursions that can be expected when it is set to produce a pulse whose low level is 1 V and whose high level is 15 V?

Solution.   Using the decimal form of equation (6.16), we have

$$0.05 = \frac{\text{overshoot}}{15 \text{ V} - 1 \text{ V}}$$

$$\text{overshoot} = (0.05)(14 \text{ V}) = 0.7 \text{ V}$$

Thus, the maximum output voltage that could be expected is

$$V_o(\text{max}) = V_{\text{HI}} + \text{overshoot} = 15 \text{ V} + 0.7 \text{ V} = 15.7 \text{ V}$$

Similarly, from equation (6.17),

$$0.01 = \frac{\text{undershoot}}{15 \text{ V} - 1 \text{ V}}$$

$$\text{undershoot} = (0.01)(14 \text{ V}) = 0.14 \text{ V}$$

$$V_o(\text{min}) = V_{\text{LO}} - \text{undershoot} = 1 \text{ V} - 0.14 \text{ V} = 0.86 \text{ V}$$

**Drill
6–7**

What are the percent overshoot and undershoot of a pulse whose nominal low and high levels are −5 V and 0 V, respectively, if it overshoots to 0.4 V and undershoots to −5.8 V?

Answer: 8% overshoot; 16% undershoot.

DC and AC Noise

Broadly speaking, noise is any alteration of a signal that makes the signal more difficult to interpret or detect. Under that definition, ringing and droop are forms of noise, since they cause digital signals to be different from ideal pulses. DC noise, the shifting of the dc level of a pulse, does not create distortion in the sense that droop and ringing do, but it can be responsible for misinterpretation of signal levels.

DC noise is possible when different parts of a digital system have different power supplies. If the dc voltage in one part of the system *drifts* (changes) and the dc voltage in another part does not or if it drifts in the opposite direction, then pulses produced in the first part may be *offset* (shifted) with respect to the common circuit ground in such a way that a low level is interpreted by the second part as a high or vice versa. The same may occur if the ground levels in different parts of a system drift with respect to each other—that is, if a voltage difference is developed between circuit commons. The problem of dc noise is compounded by the fact that a gate typically will interpret an input as high even when that input is smaller than the nominal high level and will interpret an input as low even when that input is greater than the nominal low level. If dc noise shifts a pulse so that its high level is in the range between the maximum level that is interpreted as low and the minimum level that is interpreted as high, then the response of a gate is unpredictable. The next example illustrates such a case.

Example 6.8   The gates in a certain digital system respond to any voltage above 3.4 V as a high and to any voltage less than 1 V as a low. Due to current flowing between grounds, there is a voltage drop that makes the ground level in Part B of the system 0.9 V higher than the

Figure 6.17
Because of dc noise, the high produced in Part A of the system is 3.3 V above the ground level in Part B, and is therefore 0.1 V below the minimum voltage (3.4 V) that will be interpreted as a high in Part B (Example 6–8).

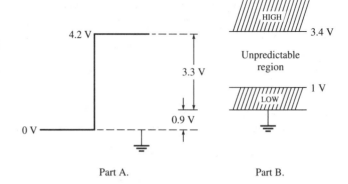

Part A.   Part B.

ground level in Part A. The supply voltage in Part A has drifted to 4.2 V. Will a pulse produced in Part A of the system be interpreted properly by a gate in Part B?

Solution. Figure 6.17 is a diagram of the voltage levels in the two parts of the system. Notice that a pulse produced in Part A has a high level of 4.2 V, but since its ground level is 0.9 V below the ground level in Part B, the input to a gate in Part B rises only 4.2 − 0.9 = 3.3 V above the ground level in Part B. Consequently, the high level of the pulse is in the unpredictable region, and the response of the gate is not dependable.

Drill 6–8

If the power supply voltage in Part A of the system in Example 6–8 drifts up to 5.4 V, what is the maximum voltage that the ground in Part B can rise above the ground in Part A without creating difficulty in interpreting a high level?
Answer: 2 V.

AC noise in a digital system typically appears in the form of very narrow pulses called *spikes* that are superimposed on a signal line. These spikes can be positive or negative; negative spikes superimposed on a positive pulse create false lows and positive spikes superimposed on a low level create false highs. Spikes are *induced* in a signal line through electromagnetic coupling, similar to transformer action. When large currents are switched on and off, rapidly changing magnetic fields are created, and any conductors in the presence of such fields will generate voltages in response to them. Power supplies are also responsible for transmitting ac noise to digital circuitry because they are connected to many different parts of a system. If the impedance of a power supply is not small, noise generated in one part of a circuit is developed across the impedance of the supply and is thereby coupled to other components of the circuit.

AC noise is minimized when

1. Gates have low output impedances;
2. Signal and power lines are shielded;
3. Power supplies have low output impedances (*decoupling* capacitors are often connected between power supply outputs and ground to provide a low-impedance path that shunts high-frequency signals to ground);
4. Shunt capacitance across signal lines is increased (at the expense of greater propagation delays).

Figure 6.18
Contact bounce in mechanical switches creates narrow pulses at the leading or trailing edges.

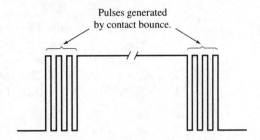

Pulses generated
by contact bounce.

Contact Bounce

When a mechanical switch, such as a push button on a keyboard, is used to produce a pulse for input to a digital system, the switch contacts may open and close several times before they settle into a permanently open or closed position. This behavior is called contact *bounce,* and, depending on how the switch is used in a circuit, it may create a series of narrow pulses at the leading or trailing edge of a pulse. Figure 6.18 illustrates the effect of contact bounce. Bouncing typically lasts 5–20 ms. In some circuits, the presence of a few on-off pulses before or after the closing or opening of a switch has no detrimental effects. In others, contact bounce is not tolerable, and special circuitry, called *hardware debouncing* must be used to "shield" the input from the bounces. A specific debouncing circuit is discussed in Section 17–3 (Figure 17.18). In the case of computers, programming techniques can also be used to eliminate the effects of contact bounce. Such techniques, called *software debouncing,* are also discussed in Section 17–3.

6–5   SQUARE WAVES AND RECTANGULAR WAVEFORMS

A square wave is a series of recurring pulses whose pulsewidths are equal to the time-intervals between them. In other words, it is a waveform that is alternately high and low for equal intervals of time, as illustrated in Figure 6.19. A square wave is an example of a *periodic* waveform, one that repeats the same pattern of values at regular intervals. The *period T* of such a waveform is the time between repetitions—that is, the time required for one complete *cycle* of values. The *frequency* is the number of cycles that occur in 1 s and is related to the period by

$$f = \frac{1}{T} \text{ hertz (Hz)} \qquad \textbf{(6.18)}$$

Square waves are present in many digital systems, where they are used to *synchronize* logic operations, or to perform such operations at prescribed instants of time. In digital

Figure 6.19
Example of a square wave.

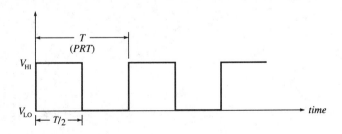

computer applications, synchronizing waveforms such as these are called *clocks*. Examples of sychronized logic operations are given in Section 6–6. Square waves are also used in *waveshaping* applications to produce other periodic waveforms having the same frequency, as is discussed in Chapter 7. In digital applications, the period *T* of a square wave is often called the *pulse repetition time (PRT)* and the frequency is called the *pulse repetition frequency (PRF)*, with units of *pulses per second* (PPS). Thus

$$PRF = \frac{1}{PRT} \text{ pulses/second (PPS)} \qquad \textbf{(6.19)}$$

A rectangular waveform is any recurring sequence of pulses, periodic or otherwise (*aperiodic*). An example is a square wave, a special case of a periodic, rectangular waveform. Many digital signals are periodic, rectangular waveforms having pulsewidths different from the intervals between pulses, as illustrated in Figure 6.20(*a*). For such waveforms, frequency (*PRF*) and period (*PRT*) are defined in the same way as they are for square waves. A *pulsetrain* is any sequence of pulses, recurring or otherwise. A sequence of eight pulses that occur in a 1-s interval and then ceasc is an example of a non-recurrent pulsetrain. Rectangular waveforms are pulsetrains of indefinite duration.

The *duty cycle* of a periodic, rectangular waveform is the ratio of the total time it is high during one cycle (period) to the period, often expressed as a percent:

$$\% \text{ duty cycle} = \frac{\text{time high in one cycle}}{\text{period}} \times 100\% = \frac{PW}{T} \times 100\% \qquad \textbf{(6.20)}$$

where *PW* is the pulsewidth of one pulse in the waveform. From this definition, we see that a square wave has a duty cycle of 50%.

Recall that the *average* value of any waveform, also called the *dc* value, is computed by finding the net area of the waveform over one cycle and dividing by the period. For a rectangular waveform that alternates between 0 V and V_{HI}, it is easy to see from the definition that the average value, V_{AVG}, is

Figure 6.20
Examples of rectangular waveforms.

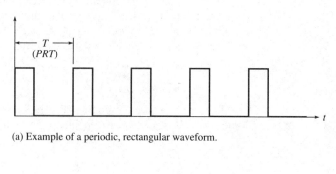

(a) Example of a periodic, rectangular waveform.

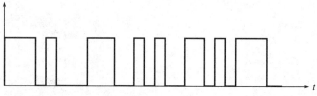

(b) Example of an aperiodic, rectangular waveform.

$$V_{AVG} = (\text{duty cycle})V_{HI} \qquad (6.21)$$

where the duty cycle is expressed in decimal form.

In the field of digital data communications, where information is transmitted in digital form, the high level of a digital signal is called a *mark,* and the low level is called a *space.* The mark-to-space *(M/S)* ratio is defined to be the ratio of the time interval that the waveform is high (the pulsewidth, *PW*) to the time interval that it is low *(T − PW)*. As a percent,

$$\% \ M/S = \frac{PW}{T - PW} \times 100\% \qquad (6.22)$$

Example 6–9

A rectangular waveform alternates between 0 V and +7.5 V. It has a duty cycle of 20% and a pulse repetition frequency of 2×10^4 PPS. Find

a. The pulse repetition period;
b. The average value;
c. The percent *M/S*.

Solution

a. From equation (6.20),

$$PRT = \frac{1}{PRF} = \frac{1}{2 \times 10^4} = 0.5 \times 10^{-4} \text{ s}, \quad \text{or} \quad 50 \ \mu\text{s}$$

b. From equation (6.22),

$$V_{AVG} = (\text{duty cycle})V_{HI} = (0.2)(7.5 \text{ V}) = 1.5 \text{ V}$$

c. Since the duty cycle is 0.2, the pulsewidth is

$$PW = 0.2T = 0.2(50 \ \mu\text{s}) = 10 \ \mu\text{s}$$

Then

$$T - PW = 50 \ \mu\text{s} - 10 \ \mu\text{s} = 40 \ \mu\text{s}$$

From equation (6.23),

$$\% \ M/S = \frac{PW}{T - PW} \times 100\% = \frac{10 \ \mu\text{s}}{40 \ \mu\text{s}} \times 100\% = 25\%$$

Drill 6–9

A rectangular waveform is high for 0.25 ms during each period. If it has a duty cycle of 40%, what is its *PRF?*
Answer: 1600 PPS.

6–6 TIMING DIAGRAMS AND SYNCHRONOUS LOGIC

A *timing diagram* is a diagram that shows the waveforms appearing at several different points in a digital system. A separate set of axes is used for each waveform, and each horizontal (time) axis is aligned with all the others. Thus, it is possible to select a time point and readily determine the state (high or low) of each waveform at that same instant

Figure 6.21

A clocked AND gate and a typical timing diagram.

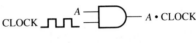

(a) Clocked AND gate.

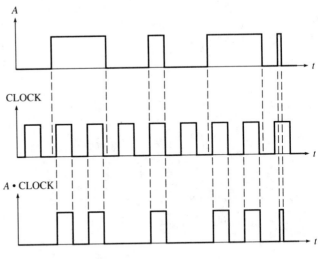

(b) Timing diagram.

of time. Dashed vertical lines are often drawn through the waveforms to show states, or changes of states, at important time points. In Example 6–5 (Figure 6.10), we used a timing diagram to examine propagation delays.

Timing diagrams are particularly useful in *synchronous logic* circuits, those in which logic operations are controlled by a rectangular waveform, or clock. In these systems, gate outputs are considered *valid*, or useful, only when the clock waveform is high. The gates are said to be *pulsed*, or *clocked*. Figure 6.21(*a*) shows an AND gate, one input of which is the digital variable *A* and the other input of which is a clock waveform. In general, *A* may change *asynchronously* (at any time), but the output of the gate equals *A* only when the clock is high. This follows from the definition of an AND gate. Note that the output is always low when CLOCK is low but equals *A* when CLOCK is high. We say that *A* is *gated* by the clock and write the output as *A* · CLOCK. Part (*b*) of the figure is an example of a timing diagram, showing some arbitrary, asynchronous changes in *A*, and showing CLOCK, and *A* · CLOCK.

Example 6–10

Construct a timing diagram to determine the waveform at point ④ in the logic circuit shown in Figure 6.22, given that variables *A* and *B* and the CLOCK have the waveforms shown in the figure.

Solution. Figure 6.22 contains a timing diagram showing the waveforms at points ①, ②, ③, and ④. Notice that the waveform at point ① is labeled *A* · CLOCK and that it is high only when both *A* and CLOCK are high. The output of the inverter, at point ②, is $\overline{A \cdot \text{CLOCK}}$. This waveform is the complement of that at point ①: It is high when *A* · CLOCK is low and vice versa. The waveform at point ③ is *B* + CLOCK, and it is high when either *B* or CLOCK is high. Finally, the waveform at point ④ is high when

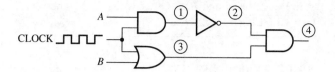

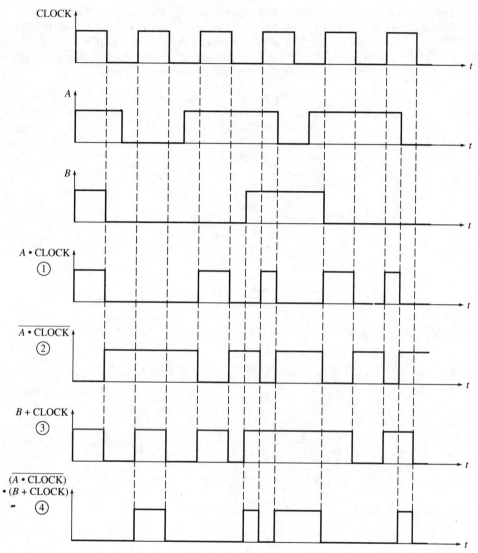

Figure 6.22
Example 6–10.

both $\overline{A \cdot \text{CLOCK}}$ and $B + \text{CLOCK}$ are high. The waveform is labeled
$(\overline{A \cdot \text{CLOCK}})(B + \text{CLOCK})$.

6–7   SPICE EXAMPLES

Example 6–11

A single pulse whose minimum value is 0 V and whose maximum value is +5 V has a rise-time, t_r, of 20 ns and a fall-time, t_f, of 40 ns. The pulsewidth is 100 ns. It is produced by a source that has internal resistance 500 Ω. Use SPICE to obtain a plot of the pulse developed across a 2-kΩ load connected to the source. Assume that the leading and trailing edges of the pulse are linear ramp-type voltages.

Solution.   In SPICE, the rise-time, TR, and fall-time, TF, are defined to be the *total* times required for a pulse to change between its minimum and maximum values. Since the rise-time, t_r, and fall-time, t_f, of a pulse are actually the times required for it to change between 10% and 90% of its final value, we must calculate the values of TR and TF based

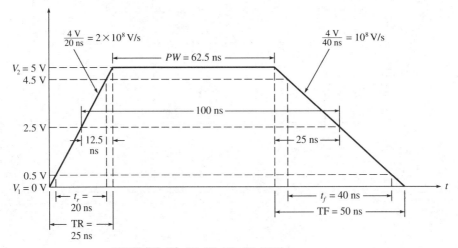

```
PULSE(V1 V2 TD TR TF PW PER)
PULSE(0 5 0 25NS 50NS 62.5NS)   Note: PER defaults to 150 ns.
```
(a)

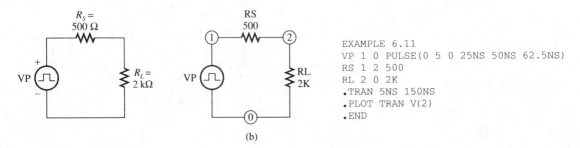

```
EXAMPLE 6.11
VP 1 0 PULSE(0 5 0 25NS 50NS 62.5NS)
RS 1 2 500
RL 2 0 2K
.TRAN 5NS 150NS
.PLOT TRAN V(2)
.END
```
(b)

Figure 6.23
Example 6–11.

on the given values of t_r and t_f. This calculation is illustrated in Figure 6.23(a). Since we are assuming linear edges, the slope of the leading edge is

$$\text{slope (leading)} = \frac{(0.9)(5 \text{ V}) - (0.1)(5 \text{ V})}{20 \text{ ns}} = \frac{(4.5 - 0.5) \text{ V}}{20 \times 10^{-9} \text{ s}} = 2 \times 10^8 \text{ V/s}$$

Therefore, the total time, TR, required for the pulse to change from 0 V to 5 V is

$$\text{TR} = \frac{5 \text{ V}}{2 \times 10^8 \text{ V/s}} = 25 \text{ ns}$$

EXAMPLE 6.11

```
****      TRANSIENT ANALYSIS                 TEMPERATURE =    27.000 DEG C


*********************************************************************************



      TIME        V(2)

                       0.000D+00    1.000D+00    2.000D+00    3.000D+00   4.000D+00
                       - - - - - - - - - - - - - - - - - - - - - - - - - - - - - -
 0.000D+00   0.000D+00 *            .            .            .           .
 5.000D-09   8.000D-01 .        *   .            .            .           .
 1.000D-08   1.600D+00 .            .        *   .            .           .
 1.500D-08   2.400D+00 .            .            .        *   .           .
 2.000D-08   3.200D+00 .            .            .            .    *      .
 2.500D-08   4.000D+00 .            .            .            .           *
 3.000D-08   4.000D+00 .            .            .            .           *
 3.500D-08   4.000D+00 .            .            .            .           *
 4.000D-08   4.000D+00 .            .            .            .           *
 4.500D-08   4.000D+00 .            .            .            .           *
 5.000D-08   4.000D+00 .            .            .            .           *
 5.500D-08   4.000D+00 .            .            .            .           *
 6.000D-08   4.000D+00 .            .            .            .           *
 6.500D-08   4.000D+00 .            .            .            .           *
 7.000D-08   4.000D+00 .            .            .            .           *
 7.500D-08   4.000D+00 .            .            .            .           *
 8.000D-08   4.000D+00 .            .            .            .           *
 8.500D-08   4.000D+00 .            .            .            .           *
 9.000D-08   3.800D+00 .            .            .            .        *  .
 9.500D-08   3.400D+00 .            .            .            .    *      .
 1.000D-07   3.000D+00 .            .            .            .*          .
 1.050D-07   2.600D+00 .            .            .        *   .           .
 1.100D-07   2.200D+00 .            .            .    *       .           .
 1.150D-07   1.800D+00 .            .        *   .            .           .
 1.200D-07   1.400D+00 .            .    *       .            .           .
 1.250D-07   1.000D+00 .            *            .            .           .
 1.300D-07   6.000D-01 .        *   .            .            .           .
 1.350D-07   2.000D-01 .    *     . .            .            .           .
 1.400D-07   0.000D+00 *          .              .            .           .
 1.450D-07   0.000D+00 *            .            .            .           .
 1.500D-07   0.000D+00 *            .            .            .           .
                       - - - - - - - - - - - - - - - - - - - - - - - - - - - - - -
                                           (c)
```

Figure 6.23
(Continued)

Similarly, for the trailing edge, the magnitude of the slope is

$$\text{slope (trailing)} = \frac{4 \text{ V}}{40 \times 10^{-9} \text{ s}} = 10^8 \text{ V/s}$$

and

$$\text{TF} = \frac{5 \text{ V}}{10^8 \text{ V/s}} = 50 \text{ ns}$$

Furthermore, SPICE defines *PW* to be the total time that the pulse is high rather than the total time between the 50% points of the leading and trailing edges. As illustrated in Figure 6.23(*a*), the value we must specify for *PW* in order to obtain 100 ns between the 50% points is

$$PW = 100 \text{ ns} - 0.5(25 \text{ ns}) - 0.5(50 \text{ ns}) = 62.5 \text{ ns}$$

In general, the SPICE parameters for a PULSE specification can be determined from conventionally defined parameters by

$$\text{TR} = \frac{\text{V2} - \text{V1}}{S_L} \quad \text{and} \quad \text{TF} = \frac{\text{V2} - \text{V1}}{S_T}$$

where

$$S_L = \frac{0.8(\text{V2} - \text{V1})}{t_r} \quad \text{and} \quad S_T = \frac{0.8(\text{V2} - \text{V1})}{t_f}$$

$$PW = PW(50\% \text{ points}) - 0.5 \,(\text{TR} + \text{TF})$$

where *PW*(50% points) is the pulsewidth measured between the 50% points on the leading and trailing edges.

Figure 6.23(*b*) shows the circuit and its redrawn version for analysis by SPICE, with node numbers and component designations. Also shown is the SPICE input file needed to obtain a *transient* (TRAN) analysis and plot. Note that we have allowed the period of the pulse to default to TSTOP = 150 ns, since only one pulse is generated.

Figure 6.23(*c*) shows the plot obtained from a SPICE program run. We see that voltage division between the source and load resistances has caused the maximum pulse amplitude to be reduced to 4 V.

Example 6–12

A 0–5 V pulse having width 0.1 μs is produced by a source whose internal resistance is 1.2 kΩ. When the source is not loaded, the pulse, for all practical purposes, is ideal. Use SPICE to find the rise- and fall-times of the pulse when the source is connected to a load whose input is equivalent to 4.8 kΩ shunted by 12 pF of capacitance.

Solution. Figure 6.24(*a*) shows the circuit, the redrawn circuit, and the SPICE input file. Although the pulse is ideal, note that the PULSE specification does *not* show zero rise- and fall-times (TR and TF). If zero time is specified for either or both of those parameters, SPICE assigns them a default value equal to TSTEP in the .TRAN statement. In some cases, that default value may be large enough to affect the accuracy of the computations significantly. Therefore, TR and TF in the example are set equal to the very small value 1 fs (10^{-15} s). Since the transient analysis must last long enough for us to view

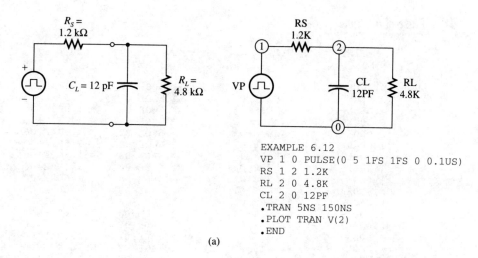

EXAMPLE 6.12
VP 1 0 PULSE(0 5 1FS 1FS 0 0.1US)
RS 1 2 1.2K
RL 2 0 4.8K
CL 2 0 12PF
.TRAN 5NS 150NS
.PLOT TRAN V(2)
.END

(a)

EXAMPLE 6.12

****   TRANSIENT ANALYSIS   TEMPERATURE =   27.000 DEG C

****

```
      TIME        V(2)

                         0.000D+00     1.000D+00     2.000D+00     3.000D+00   4.000D+00
                    - - - - - - - - - - - - - - - - - - - - - - - - - - - - - - - - - -
  0.000D+00    0.000D+00  *            .             .             .           .
  5.000D-09    1.390D+00  .            .     *       .             .           .
  1.000D-08    2.324D+00  .            .             .     *       .           .
  1.500D-08    2.911D+00  .            .             .             .*          .
  2.000D-08    3.296D+00  .            .             .             .   *       .
  2.500D-08    3.548D+00  .            .             .             .       *   .
  3.000D-08    3.706D+00  .            .             .             .         * .
  3.500D-08    3.810D+00  .            .             .             .          *.
  4.000D-08    3.878D+00  .            .             .             .           * .
  4.500D-08    3.921D+00  .            .             .             .           *.
  5.000D-08    3.949D+00  .            .             .             .           *.
  5.500D-08    3.967D+00  .            .             .             .           .*
  6.000D-08    3.979D+00  .            .             .             .           .*
  6.500D-08    3.986D+00  .            .             .             .           .*
  7.000D-08    3.991D+00  .            .             .             .           .*
  7.500D-08    3.994D+00  .            .             .             .           .*
  8.000D-08    3.996D+00  .            .             .             .           .*
  8.500D-08    3.998D+00  .            .             .             .           .*
  9.000D-08    3.998D+00  .            .             .             .           .*
  9.500D-08    3.999D+00  .            .             .             .           .*
  1.000D-07    3.999D+00  .            .             .             .           .*
  1.050D-07    3.248D+00  .            .             .             . *         .
  1.100D-07    2.112D+00  .            .             . *           .           .
  1.150D-07    1.366D+00  .            .     *       .             .           .
  1.200D-07    8.858D-01  .            *             .             .           .
  1.250D-07    5.702D-01  .         *  .             .             .           .
  1.300D-07    3.687D-01  .       *    .             .             .           .
  1.350D-07    2.391D-01  .     *      .             .             .           .
  1.400D-07    1.539D-01  . *          .             .             .           .
  1.450D-07    9.953D-02  .*           .             .             .           .
  1.500D-07    6.404D-02  .*           .             .             .           .
                    - - - - - - - - - - - - - - - - - - - - - - - - - - - - - - - - - -
```

(b)

Figure 6.24
Example 6–12.

TIME	V(2)	TIME	V(2)
0.000E+00	0.000E+00	2.500E-08	3.543E+00
2.000E-10	6.882E-02	2.520E-08	3.551E+00
4.000E-10	1.365E-01	2.540E-08	3.559E+00
6.000E-10	2.030E-01	2.560E-08	3.567E+00
8.000E-10	2.683E-01	2.580E-08	3.574E+00
1.000E-09	3.326E-01	2.600E-08	3.581E+00
1.200E-09	3.957E-01	2.620E-08	3.589E+00
1.400E-09	4.577E-01	2.640E-08	3.596E+00
1.600E-09	5.187E-01	2.660E-08	3.603E+00
1.800E-09	5.786E-01	2.680E-08	3.609E+00
2.000E-09	6.375E-01	2.700E-08	3.616E+00
2.200E-09	6.954E-01	2.720E-08	3.623E+00
2.400E-09	7.522E-01	2.740E-08	3.629E+υ0
2.600E-09	8.081E-01	2.760E-08	3.636E+00
2.800E-09	8.631E-01	2.780E-08	3.642E+00
3.000E-09	9.171E-01	2.800E-08	3.648E+00
3.200E-09	9.701E-01	2.820E-08	3.654E+00
3.400E-09	1.022E+00	2.840E-08	3.660E+00
3.600E-09	1.074E+00	2.860E-08	3.666E+00
3.800E-09	1.124E+00	2.880E-08	3.672E+00
4.000E-09	1.173E+00	2.900E-08	3.677E+00
4.200E-09	1.222E+00	2.920E-08	3.683E+00
4.400E-09	1.270E+00	2.940E-08	3.688E+00
4.600E-09	1.317E+00	2.960E-08	3.694E+00
4.800E-09	1.363E+00	2.980E-08	3.699E+00
5.000E-09	1.408E+00	3.000E-08	3.704E+00

(c)

Figure 6.24
(Continued)

the decaying trailing edge, we make the value of TSTOP in the .TRAN statement equal to 0.15 μs.

Figure 6.24(b) shows the voltage plot produced by SPICE. Note that the maximum value reached by the pulse is 4 V (not 5 V) because of the loading that occurs. The printed values of the times and voltages in the plot appear in the left-hand columns. We see that the resolution is not sufficient to determine the exact times at which the voltage rises to 0.1(4 V) = 0.4 V and to 0.9(4 V) = 3.6 V. However, it is clear that the first time is somewhere between 0 and 5 ns and that the second time is between 25 ns and 30 ns. To increase resolution, we can rerun the program to obtain analysis points in the interval from 0 to 5 ns and run it again to obtain points in the interval from 25 to 30 ns. The results of such runs, using a time increment (TSTEP) of 0.2 ns and .PRINT statements instead of .PLOT statements, are shown in Figure 6.24(c). Note that we must use a TSTART specification of 25 ns to obtain points between 25 and 30 ns: .TRAN 0.2NS 30NS 25NS. We see that the voltage reaches 0.3957 V at $t = 1.2$ ns and 3.603 V at $t = 26.6$ ns, which we will accept as sufficiently accurate. Thus, the rise-time is approximately

$$t_r = 26.6 \text{ ns} - 1.2 \text{ ns} = 25.4 \text{ ns}$$

By a similar analysis, we find that the fall-time is also 25.4 ns.

Example 6–13

An ideal, 0–10 V pulse having width 1 μs is coupled to a 2.2-kΩ load through a 1000-pF capacitor. Use SPICE to determine the percent droop that occurs as a result of this coupling.

Solution.   Figure 6.25(*a*) shows the circuit in its SPICE format and the SPICE input file. The resulting plot is shown in Figure 6.25(*b*). Although the value of TSTEP is too large to yield an accurate value for the output at $t = 0$, another run using a very small TSTEP would show that the initial value of the output is $V_{HI}(\text{max}) = 10$ V. Figure 6.25(*b*) shows that the pulse droops to $V_{HI}(\text{min}) \approx 6.35$ V. Therefore, the average pulse amplitude is

$$\frac{V_{HI}(\text{max}) + V_{HI}(\text{min})}{2} = \frac{10 \text{ V} + 6.35 \text{ V}}{2} = 8.175 \text{ V}$$

By equation (6.13), the percent droop is

$$\frac{10 \text{ V} - 6.35 \text{ V}}{8.175 \text{ V}} \times 100\% = 44.6\%$$

Example 6–14

Use SPICE to generate and plot a rectangular waveform that alternates between 0 V and 5 V with a frequency of 1 MHz (10^6 PPS) and a 25% duty cycle. Assume an ideal waveform. The plot should show two complete cycles.

Solution.   The value of the period in the PULSE specification will be

$$T = \frac{1}{10^6} = 1 \text{ μs}$$

Since the waveform is to have a 25% duty cycle, the pulsewidth specification is 0.25(1 μs) = 0.25 μs. Two complete periods require 2 μs, so the stopping time (TSTOP) specified in the .TRAN statement is 2 μs. By setting the step increment in the .TRAN statement to 50 ns, we obtain 2 μs/50 ns + 1 = 41 points in the plot. The circuit and the resulting plot are shown in Figure 6.26. Note that it is necessary to specify a load ($R = 1$ kΩ) across the terminals of the source, since SPICE requires that there be at least two connections to every node. In this example, we allow TR and TF to default to TSTEP (50 ns).

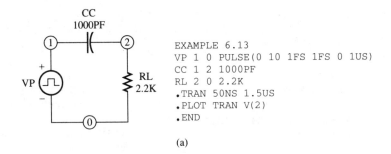

EXAMPLE 6.13
VP 1 0 PULSE(0 10 1FS 1FS 0 1US)
CC 1 2 1000PF
RL 2 0 2.2K
.TRAN 50NS 1.5US
.PLOT TRAN V(2)
.END

(a)

EXAMPLE 6.13

****   TRANSIENT ANALYSIS   TEMPERATURE =   27.000 DEG C

****


```
     TIME      V(2)

                     -5.000D+00   0.000D+00   5.000D+00   1.000D+01  1.500D+01
                     - - - - - - - - - - - - - - - - - - - - - - - - - - - -
0.000D+00   0.000D+00 .              *              .              *              .
5.000D-08   9.776D+00 .              .              .              *.             .
1.000D-07   9.556D+00 .              .              .              *.             .
1.500D-07   9.341D+00 .              .              .              *  .           .
2.000D-07   9.131D+00 .              .              .              *  .           .
2.500D-07   8.926D+00 .              .              .             *   .           .
3.000D-07   8.725D+00 .              .              .            *    .           .
3.500D-07   8.529D+00 .              .              .            *    .           .
4.000D-07   8.338D+00 .              .              .           *     .           .
4.500D-07   8.150D+00 .              .              .           *     .           .
5.000D-07   7.967D+00 .              .              .          *      .           .
5.500D-07   7.788D+00 .              .              .          *      .           .
6.000D-07   7.613D+00 .              .              .         *       .           .
6.500D-07   7.442D+00 .              .              .         *       .           .
7.000D-07   7.275D+00 .              .              .        *        .           .
7.500D-07   7.111D+00 .              .              .        *        .           .
8.000D-07   6.952D+00 .              .              .       *         .           .
8.500D-07   6.795D+00 .              .              .       *         .           .
9.000D-07   6.643D+00 .              .              .       *         .           .
9.500D-07   6.493D+00 .              .              .      *          .           .
1.000D-06   6.347D+00 .              .              .      *          .           .
1.050D-06  -3.682D+00 .       *      .              .                 .           .
1.100D-06  -3.599D+00 .       *      .              .                 .           .
1.150D-06  -3.519D+00 .       *      .              .                 .           .
1.200D-06  -3.440D+00 .       *      .              .                 .           .
1.250D-06  -3.362D+00 .        *     .              .                 .           .
1.300D-06  -3.287D+00 .        *     .              .                 .           .
1.350D-06  -3.213D+00 .        *     .              .                 .           .
1.400D-06  -3.141D+00 .        *     .              .                 .           .
1.450D-06  -3.070D+00 .        *     .              .                 .           .
1.500D-06  -3.001D+00 .         *    .              .                 .           .
                     - - - - - - - - - - - - - - - - - - - - - - - - - - - -
```

(b)

Figure 6.25
Example 6–13.

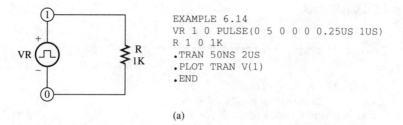

```
EXAMPLE 6.14
VR 1 0 PULSE(0 5 0 0 0 0.25US 1US)
R 1 0 1K
.TRAN 50NS 2US
.PLOT TRAN V(1)
.END
```

(a)

```
EXAMPLE 6.14

****        TRANSIENT ANALYSIS              TEMPERATURE =    27.000 DEG C

***************************************************************************


     TIME        V(1)

                        0.000D+00   2.000D+00    4.000D+00     6.000D+00  8.000D+00
                     - - - - - - - - - - - - - - - - - - - - - - - - - - - - - - - -
0.000D+00   0.000D+00 *          .            .          .            .
5.000D-08   5.000D+00 .          .            .          *  .            .
1.000D-07   5.000D+00 .          .            .          *  .            .
1.500D-07   5.000D+00 .          .            .          *  .            .
2.000D-07   5.000D+00 .          .            .          *  .            .
2.500D-07   5.000D+00 .          .            .          *  .            .
3.000D-07   5.000D+00 .          .            .          *  .            .
3.500D-07  -3.990D-17 *          .            .          .            .
4.000D-07   0.000D+00 *          .            .          .            .
4.500D-07   0.000D+00 *          .            .          .            .
5.000D-07   0.000D+00 *          .            .          .            .
5.500D-07   0.000D+00 *          .            .          .            .
6.000D-07   0.000D+00 *          .            .          .            .
6.500D-07   0.000D+00 *          .            .          .            .
7.000D-07   0.000D+00 *          .            .          .            .
7.500D-07   0.000D+00 *          .            .          .            .
8.000D-07   0.000D+00 *          .            .          .            .
8.500D-07   0.000D+00 *          .            .          .            .
9.000D-07   0.000D+00 *          .            .          .            .
9.500D-07   0.000D+00 *          .            .          .            .
1.000D-06   0.000D+00 *          .            .          .            .
1.050D-06   5.000D+00 .          .            .      *      .            .
1.100D-06   5.000D+00 .          .            .      *      .            .
1.150D-06   5.000D+00 .          .            .      *      .            .
1.200D-06   5.000D+00 .          .            .      *      .            .
1.250D-06   5.000D+00 .          .            .      *      .            .
1.300D-06   5.000D+00 .          .            .      *      .            .
1.350D-06  -5.360D-16 *          .            .          .            .
1.400D-06   0.000D+00 *          .            .          .            .
1.450D-06   0.000D+00 *          .            .          .            .
1.500D-06   0.000D+00 *          .            .          .            .
1.550D-06   0.000D+00 *          .            .          .            .
1.600D-06   0.000D+00 *          .            .          .            .
1.650D-06   0.000D+00 *          .            .          .            .
1.700D-06   0.000D+00 *          .            .          .            .
1.750D-06   0.000D+00 *          .            .          .            .
1.800D-06   0.000D+00 *          .            .          .            .
1.850D-06   0.000D+00 *          .            .          .            .
1.900D-06   0.000D+00 *          .            .          .            .
1.950D-06   0.000D+00 *          .            .          .            .
2.000D-06   0.000D+00 *          .            .          .            .
                     - - - - - - - - - - - - - - - - - - - - - - - - - - - - - - - -
```

(b)

Figure 6.26
Example 6–14.

EXERCISES

Section 6–1

6.1 An electronic switch has a resistance of 25 Ω when it is ON and a resistance of 750 kΩ when it is OFF. The switch is connected through 1.2 kΩ of internal resistance to a 5-V supply voltage. Find the low and high output voltages.

6.2 Repeat Exercise 6.1 if the switch is connected through 600 Ω of internal resistance to a -10-V supply voltage.

6.3 An electronic switch has $R_{ON} = 20$ Ω and $R_{OFF} = 1.5$ MΩ. It is connected through 400 Ω of internal resistance to a $+5$-V supply voltage. The switch drives a load of 1.4 kΩ. Find the low and high output voltages.

6.4 The output of a certain logic gate can be represented as a switch having an internal resistance of 600 Ω connected to a $+5$-V supply voltage. The off-resistance of the switch is 950 kΩ. The *input resistance* of the gate (the effective resistance to ground at each of its input terminals) is 10 kΩ. What is the maximum number of parallel gate inputs identical to its own that the output of the gate can drive when it is high, without the output falling below 3.5 V?

6.5 **(a)** Write the equation for the capacitor voltage $v_C(t)$ after the switch in Figure 6.27 is placed in position 1 at $t = 0$.
(b) Find the time required for the voltage to reach 4 V.
(c) Assuming that the capacitor has fully charged and that the switch is placed in position 2 at a new $t = 0$, write the equation for $v_C(t)$.
(d) Find the time required for the capacitor to discharge to 2 V.

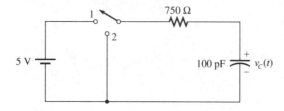

Figure 6.27
Exercise 6.5.

6.6 Repeat Exercise 6.5 for the circuit shown in Figure 6.28.

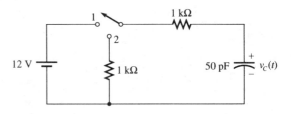

Figure 6.28
Exercise 6.6.

6.7 A logic gate has an effective source resistance of 420 Ω and an OFF resistance of 1.8 MΩ. The interelectrode capacitance at its output is 22 pF. The load is equivalent to 2.5 kΩ shunted by 60 pF of capacitance. If $V_{CC} = +5$ V and $V_{LO} = 0.2$ V, how long does it take for the gate to switch from V_{LO} to 50% of its high output level?

6.8 How long does it take the gate in Exercise 6.7 to switch from 10% of its high output level to 90% of its high output level?

6.9 A logic gate is equivalent to a switch connected through 4 kΩ of internal resistance to a $+15$-V supply. The ON resistance of the gate is 100 Ω, the OFF resistance is 2 MΩ, and it drives a 6-kΩ load. The total capacitance shunting the output is 65 pF.
(a) Draw the equivalent circuit of the gate when it switches from a high output to a low output.
(b) Assuming the load capacitance is fully charged, write the equation for the output voltage when it switches from high to low.
(c) How long does it take the output voltage to decay to 5 V?

6.10 Find the time required for the gate in Exercise 6.9 to switch from 90% of its high output level to 15% of its high output level.

Section 6–2

6.11 The leading and trailing edges of the pulse shown in Figure 6.29 are closely

approximated by the *ramps* (straight, sloped lines) shown in the figure. Find each of the following.
(a) Rise-time
(b) Fall-time
(c) Amplitude
(d) Pulsewidth

(*Hint:* The voltage of a positive-going ramp at any point in time is directly proportional to time.)

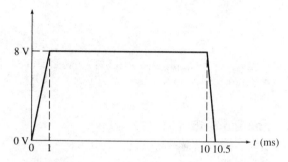

Figure 6.29
Exercise 6.11.

6.12 The leading edge of a certain pulse obeys the charging equation $v_o(t) = 10(1 - e^{-t/8 \times 10^{-9}})$ V. The trailing edge obeys the discharge equation $v_o(t) = 10e^{-t/3.5 \times 10^{-9}}$ V (beginning at a new $t = 0$). Find for the pulse:
(a) Amplitude
(b) Rise-time
(c) Fall-time

6.13 A pulse is applied to the first of three logic gates connected in cascade (series). The logic is such that the pulse is propagated through all three gates. Each gate delays the 50% point on the leading edge by 40 ns and the 50% point on the trailing edge by 15 ns. If the input pulse has pulsewidth 500 ns, what is the pulsewidth of the output pulse?

6.14 The leading edge of a positive pulse appearing at the output of a certain gate obeys the equation $v_o(t) = 5(1 - e^{-t/20 \times 10^{-9}})$ V. In order to reduce the time required for the output to reach 3.5 V, it was decided to increase the 5-V supply voltage. To what value should the voltage be increased if the output must reach 3.5 V in 9.85 ns?

Section 6–3

6.15 Figure 6.30 shows the input pulse to a certain gate and the output pulse that results. Note that the leading and trailing edges of the input are voltage *ramps* (see Exercise 6.11). Find the values of t_{pLH} and t_{pHL} for the gate.

6.16 When the input of a certain inverter is an ideal pulse, the leading edge of the output is $v_o(t) = 5e^{-t/6 \times 10^{-9}}$ V and the trailing edge of the output is $v_o(t) = 5(1 - e^{-t/9 \times 10^{-9}})$ V. Find the values of t_{pLH} and t_{pHL} for the inverter.

Section 6–4

6.17 Find the percent droop in each of the pulses shown in Figure 6.31.

6.18 The maximum voltage of a certain pulse is 8 V and it has 12% droop. What is the minimum value of its high level?

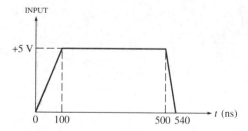

Figure 6.30
Exercise 6.15.

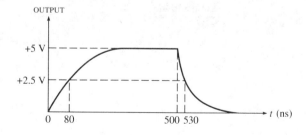

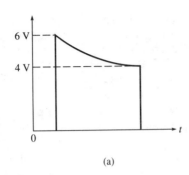

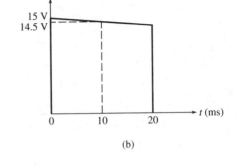

Figure 6.31
Exercise 6.17.

6.19 With reference to the pulse shown in Figure 6.32, find:
(a) Percent overshoot
(b) Percent undershoot
(c) Frequency of the ringing

6.20 A negative pulse has nominal levels of 0 V and $-E$ volts. The pulse has 25% undershoot and is applied to a logic gate that will break down if the input voltage is more negative than -15 V. What is the maximum permissible (absolute) value of E?

6.21 The gates in a certain digital system respond to any voltage above 11.5 V as a high and to any voltage less than 1.5 V as a low. The nominal low and high levels are 0 V and 15 V, respectively. The ground level in Part A of the system has drifted so that it is 2 V above the ground level in Part B. Will a 15-V pulse generated in Part A be interpreted properly by a gate in Part B? Explain.

6.22 List two ways that ac noise is created and transmitted in a digital system. List a remedy for each.

Section 6–5

6.23 Each pulse of a square wave has pulsewidth 800 ns. What is the pulse repetition frequency?

6.24 A square wave has $PRT = 1.25$ μs. How many pulses occur in 1 ms?

6.25 A rectangular waveform has a pulse repetition frequency of 250×10^3 PPS and a 20% duty cycle. What is the pulsewidth of each pulse?

6.26 What is the pulse repetition time of a rectangular waveform if it is high for 0.1 ms during each period and has a 40% duty cycle?

6.27 Find the average value of a rectangular waveform that alternates between 0 V and

Figure 6.32
Exercise 6.19.

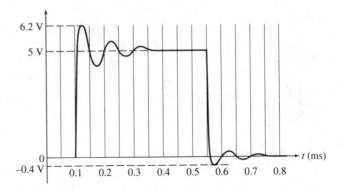

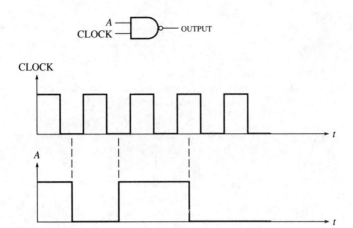

Figure 6.33
Exercise 6.31.

+9 V, given that it is high for half as long as it is low during each cycle.

6.28 What is the average value of a square wave that alternates between +5 V and −5 V?

6.29 A rectangular waveform has a 20% *M/S* ratio and a *PRF* of 50 × 10³ PPS. What is its duty cycle?

6.30 What is the percent *M/S* ratio of a rectangular waveform whose duty cycle is 25%?

Figure 6.34
Exercise 6.32.

Section 6–6

6.31 Complete the timing diagram shown in Figure 6.33 to determine the OUTPUT waveform from the NAND gate, given the waveforms for CLOCK and A shown in the figure.

6.32 Complete the timing diagram shown in Figure 6.34 to determine the OUTPUT waveform, given the waveforms for *A, B,* and *C* shown in the figure.

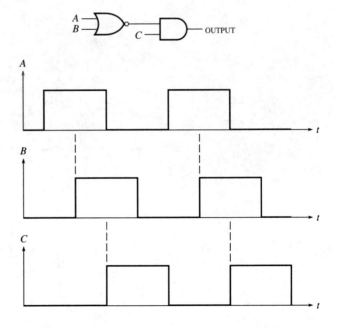

CHALLENGING EXERCISES

6.33 The switch in Figure 6.35 is placed in position 1 at $t = 0$. Twenty nanoseconds later, it is placed in position 2. Write the equations for $v_C(t)$ (assuming in each case that switching occurs at $t = 0$), and sketch $v_C(t)$ versus t.

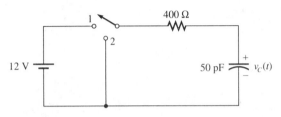

Figure 6.35
Exercise 6.33.

6.34 Show that the rise time t_r of the output $v_o(t)$ across the capacitor in Figure 6.36 can be calculated from $t_r = 2.197\tau$, where $\tau = RC$.

6.35 Assuming that the capacitor in Figure 6.37 is fully charged to E volts, show that the fall time t_f of the voltage $v_o(t)$ across the capacitor can be calculated from $t_f = 2.197\tau$, where $\tau = RC$.

6.36 Show that the average value of the rectangular waveform shown in Figure 6.38 can be calculated from $V_{AVG} = V_1 + $ (duty cycle) $\times$ $(V_2 - V_1)$. (Note that duty cycle $= T_1/T_2$.)

6.37 Show that the duty cycle d and M/S ratios of a rectangular waveform are related by the following equations:

$$dC = \frac{M/S}{M/S + 1}$$

$$M/S = \frac{dC}{1 - dC}$$

Figure 6.36
Exercise 6.34.

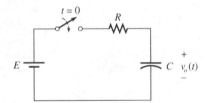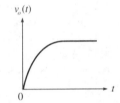

Figure 6.37
Exercise 6.35.

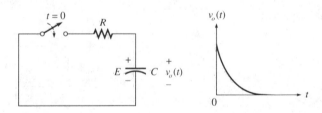

Figure 6.38
Exercise 6.36.

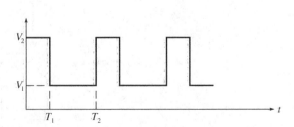

SPICE EXERCISES

6.38 A single pulse whose minimum value is 0 V and whose maximum value is +9 V has a rise-time of 30 ns and a fall-time of 24 ns. The pulsewidth is 0.15 μs. It is produced by a source that has an internal resistance of 1.2 kΩ. Use SPICE to obtain a plot of the pulse when the source drives a gate whose input resistance is 1.5 kΩ. If the gate's response to any input in the range from 4 V to 6 V is unpredictable, will it respond reliably to the input pulse?

6.39 Use SPICE to generate the pulse shown in Figure 6.39. Obtain a plot of the pulse and determine its rise and fall times. (*Hint:* Construct an RC network whose time-constant is that shown by the equations in the figure, and drive the network with a pulse.)

6.40 Figure 6.40 shows the source of a single pulse driving three loads in parallel. Without the loads connected, the pulse can be considered ideal and has pulsewidth 1 μs. Its minimum and maximum values are 0 V and 15 V. Use SPICE to obtain a plot of the pulse when the loads are connected and to determine its rise- and fall-times.

6.41 A single, ideal pulse having width 2 μs is produced by a source whose internal resistance is 50 Ω. The pulse switches between 0 V and 5 V. Use SPICE to determine the percent droop in the pulse when the source is connected through a 0.02-μF capacitor to a 200-Ω load. The load is shunted by (in parallel with) 200 pF of capacitance.

Figure 6.39
Exercise 6.39.

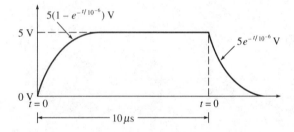

Figure 6.40
Exercise 6.40.

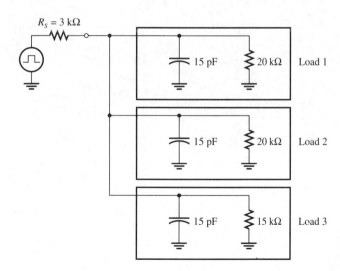

6.42 Use SPICE to obtain a plot of a 500×10^3 PPS rectangular waveform that alternates between -5 V and $+5$ V with a 75% duty cycle. The plot should show two complete cycles of the waveform.

6.43 A step input (leading edge) rising from 0 V to 5 V is applied to the circuit shown in Figure 6.41. Find the percent overshoot of the waveform across the 10-kΩ resistor that results. (*Hint:* The frequency of the ringing is approximately 10 MHz.)

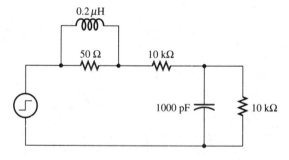

Figure 6.41
Exercise 6.43.

Chapter 7

WAVESHAPING AND WAVEFORM ANALYSIS

7-1   THE TIME AND FREQUENCY DOMAINS

Waveshaping is the alteration of a given waveform to produce a new waveform having specific characteristics. The alteration process can be studied from two viewpoints.

The Time Domain

A waveform is a voltage or current whose values change in a prescribed way with the passage of *time*. A new waveform is generated by applying the original waveform to various kinds of electronic circuits. As the original waveform changes with time, it creates new time-varying voltages and currents in those circuits. For example, when the output of a square wave generator is connected to an RC network, the capacitor charges and discharges during each cycle and thus produces an altered waveform. In this case, we can predict the new waveshape because we know how the voltage across a capacitor changes with time. This perspective of waveshaping is said to be in the *time domain*.

The Frequency Domain

Every periodic waveform, regardless of its shape, can be regarded as the sum of (an infinite number of) *sinewaves*. The frequencies, phase angles, and amplitudes of those sinewaves determine the shape of the waveform. (We study these concepts in more detail in a later paragraph.) Waveshaping can then be regarded as altering the *frequency* content of a waveform. *Filtering* is one example, where sinewaves having certain frequencies are suppressed, thus changing the frequency content and therefore the shape of a waveform. Also, new frequencies can be added to a given waveform (by *non-linear* circuits) to alter its shape. When we study waveshaping from the perspective of altering the frequency content of waveforms, we are in the *frequency domain*.

   In digital systems, waveshaping is most often studied in the time domain. *Clipping* is a common example, and, as we shall see, clipping circuits are most easily understood in terms of time-changing voltages. However, there are many good reasons for understanding how the frequency content of a waveform is altered when its shape is changed. Among those reasons is the importance of knowing what *bandwidth* is required

159

in the circuitry through which a signal must pass, that is, the frequency range that the circuitry can accommodate without distorting a signal by further altering its frequency content.

In general, waveforms that undergo abrupt changes in level, such as square waves, "spikes," and others having jagged appearances or sharp transitions, contain a broad band of frequencies and require *wideband* circuits to minimize distortion. Thus, for example, if a sinewave is shaped so that it becomes a flat-topped, nearly square wave, the circuitry required to pass the altered waveform without distortion must have a much wider bandwidth than the circuitry required for the original sinewave. In this case, new frequencies are added to the original waveform by the non-linear process called clipping. On the other hand, if a square wave is shaped into a nearly sinusoidal waveform by filtering, the new waveform can be passed by circuitry having a bandwidth much narrower than that required for the original square wave. In this case, the abrupt transitions of the square wave are replaced by the smooth variations of a sinewave. Filtering out high frequencies that cause sharp transitions in level is often called *smoothing*.

7–2   HARMONIC ANALYSIS

As discussed earlier, every periodic waveform can be represented as the sum of an infinite number of sinewaves. The frequencies, amplitudes, and phase angles of these sinewaves determine the shape of the waveform. The lowest-frequency sinewave is called the *fundamental*, and it has the same frequency as the waveform itself. All other frequencies are integer multiples of the fundamental frequency and are called *harmonics*. For example, a 1-kHz waveform has a 1-kHz fundamental, a 2-kHz *second harmonic*, a 3-kHz *third harmonic*, and so forth. Every harmonic frequency may not be present in every periodic waveform; that is, the amplitudes of some harmonics may be zero. For example, a square wave can be represented as the sum of a fundamental and all *odd* harmonics (third, fifth, etc.), so the amplitude of every even harmonic is zero. Of course, a pure sinewave consists of just one sinewave (itself, the fundamental), and *all* harmonics are zero. The process of determining the amplitudes and phase angles of all the sinewaves comprising a particular waveform is called *harmonic analysis*.

Electronic instruments, such as *distortion analyzers,* are available to perform harmonic analysis of waveforms. The analysis can also be performed mathematically, provided a mathematical expression is known, or can be determined for the periodic waveform to be analyzed. The mathematical representation of the infinite sum is called a *Fourier series*. A periodic voltage waveform, $v(t)$, is expressed in a Fourier series by writing

$$v(t) = V_{AVG} + A_1\sin(\omega t + \phi_1) + A_2\sin(2\omega t + \phi_2)$$
$$+ A_3\sin(3\omega t + \phi_3) + \cdots \tag{7.1}$$

where

V_{AVG} is the average value (dc value, or offset) or the waveform; A_1, A_2, . . . are the amplitudes (peak values) of the fundamental (A_1) and the harmonics (A_2, A_3, . . .);

ϕ_1, ϕ_2, . . . are the phase angles of the fundamental (ϕ_1) and the harmonics (ϕ_2, ϕ_3, . . .);

ω is the fundamental frequency in radians per second ($\omega = 2\pi f$, where f is the fundamental frequency in hertz).

Of course, a periodic current $i(t)$ can be expressed in a similar way.

Figure 7.1
Fourier analysis of a square wave.

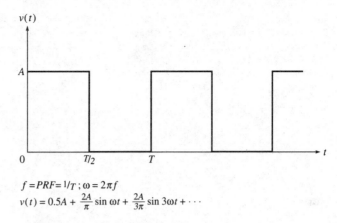

$$f = PRF = 1/T \, ; \, \omega = 2\pi f$$
$$v(t) = 0.5A + \frac{2A}{\pi} \sin \omega t + \frac{2A}{3\pi} \sin 3\omega t + \cdots$$

Harmonic Analysis of Rectangular Waveforms

Figure 7.1 shows a square wave having peak value A. Since the duty cycle is 0.5, the average value is, from equation 6.19, $V_{AVG} = 0.5\ A$. It can be shown* that the Fourier series for this square wave is

$$v(t) = 0.5A + \frac{2A}{\pi} \sin \omega t + \frac{2A}{3\pi} \sin 3\omega t + \cdots + \frac{2A}{n\pi} \sin n\omega t + \cdots \quad (7.2)$$

where n is any odd number and $\omega = 2\pi f$ is the fundamental frequency. f is the pulse repetition frequency *(PRF)* of the square wave, as discussed in Chapter 6. Comparing (7.2) with the general form of the Fourier series (equation (7.1)), we see that the amplitude of the fundamental is $A_1 = 2A/\pi$ and the amplitude of the third harmonic is $A_3 = 2A/3\pi$. All even harmonics have zero amplitude. The amplitude of the nth odd harmonic is $2A/n\pi$. All phase angles are seen to be zero.

Note that the amplitudes of the odd harmonics of a square wave are inversely proportional to n; that is, the higher the harmonic, the smaller its amplitude. The amplitude of the 25th harmonic is only $A_{25} = 2A/25\pi = 0.0255A$, or about 2.6% of the amplitude of the square wave. From a practical standpoint, this means that higher-frequency harmonics can be neglected when selecting or designing circuitry necessary to pass the square wave. (The Fourier series shows there are an *infinite* number of frequencies present, but of course we cannot design any circuit to have infinite bandwidth.) Figure 7.2 shows the sum of the fundamental, third, and fifth harmonics. We see that adding just these three frequencies gives a sum that begins to resemble a square wave.

If a square wave alternates between V_{LO} and V_{HI} with a 50% duty cycle, its average value is

$$V_{AVG} = \frac{V_{LO} + V_{HI}}{2} \quad (7.3)$$

---

*In the interest of conserving space, many of the equations related to topics that are usually considered analog in nature are given in this chapter without derivation or proof. They are included for reference and for the sake of completeness. Readers interested in derivations can consult other electronics devices books, including *Electronic Devices and Circuits* (Theodore F. Bogart, Jr., Merrill Publishing Company, 2d ed., 1990).

Figure 7.2

The sum of the fundamental and the third and fifth harmonics of a square wave begins to resemble a square wave. The more odd harmonics that are added, the closer the sum approaches a true square wave.

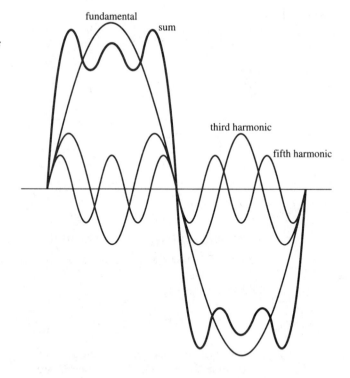

Either V_{LO} or both V_{LO} and V_{HI} may be negative. This square wave has the same odd harmonics as that previously described, except the amplitudes of each harmonic are found by substituting the following value for A in equation (7.2):

$$A = V_{HI} - V_{LO} \tag{7.4}$$

Note that (7.4) gives the *peak-to-peak* value of the waveform.

Example 7–1

Refer to the square wave shown in Figure 7.3.

a. Find the first four non-zero terms (including V_{AVG}) of the Fourier series.
b. What should be the bandwidth of an amplifier designed to pass the square wave if the largest frequency it must pass is the first non-zero harmonic whose amplitude is no greater than 10% of the fundamental?

Solution

a. From equation (7.3), the average value is

$$V_{AVG} = \frac{V_{HI} + V_{LO}}{2} = \frac{5\text{ V} + (-1\text{ V})}{2} = 2\text{ V}$$

The period of the square wave is clearly $T = 0.4$ ms, so the frequency is

$$f = PRF = \frac{1}{T} = \frac{1}{0.4 \times 10^{-3}\text{ s}} = 2.5 \times 10^3\text{ Hz}$$

Figure 7.3
Example 7–1.

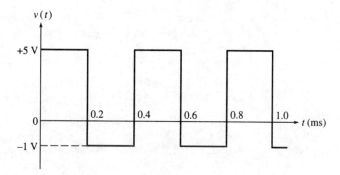

From equation (7.4),

$$A = V_{HI} - V_{LO} = 5 \text{ V} - (-1 \text{ V}) = 6 \text{ V}$$

From equation (7.2), the amplitudes of the fundamental, third, and fifth harmonics are

$$A_1 = \frac{2A}{\pi} = \frac{2(6 \text{ V})}{\pi} = 3.82 \text{ V}$$

$$A_3 = \frac{2A}{3\pi} = \frac{2(6 \text{ V})}{3\pi} = 1.27 \text{ V}$$

$$A_5 = \frac{2A}{5\pi} = \frac{2(6 \text{ V})}{5\pi} = 0.76 \text{ V}$$

Thus,

$$v(t) = 2 + 3.82 \sin(2\pi \times 2.5 \times 10^3 t) + 1.27 \sin(2\pi \times 7.5 \times 10^3 t)$$
$$+ 0.76 \sin(2\pi \times 12.5 \times 10^3 t) + \cdots \text{ V}$$

b. The ratio of the amplitude of the nth harmonic to the fundamental is

$$\frac{A_n}{A_1} = \frac{2A/n\pi}{2A/\pi} = \frac{1}{n}$$

Since the amplitude of the largest harmonic must be no greater than 10% of A_1, we require

$$\frac{A_n}{A_1} = \frac{1}{n} \leq 0.1$$

or $n \geq 10$.

Because the square wave contains only odd harmonics, the amplifier must be capable of passing the eleventh harmonic—that is,

$$f = 11(2.5 \text{ kHz}) = 27.5 \text{ kHz}$$

(Note that the ratio A_{11}/A_1 is $\frac{1}{11}$, or 9.09%, whereas the ratio A_9/A_1 is $\frac{1}{9}$, or 11.1%). Since the amplifier must also pass the dc value, its frequency range must extend from 0 to 27.5 kHz. Thus, the required bandwidth is 27.5 kHz.

Drill
7–1

In the square wave of Example 7–1, what is the amplitude of the first non-zero harmonic whose amplitude is no greater than 1% of the peak-to-peak value of the square wave? *Answer:* 58.76 mV.

Figure 7.4 shows a general rectangular waveform that alternates between V_{LO} and V_{HI} and has a duty cycle of d ($0 < d < 1$). The average value of this waveform is

$$V_{AVG} = V_{HI}d + V_{LO}(1 - d) \tag{7.5}$$

where $d = T_1/T$. Unlike the special case of a square wave (where $d = 0.5$), this waveform in general contains both even and odd harmonics, and their phase angles are non-zero. However, the nth harmonic is zero when any combination of n and d makes the product nd equal to an integer. For example, if $d = 0.1$, then the 20th, 30th, . . . harmonics have zero amplitudes, since $10(0.1) = 1$, $20(0.1) = 2$, $30(0.1) = 3$, and so on. In general, the amplitude of the nth harmonic is

$$A_n = \frac{\sqrt{2}\,A}{n\pi}\sqrt{1 - \cos(2\pi nd)} \tag{7.6}$$

where $A = V_{HI} - V_{LO}$. The phase angle of the nth harmonic is

$$\phi_n = \tan^{-1}\left[\frac{\sin(2\pi nd)}{1 - \cos(2\pi nd)}\right] \tag{7.7}$$

Table 7.1 shows *normalized* amplitudes (values of A_n for $A = 1$) of the first 25 harmonics, corresponding to several values of d. As can be seen in the table, *the harmonic content of a waveform having duty cycle d is identical to that of a waveform having duty cycle 1 − d*. Of course, the average value, as given by equation (7.5), differs according to the value of d, but the ac (harmonic) content is the same for d and $1 - d$.

Examination of Table 7.1 shows that when d is a value different from 0.5, the amplitudes of the harmonics are not necessarily inversely proportional to n. For example, when $d = 0.1$, the amplitude of the 15th harmonic is more than twice the amplitude of the 9th harmonic. This fact must be kept in mind when considering the bandwidth required to pass a rectangular waveform.

Figure 7.4
Harmonic analysis of a rectangular waveform.

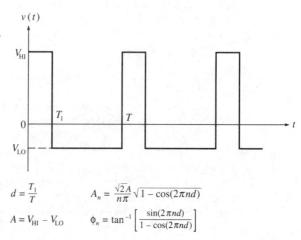

Table 7.1

n	$d = 0.1, 0.9$ A_n	$d = 0.2, 0.8$ A_n	$d = 0.3, 0.7$ A_n	$d = 0.4, 0.6$ A_n	$d = 0.5$ A_n
1	0.1967	0.3742	0.5150	0.6055	0.6366
2	0.1871	0.3027	0.3027	0.1871	0
3	0.1717	0.2018	0.0656	0.1247	0.2122
4	0.1514	0.0935	0.0935	0.1514	0
5	0.1273	0	0.1273	0	0.1273
6	0.1009	0.0624	0.0624	0.1009	0
7	0.0736	0.0865	0.0281	0.0535	0.0909
8	0.0468	0.0757	0.0757	0.0468	0
9	0.0219	0.0416	0.0572	0.0673	0.0707
10	0	0	0	0	0
11	0.0179	0.0340	0.0468	0.0550	0.0579
12	0.0312	0.0505	0.0505	0.0312	0
13	0.0396	0.0466	0.0151	0.0288	0.0490
14	0.0432	0.0267	0.0267	0.0432	0
15	0.0424	0	0.0424	0	0.0424
16	0.0378	0.0234	0.0234	0.0378	0
17	0.0303	0.0356	0.0157	0.0220	0.0374
18	0.0208	0.0336	0.0336	0.0208	0
19	0.0104	0.0197	0.0271	0.0319	0.0335
20	0	0	0	0	0
21	0.0094	0.0178	0.0245	0.0288	0.0303
22	0.0170	0.0275	0.0275	0.0170	0
23	0.0224	0.0263	0.0086	0.0163	0.0277
24	0.0252	0.0156	0.0156	0.0252	0
25	0.0255	0	0.0255	0	0.0255

Normalized amplitudes (A_n) of harmonics in rectangular waveforms with duty cycles d ($A = 1$).

Example 7-2

Find the average value and the amplitudes of the first four harmonics (including the fundamental) of the waveform shown in Figure 7.5. Also find the frequencies of the first four harmonics and their phase angles.

Solution.   The duty cycle is

$$d = \frac{T_1}{T} = \frac{20 \ \mu s}{80 \ \mu s} = 0.25$$

Figure 7.5
Example 7-2.

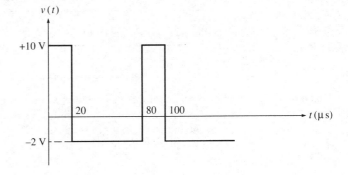

From equation (7.5),

$$V_{AVG} = (+10 \text{ V})(0.25) + (-2 \text{ V})(0.75) = 1.0 \text{ V}$$

Since Table 7.1 does not contain values of A_n for $d = 0.25$, we must use equation (7.6):

$$A = V_{HI} - V_{LO} = 10 \text{ V} - (-2 \text{ V}) = 12 \text{ V}$$

$$A_1 = \frac{\sqrt{2}(12)}{\pi}\sqrt{1 - \cos[2\pi(0.25)]} = 5.40 \text{ V}$$

$$A_2 = \frac{\sqrt{2}(12)}{2\pi}\sqrt{1 - \cos[2\pi(2)(0.25)]} = 3.82 \text{ V}$$

$$A_3 = \frac{\sqrt{2}(12)}{3\pi}\sqrt{1 - \cos[2\pi(3)(0.25)]} = 1.80 \text{ V}$$

$$A_4 = \frac{\sqrt{2}(12)}{4\pi}\sqrt{1 - \cos[2\pi(4)(0.25)]} = 0 \text{ V}$$

(Note that the preceding computations involve calculating the cosine function of an angle in *radians*.) A_4 is zero because $nd = 4(0.25) = 1$ and $\cos(2\pi nd) = \cos(2\pi) = 1$.

From Figure 7.5, $T = 80 \ \mu s$. Therefore, the fundamental frequency is

$$f = \frac{1}{T} = \frac{1}{80 \times 10^{-6} \text{ s}} = 12.5 \text{ kHz}$$

The second, third, and fourth harmonic frequencies are $2f = 25$ kHz, $3f = 37.5$ kHz, and $4f = 50$ kHz.

From equation (7.7), the phase angles are

$$\phi_1 = \tan^{-1}\left\{\frac{\sin[(2\pi)(0.25)]}{1 - \cos[(2\pi)(0.25)]}\right\} = \tan^{-1}(1) = 45°$$

$$\phi_2 = \tan^{-1}\left\{\frac{\sin[(2\pi)(2)(0.25)]}{1 - \cos[(2\pi)(2)(0.25)]}\right\} = \tan^{-1}(0) = 0°$$

$$\phi_3 = \tan^{-1}\left\{\frac{\sin[(2\pi)(3)(0.25)]}{1 - \cos[(2\pi)(3)(0.25)]}\right\} = \tan^{-1}(-1) = -45°$$

Equation (7.7) gives an indeterminate result for ϕ_4, but that is irrelevant, since $A_4 = 0$. Using the values calculated, we may write the Fourier series as

$$v(t) = 1 + 5.4 \sin(2\pi \times 12.5 \times 10^3 t + 45°) + 3.82 \sin(2\pi \times 25 \times 10^3 t)$$
$$+ 1.8 \sin(2\pi \times 37.5 \times 10^3 t - 45°) + \cdots \text{V}$$

Drill 7–2

What is the next harmonic after 50 kHz that has zero amplitude in the waveform of Example 7–2?

Answer: 100 kHz.

Example 7–3

A rectangular waveform has a pulse repetition frequency of 1.0×10^5 PPS and a duty cycle of 0.8. What minimum bandwidth is required to pass it if the bandwidth must extend from 0 to the highest harmonic frequency whose amplitude is no greater than 3% of the peak-to-peak value of the waveform?

Solution.   Examination of Table 7.1 for $d = 0.8$ shows that the amplitude of the 14th harmonic falls below 3% of A (0.0267), but the amplitudes of some succeeding harmonics rise above 3%. The 19th harmonic is the first whose amplitude falls below $0.03A$ and the first for which succeeding harmonics stay below $0.03A$. Therefore, the required bandwidth is 19 times the fundamental frequency of 10^5 Hz, or

$$19 \times 10^5 \text{ Hz} = 1.9 \text{ MHz}$$

Drill 7–3   Repeat Example 7–3 if the duty cycle is changed to 40%.
Answer: 2 MHz.

Harmonic Analysis of Other Waveforms

Figure 7.6 shows a *sawtooth*, or *sweep*, waveform. The Fourier series for this waveform is

$$v(t) = \frac{A}{2} - \frac{A}{\pi} \sin \omega t - \frac{A}{2\pi} \sin(2\omega t) - \frac{A}{3\pi} \sin(3\omega t)$$
$$- \cdots - \frac{A}{n\pi} \sin(n\omega t) - \cdots \tag{7.8}$$

where A is the peak value of the sawtooth and $\omega = 2\pi f = 2\pi/T$ is the fundamental frequency. Note that *all* harmonics are present and that the amplitude of A_n is inversely proportional to n. The significance of the minus sign in front of each harmonic is the same as stating that each harmonic is added to the sum with a 180° phase shift. That is,

$$-\frac{A}{n\pi} \sin(n\omega t) = \frac{A}{n\pi} \sin(n\omega t + 180°) \tag{7.9}$$

As usual, the average value can be adjusted if the sawtooth is shifted up or down, and A in (7.8) becomes the peak-to-peak value of the sawtooth. For a sawtooth that alternates between V_{LO} and V_{HI},

$$V_{AVG} = \tfrac{1}{2}(V_{HI} + V_{LO}) \tag{7.10}$$

and the peak-to-peak value is $A = V_{HI} - V_{LO}$.

Figure 7.6
Harmonic analysis of a sawtooth waveform.

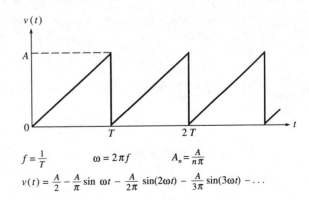

$$f = \frac{1}{T} \qquad \omega = 2\pi f \qquad A_n = \frac{A}{n\pi}$$

$$v(t) = \frac{A}{2} - \frac{A}{\pi} \sin \omega t - \frac{A}{2\pi} \sin(2\omega t) - \frac{A}{3\pi} \sin(3\omega t) - \cdots$$

Figure 7.7
Harmonic analysis of a triangular waveform.

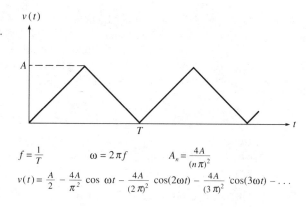

$$f = \frac{1}{T} \qquad \omega = 2\pi f \qquad A_n = \frac{4A}{(n\pi)^2}$$

$$v(t) = \frac{A}{2} - \frac{4A}{\pi^2}\cos \omega t - \frac{4A}{(2\pi)^2}\cos(2\omega t) - \frac{4A}{(3\pi)^2}\cos(3\omega t) - \cdots$$

Figure 7.7 shows a *triangular* waveform. The Fourier series is

$$v(t) = \frac{A}{2} - \frac{4A}{\pi^2}\cos \omega t - \frac{4A}{(2\pi)^2}\cos(2\omega t) - \frac{4A}{(3\pi)^2}\cos(3\omega t) - \cdots$$
$$- \frac{4A}{(n\pi)^2}\cos(n\omega t) - \cdots \qquad (7.11)$$

where A is the peak value and $\omega = 2\pi f = 2\pi/T$ is the fundamental frequency of the triangular wave. Note again that all harmonics are present, but in this case the amplitude of each is inversely proportional to n^2. The significance of each negative cosine term is the same as stating that each harmonic can be expressed as a sine term with a phase shift of $-90°$. That is,

$$\frac{-4A}{(n\pi)^2}\cos(n\omega t) = \frac{4A}{(n\pi)^2}\sin(n\omega t - 90°) \qquad (7.12)$$

If the triangular waveform is shifted so that it alternates between V_{LO} and V_{HI}, its average value becomes

$$V_{AVG} = \tfrac{1}{2}(V_{HI} + V_{LO}) \qquad (7.13)$$

and A in equation (7.11) becomes the peak-to-peak value $V_{HI} - V_{LO}$.

Example 7–4

Find the first four terms of the Fourier series for the waveform shown in Figure 7.8.

Solution. Since the waveform is symmetrical about the horizontal (time) axis, it is clear that its average value is zero. Equation (7.13) confirms this fact:

$$V_{AVG} = \tfrac{1}{2}[10 \text{ V} + (-10 \text{ V})] = 0 \text{ V}$$

The period T is 50 μs, so the fundamental frequency is

$$f = \frac{1}{50 \times 10^{-6} \text{ s}} = 20 \text{ kHz}$$

The peak-to-peak value is $A = 10$ V $- (-10$ V$) = 20$ V. From equation (7.11),

$$A_1 = \frac{4(20)}{\pi^2} = 8.11 \text{ V}$$

Figure 7.8
Example 7–4.

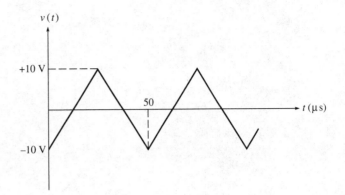

$$A_2 = \frac{4(20)}{(2\pi)^2} = 2.03 \text{ V}$$

$$A_3 = \frac{4(20)}{(3\pi)^2} = 0.90 \text{ V}$$

$$A_4 = \frac{4(20)}{(4\pi)^2} = 0.51 \text{ V}$$

Thus,

$$v(t) = -8.11 \cos(2\pi \times 20 \times 10^3 t) - 2.03 \cos(2\pi \times 40 \times 10^3 t)$$
$$- 0.9 \cos(2\pi \times 60 \times 10^3 t) - 0.51 \cos(2\pi \times 80 \times 10^3 t) - \cdots$$

Drill
7–4

What is the frequency of the first harmonic whose amplitude is less than 1% of the amplitude of the fundamental of the triangular wave in Example 7–4?
Answer: 220 kHz.

7–3   RC SHAPING CIRCUITS

The Low-Pass RC Filter

A waveform applied to the input of the RC circuit shown in Figure 7.9(a) is altered because the capacitive reactance of the capacitor (X_C) is different at every harmonic frequency contained in the input. The resistor and capacitor form a voltage divider whose output, across the capacitor, decreases as frequency increases. This follows from the fact that capacitive reactance is inversely proportional to frequency: $X_C = 1/\omega C = 1/(2\pi f C)$. Thus, high-frequency harmonics are reduced *(attenuated)*, whereas low-frequency harmonics are "passed." This is a simple example of a *low-pass filter*, one that shapes a waveform by reducing the amplitudes of high-frequency harmonics.

Figure 7.9(b) shows how the amplitude of the output varies as the frequency of the input signal is changed. The frequency at which the magnitude of the output falls to $\sqrt{2}/2 \approx 0.707$ times the magnitude of the input is called the (upper) *cutoff frequency*, f_2.

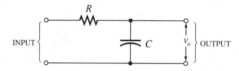

(a) The low-pass RC filter.

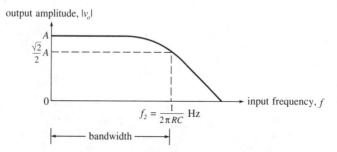

(b) Variation of output amplitude as input frequency is changed.
INPUT = $A \sin(2\pi ft)$.

Figure 7.9
The low-pass RC filter and its frequency response.

The input amplitude is assumed to be A at every frequency in Figure 7.9(b), so it can be seen that the cutoff frequency is

$$f_2 = \frac{1}{2\pi RC} \text{ hertz} \qquad (7.14)$$

or,

$$\omega_2 = 2\pi f_2 = \frac{1}{RC} \text{ radians/second}$$

Since the bandwidth of a low-pass filter extends from 0 to its cutoff frequency, we see that the bandwidth is the same as the cutoff frequency. The amplitude of the output when the input is any sinusoidal signal having frequency f can be found from

$$|v_o| = \frac{A}{\sqrt{(f/f_2)^2 + 1}} \qquad (7.15)$$

where $|v_o|$ is the output amplitude and A is the input amplitude. The way the output of a filter varies as the frequency of the input changes is called the *frequency response* of the filter. Thus, Figure 7.9(b) is a plot of the frequency response of the low-pass RC filter.

The phase angle of the output relative to the input is given by

$$\theta = -\tan^{-1}\left(\frac{f}{f_2}\right) \qquad (7.16)$$

The fact that this angle is always negative means that the output lags the input at every frequency. It is easy to see from (7.16) that the *output lags the input by 45° at the cutoff frequency* (where $f = f_2$).

Example 7–5

For the RC filter in Figure 7.9(a), $R = 1$ kΩ and $C = 0.01$ μF. The input signal is a 10-kHz square wave that alternates between 0 V and 9 V.

a. Find the bandwidth of the filter.
b. Find the amplitudes of the fundamental and third harmonic in the output waveform.

Solution

a. The bandwidth (BW) is the same as the cutoff frequency:

$$BW = f_2 = \frac{1}{2\pi RC} = \frac{1}{2\pi(10^3\ \Omega)(10^{-8}\ \text{F})} = 15{,}915.5\ \text{Hz}$$

b. The amplitude of the fundamental in the input square wave is

$$A_1 = \frac{2A}{\pi} = \frac{2(9\ \text{V})}{\pi} = 5.73\ \text{V}$$

Then we may calculate the effect of the filter acting on a 10-kHz sinewave with peak value 5.73 V. From equation (7.15),

$$|v_o| = \frac{A_1}{\sqrt{\left(\dfrac{f}{f_2}\right)^2 + 1}} = \frac{5.73\ \text{V}}{\sqrt{\left(\dfrac{10{,}000}{15{,}915.5}\right)^2 + 1}} = 4.85\ \text{V}$$

The amplitude of the third harmonic, $f_3 = 30$ kHz, in the input square wave is

$$A_3 = \frac{2A}{3\pi} = \frac{2(9\ \text{V})}{3\pi} = 1.91\ \text{V}$$

The output of the filter when the input is a 30-kHz sinewave with peak value 1.91 V is

$$|v_o| = \frac{1.91\ \text{V}}{\sqrt{\left(\dfrac{30{,}000}{15{,}915.5}\right)^2 + 1}} = 0.896\ \text{V}$$

Note that the filter has reduced the amplitude of the fundamental by the factor $4.85/5.73 = 0.846$, whereas the amplitude of the third harmonic has been reduced by the factor $0.896/1.91 = 0.469$. Thus the output contains a much smaller proportion of third-harmonic signal than does the input, so we know the waveshape has been altered between input and output.

Drill 7–5

Find the ratio of the amplitude of the fifth harmonic in the output to the fifth harmonic in the input in Example 7–5. Also find the phase angle of the fifth harmonic in the output, relative to the fifth harmonic in the input.
Answer: 0.303, −72.34°

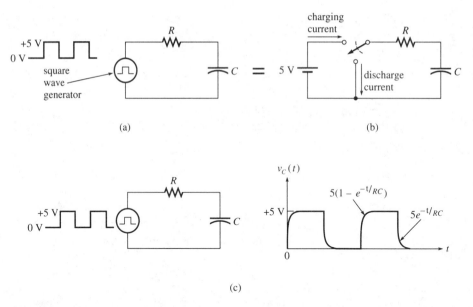

Figure 7.10
The smoothing of a square wave by a low-pass RC filter.

Recall that abruptly changing waveforms have substantial high-frequency content. Therefore, we would expect a low-pass filter to ''smooth'' such a waveform by reducing the amplitudes of its high-frequency harmonics. This fact is illustrated in Example 7–5 and in Figure 7.10(c), which shows how a square wave is smoothed by a low-pass RC filter. Note that the input alternates between 0 V and 5 V, which is the same as switching the input between 5 V and *ground* (Figure 7.10(b)). The capacitor charges with its characteristic charging curve when the input is at +5 V and discharges back through R to ground when the input is at 0 V. The smoothed output—i.e., the charge and discharge curves shown in Figure 7.10(c)—are predicted in this case by analysis in the *time domain*.

Figure 7.11 shows how the value of the RC time-constant affects the waveshape. When the time constant is small in comparison to the period T of the square wave, the capacitor charges and discharges rapidly and the square wave is only slightly rounded, as shown in Figure 7.11(b). This is a *time*-based conclusion. From a frequency domain standpoint, consider that equation (7.14) shows the cutoff frequency to be inversely proportional to RC. Thus, a small value of RC means a high cutoff frequency, which in turn means that more high-frequency harmonics are passed. Consequently, there is less smoothing. On the other hand, when the time constant is long, it takes longer for the capacitor to charge and discharge. In fact, it may not fully charge before it begins to discharge, as shown in Figure 7.11(d). In this case, the cutoff frequency is small, so a substantial number of high-frequency harmonics are attenuated and considerable smoothing results. (In all cases, the average value of the capacitor voltage equals the average value of the input square wave, $V_{HI}/2$.)

Figure 7.11(c) shows that when RC is large, the shaped square wave is nearly triangular (with dc value $V_{HI}/2$). This result is attributable to the fact that the exponential term $e^{-t/RC}$ in the charge and discharge equations is nearly *linear* for small values of t. In fact, $e^x \approx 1 + x$ for small x, so, with $x = -t/RC$ and t/RC very small, we have

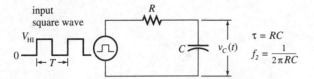

(a) *RC* low-pass filter driven by a square wave having period *T*.

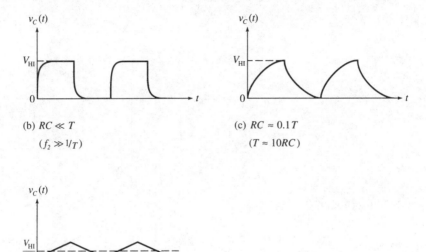

(b) $RC \ll T$

($f_2 \gg 1/T$)

(c) $RC \approx 0.1T$

($T \approx 10RC$)

(d) $RC \gg T$

($f_2 \ll 1/T$)

Figure 7.11
The effect of time-constant (and cutoff frequency) on the output waveform from a low-pass RC filter.

$$E(1 - e^{-t/RC}) \approx E\left[1 - \left(1 - \frac{t}{RC}\right)\right] = \frac{Et}{RC}$$

and

(7.17)

$$Ee^{-t/RC} \approx E\left(1 - \frac{t}{RC}\right)$$

These approximations are illustrated in Figure 7.12(*a*) and (*b*). Figure 7.12(*c*) shows that when *RC* is large compared to the period *T* of the square wave, the capacitor has but a small amount of time to charge before it must begin to discharge (due to the square wave suddenly returning to zero). Similarly, it has just begun to discharge when it must begin to charge again. Thus, charging and discharging times are small and the waveform segments are nearly linear. Note that the slopes of the straight-line approximations are *E/RC* and −*E/RC* volts/second.

Figure 7.12
Exponential charge and discharge curves are approximately linear when t is small.

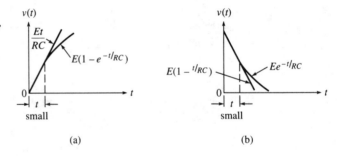

(a)   (b)

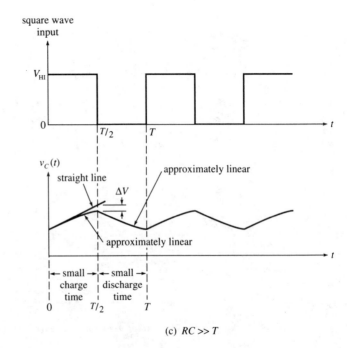

(c) $RC \gg T$

The maximum deviation of the capacitor voltage from a true straight line occurs at $t = T/2$ and is identified in Figure 7.12(c) as ΔV. The maximum percent deviation D from a straight line is then

$$\%D = \frac{\Delta V \text{ at } t = T/2}{\text{straight-line value at } t = T/2} \times 100\%$$

$$= \frac{\dfrac{V_0(T/2)}{RC} - V_0(1 - e^{-T/2RC})}{\dfrac{V_0(T/2)}{RC}} \times 100\%$$

where V_0 is the initial capacitor voltage. Simplifying, we obtain

$$\%D = \frac{\dfrac{T}{2RC} + e^{-T/2RC} - 1}{T/2RC} \times 100\% \qquad (7.18)$$

Example
7–6

A low-pass RC filter used to produce a triangular waveform from a square wave has $R =$ 100 kΩ and $C = 0.2$ µF. If the frequency of the square wave is 500 Hz, find the maximum percent deviation of the waveform from a true triangular wave.

Solution

$$RC = (10^5 \ \Omega)(0.2 \times 10^{-6} \ F) = 0.02 \ s$$

$$T = \frac{1}{f} = \frac{1}{500 \ Hz} = 2 \times 10^{-3} \ s$$

From equation (7.18),

$$\%D = \frac{\dfrac{2 \times 10^{-3}}{2(0.02)} + e^{-2 \times 10^{-3}/2(0.02)} - 1}{\dfrac{2 \times 10^{-3}}{2(0.02)}} \times 100\%$$

$$= \frac{0.05 - e^{-0.05} - 1}{0.05} \times 100\%$$

$$= 2.46\%$$

In this example, we see that the maximum deviation is only about 2.5% when the time constant is ten times greater than the period of the square wave: $RC = 10T$.

Drill
7–6

Find the maximum percent deviation of the capacitor voltage from the triangular wave in Example 7–6 if the frequency of the square wave is changed so that its period is 20 times the RC time constant.
Answer: 1.24%.

From a practical standpoint, the problem with obtaining a triangular waveform from an RC filter is that the peak-to-peak variation of the triangular wave is quite small. For example, when $RC = 10T$, the peak-to-peak value of the triangular waveform is only about 2.5% of the peak-to-peak value of the input square wave. Increasing the time constant to improve linearity further reduces the total variation of the triangle. The peak-to-peak value of the triangular waveform can be found from

$$V_{PP} = (V_{HI} - V_{LO})\frac{1 - e^{-T/2RC}}{1 + e^{-T/2RC}} \tag{7.19}$$

where V_{HI} and V_{LO} are the high and low voltages between which the input square wave alternates. In a later discussion, we will learn of an improved method for generating triangular waveforms, using *active* circuitry (operational amplifiers).

Since the average value of the capacitor voltage equals the average value of the input square wave, it follows that shifting the square wave causes the same shift in the output waveform. Figure 7.13 shows the waveforms produced when the input square wave alternates between $+E$ and $-E$ volts. In each case, the average value is zero.

Calculation of Rise-Time

Recall from Chapter 6 that the *rise-time* t_r of a pulse is defined to be the time required for it to change from 10% of its maximum value to 90% of its maximum value. Figure 7.14

Figure 7.13
Shifting the input square wave shifts the output waveform by the same amount. In the example shown, the dc values of input and output are 0 V. Compare with Figure 7.11.

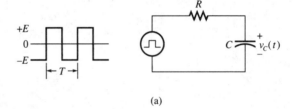

(a)

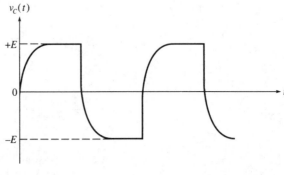

(b) $RC \ll T$

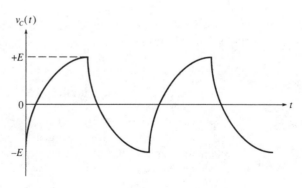

(c) $RC \approx 0.1T$

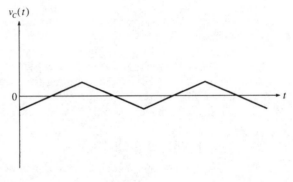

(d) $RC \gg T$

Figure 7.14

Calculating the rise time of the output of a low-pass RC filter.

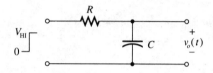

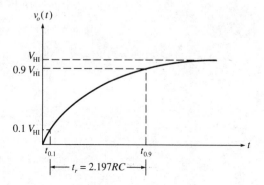

shows the rising voltage at the output of a low-pass RC circuit when the input is a square wave that alternates between 0 V and V_{HI}. We assume that the RC time constant is short enough for the capacitor to charge fully during one half-cycle of input. The 10% and 90% points are identified in the figure. The corresponding time points, $t_{0.1}$ and $t_{0.9}$, can be found by solving the charging equation as follows:

$$0.1\ V_{HI} = V_{HI}(1 - e^{-t_{0.1}/RC})$$
$$0.9 = e^{-t_{0.1}/RC}$$
$$t_{0.1} = -RC\ \ln(0.9) = 0.10536RC \tag{7.20}$$

$$0.9\ V_{HI} = V_{HI}(1 - e^{-t_{0.9}/RC})$$
$$0.1 = e^{-t_{0.9}/RC}$$
$$t_{0.9} = -RC\ \ln(0.1) = 2.30259RC \tag{7.21}$$

Therefore, the rise-time is

$$t_r = t_{0.9} - t_{0.1} = 2.197RC \tag{7.22}$$

Since the cutoff frequency of the RC network is $f_2 = 1/(2\pi RC)$ hertz, we have $RC = 1/(2\pi f_2)$ second, and (7.22) can then be written as

$$t_r = 2.197\left(\frac{1}{2\pi f_2}\right) = \frac{0.35}{f_2} \tag{7.23}$$

Note that the *time-domain* parameter t_r is expressed in terms of the *frequency-domain* parameter f_2, the cutoff frequency. This is an example of a very important relationship that is true in general: rise-time and bandwidth are inversely proportional to each other. In fact, equation (7.23) is often used as an approximation to find t_r or f_2 in more complex circuits containing a low-pass RC network as the dominant filter type. It is an exercise at the end of this chapter to show that the fall-time t_r of the output of a low-pass RC filter is also $2.197RC$.

Figure 7.15
The high-pass RC filter and its frequency response.

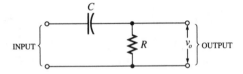

(a) The high-pass RC filter.

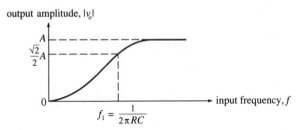

(b) Variation of output amplitude as input frequency is changed.
INPUT = $A \sin(2\pi ft)$.

The High-Pass RC Filter

Figure 7.15(*a*) shows a high-pass RC filter. Note that the output is taken across the resistor. The higher the frequency of the input, the smaller the reactance of the capacitor and the smaller the voltage drop across it. Therefore, the amplitude of the output increases as frequency increases, which accounts for the name *high-pass*. The variation in output amplitude versus frequency is shown in Figure 7.15(*b*). The frequency at which the output amplitude is $\sqrt{2}/2$ times the input amplitude is called the (lower) cutoff frequency f_1. This frequency is calculated in exactly the same way as the cutoff frequency of a low-pass RC filter:

$$f_1 = \frac{1}{2\pi RC} \text{ hertz} \tag{7.24}$$

or

$$\omega_1 = \frac{1}{RC} \text{ radians/second} \tag{7.25}$$

The magnitude and phase angle of the output when the input is a sinewave signal having amplitude A and phase angle $0°$ can be found from

$$|v_o| = \frac{A}{\sqrt{(f_1/f)^2 + 1}} \tag{7.26}$$

$$\theta = \tan^{-1}\left(\frac{f_1}{f}\right) \tag{7.27}$$

Note that the output leads the input at every frequency and that $\theta = 45°$ at the cutoff frequency ($f = f_1$).

From a frequency-domain viewpoint, we can expect the output of a high-pass filter to be a more abruptly changing waveform than the input, because low-frequency harmonics are suppressed in favor of higher frequency harmonics. Figure 7.16 shows

Figure 7.16
The effect of the time-constant (and cutoff frequency) on the output waveform from a high-pass RC filter.

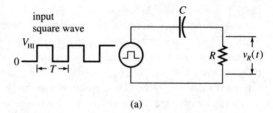

(a)

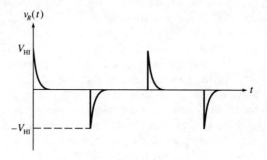

(b) $RC \ll T$ $(f_1 \gg 1/_T)$

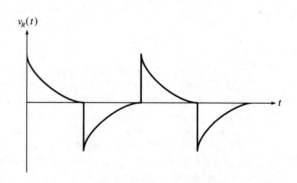

(c) $RC \approx 0.1T$ $(T \approx 10RC)$

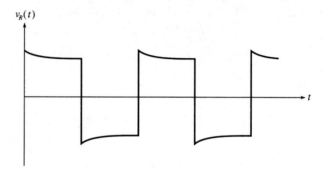

(d) $RC \gg T$ $(f_1 \ll 1/_T)$

output waveforms that result when the input is a square wave. As in the case of the low-pass filter, the output waveshape depends heavily on the value of the cutoff frequency of the filter. Note that the output is a series of very abrupt "spikes" when the cutoff frequency is much greater than the frequency of the square wave ($f_1 \gg 1/T$). In this case, only the very high frequency harmonics of the square wave are passed by the filter. At the other extreme, when the cutoff frequency is small, the fundamental and the low-frequency harmonics as well as high-frequency harmonics are passed. The output then undergoes very little alteration, as shown in Figure 7.16(d). From a time-domain viewpoint, a small RC time-constant means that the capacitor charges and discharges rapidly, so the output across the resistor consists of transients having very short durations. On the other hand, when the time-constant is large in comparison to the period T of the square wave, the capacitor voltage changes very little and the output waveform resembles the input.

Note that the average value of every output waveform shown in Figure 7.16 is 0 V. The output of *every* high-pass filter has zero average value, regardless of input waveshape and irrespective of the minimum and maximum values (positive and/or negative peaks) of the input. From a frequency-domain viewpoint, this result follows from the fact that a high-pass filter blocks the zero-frequency (dc) component of the input. From a time-domain viewpoint, the capacitor charges to the value of any dc component in the input, so no dc current can flow in the resistor. In any event, the output waveform always has equal areas below and above the horizontal (time) axis.

Note that the peak-to-peak value of the output for the case $RC \ll T$, as shown in Figure 7.16(b), equals twice the peak-to-peak value of the input. Such is always the case when the time constant is so small that the capacitor can fully charge and discharge during each cycle of the input. Whether the input has zero, positive, or negative average value, the peak-to-peak value of the output in those cases is twice the peak-to-peak value of the input. Of course, the positive and negative peak values of the output may not equal those of the input because the output waveform must always be shifted to have equal areas below and above the time axis. The positive peak of the output will equal the peak-to-peak value of the input, as will the negative peak.

The series of spikes resulting from the shaping of a square wave by a short time-constant filter is often used to *synchronize* or *trigger* other circuits, so that they respond in unison with the transitions of the square wave. The smaller the time-constant, the narrower the spikes. The next example illustrates that the width of the spikes can be predicted, since the capacitor for all practical purposes fully charges and discharges in five time-constants.

Example 7–7

Sketch the output waveform produced by the high-pass filter shown in Figure 7.17(a) when the input is a square wave that alternates between -5 V and $+10$ V with a frequency of 8×10^4 PPS.

Solution. The period T of the input square wave is

$$T = \frac{1}{8 \times 10^4 \text{ PPS}} = 1.25 \times 10^{-5} \text{ s} = 12.5 \text{ } \mu\text{s}$$

The RC time-constant of the filter is

$$RC = (2 \times 10^3 \text{ } \Omega)(250 \times 10^{-12} \text{ F}) = 500 \times 10^{-9} \text{ s} = 0.5 \text{ } \mu\text{s}$$

We see that $RC = 0.04T$, so the condition $RC \ll T$ is satisfied, and the output will consist

Figure 7.17
Example 7–7.

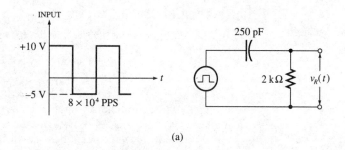

(a)

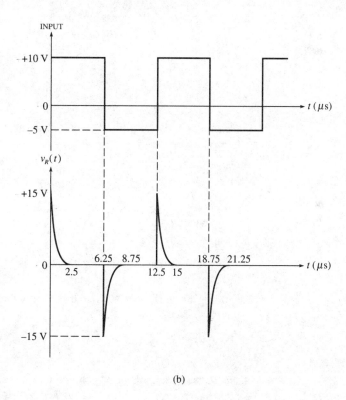

(b)

of alternating positive and negative spikes. Assuming the capacitor is fully charged in 5 time-constants, the spikes will return to zero $5(0.5 \ \mu s) = 2.5 \ \mu s$ after each peak occurs. The positive peak value of each spike is the peak-to-peak value of the input: $10 \ V - (-5 \ V) = 15 \ V$, and the negative peak is $-15 \ V$. The output waveform is sketched in Figure 7.17(*b*). Note that its peak-to-peak value is 30 V, twice the peak-to-peak value of the input, and that the output has zero average value.

**Drill
7–7**

To what value should the capacitance in Example 7–7 be changed if it is necessary to make each output spike return to zero in (essentially) 1.5 μs?
Answer: 150 pF

Calculation of Droop

Recall from Chapter 6 that the percent droop, or tilt, of a rectangular pulse is defined by

$$\% \text{ droop} = \frac{\text{change in pulse amplitude}}{\text{average pulse amplitude}} \times 100\% \tag{7.28}$$

We can apply this definition to compute the percent droop in each half-cycle of a square wave after it has been shaped by a high-pass RC filter. The following equations can be used to compute the various quantities shown in Figure 7.18:

$$V_{HI}(\text{max}) + V_{HI}(\text{min}) = V_{HI}(\text{input}) \tag{7.29}$$

$$V_{HI}(\text{max}) = \left(\frac{1}{1 + e^{-T/2RC}}\right) V_{HI}(\text{input}) \tag{7.30}$$

$$V_{HI}(\text{min}) = \left(\frac{e^{-T/2RC}}{1 + e^{-T/2RC}}\right) V_{HI}(\text{input}) \tag{7.31}$$

Subtracting (7.31) from (7.30), we find the change in pulse amplitude during the positive half-cycle to be

$$V_{HI}(\text{max}) - V_{HI}(\text{min}) = \left(\frac{1 - e^{-T/2RC}}{1 + e^{-T/2RC}}\right) V_{HI}(\text{input}) \tag{7.32}$$

The average amplitude of the positive pulse is

$$V_{AVG}(\text{positive}) = \frac{2RC}{T}\left(\frac{1 - e^{-T/2RC}}{1 + e^{-T/2RC}}\right) V_{HI}(\text{input}) \tag{7.33}$$

According to equation (7.28), the droop is then found by dividing equation (7.32) by equation (7.33):

$$\% \text{ droop} = \frac{T}{2RC} \times 100\% \tag{7.34}$$

As expected, equation (7.34) shows that the shorter the time constant, the greater the droop. In terms of the cutoff frequency f_1 of the filter and the frequency f of the square wave, the percent droop is

$$\% \text{ droop} = \pi\left(\frac{f_1}{f}\right) \times 100\% \tag{7.35}$$

Example 7–8

A square wave that alternates between 0 V and 5 V is *capacitor-coupled* to an amplifier whose input resistance is 10 kΩ. The capacitor and the resistance form a high-pass filter. If the square wave has frequency 50 kHz, what is the smallest capacitor that can be used to ensure that the droop does not exceed 2%?

Solution. The period of the square wave is

$$T = \frac{1}{50 \times 10^3 \text{ Hz}} = 0.2 \times 10^{-4} \text{ s}$$

Figure 7.18
Computing the percent droop in the output of a high-pass RC filter.

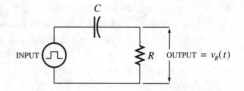

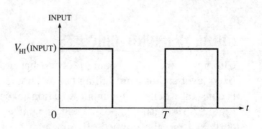

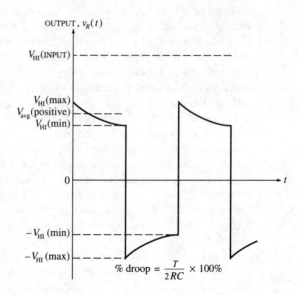

$$\% \text{ droop} = \frac{T}{2RC} \times 100\%$$

From equation (7.34), we require that

$$\frac{T}{2RC} \leq 0.02$$

$$\frac{0.2 \times 10^{-4}}{(2 \times 10^4)C} \leq 0.02$$

$$C \geq \frac{0.2 \times 10^{-4}}{(2 \times 10^4)(0.02)} = 0.05 \ \mu\text{F}$$

**Drill
7–8**

What would be the percent droop in Example 7–8 if the required value of capacitance were used but the frequency of the square wave were halved? Doubled?
Answer: 4%, 1%

We have seen that low-pass and high-pass filters are in many respects "opposites" of one another. Indeed, when a waveform produced by a filter of one type is passed through a filter of the opposite type, the original (unshaped) waveform is recovered. In a sense, one filter "undoes" what the other filter does to frequency content. For example, when a triangular wave produced by shaping a square wave with a low-pass filter is then applied to a high-pass filter, the output resembles the original square wave. (In reality this process only works perfectly for *ideal* filters, which RC filters are not.)

7-4   DIODE CLIPPING CIRCUITS

Clipping is the waveshaping process that prevents a waveform from exceeding a certain fixed level and/or from falling below another fixed level. The output of a clipping circuit is constant whenever the input variations exceed and/or fall below specific clipping levels. Figure 7.19 illustrates positive and negative clipping of a sinewave. The term *clipping* is derived from the clipped-off appearance of the peaks. In analog circuits, clipping is considered a form of distortion. In digital circuits, it is used to shape waveforms into nearly square waves and to prevent large voltage variations from damaging semiconductor devices. Clipping is a *non-linear* process that causes new frequencies to be added to a waveform, but the process is best understood by studying it in the time domain.

Semiconductor diodes are widely used in clipping circuits because of their ability to serve as *voltage-controlled switches:* they conduct heavily when *forward-biased* by a voltage having one polarity and are essentially open circuits when *reverse-biased* by a voltage of the opposite polarity. Figure 7.20 shows a typical *characteristic curve* for a silicon diode: a plot of the current through it versus the voltage across it. When the anode is positive with respect to the cathode (forward-biased) the diode conducts current from anode to cathode. When that voltage is greater than about 0.7 V for a silicon diode (about 0.3 V for a germanium diode), the conduction is quite heavy, as evident from the steepness of the curve in that region. When the anode-to-cathode voltage is negative (reverse-biased), the current is essentially cut off. Thus, the diode is like a switch that closes when the anode-to-cathode voltage is positive and opens when the voltage is negative.

Figure 7.21 shows how a diode can be used to clip essentially the entire positive or the entire negative variation from a waveform. In 7.21(a), the diode is forward-biased when the input is positive, so the diode behaves like a closed switch, or short circuit,

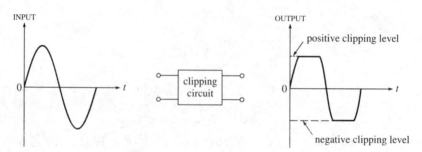

Figure 7.19
Example of a clipped waveform.

Figure 7.20
Characteristic curve of a silicon diode.

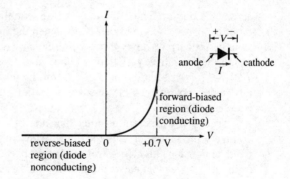

Figure 7.21
Clipping positive and negative peaks.

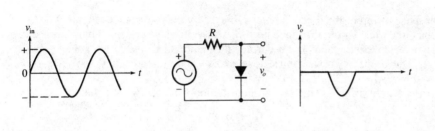

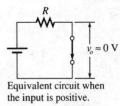

Equivalent circuit when
the input is positive.

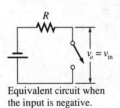

Equivalent circuit when
the input is negative.

(a) Clipping the positive peaks from a waveform.

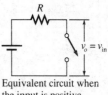

Equivalent circuit when
the input is positive.

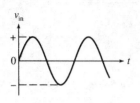

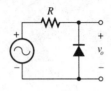

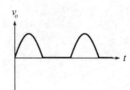

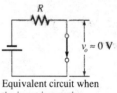

Equivalent circuit when
the input is negative.

(b) Reversing the diode clips the negative peaks.

across the output. Thus, the output is essentially zero (actually, about 0.7 V) when the input is positive. On the other hand, when the input is negative, the diode is like an open switch, no current flows through R, there is no voltage drop across R, and the output equals the negative variations of the input. Reversing the diode, as shown in Figure 7.21(b), forward-biases the diode when the input is negative and reverse-biases it when the input is positive, so the negative peaks are clipped off. In each case, the clipping level is essentially 0 V.

To clip a waveform at other than the 0-V level, it is necessary to connect a fixed voltage source in series with the diode. This combination is called a *biased diode,* and the fixed source can either forward- or reverse-bias the diode. Clipping occurs over the range of input voltages that cause the diode to be forward-biased, and the output level over that range equals the value of the fixed source. The next example demonstrates how clipping levels can be calculated using Kirchhoff's voltage law.

Example 7–9

Sketch the output waveform produced by the biased-diode clipping circuits shown in Figure 7.22(a) and (b) when the input waveforms are as shown. Assume each diode conducts when its anode-to-cathode voltage is positive.

Solution

a. In the absence of input, the diode is reverse-biased by the 6-V source. Figure 7.23(a) shows the circuit just before the diode becomes forward-biased by the input. Since the diode is reverse-biased, there is no current flow and, therefore, zero voltage drop across R. The diode is forward-biased when its anode-to-cathode voltage, V_D, is positive—i.e., when $V_D > 0$.

Writing Kirchhoff's voltage law around the loop shown, we see that

$$v_{in} = V_D + 6 \text{ V}$$

or

$$V_D = V_{in} - 6 \text{ V}$$

Figure 7.22
Example 7–9.

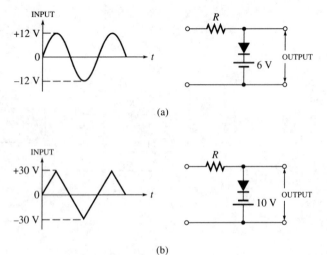

(a)

(b)

For V_{in} to forward-bias the diode,

$$V_D > 0 \implies v_{in} - 6\text{ V} > 0$$
$$\implies v_{in} \qquad > 6\text{ V}$$

Thus, the diode becomes forward-biased when the input exceeds 6 V. Under those circumstances, the diode is like a closed switch, as shown in Figure 7.23(b), and the

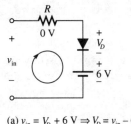

(a) $v_{in} = V_D + 6\text{ V} \Rightarrow V_D = v_{in} - 6\text{ V}$

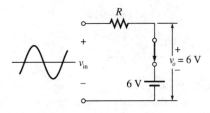

(b) Equivalent circuit when $v_{in} > 6$ V.

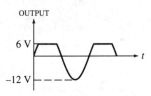

(c) Output waveform.

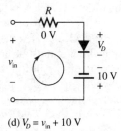

(d) $V_D = v_{in} + 10$ V

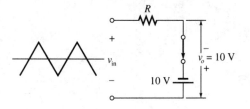

(e) Equivalent circuit when $v_{in} > -10$ V.

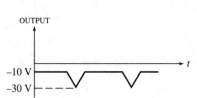

(f) Output waveform.

Figure 7.23
Example 7–22.

output remains fixed at the dc source voltage, 6 V. When the input falls below 6 V, the diode is reverse-biased, and the output is the same as the input. The output waveform is shown in Figure 7.23(c).

b. Figure 7.23(d) shows the circuit before the diode is forward-biased, and the voltage drop across R is 0 V. Writing Kirchhoff's voltage law around the loop, we find

$$V_D = v_{in} + 10 \text{ V}$$

For v_{in} to forward-bias the diode,

$$V_D > 0 \implies v_{in} + 10 \text{ V} > 0$$
$$\implies v_{in} > -10 \text{ V}$$

Thus, the diode is forward-biased when the input is greater (more positive) than -10 V. Under those circumstances, the diode is like a closed switch, shown in Figure 7.23(e), and the output remains fixed at -10 V. When the input is more negative than -10 V, the diode is open, and the output is the same as the input. Figure 7.23(f) shows the output waveform.

Drill
7-9

Find the range of input voltages for which clipping occurs in (a) and (b) of Example 7-9 when each diode is reversed.
Answer: (a) $v_{in} < 6$ V; (b) $v_{in} < -10$ V.

A waveform can be clipped at both positive and negative levels by connecting two biased diodes in parallel. Figure 7.24 shows an example, in which a 10-V peak sinewave is clipped at $+5$ V and -3 V. Note that only diode D_1 conducts when $v_{in} > 5$ V, neither diode conducts when $-3 \text{ V} < v_{in} < 5$ V, and only diode D_2 conducts when $v_{in} < -3$ V.

One of the practical disadvantages of diode clippers is that clipping levels are not precise because of the voltage drop across a forward-biased diode (about 0.7 V for silicon). This drop can be neglected if signal levels are large in comparison to 0.7 V but is significant if precision clipping must be achieved in low-level signals. If it is necessary to account for the forward drop, the procedures illustrated in Example 7-9 can be modified to include it when calculating clipping levels. For example, if we assume that the diode does not conduct until $V_D > 0.7$ V, then in Example 7-9(a), clipping occurs when $v_{in} - 6 \text{ V} > 0.7$ V, or $v_{in} > 6.7$ V. The output voltage when clipping occurs is then $6 \text{ V} + 0.7 \text{ V} = 6.7$ V instead of 6 V. In Example 7-9(b), clipping occurs when $v_{in} > -9.3$ V.

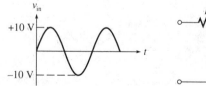
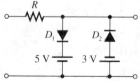
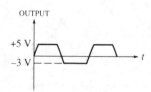

Figure 7.24
An example of positive and negative clipping by two biased diodes connected in parallel.

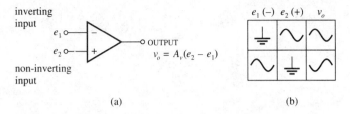

Figure 7.25
Schematic symbol and phase relations for an operational amplifier.

7–5 OPERATIONAL AMPLIFIER FUNDAMENTALS

An operational amplifier is a very high gain voltage amplifier having a single output and two inputs: an *inverting* input and a *non-inverting* input. Figure 7.25(a) shows the schematic symbol for an operational amplifier, with the inverting and non-inverting inputs labeled − and +, respectively. If the + input is grounded (at 0 V), the output is 180° out of phase with the input connected to the − input, and if the − input is grounded, the output is in phase with the input connected to the + input. These relationships are summarized graphically in Figure 7.25(b). When signals e_1 and e_2 are applied to the inverting and non-inverting inputs, respectively, the output is the amplified *difference voltage* $A_v(e_2 - e_1)$, where A_v is the voltage gain of the amplifier.

An operational amplifier is a *dc* amplifier, in the sense that it is direct-coupled (no coupling capacitors), and direct-current. Thus, it can be used to amplify slowly varying or very low frequency input signals. In most applications, there is no dc bias level, or offset voltage, at the output, so the output can go both positive and negative. Positive and negative power supplies must be used with the amplifier for it to function in this manner. Ideally, the output should be exactly 0 V when both inputs are grounded. Virtually all modern operational amplifiers are fabricated as integrated circuits.

The *ideal* operational amplifier has infinite voltage gain, infinite input impedance, and zero output impedance. Practical amplifiers have voltage gains on the order of 10,000 or more, input impedances of at least several megohms, and output impedances of a few ohms. However, the amplifier is rarely used with its full gain capability. Instead, external circuitry is connected between the output and input(s) to reduce the overall gain and to achieve certain other characteristics. Figure 7.26(a), (b), and (c) shows examples. Assuming an ideal amplifier, the output in 7.26(a) is

$$v_o = -\frac{R_f}{R_1} e_{\text{in}} \tag{7.36}$$

R_f is called the *feedback* resistance. We see that the output is inverted (out of phase with the input), as signified by the minus sign, and that the voltage gain is the ratio R_f/R_1 of resistance values. Unlike conventional amplifiers, the voltage gain is independent of amplifier characteristics and can be set as precisely as the resistance values can be obtained. Figure 7.26(b) shows how a non-inverted output can be obtained. In this case,

$$v_o = \left(\frac{R_f}{R_1} + 1\right) e_{\text{in}} \tag{7.37}$$

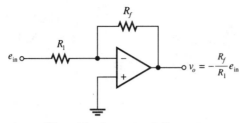

(a) The inverting amplifier, with voltage gain $-R_f/R_1$.

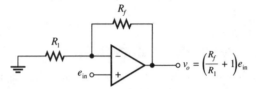

(b) The non-inverting amplifier, with voltage gain $(R_f/R_1) + 1$.

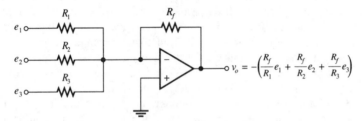

(c) The summing amplifier, with phase inversion.

Figure 7.26
Examples of operational amplifier applications.

The advantage of this configuration is that the impedance presented to e_{in} at the + input is very large, whereas the impedance to e_{in} in Figure 7.26(a) is just the value of R_1. Figure 7.26(c) shows how the amplifier can be connected to obtain an output proportional to the (inverted) *sum* of several input signals:

$$v_o = -\left(\frac{R_f}{R_1}e_1 + \frac{R_f}{R_2}e_2 + \frac{R_f}{R_3}e_3\right) \tag{7.38}$$

The number of inputs can be extended in an obvious way to increase the number of signals that are summed. This example demonstrates the origin of the name *operational* amplifier: It is often used to perform mathematical operations on signal voltages. Although the relationships expressed by equations (7.36), (7.37), and (7.38) assume ideal amplifiers, modern amplifiers are close enough to ideal to make the relations valid in most practical applications.

Example 7–10

Design an operational amplifier circuit that sums three equal-amplitude sinusoidal voltages having the frequencies of the fundamental, third, and fifth harmonics of a 500-Hz square wave. Each voltage should be summed so that its proportion in the output is the same as that in the harmonic content of a square wave. The gain for the fundamental component should be 15.

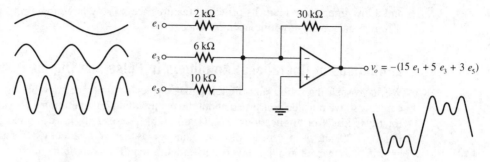

Figure 7.27
Example 7–10.

Solution.   The amplitudes of the fundamental, third, and fifth harmonics of a square wave are $2A/\pi$, $2A/3\pi$, and $2A/5\pi$, so their proportions in the output should be 1, $\frac{1}{3}$, and $\frac{1}{5}$. Since the gain for the fundamental component is to be 15, the gain for the third harmonic will be $\frac{15}{3} = 5$ and the gain for the fifth harmonic will be $\frac{15}{5} = 3$. We use the inverting summing amplifier to produce $v_o = -(15e_1 + 5e_3 + 3e_5)$, where e_1, e_3, and e_5 represent the sinusoidal inputs at the three frequencies.

With reference to Figure 7.26(c), we may arbitrarily select $R_f = 30$ kΩ. Then

$$\frac{R_f}{R_1} = 15 \implies \frac{30 \text{ k}\Omega}{R_1} = 15 \implies R_1 = 2 \text{ k}\Omega$$

$$\frac{R_f}{R_2} = 5 \implies \frac{30 \text{ k}\Omega}{R_2} = 5 \implies R_2 = 6 \text{ k}\Omega$$

$$\frac{R_f}{R_3} = 3 \implies \frac{30 \text{ k}\Omega}{R_3} = 3 \implies R_3 = 10 \text{ k}\Omega$$

The circuit is shown in Figure 7.27.

Drill 7–10   If the seventh harmonic is to be included in the output in Example 7–10, what should be the value of the input resistor to which it is connected (assuming it has the same amplitude as the other inputs)?
Answer: 14 kΩ.

Figure 7.28 shows another useful application of an operational amplifier. Note that the output is connected directly to the inverting input and the input signal is connected directly to the non-inverting input. This configuration is called a *voltage follower* because the output is identical to (follows) the input. The gain is one, and there is no phase inversion. The voltage follower is often used as a *buffer* between a high-impedance source

Figure 7.28
The voltage follower.

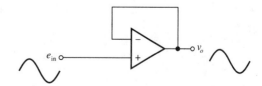

and a low-impedance load. Its principal features are its very large input impedance at the + input and a very low output impedance.

Limitations of Operational Amplifiers in Pulse and Digital Systems

We have seen that pulse and digital signals typically have considerable high-frequency content, so we must be concerned about the bandwidth of an operational amplifier selected for those kinds of applications. High-gain amplifiers usually have limited bandwidths. Figure 7.29 shows a typical plot of gain versus cutoff frequency for an operational amplifier connected in a non-inverting configuration. The *open-loop gain* is the very large voltage gain that the amplifier exhibits when no circuitry is connected between its output and input(s). Note in the example shown in the figure that the cutoff frequency, or bandwidth, when the gain is open-loop is only $f_c = 10$ Hz. When external circuitry is connected to reduce the gain (closed loop), the figure shows that the bandwidth increases. When the closed-loop gain is reduced to unity, the bandwidth is increased to the value labeled f_T in the figure, 1 MHz. These observations lead us to believe that the gain-bandwidth product is *constant* (the greater one, the smaller the other and vice versa). This is indeed the case.

The *feedback factor* β of a closed-loop amplifier is the proportion of output voltage fed back through external circuitry to the input, and its value influences both the value of the closed-loop gain and the bandwidth: $0 \leq \beta \leq 1$. For the inverting and non-inverting amplifiers shown in Figure 7.26(a) and (b),

$$\beta = \frac{R_1}{R_1 + R_f} \tag{7.39}$$

For the summing amplifier shown in Figure 7.26(c),

$$\beta = \frac{R_p}{R_p + R_f} \tag{7.40}$$

where $R_p = R_1 \parallel R_2 \parallel R_3$.

Figure 7.29
Typical gain versus frequency plots for a non-inverting operational amplifier.

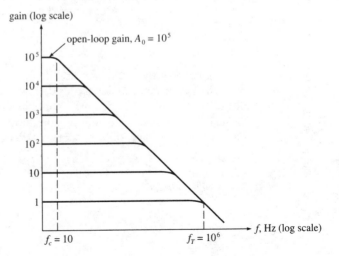

For the voltage follower, $\beta = 1$. The closed-loop bandwidth, BW_{CL}, of an operational amplifier is closely approximated by

$$BW_{CL} = f_T\beta = A_o f_c \beta \qquad (7.41)$$

where f_T is the *unity-gain frequency*, identified in Figure 7.29, β is the feedback factor, A_o is the open-loop gain, and f_c is the open-loop cutoff frequency, at which the output equals $\sqrt{2}/2$ times its value at dc (see Figure 7.29).

It is apparent from equation (7.41) that the unity-gain frequency f_T equals $A_o f_c$, so it is also called the *gain-bandwidth* product, the constant we mentioned earlier. It is often referred to by that name in manufacturers' specifications.

Example 7–11

An inverting amplifier is to be used to amplify a 1-kHz square wave. The gain-bandwidth product of the amplifier is 10^6. If the amplifier must pass the 15th harmonic without distortion, what is the maximum closed-loop gain the amplifier can have?

Solution.   The closed-loop bandwidth of the amplifier must be at least 15×1 kHz $= 15$ kHz. From equation (7.41),

$$f_T\beta = BW_{CL}$$

so we require that

$$10^6\beta \geq 15 \times 10^3$$

or

$$\beta \geq \frac{15 \times 10^3}{10^6} = 0.015$$

From equation (7.39),

$$\beta = \frac{R_1}{R_1 + R_f} \geq 0.015$$

Therefore

$$R_1 \geq 0.015R_1 + 0.015R_f$$
$$0.985R_1 \geq 0.015R_f$$
$$\frac{R_f}{R_1} \leq 65.67$$

Since the magnitude of the gain of the inverting amplifier is R_f/R_1, we see that that value cannot exceed 65.67.

Drill 7–11

What maximum gain can the amplifier in Example 7–11 have if it is used in a non-inverting configuration?
Answer: 66.67.

Slew Rate

Another limitation of practical operational amplifiers used in digital systems is that the *rate* at which their output voltage changes cannot exceed a specified value. The maximum rate of change of output voltage, in volts/second (V/s), is called the maximum

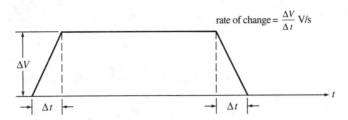

Figure 7.30
Approximation of a real pulse and calculation of its rate of change in volts per second.

slew rate of the amplifier. A typical value is 10^6 V/s, which can be expressed equivalently as 1 V/μs. If it were possible for pulses or rectangular waves to change levels *instantaneously* (in zero time), then their rates of change would be infinite and no practical amplifier could accommodate them. However, every real pulse requires a non-zero time to change levels, and the leading and trailing edges are often approximated as sloped lines. Figure 7.30 shows how a real pulse can be approximated. The rate of change of the leading or trailing edge is simply its slope $\Delta V/\Delta t$ in volts per second.

Example 7–12

The operational amplifier shown in Figure 7.31 has a specified maximum slew rate of 1 V/μs. What is the maximum permissible value of R_f when the input is the pulse shown?

Solution. Let A_v represent the closed-loop gain of the amplifier: $|A_v| = R_f/R_1$. As shown in Figure 7.31, the inverting amplifier will then cause the output pulse to alternate between 0 V and $-5A_v$ volts. The leading and trailing edges of the output pulse have slopes

$$\frac{\Delta V}{\Delta t} = \frac{-5A_v - 0}{25 \ \mu s - 0} = -(2 \times 10^5)A_v \ \text{V/s} \quad \text{(leading edge)}$$

$$\frac{\Delta V}{\Delta t} = \frac{0 - (-5A_v)}{125 \ \mu s - 100 \ \mu s} = (2 \times 10^5)A_v \ \text{V/s} \quad \text{(trailing edge)}$$

We see that the magnitude of the output rate of change in both cases is $2 \times 10^5 A_v$ V/s. This rate must not exceed the specified slew rate of 1 V/μs $= 10^6$ V/s:

$$2 \times 10^5 \ A_v \leq 10^6 \ \text{V/s}$$

$$A_v \leq \frac{10^6}{2 \times 10^5} = 5$$

Thus,

$$\frac{R_f}{R_1} \leq 5 \implies \frac{R_f}{2 \ \text{k}\Omega} \leq 5$$

$$R_f \leq 5(2 \ \text{k}\Omega) = 10 \ \text{k}\Omega$$

Drill 7–12

If both resistors R_1 and R_f in Figure 7.31 were made equal to 10 kΩ, what is the shortest interval of time that the input pulse could rise to 5 V without exceeding the amplifier's slew rate?
Answer: 5 μs.

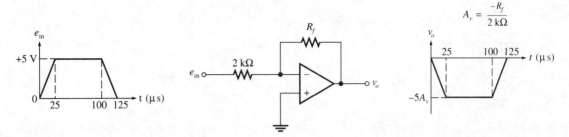

Figure 7.31
Example 7-12.

Of course, the output of an operational amplifier also undergoes a rate of change when the input is a sinusoidal signal. The greater the amplitude and the higher the frequency of the signal, the greater the rate of change of output voltage. For a given slew rate S, the maximum sinusoidal frequency the amplifier can accommodate is found from

$$f_{(max)} \leqslant \frac{S}{2\pi A} \tag{7.42}$$

where A is the peak value of the output sinewave. The maximum frequency specified by (7.42) may be quite different from (less than or greater than) the closed-loop bandwidth determined by the feedback factor. In other words, both slew rate and bandwidth must be considered when determining the maximum sinusoidal frequency that an amplifier can accommodate without introducing distortion.

7-6 ELECTRONIC INTEGRATION AND DIFFERENTIATION

Integration

Integration of a waveform produces another waveform whose value at any instant of time t equals the net *area* under the original waveform up to that time. Figure 7.32 shows examples. In 7.32(*a*), the input "waveform" is simply a dc level (constant) applied to an ideal integrator. The integrator is represented by the standard mathematical symbol $\int dt$. We assume the integration process begins at $t = 0$, when dc level E is first applied to the input of the integrator. As time passes, the total area under the constant level increases *linearly*, so the output is the *ramp* voltage Et shown in the figure. At any particular instant of time t_1, the output equals the area $E \times t_1 = Et_1$ volts. Of course, in a practical integrator, this process cannot continue indefinitely because the output voltage grows without limit. Figure 7.32(*b*) shows integration of a square wave that alternates between $+E$ and $-E$. In this case, the output is initially the same ramp voltage Et, but at $t = T/2$, the input becomes negative. Area beyond that instant is *negative*, so negative area is added to the total positive area accumulated up to that point. Therefore, the total (net) area begins to *decrease*, and the output voltage decreases linearly. At $t = T$, the total negative area has completely cancelled the total positive area, the net area is zero, and the output is 0 V. The process then repeats itself. We see that integration of a square wave produces a triangular wave.

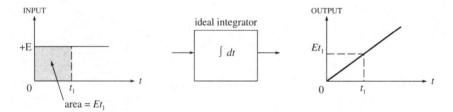

(a) Integration of dc level E produces a ramp output whose value at any instant t_1 equals the area Et_1 up to that time.

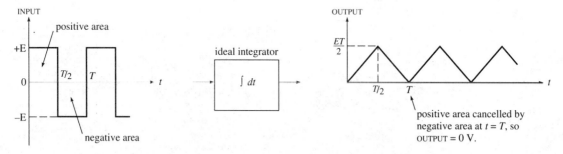

(b) Integration of a square wave adds negative area to positive area; so the output is 0 V after one complete cycle.

Figure 7.32
Examples of waveform integration by an ideal integrator.

Several other examples of waveshapes that can be produced by waveform integration are shown in Figure 7.33. Study each example and verify that the output at any instant of time t is the net area under the input waveform up to that time.

Figure 7.34 shows how an operational amplifier can be used to construct an electronic integrator. As shown in the figure, the output is

$$v_o = -\frac{1}{R_1 C} \int e_{\text{in}} \, dt \tag{7.43}$$

where $\int e_{\text{in}} \, dt$ means the integral of the input voltage, and the minus sign means phase inversion: The output is the negative of that which would be produced by an ideal integrator. For example, if the input were the constant level E, the output would be

$$v_o = -\frac{1}{R_1 C} \int E \, dt = -\frac{Et}{R_1 C} \tag{7.44}$$

If the amplifier were ideal, resistors R_f and R_c would not be required; their purpose is simply to prevent any dc offset in the amplifier from being integrated and producing an unlimited ramp. For optimum performance, the *compensating* resistance R_c is made equal to the parallel equivalent of R_1 and R_f.

Figure 7.35 shows the output waveform produced by the integrator in Figure 7.34 when the input is a square wave that alternates between $+E$ and $-E$ volts. Note that it differs from the output of an ideal integrator in several respects. The phase inversion caused by the amplifier is apparent from the fact that the output is increasing when the

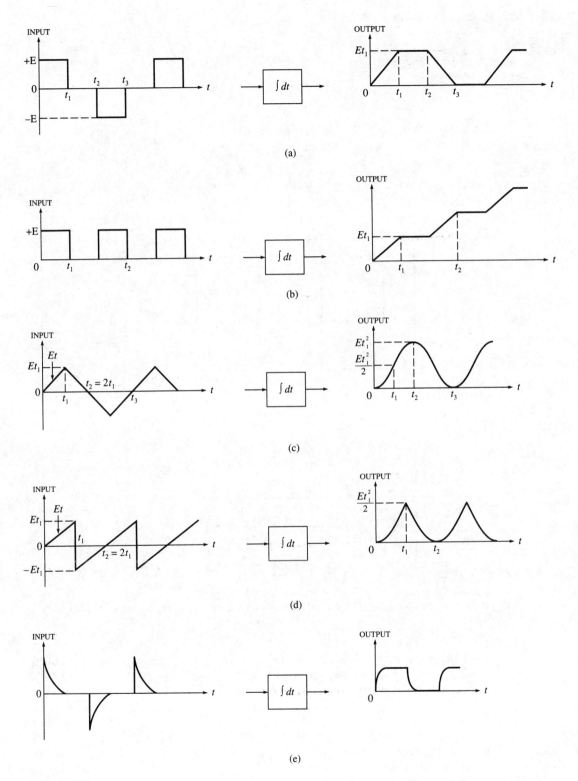

Figure 7.33
Examples of waveshaping by waveform integration.

Figure 7.34
Construction of an electronic integrator using an operational amplifier.

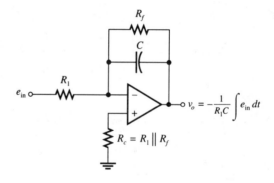

$$v_o = -\frac{1}{R_1C}\int e_{in}\,dt$$

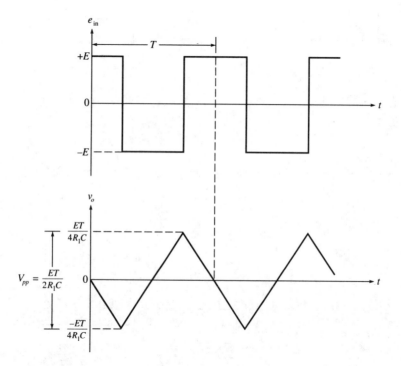

Figure 7.35
Triangular waveform produced by the integrator in Figure 7.33 when the input is a square wave.

input area is negative and decreasing when the input area is positive. The slope of the ramps is $\pm E/R_1C$ volts/second. The positive and negative peaks are therefore $\pm ET/4R_1C$ volts and the peak-to-peak value of the triangular waveform is

$$V_{pp} = \frac{ET}{2R_1C} \text{ volts} \tag{7.45}$$

Example 7–13

a. Sketch the output waveform produced by the integrator in Figure 7.36(a) when the input is the square wave shown.

b. What minimum slew rate must the amplifier have when used in this application?

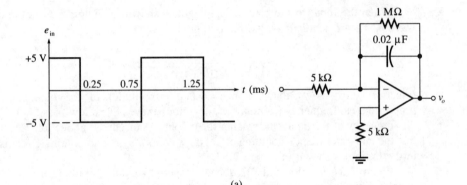

(a)

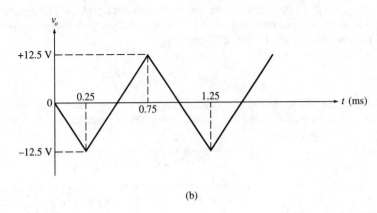

(b)

Figure 7.36
Example 7–13.

Solution

a. From the figure, the period of the input square wave is seen to be $T = (1.25 \text{ ms}) - (0.25 \text{ ms}) = 1 \text{ ms}$. Therefore, the positive and negative peaks of the triangular output are

$$\pm \frac{ET}{4R_1C} = \pm \frac{(5 \text{ V})(10^{-3} \text{ s})}{4(5 \times 10^3 \ \Omega)(0.02 \times 10^{-6} \text{ F})} = \pm 12.5 \text{ V}$$

The output waveform is shown in Figure 7.36(b).

b. The slew rate of the amplifier must be at least as great as the magnitude of the slopes of the triangular waveform:

$$\frac{E}{R_1C} = \frac{5 \text{ V}}{(5 \times 10^3 \ \Omega)(0.02 \times 10^{-6} \text{ F})} = 5 \times 10^4 \text{ V/s}$$

Drill
7–13

To what value should the 5-kΩ resistor in Figure 7.36(a) be changed if it is necessary to reduce the peak-to-peak value of the triangular output to 20 V?
Answer: 6.25 kΩ.

When the input to the electronic integrator shown in Figure 7.34 is the sinusoidal wave $e_{in}(t) = E_p\sin(2\pi ft)$ volts, the output is

$$v_o(t) = \frac{E_p}{2\pi fR_1C}\cos(2\pi ft) \text{ volts} \tag{7.46}$$

Equation 7.46 shows that the peak value (amplitude) of the output is *inversely* proportional to frequency f. Thus, the integrator behaves like a low-pass filter: The higher the frequency of the input, the smaller the output. As discussed in Section 7.3, this means that the integrator has a smoothing effect on waveforms. The smoothing effect is apparent in the examples shown in Figure 7.33.

Figure 7.37 is a plot of the frequency response (output amplitude versus frequency), on logarithmic scales, for the integrator shown in Figure 7.34. If the integrator were ideal, the plot would be a straight line over all frequency ranges. The response of the practical integrator departs from the ideal line in the vicinity of a *break* frequency, f_b, caused by the presence of R_f. Its value is given by

$$f_b = \frac{1}{2\pi R_fC} \text{ hertz} \tag{7.47}$$

For accurate integration, the lowest frequency component of the input waveform—i.e., the fundamental—should be at least 10 times greater than f_b. In Example 7–13,

$$f_b = \frac{1}{2\pi(10^6 \ \Omega)(0.02 \times 10^{-6} \ F)} = 7.96 \text{ Hz}$$

Since the fundamental frequency of the square wave in the example is 1 kHz, we see that accurate integration will occur. The break frequency can be made smaller by increasing the value of R_f, but that remedy increases the amplification of any dc offset. The voltage gain at dc is $-R_f/R_1$.

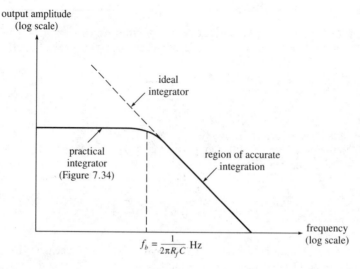

Figure 7.37
Frequency response of ideal and practical integrators.

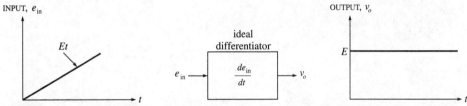

(a) Differentiation of a ramp produces a constant (dc) output, because the slope (rate of change) of the ramp is the constant value E V/s.

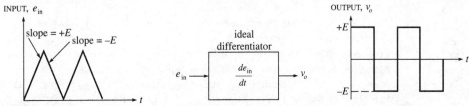

(b) Differentiation of a triangular wave produces a square wave, because the input slopes are alternately $+E$ V/s and $-E$ V/s.

Figure 7.38
Examples of waveform differentiation by an ideal differentiator.

Differentiation

Differentiation of a waveform produces another waveform whose value at any instant of time t equals the *rate of change* (slope) of the input waveform at that instant. When the input is any constant (dc) level, the input rate of change is clearly zero (a dc level does not change), so the output is 0 V. Figure 7.38(*a*) shows differentiation of a ramp waveform by an ideal differentiator. The differentiator is represented by the standard mathematical symbol $d(e_{in})/dt$. The slope of a ramp is everywhere the same—that is, its rate of change is constant. In the figure, the input is Et, so its slope is the constant value E volts/second. Consequently, the output is the dc level E. Figure 7.38(*b*) shows differentiation of a triangular waveform. When the triangular input is the increasing ramp Et, having slope E volts/second, the output is the dc level E volts. When the triangular wave is *decreasing,* its slope is $-E$ volts/second, so the output is the dc level $-E$ volts. We see that differentiation of a triangular waveform produces a square wave.

Comparing Figure 7.38 with the integrator waveforms in Figure 7.32, we note that integration of a square wave produces a triangular wave, whereas differentiation of a triangular wave produces a square wave. In other words, integration and differentiation are the *inverse* of each other, in the sense that making the output of one equal the input of the other reproduces the original waveform. All the integrator examples shown in Figure 7.33 can be "turned around" so that they also serve as examples of differentiation: If the output waveform in each case is made the input to a differentiator, then the output of the differentiator is the input waveform shown with each integrator.

Figure 7.39 shows how a practical differentiator can be constructed using an operational amplifier. As shown in the figure, the output is

$$v_o = -R_f C \frac{de_{in}}{dt} \tag{7.48}$$

Figure 7.39
Construction of a practical differentiator using an operational amplifier.

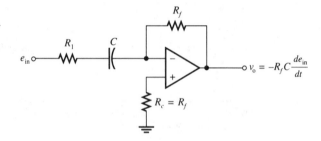

As before, the minus sign means that the output is inverted. Resistor R_1 is used to prevent differentiation of signals having frequencies beyond a specific value. As we shall presently learn, differentiation effectively amplifies high-frequency components, including those present in random electrical noise, so it is necessary to inhibit differentiation above a certain frequency. Resistor R_c is a compensating resistor used to minimize the effects of dc offset in the amplifier. Its value should be the same as R_f.

The slew-rate limitation of an operational amplifier imposes serious restrictions on its use as a practical differentiator. For example, when the input is a triangular wave, the square wave output must theoretically change levels in zero time. Since instantaneous level changes require infinite slew rates, it is clear that differentiation is not a practical way to generate square waves. Waveshaping by differentiation should be confined to input waveforms that are relatively smooth.

When the input to the differentiator shown in Figure 7.38 is the sinusoidal wave $e_{in}(t) = E_p\sin(2\pi ft)$, the output is

$$v_o(t) = E_p(2\pi f)R_fC \sin(2\pi ft - 90°) \tag{7.49}$$

Equation (7.49) shows that the peak value of the output is directly proportional to the frequency f of the input. The output is zero at dc ($f = 0$) and increases as frequency increases. Thus, the differentiator behaves as a high-pass filter. A practical difficulty with this behavior is that high-frequency noise in the input is effectively amplified in the output. Figure 7.40 shows the frequency responses of the ideal and practical

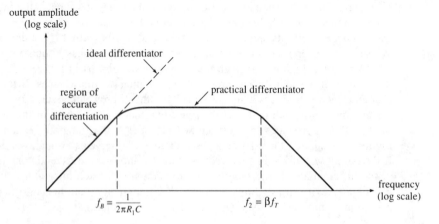

Figure 7.40
Frequency response of ideal and practical differentiators.

differentiators. Note that the response of the practical differentiator is made to depart from that of the ideal near a break frequency determined by R_1 and C:

$$f_B = \frac{1}{2\pi R_1 C} \text{ hertz} \tag{7.50}$$

This break suppresses high-frequency amplification and should be set much higher than the highest frequency at which accurate differentiation is required. Another break occurs at f_2, the cutoff frequency of the amplifier:

$$f_2 = \beta f_T \tag{7.51}$$

where

$$\beta = \frac{R_1}{R_1 + R_f}$$

and f_T is the unity-gain frequency of the amplifier, or its gain-bandwidth product.

Example 7–14

For the differentiator shown in Figure 7.39, $R_f = 100$ kΩ and $C = 0.01$ μF. The input waveform has frequency components ranging from 10 Hz at 1 V pk to 1 kHz at 0.4 V pk. Suppose it is desired to differentiate this waveform accurately.

a. Find the value of R_1 that should be used.
b. Find the peak values of the 10-Hz and 1-kHz components in the output.
c. Find the cutoff frequency of the amplifier if its gain-bandwidth product is 10^6.

Solution

a. The break frequency f_B should be set well above the highest frequency (1 kHz) at which accurate differentiation is required. Let $f_B = 10$ kHz. Then from equation (7.50),

$$f_B = \frac{1}{2\pi R_1 C} \implies R_1 = \frac{1}{2\pi f_B C} = \frac{1}{2\pi(10^4 \text{ Hz})(10^{-8} \text{ F})} = 1592 \ \Omega$$

b. From equation (7.49), at $f = 10$ Hz,

$$|v_o| = E_p(2\pi f)R_f C = (1 \text{ V})2\pi(10)(10^5)(10^{-8}) = 62.83 \text{ mV pk}$$

At $f = 1$ kHz,

$$|v_o| = (0.4 \text{ V})(2\pi)(10^3)(10^5)(10^{-8}) = 2.51 \text{ V pk}$$

c.

$$\beta = \frac{R_1}{R_1 + R_f} = \frac{1592}{1592 + 10^5} = 16.18 \times 10^{-3}$$

From equation (7.51),

$$f_2 = \beta f_T = (16.18 \times 10^{-3})10^6 = 16.18 \text{ kHz}$$

Drill 7–14

If the input to the amplifier in Example 7–14 is a 2-kHz sinewave having peak value 0.5 V, what minimum slew rate must the amplifier have?
Answer: 39.477×10^3 V/s.

7–7   VOLTAGE COMPARATORS AND SCHMITT TRIGGERS

Voltage Comparators

A voltage comparator effectively compares two input voltages and produces a low-to-high (or high-to-low) output when one of the inputs reaches or exceeds the level of the other. In a typical application, one of the inputs is a fixed voltage level and the other input varies below and above that level. When the variable input reaches or exceeds the level of the fixed input, the output of the comparator changes from low to high or from high to low, depending on its design. When the variable input falls below the fixed input level, the output reverts back to its original state.

An operational amplifier can serve as a voltage comparator by operating it with its very large open-loop gain, that is, with no feedback circuitry between output and input(s). Recall that $v_o = A_v(e_{in}^+ - e_{in}^-)$, where e_{in}^- in is the voltage at the inverting input and e_{in}^+ is the voltage at the non-inverting input. The signals to be compared are connected to these inputs, and, because of the very high gain A_v, the output rapidly switches to its maximum positive value when e_{in}^+ is slightly greater than e_{in}^-. When e_{in}^+ falls slightly below e_{in}^-, the output quickly switches to its maximum negative value. In practice, general-purpose operational amplifiers are not used in exactly this way because of stability problems (a tendency to oscillate) and slew-rate limitations. Specially designed operational amplifiers, marketed as voltage comparators, are used instead. Often the output levels of these devices can be set to values such as 0 V and +5 V rather than the maximum positive and negative values of the amplifier's output range. Nevertheless, comparators operate on essentially the same principles as open-loop operational amplifiers.

Figure 7.41 shows four examples of waveshapes produced by voltage comparators having sinusoidal voltages as their variable inputs. When $(e_{in}^+ - e_{in}^-)$ is positive, the amplifier output, $A_v(e_{in}^+ - e_{in}^-)$, is positive. Therefore, in each example, the output switches to its maximum positive output whenever the condition $e_{in}^+ - e_{in}^- > 0$ is satisfied. In Figure 7.41(a), $e_{in}^- = +5$ V, so the output switches high when $e_{in}^+ - 5$ V > 0—i.e., when $e_{in}^+ > +5$ V. In 7.41(b), the output switches high when 5 V $- e_{in}^- > 0$—i.e., when $e_{in}^- < 5$ V. In this example, we assume a voltage comparator whose output low is 0 V. In 7.41(c), the output switches high when $e_{in}^+ - (-5$ V$) > 0$—i.e., when $e_{in}^+ > -5$ V.

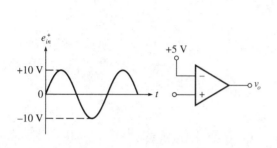

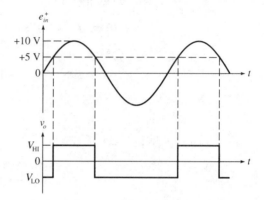

(a) The output switches high when $e_{in}^+ > 5$ V.

Figure 7.41
Examples of waveshaping by voltage comparators. *(Continued on page 205.)*

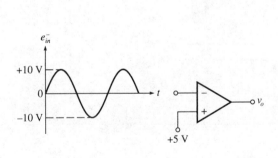

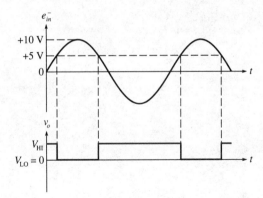

(b) The output switches high when $e_{in}^- < 5$ V. In this example, $V_{LO} = 0$ V.

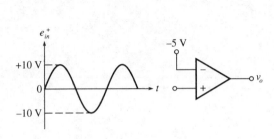

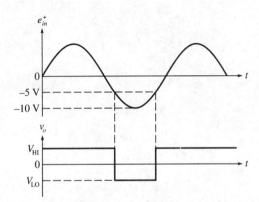

(c) The output switches high when $e_{in}^+ > -5$ V.

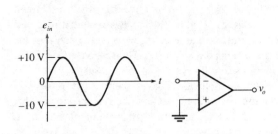

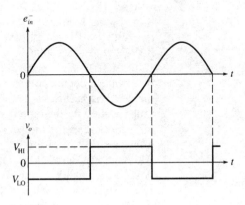

(d) The output switches high when $e_{in}^- < 0$ V.

Figure 7.41
(Continued)

Figure 7.42

Response time and rise-time of a voltage comparator.

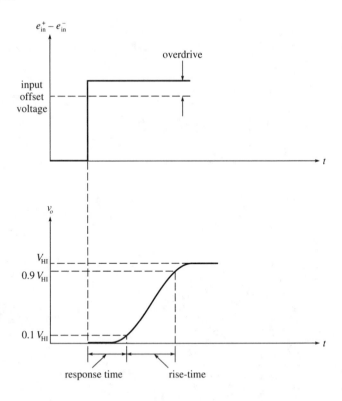

Finally, 7.41(*d*) shows an example where one input, e_{in}^+, is grounded. Thus, $e_{in}^+ = 0$, and the output switches high when $0 - e_{in}^- > 0$—i.e., when $e_{in}^- < 0$ V.

The *input offset voltage* of a voltage comparator is the smallest value of difference voltage $(e_{in}^+ - e_{in}^-)$ that will cause the comparator output to switch levels. For example, if the output high level is $+5$ V and the gain A_v is 50,000, then the input offset voltage is 5 V/$(5 \times 10^4) = 0.1$ mV. Thus, when e_{in}^+ is 0.1 mV greater than e_{in}^-, we have $v_o = A_v(e_{in}^+ - e_{in}^-) = 50,000(0.1$ mV$) = 5$ V. *Overdrive* is the amount by which the actual input value of $e_{in}^+ - e_{in}^-$ exceeds the input offset voltage. Figure 7.42 shows a typical response of a voltage comparator to a sudden change in the value of $e_{in}^+ - e_{in}^-$. We see that the output delay is characterized by a *response time* and a *rise-time*. The response time is the time required for the output to change from its low level (0 V in the figure) to 10% of its high level, and the rise-time is as previously defined. The response time is heavily influenced by the overdrive: the greater the overdrive, the smaller the response time. Although it is desirable in many applications to have an input offset voltage as small as possible, a small value adversely affects the rise time. To understand this fact, we must recognize that a small input offset voltage is achieved by having a very large voltage gain, A_v. But a large voltage gain means a small bandwidth because the gain-bandwidth product is constant. Recalling our previous discussion of filter theory, we know that a small bandwidth means a long rise-time.

Schmitt Triggers

A Schmitt trigger operates much like a voltage comparator, except the value of $e_{in}^+ - e_{in}^-$ that causes the output to switch from low to high is different from the value that causes

Figure 7.43
An inverting Schmitt trigger—a voltage comparator with hysteresis.

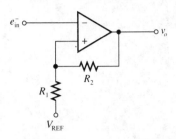

it to switch from high to low. Thus, there is a *range* of values between the low and high switching levels where no switching occurs. This range is called the *hysteresis* of the Schmitt trigger. Hysteresis is useful in applications where electrical noise could otherwise cause random switching of a comparator. For example, if the input offset voltage is 0.1 mV, then noise voltages of that order of magnitude superimposed on e_{in}^+ and/or e_{in}^- would cause unpredictable switching when $e_{in}^+ - e_{in}^-$ is near zero. A common application for voltage comparators and Schmitt triggers is in *level shifting*, to make the logic levels of one digital system compatible with another. For example, if the logic levels in one system are $V_{LO} = 0$ V and $V_{HI} = +5$ V and if the outputs are delivered to a system whose logic levels are $V_{LO} = -10$ V and $V_{HI} = 0$ V, then the low level must be shifted from 0 V to -10 V and the high level shifted from $+5$ V to 0 V. Since the inputs to the level shifter switch from 0 V to $+5$ V, it is possible to have a range of values—i.e., a dead zone, or hysteresis—between those two values that does not affect the level-shifting function.

Figure 7.43 shows how a Schmitt trigger can be constructed with an operational amplifier or voltage comparator. V_{REF} is a dc reference voltage, which may be zero (ground). This configuration is called an *inverting* Schmitt trigger because the output switches to V_{HI} when the input, e_{in}^-, *falls below* a lower trigger level *(LTL)* and switches to V_{LO} when e_{in}^- *rises above* an upper trigger level *(UTL)*. The trigger levels can be calculated as follows:

$$LTL = \frac{R_2}{R_1 + R_2} V_{REF} + \frac{R_1}{R_1 + R_2} V_{LO} \qquad (7.52)$$

$$UTL = \frac{R_2}{R_1 + R_2} V_{REF} + \frac{R_1}{R_1 + R_2} V_{HI} \qquad (7.53)$$

The hysteresis is

$$UTL - LTL = \frac{R_1}{R_1 + R_2} V_{HI} - \frac{R_1}{R_1 + R_2} V_{LO}$$

$$= \left(\frac{R_1}{R_1 + R_2} \right) (V_{HI} - V_{LO}) \qquad (7.54)$$

Example 7-15

Sketch the output of the Schmitt trigger shown in Figure 7.44 when the input is the sinewave shown.

a. Assume $V_{LO} = -5$ V, $V_{HI} = +5$ V, and $V_{REF} = 3$ V.
b. Assume $V_{LO} = 0$ V, $V_{HI} = +5$ V, and $V_{REF} = 3$ V.
c. Assume $V_{LO} = -5$ V, $V_{HI} = +5$ V, and $V_{REF} = 0$ V.

Find the hysteresis in each case.

Figure 7.44
Example 7–15.

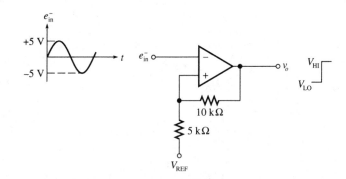

Solution

a. From equations (7.52) and (7.53),

$$LTL = \frac{R_2}{R_1 + R_2}V_{REF} + \frac{R_1}{R_1 + R_2}V_{LO}$$

$$= \left(\frac{10 \text{ k}\Omega}{5 \text{ k}\Omega + 10 \text{ k}\Omega}\right)3 \text{ V} + \left(\frac{5 \text{ k}\Omega}{5 \text{ k}\Omega + 10 \text{ k}\Omega}\right)(-5 \text{ V})$$

$$= 2 \text{ V} - 1.67 \text{ V} = 0.33 \text{ V}$$

$$UTL = \frac{R_2}{R_1 + R_2}V_{REF} + \frac{R_1}{R_1 + R_2}V_{HI}$$

$$= \left(\frac{10 \text{ k}\Omega}{5 \text{ k}\Omega + 10 \text{ k}\Omega}\right)3 \text{ V} + \left(\frac{5 \text{ k}\Omega}{5 \text{ k}\Omega + 10 \text{ k}\Omega}\right)(5 \text{ V})$$

$$= 2 \text{ V} + 1.67 \text{ V} = 3.67 \text{ V}$$

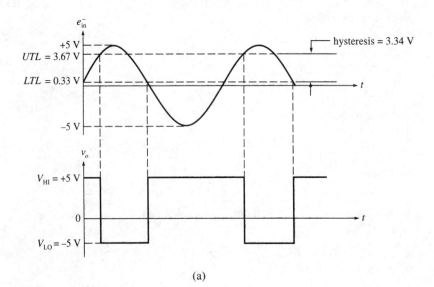

(a)

Figure 7.45
Example 7–15.

The hysteresis is

$$UTL - LTL = 3.67 \text{ V} - 0.33 \text{ V} = 3.34 \text{ V}$$

The output waveform is shown in Figure 7.45(a).

b. $$LTL = \left(\frac{10 \text{ k}\Omega}{5 \text{ k}\Omega + 10 \text{ k}\Omega}\right)(3 \text{ V}) + \left(\frac{5 \text{ k}\Omega}{5 \text{ k}\Omega + 10 \text{ k}\Omega}\right)(0 \text{ V}) = 2 \text{ V}$$

$$UTL = \left(\frac{10 \text{ k}\Omega}{5 \text{ k}\Omega + 10 \text{ k}\Omega}\right)(3 \text{ V}) + \left(\frac{5 \text{ k}\Omega}{5 \text{ k}\Omega + 10 \text{ k}\Omega}\right)(5 \text{ V}) = 3.67 \text{ V}$$

$$\text{hysteresis} = 3.67 \text{ V} - 2 \text{ V} = 1.67 \text{ V}$$

The output is shown in Figure 7.45(b).

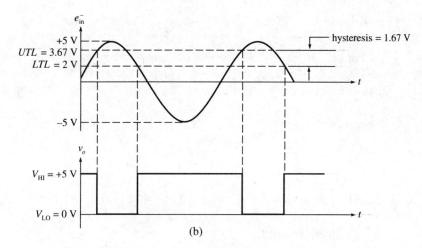

(b)

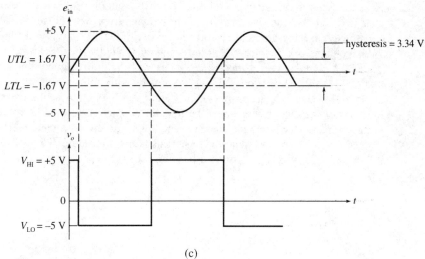

(c)

Figure 7.45
(Continued)

c.
$$LTL = \left(\frac{10\ k\Omega}{5\ k\Omega\ +\ 10\ k\Omega}\right)(0\ V) + \left(\frac{5\ k\Omega}{5\ k\Omega\ +\ 10\ k\Omega}\right)(-5\ V) = -1.67\ V$$

$$UTL = \left(\frac{10\ k\Omega}{5\ k\Omega\ +\ 10\ k\Omega}\right)(0\ V) + \left(\frac{5\ k\Omega}{5\ k\Omega\ +\ 10\ k\Omega}\right)(5\ V) = 1.67\ V$$

hysteresis $= 1.67\ V - (-1.67\ V) = 3.34\ V$

The output waveform is shown in Figure 7.45(c).

Drill 7–15   Assume that the outputs of the Schmitt trigger in Example 7–15 are $V_{HI} = +10\ V$ and $V_{LO} = -10\ V$ and that $V_{REF} = -2\ V$.

a. To what value should R_2 be changed if it is desired to have a hysteresis of 5 V?
b. What then are the values of *LTL* and *UTL*?

Answer: (a) $R_2 = 15\ k\Omega$; (b) $LTL = -4\ V$, $UTL = +1\ V$.

Figure 7.46 shows a *non-inverting* Schmitt trigger. In this case, the output switches to V_{HI} when the input, e_{in}, rises above an upper trigger level and switches to V_{LO} when the input falls below a lower trigger level. The trigger levels can be found from

$$LTL = -\frac{R_1}{R_2}V_{HI} \tag{7.55}$$

$$UTL = \frac{R_1}{R_2}|V_{LO}| \tag{7.56}$$

where $|V_{LO}|$ is the absolute value of V_{LO}.

TTL Schmitt Triggers

Because Schmitt triggers provide sharp, fast-changing outputs in response to slowly changing inputs and because their hysteresis improves noise immunity, they are constructed in integrated-circuit form for use with the TTL family of logic circuitry. The 7414 contains six inverting Schmitt triggers (called a hex Schmitt trigger inverter). The 7413 contains two 4-input NAND gates with Schmitt triggers built in at the inputs, and the 74132 contains four 2-input NAND gates with built-in Schmitt triggers. The logic symbols for these devices, shown in Figure 7.47, contain a boxlike symbol called a *hysteresis loop,* which represents the transfer characteristic (output voltage versus input voltage) of a device having hysteresis. The lower and upper trigger levels of TTL Schmitt triggers are fixed by design so the hysteresis is not adjustable. Typical values are $LTL = 0.9\ V$ and $UTL = 1.7\ V$, giving a hysteresis of 0.8 V.

Figure 7.46
The non-inverting Schmitt trigger.

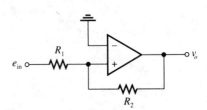

Figure 7.47

Logic symbols used for integrated-circuit Schmitt triggers.

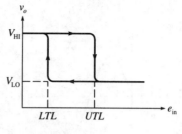

(a) Hysteresis loop.

(b) Schmitt trigger (7414).

(c) 4-input NAND Schmitt trigger (7413).

(d) 2-input NAND Schmitt trigger (74132).

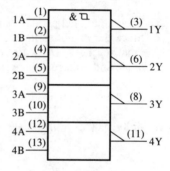

(e) ANSI/IEEE standard logic symbol for the 74132 quad 2-input NAND Schmitt trigger.

7–8   CLIPPING WITH OPERATIONAL AMPLIFIERS

As discussed in Section 7–4, the practical difficulty with diode clippers is that a small signal cannot be clipped accurately due to the voltage drop across the diode. For example, it is not feasible to clip a 1-V pk sinewave at its 50% level (0.5 V) because the drop across a (silicon) diode alone is 0.7 V. The advantages of operational amplifier clippers are (1) they can amplify small signals so that diode drops are not so significant, and (2) they provide a low-impedance source having power gain, so they can drive heavier loads than passive diode clippers. Furthermore, load resistance does not affect the clipping level, as it does in diode clippers.

Figure 7.48 shows an operational amplifier having a biased diode in parallel with feedback resistor R_f. In an operational amplifier having feedback, it can be shown that the point labeled G in the figure is held at essentially 0 V (called *virtual ground*). Therefore, the voltage across R_f has the same value as the output voltage, v_o, with respect to ground. When this voltage reaches the bias value E, the diode becomes forward-biased, and the

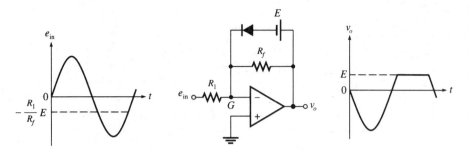

Figure 7.48
Clipping circuit using a biased diode in the feedback path of an operational amplifier.

voltage across R_f is held at E volts. This action is the same as that described for biased diodes in Section 7–4. Since the voltage across R_f is held at E volts, the output is clipped at E volts. Note the following important points:

1. The output is clipped at *output* level E, not input level E.
2. Since the amplifier inverts and has voltage gain $-R_f/R_1$, clipping occurs when the input reaches $-(R_1/R_f)E$ volts. Negative clipping can be achieved by reversing the diode and voltage source, as illustrated in the next example.

Example
7–16

Sketch the output waveform produced by the circuit shown in Figure 7.49(a).

Solution. The voltage gain of the inverting amplifier is

$$\frac{-R_f}{R_1} = -\frac{100 \text{ k}\Omega}{10 \text{ k}\Omega} = -10$$

Therefore, if the output were not clipped, it would be an inverted sinewave having peak value 10 V. As shown in Figure 7.49(b), the onset of clipping occurs when the voltage across R_f reaches 5 V—i.e., when the output voltage is -5 V (neglecting the drop across the diode). The output waveform is shown in Figure 7.49(c). Note that clipping begins when

$$\left(\frac{R_f}{R_1}\right)e_{\text{in}} = 5 \text{ V}$$

i.e., when

$$e_{\text{in}} = 5 \text{ V}\left(\frac{R_1}{R_f}\right) = 5 \text{ V}\left(\frac{10 \text{ k}\Omega}{100 \text{ k}\Omega}\right) = 0.5 \text{ V}$$

Drill
7–16

What is the output clipping level in Example 7–16 if the 0.7-V drop across the diode is taken into account?
Answer: -5.7 V.

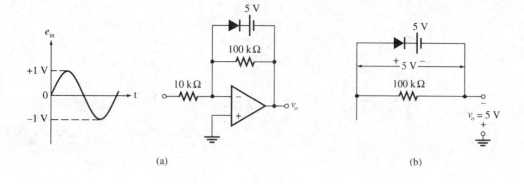

(a)   (b)

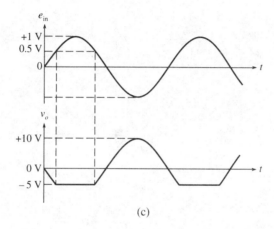

(c)

Figure 7.49
Example 7–16.

The disadvantage of the clipping circuits shown in Figures 7.48 and 7.49 is the need for a *floating* voltage source: one that has neither terminal connected to ground. Figure 7.50(a), (b), and (c) shows how clipping can be achieved using *zener diodes* instead of biased diodes in the feedback path. A zener diode behaves much like a biased diode, in that it conducts heavily when the (reverse) biasing voltage across it reaches a specific value, V_Z. It is then like a closed switch in series with a voltage source, so it holds the voltage across it essentially constant at the value V_Z. In Figure 7.50(a), the zener diode conducts when the output voltage reaches $V_Z + V_D$, where $V_D \approx 0.7$ V for silicon. The purpose of the silicon diode is to prevent the feedback path from being a short circuit when the output is negative, since a zener diode behaves like a conventional diode when it is forward-biased. Figure 7.50(b) shows that negative clipping can be achieved by reversing the connections to the zener and silicon diodes. Finally, Figure 7.50(c) shows that positive and negative clipping can be achieved by connecting two zener diodes "back-to-back" in the feedback path. In this case the forward voltage drop V_D is the drop across a forward-biased zener diode, which is the same as that of a conventional diode.

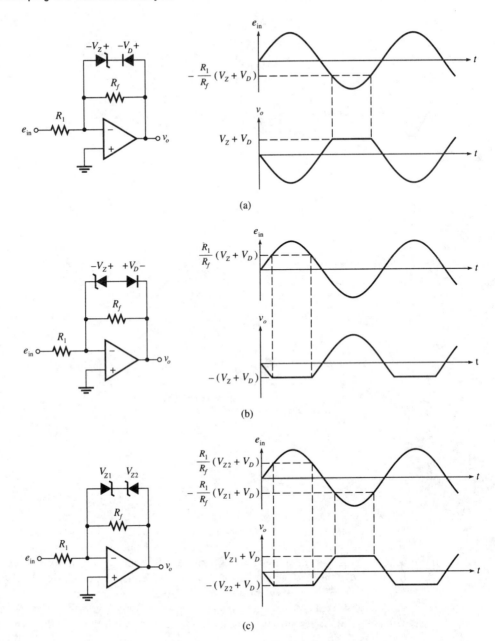

Figure 7.50
Clipping circuits utilizing zener diodes.

7–9   SPICE EXAMPLES

Example
7–17

Use SPICE to obtain a Fourier analysis of the rectangular waveform in Figure 7.5 (Example 7–2).

Solution.   The waveform is redrawn in Figure 7.51(*a*). We see that the pulsewidth to be used in the PULSE specification is 20 μs, the period is 80 μs, and the frequency is 1/80 μs = 12.5 kHz. To obtain a Fourier analysis from SPICE, we must perform a transient (.TRAN) analysis and include either a .PRINT TRAN or a .PLOT TRAN control statement in addition to a .FOUR control statement. The greater the number of analysis points (the smaller of size of TSTEP in the .TRAN statement), the more accurate will be the analysis. Making TSTEP smaller has the effect of reducing the rise and fall times of the waveform, since SPICE defaults TR and TF to TSTEP when those values are specified to be 0 in the PULSE definition. We can also improve accuracy by specifying very small values for TR and TF, thus making the waveform closer to ideal. Still another way to improve accuracy is to specify a very small value for TMAX (an optional specification in the .TRAN statement), such as TSTEP/100. If we make TSTEP very small, we may exceed the limit of 201 points that can be plotted or printed. If we make TMAX very small, we may exceed the limit on the number of iterations that SPICE will perform to obtain a solution. Both of these limits can be overridden using an .OPTIONS control statement.

Figure 7.51(*b*) shows the SPICE circuit and an input data file used to obtain a Fourier analysis. Note that we set TSTOP in the .TRAN statement to 80 μs, which is exactly one period. Maximum accuracy is obtained by making all the points that generate the waveform occur in one cycle. In this case, there are 81 points (80 μs/1 μs + 1). The results of the Fourier analysis are shown in Figure 7.51(*c*). Table 7.2, p. 217, compares the computed values of the "Fourier components" and "phase" with the theoretical values computed in Example 7–2. We see that the SPICE computations are somewhat inaccurate when the waveform is represented by 81 points. Figure 7.51(*d*) shows the results from an input data file used to obtain greater accuracy by increasing the number of points in one cycle (to 80 μs/0.2 μs + 1 = 401). In this case, it is necessary to use an .OPTIONS statement to override the limit of 201 points that can be printed. Comparing the results of this Fourier analysis with the theoretical computation (Table 7.2), we see that we have achieved greater accuracy. It would be an instructive exercise for the reader to experiment with different values of TF, TR, TSTEP, TSTOP, and TMAX to determine the effects of those parameters on the accuracy of the Fourier analysis.

The total harmonic distortion (*THD*) computed by SPICE in the Fourier analysis is the distortion that the rectangular waveform represents when compared to a pure 12.5-kHz sinewave: $THD = \sqrt{N_2^2 + N_3^2 + \cdots + N_9^2}$, where $N_2, N_3, \cdots, N_9$ are the normalized components printed in the analysis.

Example
7–18

Use SPICE to obtain (a) a plot of the waveform produced by the low-pass RC filter in Example 7–5, (b) a plot of the frequency response of the filter, and (c) a Fourier analysis of the output waveform when the input is the 10 kHz square wave of Example 7–5.

Solution

a. The circuit is redrawn in Figure 7.52, p. 218. The input square wave has frequency 10 kHz, so the period in the PULSE specification is $1/10^4 = 0.1$ ms and the pulsewidth is one-half the period, 0.05 ms. The .TRAN control statement will cause the plot to

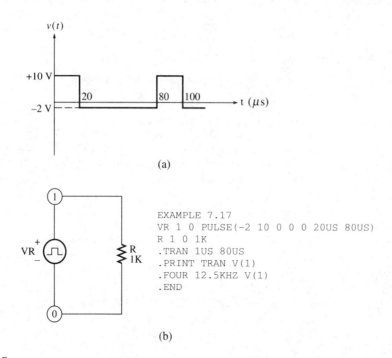

(a)

EXAMPLE 7.17
VR 1 0 PULSE(-2 10 0 0 0 20US 80US)
R 1 0 1K
.TRAN 1US 80US
.PRINT TRAN V(1)
.FOUR 12.5KHZ V(1)
.END

(b)

EXAMPLE 7.17

```
****      FOURIER ANALYSIS                    TEMPERATURE =    27.000 DEG C

*****************************************************************************

FOURIER COMPONENTS OF TRANSIENT RESPONSE V(1)

DC COMPONENT =    1.150D+00
```

HARMONIC NO	FREQUENCY (HZ)	FOURIER COMPONENT	NORMALIZED COMPONENT	PHASE (DEG)	NORMALIZED PHASE (DEG)
1	1.250D+04	5.609D+00	1.000000	40.466	.000
2	2.500D+04	3.807D+00	.678764	-9.033	-49.498
3	3.750D+04	1.577D+00	.281080	-58.371	-98.836
4	5.000D+04	2.973D-01	.052999	70.308	29.843
5	6.250D+04	1.268D+00	.225985	22.161	-18.305
6	7.500D+04	1.236D+00	.220304	-26.931	-67.397
7	8.750D+04	5.336D-01	.095138	-75.158	-115.623
8	1.000D+05	2.920D-01	.052058	50.758	10.292
9	1.125D+05	7.658D-01	.136533	3.883	-36.583

```
     TOTAL HARMONIC DISTORTION =      82.008576   PERCENT
```

(c)

Figure 7.51
Fourier analysis of the rectangular waveform in Example 7–2 (Example 7–17). (Continued on p. 217.)

EXAMPLE 7.17

```
****      FOURIER ANALYSIS                   TEMPERATURE =    27.000 DEG C


*************************************************************************

FOURIER COMPONENTS OF TRANSIENT RESPONSE V(1)

DC COMPONENT =    1.030D+00

HARMONIC    FREQUENCY     FOURIER      NORMALIZED     PHASE      NORMALIZED
  NO          (HZ)       COMPONENT     COMPONENT      (DEG)     PHASE  (DEG)


   1        1.250D+04    5.444D+00     1.000000      44.099        .000

   2        2.500D+04    3.819D+00      .701532      -1.802      -45.900

   3        3.750D+04    1.758D+00      .322865     -47.696      -91.795

   4        5.000D+04    5.995D-02      .011012      86.061       41.963

   5        6.250D+04    1.122D+00      .206073      40.481       -3.618

   6        7.500D+04    1.272D+00      .233602      -5.404      -49.502

   7        8.750D+04    7.281D-01      .133742     -51.269      -95.368

   8        1.000D+05    5.991D-02      .011004      82.124       38.025

   9        1.125D+05    6.410D-01      .117739      36.847       -7.251


   TOTAL HARMONIC DISTORTION =        85.171365   PERCENT
```
(d)

Figure 7.51
(Continued)

Table 7.2
Comparison of Theoretical and SPICE Computations in Fourier Analysis (Example 7–17).

	Theoretical	**SPICE (81 points)**	**SPICE (401 points)**
DC component (average value)	1.00 V	1.15 V	1.03 V
Fundamental (A_1)	5.40 V	5.61 V	5.44 V
Second harmonic (A_2)	3.82 V	3.81 V	3.82 V
Third harmonic (A_3)	1.80 V	1.58 V	1.76 V
Fourth harmonic (A_4)	0 V	0.297 V	0.060 V
Phase (ϕ_1)	45°	40.47°	44.1°
Phase (ϕ_2)	0°	−9.03°	−1.80°
Phase (ϕ_3)	−45°	−58.37°	−47.70°
Phase (ϕ_4)	Indeterminate	70.38°	86.06°

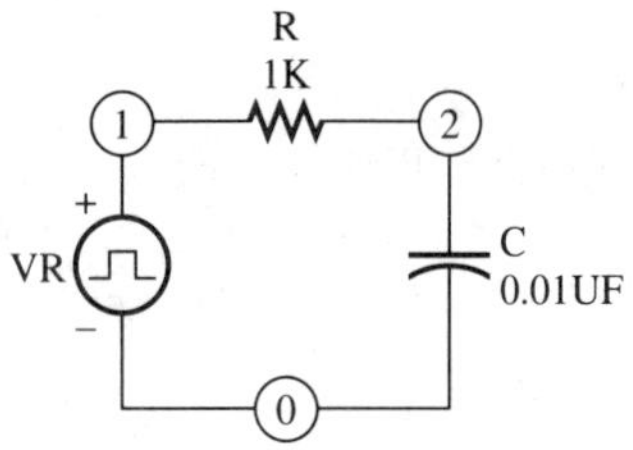

```
                                    EXAMPLE 7.18A
                                    VR 1 0 PULSE(0 9 0 0 0 0.05MS 0.1MS)
                                    R 1 2 1K
                                    C 2 0 0.01UF
                                    .TRAN 5US 0.2MS
                                    .PLOT TRAN V(2)
                                    .END
```

(a)

```
EXAMPLE 7.18A

****      TRANSIENT ANALYSIS                    TEMPERATURE =    27.000 DEG C

*****************************************************************************

      TIME        V(2)

                          -5.000D+00    0.000D+00    5.000D+00    1.000D+01 1.500D+01
                        - - - - - - - - - - - - - - - - - - - - - - - - - - - - -
   0.000D+00   0.000D+00 .              *            .            .            .
   5.000D-06   1.909D+00 .              .      *     .            .            .
   1.000D-05   4.694D+00 .              .            *.           .            .
   1.500D-05   6.380D+00 .              .            .      *     .            .
   2.000D-05   7.410D+00 .              .            .         *  .            .
   2.500D-05   8.044D+00 .              .            .           *.            .
   3.000D-05   8.432D+00 .              .            .            *            .
   3.500D-05   8.655D+00 .              .            .            *            .
   4.000D-05   8.791D+00 .              .            .            .*           .
   4.500D-05   8.874D+00 .              .            .            .*           .
   5.000D-05   8.925D+00 .              .            .            . *          .
   5.500D-05   8.955D+00 .              .            .            . *          .
   6.000D-05   7.061D+00 .              .            .         *  .            .
   6.500D-05   4.291D+00 .              .            *.           .            .
   7.000D-05   2.619D+00 .              .       *    .            .            .
   7.500D-05   1.584D+00 .              .   *        .            .            .
   8.000D-05   9.483D-01 .              .  *         .            .            .
   8.500D-05   5.651D-01 .              . *          .            .            .
   9.000D-05   3.448D-01 .              .*           .            .            .
   9.500D-05   2.086D-01 .              .*           .            .            .
   1.000D-04   1.233D-01 .              *            .            .            .
   1.050D-04   1.987D+00 .              .      *     .            .            .
   1.100D-04   4.720D+00 .              .            *.           .            .
   1.150D-04   6.394D+00 .              .            .      *     .            .
   1.200D-04   7.428D+00 .              .            .         *  .            .
   1.250D-04   8.062D+00 .              .            .           *.            .
   1.300D-04   8.436D+00 .              .            .            *            .
   1.350D-04   8.657D+00 .              .            .            *            .
   1.400D-04   8.793D+00 .              .            .            .*           .
   1.450D-04   8.876D+00 .              .            .            .*           .
   1.500D-04   8.926D+00 .              .            .            . *          .
   1.550D-04   8.956D+00 .              .            .            . *          .
   1.600D-04   7.061D+00 .              .            .         *  .            .
   1.650D-04   4.292D+00 .              .            *.           .            .
   1.700D-04   2.619D+00 .              .       *    .            .            .
   1.750D-04   1.584D+00 .              .   *        .            .            .
   1.800D-04   9.483D-01 .              .  *         .            .            .
   1.850D-04   5.652D-01 .              . *          .            .            .
   1.900D-04   3.448D-01 .              .*           .            .            .
   1.950D-04   2.086D-01 .              .*           .            .            .
   2.000D-04   1.233D-01 .              *            .            .            .
                        - - - - - - - - - - - - - - - - - - - - - - - - - - - - -
```

(b)

Figure 7.52
Output waveform in a low-pass RC filter when the input is a square wave (Example 7–18(a)).

extend over two full cycles (0.2 ms) and will generate 0.2 ms/5 μs + 1 = 41 points. The resulting plot is shown in Figure 7.52.

b. To obtain a plot of the frequency response, we must change the input to an ac sinewave generator. See Figure 7.53, p. 220. Note that we do not specify an amplitude or phrase angle for V1, so SPICE defaults those values to 1 V and 0°, respectively. It is necessary to identify the source as AC for SPICE to generate a plot of the frequency response. The .AC control statement causes SPICE to plot the output magnitude at frequencies ranging from 160 Hz (about 2 decades below the cutoff frequency: 15.915 kHz/100 = 159.15 Hz) through 1.6 MHz (about 2 decades above the cutoff frequency: 15.915 kHz × 100 = 1.5915 MHz). The DEC 10 specification in the .AC control statement means that we will obtain output at 10 frequencies within each decade and that the output will be plotted on logarithmically scaled axes (a log-log plot). The resulting plot is shown in Figure 7.53. Note the logarithmically scaled axes. Also note that the output is close to 0.707 (0.7052) at a frequency (16 kHz) close to the cutoff frequency: 15.915 kHz.

c. The Fourier analysis of the shaped waveform is shown in Figure 7.54, p. 221. The values of one cycle of the waveform are printed at 501 points to improve accuracy. (Printed output values are not shown.) An .OPTIONS statement is used to override the limit of 201 points of output. Since the square wave input has zero even-harmonic content, the output should theoretically have zero magnitude at all even harmonics. We see from the SPICE analysis that the computed outputs are near zero at the even harmonics. In Example 7-5, we computed the fundamental and third harmonic magnitudes to be 4.85 V and 0.896 V, respectively. The corresponding values computed by SPICE are seen to be 4.855 V and 0.8958 V, in close agreement with the theoretical.

Example 7-19   Use SPICE to obtain plots of the phase angle and magnitude of the output of the high-pass RC filter in Figure 7.55(a), p. 221, versus frequency. Both plots should appear on the same frequency axis and should extend from 2 decades below the cutoff frequency $(f_1/100)$ to 2 decades above the cutoff frequency $(100f_1)$.

Solution.   The cutoff frequency of the filter is

$$f_1 = \frac{1}{2\pi RC} = \frac{1}{2\pi(3183 \ \Omega)(10^{-8} \ F)} = 5 \ kHz$$

Thus, the frequency range specified in the .AC control statement is from 5 kHz/100 = 50 Hz through 100(5 kHz) = 500 kHz. Note that one .PLOT statement is used to obtain plots of two variables: VP(2) (phase) and V(2) (magnitude). Thus, both plots will appear on the same frequency axis, as shown in Figure 7.55(b), p. 222. Since VP(2) is the first variable specified in the .PLOT statement, its values only are printed down the left-hand column. Note that both axes of the magnitude plot are logarithmically scaled, but VP(2) is plotted on a linear axis versus a logarithmic frequency axis. We see that the phase angle is 45° at the cutoff frequency (5 kHz).

```
EXAMPLE 7.18B
VAC 1 0 AC
R 1 2 1K
C 2 0 0.01UF
.AC DEC 10 160HZ 1.6MEGAHZ
.PLOT AC V(2)
.END
```

(a)

EXAMPLE 7.18B

****   AC ANALYSIS   TEMPERATURE =   27.000 DEG C

****

```
      FREQ        V(2)

                   1.000D-03    1.000D-02    1.000D-01    1.000D+00  1.000D+01
                  - - - - - - - - - - - - - - - - - - - - - - - - - - - - - -
   1.600D+02   9.999D-01 .            .            .            *          .
   2.014D+02   9.999D-01 .            .            .            *          .
   2.536D+02   9.999D-01 .            .            .            *          .
   3.192D+02   9.998D-01 .            .            .            *          .
   4.019D+02   9.997D-01 .            .            .            *          .
   5.060D+02   9.995D-01 .            .            .            *          .
   6.370D+02   9.992D-01 .            .            .            *          .
   8.019D+02   9.987D-01 .            .            .            *          .
   1.010D+03   9.980D-01 .            .            .            *          .
   1.271D+03   9.968D-01 .            .            .            *          .
   1.600D+03   9.950D-01 .            .            .            *          .
   2.014D+03   9.921D-01 .            .            .            *          .
   2.536D+03   9.875D-01 .            .            .            *          .
   3.192D+03   9.805D-01 .            .            .            *          .
   4.019D+03   9.696D-01 .            .            .            *          .
   5.060D+03   9.530D-01 .            .            .            *          .
   6.370D+03   9.284D-01 .            .            .            *          .
   8.019D+03   8.930D-01 .            .            .           *.          .
   1.010D+04   8.444D-01 .            .            .           *.          .
   1.271D+04   7.814D-01 .            .            .           *.          .
   1.600D+04   7.052D-01 .            .            .          * .          .
   2.014D+04   6.200D-01 .            .            .         *  .          .
   2.536D+04   5.316D-01 .            .            .        *   .          .
   3.192D+04   4.462D-01 .            .            .       *    .          .
   4.019D+04   3.682D-01 .            .            .      *     .          .
   5.060D+04   3.001D-01 .            .            .     *      .          .
   6.370D+04   2.424D-01 .            .            .   *        .          .
   8.019D+04   1.947D-01 .            .            .  *         .          .
   1.010D+05   1.557D-01 .            .            . *          .          .
   1.271D+05   1.243D-01 .            .            .*           .          .
   1.600D+05   9.898D-02 .            .            *            .          .
   2.014D+05   7.877D-02 .            .          *.             .          .
   2.536D+05   6.264D-02 .            .        *                .          .
   3.192D+05   4.979D-02 .            .      *                  .          .
   4.019D+05   3.957D-02 .            .    *                    .          .
   5.060D+05   3.144D-02 .            .   *                     .          .
   6.370D+05   2.498D-02 .            . *                       .          .
   8.019D+05   1.984D-02 .            *                         .          .
   1.010D+06   1.576D-02 .          *                           .          .
   1.271D+06   1.252D-02 .        .*                            .          .
   1.600D+06   9.947D-03 .        *                             .          .
                  - - - - - - - - - - - - - - - - - - - - - - - - - - - - - -
```

(b)

Figure 7.53
Frequency response of a low-pass RC filter (Example 7–18(b)).

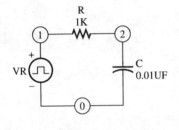

```
EXAMPLE 7.18C
VR 1 0 PULSE(0 9 0 0 0 0.05MS 0.1MS)
R 1 2 1K
C 2 0 0.01UF
.TRAN 0.2US 0.1MS
.OPTIONS LIMPTS = 501
.PRINT TRAN V(2)
.FOUR 10KHZ V(2)
.END
```

(a)

EXAMPLE 7.18C

****   FOURIER ANALYSIS   TEMPERATURE =   27.000 DEG C

****

FOURIER COMPONENTS OF TRANSIENT RESPONSE V(2)

DC COMPONENT =   4.512D+00

HARMONIC NO	FREQUENCY (HZ)	FOURIER COMPONENT	NORMALIZED COMPONENT	PHASE (DEG)	NORMALIZED PHASE (DEG)
1	1.000D+04	4.855D+00	1.000000	-33.016	.000
2	2.000D+04	1.467D-02	.003022	35.758	68.774
3	3.000D+04	8.958D-01	.184510	-64.839	-31.822
4	4.000D+04	8.715D-03	.001795	16.019	49.035
5	5.000D+04	3.470D-01	.071468	-77.005	-43.988
6	6.000D+04	6.039D-03	.001244	6.339	39.355
7	7.000D+04	1.805D-01	.037169	-83.637	-50.620
8	8.000D+04	4.591D-03	.000946	.001	33.018
9	9.000D+04	1.097D-01	.022605	-88.091	-55.074

TOTAL HARMONIC DISTORTION =   20.262995 PERCENT

(b)

Figure 7.54
Fourier analysis of the waveform in Figure 7.52 (Example 7–18(c)).

Figure 7.55
Frequency response of a high-pass RC filter (Example 7–19). (Continued on p. 222.)

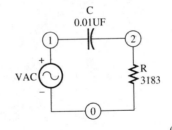

```
EXAMPLE 7.19
VAC 1 0 AC
C 1 2 0.01UF
R 2 0 3183
.AC DEC 10 50HZ 500KHZ
.PLOT AC VP(2) V(2)
.END
```

(a)

EXAMPLE 7.19

```
****      AC ANALYSIS                      TEMPERATURE =    27.000 DEG C


**************************************************************************


LEGEND:

*: VP(2)
+: V(2)

     FREQ       VP(2)

*)------------- -5.000D+01      0.000D+00      5.000D+01      1.000D+02  1.500D+02
                 - - - - - - - - - - - - - - - - - - - - - - - - - - - - -

+)------------- 1.000D-03      1.000D-02      1.000D-01      1.000D+00  1.000D+01
                 - - - - - - - - - - - - - - - - - - - - - - - - - - - - -
  5.000D+01   8.943D+01 .             +            .            *   .            .
  6.295D+01   8.928D+01 .            .+            .            *   .            .
  7.924D+01   8.909D+01 .             . +          .            *   .            .
  9.976D+01   8.886D+01 .             .   +        .            *   .            .
  1.256D+02   8.856D+01 .             .     +      .            *   .            .
  1.581D+02   8.819D+01 .             .      +     .            *   .            .
  1.991D+02   8.772D+01 .             .        +   .            *   .            .
  2.506D+02   8.713D+01 .             .          + .            *   .            .
  3.155D+02   8.639D+01 .             .           +.            *   .            .
  3.972D+02   8.546D+01 .             .          +.             *   .            .
  5.000D+02   8.429D+01 .             .            +            *   .            .
  6.295D+02   8.282D+01 .             .            .+           *   .            .
  7.924D+02   8.099D+01 .             .            . +         *    .            .
  9.976D+02   7.872D+01 .             .            .   +     *      .            .
  1.256D+03   7.590D+01 .             .            .     + *        .            .
  1.581D+03   7.245D+01 .             .            .       *+       .            .
  1.991D+03   6.829D+01 .             .            .      *    +    .            .
  2.506D+03   6.338D+01 .             .            .    *        +  .            .
  3.155D+03   5.775D+01 .             .            .  *            + .            .
  3.972D+03   5.154D+01 .             .            .*              + .            .
  5.000D+03   4.500D+01 .             .          *.              +  .            .
  6.295D+03   3.846D+01 .             .        *   .            +.  .            .
  7.924D+03   3.225D+01 .             .      *     .            +.  .            .
  9.976D+03   2.662D+01 .             .     *      .            +.  .            .
  1.256D+04   2.171D+01 .             .    *       .            +   .            .
  1.581D+04   1.755D+01 .             .   *        .            +   .            .
  1.991D+04   1.410D+01 .             .  *         .            +   .            .
  2.506D+04   1.128D+01 .             . *          .            +   .            .
  3.155D+04   9.006D+00 .             . *          .            +   .            .
  3.972D+04   7.176D+00 .             .*           .            +   .            .
  5.000D+04   5.711D+00 .             .*           .            +   .            .
  6.295D+04   4.542D+00 .             .*           .            +   .            .
  7.924D+04   3.610D+00 .             .*           .            +   .            .
  9.976D+04   2.869D+00 .             .*           .            +   .            .
  1.256D+05   2.280D+00 .             .*           .            +   .            .
  1.581D+05   1.811D+00 .             .*           .            +   .            .
  1.991D+05   1.439D+00 .             *            .            +   .            .
  2.506D+05   1.143D+00 .             *            .            +   .            .
  3.155D+05   9.080D-01 .             *            .            +   .            .
  3.972D+05   7.213D-01 .             *            .            +   .            .
  5.000D+05   5.730D-01 .             *            .            +   .            .
                 - - - - - - - - - - - - - - - - - - - - - - - - - - - - -
```

(b)

Figure 7.55
(Continued)

Example 7-20

Use SPICE to obtain a plot of the waveform at the output of the clipping circuit shown in Figure 7.56(a). What are the clipping levels of the circuit?

Solution

The circuit is redrawn in its SPICE format in Figure 7.56(b). To obtain a plot of the output versus time, we must use a SIN source and a .TRAN analysis, rather than an AC source and an .AC analysis. The period of the sinewave is 1/10 kHz = 0.1 ms, so the .TRAN analysis produces 0.1 ms/2 μs + 1 = 51 points on a plot that extends through one complete cycle. Note that we allow all diode parameters to have their default values, since none are specified in the .MODEL statement. Examination of the results, shown in Figure 7.56(c), p. 224, reveals that the clipping levels are approximately 6.83 V and − 10.81 V. These results reflect the voltage drops across the diodes.

Example 7-21

Use SPICE to obtain plots of the input and output waveforms (on the same set of axes) in the circuit shown in Figure 7.57(a), p. 225. The operational amplifier has an input resistance of 100 MΩ and a voltage gain of 10^6.

Solution. Figure 7.57(b), p. 225, shows how the operational amplifier can be modeled as a voltage-controlled voltage source. The phase inversion is implemented by grounding the positive node (N+) of the controlled source, EOP. Resistor RIN, with value 100 MΩ, is used to model the very large input resistance of the operational amplifier. In reality, it could be omitted entirely, leaving an open circuit between nodes 0 and 2 and thus simulating an ideal operational amplifier with infinite input resistance. The results of the computations are the same either way. The triangular wave input is approximated by a PULSE waveform whose pulsewidth is very narrow in comparison to its rise and fall times (TR and TF). (If the pulsewidth is set to 0, SPICE assigns it the default value TSTEP.) In this example, we make the pulsewidth 1 μs, which is very small in comparison to the 5-ms values of TR and TF.

The plot is shown in Figure 7.57(c), p. 225. Note the inversion of the output waveform, with respect to the input. The peak-to-peak variation of the output is 50 V, and that of the input is 10 V. Thus, the closed-loop gain is −50/10 = −5, in agreement with the theoretical: $-R_f/R_1$ = −50 kΩ/10 kΩ = −5. Note that SPICE prints an X where the two plots intersect (at 0 V).

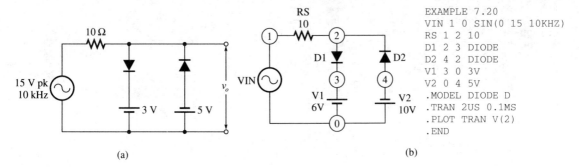

(a)   (b)

Figure 7.56
Output of a clipping circuit (Example 7−20). (Continued on p. 224.)

```
****      TRANSIENT ANALYSIS              TEMPERATURE =   27.000 DEG C
  TIME    V(2)      -2.000D+01     -1.000D+01     0.000D+00      1.000D+01 2.000D+01
                     -- - - - - - - - - - - - - - - - - - - - - - - -- - - - -
0.000D+00 -4.000D-11 .                 .             *            .             .
2.000D-06  1.879D+00 .                 .             .   *        .             .
4.000D-06  3.725D+00 .                 .             .      *     .             .
6.000D-06  5.513D+00 .                 .             .          * .             .
8.000D-06  6.558D+00 .                 .             .         *  .             .
1.000D-05  6.790D+00 .                 .             .          * .             .
1.200D-05  6.806D+00 .                 .             .          * .             .
1.400D-05  6.814D+00 .                 .             .          * .             .
1.600D-05  6.820D+00 .                 .             .          * .             .
1.800D-05  6.823D+00 .                 .             .          * .             .
2.000D-05  6.826D+00 .                 .             .          * .             .
2.200D-05  6.828D+00 .                 .             .          * .             .
2.400D-05  6.828D+00 .                 .             .          *  .            .
2.600D-05  6.828D+00 .                 .             .          * .             .
2.800D-05  6.828D+00 .                 .             .          * .             .
3.000D-05  6.826D+00 .                 .             .          * .             .
3.200D-05  6.823D+00 .                 .             .          * .             .
3.400D-05  6.820D+00 .                 .             .          * .             .
3.600D-05  6.814D+00 .                 .             .          * .             .
3.800D-05  6.806D+00 .                 .             .          * .             .
4.000D-05  6.791D+00 .                 .             .          * .             .
4.200D-05  6.725D+00 .                 .             .         *  .             .
4.400D-05  5.496D+00 .                 .             .        *   .             .
4.600D-05  3.724D+00 .                 .             .      *     .             .
4.800D-05  1.877D+00 .                 .             .    *       .             .
5.000D-05 -4.399D-04 .                 .             *            .             .
5.200D-05 -1.877D+00 .                 .          *  .            .             .
5.400D-05 -3.725D+00 .                 .       *     .            .             .
5.600D-05 -5.513D+00 .                 .   *         .            .             .
5.800D-05 -7.215D+00 .                 . *           .            .             .
6.000D-05 -8.803D+00 .               * .             .            .             .
6.200D-05 -1.025D+01 .             *   .             .            .             .
6.400D-05 -1.074D+01 .          *.      .            .            .             .
6.600D-05 -1.079D+01 .          *.      .            .            .             .
6.800D-05 -1.080D+01 .          *.      .            .            .             .
7.000D-05 -1.081D+01 .          *.      .            .            .             .
7.200D-05 -1.081D+01 .          *.      .            .            .             .
7.400D-05 -1.081D+01 .          *.      .            .            .             .
7.600D-05 -1.081D+01 .          *.      .            .            .             .
7.800D-05 -1.081D+01 .          *.      .            .            .             .
8.000D-05 -1.081D+01 .          *.      .            .            .             .
8.200D-05 -1.080D+01 .          *.      .            .            .             .
8.400D-05 -1.079D+01 .          *.      .            .            .             .
8.600D-05 -1.076D+01 .          *.      .            .            .             .
8.800D-05 -1.012D+01 .          *       .            .            .             .
9.000D-05 -8.802D+00 .            *     .            .            .             .
9.200D-05 -7.214D+00 .             . *          .            .             .
9.400D-05 -5.513D+00 .             .    *       .            .             .
9.600D-05 -3.724D+00 .             .       *    .            .             .
9.800D-05 -1.877D+00 .             .          *  .            .             .
1.000D-04 -4.000D-11 .             .             *            .             .
```

(c)

Figure 7.56
(Continued)

Figure 7.57
Example 7-21.

(a)

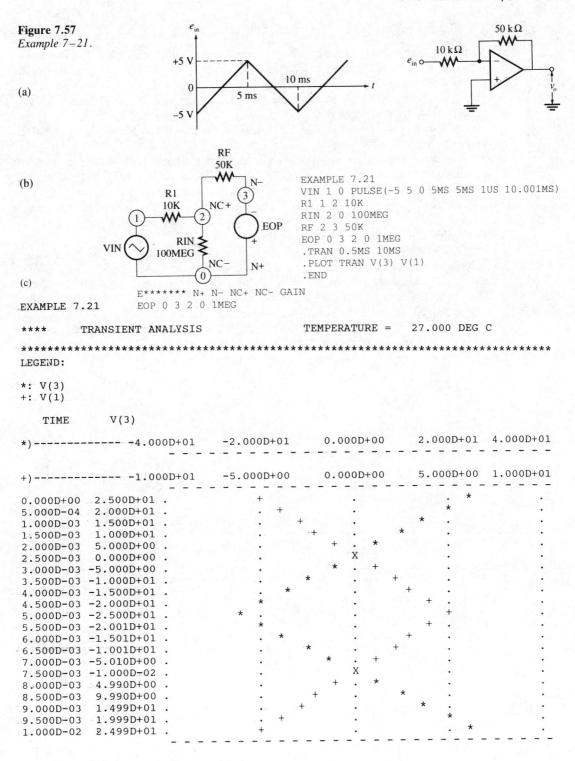

(b)

(c)

EXAMPLE 7.21

```
EXAMPLE 7.21
VIN 1 0 PULSE(-5 5 0 5MS 5MS 1US 10.001MS)
R1 1 2 10K
RIN 2 0 100MEG
RF 2 3 50K
EOP 0 3 2 0 1MEG
.TRAN 0.5MS 10MS
.PLOT TRAN V(3) V(1)
.END
```

```
E******* N+ N- NC+ NC- GAIN
EOP 0 3 2 0 1MEG
```

```
****     TRANSIENT ANALYSIS                TEMPERATURE =    27.000 DEG C

***************************************************************************

LEGEND:

*: V(3)
+: V(1)

    TIME       V(3)

*)------------- -4.000D+01    -2.000D+01     0.000D+00     2.000D+01  4.000D+01
- - - - - - - - - - - - - - - - - - - - - - - - - - - - - - - - - - - - - - -

+)------------- -1.000D+01    -5.000D+00     0.000D+00     5.000D+00  1.000D+01
- - - - - - - - - - - - - - - - - - - - - - - - - - - - - - - - - - - - - - -

0.000D+00  2.500D+01 .              +           .            .  *       .
5.000D-04  2.000D+01 .                 +        .            .*         .
1.000D-03  1.500D+01 .                     +    .        *   .          .
1.500D-03  1.000D+01 .                        +.      *     .          .
2.000D-03  5.000D+00 .                      +  . *         .           .
2.500D-03  0.000D+00 .                         X           .           .
3.000D-03 -5.000D+00 .                     *   . +         .           .
3.500D-03 -1.000D+01 .                 *       .        +  .           .
4.000D-03 -1.500D+01 .              *          .           +           .
4.500D-03 -2.000D+01 .           *             .            +          .
5.000D-03 -2.500D+01 .      *     .             .            +         .
5.500D-03 -2.001D+01 .        *   .             .         +            .
6.000D-03 -1.501D+01 .          *  .            .        +             .
6.500D-03 -1.001D+01 .              *           .       +              .
7.000D-03 -5.010D+00 .                    *     . +                    .
7.500D-03 -1.000D-02 .                           X                     .
8.000D-03  4.990D+00 .                      +  . *                     .
8.500D-03  9.990D+00 .                        + .        *             .
9.000D-03  1.499D+01 .                     +    .           *          .
9.500D-03  1.999D+01 .               +         .             *         .
1.000D-02  2.499D+01 .            +             .                *      .
- - - - - - - - - - - - - - - - - - - - - - - - - - - - - - - - - - - - - - -
```

Example 7-22

Use SPICE to obtain plots of the waveforms for v_{01} and v_{02} in Figure 7.58(a).

Solution. As shown in Figure 7.58(b), the operational amplifiers E0P1 and E0P2 are modeled in the same way as the operational amplifier in Example 7-21. The triangular pulses at the inputs are approximated by PULSE specifications that make the pulsewidths (1 μs) very small in comparison to TR and TF, also as in Example 7-21. Note that VP2 is delayed by the total width of the first pulse produced by VP1: 5 ms + 5 ms + 1 μs = 10.001 ms. The period of VP1 and VP2 equals the sum of the widths of one VP1 pulse and one VP2 pulse: 20.002 ms.

As can be seen in the plot of V(4), the first operational amplifier inverts and sums the triangular pulses to produce a triangular wave that alternates between ±5 V and begins (at $t = 0$) with the value 0 V. (This waveform cannot be generated with a single PULSE specification.) The second operational amplifier integrates the triangular wave and produces the waveform shown in Figure 7.58(d). Although this output appears to be sinusoidal, it is actually parabolic. See Figure 7.33(c).

When using SPICE to perform integration, it is important that all inputs have initial value zero (as in this example). For some input waveforms, it may be necessary to use the PULSE definition in innovative ways to meet this requirement, since SPICE always sets the initial ($t = 0$) value of a PULSE to the value specified for V1. We should, for example, define a negative 5-V pulse by PULSE (0 −5 · · ·) rather than PULSE (−5 0 · · ·). It is permissible to make V2 smaller than V1. See, for example, Exercise 7.68(c) and (d), pages 228−229.

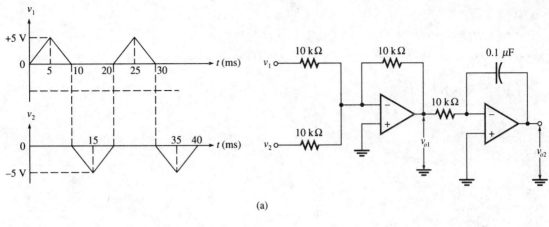

(a)

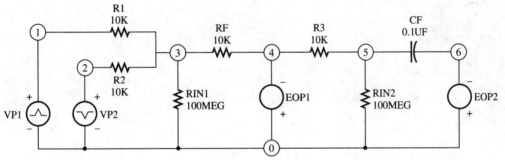

```
EXAMPLE 7.22
VP1 1 0 PULSE(0 5 0 5MS 5MS 1US 20.002MS)
VP2 2 0 PULSE(0 -5 10.001MS 5MS 5MS 1US 20.002MS)
R1 1 3 10K
R2 2 3 10K
RIN1 3 0 100MEG
RF 3 4 10K
EOP1 0 4 3 0 1MEG
R3 4 5 10K
RIN2 5 0 100MEG
CF 5 6 0.1UF
EOP2 0 6 5 0 1MEG
.TRAN 1MS 40MS
.PLOT TRAN V(4)
.PLOT TRAN V(6)
.END
```

(b)

Figure 7.58
Example 7–22. (Continued on pp. 228–229.)

EXAMPLE 7.22

```
****      TRANSIENT ANALYSIS              TEMPERATURE =    27.000 DEG C

*************************************************************************


     TIME       V(4)

                     -1.000D+01   -5.000D+00   0.000D+00   5.000D+00  1.000D+01
                      - - - - - - - - - - - - - - - - - - - - - - - - -
0.000D+00   0.000D+00 .            .            *            .           .
1.000D-03  -1.000D+00 .            .          *  .            .           .
2.000D-03  -2.000D+00 .            .        *   .            .           .
3.000D-03  -3.000D+00 .            .      *     .            .           .
4.000D-03  -4.000D+00 .          *         .            .           .
5.000D-03  -5.000D+00 .        *           .            .           .
6.000D-03  -4.001D+00 .          *         .            .           .
7.000D-03  -3.001D+00 .            .     *      .            .           .
8.000D-03  -2.001D+00 .            .       *    .            .           .
9.000D-03  -1.001D+00 .            .          * .            .           .
1.000D-02  -1.000D-03 .            .            *            .           .
1.100D-02   9.990D-01 .            .            .  *          .           .
1.200D-02   1.999D+00 .            .            .    *        .           .
1.300D-02   2.999D+00 .            .            .      *      .           .
1.400D-02   3.999D+00 .            .            .         *   .           .
1.500D-02   4.999D+00 .            .            .            *           .
1.600D-02   4.002D+00 .            .            .         *   .           .
1.700D-02   3.002D+00 .            .            .       *     .           .
1.800D-02   2.002D+00 .            .            .     *       .           .
1.900D-02   1.002D+00 .            .            .   *         .           .
2.000D-02   2.000D-03 .            .            *            .           .
2.100D-02  -9.980D-01 .            .          *  .            .           .
2.200D-02  -1.998D+00 .            .        *    .            .           .
2.300D-02  -2.998D+00 .            .      *      .            .           .
2.400D-02  -3.998D+00 .          *          .            .           .
2.500D-02  -4.998D+00 .        *            .            .           .
2.600D-02  -4.003D+00 .          *          .            .           .
2.700D-02  -3.003D+00 .            .      *      .            .           .
2.800D-02  -2.003D+00 .            .        *    .            .           .
2.900D-02  -1.003D+00 .            .          *  .            .           .
3.000D-02  -3.000D-03 .            .            *            .           .
3.100D-02   9.970D-01 .            .            .  *          .           .
3.200D-02   1.997D+00 .            .            .    *        .           .
3.300D-02   2.997D+00 .            .            .      *      .           .
3.400D-02   3.997D+00 .            .            .         * .            .
3.500D-02   4.997D+00 .            .            .            *           .
3.600D-02   4.004D+00 .            .            .         * .            .
3.700D-02   3.004D+00 .            .            .       *     .           .
3.800D-02   2.004D+00 .            .            .     *       .           .
3.900D-02   1.004D+00 .            .            .  *          .           .
4.000D-02   4.000D-03 .            .            *            .           .
                      - - - - - - - - - - - - - - - - - - - - - - - - -
```

(c)

Figure 7.58
(Continued)

EXAMPLE 7.22

****   TRANSIENT ANALYSIS   TEMPERATURE =   27.000 DEG C

****

```
   TIME      V(6)

             -1.000D+01    0.000D+00    1.000D+01    2.000D+01  3.000D+01
            - - - - - - - - - - - - - - - - - - - - - - - - - - - - - - -
0.000D+00  0.000D+00 .            *            .            .          .
1.000D-03  5.584D-01 .           .*            .            .          .
2.000D-03  2.057D+00 .           . *           .            .          .
3.000D-03  4.506D+00 .           .      *      .            .          .
4.000D-03  8.063D+00 .           .         *   .            .          .
5.000D-03  1.250D+01 .           .            .*            .          .
6.000D-03  1.695D+01 .           .            .        *    .          .
7.000D-03  2.050D+01 .           .            .            .*          .
8.000D-03  2.294D+01 .           .            .            .   *       .
9.000D-03  2.442D+01 .           .            .            .       *   .
1.000D-02  2.500D+01 .           .            .            .         * .
1.100D-02  2.446D+01 .           .            .            .        *  .
1.200D-02  2.300D+01 .           .            .            .     *     .
1.300D-02  2.044D+01 .           .            .            .*          .
1.400D-02  1.692D+01 .           .            .        *    .          .
1.500D-02  1.250D+01 .           .            .*           .          .
1.600D-02  8.049D+00 .           .         *   .            .          .
1.700D-02  4.503D+00 .           .      *      .            .          .
1.800D-02  2.060D+00 .           . *           .            .          .
1.900D-02  5.786D-01 .           .*            .            .          .
2.000D-02 -2.817D-03 .          *             .            .          .
2.100D-02  5.010D-01 .           .*            .            .          .
2.200D-02  2.047D+00 .           . *           .            .          .
2.300D-02  4.570D+00 .           .      *      .            .          .
2.400D-02  8.053D+00 .           .         *   .            .          .
2.500D-02  1.249D+01 .           .            .*           .          .
2.600D-02  1.694D+01 .           .            .        *    .          .
2.700D-02  2.049D+01 .           .            .            .*          .
2.800D-02  2.293D+01 .           .            .            .   *       .
2.900D-02  2.442D+01 .           .            .            .       *   .
3.000D-02  2.500D+01 .           .            .            .         * .
3.100D-02  2.445D+01 .           .            .            .        *  .
3.200D-02  2.300D+01 .           .            .            .     *     .
3.300D-02  2.044D+01 .           .            .            .*          .
3.400D-02  1.693D+01 .           .            .        *    .          .
3.500D-02  1.251D+01 .           .            .*           .          .
3.600D-02  8.053D+00 .           .         *   .            .          .
3.700D-02  4.506D+00 .           .      *      .            .          .
3.800D-02  2.061D+00 .           . *           .            .          .
3.900D-02  5.771D-01 .           .*            .            .          .
4.000D-02 -6.858D-03 .          *             .            .          .
            - - - - - - - - - - - - - - - - - - - - - - - - - - - - - - -
```

(d)

Figure 7.58
(Continued)

Example 7-23

Design an operational amplifier circuit whose output reaches +9 V after 3 pulses of a 500-kHz square wave have occurred at its input. The square wave alternates between 0 V and +5 V. Verify your design using SPICE.

Solution. As shown in Figure 7.33(b), the output of an ideal integrator is a *staircase* waveform that increases by Et_1 volts after the occurrence of each new pulse, where E is the high level and t_1 is the width of the pulse. Since the output of an operational-amplifier integrator is

$$v_o = -\frac{1}{RC} \int e_{in}\, dt$$

we must use a second amplifier to obtain a positive output. The gain of the second amplifier will be $-R_f/R_1$, so the overall gain will be

$$\text{gain} = \left(-\frac{1}{RC}\right)\left(-\frac{R_f}{R_1}\right) = \frac{R_f}{R_1 RC}$$

In our example, we want the output to be +9 V after 3 pulses, so we require that

$$3Et_1 \frac{R_f}{R_1 RC} = 9$$

where E = +5 V and t_1 is one-half the period of the 500 kHz square wave:

$$t_1 = 0.5 \left(\frac{1}{500 \times 10^3}\right) = 10^{-6}\ \text{s. Thus,}$$

$$3(5)(10^{-6}) \frac{R_f}{R_1 RC} = 9$$

or

$$\frac{R_f}{R_1 RC} = 9/(15 \times 10^{-6}) = 6 \times 10^5$$

Letting C = 0.001 μF = 10^{-9} and $R = R_1$ = 10 kΩ = 10^4, we have

$$R_f = (6 \times 10^5)(10^{-9})(10^4)(10^4) = 60\ \text{kΩ}$$

The complete design is shown in Figure 7.59(a).

The SPICE circuit and input file are shown in Figure 7.59(b). The .TRAN statement makes the analysis last for the duration of three positive pulses, or 2½ periods of the square wave: $(2.5) \times 1/(500 \times 10^3)$ = 5 μs. Note that we set TR and TF in the pulse specification to 1 ns. If these were made 0, SPICE would assign them the default value TSTEP = 0.1 μs. The accuracy of the integration would be affected by such relatively long rise and fall times, since the output of the integrator is proportional to the total area under the pulse waveform. The plot of the output is shown in Figure 7.59(c), p. 232, and we see that the output reaches +9.004 V in 5 s.

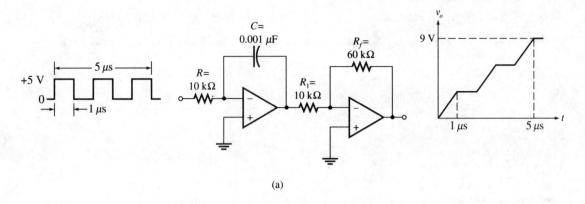

(a)

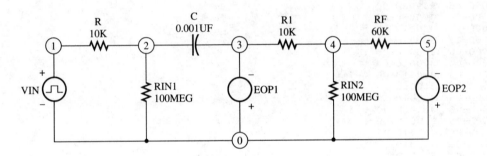

```
EXAMPLE 7.23
VIN 1 0 PULSE(0 5 0 1NS 1NS 1US 2US)
R 1 2 10K
RIN1 2 0 100MEG
C 2 3 0.001UF
EOP1 0 3 2 0 1MEG
R1 3 4 10K
RIN2 4 5 60K
EOP2 0 5 4 0 1MEG
.TRAN 0.1US 5US
.PLOT TRAN V(5)
.END
```

(b)

Figure 7.59
Example 7–23. *(Continued on p. 232.)*

EXAMPLE 7.23

```
****     TRANSIENT ANALYSIS              TEMPERATURE =   27.000 DEG C

***************************************************************************

     TIME     V(5)

                   -5.000D+00    0.000D+00    5.000D+00   1.000D+01  1.500D+01
                - - - - - - - - - - - - - - - - - - - - - - - - - - - - - - -
 0.000D+00   0.000D+00 .              *              .            .          .
 1.000D-07   2.985D-01 .              .*             .            .          .
 2.000D-07   5.985D-01 .              . *            .            .          .
 3.000D-07   8.985D-01 .              .  *           .            .          .
 4.000D-07   1.198D+00 .              .  *           .            .          .
 5.000D-07   1.498D+00 .              .   *          .            .          .
 6.000D-07   1.798D+00 .              .    *         .            .          .
 7.000D-07   2.098D+00 .              .     *        .            .          .
 8.000D-07   2.398D+00 .              .      *       .            .          .
 9.000D-07   2.698D+00 .              .       *      .            .          .
 1.000D-06   2.998D+00 .              .       *      .            .          .
 1.100D-06   3.001D+00 .              .       *      .            .          .
 1.200D-06   3.001D+00 .              .       *      .            .          .
 1.300D-06   3.001D+00 .              .       *      .            .          .
 1.400D-06   3.001D+00 .              .       *      .            .          .
 1.500D-06   3.001D+00 .              .       *      .            .          .
 1.600D-06   3.001D+00 .              .       *      .            .          .
 1.700D-06   3.001D+00 .              .       *      .            .          .
 1.800D-06   3.001D+00 .              .       *      .            .          .
 1.900D-06   3.001D+00 .              .       *      .            .          .
 2.000D-06   3.001D+00 .              .       * .            .          .
 2.100D-06   3.301D+00 .              .        *    .            .          .
 2.200D-06   3.601D+00 .              .         *   .            .          .
 2.300D-06   3.901D+00 .              .          *  .            .          .
 2.400D-06   4.201D+00 .              .           * .            .          .
 2.500D-06   4.501D+00 .              .           *.            .          .
 2.600D-06   4.801D+00 .              .            *.            .          .
 2.700D-06   5.101D+00 .              .             .*           .          .
 2.800D-06   5.401D+00 .              .             . *          .          .
 2.900D-06   5.701D+00 .              .             .  *         .          .
 3.000D-06   6.001D+00 .              .             .   *        .          .
 3.100D-06   6.004D+00 .              .             .   *        .          .
 3.200D-06   6.004D+00 .              .             .   *        .          .
 3.300D-06   6.004D+00 .              .             .   *        .          .
 3.400D-06   6.004D+00 .              .             .   *        .          .
 3.500D-06   6.004D+00 .              .             .   *        .          .
 3.600D-06   6.004D+00 .              .             .   *        .          .
 3.700D-06   6.004D+00 .              .             .   *        .          .
 3.800D-06   6.004D+00 .              .             .   *        .          .
 3.900D-06   6.004D+00 .              .             .   *        .          .
 4.000D-06   6.004D+00 .              .             .   *        .          .
 4.100D-06   6.304D+00 .              .             .    *       .          .
 4.200D-06   6.604D+00 .              .             .     *      .          .
 4.300D-06   6.904D+00 .              .             .      *     .          .
 4.400D-06   7.204D+00 .              .             .       *    .          .
 4.500D-06   7.504D+00 .              .             .        *   .          .
 4.600D-06   7.804D+00 .              .             .         *  .          .
 4.700D-06   8.104D+00 .              .             .          * .          .
 4.800D-06   8.404D+00 .              .             .           *.          .
 4.900D-06   8.704D+00 .              .             .           * .          .
 5.000D-06   9.004D+00 .              .             .            *.          .
                - - - - - - - - - - - - - - - - - - - - - - - - - - - - - - -
```

(c)

Figure 7.59
(Continued)

EXERCISES

Section 7–2

7.1 A square wave alternates between 0 V and +5 V and has a period of 0.25 ms. Find each of the following.
(a) The average value.
(b) The amplitude and frequency of the fundamental and of the second and third harmonics.

7.2 Write the first four non-zero terms of the Fourier series for the square wave shown in Figure 7.60.

7.3 A square wave has a pulse repetition frequency of 2×10^5 PPS. What is the frequency of the first non-zero harmonic whose amplitude is less than 1% of the amplitude of the fundamental?

7.4
(a) Write the first four non-zero terms of the Fourier series for the square wave shown in Figure 7.61.
(b) What bandwidth is required for an amplifier designed to pass the square wave

Figure 7.60
Exercise 7.2.

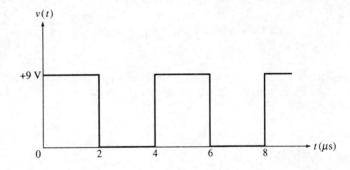

Figure 7.61
Exercise 7.4.

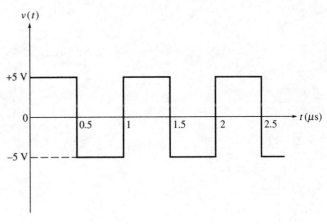

Figure 7.62
Exercise 7.5.

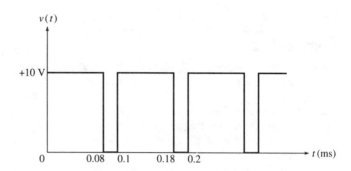

if its frequency must extend from zero to the first non-zero harmonic whose amplitude is no greater than 2% of the peak-to-peak value of the square wave?

7.5 Find the average value, the amplitude of the fundamental, and the amplitudes of the second and third harmonics of the waveform shown in Figure 7.62.

7.6

(a) Write the first four non-zero terms of the Fourier series for the waveform shown in Figure 7.63.

(b) What is the minimum bandwidth of an amplifier designed to pass the waveform if its maximum frequency must equal the maximum harmonic frequency whose amplitude is no greater than 5% of the peak-to-peak value of the waveform?

7.7 Write the first four non-zero terms of the Fourier series for the waveform shown in Figure 7.64.

7.8 Write the first five terms of the Fourier series for the waveform shown in Figure 7.65.

Figure 7.63
Exercise 7.6.

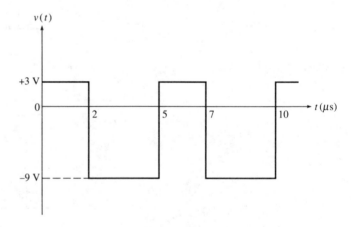

Figure 7.64
Exercise 7.7.

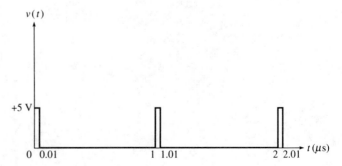

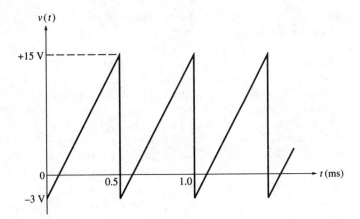

Figure 7.65
Exercise 7.8.

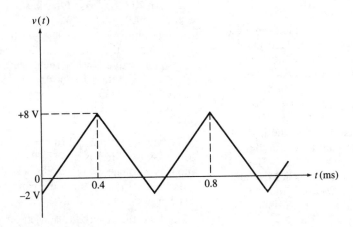

Figure 7.66
Exercise 7.10.

7.9 A certain filter is required to extract a 12-kHz third harmonic from a sawtooth waveform whose values range from 0 V to V_{HI}. If the third harmonic must have an amplitude of 0.75 V, what should be the frequency of the sawtooth and the value of V_{HI}?

7.10 Write the first four terms of the Fourier series for the waveform shown in Figure 7.66.

Section 7–3

7.11 The sinusoidal input to the circuit shown in Figure 7.67 is $v(t) = 20 \sin(2\pi ft)$ V. Write the sinusoidal expression for the capacitor

voltage $v_c(t)$ when (a) $f = 100$ Hz, (b) $f = 1$ kHz, and (c) $f = 10$ kHz.

7.12 The input to the circuit shown in Figure 7.67 is a square wave whose frequency is 1

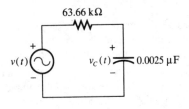

Figure 7.67
Exercises 7.11 and 7.12.

kHz. Find the ratio of the amplitude of the fifth harmonic to the amplitude of the fundamental
(a) In the input square wave;
(b) In the waveform developed across the capacitor.

7.13 A low-pass RC filter is to be designed so that its bandwidth equals the third-harmonic frequency of a triangular waveform whose period is 2.5 ms. If the resistor value is 3.3 kΩ, what should be the value of the capacitance?

7.14 A low-pass RC filter is to be designed so that it reduces the amplitude A of a 25-kHz sinusoidal signal to $0.1A$.
(a) What should be the cutoff frequency of the filter?
(b) What should be the value of its RC time constant?

7.15 Figure 7.68 shows the equivalent circuit of a square wave generator that produces a digital clock signal. Note that the source resistance R_S and the load capacitance C_L form a low-pass filter. To minimize distortion, it was determined that the R_SC_L time constant should not exceed 2% of the period of the clock.
(a) When that criterion is met, find the highest harmonic of the clock that is within the bandwidth of the filter.
(b) If the frequency of the clock is 1 MHz and the load capacitance is 200 pF, find the maximum permissible source resistance.

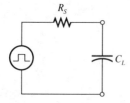

Figure 7.68
Exercises 7.15 and 7.16.

7.16 Repeat Exercise 7.15 if the R_SC_L time-constant cannot exceed 5% of the clock period and the clock frequency is changed to 2 MHz.

7.17 A low-pass RC filter used to produce a triangular waveform from a square wave has $R = 20$ kΩ and $C = 1.5$ μF. If the frequency of the square wave is 1 kHz, find the maximum

percent deviation of the waveform from a true triangular wave.

7.18 Find the maximum percent deviation of the output of a low-pass RC filter from a true triangular wave if the cutoff frequency of the filter is 2% of the frequency of the input square wave.

7.19 Find the average value and the peak-to-peak value of the triangular waveform produced by the filter in Exercise 7.17, assuming the square wave alternates between 0 V and 15 V.

7.20 For an RC low-pass filter, $R = 500$ Ω and $C = 1$ μF. Find the peak-to-peak value of its triangular output when the input is a 20-kHz square wave that alternates between $+10$ V and 5 V. Sketch the output waveform.

7.21 An RC low-pass filter has $R = 470$ Ω and $C = 0.015$ μF.
(a) Find the rise time of the positive-going output of the filter when the input is a square wave that alternates between 0 V and 5 V with a frequency of 2 kHz.
(b) Find the length of time required for the output to rise to 10% of its maximum value.

7.22 Show that the fall time of the output of a low-pass RC filter is $t_f = 2.197RC$ when the input is a square wave whose period is sufficiently long to allow the capacitor to fully discharge during one half-cycle.

7.23 Find the amplitude and phase angle of the output of the circuit shown in Figure 7.69 when the input is a sinewave having
(a) Amplitude 12 V, frequency 1 kHz, and phase 0°;
(b) Amplitude 2 V, frequency 3×10^4 rad/s, and phase 30°;

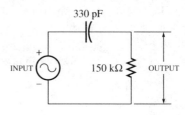

Figure 7.69
Exercises 7.23 and 7.24.

(c) Amplitude 25 mV, phase $-65°$, and frequency equal to 10 times the cutoff frequency of the circuit.

7.24 The input to the circuit in Figure 7.60 is a square wave having pulse repetition frequency 1000 PPS. Find the ratio of the amplitude of the fundamental to the amplitude of the fifth harmonic
(a) In the input;
(b) In the output.

7.25 A 0.5-MHz square wave clock signal that alternates between 0 V and 5 V is coupled through a 0.005-μF capacitor to a 5-kΩ load.
(a) Find the percent droop in the load voltage.
(b) Find the maximum positive value of the load voltage.
(c) What is the minimum frequency to which the clock signal can be reduced without the droop exceeding 10%?

7.26 A 50-kHz square wave that alternates between 0 V and 15 V is the input to a high-pass RC filter having $R = 1$ kΩ and $C = 0.01$ μF.
(a) Sketch the filter's output waveform. Label all voltage levels.
(b) Using the voltage levels found in (a), calculate the percent droop.
(c) Calculate the percent droop using equation (7.34).

Section 7–4

7.27 Sketch the output waveforms produced by each of the circuits shown in Figure 7.70 when the inputs are as shown. Assume each diode conducts when its anode-to-cathode voltage is positive. Label the clipping levels on the sketches.

7.28 Repeat Exercise 7.27 when
(a) Each of the diodes in Figure 7.70 is reversed;
(b) Each of the diodes *and* each of the voltage sources in Figure 7.70 is reversed.

7.29 Design one clipping circuit that clips an input waveform at $+6$ V and -12 V.

7.30 Repeat Exercise 7.27 under the assumption that each diode conducts when its anode-to-cathode voltage exceeds 0.7 V.

Section 7–5

7.31 Design operational amplifier circuits that produce each of the following outputs:
(a) $v_o = -4e_{in}$
(b) $v_o = -0.2e_{in}$
(c) $v_o = 12e_{in}$
(d) $v_o = -(2.5e_1 + e_2 + 5e_3)$

7.32 Design an operational amplifier circuit that sums four equal-amplitude sinusoidal voltages having the frequencies of the fundamental, second, third, and fourth harmonics of a 10-kHz triangular wave. Each voltage should be summed so that its proportion in the output is the same as that in the harmonic content of a triangular wave. The gain for the fundamental component should be 7.2.

7.33 Find the value of the feedback factor in each of the circuits designed in Exercise 7.31.

7.34 Assuming the unity-gain frequency of the amplifier used in Exercise 7.32 is 800 kHz, find the closed-loop bandwidth of the circuit. Is the amplifier capable of passing all the input signals without distortion?

7.35 An operational amplifier has a gain-bandwidth product of 1.2×10^6. It is to be used in a non-inverting configuration and must pass the fifth harmonic of a 10-kHz square wave. If the feedback resistor R_f used in the circuit is to be 100 kΩ, what is the minimum permissible value of R_1?

7.36 An operational amplifier is to be selected for a design in which it must pass the 11th harmonic of a rectangular waveform having a *PRF* of 6×10^3 PPS. The required gain is 12. What is the minimum unity-gain frequency that the amplifier must have? Assume the amplifier will be used in an inverting configuration.

7.37 A non-inverting operational amplifier circuit has $R_1 = 5$ kΩ and $R_f = 40$ kΩ. What slew rate must it have if it is required to amplify without distortion a pulse that changes from -0.5 V to $+1.2$ V in 2 μs?

7.38 An operational amplifier circuit has a closed-loop gain of 4.8. The maximum slew rate of the amplifier is 2.5 V/μs. A square wave input alternates between $-E$ and $+E$ volts, each transition requiring 0.8 μs. What is

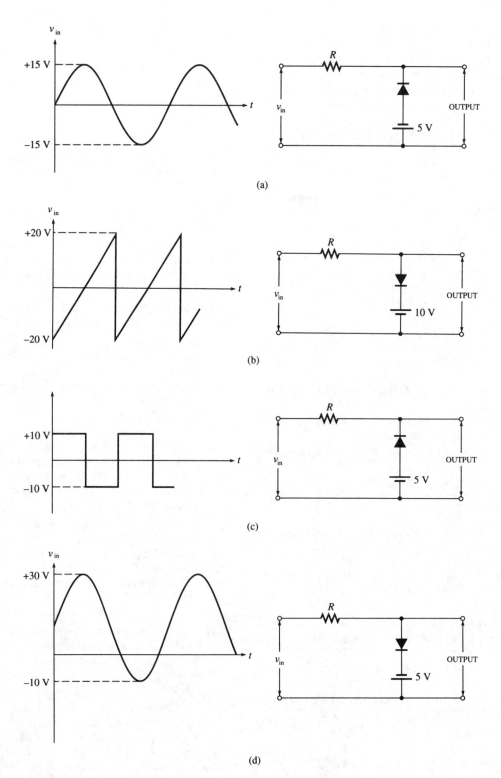

Figure 7.70
Exercises 7.27, 7.28, and 7.30.

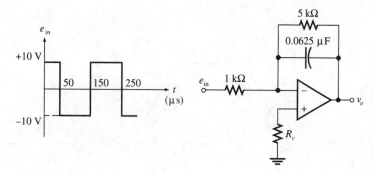

Figure 7.71
Exercises 7.39 and 7.41.

the maximum permissible value of E if the specified slew rate cannot be exceeded?

Section 7–6

7.39

(a) Sketch the output waveform produced by the operational amplifier circuit shown in Figure 7.71.
(b) What should be the value of R_c?
(c) What minimum slew rate must the amplifier have when used in this application?

7.40 Design an integrator that will produce a triangular waveform having peak-to-peak value 20 Vpp when the input is a 10-kHz square wave that alternates between ±5 V. The input

resistance of the circuit should be at least 4 kΩ.

7.41 The fundamental component of a square wave has a peak-to-peak value of 6 Vpp. The square wave has a *PRF* of 2.5×10^3 PPS and is applied to the input of the integrator shown in Figure 7.71. What is the peak-to-peak value of the fundamental component in the output waveform?

7.42 Show that the dc gain, A_{dc}, of the integrator in Figure 7.34 can be expressed as

$$A_{dc} = \frac{-1}{2\pi f_B R_1 C}$$

where f_B is the break frequency caused by the parallel combination of R_f and C.

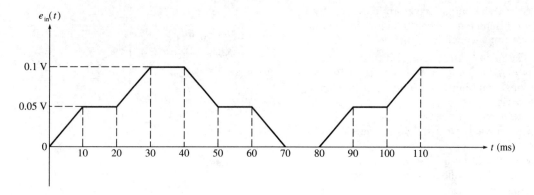

(a)

Figure 7.72
Exercise 7.43.

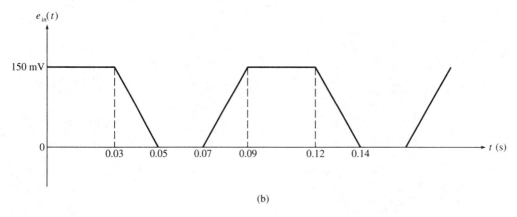

(b)

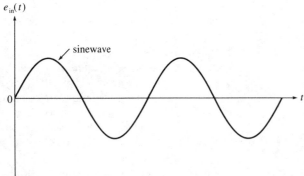

(Hint: Where are the slopes maximum positive, maximum negative, and zero?)

(c)

Figure 7.72
(Continued)

7.43 Sketch the output waveform produced by an *ideal* differentiator when the input is each of the waveforms shown in Figure 7.72.

7.44 The input to the circuit shown in Figure 7.73 is $e_{in}(t) = 3 \sin(2\pi \times 1.5 \times 10^3 t)$ V. The operational amplifier has a unity-gain frequency of 2.5 MHz.

(a) If the input is to be accurately differentiated, what should be the value of R_1?

(b) What should be the value of R_c?

(c) Write the sinusoidal expression for the output.

(d) What minimum slew rate should the amplifier have?

(e) What is the upper cutoff frequency of the circuit?

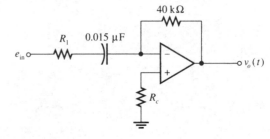

Figure 7.73
Exercise 7.44.

Section 7–7

7.45 Sketch the output waveform produced by each of the circuits in Figure 7.41 when the

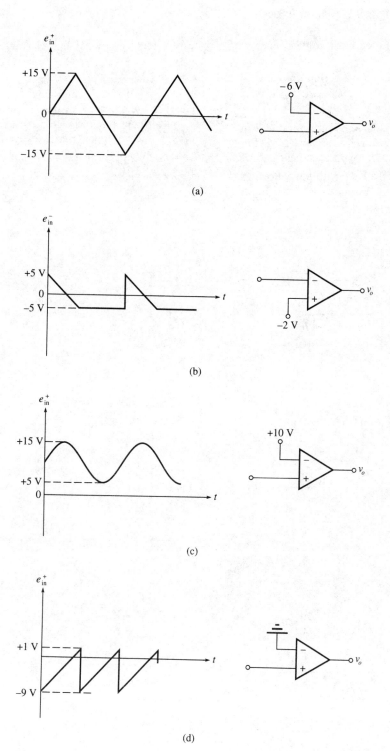

Figure 7.74
Exercise 7.46.

input waveform and the fixed voltage input are interchanged.

7.46 Sketch the output waveform produced by each of the circuits in Figure 7.74. In each comparator, $V_{LO} = -5$ V and $V_{HI} = +5$ V.

7.47 Sketch the output waveform produced by each of the circuits shown in Figure 7.75. Find the value of the hysteresis in each case.

7.48 Repeat Exercise 7.47 with the following changes: (a) $V_{LO} = 0$ V, $V_{HI} = +5$ V; (b) 2-kΩ

resistor replaced by a 1-kΩ resistor; (c) $V_{HI} = 0$ V.

7.49 A voltage comparator has $V_{LO} = -5$ V and $V_{HI} = +10$ V. Using the comparator and a reference voltage of 0 V, design an inverting Schmitt trigger that has a lower trigger level of -2.5 V and an upper trigger level of $+5$ V.

7.50 A voltage comparator has $V_{LO} = -10$ V and $V_{HI} = +10$ V.

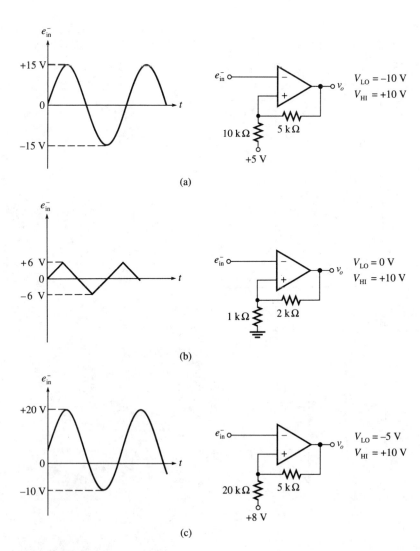

Figure 7.75
Exercise 7.47.

(a) Design a non-inverting Schmitt trigger that has a lower trigger level of −4 V and an upper trigger level of +4 V.

(b) Sketch the output of the trigger when the input is a 10-V pk sinewave.

Section 7–8

7.51 Sketch the output waveform produced by each of the circuits shown in Figure 7.76. Neglect the voltage drop across the diode. In

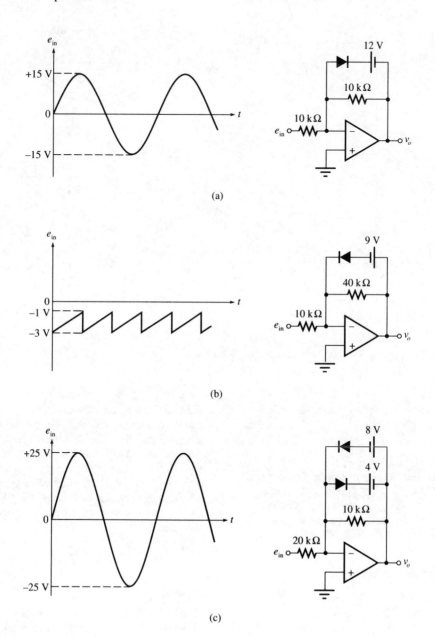

(a)

(b)

(c)

Figure 7.76
Exercises 7.51 and 7.52.

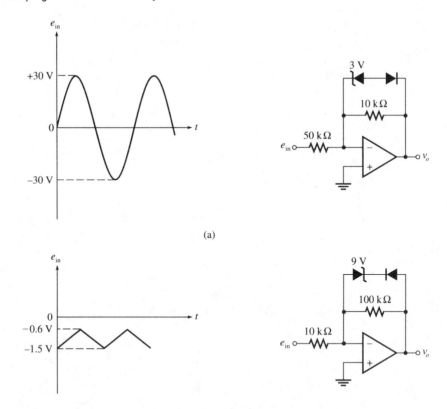

Figure 7.77
Exercise 7.53.

each case, indicate the input level at which output clipping occurs.

7.52 Repeat Exercise 7.51 with each of the diodes and each of the dc sources reversed.

7.53 Sketch the output of each of the circuits shown in Figure 7.77. Assume each of the silicon diodes has a forward-biased voltage drop of 0.7 V. In each case, indicate the input level at which output clipping occurs.

7.54 Design a zener diode clipping circuit using an operational amplifier whose voltage gain is -10. The input to the circuit is $e_{in}(t) = 1.2 \sin \omega t$ V and the output should be clipped as follows: negative peak clipped when ωt is between 30° and 150°; positive peak clipped when ωt is between 225° and 315°. Assume the forward-biased voltage drop across each zener diode is 0.7 V.

CHALLENGING EXERCISES

7.55 Write the Fourier series for the waveform shown in Figure 7.78.

7.56 Write several terms of the Fourier series for the waveform shown in Figure 7.79. (*Hint:* Find two waveforms whose sum equals $v(t)$.)

7.57 To approximate a square wave, a sinusoidal voltage having frequency 100 kHz is clipped at 0 V and at 50% of its peak positive value. Find the rise-time of the clipped waveform.

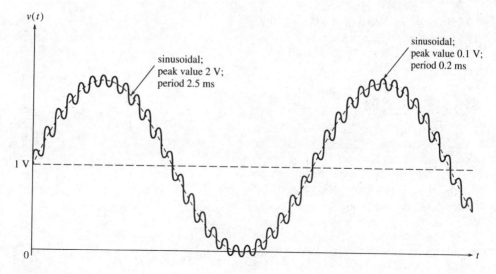

Figure 7.78
Exercise 7.55.

Figure 7.79
Exercise 7.56.

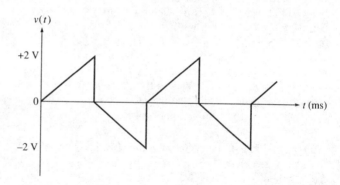

Figure 7.80
Exercise 7.59.

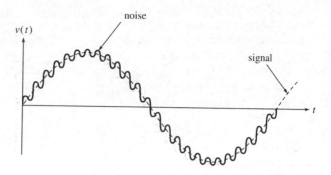

7.58 To approximate a square wave, a triangular voltage that alternates between +20 V and −20 V with a frequency of 1 MHz is clipped at 0 V and at +C volts. What should the clipping level C be if the rise-time of the clipped waveform is to be 40 ns?

7.59 Figure 7.80 shows a 1-kHz sinusoidal signal with peak value 1 V contaminated by a

20-kHz noise voltage with peak value 0.2 V. The ratio of the signal voltage to the noise voltage is thus 1 V/0.2 V = 5. Find the ratio of the signal-to-noise voltage

(a) After the waveform is integrated;
(b) After the waveform is differentiated.

SPICE EXERCISES

7.60 Use SPICE to obtain a Fourier analysis of a square wave with frequency 5×10^3 PPS. The square wave alternates between -5 V and $+5$ V. Compare the theoretical values of V_{AVG}, $A_1, A_2, \cdots, A_9$ with those computed by SPICE.

7.61 Use SPICE to obtain a Fourier analysis of the rectangular waveform shown in Figure 7.81. Compute the percent error, with respect to the theoretical exact values, in each of the values computed by SPICE for V_{AVG}, A_1, A_2, A_3, A_4, A_5, A_6, A_7, A_8, and A_9.

7.62 Use SPICE to obtain plots of the output waveform from a high-pass RC filter having $R = 40$ kΩ and $C = 250$ pF when the input is a 0–5 V square wave with frequency (a) 10^4 PPS and (b) 2×10^4 PPS. The plots should cover one complete cycle of the output waveform. What is the peak-to-peak voltage of the output in each case? What is the theoretical peak-to-peak value of the output? How could the accuracy of the SPICE solution be improved?

7.63 Use SPICE to determine the cutoff frequency (the frequency at which the magnitude of the output is 0.707 times the magnitude of the input) of the circuit shown in Figure 7.82. (*Note:* It will be necessary to perform trial-and-error runs to narrow the frequency range to one that includes the cutoff frequency.) What is the phase angle of the output at the cutoff frequency?

7.64 By trial-and-error runs of SPICE programs, find the value of capacitance

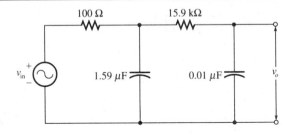

Figure 7.82
Exercise 7.63.

required in a low-pass RC filter having $R = 15$ kΩ in order that the magnitude of the output be one-half the magnitude of the input when the input is a sinewave with frequency 5 kHz.

7.65
(a) Use SPICE to obtain a plot of the output of the circuit shown in Figure 7.83.
(b) Repeat with the polarity of the 6-V source reversed.
(c) Repeat with the 6-V source connected as in (a) but with the diode terminals reversed.

Each plot should show one full cycle of output. Allow the diode to have its default parameter values.

7.66 Use SPICE to obtain a plot of the output of the circuit shown in Figure 7.84. The plot should show one full cycle of output. Allow the diodes to have their default parameter values.

7.67 Use SPICE to obtain a plot of one cycle of a waveform equal to the sum of the

Figure 7.81
Exercise 7.61.

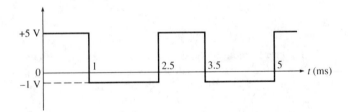

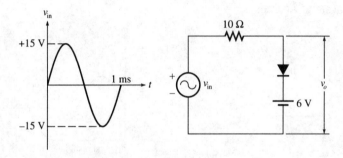

Figure 7.83
Exercise 7.65.

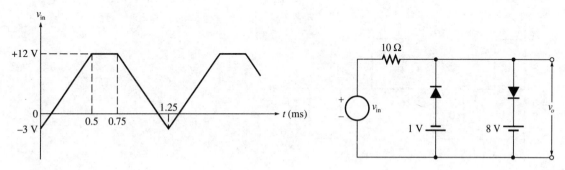

Figure 7.84
Exercise 7.66.

Figure 7.85
Exercise 7.67.

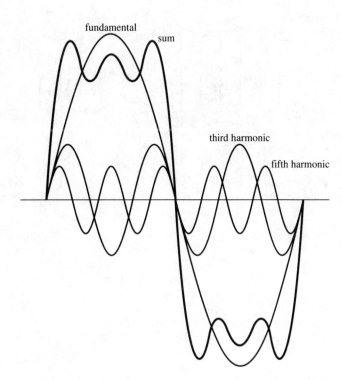

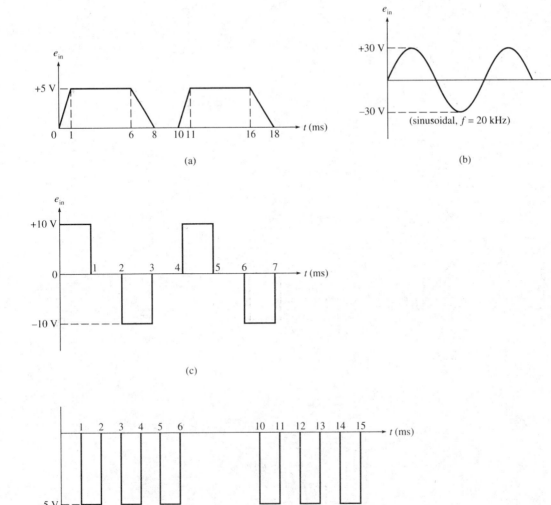

Figure 7.86
Exercise 7.68.

fundamental, third, and fifth harmonics of a 1-kHz square wave that alternates between −5 V and +5 V (Figure 7.85). The harmonics should be added in the same proportions as their content in the square wave, with the gain of the fundamental being 15. Use an operational amplifier having input resistance 100 MΩ and open-loop gain 10^6 to perform the summation.

7.68 Use SPICE to plot the integral of one cycle of each of the waveforms shown in Figure 7.86. The integrations should be performed with an operational amplifier having $1/RC = 100$. In each case, compare the output level produced by SPICE at the end of the cycle to the theoretical value. (*Hint:* See the remarks in Example 7–22).

Chapter 8

PULSE AND SIGNAL GENERATION

8–1   MONOSTABLE MULTIVIBRATORS

A monostable multivibrator produces a pulse of predetermined width in response to a *trigger* input. The trigger may itself be a pulse, whose low-to-high or high-to-low transition (depending on design) initiates the output pulse. The width of the output pulse is usually determined by the resistance and capacitance values in an RC network called a *timing circuit* connected to the device. A monostable multivibrator is often called simply a monostable, a *one-shot*, or a *single-shot*, because it produces a single pulse in response to a trigger input. The word *monostable* means it has *one* stable state. Typically, the output is low and remains in that (stable) state until the device goes high for a predetermined time, ultimately reverting back to the stable state. Figure 8.1 shows the schematic symbol* for a monostable and typical outputs produced in response to trigger inputs. Note that the output is labeled Q; many monostables also have a $\overline{Q}$ output, the complement of Q, which goes low when Q goes high and vice versa.

A *retriggerable* monostable is one that will accept a new trigger input while the output pulse produced by the previous trigger is still in progress. The new trigger initiates a new timing cycle, so the pulse is extended, beginning where the new trigger occurred, a length of time equal to the monostable's full-output pulsewidth. In other words, regardless of how long an output pulse has been high, a new trigger input effectively restarts time and superimposes a new pulse beginning where the trigger occurred. A non-retriggerable monostable simply ignores any new trigger that occurs while a pulse output is in progress. Figure 8.2 illustrates these ideas by showing the response of retriggerable and non-triggerable monostables to the same sequence of trigger inputs. Note that three trigger inputs create two output pulses from the non-retriggerable device because the second trigger occurs while the output is still high. However, the retriggerable monostable produces a single "stretched" pulse because it responds to each new trigger.

---

*Section 8.6 shows ANSI/IEEE standard symbols.

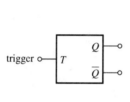

(a) Symbol.

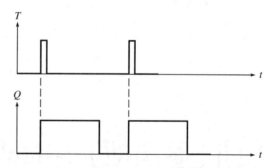

(b) Typical output produced in response to the low-to-high transitions of a trigger input.

Figure 8.1
The monostable multivibrator.

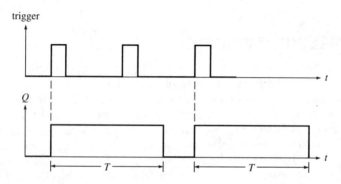

(a) The non-retriggerable monostable ignores the second trigger pulse.

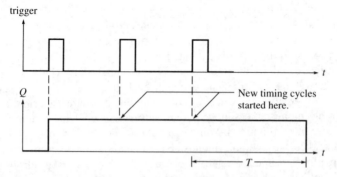

(b) The retriggerable monostable starts timing a new pulse having width *T* each time it is triggered. Only after the third trigger does it fully time out.

Figure 8.2
Response of retriggerable and non-retriggerable monostables to a sequence of three trigger pulses.

7400-Series Monostables

A widely used integrated-circuit version of the non-retriggerable monostable is the 74121, from the 7400 TTL family introduced in Chapter 3. The 74121 is constructed with logic gates at its trigger input. Depending on how external signal inputs are connected to these gates, the monostable can be triggered either by a low-to-high or by a high-to-low level transition. Figure 8.3 shows the wiring diagram of the 74121 with connections to the external RC timing circuit. Also shown is the truth table governing operation and triggering of the device. We see that it can be triggered by a high-to-low transition on input A_2 when the other inputs (A_1 and B) are high or by a high-to-low transition on A_1 when A_2 and B are high. It can also be triggered by a low-to-high transition on B when either A_1 or A_2 is low. Note that the AND gate incorporates a Schmitt trigger to provide sharp triggering from slowly varying inputs.

Figure 8.3
The 74121 non-retriggerable monostable multivibrator.

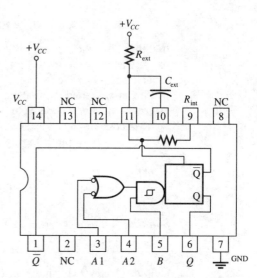

(a) Wiring diagram showing connection of the external RC timing circuit (R_{ext}, C_{ext}).

A_1	A_2	B	Q	$\bar{Q}$
L	x	H	L	H
x	L	H	L	H
X	X	L	L	H
H	H	x	L·	H
H	↓	H	⊓	⊔
↓	H	H	⊓	⊔
↓	↓	H	⊓	⊔
L	x	↑	⊓	⊔
x	L	↑	⊓	⊔

(b) Truth table. L = low, H = high, x = *don't care*,
↑ = low-to-high transition, ↓ = high-to-low transition.

The width of the output pulse produced by the 74121 is given by

$$PW = (\ln 2)R_{ext}C_{ext} \approx 0.69 R_{ext}C_{ext} \qquad \textbf{(8.1)}$$

When a monostable is repeatedly triggered, the output is a series of pulses. A variation in output pulse widths due to failure of all pulses to return low in exactly the same time intervals is called trailing-edge *jitter*. To prevent trailing-edge jitter in the 74121, the external timing components must be in the range

$$2 \text{ k}\Omega < R_{ext} < 40 \text{ k}\Omega$$
$$10 \text{ pF} < C_{ext} < 10 \text{ }\mu\text{F} \qquad \textbf{(8.2)}$$

If the exact time-point where the output pulse returns low is not critical, the component range can be extended:

$$1.4 \text{ k}\Omega < R_{ext} < 40 \text{ k}\Omega$$
$$10 \text{ pF} < C_{ext} < 1000 \text{ }\mu\text{F} \qquad \textbf{(8.3)}$$

In either case, there is internal capacitance of approximately 30 pF in parallel with C_{ext}, so its (additive) effect must be considered when using small values of external capacitance. Over the full range of component values specified by (8.3), the output pulse width can be varied from about 40 ns to 28 s. The 74121 can also be operated without external timing components by leaving pins 10 and 11 open and connecting V_{CC} directly to the internal timing resistance (R_{int}) at pin 9. The internal resistance and capacitance then produce an output pulse having a width of about 35 ns.

Example 8-1   A 74121 has $R_{ext} = 14.43$ kΩ and $C_{ext} = 1$ μF. Sketch Q and $\overline{Q}$ when A_1, A_2, and B are the waveforms shown in the timing diagram in Figure 8.4(a).

Solution.   From equation (8.1), the output pulsewidth is

$$PW = (\ln 2)(14.43 \times 10^3 \text{ }\Omega)(1 \times 10^{-6} \text{ F}) = 10 \text{ ms}$$

Outputs Q and $\overline{Q}$ are shown in Figure 8.4(b). At $t = 5$ ms, A_1 goes low while A_2 and B are high. Referring to the truth table in Figure 8.3(b), we see that this action triggers a 10-ms pulse. The low-to-high transition of A_1 at $t = 10$ ms has no effect. The following table summarizes the responses of the monostable:

t (ms)	A_1	A_2	B	Q
5	↓	H	H	⎍
10	↑	H	H	No change
30	H	↓	H	⎍
35	↓	L	H	No change
45	L	L	↓	No change
50	L	↑	L	No change
55	L	H	↑	⎍
60	L	↓	H	No change
70	L	L	↓	No change
80	L	L	↑	⎍
100	L	L	↓	No change

Figure 8.4
Example 8–1.

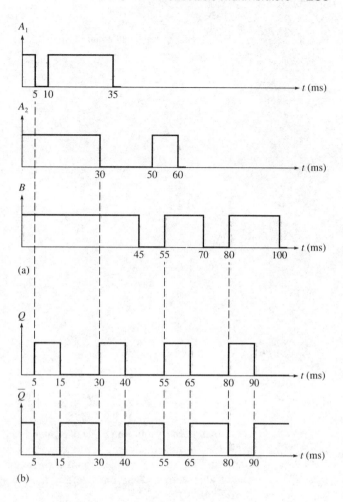

(a)

(b)

*Drill
8–1*

To what value would C_{ext} in Example 8–1 be changed if it is desired to reduce the pulsewidth to 7.5 ms?
Answer: 0.75 μF.

Figure 8.5 shows the wiring diagram and truth table for one-half of the dual 74123 *retriggerable* monostable multivibrator. Notice that this version has a clear input (CLR), which, when made low, will cause any output pulse already in progress to be terminated. When $C_{ext} > 1000$ pF, the output pulsewidth is approximated by

$$PW \approx 0.28 R_{ext} C_{ext}\left(1 + \frac{6.7}{R_{ext}}\right) \tag{8.4}$$

If C_{ext} is an electrolytic capacitor or if the clear function is used, the manufacturer recommends that a diode be inserted between R_{ext} and pin 15 (cathode to pin 15), and the

Figure 8.5
The 74123 dual, retriggerable monostable multivibrator.

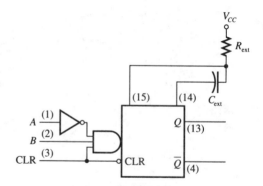

(a) One-half of a 74123 dual monostable.
(Pin numbers in parentheses.)

CLR	A	B	Q	$\overline{Q}$
L	x	x	L	H
x	H	x	L	H
x	x	L	L	H
H	L	↟	⎍	⎎
H	↓	H	⎍	⎎
↟	L	H	⎍	⎎

(b) Truth table.

coefficient 0.28 in equation (8.4) is changed to 0.25. When $C_{ext} \leq 1000$ pF, a nomograph available in manufacturer's product literature must be consulted to determine *PW*. Another TTL version of the monostable is the 74122, which contains one retriggerable monostable with a clear input and extensive input trigger logic. (See Figure 8.24 in Section 8–6.)

Example 8–2

The CLR and *B* inputs of a 74123 are held high and a pulse train consisting of three 1-μs pulses occurring at 8-μs intervals is applied to the *A* input. If $C_{ext} = 3571$ pF and $R_{ext} = 10$ kΩ, what is the width of the output pulse produced by the 74123?

Solution. From equation (8.4), the pulsewidth with no retriggering is approximately

$$0.28(10^4 \ \Omega)(3571 \times 10^{-12} \ \text{F})\left(1 + \frac{0.7}{10^4 \ \Omega}\right) = 10 \ \mu s$$

The truth table in Figure 8.5(*b*) shows that the monostable will trigger (or retrigger) on the trailing edges of the pulses applied to input *A*, since CLR and *B* are high. Figure 8.6 shows the timing diagram. The output pulse begins at the trailing edge of the first trigger pulse and is retriggered at the trailing edges of the second and third pulses. Thus, the full 10-μs timing cycle does not occur until after the trailing edge of the third pulse, and the total duration of the output pulse is 8 μs + 8 μs + 10 μs = 26 μs.

Figure 8.6
Example 8.2.

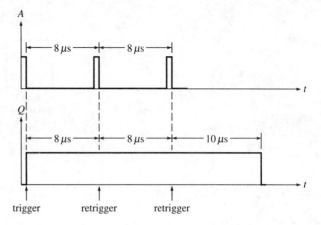

What would be the width of the output pulse in Example 8–2 if the value of C_{ext} were doubled and the interval between the trigger pulses were reduced to 5 μs? *Answer:* 30 μs.

8–2 APPLICATIONS OF MONOSTABLE MULTIVIBRATORS

Gating

In many digital systems, it is necessary to enable a logic gate to permit the passage, or *gating*, of digital signals to another part of the system for a prescribed period of time. For example, it is often necessary to deliver a certain number of pulses from a fixed-frequency clock to other logic circuitry for counting or synchronizing purposes. A monostable can be used to enable (or inhibit) logic gates for the necessary times, as illustrated in Figure 8.7. In this example, a 1-MHz clock signal is normally delivered to system *B* while system *A* is idle. When the monostable is triggered, it produces an 8-μs pulse, causing *Q* to go high for 8 μs and $\overline{Q}$ to go low for 8 μs. AND gates 1 and 2 are thus enabled and inhibited, respectively, allowing exactly 8 pulses to be delivered to system *A* and idling system *B* for 8 μs.

Time Delays

Monostables are widely used to deliver a pulse a certain time *after* the occurrence of another pulse, i.e., to create a prescribed *time delay* in the delivering of a pulse. Figure 8.8 shows an example in which two monostables are used to delay a 1-μs pulse by 1 ms. Each monostable is assumed to trigger on a low-to-high transition at its trigger input. As can be seen in the timing diagram, the input pulse produces an output which causes $\overline{Q}_1$ to go from low to high 1 ms after the trigger. This low-to-high transition triggers the second monostable, which produces a 1-μs pulse. In effect, the original 1-μs pulse has been delayed by 1 ms.

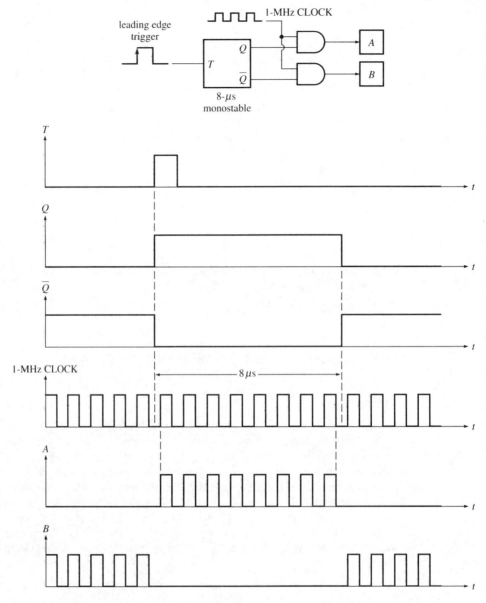

Figure 8.7
Using a monostable to route pulses through logic gates.

Synchronizing

Digital computer operations are often synchronized by sequences of pulses that occur on different control lines at different times. In a common example, each of several control lines goes high in a regular sequence so that one and only one is high at any given time. Figure 8.9(*a*) shows how 74121 monostables can be used to generate a sequence of three

Figure 8.8
An example of using monostables to delay a pulse.

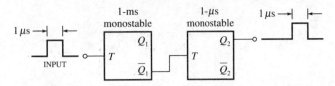

(a) Connecting two monostables to delay a 1-μs pulse by 1 ms. Both are assumed to trigger on a low-to-high transition.

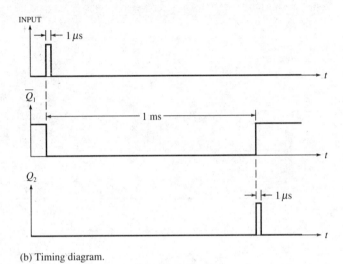

(b) Timing diagram.

such pulses. Referring to the truth table in Figure 8.3, we see that each monostable is connected to trigger on a low-to-high transition at its B input. Note that the $\overline{Q}_3$ output is *fed back* to the trigger input of the first monostable. Thus, every time Q_3 times out, making $\overline{Q}_3$ go from low to high, Q_1 is triggered. As shown in the timing diagram, $\overline{Q}_1$ triggers Q_2 and $\overline{Q}_2$ triggers Q_3. The process continues indefinitely and is an example of an *astable* circuit, about which we have more to say later. Note particularly the timing of the Q_1, Q_2, and Q_3 outputs: At any instant of time, one and only one output is high. The width of each pulse is given by equation (8.1): $PW = (\ln 2)R_{ext}C_{ext}$. The circuit can be expanded in an obvious way to create four or more timing pulses.

Detection of a Missing Pulse

A retriggerable monostable can be used to detect a missing pulse or the cessation of pulses in a pulsetrain that is supposed to consist of a sequence of regularly recurring pulses. One example where such detection is important is in a pulsetrain representing heartbeats. The pulsewidth of the retriggerable monostable is set to between one and two periods of the pulsetrain. The pulsetrain continually retriggers the monostable, which never times out unless a pulse is missing or the pulsetrain ceases. Figure 8.10 illustrates the concept. Here, a missing pulse is detected by the occurrence of a high-to-low transition on the Q output, which can be used to set an alarm or to trigger some other indicating device.

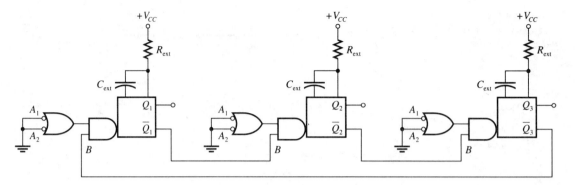

(a) Logic diagram.

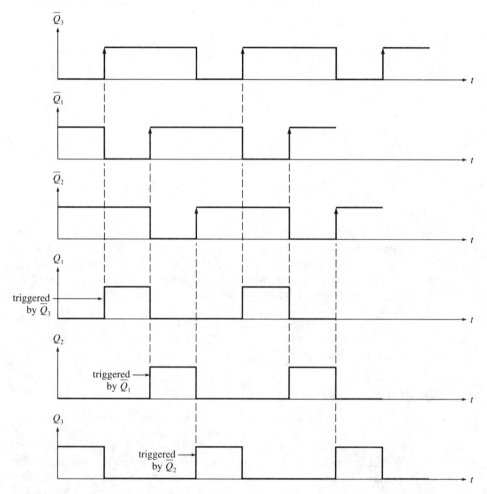

(b) Timing diagram.

Figure 8.9
Using monostables to produce continuous sequences of timing pulses (Q_1, Q_2, Q_3).

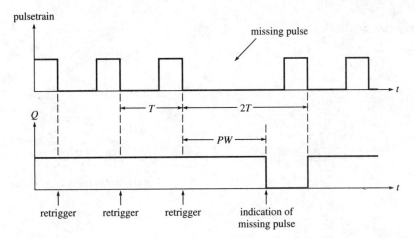

Figure 8.10
Use of a retriggerable monostable to detect a missing pulse: T < PW < 2T.

8–3   ASTABLE MULTIVIBRATORS

Astable means *not* stable, and an astable multivibrator has zero stable states. Its output continually alternates between low and high, so it is in fact a square (or rectangular) wave generator. An astable multivibrator is also called a *free-running* multivibrator.

In Section 8–2, we saw how an astable multivibrator can be constructed using monostable multivibrators. Figure 8.11(*a*) shows how one can be constructed using a voltage comparator. Assuming the maximum positive and negative outputs of the comparator are $\pm V_{\text{max}}$, capacitor C continually charges and discharges *toward* those values, as shown in Figure 8.11(*b*). The voltage fed back to the noninverting input is $\pm\beta V_{\text{max}}$, where $\beta = R_1/(R_1 + R_2)$. Therefore, when the capacitor charges or discharges to one of those levels, $v_{\text{in}}^+ = v_{\text{in}}^-$ and the output switches state. It can be shown that the period of the output square wave is

$$T = 2RC \ln\left(\frac{1 + \beta}{1 - \beta}\right) \text{ seconds} \tag{8.5}$$

where

$$\beta = \frac{R_1}{R_1 + R_2}$$

Example 8–3   For the astable multivibrator shown in Figure 8.11(*a*), $R = 10 \text{ k}\Omega$, $C = 0.001$ μF, and $R_1 = 10 \text{ k}\Omega$. It is desired to make the frequency of the output square wave adjustable from 20 kHz through 100 kHz by making R_2 adjustable. Through what range of values should R_2 be adjustable to obtain the required frequency range?

Solution.   The period of the output must range from $T = 1/(20 \text{ kHz}) = 5 \times 10^{-5}$ s through $T = 1/(100 \text{ kHz}) = 10^{-5}$ s. Applying equation (8.5) for the case $T = 10^{-5}$ s, we have

Figure 8.11
An astable multivibrator.

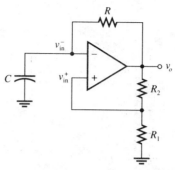

(a) Construction of an astable multivibrator using a voltage comparator.

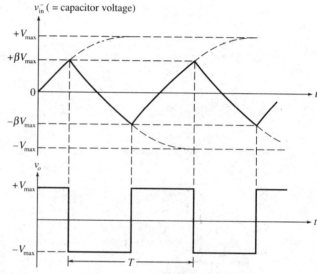

(b) Timing diagram. $T = 2RC \ln\left(\dfrac{1+\beta}{1-\beta}\right)$ where $\beta = \dfrac{R_1}{R_1+R_2}$.

$$10^{-5} = 2(10^4 \ \Omega)(10^{-9} \ \text{F}) \ln\left(\frac{1+\beta}{1-\beta}\right)$$

$$0.5 = \ln\left(\frac{1+\beta}{1-\beta}\right)$$

$$e^{0.5} = 1.6487 = \frac{1+\beta}{1-\beta}$$

$$2.6487\beta = 0.6487$$

$$\beta = 0.25$$

Since

$$\beta = \frac{R_1}{R_1+R_2}$$

we have

$$0.25 = \frac{10 \text{ k}\Omega}{10 \text{ k}\Omega + R_2}$$

Solving for R_2 gives $R_2 = 30$ kΩ.

Setting equation (8.5) equal to 5×10^{-5} s and following this same procedure to solve for R_2, we find $R_2 = 1.79$ kΩ. Thus, R_2 must be adjustable from about 1.8 kΩ through 30 kΩ.

Drill
8–3

If C in Example 8–3 is changed to 0.002 μF, what range of frequencies can be obtained by adjusting R_2 from 5 kΩ through 20 kΩ?
Answer: 15.5 kHz through 36.1 kHz.

Figure 8.12 shows an astable multivibrator constructed with the 74HC04 hex inverter. The 74HC00-series family is logically equivalent to the 7400-series TTL family but is constructed using *CMOS* technology, which is discussed in the next chapter. The advantage of using CMOS devices in astable multivibrators is that they have a much higher input impedance than their TTL counterparts. This characteristic makes the performance of CMOS multivibrators more predictable than TTL designs, in the sense that they are less sensitive to variations in device characteristics. The oscillation frequency of the circuit in Figure 8.12 is approximately

$$f \approx \frac{1}{1.8 \, RC} \text{ hertz} \tag{8.6}$$

The circuit can be operated with R in the range from 2.7 kΩ to 2.7 MΩ and with C from 50 pF to 10 μF.

Figure 8.13 shows two more astable designs using 74HC00-series devices. Although these designs have been found to operate reliably, their frequencies are not easily predictable. Experience shows that frequencies calculated using equations given in manufacturers' product literature may deviate by 10 to 40% from actual frequencies. In applications, it is necessary to adjust resistor and capacitor values if precise frequency control is required.

Figure 8.12
An astable multivibrator constructed with 74HC04 CMOS inverters.

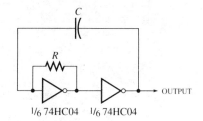

Figure 8.13
Astable multivibrator designs using 74HC00-series devices.

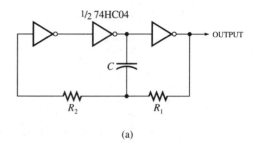

(a)

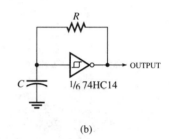

(b)

8–4   CRYSTAL-CONTROLLED CLOCK GENERATORS

A clock generator is a source for the square wave that synchronizes operations in a digital system. Some clock generators produce a reasonably sharp square wave, such as an astable multivibrator, and others produce sinusoidal or rounded outputs that can be shaped by Schmitt triggers or clipping circuits. The term astable multivibrator is usually reserved for circuits whose operation depends on the charging and discharging of a capacitor, such as those discussed in Section 8–3, whereas an *oscillator* refers to any unstable system whose output continually changes value. All oscillators have feedback paths from output to input. The feedback is said to be *positive* when the gain in the feedback path is 1 and the phase shift is zero (or a multiple of 360°). These are the criteria, called the Barkhausen criteria, for oscillation to occur. By connecting frequency-sensitive components in the feedback path (capacitors, inductors and/or combinations thereof), we can provide a path that has unity gain and zero-phase shift at *some* frequency. The frequency of oscillation is that one frequency at which the Barkhausen criteria are satisfied by a signal passing through the feedback path.

Many clock generators employ a crystal as the frequency-sensitive component. A crystal behaves like an LC network in that there is one frequency at which it is resonant (*tuned*). Crystals are available in a wide range of frequencies and are more stable than inductor/capacitor networks, in the sense that their characteristics are less likely to change with temperature and age. Thus, a crystal-controlled oscillator produces a signal whose frequency is less likely to *drift*. This frequency stability is particularly important in applications where frequency serves as a time reference, as, for example, in digital watches. Also, since crystal-controlled oscillators produce signals whose frequencies equal their crystal frequency, they are very *predictable*. By contrast, multivibrators whose operation depends on the charge and discharge cycles of an RC network are often sensitive to variations in integrated-circuit characteristics and dc supply voltage.

Figure 8.14

A crystal-controlled clock generator using TTL inverters.

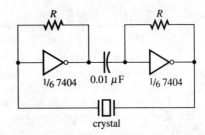

Oscillators used as clock generators are often designed with enough voltage gain to make the oscillations drive the amplifiers in the circuit between their extreme high and low voltage limits, producing essentially square outputs. This overdrive is a form of clipping. Sinusoidal oscillations can also be shaped into square waves using a Schmitt trigger. In applications, the crystal-controlled clock frequency is often *divided down* using flip-flops, a technique that is discussed in a later chapter, so the crystal frequency is an integer multiple of the actual clock frequency. One advantage of frequency division by flip-flops is further improvement of the waveform.

Figure 8.14 shows a popular crystal-controlled clock generator constructed with inverters from a TTL 7404 hex inverter. Although the oscillations produced by this design are not perfectly square, the output is generally adequate for many TTL synchronizing and triggering functions. The 7414 hex Schmitt trigger inverter can also be used. The value of R in the circuit controls the gain and is usually between 300 Ω and 1.5 kΩ. The optimum value depends on the type of crystal used and its frequency. If R is set too low or too high, the generator may oscillate at a harmonic of the crystal frequency and have smaller amplitude. The design has been used to produce clock frequencies from about 1 MHz to 20 MHz.

Figure 8.15 shows two crystal-controlled oscillators constructed with inverters from 74HC04 CMOS hex inverters. As in the TTL circuit of Figure 8.14, the oscillation frequencies may be sensitive to the values of R used. A typical value for R in Figure 8.15(*a*) is 100 kΩ, but it may have to be specially selected to prevent oscillation at a harmonic frequency. The 100-pF capacitor suppresses spurious high-frequency oscillations in the 30-MHz to 50-MHz range. The resistor in Figure 8.15(*b*) is on the order of 1 to 5 MΩ, and the circuit will oscillate for crystal frequencies up to about 9 MHz.

Figure 8.15

Crystal-controlled CMOS oscillators.

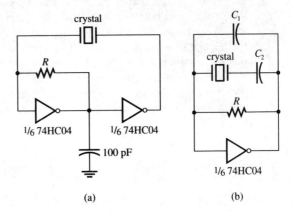

(a)   (b)

8–5   THE 555 TIMER

The 555 timer is a widely used integrated circuit having considerable versatility: It can be operated as a monostable or as an astable multivibrator as well as perform many other special functions. The principal components of the circuit are two voltage comparators, a flip-flop, and a transistor. As we will see later, a flip-flop is a device whose Q output can be triggered into a high state (and $\overline{Q}$ into a low state) by a pulse applied to its *set* input, and whose Q and $\overline{Q}$ outputs go low and high, respectively, when a pulse is applied to its *reset* input. These actions are summarized in Figure 8.16. In the 555 timer, only the $\overline{Q}$ output of the flip-flop is used. Thus, for our purposes now, it is necessary simply to remember that a positive edge applied to S causes $\overline{Q}$ to go low and a positive edge applied to R causes $\overline{Q}$ to go high.

Monostable Operation

Figure 8.17 shows a block diagram of the 555 timer connected with external timing components R and C for operation as a monostable. The control-voltage input (pin 5) is not used, and a 0.01-μF capacitor is connected to it to act as a noise filter. Note the voltage divider consisting of three 5-kΩ resistors connected to V_{CC}. The voltage at the − input of the threshold comparator is

$$\left(\frac{5\ k\Omega + 5\ k\Omega}{5\ k\Omega + 5\ k\Omega + 5\ k\Omega}\right)V_{CC} = \frac{2}{3}\,V_{CC}$$

and the voltage at the + input of the trigger comparator is

$$\left(\frac{5\ k\Omega}{5\ k\Omega + 5\ k\Omega + 5\ k\Omega}\right)V_{CC} = \frac{1}{3}V_{CC}$$

In the stable state (no pulse output), the trigger input at pin 2 is held high, typically at V_{CC} volts, so the output of the trigger comparator is low. The flip-flop is in its reset state, $\overline{Q}$ is high, and the output at pin 3 is low. The high voltage at $\overline{Q}$ applied to the base of the discharge transistor keeps it *on*—i.e., conducting, like a closed switch. The closed switch shorts the external capacitor, so it cannot charge. Since the voltage across the capacitor is therefore low, the threshold input at pin 6 is low, and the output of the threshold comparator is low. Thus, both the set and reset inputs of the flip-flop are low in the stable state.

   To trigger an output pulse, a negative-going pulse is applied to the trigger input. When that input falls below $(\frac{1}{3})V_{CC}$, the output of the trigger comparator goes high. The low-to-high transition sets the flip-flop, which makes $\overline{Q}$ go low. The output at pin 3 is then the leading edge of the output pulse. The low at $\overline{Q}$ turns off the discharge transistor, so it is like a switch that is opened, and the external capacitor C can begin to charge through R. When the capacitor voltage rises above $(\frac{2}{3})V_{CC}$, the output of the threshold comparator switches high. This low-to-high transition resets the flip-flop, and $\overline{Q}$ goes high. The output at pin 3 returns low. The high voltage at $\overline{Q}$ applied to the base of the discharge transistor

Figure 8.16
Behavior of the flip-flop in the 555 timer in response to pulses at its set and reset inputs.

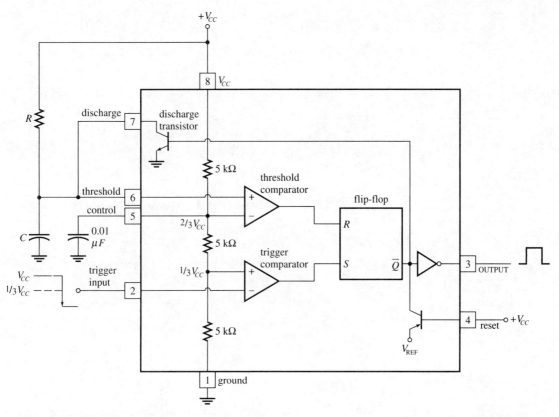

Figure 8.17
Block diagram of the 555 timer connected for operation as a monostable multivibrator.

turns it on again, and the capacitor discharges through the transistor. Note that external capacitor C charges according to

$$v_C(t) = V_{CC}(1 - e^{-t/RC}) \text{ volts} \tag{8.7}$$

The output pulse has width PW equal to the time required for $v_C(t)$ to rise from 0 V to $(2/3)V_{CC}$ volts. Setting $v_C(t) = (2/3)V_{CC}$ and $t = PW$, we find

$$2/3 V_{CC} = V_{CC}(1 - e^{-PW/RC})$$
$$2/3 = 1 - e^{-PW/RC}$$
$$1/3 = e^{-PW/RC}$$
$$PW = -RC \ln(1/3) = RC \ln 3$$

Thus, the output pulsewidth is

$$PW = RC \ln 3 \approx 1.1 \, RC \text{ seconds} \tag{8.8}$$

Figure 8.18 is a timing diagram that summarizes the operations we have described. After the output pulse is initiated, the flip-flop is insensitive to negative-going pulses at its S and R inputs. Thus, when the trigger input is returned high at a later time, the flip-flop and the

Figure 8.18

Timing diagram showing operation of the 555 timer as a monostable.

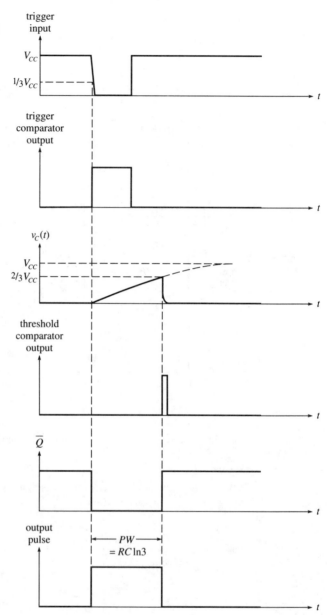

timing cycle are not affected. The manufacturer recommends that the trigger input be returned high before the end of the timing cycle.

Monostable operation of the 555 is nonretriggerable: Application of new trigger pulses during the timing cycle has no effect. However, the reset input at pin 4 can be used to terminate an output pulse during the timing cycle if desired. When a negative-going pulse applied to that input falls to about 0.5 V, the output pulse is terminated. A new trigger input will then initiate a new timing cycle. The reset input should be connected to V_{CC} when not used. The 555 timer is also available in a dual version: two timers in a single integrated-circuit package, designated a 556 timer.

Example 8–4

When the discharge transistor in the 555 timer is on, the voltage across the external capacitor is not exactly 0 V but is instead the small saturation voltage of the transistor: $V_{CE(\text{sat})} \approx 150$ mV. Thus, when the 555 is triggered, the capacitor actually charges according to the equation

$$v_C(t) = V_{CC} + (V_{CE(\text{sat})} - V_{CC})e^{-t/RC} \tag{8.9}$$

Assuming $V_{CC} = 10$ V, $RC = 10^{-3}$ s, and $V_{CE(\text{sat})} = 150$ mV, find the percent error in the calculation of the output pulsewidth using equation (8.8) instead of equation (8.9).

Solution. The actual pulsewidth is found by solving equation (8.9) for $t = PW$ when $v_C(t) = (\frac{2}{3})V_{CC}$:

$$\frac{2}{3}V_{CC} = V_{CC} + (V_{CE(\text{sat})} - V_{CC})e^{-PW/RC}$$

$$\frac{\frac{1}{3}V_{CC}}{V_{CC} - V_{CE(\text{sat})}} = e^{-PW/RC} \tag{8.10}$$

$$PW = -RC \ln\left[\frac{V_{CC}}{3(V_{CC} - V_{CE(\text{sat})})}\right]$$

$$= RC \ln\left[\frac{3(V_{CC} - V_{CE(\text{sat})})}{V_{CC}}\right]$$

Thus,

$$PW = 10^{-3}\ln\left[\frac{3(10 \text{ V} - 0.15 \text{ V})}{10 \text{ V}}\right] = 1.0835 \text{ ms}$$

By equation (8.8),

$$PW = 10^{-3}\ln 3 = 1.0986 \text{ ms}$$

The percent error is then

$$\frac{1.0986 \text{ ms} - 1.0835 \text{ ms}}{1.0835 \text{ ms}} \times 100\% = 1.39\%$$

This example shows that there is very little error in using equation (8.8) to calculate the pulsewidth. Note also that equation (8.10) shows that PW is somewhat dependent on V_{CC}. Over the manufacturer's specified range of V_{CC}, 4.5 V to 18 V, the variation is only about 2%.

Drill 8–4

A 1-kHz square wave is applied to the trigger input of the circuit in Figure 8.17. If $C = 0.1$ μF, what should be the value of R if the output at pin 3 is to have a 75% duty cycle? *Answer:* 6827 Ω.

Astable Operation

Figure 8.19 shows how the 555 timer can be connected to external timing components for operation as an astable multivibrator. R_1, R_2, and C determine the frequency of oscillation

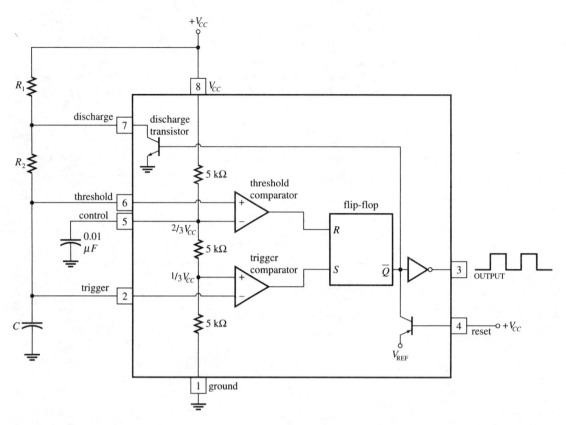

Figure 8.19
Block diagram of the 555 timer connected for operation as an astable multivibrator.

and the duty cycle. The capacitor charges through R_1 and R_2 until it reaches $(\frac{2}{3})V_{CC}$, at which time the threshold comparator resets the flip-flop. The high output at $\overline{Q}$ turns on the discharge transistor, and the capacitor discharges through R_2 only. When the capacitor voltage falls to $(\frac{1}{3})V_{CC}$, the trigger comparator sets the flip-flop, $\overline{Q}$ goes low, the discharge transistor turns off, and the cycle starts over again.

As shown in the timing diagram in Figure 8.20, the capacitor continually charges and discharges between $(\frac{1}{3})V_{CC}$ and $(\frac{2}{3})V_{CC}$ volts. Since it charges through both R_1 and R_2, the charging equation is

$$v_C(t) = V_{CC} + (\tfrac{1}{3}V_{CC} - V_{CC})e^{-t/(R_1+R_2)C} \text{ volts} \qquad (8.11)$$

The time T_1 for the capacitor to charge to $(\frac{2}{3})V_{CC}$ is found by setting $v_C(t) = (\frac{2}{3})V_{CC}$ and $t = T_1$:

$$\frac{2}{3}V_{CC} = V_{CC} + \left(\frac{1}{3}V_{CC} - V_{CC}\right)e^{-T_1/(R_1+R_2)C}$$

$$\frac{\frac{2}{3}V_{CC} - V_{CC}}{\frac{1}{3}V_{CC} - V_{CC}} = \frac{1}{2} = e^{-T_1/(R_1+R_2)C} \qquad (8.12)$$

$$T_1 = (R_1 + R_2)C \ln 2 \text{ seconds}$$

Figure 8.20
Timing diagram showing the external capacitor voltage and the output of the 555 timer operated as an astable multivibrator.

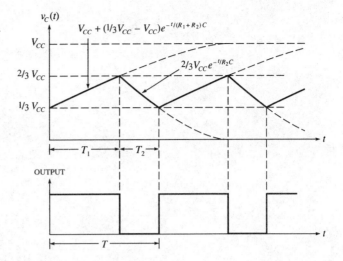

The capacitor discharges through R_2 only, so the discharge equation is

$$v_C(t) = \tfrac{2}{3}V_{CC}\, e^{-t/R_2 C} \tag{8.13}$$

The time T_2 for the capacitor to discharge to $(\tfrac{1}{3})V_{CC}$ is found by setting $v_C(t) = (\tfrac{1}{3})V_{CC}$ and $t = T_2$:

$$\tfrac{1}{3}V_{CC} = \tfrac{2}{3}V_{CC}\, e^{-T_2/R_2 C}$$
$$\tfrac{1}{2} = e^{-T_2/R_2 C} \tag{8.14}$$
$$T_2 = R_2 C \ln 2 \text{ seconds}$$

Thus, the total period of the oscillation is

$$T = T_1 + T_2 = (R_1 + R_2)C \ln 2 + R_2 C \ln 2$$
$$T = (R_1 + 2R_2)C \ln 2 \text{ seconds} \tag{8.15}$$

and the frequency is

$$f = \frac{1}{T} = \frac{1}{(R_1 + 2R_2)C \ln 2} \approx \frac{1.44}{(R_1 + 2R_2)C} \text{ hertz} \tag{8.16}$$

The duty cycle d is

$$d = \frac{T_1}{T_1 + T_2} = \frac{(R_1 + R_2)C \ln 2}{(R_1 + 2R_2)C \ln 2}$$
$$d = \frac{R_1 + R_2}{R_1 + 2R_2} \tag{8.17}$$

Note that the duty cycle must be greater than 0.5 because the capacitor always takes longer to charge than to discharge.

Example 8–5   An astable 555 circuit is required to produce a 20-kHz waveform with a 75% duty cycle. Find values for R_1, R_2, and C.

Solution. From Equation (8.17),

$$d = 0.75 = \frac{R_1 + R_2}{R_1 + 2R_2}$$

Letting $R_1 = 10$ kΩ, we find

$$0.75 = \frac{10 \text{ k}\Omega + R_2}{10 \text{ k}\Omega + 2R_2}$$

$$7.5 \text{ k}\Omega + 1.5R_2 = 10 \text{ k}\Omega + R_2$$

$$R_2 = 5 \text{ k}\Omega$$

From equation (8.16),

$$f = \frac{1}{(R_1 + 2R_2)C \ln 2} = \frac{1}{(10 \text{ k}\Omega + 10 \text{ k}\Omega)C \ln 2} = 20 \times 10^3 \text{ Hz}$$

$$C = \frac{1}{(20 \times 10^3 \text{ }\Omega)(20 \times 10^3 \text{ Hz}) \ln 2} = 3607 \text{ pF}$$

Drill 8–5

What will be the frequency and duty cycle of the output of the 555 timer in Example 8–5 if the resistor values of R_1 and R_2 are interchanged?
Answer: $f = 16$ kHz, $d = 60\%$.

If the anode of a diode is connected to pin 7 in Figure 8.19 and its cathode to pin 6, then the diode will be forward-biased when the capacitor is charging and will act as a closed switch that shorts out R_2. In that case, the time-constant during the charge cycle will be $R_1 C$ seconds. Since the diode is like an open switch when the capacitor is discharging, the capacitor still discharges through R_2 only, with time constant $R_2 C$ seconds. Thus, the capacitor charges through R_1 only and discharges through R_2 only. The charge and discharge times in that situation are

$$T_1 = R_1 C \ln 2 \text{ seconds} \tag{8.18}$$

$$T_2 = R_2 C \ln 2 \text{ seconds} \tag{8.19}$$

The frequency and duty cycle are then

$$f = \frac{1}{(R_1 + R_2)C \ln 2} \approx \frac{1.44}{(R_1 + R_2)C} \text{ hertz} \tag{8.20}$$

$$d = \frac{R_1}{R_1 + R_2} \tag{8.21}$$

We see that with the shorting diode connected, we can obtain duty cycles less than, equal to, or greater than 0.5.

Pulsewidth Modulation

Modulation means changing one characteristic of a voltage or current in response to changes that occur in another voltage or current. In pulsewidth modulation, the pulsewidths of a rectangular waveform change as the amplitude of another waveform (the modulating voltage) changes. When the modulating voltage is large, the pulsewidths are long, and as the modulating voltage decreases, the pulsewidths become narrower. Figure

Figure 8.21
An example of pulsewidth modulation.

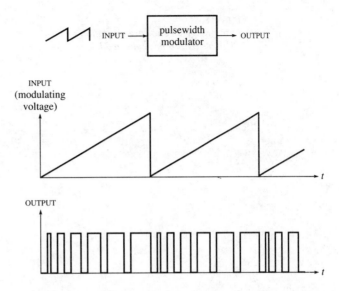

8.21 shows an example. The modulating voltage in this case is a sawtooth waveform. Note how the pulsewidths of the modulator output increase as the sawtooth increases.

Pulsewidth modulators are used in linear and digital systems, including so-called class-D digital amplifiers and digital data communications systems. The 555 timer can be operated as a pulsewidth modulator by applying the modulating voltage to the control input at pin 5. See Figure 8.17. As the modulating voltage changes, the voltage at the − input to the threshold comparator varies above and below $(\frac{2}{3})V_{CC}$. In this operation, the 555 is connected as a monostable and is continually triggered by a series of trigger pulses. When the modulating voltage increases, capacitor C must charge to a voltage *higher* than $(\frac{2}{3})V_{CC}$ before the threshold comparator can change state, which requires a longer time, so the output pulse is wider. Similarly, if the modulating voltage decreases, the capacitor charges to a smaller voltage and the output pulse is narrower. This behavior is illustrated in the timing diagram shown in Figure 8.22.

Note the following points in connection with operation of the 555 as a pulsewidth modulator:

1. There is a dc voltage $(\frac{2}{3}V_{CC})$ established by the voltage divider at the control input, so the modulating voltage should be connected to pin 5 through a *coupling capacitor* to block the flow of dc current.
2. The ac impedance at the control input is 5 kΩ ∥ 10 kΩ ≈ 3.3 kΩ, so the coupling capacitor, C_c, should be large enough to make $X_{C_c} = 1/(2\pi fC_c)$ small in comparison to 3.3 kΩ at the lowest frequency f of the modulating voltage.
3. The modulation process is *non-linear* in the sense that pulsewidths do not increase in direct, linear proportion to increases in the modulating voltage. This characteristic is due to the fact that the charging time of the capacitor is not directly proportional to the voltage to which it charges.
4. The pulse-repetition period of the trigger input must be greater than the width of the widest output pulse that the modulator is expected to produce.
5. In order for the trigger input to return low before the termination of any timing cycle, the trigger pulses must be narrower than the width of the smallest output pulse that the modulator is expected to produce.

Figure 8.22
Timing diagram showing operation of the 555 timer as a pulsewidth modulator.

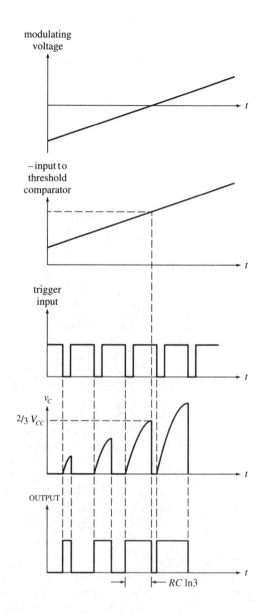

8-6   ANSI/IEEE STANDARD SYMBOLS

The qualifying symbol for a nonretriggerable monostable is a 1 followed by a single pulse: 1⎍. The qualifying symbol for a retriggerable monostable is a pulse: ⎍. Figure 8.23 shows the ANSI/IEEE logic symbols for the 74221 dual nonretriggerable monostable and the 74123 dual retriggerable monostable. Note that the trigger inputs are labeled 1A, 1B, 1$\overline{\text{CLR}}$ and 2A, 2B, 2$\overline{\text{CLR}}$ and that each set of inputs is connected to an AND block. A small triangle appears on the output side of each AND block. This symbol signifies that an output pulse is generated when there is a 0-to-1 transition at the output of the AND block. For example, if 1A (the active-low input) is low, 1$\overline{\text{CLR}}$ is high, and 1B goes from low to high, the output of the AND block goes from low to high and an output

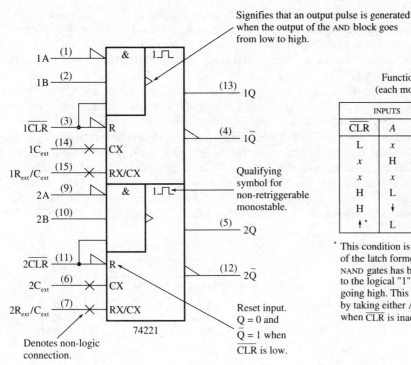

Signifies that an output pulse is generated when the output of the AND block goes from low to high.

Qualifying symbol for non-retriggerable monostable.

Reset input. Q = 0 and Q̄ = 1 when CLR is low.

Denotes non-logic connection.

(a) The 74221 dual non-retriggerable monostable.

Function Table
(each monostable)

INPUTS			OUTPUTS	
CLR	A	B	Q	Q̄
L	x	x	L	H
x	H	x	L	H
x	x	L	L	H
H	L	↑	⊓	⊔
H	↓	H	⊓	⊔
↑*	L	H	⊓	⊔

* This condition is true only if the output of the latch formed by the two NAND gates has been conditioned to the logical "1" state prior to CLR going high. This latch is conditioned by taking either A high or B low when CLR is inactive.

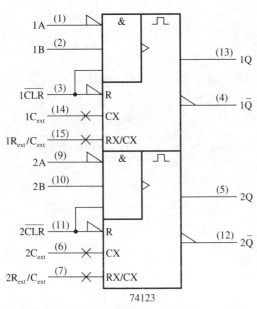

(b) The 74123 dual retriggerable monostable.

Function Table

INPUTS			OUTPUTS	
CLR	A	B	Q	Q̄
L	x	x	L	H
x	H	x	L †	H †
x	x	L	L †	H †
H	L	↑	⊓	⊔
H	↓	H	⊓	⊔
↑	L	H	⊓	⊔

† These lines of the function table assume that the indicated steady-state conditions at the A and B inputs have been set up long enough to complete any pulse started before the set up.

Figure 8.23
ANSI/IEEE symbols for the 74221 and 74123 monostables.

pulse is generated. See the function tables for other examples. The $\overline{\text{CLR}}$ input is also connected through an active-low symbol to the letter R. R stands for *reset,* which means that the Q output is held low, regardless of the states of A and B, when $\overline{\text{CLR}}$ is low. The CX and RX/CX characters inside the rectangle designate connections for external resistors and capacitors—i.e., the timing components for the monostable. The × drawn on the connection lines for these components means that the connections are not to logic variables. RX/CX is the connection point where the external resistor and external capacitor are joined, corresponding to pin 11 in Figure 8.3 and pin 15 in Figure 8.5.

Figure 8.24(*a*) shows the logic symbol for the 74122 retriggerable monostable. In this case, there is an OR block followed by an AND block. Active-low inputs A1 and A2 are ORed and the result is ANDed with B1, B2, and $\overline{\text{CLR}}$. Again, the triangle at the output of the AND block means that any condition creating a low-to-high transition at the output of the AND block will generate an output pulse. The equivalent logic diagram for the trigger input is shown using traditional logic symbols in Figure 8.24(*b*).

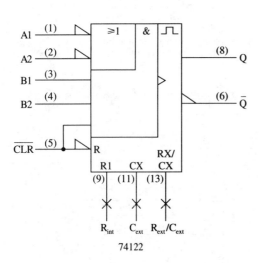

(a) ANSI/IEEE logic symbol.

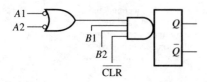

(b) Equivalent logic diagram for the trigger input drawn with traditional symbols.

Figure 8.24
The 74122 retriggerable monostable.

Function Table

	INPUTS				OUTPUTS	
$\overline{\text{CLR}}$	A1	A2	B1	B2	Q	$\overline{\text{Q}}$
L	x	x	x	x	L	H
x	H	H	x	x	L†	H†
x	x	x	L	x	L†	H†
x	x	x	x	L	L†	H†
H	L	x	↑	H	⊓	⊔
H	L	x	H	↑	⊓	⊔
H	x	L	↑	H	⊓	⊔
H	x	L	H	↑	⊓	⊔
H	H	↓	H	H	⊓	⊔
H	↓	↓	H	H	⊓	⊔
H	↓	H	H	H	⊓	⊔
↑	L	x	H	H	⊓	⊔
↑	x	L	H	H	⊓	⊔

† These lines of the function table assume that the indicated steady-state conditions at the A and B inputs have been set up long enough to complete any pulse started before the set up.

Figure 8.25
Example 8–6.

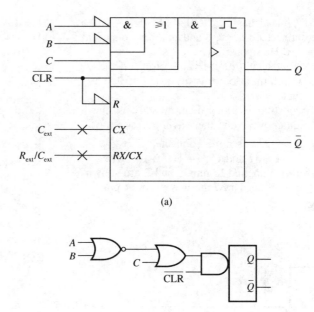

(a)

(b)

***Example
8–6***

Using traditional logic symbols, draw a logic diagram that is equivalent to the logic implemented by the circuit shown in Figure 8.25(a) for the trigger input on the monostable.

Solution.   The AND block with active-low inputs implements $\overline{A}\,\overline{B} = \overline{A + B}$. The output of the AND block is ORed with C to produce $\overline{A + B} + C$. Finally, $\overline{\text{CLR}}$ is ANDed with the output of the OR block to produce $(\overline{A + B} + C)\overline{\text{CLR}}$. The equivalent logic diagram is shown in Figure 8.25(b).

EXERCISES

8.1 In response to the leading edge of a trigger input, a monostable multivibrator produces a 0.75-ms pulse. If the trigger input is a square wave having frequency 10^3 PPS, what is the time interval between the trailing and leading edges of successive output pulses? Draw a timing diagram.

8.2 In response to the trailing edge of a trigger input, a monostable multivibrator produces a 20-μs pulse. If the trigger input is a square wave having frequency 12.5×10^3 PPS, what is the duty cycle of the monostable's output? Draw a timing diagram.

8.3 In response to the trailing edge of a trigger input, a retriggerable monostable multivibrator produces a 90-ns pulse. If the trigger input is the first five pulses of a square wave having frequency 12.5×10^6 PPS, what is the width of the output pulse? Draw a timing diagram.

8.4 A retriggerable monostable produces an output pulse having width 3 μs. How many pulses from a 500-kHz square wave should be applied to its trigger input to produce an output pulse having width 27 μs?

8.5 A 74121 has $R_{ext} = 22$ kΩ and $C_{ext} = 0.5$ μF. Pin 4 is connected to ground and pin 5 is connected to a 100-Hz square wave. Construct a timing diagram showing the square wave and the waveforms at pins 6 and 1.

8.6 A 74121 has $C_{ext} = 0.1$ μF. Pins 4 and 5 are connected to $+V_{CC}$ and pin 3 is connected to a 100-Hz square wave.
(a) What should be the value of R_{ext} if the waveform at pin 6 is to have a 40% duty cycle?
(b) Construct a timing diagram showing the square wave and the waveform at pin 6.

8.7 A 74123 has pin 3 connected to $+V_{CC}$. $R_{ext} = 7143$ Ω and $C_{ext} = 0.005$ μF. The waveforms applied to pins 1 and 2 are shown in Figure 8.26. Draw the waveform at pin 13.

pin 1

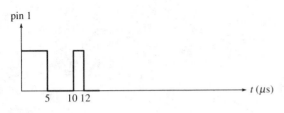

pin 2

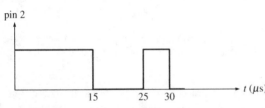

Figure 8.26
Exercise 8.7.

8.8 Repeat Exercise 8.7 when R_{ext} is changed to 14,286 Ω.

Section 8–2

8.9 A 74121 monostable is to be used to enable an AND gate so that 4096 pulses of a 2.5-MHz clock can be passed when the monostable is triggered. If the external timing capacitance is 0.1 μF, what value of external timing resistance should be used?

8.10 A 74121 monostable has $R_{ext} = 12$ kΩ and $C_{ext} = 0.5$ μF. Its $\overline{Q}$ output is connected to one input of a NOR gate whose other input is a 1202-Hz clock. The monostable is triggered on the leading edge of one of the clock pulses. Draw a timing diagram showing $\overline{Q}$, the clock, and the output of the NOR gate.

8.11 A 15-ms pulse is to be delayed 200 ms. Design a circuit using 74121 monostables that performs the delay.

8.12 A digital circuit is required to deliver the trailing edge of 50-μs pulse 0.25 ms after the occurrence of the leading edge of a 100-μs pulse. Design a circuit using 74121 monostables that performs this function.

8.13 Design a digital circuit that produces sequences of 1-ms synchronizing pulses on four different lines so that one and only one line is high at every instant of time. Use 74121 monostable multivibrators.

8.14 Design a digital circuit using 74121 monostables that produces the waveforms shown in Figure 8.27.

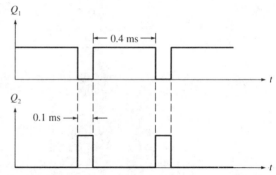

Figure 8.27
Exercise 8.14.

8.15 Design a digital circuit using the 74123 monostable that produces a high-to-low transition if and only if one pulse is lost from a pulsetrain having frequency 8×10^3 PPS.

8.16 Design a digital circuit using 74123 monostables that produces a low-to-high transition if and only if one pulse is lost from a 125-kHz pulsetrain and produces a high-to-low transition (on a different line) if and only if two consecutive pulses are lost.

Section 8–3

8.17 (a) Find the frequency of the square wave produced by the circuit shown in Figure 8.28.

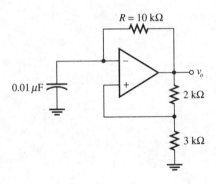

Figure 8.28
Exercise 8.17.

(b) To what value should feedback resistor R be changed if it is desired to double the frequency?

8.18 The astable multivibrator shown in Figure 8.11(*a*) has $C = 0.002$ μF, $R = 22$ kΩ, and $R_2 = 1$ kΩ. What value of R_1 should be used to obtain a frequency of 10 kHz?

8.19 Through what range of output frequencies can the astable multivibrator in Figure 8.12 be operated?

8.20 Design an astable multivibrator having an adjustable resistance that allows the period of the output to be adjusted from 0.5 ms to 5 ms.

Section 8–4

8.21 The period of the clock signal for a digital system is to be 0.4 μs. The clock is to be obtained by using a flip-flop to divide the output of a crystal-controlled oscillator in half. What should be the crystal frequency?

8.22 The outputs of two crystal-controlled clock generators are connected to the inputs of a 2-input AND gate. The output of the AND gate is required to be 100 pulses in a 2-μs time interval, zero pulses in the next 2-μs interval, followed by another 100 pulses in the next 2-μs interval, and so forth. What should be the crystal frequencies of the two oscillators?

Section 8–5

8.23 Design a monostable multivibrator using a 555 timer that produces a 1-s-wide output pulse.

8.24 The trigger input to a 555 timer operated as a monostable is shown in Figure 8.29. If the external timing components are $R = 4.7$ kΩ and $C = 0.1$ μF, what is the duty cycle of the output of the 555? Draw a timing diagram showing the trigger input and the 555 output.

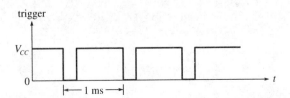

Figure 8.29
Exercise 8.24.

8.25 Design an astable multivibrator using a 555 timer that produces a 5-kHz waveform with an 80% duty cycle.

8.26 (a) Show that the frequency f and duty cycle d of the output of the astable multivibrator in Figure 8.19 are related by

$$f = \frac{d}{(R_1 + R_2)C \, \ln 2}$$

(b) If $R_1 = R_2 = R$, show that the duty cycle is 0.667 and that

$$f = \frac{1}{3RC \, \ln 2}$$

8.27 Design an astable multivibrator using a 555 timer that produces a rectangular waveform whose period can be adjusted from 0.5 ms through 10 ms. The adjustment should be performed by adjusting the value of R_1 in Figure 8.19. What range of duty cycles occur over the range through which the period is adjusted?

8.28 Design an astable multivibrator using a 555 timer that produces a 15-kHz rectangular output whose duty cycle is 20%. Draw the schematic diagram, showing all external components and their pin connections to the 555 timer.

8.29 A 555 timer operated as a pulsewidth modulator is required to produce output pulses that range in width from 1 ms to 5 ms.

(a) The pulse repetition frequency of the trigger input should be less than what value?

(b) At the frequency found in (a), what is the minimum duty cycle that the trigger input should have?

8.30 A 555 timer operated as a pulsewidth modulator has $R = 20$ kΩ and $C = 0.02$ μF. If $V_{CC} = 10$ V, what is the output pulsewidth when the $-$ input to the threshold comparator at the end of a timing cycle is 8 V?

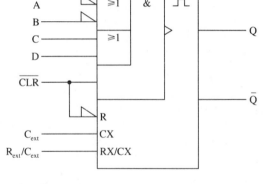

Figure 8.30
Exercise 8.31.

Section 8–6

8.31 Using traditional logic symbols, draw a logic diagram that is equivalent to the logic implemented by the circuit shown in Figure 8.30 for the trigger input of the monostable.

8.32 Draw the ANSI/IEEE standard symbol for a nonretriggerable monostable that generates an output pulse when a low-to-high transition occurs in the function $(A + B + \overline{CD})\overline{CLR}$.

CHALLENGING EXERCISES

8.33 Draw a timing diagram showing the trigger input and the system output in Figure 8.31. Be certain to show the time interval between the trigger input and the system output.

8.34 Derive equation (8.5) for the period of the astable multivibrator shown in Figure 8.11(a):

$$T = 2RC \ln\left(\frac{1 + \beta}{1 - \beta}\right)$$

where

$$\beta = \frac{R_1}{R_1 + R_2}$$

8.35 Find the duty cycle of the system output shown in Figure 8.32.

8.36 To what value should the 13.65-kΩ resistor in Figure 8.32 be changed if the duty cycle of the output is to be changed to 20%?

8.37 The voltage at the $-$ input of the threshold comparator of a 555 timer operated as a pulsewidth modulator is

$$v^- = v_{control} + (\tfrac{2}{3})V_{CC}$$

If $V_{CC} = 15$ V and the external timing components are $R = 22$ kΩ and $C = 0.015$ μF, what value of $v_{control}$ is necessary to produce an output pulsewidth of 0.33 ms? (Assume that v^- is the voltage at the end of a timing cycle.)

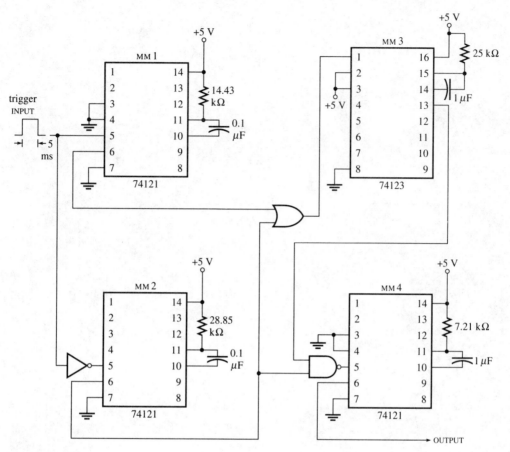

Figure 8.31
Exercise 8.33.

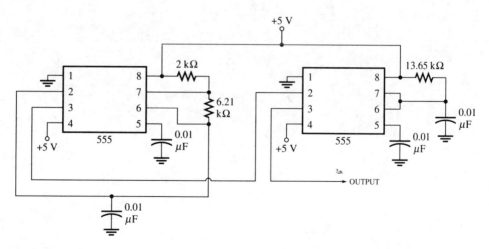

Figure 8.32
Exercise 8.35.

Chapter 9

LOGIC CIRCUIT FUNDAMENTALS

9–1 STATIC ANALYSIS OF LOGIC GATES

Static (or dc) analysis of an electronic logic gate means determining whether an output is low or high, based on a knowledge of the dc voltages and currents that result when dc inputs are applied. The static analysis of most gates can be performed by simply regarding the individual semiconductor devices in the gates as *voltage-controlled switches* and by knowing what voltages cause those switches to open and close. For reference purposes and as an aid in understanding the behavior of the logic families we study later, we present now a summary of the static characteristics of semiconductor devices, viewed as voltage-controlled switches.

Diodes

We have already discussed the behavior of a PN diode as a voltage-controlled switch (see Section 7–4 and Figure 7.20). Recall that making the anode positive *with respect to* the cathode forward-biases the diode and causes it to conduct like a closed switch. When the anode is negative with respect to the cathode, the diode is reverse-biased and behaves like an open switch. Figure 9.1 summarizes this behavior.

The key words in understanding the behavior of a diode as a voltage-controlled switch are *with respect to:* For the switch to close, the anode must be positive *with respect to* the cathode. Indeed, those are the key words in understanding the switching operation of all semiconductor devices. Always consider the *difference* in voltage *across* the terminals that control the device. For example, it may well be that the anode voltage is negative with respect to ground but that the diode is switched on because the cathode is more negative than the anode, with respect to ground. Remember that the voltage at the anode, A, with respect to the cathode, K, is written V_{AK} and that

$$V_{AK} = V_A - V_k \qquad (9.1)$$

See Figure 9.1(*c*). It is always true that

$$V_{AK} = -V_{KA} \quad \text{and} \quad V_{KA} = -V_{AK} \qquad (9.2)$$

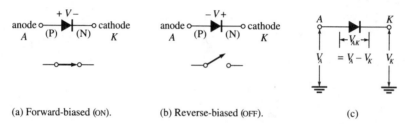

(a) Forward-biased (ON). (b) Reverse-biased (OFF). (c)

Figure 9.1
Behavior of a diode as a voltage-controlled switch.

Example 9–1

Determine which of the diodes in Figure 9.2 are forward-biased and which are reverse-biased. (A resistor is connected in series with each diode to limit current flow when each is forward-biased.)

Solution

a. The anode-to-cathode voltage (V_{AK}) is −5 V, or, equivalently, the cathode is 5 V positive with respect to the anode. Therefore, the diode is reverse-biased.

b. The anode side of the circuit is at +10 V and the cathode side is at +5 V. The anode is therefore positive with respect to the cathode, and the diode is forward-biased. (The actual voltage across the forward-biased diode itself is about 0.7 V; 4.3 V is dropped across the resistor.)

c. The cathode side of the circuit is 5 V more negative than the anode side; equivalently, the anode side is 5 V more positive. The diode is forward-biased. $V_{AK} \approx 0.7$ V.

d. D_1 and D_3 are forward-biased; D_2 is reverse-biased. The anode-to-cathode voltage of D_2 is $V_{AK} = V_A - V_K = -5$ V − (+5 V) = −10 V.

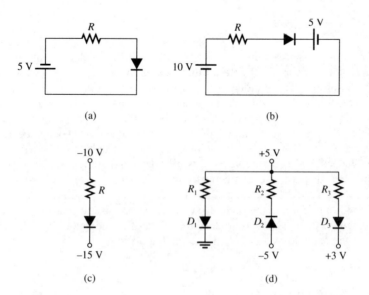

Figure 9.2
Example 9–1.

Drill 9–1

What are the voltage drops

a. Across R in Figure 9.2(c)?
b. Across R_1 in Figure 9.2(d)?
c. Across R_3 in Figure 9.2(d)?

Answer: (a) 4.3 V; (b) 4.3 V; (c) 1.3 V.

Bipolar Junction Transistors (BJTs)

The switch in a bipolar junction transistor is between the collector and emitter terminals. The base-to-emitter voltage, $V_{BE} = V_B - V_E$, controls the switch. In an NPN transistor, the switch closes and current flows from collector to emitter when the base is made positive with respect to the emitter. When the base is made zero or negative with respect to the emitter, the switch opens. See Figure 9.3(a). In a PNP transistor, the switch closes and current flows from emitter to collector when the base is made negative with respect to the emitter. The switch opens when the base is made zero or positive with respect to the emitter. See Figure 9.3(b). One way to remember these facts is to remember that the base-to-emitter circuit is a PN junction, like a diode. In both types of transistors, the switch closes when that PN junction is forward-biased by about 0.7 V (*P* positive, *N* negative) and opens when it is reverse-biased:

(C) (B) (E)
N P N
  + –   ⟹   base-to-emitter forward-biased   (switch ON)
  – +   ⟹   base-to-emitter reverse-biased   (switch OFF)

(C) (B) (E)
P N P
  + –   ⟹   base-to-emitter reverse-biased   (switch OFF)
  – +   ⟹   base-to-emitter forward-biased   (switch ON)

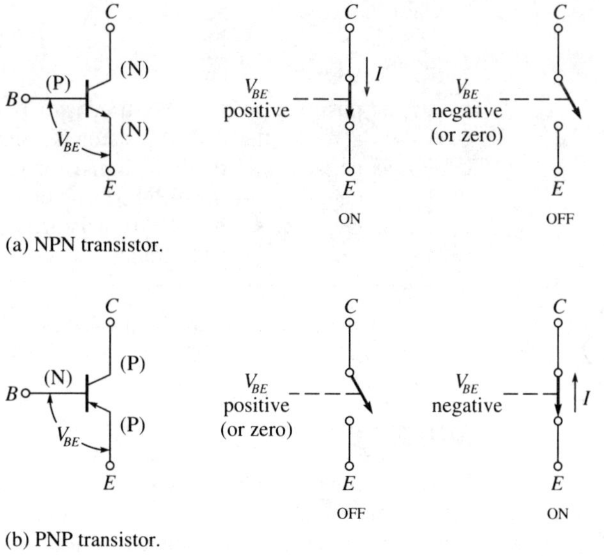

Figure 9.3
Behavior of BJTs as voltage-controlled switches.

(a) NPN transistor.

(b) PNP transistor.

Figure 9.4
Example 9-2.

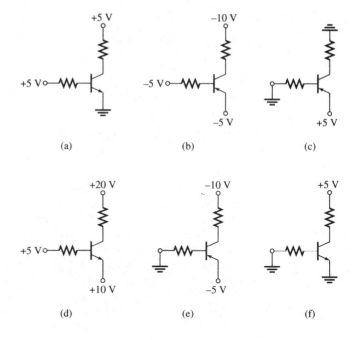

(a)   (b)   (c)

(d)   (e)   (f)

Successful switching of BJTs actually depends on the resistor values used in the collector and/or base circuits. These, in turn, depend on individual transistor characteristics, which are discussed in a later section. When a BJT is switched ON, it is said to be in *saturation,* and when it is switched OFF, it is said to be *cutoff.* If the BJT were an ideal switch, its collector-to-emitter voltage would be zero volts when saturated. In reality that value, called $V_{CE(sat)}$, is typically about 100 mV.

Example 9-2

Determine which of the transistors in Figure 9.4 are ON and which are OFF. Assume resistor values are such that a transistor is in saturation (ON) if its base-to-emitter junction is forward-biased.

Solution

a. V_{BE} is positive, so the NPN transistor is ON.
b. V_{BE} is zero, so the PNP transistor is OFF.
c. V_{BE} is negative, so the PNP transistor is ON.
d. V_{BE} is negative, so the NPN transistor is OFF.
e. V_{BE} is positive, so the PNP transistor is OFF.
f. V_{BE} is zero, so the NPN transistor is OFF.

Drill 9-2

Assuming the transistors in Figure 9.4 are all silicon, find each value of V_{BE}.
Answer: (a) 0.7 V; (b) 0 V; (c) −0.7 V; (d) −5 V; (e) 5 V; (f) 0 V.

MOSFETs

MOSFET (metal-oxide-semiconductor field-effect transistors) devices are widely used in digital integrated circuits. The most common types are *enhancement mode* transistors. These are completely analogous to NPN and PNP transistors in terms of their static

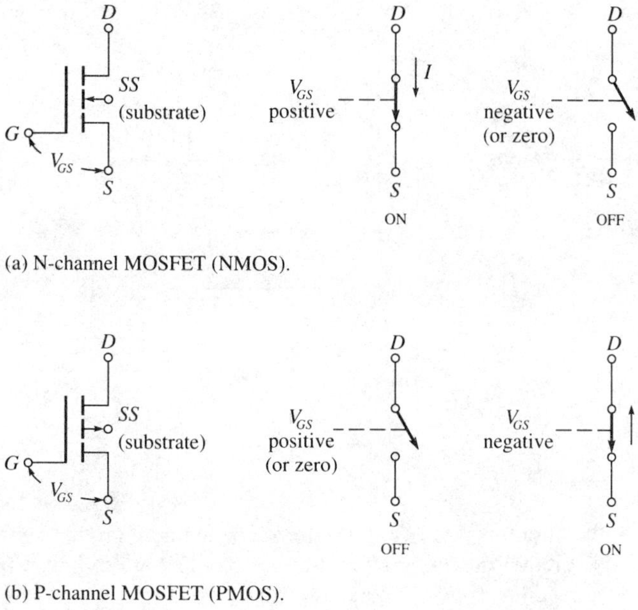

(a) N-channel MOSFET (NMOS).

(b) P-channel MOSFET (PMOS).

Figure 9.5
Behavior of MOSFETs as voltage-controlled switches.

behavior. Only the terminal names are different. Hereafter, we assume enhancement mode devices.

The switch in a MOSFET is between the drain and source terminals. The gate-to-source voltage, V_{GS}, controls the switch. In an N-channel MOSFET, called an NMOS transistor, the switch closes and current flows from drain to source when the gate is positive with respect to the source. When the gate is zero or negative with respect to the source, the switch opens. See Figure 9.5(*a*). In a P-channel MOSFET (PMOS transistor), the switch closes and current flows from source to drain when the gate is negative with respect to the source. The switch opens when the gate is zero or positive with respect to the source. See Figure 9.5(*b*).

To switch an NMOS transistor ON, it is not strictly sufficient to make V_{GS} positive. In reality, V_{GS} must be more positive than a *threshold voltage*, V_T, typically on the order of 2 V. Similarly, in a PMOS transistor, V_{GS} must be more negative than a threshold voltage of about -2 V. In practice, the voltages used to switch MOSFET transistors ON are usually much greater than threshold values.

9–2   DIODE LOGIC

Although diode logic is not widely used for performing complex logic operations, a thorough understanding of its principles will greatly aid the understanding of other logic families that incorporate PN junctions. Of particular importance is the way that PN junctions in the same circuit interact with each other, that is, how the biasing of one junction affects the others.

Figure 9.6(*a*) shows a 2-input diode AND gate. The two inputs, *A* and *B*, and the output have the same ground reference (common). To facilitate analysis and improve

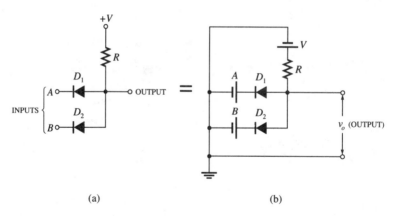

Figure 9.6
The diode AND gate

comprehension, the circuit is redrawn in Figure 9.6(b) with all ground connections shown explicitly. The dc voltage sources shown as inputs may be high or low, where low can be 0 V (ground) or a negative voltage. V is a dc supply voltage equal to or greater than the high level of the inputs.

Analysis of the diode AND gate is best understood using an example. Let us suppose that the logic levels are $0 = 0$ V (low) and $1 = +5$ V (high) and that $V = +10$ V. Figure 9.7(a) shows the circuit when $A = B = 0$. The 10-V supply forward-biases both diodes, so both act as closed switches, as shown in Figure 9.7(b). It is clear that the output voltage is 0 V, or logic 0. (In fact, the output is the forward drop across the parallel diodes: $V_o \approx 0.7$ V.) Figure 9.7(c) shows the circuit when $A = 0$ V and $B = +5$ V. At first, it might seem that both diodes are forward-biased, since the anode side of each is more positive than its cathode side. In reality, only D_1 is forward-biased. It therefore acts like a closed switch that connects 0 V to the output *and to the anode side of D_2*. See Figure 9.7(d). Thus, D_2 is *reverse*-biased. If D_1 and D_2 were both forward-biased, we would have closed switches connecting both 5 V and 0 V to the output, a clear contradiction. One way to remember which diode is truly forward-biased is to determine which has the greatest forward-biasing voltage from its anode side to its cathode side. In this example, that voltage is 10 V for D_1, whereas it is only $10 V - 5 V = 5$ V for D_2. D_1 is therefore forward-biased. When $A = +5$ V and $B = 0$ V, D_2 is forward-biased and D_1 is reverse-biased. The output is again 0. Finally, when A and B are both $+5$ V, both diodes are forward-biased and both connect the output to $+5$ V, as shown in Figures 9.7(e) and (f). The *voltage* truth table is shown in Figure 9.7(g) and the corresponding logic truth table in 9.7(h). The truth table confirms that the circuit implements the logical AND operation.

It is easy to see how the 2-input AND gate can be expanded to become a 3- or more input AND gate: Simply connect additional inputs through diodes configured like those shown in Figure 9.6. Once again, the key to analyzing diode logic is to remember that whichever diode(s) are "most" forward-biased are truly forward-biased, regardless of the number of inputs or the voltage levels corresponding to 0 and 1. For example, in a 3-input AND gate having $V = +5$ V, $A = 0$ V, $B = -5$ V, and $C = 0$ V, the diode connected to the B input is most forward-biased [by $5 V - (-5 V) = 10$ V] so it is truly forward-biased, and the other diodes are reverse-biased.

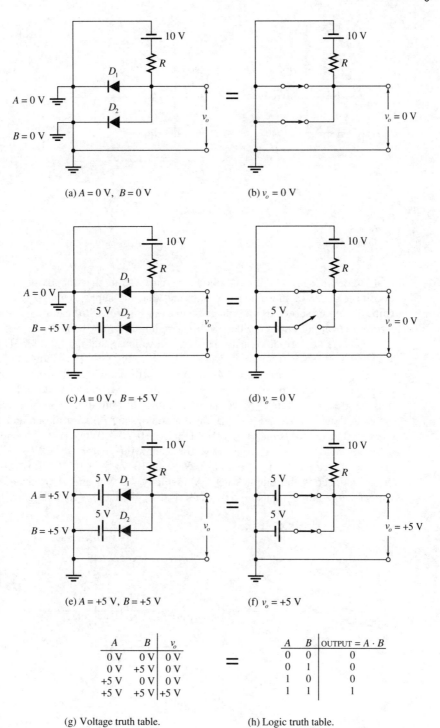

(a) $A = 0$ V, $B = 0$ V

(b) $v_o = 0$ V

(c) $A = 0$ V, $B = +5$ V

(d) $v_o = 0$ V

(e) $A = +5$ V, $B = +5$ V

(f) $v_o = +5$ V

A	B	v_o
0 V	0 V	0 V
0 V	+5 V	0 V
+5 V	0 V	0 V
+5 V	+5 V	+5 V

(g) Voltage truth table.

A	B	OUTPUT $= A \cdot B$
0	0	0
0	1	0
1	0	0
1	1	1

(h) Logic truth table.

Figure 9.7
Analysis of the diode AND *gate.*

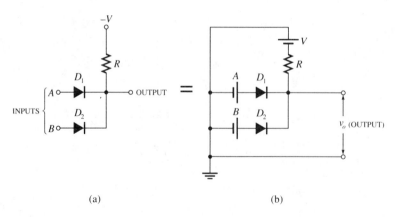

(a) (b)

Figure 9.8
The diode OR *gate.*

Figure 9.8 shows a 2-input diode OR gate. Note that the diode orientations are opposite to those of the diode AND gate and that the supply voltage is negative. Once again, the inputs shown as dc voltage sources in Figure 9.8(b) can be positive, negative, or zero, depending on the voltages corresponding to logic 0 and 1.

To analyze the diode OR gate, let us again suppose that the logic levels are $0 = 0$ V (low) and $1 = +5$ V (high) and that $V = -10$ V. Figure 9.9(a) shows the circuit when $A = B = 0$. In this example, we dispense with showing all ground connections to encourage development of the important ability to visualize voltage levels having a common ground reference. Since the anode sides of both diodes are positive (0 V) with respect to their cathode sides (-10 V), both diodes are forward-biased and act as closed switches. The output is therefore 0 V (actually -0.7 V). When $A = 0$ V and $B = +5$ V, as shown in Figure 9.9(c), the forward-biasing voltage for D_2 is 5 V $- (-10$ V$) = +15$ V, whereas that for D_1 is only 0 V $- (-10$ V$) = +10$ V. Thus, D_2 is forward-biased, connecting the output to $+5$ V (actually 4.3 V), and D_1 is reverse-biased. By symmetry, the output is also $+5$ V when $A = +5$ V and $B = 0$ V. Figure 9.9(e) and (f)

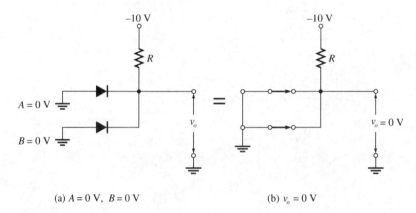

(a) $A = 0$ V, $B = 0$ V (b) $v_o = 0$ V

Figure 9.9
Analysis of the diode OR *gate. (Continued on 289.)*

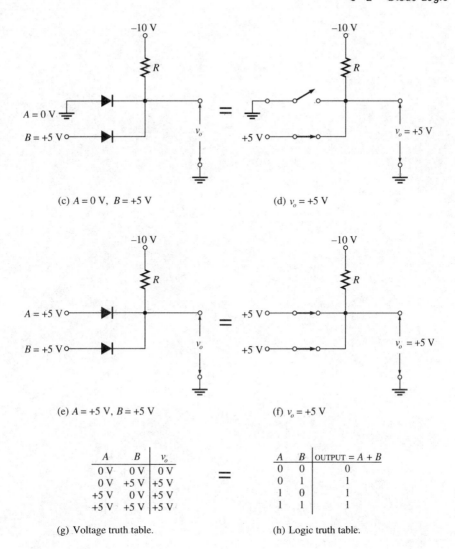

(c) $A = 0$ V, $B = +5$ V

(d) $v_o = +5$ V

(e) $A = +5$ V, $B = +5$ V

(f) $v_o = +5$ V

A	B	v_o
0 V	0 V	0 V
0 V	+5 V	+5 V
+5 V	0 V	+5 V
+5 V	+5 V	+5 V

A	B	OUTPUT = A + B
0	0	0
0	1	1
1	0	1
1	1	1

(g) Voltage truth table.

(h) Logic truth table.

Figure 9.9
(Continued)

show that both diodes are forward-biased when $A = B = +5$ V. The truth tables in Figure 9.9(g) and (h) confirm that the circuit implements the logical OR operation.

Example 9-3
Construct voltage and logic truth tables for the circuits shown in Figure 9.10(a) and (b). Determine the logic operation performed by each. (Note the logic levels defined in the figure.) Assume ideal diodes; that is, neglect the voltage drop across each diode when it is forward-biased.

Solution

a. When $A = B = C = -5$ V, all three diodes are forward-biased, because their anode sides are each 5 V positive with respect to their cathode sides $[-5$ V $- (-10$ V$) =$

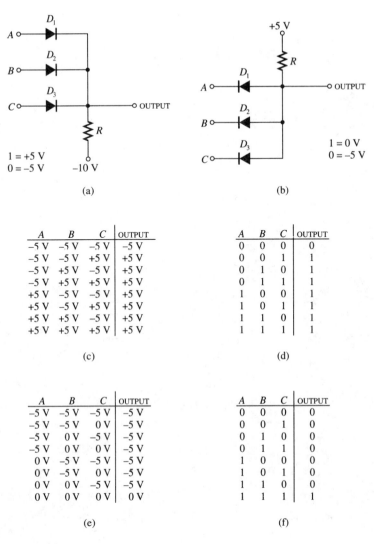

Figure 9.10
Example 9–3.

+5 V]. The output is therefore −5 V. If $A = B = −5$ V and $C = +5$ V, then D_3 is forward-biased because the total forward-biasing voltage across it is +5 V − (−10 V) = +15 V, whereas the forward-biasing voltage across D_1 and D_2 is only −5 V − (−10 V) = +5 V. D_3 therefore acts as a closed switch that connects the output to +5 V. Since +5 V then appears at the cathodes of D_1 and D_2, they are each reverse-biased. Similarly, if *any* one or two inputs are +5 V, the corresponding diodes are forward-biased, the output is +5 V, and the other diode(s) are reverse-biased. If $A = B = C = +5$ V, all three diodes are forward-biased, and the output is again +5 V. The voltage truth table and corresponding logic truth table are shown in Figure 9.10(c) and (d). We see that the circuit performs the logical OR operation.

b. When $A = B = C = -5$ V, all diodes are forward-biased, and the output is -5 V. If $A = B = -5$ V and $C = 0$ V, then D_1 and D_2 are forward-biased because the forward-biasing voltage across each is 5 V $-$ $(-5$ V$) = 10$ V, whereas the forward-biasing voltage across D_3 is only 5 V $- 0$ V $= 5$ V. Similarly, if *any* one or two inputs are at -5 V, the corresponding diodes are forward-biased, the output is -5 V, and the other diode(s) are reverse-biased. If $A = B = C = 0$ V, then all three diodes are forward-biased and the output is 0 V. Figure 9.10*(e)* and *(f)* show the voltage and logic truth tables. We see that the circuit performs the logical AND operation.

Drill 9–3

Assume that all diodes in Figure 9.10 are silicon.

a. Find the actual output voltage in Figure 9.10*(a)* when $ABC = 000$ and when $ABC = 001$.

b. Find the actual output voltage in Figure 9.10*(b)* when $ABC = 000$ and when $ABC = 111$.

Answer: (a) -5.7 V; $+4.3$ V; (b) -4.3 V; 0.7 V

9–3   BJT INVERTERS

Figure 9.11*(a)* shows how a logical *inverter* is constructed with an NPN transistor. When the input to the base circuit is high (typically V_{CC}), the base-to-emitter junction is forward-biased, and, as discussed in Section 9–1, the collector-to-emitter switch closes. As shown in Figure 9.11*(b)*, the output is then 0 V, or low. When the input is low, the switch is open, and since there is no voltage drop across collector resistor R_C, the output is V_{CC} (high). There are two points to remember:

1. If there is a load R_L connected to the output, then the high output voltage is reduced by voltage division, as discussed in Section 6–1.
2. The low output voltage is actually $V_{CE(\text{sat})}$, on the order of 100 mV.

Although NPN inverters are found in some integrated circuits, their most common application is in discrete form, where heavy currents and/or large power dissipations are involved. An example is an inverter used as a *driver* for a light-emitting diode (LED).

In order for the output to be zero when the input is high, resistors R_B and R_C must be selected so that V_{CC} volts are dropped across R_C. In other words, the current through R_C when the input is high must be the *saturation* value

$$I_{C(\text{sat})} = \frac{V_{CC}}{R_C} \tag{9.3}$$

See Figure 9.11*(b)*. The base current when the input is high is

$$I_{B(\text{ON})} = \frac{V_{\text{HI}} - V_{BE}}{R_B} = \frac{V_{CC} - 0.7 \text{ V}}{R_B} \tag{9.4}$$

The current gain of the transistor is

$$\beta = \frac{I_C}{I_B} \tag{9.5}$$

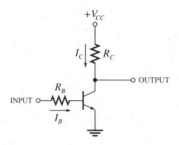

(a) The BJT inverter.

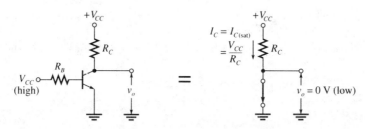

(b) The output is low when the input is high.

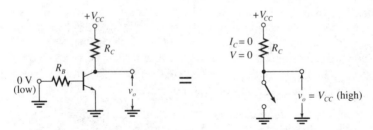

(c) The output is high when the input is low.

INPUT	OUTPUT
0 V	V_{CC}
V_{CC}	0 V

(d) Voltage truth table.

INPUT	OUTPUT
0	1
1	0

(e) Logic truth table.

Figure 9.11
Analysis of a BJT inverter.

Thus, we require

$$\beta I_{B(\text{ON})} = I_{C(\text{sat})}$$

or

$$\beta \left(\frac{V_{CC} - 0.7}{R_B} \right) = \frac{V_{CC}}{R_C} \tag{9.6}$$

Simplifying, we obtain the following relationship between β, R_B, and R_C:

$$\frac{R_B}{R_C} = \beta\left(1 - \frac{0.7}{V_{CC}}\right) \qquad (9.7)$$

If $V_{CC} \gg 0.7$ V, a good approximation is

$$\frac{R_B}{R_C} \approx \beta \qquad (9.8)$$

Equation (9.8) shows that R_C must be approximately R_B/β. Since the value of β depends heavily on temperature and also varies among transistors (even among those of the same type), it is good practice to make R_C *larger* than R_B/β (or R_B smaller than βR_C). This practice ensures that the transistor will saturate when β is smaller than anticipated. If the minimum possible value of β, β_{min}, is known, then we should have

$$R_C \geq \frac{R_B}{\beta_{min}} \qquad (9.9)$$

The problem with making R_C too large or R_B too small is that it *overdrives* the transistor. Because of inherent transistor properties, overdrive increases the delay in switching it from ON to OFF.

Example 9-4

The range of β values for a certain NPN transistor is 50–200. It is to be used in inverter circuits with $V_{CC} = +5$ V. The logic levels are $1 = +5$ V and $0 = 0$ V.

a. Find R_B if R_C is to be 1 kΩ.
b. Find the base current when the input is high. Assume $V_{BE} = 0.7$ V.
c. Find the output voltage when the input is low, if a load of 2 kΩ is connected to the output.

Solution

a. From equation (9.9), $R_B \leq \beta_{min}R_C = 50(1\ \text{k}\Omega) = 50$ kΩ. A good practical value for R_B is 47 kΩ.
b. From equation (9.4),

$$I_{B(ON)} = \frac{5\ \text{V} - 0.7\ \text{V}}{47\ \text{k}\Omega} = 91.5\ \mu\text{A}$$

c. By voltage division,

$$v_o = \left(\frac{R_L}{R_C + R_L}\right)V_{CC} = \left(\frac{2\ \text{k}\Omega}{1\ \text{k}\Omega + 2\ \text{k}\Omega}\right)5\ \text{V} = 3.33\ \text{V}$$

Drill 9-4

The output of the inverter in Example 9–4 (inverter 1) is connected to the inputs of two inverters identical to itself. Assuming that the base-to-emitter voltage of each of the two load inverters is 0.7 V, find the output voltage of inverter 1 when its input is low. *Answer:* 4.824 V.

9–4 NMOS INVERTERS

Most modern MOSFET circuitry is constructed using NMOS devices because they operate at about three times the speed of their PMOS counterparts. (However, PMOS devices are found in CMOS logic, which is discussed in the next section.) The underlying principle of an NMOS inverter is identical to that of a BJT inverter: It is a voltage-controlled switch that (1) closes and makes the output low when the input is high and (2) opens and makes the output high when the input is low. Figure 9.12 shows an NMOS inverter having resistance R_D in its drain circuit. This is the resistance across which the drain supply voltage, V_{DD}, must be dropped to make the output low when the inverter is on:

$$I_{D(\text{sat})} = \frac{V_{DD}}{R_D} \tag{9.10}$$

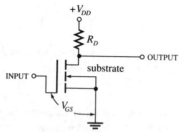

(a) The NMOS inverter.

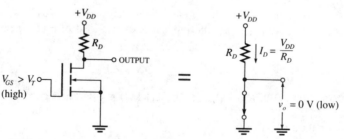

(b) The output is low when the input is high.

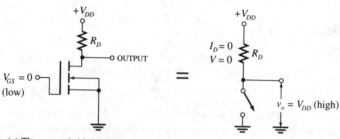

(c) The output is high when the input is low.

Figure 9.12
Analysis of an NMOS inverter having a fixed-load resistor R_D.

As we shall presently discover, NMOS *integrated circuits* do not have a fixed drain resistance. In any case, we can still regard the NMOS inverter as a voltage-controlled switch, where V_{GS} is the controlling voltage. Recall that the transistor conducts when V_{GS} is made more positive than a threshold voltage, V_T, of about 2 V.

NMOS transistors are easier to construct in integrated circuits than are resistors. For that reason, NMOS transistors are used to serve the same purpose as resistors. Figure 9.13(a) shows an NMOS transistor connected for use as a resistor. Note that the gate is connected directly to the drain, so $V_{GS} = V_{DS}$ at all times. Figure 9.13(b) is a typical set of *drain characteristics* showing how the drain current, I_D, of an NMOS transistor varies with V_{DS}, for different values of V_{GS}. Superimposed on the characteristics is a plot of the line $V_{GS} = V_{DS}$, the special condition that holds when the device is operated as a resistor. For any value of voltage V_{DS} across the "resistor," the corresponding current through it

Figure 9.13
The NMOS transistor as a resistor.

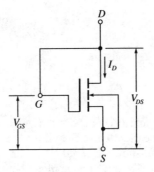

(a) An NMOS transistor connected as a resistor. Note that $V_{GS} = V_{DS}$.

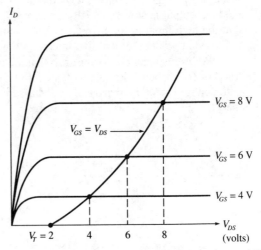

(b) A plot of $V_{GS} = V_{DS}$ on a set of NMOS drain characteristics. The plot shows how the current, I_D, through the resistor-connected NMOS varies with the voltage, V_{DS}, across it. The variation is clearly nonlinear.

is the value found on the plot of $V_{GS} = V_{DS}$. We see that the voltage-versus-current plot is *non-linear*, unlike that of a conventional, fixed resistor. In other words, the value of resistance presented by a resistor-connected NMOS device *depends* on the current through it.

The drain current I_D in an NMOS transistor depends on its gate-to-source voltage V_{GS} according to

$$I_D = 0.5\beta(V_{GS} - V_T)^2 \tag{9.11}$$

where β is a device parameter related to the physical dimensions of the transistor. The larger the value of β, the smaller the resistance of the region between drain and source. The units of β are A/V^2. Equation (9.11) is applicable only when I_D is in the "flat" region of the characteristic curves. The ac, or *dynamic*, resistance of a resistor-connected NMOS transistor in that region can be found from

$$r_{ac} = \frac{\Delta V}{\Delta I} = \frac{1}{\sqrt{2\beta I}} \text{ ohms} \qquad (V > V_T) \tag{9.12}$$

and its dc, or *static*, resistance is

$$R_{dc} = \frac{V}{I} = \sqrt{\frac{2}{\beta I}} + \frac{V_T}{I} \text{ ohms} \qquad (V > V_T) \tag{9.13}$$

Equations (9.12) and (9.13) show the dependence of resistance on current and clearly reveal the non-linear nature of the NMOS resistor.

Figure 9.14(b) shows the *load line* of an NMOS transistor having fixed resistance R_D in its drain circuit. The load line is plotted on a set of drain characteristics and represents all possible combinations of drain current and drain-to-source voltage that the transistor undergoes when switching between ON and OFF. Note that I_D is large and V_{DS} is small (near 0) when the device is ON and I_D is small and V_{DS} large (V_{DD}) when the device is OFF. Figure 9.14(c) shows an NMOS inverter having a resistor-connected NMOS transistor in place of R_D. Figure 9.14(d) shows that the load line in this case is non-linear. Note that the high output is now $V_{DD} - V_{T2}$, where V_{T2} is the threshold voltage of Q_2. We assume that the high input to the inverter is from another identical inverter, so the output is low at the point where the load line intersects the curve $V_{GS} = V_{DD} - V_{T2}$.

NMOS inverters are designed so that the β of Q_2 in Figure 9.14(c) is about one-tenth the β of Q_1. As a result, the resistance of Q_2 is about ten times the resistance of Q_1 when conducting. Under the reasonably good assumption that $V_{T1} = V_{T2} = V_T$, it can be shown that

$$V_{LO} = \frac{-B - \sqrt{B^2 - 4AC}}{2A} \tag{9.14}$$

where

$$A = 1 + \frac{\beta_1}{\beta_2}$$

$$B = -2\left[(V_{DD} - V_T) + \frac{\beta_1}{\beta_2}(V_{DD} - 2V_T)\right]$$

$$C = (V_{DD} - V_T)^2$$

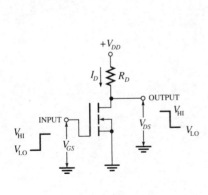

(a) NMOS inverter with fixed load R_D.

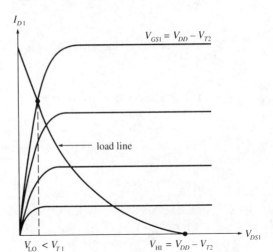

(b) Load line for (a).

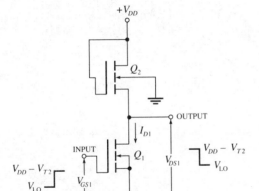

(c) NMOS inverter with NMOS load.

(d) Load line for (c).

Figure 9.14
Load lines of NMOS inverters showing low and high output voltages.

and β_1, β_2 are the β-values of Q_1 and Q_2. This equation is derived* under the assumption that $V_{HI} = V_{DD} - V_T$.

***Example
9–5***   The NMOS inverter in Figure 9.14(c) has $\beta_1 = 0.25 \times 10^{-3}$ A/V^2, $\beta_2 = 0.025 \times 10^{-3}$ A/V^2, $V_{T1} = V_{T2} = 2$ V, and $V_{DD} = 10$ V.

---

*See pages 314–315 in Bogart, *Electronic Devices and Circuits*, 2d ed., Merrill Publishing Company, 1990.

a. Find V_{LO} when the input is $V_{DD} - V_T$.
b. Find V_{HI} when the input is V_{LO}.

Solution

a. With reference to equation (9.14),

$$A = 1 + \frac{\beta_1}{\beta_2} = 1 + \frac{0.25 \times 10^{-3}}{0.025 \times 10^{-3}} = 11$$

$$B = -2\left[(V_{DD} - V_T) + \frac{\beta_1}{\beta_2}(V_{DD} - 2V_T)\right]$$

$$= -2[(10 - 2) + 10(10 - 4)] = -136$$

$$C = (V_{DD} - V_T)^2 = (10 - 2)^2 = 64$$

Then

$$V_{LO} = \frac{-B - \sqrt{B^2 - 4AC}}{2A} = \frac{136 - \sqrt{(136)^2 - 4(11)(64)}}{2(11)} = 0.49 \text{ V}$$

Note that V_{LO} is less than V_T, a necessary condition for successful operation.
b. Since $V_{LO} < V_T$, Q_1 is off when the input is V_{LO}, and $V_{HI} = V_{DD} - V_T = 10 \text{ V} - 2 \text{ V} = 8 \text{ V}$.

Drill
9–5

If both NMOS devices in Example 9–5 had the same value of β, would the value of V_{LO} be acceptable? Explain.
Answer: No; $V_{LO} > V_T$.

9–5 CMOS INVERTERS

Recall from Chapter 6 that shunt capacitance limits the speed at which logic circuits can be switched on and off. We saw that a small RC time constant is desirable and that one way to minimize it is to reduce the effective source resistance through which the capacitance charges and discharges. Figure 9.15 shows that the time constants governing the charge and discharge cycles of an NMOS inverter are quite different. When Q_1 is turned off, the capacitance must charge through the large resistance of Q_2, but when Q_1 turns on, the capacitance discharges through the small resistance of Q_1. As a consequence, there is a significant time delay in switching the output from low to high.

It is clear from Figure 9.15 that it would be desirable if a switching circuit could both discharge *and* charge its load capacitance through a small resistance. CMOS (*complementary* metal-oxide-semiconductor) circuitry was developed to meet that need. CMOS logic contains both NMOS and PMOS devices and provides low-resistance paths through which capacitance can both charge and discharge. It is widely used for general-purpose logic circuitry: circuitry containing individually accessible gates that can be connected externally to implement various logic operations. An example mentioned in Chapter 8 is the 74HC00 series that mimics 7400-series TTL logic. Such general-purpose logic circuitry must be capable of driving heavy capacitive loads associated with external wiring. Another characteristic of CMOS logic is its extremely low power consumption, a feature that makes it useful for applications such as watches and calculators.

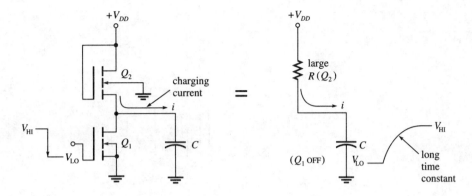

(a) The time required for the output to switch from low to high is long, because the time constant is large when the capacitance charges through Q_2.

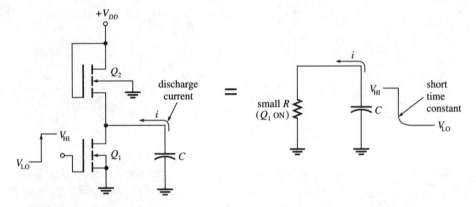

(b) The time required for the output to switch from high to low is short, because the time constant is small when the capacitance discharges through Q_1.

Figure 9.15
Comparison of switching times in an NMOS inverter.

Figure 9.16(a) shows a CMOS inverter, consisting of NMOS transistor Q_1 and PMOS transistor Q_2. Note the following points:

1. The input is connected to the gate of both devices; the output is the drain of both devices.
2. The positive supply voltage is connected to the *source* of PMOS transistor Q_2.

When the input is low (0 V), the *gate-to-source* voltage of Q_2 is *negative*: $V_{GS2} = -V_{DD}$, whereas $V_{GS1} = 0$ V. Thus, Q_2 is on and Q_1 is off, as illustrated in Figure 9.16(b). When the input is high ($+V_{DD}$), V_{GS2} is zero, which is more positive than its (negative) threshold voltage, so Q_2 is off. (Recall that the gate-to-source voltage of a PMOS transistor must be more negative than its threshold for it to turn on.) However, $V_{GS1} = +V_{DD}$, so Q_1 is on. See Figure 9.16(c).

Figure 9.16 shows that Q_2 is on when the output of the CMOS inverter switches high and that Q_1 is on when the output switches low. Thus, the load capacitance charges

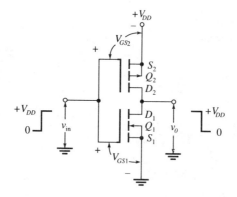

(a) The CMOS inverter.

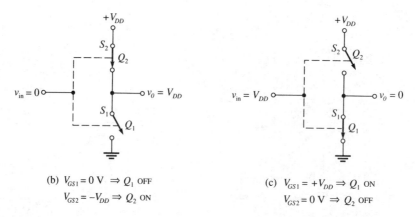

(b) $V_{GS1} = 0$ V $\Rightarrow Q_1$ OFF
$V_{GS2} = -V_{DD} \Rightarrow Q_2$ ON

(c) $V_{GS1} = +V_{DD} \Rightarrow Q_1$ ON
$V_{GS2} = 0$ V $\Rightarrow Q_2$ OFF

Figure 9.16
The CMOS inverter and its representation as voltage-controlled switches.

through the small resistance of Q_2 and discharges through the small resistance of Q_1. The charge and discharge times are both small because both are governed by the small resistance of an ON transistor.

9–6   POSITIVE AND NEGATIVE LOGIC

In previous examples and discussions, we have seen that *logic levels* were defined so that the high, or most positive, voltage corresponded to logical 1 and that the low, or least positive, voltage corresponded to logical 0. This assignment of logic levels is called *positive* logic. *Negative* logic is the assignment of logical 0 to the most positive voltage and 1 to the least positive.

Figure 9.17(*a*) shows the *voltage* truth table of one logic gate. Figure 9.17(*b*) and (*c*) show the corresponding logical truth tables using positive and negative logic, respectively. We see that a positive-logic AND gate is equivalent to a negative-logic OR gate. Note that both the inputs *and* outputs in the voltage truth table are "translated" to negative

Figure 9.17
Positive-logic and negative-logic truth tables derived from a voltage truth table.

A	B	OUTPUT
0 V	0 V	0 V
0 V	+5 V	0 V
+5 V	0 V	0 V
+5 V	+5 V	+5 V

(a) Voltage truth table.

A	B	OUTPUT
0	0	0
0	1	0
1	0	0
1	1	1

(b) Positive-logic truth table corresponding to (a). +5 V = 1, 0 V = 0. OUTPUT = AB.

A	B	OUTPUT
1	1	1
1	0	1
0	1	1
0	0	0

(c) Negative-logic truth table corresponding to (a). +5 V = 0, 0 V = 1. OUTPUT = A + B

logic to produce the truth table of Figure 9.17(c). It is also very important to realize that the voltages produced by a logic gate are independent of how logic levels are assigned. Positive and negative logic are merely different ways for a user to *interpret* a given set of voltages. It would be absurd to say that a gate was *itself* a positive or negative logic gate or to say that a certain computer was built with positive or negative logic gates.

Example 9-6
Find the logical operations performed by a gate whose voltage truth table is

A	B	OUTPUT
−5 V	−5 V	0 V
−5 V	0 V	−5 V
0 V	−5 V	−5 V
0 V	0 V	−5 V

Solution.   Under positive logic, 0 V = 1 and −5 V = 0. Translating the voltage truth table, we see that the result is a NOR gate:

A	B	OUTPUT = $\overline{A + B}$
0	0	1
0	1	0
1	0	0
1	1	0

Under negative logic, 0 V = 0 and −5 V = 1. Translating the voltage truth table, we see that the result is a NAND gate:

A	B	OUTPUT $= \overline{AB}$
1	1	0
1	0	1
0	1	1
0	0	1

Drill 9–6

In the voltage truth table of Example 9–6, replace 0 V by −5 V and −5 V by 0 V in the output *only*. What logic operation is then performed under positive logic? Under negative logic?

Answer: OR; AND.

Regardless of actual voltage levels, a voltage truth table that represents an AND gate under positive logic always becomes an OR gate under negative logic and vice versa. Similarly, a positive-logic NAND gate always becomes a negative-logic NOR gate and vice versa. A positive-logic truth table can be translated to a negative-logic truth table by replacing every 0 by a 1 and every 1 by a 0 in *both* the input and output columns of the table.

9–7 NOISE IMMUNITY

In Chapter 6 we described several sources of noise that can distort pulse and digital signals. When noise is present, digital systems can fail to operate properly for two reasons:

1. Voltage levels may be shifted so that a voltage that is supposed to represent logical 0 is interpreted by a gate as logical 1 and/or a logical 1 is interpreted as 0.

2. Voltage levels may be shifted so that either or both of the 0 and 1 levels are in the range where a gate's response is unpredictable—that is, in the range between the maximum level it interprets as low and the minimum level it interprets as high.

Clearly, the design of a logic gate and the way it responds to different voltage levels affect its ability to perform satisfactorily in the presence of noise. Thus, gates from the various logic families (TTL, CMOS, and so on, discussed in the next chapter) can be expected to have varying degrees of sensitivity to noise. One family, called high-threshold logic (HTL), is specially designed for use in noisy environments and is characterized by a large *difference* between its low and high voltage levels.

Noise *immunity* is a measure of a gate's insensitivity to noise. To obtain a numerical value for this measure, we always assume that the input to a gate of a given type is obtained from the output of another gate of identical type. Thus, we must consider the possible range of *output* voltages produced by a gate (due to loading, manufacturing tolerances, and other factors) as well as the ranges of input voltages it interprets as low and high. Before giving an equation that can be used to calculate noise immunity, let us consider a specific example to help refine the notion. Suppose that a gate interprets any input voltage in the range from 0 V to 1.5 V as a low, and that it produces a low output that may range from 0.1 V to 1 V. When the input to the gate (from an identical gate) is the maximum that can be expected, 1 V, noise superimposed on that level can raise it to 1.5 V and it will still be interpreted as a low input. (If the noise causes the 1 V level to exceed 1.5 V, the input signal will be in the unpredictable range.) In this example, the noise that can be tolerated is 1.5 V − 1 V = 0.5 V, the difference between the maximum input level recognized as a low and the maximum output level representing a low. Thus, we define noise immunity when the input is low, N_L, to be

$$N_L = V_{IL}(\text{max}) - V_{OL}(\text{max}) \tag{9.15}$$

where

$$V_{IL}(\text{max}) = \text{maximum input level interpreted as low}$$
$$V_{OL}(\text{max}) = \text{maximum output level representing a low}$$

Values of $V_{IL}(\text{max})$ and $V_{OL}(\text{max})$ can usually be obtained from manufacturers' specifications. Noise immunity when the input is high, N_H, is defined by

$$N_H = V_{OH}(\text{min}) - V_{IH}(\text{min}) \tag{9.16}$$

where

$$V_{OH}(\text{min}) = \text{minimum output level representing a high}$$
$$V_{IH}(\text{min}) = \text{minimum input level interpreted as high}$$

Noise immunity, as defined by these equations, is also called noise *margin*.

Example 9–7   The specifications for a 7400 NAND gate state that the minimum and maximum input voltages interpreted as high and low are 2 V and 0.8 V, respectively. The minimum and maximum output voltages representing high and low are 2.4 V and 0.4 V, respectively. Find N_L and N_H.

Solution

$$V_{IL}(\text{max}) = 0.8 \text{ V}$$
$$V_{OL}(\text{max}) = 0.4 \text{ V}$$

From equation (9.15),

$$N_L = 0.8 \text{ V} - 0.4 \text{ V} = 0.4 \text{ V}$$
$$V_{OH}(\text{min}) = 2.4 \text{ V}$$
$$V_{IH}(\text{min}) = 2 \text{ V}$$

From equation (9.16),

$$N_H = 2.4 \text{ V} - 2 \text{ V} = 0.4 \text{ V}$$

Figure 9.18, p. 304, is a diagram showing the input and output voltage ranges for the gate. Note how N_L and N_H can be identified in such a diagram.

Drill 9–7   The output of a gate (gate 1) is the input to another gate (gate 2) of the same type. Both gates have the specifications given in Example 9–7. Under worst-case conditions, what noise voltage added to a low input to gate 2 will cause it to be interpreted as a high? *Answer:* 1.6 V.

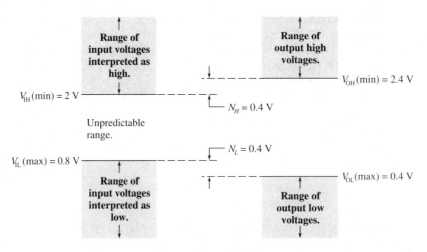

Figure 9.18
Example 9–7.

9–8   SPICE EXAMPLES

Example 9–8

Use SPICE to obtain a truth table for the diode logic gate shown in Figure 9.19(*a*). Logic levels are 1 = +5 V and 0 = 0 V. Allow all diode parameters to have their default values.

Solution.   Figure 9.19(*b*) shows the circuit redrawn in its SPICE format and the SPICE input file. The statement .DC VB 0 5 5 VA 0 5 5 sets input VA to 0 V and steps input VB from 0 V to 5 V and then sets VA to 5 V and again steps VB from 0 V to 5 V. Thus, output voltages are obtained when the inputs are stepped through the sequence AB = 00, 01, 10, 11, as required to generate a truth table.

The computed output, identified as "DC TRANSFER CURVES," is shown in Figure 9.19(*c*). Note that only the values of VB are printed. The first two values of VB (0 V and 5 V) are for VA = 0 V, and the second two values of VB are for VA = 5 V. The voltage truth table is, therefore,

VA	VB	OUTPUT $V(3)$
0 V	0 V	0.6949 V
0 V	5 V	0.7127 V
5 V	0 V	0.7127 V
5 V	5 V	5.675 V

The output voltages reflect the voltage drops across the diodes. Replacing each high voltage in the table by logical 1 and each low voltage by logical 0, we see that the gate performs the logical AND operation.

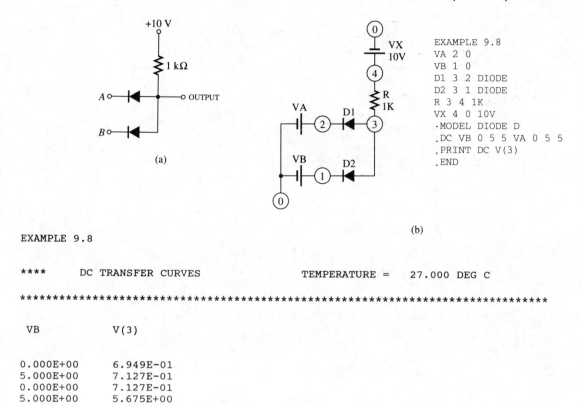

EXAMPLE 9.8
VA 2 0
VB 1 0
D1 3 2 DIODE
D2 3 1 DIODE
R 3 4 1K
VX 4 0 10V
.MODEL DIODE D
.DC VB 0 5 5 VA 0 5 5
.PRINT DC V(3)
.END

(b)

EXAMPLE 9.8

****   DC TRANSFER CURVES   TEMPERATURE =   27.000 DEG C

****

   VB   V(3)

0.000E+00   6.949E-01
5.000E+00   7.127E-01
0.000E+00   7.127E-01
5.000E+00   5.675E+00

(c)

Figure 9.19
Example 9–8.

Example   **a.** Use SPICE to find the output low and high voltages from the transistor inverter shown
9–9   in Figure 9.20(a), p. 306. The β of the transistor is 120. Allow all other parameters
             to have their default values. The input logic levels are 1 = +5 V and 0 = 0 V.
             b. Repeat, when the β of the transistor is changed to 50.

Solution.   Figure 9.20(b), p. 306, shows the circuit redrawn in its SPICE format and the
SPICE input file. Note that the β of the transistor is specified in the .MODEL statement
by BF = 120. The .DC statement causes the input to be stepped from 0 V to 5 V in one
5-V increment. The output voltage, V_{CE}, is the voltage from node 3 to ground, V(3).
   The results are shown in Figure 9.20(c), p. 306. We see that the output is high (5 V)
when VIN is 0 and low (0.2033 V) when VIN is 5 V, confirming inverter operation. When
β is changed to 50, the outputs are as shown in Figure 9.20(d). Here, the output is high
when the input is low but equals 2.897 V when the input is high. We see that reduction
in the value of β is responsible for unsuccessful operation as an inverter.

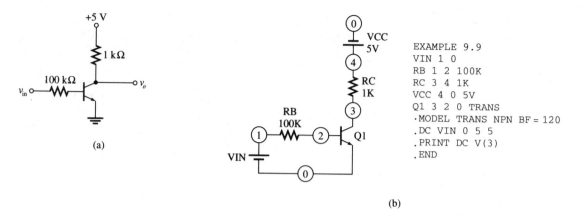

(a)

(b)

EXAMPLE 9.9

****   DC TRANSFER CURVES   TEMPERATURE =   27.000 DEG C

****


```
    VIN              V(3)


  0.000E+00        5.000E+00
  5.000E+00        2.033E-01
```

(c)


```
    VIN              V(3)


  0.000E+00        5.000E+00
  5.000E+00        2.897E+00
```

(d)

Figure 9.20
Example 9–9.

Example
9–10

Use SPICE to confirm the computations of the output voltages from the NMOS inverter in Example 9–5.

Solution.   The NMOS inverter with NMOS load is shown in its SPICE format in Figure 9.21(a), along with the SPICE input file. Note that two .MODEL statements are required because the NMOS transistors have different values of β (specified by KP). The value of VIN is set first to $V_{HI} = V_{DD} - V_T = 10\,V - 2\,V = 8\,V$. The computed output voltage, V(2), is shown in Figure 9.21(b). We see that $V_{LO} = 0.49\,V$, in exact agreement with the value computed in Example 9-5.

We next set VIN to $V_{LO} = 0.49\,V$ (not shown) and obtain the value of V_{HI} shown in Figure 9.21(c). Again, we see very close agreement with the value computed in Example 9-5 ($V_{HI} = 7.999\,V$).

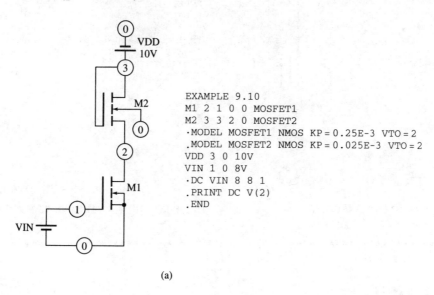

```
EXAMPLE 9.10
M1 2 1 0 0 MOSFET1
M2 3 3 2 0 MOSFET2
·MODEL MOSFET1 NMOS KP = 0.25E-3 VTO = 2
.MODEL MOSFET2 NMOS KP = 0.025E-3 VTO = 2
VDD 3 0 10V
VIN 1 0 8V
·DC VIN 8 8 1
.PRINT DC V(2)
.END
```

(a)

```
EXAMPLE 9.10


****       DC TRANSFER CURVES                 TEMPERATURE =    27.000 DEG C


***************************************************************************


    VIN            V(2)


8.000E+00      4.900E-01
```

(b)

```
    VIN            V(2)


4.900E-01      7.999E+00
```

(c)

Figure 9.21
Example 9-10.

Figure 9.22
Example 9–11.

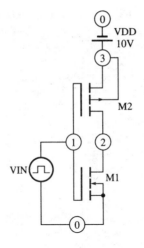

```
EXAMPLE 9.11
M1 2 1 0 0 MOSFET1
M2 2 1 3 3 MOSFET2
.MODEL MOSFET1 NMOS
.MODEL MOSFET2 PMOS
VDD 3 0 10V
VIN 1 0 PULSE(0 10 0 1PS 1PS 0.5US 1US)
.TRAN 0.05US 1US
.PLOT TRAN V(2)
.END
```

(a)

EXAMPLE 9.11

****   TRANSIENT ANALYSIS   TEMPERATURE =   27.000 DEG C

****

TIME	V(2)				
		-5.000D+00	0.000D+00	5.000D+00	1.000D+01 1.500D+01
0.000D+00	1.000D+01	.	.	.	* .
5.000D-08	6.933D-08	.	*	.	. .
1.000D-07	6.933D-08	.	*	.	. .
1.500D-07	6.933D-08	.	*	.	. .
2.000D-07	6.933D-08	.	*	.	. .
2.500D-07	6.933D-08	.	*	.	. .
3.000D-07	6.933D-08	.	*	.	. .
3.500D-07	6.933D-08	.	*	.	. .
4.000D-07	6.933D-08	.	*	.	. .
4.500D-07	6.933D-08	.	*	.	. .
5.000D-07	6.933D-08	.	*	.	. .
5.500D-07	1.000D+01	.	.	.	* .
6.000D-07	1.000D+01	.	.	.	* .
6.500D-07	1.000D+01	.	.	.	* .
7.000D-07	1.000D+01	.	.	.	* .
7.500D-07	1.000D+01	.	.	.	* .
8.000D-07	1.000D+01	.	.	.	* .
8.500D-07	1.000D+01	.	.	.	* .
9.000D-07	1.000D+01	.	.	.	* .
9.500D-07	1.000D+01	.	.	.	* .
1.000D-06	1.000D+01	.	.	.	* .

(b)

EXAMPLE 9.11

```
****      TRANSIENT ANALYSIS                 TEMPERATURE =    27.000 DEG C

*****************************************************************************


     TIME        V(2)

                      -5.000D+00      0.000D+00      5.000D+00      1.000D+01  1.500D+01
                   - - - - - - - - - - - - - - - - - - - - - - - - - - - - - - -
  0.000D+00   1.000D+01 .            .            .            *            .
  5.000D-08   5.414D+00 .            .            .*           .            .
  1.000D-07   2.413D+00 .            .      *      .            .            .
  1.500D-07   9.534D-01 .          . *          .            .            .
  2.000D-07   3.594D-01 .        .*           .            .            .
  2.500D-07   1.315D-01 .        *            .            .            .
  3.000D-07   4.812D-02 .        *            .            .            .
  3.500D-07   1.741D-02 .        *            .            .            .
  4.000D-07   6.351D-03 .        *            .            .            .
  4.500D-07   2.295D-03 .        *            .            .            .
  5.000D-07   8.238D-04 .        *            .            .            .
  5.500D-07   4.607D+00 .            .       *.            .            .
  6.000D-07   7.582D+00 .            .            .     *      .            .
  6.500D-07   9.055D+00 .            .            .         * .            .
  7.000D-07   9.640D+00 .            .            .          *.            .
  7.500D-07   9.870D+00 .            .            .            *            .
  8.000D-07   9.952D+00 .            .            .            *            .
  8.500D-07   9.983D+00 .            .            .            *            .
  9.000D-07   9.994D+00 .            .            .            *            .
  9.500D-07   9.998D+00 .            .            .            *            .
  1.000D-06   9.999D+00 .            .            .            *            .
                   - - - - - - - - - - - - - - - - - - - - - - - - - - - - - - -
```

(c)

Figure 9.22
(Continued)

Example 9–11

The input to the CMOS inverter in Figure 9.16(a) is a 1-MHz square wave that alternates between 0 V and +10 V. Use SPICE to obtain a plot of the output over one full cycle of the input. Assume all MOSFET parameters have their default values. Repeat when the output is loaded by 10 pF of capacitance.

Solution.   The CMOS inverter is shown in its SPICE format in Figure 9.22(a), along with the SPICE input file. One .MODEL statement is used to model the NMOS transistor (M1) and another to model the PMOS transistor (M2). The .TRAN statement specifies that output will be plotted over one period of the input (1 μs) at 0.05-μs intervals.

The resulting plot is shown in Figure 9.22(b). Since the input is high during the first half-cycle, we see that the output is inverted with respect to the input. In the absence of capacitive loading, there is no discernible delay in the leading or trailing edges of the output. The plot that results when 10 pF of capacitance is connected across the output is shown in Figure 9.22(c). It is clear that capacitive loading delays the rise and fall of the inverter output.

EXERCISES

Section 9–1

9.1 Determine which of the diodes in Figure 9.23 are conducting and which are non-conducting.

9.2 Assuming each diode in Figure 9.24 is silicon, find the anode-to-cathode voltage across each.

9.3 Determine which transistors in Figure 9.25 are ON and which are OFF.

9.4 Find the base-to-emitter voltage, V_{BE}, of each transistor in Figure 9.25. Assume each transistor is silicon.

9.5 Determine which of the MOSFET transistors in Figure 9.26, p. 312, are ON and which are OFF.

9.6 Determine which of the MOSFETs in Figure 9.27, p. 312, are ON and which are OFF.

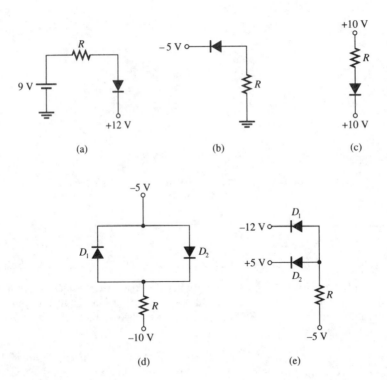

(a)   (b)   (c)

(d)   (e)

Figure 9.23
Exercise 9.1.

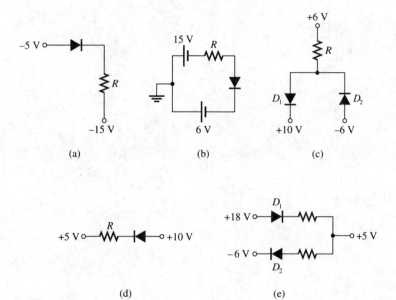

Figure 9.24
Exercise 9.2.

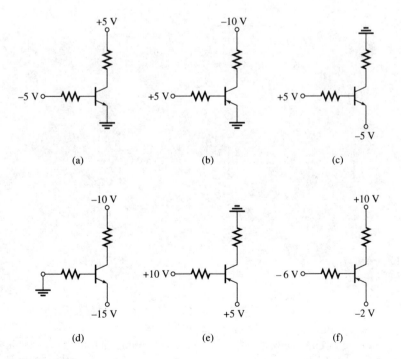

Figure 9.25
Exercises 9.3 and 9.4.

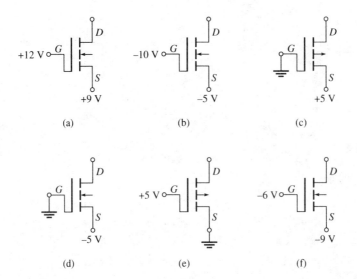

Figure 9.26
Exercise 9.5.

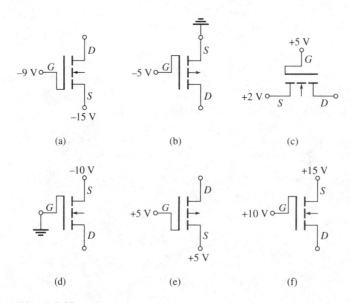

Figure 9.27
Exercise 9.6.

Section 9–2

9.7 Construct a voltage truth table and a logic truth table for each of the circuits shown in Figure 9.28. Determine the logic operation performed by each circuit. Note the logic levels defined in the figures. Assume ideal diodes.

9.8 Repeat Exercise 9.7 for the circuits shown in Figure 9.29.

9.9 Assuming that the forward-biased voltage drop across each diode in Figure 9.29 is 0.7 V, construct the voltage truth table for each circuit.

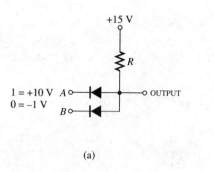

Figure 9.28
Exercise 9.7.

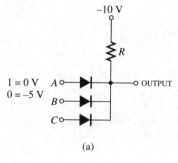

Figure 9.29
Exercise 9.8.

9.10 Assuming that the forward-biased voltage drop across each diode in Figure 9.28 is 0.7 V, construct the voltage truth table for each circuit.

Section 9–3

9.11 A BJT inverter is to be constructed using an NPN transistor whose β may range from 80 to 150. If R_B is to be 120 kΩ, what minimum value of R_C should be used? Assume $V_{CC} \gg V_{BE}$.

9.12 A BJT inverter has $R_B = 47$ kΩ and $R_C = 1.2$ kΩ. For what minimum value of β will the inverter perform satisfactorily? Assume $V_{CC} \gg V_{BE}$.

9.13 An NPN inverter has $R_C = 1.5$ kΩ and its output is directly connected to load resistance R_L. If $V_{CC} = 5$ V, what is the minimum value of R_L that can be used if the high output of the inverter cannot be less than 2 V?

9.14 The saturation voltage of an NPN transistor used in an inverter is 150 mV.
(a) What is the saturation current if $V_{CC} = 5$ V and $R_C = 2.2$ kΩ?
(b) What is the low output voltage of the inverter?

Section 9–4

9.15 An NMOS transistor connected for use as a resistor in an integrated circuit has β = 10 μA/V² and $V_T = 1.5$ V. If the current through the transistor ranges from 0.5 mA to 2 mA, find each value:
(a) Its minimum and maximum dc resistance
(b) Its minimum and maximum ac resistance

9.16 An NMOS transistor connected for use as a resistor in an integrated circuit has an ac resistance of 5 kΩ when the current through it is 0.4 mA. What value of current through the transistor will double its ac resistance?

9.17 The NMOS inverter in Figure 9.14(c) has $\beta_1 = 100$ μA/V², $\beta_2 = 12$ μA/V², $V_{T1} =$

$V_{T2} = 2.5$ V, and $V_{DD} = 15$ V. Find the output voltage when the input is:
(a) $V_{DD} - V_T$ **(b)** V_{LO}

9.18 The NMOS inverter in Figure 9.14(c) has $\beta_1 = \beta_2 = 100$ μA/V^2, $V_{T1} = V_{T2} = 1$ V, and $V_{DD} = 9$ V. Will it operate successfully as an inverter? Explain.

Section 9–5

9.19 In the circuit shown in Figure 9.30, find V_{GS1} and V_{GS2} in each case.
(a) The input is 0 V.
(b) The input is 10 V.

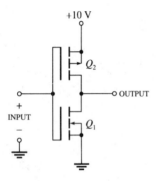

Figure 9.30
Exercise 9.19.

9.20 In the CMOS inverter shown in Figure 9.31, find V_{GS1} and V_{GS2} and determine whether the output is low or high when the input is
(a) 0 V; **(b)** -10 V.

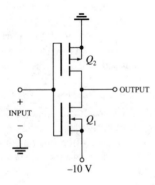

Figure 9.31
Exercise 9.20.

Section 9–6

9.21 Find the logic operation performed by each of the gates whose voltage truth tables are shown in Figure 9.32
(i) Under positive logic;
(ii) Under negative logic.

A	B	OUTPUT
0 V	0 V	+10 V
0 V	+10 V	+10 V
+10 V	0 V	+10 V
+10 V	+10 V	0 V

(a)

A	B	OUTPUT
−0.7 V	−0.7 V	−0.7 V
−0.7 V	−5.7 V	−0.7 V
−5.7 V	−0.7 V	−0.7 V
−5.7 V	−5.7 V	−5.7 V

(b)

Figure 9.32
Exercise 9.21.

9.22 Find the logic operation performed by each of the gates whose voltage truth tables are shown in Figure 9.33
(i) Under positive logic;
(ii) Under negative logic.

X_1	X_2	OUTPUT
−5 V	−5 V	−5 V
−5 V	−10 V	−10 V
−10 V	−5 V	−10 V
−10 V	−10 V	−10 V

(a)

A	B	C	OUTPUT
+3 V	+7 V	+3 V	+3 V
+7 V	+3 V	+3 V	+3 V
+7 V	+7 V	+7 V	+3 V
+3 V	+3 V	+3 V	+7 V
+7 V	+3 V	+7 V	+3 V
+3 V	+3 V	+7 V	+3 V
+7 V	+7 V	+3 V	+3 V
+3 V	+7 V	+7 V	+3 V

(b)

Figure 9.33
Exercise 9.22.

Section 9–7

9.23 The manufacturer's specifications for a certain logic gate state that the low output may range from 0.15 V to $0.2V_{CC}$ and that the high output may range from $0.75V_{CC}$ to V_{CC}. The gate will interpret as low any input in the range from 0 V to $0.25V_{CC}$ and as high any input in the range from $0.5V_{CC}$ to V_{CC}. Find the noise immunity when the input is low and when the input is high, given that $V_{CC} = +15$ V.

9.24 Describe how the gate in Exercise 9.23 will respond to inputs that alternate between $+1$ V and $+9$ V when $V_{CC} = +20$ V.

CHALLENGING EXERCISES

9.25 Figure 9.34 shows an NPN inverter used as an LED driver. For the LED to illuminate, the current through it must be 3 mA. If the voltage drop across the LED when it is

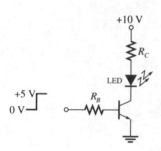

Figure 9.34
Exercise 9.25.

conducting is 2.5 V and if the β of the transistor is 100, what values of R_B and R_C should be used to ensure that the transistor is fully saturated when conducting and that the LED is illuminated? Assume $V_{CE}(\text{sat}) = 100$ mV.

9.26 An NMOS inverter with a resistor-connected NMOS load [Figure 9.14(c)] is to be operated with a $+15$-V supply. The threshold voltage of each NMOS device is 2 V. What should the ratio β_1/β_2 be if the low output voltage is to be 0.75 V?

9.27 An NMOS inverter with a resistor-connected NMOS load [Figure 9.14(c)] is to be operated so that the low output voltage is 0.25 V. If $\beta_1/\beta_2 = 12$ and the threshold voltage of each device is 1 V, what value of supply voltage, V_{DD}, should be used?

SPICE EXERCISES

Note: In each of the following exercises, assume all devices have their default parameter values, unless otherwise specified.

9.28 Use SPICE to obtain a truth table for the diode OR gate shown in Figure 9.8(a). The supply voltage is -12.5 V, $R = 2.2$ kΩ, and the logic levels are $1 = +6$ V, $0 = -6$ V.

9.29 The output of the BJT inverter in Figure 9.11(a) is connected to the input of another, identical inverter. Each inverter has $V_{CC} = +5$ V, $R_B = 150$ kΩ, and $R_C = 1.5$ kΩ. The β of each transistor is 150. Use SPICE to determine the high and low output voltages from the second inverter when the input to the first is 0 V and when it is $+5$ V.

9.30 The NMOS inverter with NMOS load shown in Figure 9.14(c) has $V_{DD} = 10$ V,

$V_{T1} = V_{T2} = 1.8$ V, $\beta_1 = 0.2 \times 10^{-3}$ A/V^2, and $\beta_2 = 0.02 \times 10^{-3}$ A/V^2. The output is loaded by 10 pF of capacitance. The input is a rectangular waveform that alternates between 0 V and $+10$ V with a frequency of 400 kHz and a pulsewidth of 0.5 μs. Use SPICE to obtain a plot of the output over an interval of time sufficient to observe the output switching from low to high and from high to low. Without actually computing the rise and fall times, comment on and explain any apparent difference in their values that can be discerned from the plot.

9.31 Use SPICE to determine the rise and fall times of the output of the CMOS inverter in Example 9–11 when the output is loaded by 10 pF.

Chapter 10

FAMILIES OF LOGIC CIRCUITS

Overview

Transistor-transistor logic, TTL, is named for its dependence on transistors alone to perform basic logic operations (unlike its obsolescent predecessors: DTL, diode-transistor logic, and RTL, resistor-transistor logic). The original version, which is now known as "standard" TTL, was developed in 1965 and is rarely used in new systems. Through the years, the basic design has been modified to improve its performance in several respects, and as a consequence, a number of subfamilies have evolved. The most significant modification was the addition of a *Schottky* diode in parallel with the base-collector junction of the saturating transistors. As we shall see in later discussions, this modification dramatically reduced propagation delays, and virtually all modern TTL devices incorporate the so-called Schottky clamp. The TTL variations include the following:

1. Standard TTL, whose basic building blocks are still an integral part of several subfamilies. Standard TTL is identified by 7400-series designations and their 5400-series counterparts (which are functionally equivalent but can be operated over wider temperature and voltage ranges, as required by military specifications). No letters are inserted between numerals in the component designations. For example, the hex inverter is identified simply as a 7404 or 5404.
2. High-speed TTL, an early modification of standard TTL, in which an additional output-driver stage was added to approximately double switching speeds. However, power consumption was also doubled. High-speed TTL designations have the letter *H* inserted between 74 (or 54) and the rest of the number. For example, the 74H04 is a high-speed hex inverter. Newer Schottky variations are superior in both speed and power consumption.
3. Low-power TTL, an early modification that reduced power consumption by increasing resistor values. Although the power consumption is about $\frac{1}{10}$ that of standard TTL, standard TTL is more than three times faster than low-power TTL. The low-power version is not available in a variety of 7400-series devices, but can be obtained in some of the 5400-series, where the letter *L* is inserted to indicate low-power. For example,

317

the 54L04 is a low-power hex inverter. Low-power Schottky and CMOS versions of the 7400-series are now widely used in lieu of low-power TTL.

4. Schottky TTL, the first version to include the Schottky clamp, mentioned earlier. Schottky TTL more than tripled the switching speed of standard TTL, at the expense of approximately doubling the power consumption. The letter S is used to designate a Schottky version. For example, the 74S04 is a Schottky hex inverter.

5. Low-power Schottky TTL, a low-power version of Schottky TTL, with switching speeds about the same as those of standard TTL but with about one-fifth the power consumption. The letters LS are used to designate low-power Schottky. For example, the 74LS04 is a low-power, Schottky hex inverter.

6. Advanced Schottky TTL, a new Schottky version that is almost twice as fast and consumes less than half as much power as the first Schottky version. The letters AS are used to designate advanced Schottky. For example, the 74AS04 is an advanced Schottky hex inverter.

7. Advanced low-power Schottky TTL, a low-power version of advanced Schottky (i.e., an advanced version of low-power Schottky) with greater speed and lower power consumption than low-power Schottky. The letters ALS are used in device designations, as, for example, the 74ALS04.

8. F (fast), the newest and fastest TTL-series, marketed by Texas Instruments as F logic. A comparable series is marketed by Fairchild and Motorola under the Fairchild trademark FAST. Devices in all of these series have the letter F inserted in their designations, as in the 74F04.

Speed-Power Product

As the foregoing comparisons suggest, there is usually a trade-off between switching speed and power consumption in the design of a logic gate: Speed is gained at the expense of increased power consumption. The *speed-power product* is a measure of overall performance that takes both of those factors into account. Actually, the speed-power product is computed as the product of propagation *delay* and average power consumption per logic gate, so the smaller the product, the better the overall performance. Notice that the product of delay in seconds and power in watts (joules per second) has the units of energy—i.e., joules (J):

$$\text{propagation delay (s)} \times \text{power (J/s)} = \text{energy (J)} \qquad \textbf{(10.1)}$$

Figure 10.1(a) summarizes the propagation delays, power consumptions, and speed-power products of the various TTL versions. Figure 10.1(b) is a plot of delay versus power for the same versions. Figure 10.1(c) compares the dc noise immunities (noise margins).

The Multiple-Emitter Transistor

Figure 10.2(a) shows a transistor configuration found at the input of many TTL circuits. Notice that Q_1 is an NPN transistor having two emitters, one for each input to the gate. This particular configuration is found in the 7400 quad two-input NAND gate, among others. Figure 10.2(b) shows the equivalent circuit, where Q_1 has been replaced by a two-transistor-equivalent, Q_A and Q_B. Note that Q_A and Q_B have their collectors joined and their bases joined.

Figure 10.3(a) shows the equivalent input circuit when $A = B = 0$ V. Since the base-emitter junctions of Q_A and Q_B are both forward-biased, Q_A and Q_B are both conducting (ON). Note that the collector-emitter saturation voltage of Q_A and Q_B, about

CIRCUIT TECHNOLOGY	FAMILY	MINIMIZING POWER				FAMILY	MINIMIZING DELAY TIME			
		PROP DELAY (ns)	PWR DISS (mW)	SPD/PWR PRODUCT (pJ)	MAXIMUM FLIP-FLOP FREQ (MHz)		PROP DELAY (ns)	PWR DISS (mW)	SPD/PWR PRODUCT (pJ)	MAXIMUM FLIP-FLOP FREQ (MHz)
Gold Doped	TTL	10	10	100	35	TTL	10	10	100	35
	L TTL	33	1	33	3	H TTL	6	22	132	50
Schottky Clamped	LS TTL	9	2	18	45	S TTL	3	19	57	125
	'ALS	4	1.2	4.8	70	'AS	1.7	8	13.6	200

(a) Typical performance characteristics.

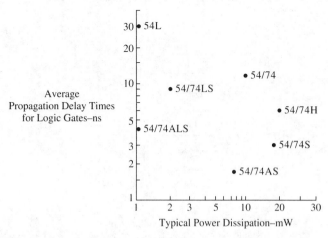

(b) Speed-power relationships.

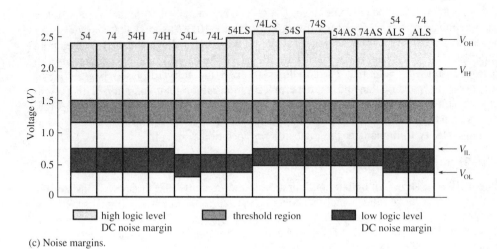

(c) Noise margins.

Figure 10.1
Comparisons of TTL subfamilies (Courtesy of Texas Instruments).

Figure 10.2

A typical TTL multiple-emitter input and its equivalent circuit.

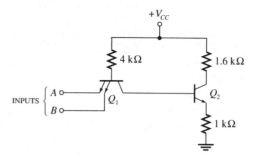

(a) Input circuit for one gate in the 7400 quad two-input
NAND gate.

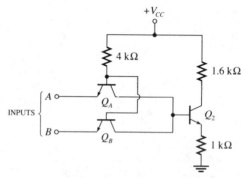

(b) A circuit equivalent to (a). The two-emitter transistor
Q_1 is equivalent to two transistors, Q_A and Q_B, having
their bases joined and their collectors joined.

0.1 V, is the same as the base voltage of Q_2. Q_2 is therefore off because its base-to-emitter voltage must be about 0.7 V for it to turn on. Figure 10.3(b) shows that Q_2 is also off when input A alone (or input B alone) is 0 V. Figure 10.3(c) shows the circuit when both inputs are +5 V. Q_A and Q_B are now both off, since the base-to-emitter junction of each is reverse-biased. Note that current flows through the forward-biased base-collector junctions of Q_A and Q_B, an unusual occurrence in transistor circuit operation. This current is sufficient to turn on Q_2. In summary, Q_2 is off when at least one of the inputs is low and is on when both inputs are high. In the next paragraph, we see how Q_2 is coupled to output circuitry to implement the logical NAND operation.

The Totem Pole Output

Figure 10.4(a) shows an output configuration found in many TTL circuits. Transistor Q_3 and Q_4 form a *totem pole*, which is shown connected to the input circuitry discussed in the previous paragraph. The complete circuit is one gate in the 7400 quad two-input NAND gate. Note that diodes D_1 and D_2 have been added at the input terminals. These diodes protect the circuit from large negative transients on input lines: If an input attempts to go more than about 1 V negative, the protective diode conducts like a short circuit to ground. Figure 10.4(b) shows the circuit under one set of conditions that makes Q_2 off (A = 0 V, B = +5 V). Since Q_2 is like an open switch, no current flows through it. Instead, current

Figure 10.3
Static analysis of the TTL multiple-emitter input circuit.

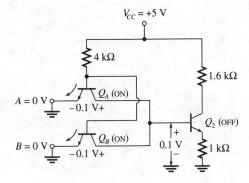

(a) When inputs *A* and *B* are both 0 V, Q_A and Q_B are both ON, so Q_2 is OFF.

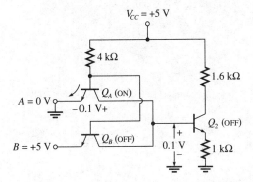

(b) When input *A* (or input *B*) is 0 V, Q_A (or Q_B) is ON, so Q_2 is OFF.

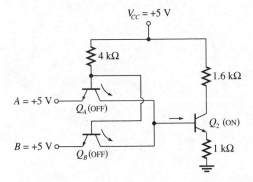

(c) When inputs *A* and *B* are both +5 V, Q_A and Q_B are both OFF. Current flows from base to collector in Q_A and Q_B, supplying base current to turn on Q_2.

flows through the 1.6-kΩ resistor and into the base of Q_3, turning it on. Q_4 remains off because there is no path through which it can receive base current. Part *(c)* of the figure shows the equivalent switches in the totem pole under these conditions. We see that the output is connected through D_3 and the 130-Ω resistor to V_{CC} (+5 V). Assuming

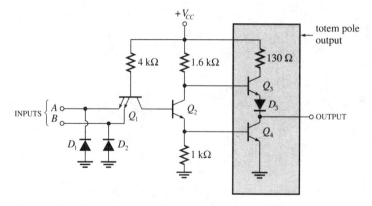

(a) A totem pole output connected to the multiple-emitter input.

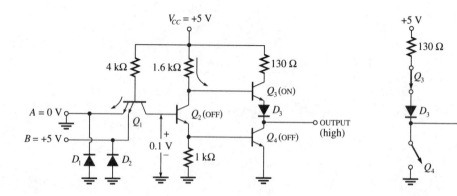

(b) Static analysis when Q_2 is off.

(c) Totem pole equivalent.

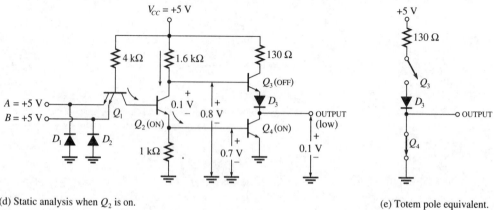

(d) Static analysis when Q_2 is on.

(e) Totem pole equivalent.

Figure 10.4
Static analysis of one gate in a 7400 quad 2-input NAND gate, including the totem pole output.

negligible load current, the output is therefore high. If load current, I_L, is taken into account, the high output voltage is

$$V_{HI} = V_{CC} - V_{CE(sat)} - V_D - I_L(130 \ \Omega) \tag{10.2}$$

where V_D is the forward drop across D_3, about 0.7 V, and $V_{CE(sat)}$ is the saturation voltage of Q_3, about 0.1 V. Note that $I_L = V_{HI}/R_L$, where R_L is the load resistance.

Figure 10.4(d) shows the circuit under the input conditions that make Q_2 on ($A = +5$ V, $B = +5$ V). Current can now be supplied to the base of Q_4, turning it on. The output is therefore $V_{CE(sat)} \approx 0.1$ V—i.e., low. Note that the voltage at the base of Q_3 equals the sum of the base-to-emitter drop of Q_4 and the saturation voltage of Q_2:

$$V_{B3} = V_{BE}(Q_4) + V_{CE(sat)} \approx 0.7 \text{ V} + 0.1 \text{ V} = 0.8 \text{ V}$$

The purpose of diode D_3 is now apparent: If it were not present, the 0.8 V on the base of Q_3 would be enough to turn Q_3 on. With D_3 in place, the base voltage of Q_3 has to rise an additional 0.7 V before Q_3 can turn on. Thus, diode D_3 ensures that Q_3 remains off when Q_4 is on. In summary, the voltage truth table shown in Figure 10.5, and its positive-logic counterpart, confirm that the gate performs the logical NAND operation.

Note that an unconnected (open) input has the same effect as if a *high* were connected to that input, since no current can flow through an open-circuited base-to-emitter junction. Thus, if B is left open, the output is $\overline{AB} = \overline{A \cdot 1} = \overline{A} + 0 = \overline{A}$. On the other hand, if input B to a 2-input TTL NOR gate is left open, the output is $\overline{A + B} = \overline{A + 1} = \overline{A} \cdot 0 = 0$. Such inputs are said to *float high*. Unused inputs should be connected high or to ground, depending on the logic operation performed.

Example 10–1

A 7400 NAND gate has $V_{CC} = +5$ V and a 5-kΩ load connected to its output. Find the output voltage (a) when both inputs are 0 V; (b) when both inputs are +5 V.

Solution

a. When both inputs are 0 V, the output is high, and V_{HI} is found from equation (10.2):

$$V_{HI} = V_{CC} - V_{CE(sat)} - V_D - I_L(130 \ \Omega)$$
$$= 5 \text{ V} - 0.1 \text{ V} - 0.7 \text{ V} - I_L(130 \ \Omega)$$
$$V_{HI} = 4.2 \text{ V} - I_L(130 \ \Omega)$$

The load current is

$$I_L = \frac{V_{HI}}{R_L} = \frac{V_{HI}}{5 \text{ k}\Omega}$$

A	B	OUTPUT
0 V	0 V	+5 V
0 V	+5 V	+5 V
+5 V	0 V	+5 V
+5 V	+5 V	0 V

(a) Voltage truth table.

A	B	OUTPUT = $\overline{AB}$
0	0	1
0	1	1
1	0	1
1	1	0

(b) Positive-logic truth table.

Figure 10.5
Voltage and logic truth tables for the TTL NAND gate.

Substituting for I_L in the previous equation, we find

$$V_{HI} = 4.2 \text{ V} - \frac{V_{HI}}{5 \text{ k}\Omega} (130 \text{ }\Omega)$$

or

$$V_{HI} = \frac{4.2 \text{ V}}{1 + \dfrac{130}{5 \times 10^3}} = 4.09 \text{ V}$$

b. When both inputs are $+5$ V, the output is low and equals the saturation voltage of Q_4:

$$V_{LO} = V_{CE(sat)} \approx 0.1 \text{ V}$$

***Drill
10–1***

What is the minimum value of R_L that can be used in Example 10.1 if the output high voltage cannot be less than 3.5 V?
Answer: 650 Ω.

Input and Output Currents—Fanout

Transistor Q_3 in the totem pole output is called a *pull-up* transistor because it supplies current to drive the load voltage up when the output is switching from low to high. Transistor Q_4 is called a *pull-down* transistor because it draws current from the load to pull the load voltage down, when the output is switching from high to low. Both transistors aid in reducing propagation delays caused by load capacitance.

Since current flows out of the totem pole when the output is high, Q_3 acts as a current *source* to a load. When the output is low, current flows *into* Q_4, and we say that Q_4 is a current *sink*. It is clear that a load connected to a TTL output must be compatible with it in the sense that the load must draw current when its input is high and must supply current when its input is low.

Figure 10.6 shows a totem pole output connected to one input of another TTL gate. In *(a)*, the totem pole output is high, so the only current it must supply is a small leakage current to the reverse-biased emitter-base junction of the input. The magnitude of this current is on the order of 50 μA. Note that I_{OH} is output current when the output is high and I_{IH} is input current when the input is high. In *(b)*, the totem pole output is low and Q_4 sinks current from the forward-biased base-emitter junction of the input. This current is much larger, on the order of 1 mA, and is designated I_{OL} at the output and I_{IL} at the input. By convention, current flowing into a device is positive and current flowing out is negative. Therefore, manufacturer's specifications list negative values for I_{OH} and I_{IL}.

Figure 10.7 shows a TTL output connected to several TTL loads. When the output is low, each load supplies current to the totem pole, so Q_4 must be capable of sinking the *sum* of these currents. Manufacturers specify a maximum value for I_{OL}, the maximum current the output can sink, and a maximum value for I_{IL}. *Fanout* is the maximum number of loads identical to itself that a logic gate can drive, so

$$\text{Fanout} = \frac{I_{OL} \text{ (max)}}{|I_{IL} \text{ (max)}|} \qquad \textbf{(10.3)}$$

Figure 10.6
TTL input and output currents.

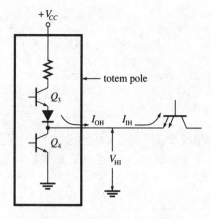

(a) Q_3 acts as a current source that supplies leakage
current to a TTL input when the output voltage is high.

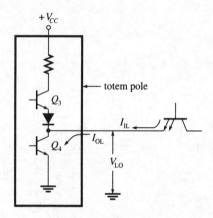

(b) Q_4 acts as a current sink that draws current from
the load when the output voltage is low.

Figure 10.7
A TTL output driving several TTL loads. I_{OL}
is the sum of the currents supplied by the
loads and Q_4 must be capable of sinking all
of this current.

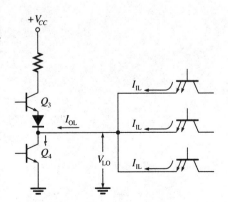

In standard TTL, typical values are $I_{OL}(max) = 20$ mA and $|I_{IL}(max)| = |-2$ mA$| = 2$ mA, so a typical fanout is 20 mA/2mA = 10. *Fan-in* is simply the number of inputs to a gate. For example, the fan-in of a 7400 NAND gate is 2.

The fanout in some logic circuits is determined by the maximum load that the output can drive when the output is *high,* rather than low. In those cases, the fanout is $I_{OH}(max)/I_{IH}(max)$. In any case, the actual fanout is always the smaller of the two computations.

High-Speed (H) and Low-Power (L) TTL

Figure 10.8(*a*) shows one gate in a 74H00 quad 2-input, high-speed NAND gate. Besides smaller resistance values, the only difference between this gate and the standard version is the additional transistor (Q_5) driving pull-up transistor Q_3. Q_3 and Q_5 form a *Darlington pair,* which has a much greater current gain than a single transistor and which speeds up switching of the output from low to high. Note that the base-emitter junctions of Q_5 and Q_3 raise the threshold level at the base of Q_5 to 1.4 V and therefore eliminate the need for a diode in the totem pole. Figure 10.8(*b*) shows one gate in the 74L00 quad 2-input

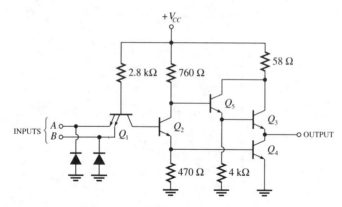

(a) 74H00 NAND gate.

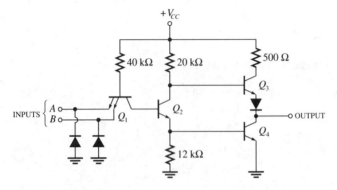

(b) 74L00 NAND gate.

Figure 10.8
Typical high-speed (H) and low-power (L) TTL gates.

low-power NAND gate. The only difference between this gate and the standard version is the much larger resistance values, which reduce power consumption at the expense of switching speed.

The Schottky Clamp

The TTL circuits we have discussed up to this point are examples of *saturating* logic: Transistors are driven into saturation when turned on. When a transistor is saturated, excess charge carriers in the base region must be removed before the transistor can be turned off. The time required to remove these carriers, called *storage time,* is responsible for a storage-time *delay* that slows down the switching of an output from low to high. One way to counter this problem is to prevent the transistor from going into deep saturation when it is turned on—i.e., to prevent its base-to-collector junction from becoming forward-biased by more than a few tenths of a volt. This can be accomplished by connecting a diode in parallel with the base-to-collector junction. If the forward voltage drop of the diode is smaller than the usual 0.7-V forward drop across a silicon PN junction, the diode will begin to conduct before the junction voltage can reach 0.7 V. When it conducts, base current is diverted around the junction and the junction is said to be *clamped* to the smaller voltage. Since the junction voltage cannot then rise above the small forward drop of the diode, the junction cannot become heavily forward-biased and the transistor is kept out of saturation.

Although germanium diodes have a smaller forward drop than silicon diodes, they cannot be fabricated in silicon integrated circuits. The Schottky diode, consisting of a silicon-metal junction, has a forward drop of $0.3 - 0.4$ V and is ideal for that purpose. Figure 10.9(*a*) and (*b*) shows the construction and schematic symbol for a Schottky diode, also called a Schottky *barrier* diode. Figure 10.9(*c*) shows the Schottky diode connected around a base-to-collector junction to form a Schottky clamp. The combination is represented in integrated circuits by the single symbol shown in Figure 10.9(*d*).

Schottky (S) and Low-Power Schottky (LS) TTL

Unlike conventional diodes, Schottky diodes have very little capacitance and have fast *recovery* times: They can be switched rapidly without storage-time delays. These characteristics enhance their performance as transistor clamps and make them desirable as

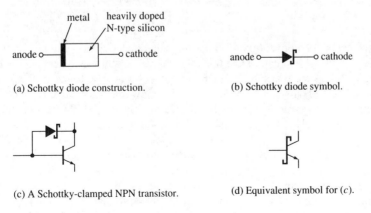

(a) Schottky diode construction.

(b) Schottky diode symbol.

(c) A Schottky-clamped NPN transistor.

(d) Equivalent symbol for (*c*).

Figure 10.9
The Schottky diode and Schottky clamp.

Figure 10.10
Typical Schottky (S) gates.

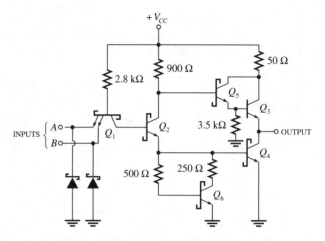

(a) One gate in the 74S00 quad two-input NAND gate.

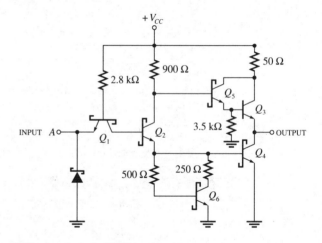

(b) One inverter in the 74S04 hex inverter.

replacements for conventional diodes in integrated circuits. Figure 10.10(*a*) shows a 74S00 Schottky NAND gate. Note that all devices are Schottky, except Q_3, which does not saturate. Also note that the Darlington driver, Q_3 and Q_5, found in the 74H00-series is also used here. Transistor Q_6 regulates current flow into the base of Q_4 and aids in turning Q_4 off rapidly. Figure 10.10(*b*) is one inverter in the 74S04 hex inverter. It can be regarded as simply a 1-input version of a 74S00 NAND gate.

Figure 10.11 shows a 74LS00 low-power Schottky NAND gate. Note that the multiple-emitter input is not used in this version. Instead, diodes D_1 and D_2 and the 20-kΩ resistor form a diode AND gate, as discussed in Section 9–2. Transistor Q_1 serves as an inverter, so the output at its collector represents the NAND function. The remaining circuitry, including the totem pole output, is similar to that in the 74S00, except for larger resistance values. The network consisting of Schottky diodes D_3 and D_4 and the 4-kΩ

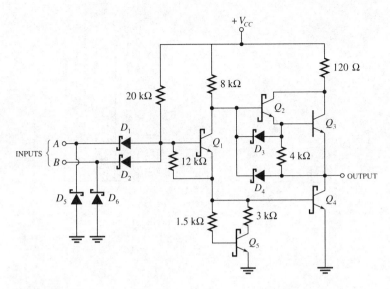

Figure 10.11
One gate in the 74LS00 quad 2-input NAND *gate. Note that Schottky diodes* D_1 *and* D_2 *and the 20-kΩ resistor form a diode* AND *gate.*

resistor is connected to the output and aids in charging and discharging load capacitance when Q_3 and Q_4 are changing states.

Advanced Low-Power Schottky (ALS) and Advanced Schottky (AS) TTL

ALS and AS technologies are recent enhancements in Schottky TTL circuitry. Among other refinements, advanced Schottky devices are fabricated with an improved *doping* technique (ion implantation instead of diffusion), and Schottky-clamped transistors have improved isolation (using oxide instead of P^+ material). These enhancements reduce capacitance and thus improve switching times. Also, a more complex circuit design uses additional active devices to speed switching, reduce power consumption, and increase fanout. Use of the advanced technologies is responsible for TTL gates that have higher speeds, lower power consumptions, and smaller speed-power products than any previous TTL versions (see Figure 10.1) and makes TTL competitive with other high-speed families to be discussed later. Figure 10.12 compares the fanouts of the various TTL versions, including advanced Schottky.

Figure 10.13(*a*) shows one gate in the 74ALS00A quad 2-input NAND gate. Instead of a multiple-emitter transistor, parallel-connected PNP transistors Q_1 and Q_2 are used at the input. These transistors reduce the current flow, I_{IL}, when the inputs are low and thus increase fanout. If inputs *A, B,* or both are low, then the respective PNP transistors turn on because their emitters are then more positive than their bases. Figure 10.13(*b*) shows a simplified equivalent circuit of the input, where Q_1 and Q_2 are represented as voltage-controlled switches. If at least one of the inputs is low, the corresponding switch is closed, making the base of Q_3 low and keeping Q_3 off. If both *A* and *B* are high, both switches are open and Q_3 turns on. Referring to part (*a*) of the figure, we see that Q_3 drives Q_4 (by emitter-follower action), and Q_4 drives the output totem pole. Schottky diodes D_3,

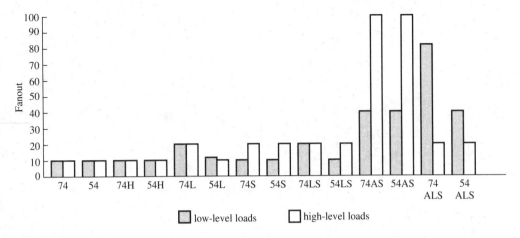

Figure 10.12
Comparison of fanouts for TTL subfamilies.

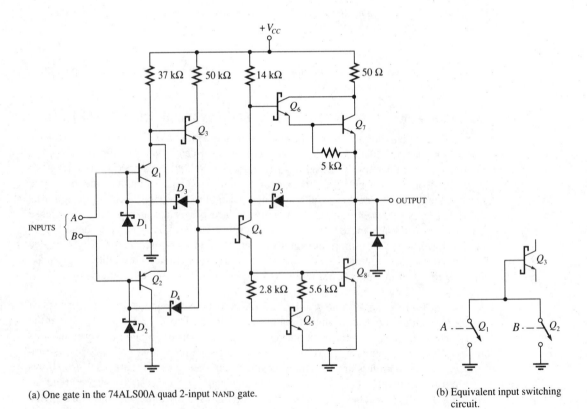

(a) One gate in the 74ALS00A quad 2-input NAND gate.

(b) Equivalent input switching circuit.

Figure 10.13
Typical advanced Schottky (AS) and advanced low-power Schottky (ALS) gates.

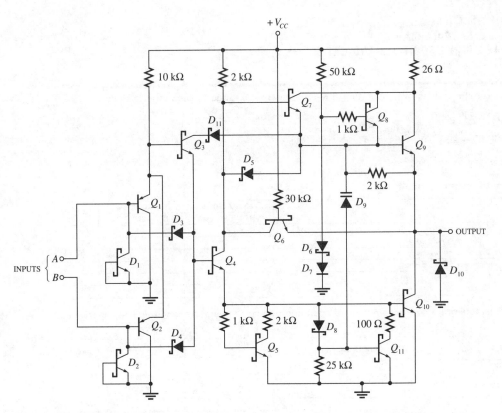

(c) One gate in the 74AS00 quad 2-input NAND gate. (D_1 and D_2 are transistors connected for operation as diodes.)

Figure 10.13
(Continued)

D_4, and D_5 are used to speed the switching and do not affect the logic. Note that the output (as well as the inputs) has a Schottky protective diode. Figure 10.13(c) shows one gate in the 74AS00. Note that the input logic circuitry is essentially the same as that in the 74ALS00 gate, as is the output totem pole. The additional circuitry between input and output improves switching speeds using sophisticated drivers and feedback networks.

Figure 10.14 shows a typical set of manufacturer's specifications for TTL devices, in this case the AS and ALS quad 2-input NAND gates. The majority of the parameters for which specifications are listed have already been defined or are self-evident. V_{IK} is the input clamping voltage, the negative voltage at which the protective diode on each input will short to ground. I_{CCH} and I_{CCL} are the total currents drawn from the power supply by the integrated circuit when all outputs are high and low, respectively. Note that some specifications show only minimum or maximum values. These correspond to worst-case conditions. For example, I_{IH}, which ideally should be very small, is specified to have a maximum value of 20 μA. Note also that most values can be expected to change with supply voltage, temperature, and other factors, so test conditions are given for each specification. The specifications shown here are referred to in some of the end-of-chapter exercises.

ALS and AS Circuits
SN54ALS00A, SN74ALS00A
QUADRUPLE 2-INPUT POSITIVE-NAND GATES

absolute maximum ratings over operating free-air temperature range (unless otherwise noted)

Supply voltage, V_{CC} . . . . 7 V
Input voltage . . . . 7 V
Operating free-air temperature range: SN54ALS00A . . . . −55 °C to 125 °C
SN74ALS00A . . . . 0 °C to 70 °C
Storage temperature range . . . . −65 °C to 150 °C

recommended operating conditions

		SN54ALS00A			SN74ALS00A			UNIT
		MIN	NOM	MAX	MIN	NOM	MAX	
V_{CC}	Supply voltage	4.5	5	5.5	4.5	5	5.5	V
V_{IH}	High-level input voltage	2			2			V
V_{IL}	Low-level input voltage			0.7			0.8	V
I_{OH}	High-level output current			−0.4			−0.4	mA
I_{OL}	Low-level output current			4			8	mA
T_A	Operating free-air temperature	−55		125	0		70	°C

electrical characteristics over recommended operating free-air temperature range (unless otherwise noted)

PARAMETER	TEST CONDITIONS		SN54ALS00A			SN74ALS00A			UNIT
			MIN	TYP[†]	MAX	MIN	TYP[†]	MAX	
V_{IK}	$V_{CC} = 4.5$ V,	$I_I = -18$ mA			−1.5			−1.5	V
V_{OH}	$V_{CC} = 4.5$ V to 5.5 V,	$I_{OH} = -0.4$ mA	$V_{CC} - 2$			$V_{CC} - 2$			V
V_{OL}	$V_{CC} = 4.5$ V,	$I_{OL} = 4$ mA		0.25	0.4		0.25	0.4	V
	$V_{CC} = 4.5$ V,	$I_{OL} = 8$ mA					0.35	0.5	
I_I	$V_{CC} = 5.5$ V,	$V_I = 7$ V			0.1			0.1	mA
I_{IH}	$V_{CC} = 5.5$ V,	$V_I = 2.7$ V			20			20	µA
I_{IL}	$V_{CC} = 5.5$ V,	$V_I = 0.4$ V			−0.1			−0.1	mA
I_O[‡]	$V_{CC} = 5.5$ V,	$V_O = 2.25$ V	−30		−112	−30		−112	mA
I_{CCH}	$V_{CC} = 5.5$ V,	$V_I = 0$ V		0.5	0.85		0.5	0.85	mA
I_{CCL}	$V_{CC} = 5.5$ V,	$V_I = 4.5$ V		1.5	3		1.5	3	mA

[†]All typical values are at $V_{CC} = 5$ V, $T_A = 25$ °C.
[‡]The output conditions have been chosen to produce a current that closely approximates one half of the true short-circuit output current, I_{OS}.

switching characteristics (see Note 1)

PARAMETER	FROM (INPUT)	TO (OUTPUT)	$V_{CC} = 5$ V, $C_L = 50$ pF, $R_L = 500 \Omega$, $T_A = 25$ °C	$V_{CC} = 4.5$ V to 5.5 V, $C_L = 50$ pF, $R_L = 500 \Omega$, $T_A = $ MIN to MAX				UNIT
			'ALS00A	SN54ALS00A		SN74ALS00A		
			TYP	MIN	MAX	MIN	MAX	
t_{PLH}	A or B	Y	7	3	16	3	11	ns
t_{PHL}	A or B	Y	5	2	13	2	8	

NOTE 1: Load circuit and voltage waveforms are shown in Section 1.

Figure 10.14
Typical manufacturer's specifications (Courtesy of Texas Instruments).

ALS and AS Circuits
SN54AS00, SN74AS00
QUADRUPLE 2-INPUT POSITIVE-NAND GATES

absolute maximum ratings over operating free-air temperature range (unless otherwise noted)

Supply voltage, V_{CC} . . . . 7 V
Input voltage . . . . 7 V
Operating free-air temperature range: SN54AS00 . . . . −55°C to 125°C
　　　　　　　　　　　　　　　　　SN74AS00 . . . . 0°C to 70°C
Storage temperature range . . . . −65°C to 150°C

recommended operating conditions

		SN54AS00			SN74AS00			UNIT
		MIN	NOM	MAX	MIN	NOM	MAX	
V_{CC}	Supply voltage	4.5	5	5.5	4.5	5	5.5	V
V_{IH}	High-level input voltage	2			2			V
V_{IL}	Low-level input voltage			0.8			0.8	V
I_{OH}	High-level output current			−2			−2	mA
I_{OL}	Low-level output current			20			20	mA
T_A	Operating free-air temperature	−55		125	0		70	°C

electrical characteristics over recommended operating free-air temperature range (unless otherwise noted)

PARAMETER	TEST CONDITIONS		SN54AS00			SN74AS00			UNIT
			MIN	TYP[†]	MAX	MIN	TYP[†]	MAX	
V_{IK}	$V_{CC} = 4.5$ V,	$I_I = −18$ mA			−1.2			−1.2	V
V_{OH}	$V_{CC} = 4.5$ V to 5.5 V,	$I_{OH} = −2$ mA	$V_{CC}-2$			$V_{CC}-2$			V
V_{OL}	$V_{CC} = 4.5$ V,	$I_{OL} = 20$ mA		0.35	0.5		0.35	0.5	V
I_I	$V_{CC} = 5.5$ V,	$V_I = 7$ V			0.1			0.1	mA
I_{IH}	$V_{CC} = 5.5$ V,	$V_I = 2.7$ V			20			20	µA
I_{IL}	$V_{CC} = 5.5$ V,	$V_I = 0.4$ V			−0.5			−0.5	mA
I_O[‡]	$V_{CC} = 5.5$ V,	$V_O = 2.25$ V	−30		−112	−30		−112	mA
I_{CCH}	$V_{CC} = 5.5$ V,	$V_I = 0$ V		2	3.2		2	3.2	mA
I_{CCL}	$V_{CC} = 5.5$ V,	$V_I = 4.5$ V		10.8	17.4		10.8	17.4	mA

[†]All typical values are at $V_{CC} = 5$ V, $T_A = 25$°C.
[‡]The output conditions have been chosen to produce a current that closely approximates one half of the true short-circuit output current, I_{OS}.

switching characteristics (see Note 1)

PARAMETER	FROM (INPUT)	TO (OUTPUT)	$V_{CC} = 4.5$ V to 5.5 V, $C_L = 50$ pF, $R_L = 50$ Ω, $T_A =$ MIN to MAX				UNIT
			SN54AS00		SN74AS00		
			MIN	MAX	MIN	MAX	
t_{PLH}	A or B	Y	1	5	1	4.5	ns
t_{PHL}	A or B	Y	1	5	1	4	

NOTE 1: Load circuit and voltage waveforms are shown in Section 1.

Figure 10.4
(Continued)

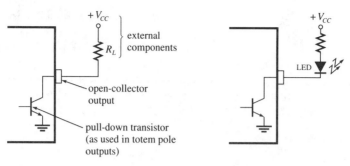

(a) An open-collector output connected to an external load. The output at the collector is low when the pull-down transistor is on.

(b) Using an open-collector output to drive a light-emitting diode (LED).

Figure 10.15
The open-collector output.

Open-Collector Outputs—The Wired AND

Many TTL gates are available in *open-collector* versions, where the pull-up transistors of the totem pole outputs are omitted. The output of such a gate is simply the collector of the pull-down transistor, which the user may connect to an external resistor and power-supply voltage. See Figure 10.15*(a)*. Any input(s) to a gate that would normally cause the output to be low also cause the pull-down transistor to be on, which allows current to flow through the external resistor and into the transistor. When the output is high, the transistor turns off and the output voltage rises. The external resistor is often called a pull-up resistor. One use for open-collector outputs is illustrated in Figure 10.15*(b)*. Here, the external load is a light-emitting diode (LED), which is turned on when the output transistor turns on. Hex inverters with open-collector outputs are often used in this way as LED drivers. Other external loads, such as relay coils, can also be activated using the method. Special versions, called buffers, are capable of operating with high-valued external supply voltages and are therefore useful for interfacing with logic families that have higher logic levels (such as CMOS).

The open-collector outputs of two or more gates can be used to logically AND the outputs of the gates. As illustrated in Figure 10.16*(a)*, the outputs are simply connected together and joined to a single pull-up resistor. The connection is called a *wired-AND* and is represented schematically by the special AND-gate symbol shown in the figure. Figure 10.16*(b)* shows the equivalent switching circuit. The output is high only when both switches are open—i.e., only when the output of each gate is high. Thus, the output is the logical AND operation of the logic functions performed by the gates. Figure 10.16*(c)* shows two 7401 open-collector NAND gates wire-ANDed to implement $\overline{AB} \cdot \overline{CD} = (\overline{A} + \overline{B})(\overline{C} + \overline{D})$. Figure 10.16*(d)* shows how three inverters can be wire-ANDed to implement a NOR operation. Figure 10.16*(e)* shows that the wired-AND operation applied to NOR gates expands the NOR operation: $(\overline{A + B})(\overline{C + D}) = \overline{ABCD} = \overline{A + B + C + D}$. Because of the absence of pull-up transistors, wired-AND connections significantly reduce switching speeds. However, they are useful in reducing the *chip count* of a system when speed is not a consideration.

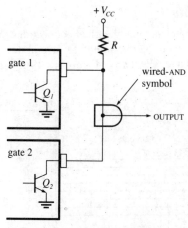

(a) Connecting open-collector outputs to perform a wired-AND operation. Each transistor is off if the output of its respective gate is high (logical 1).

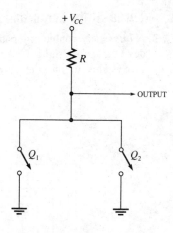

(b) Switching equivalent of (a). The output is high only if Q_1 and Q_2 are both open; i.e., only if the outputs of gates 1 and 2 are both high.

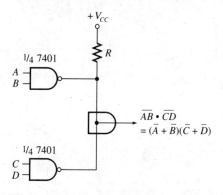

(c) Wire-ANDed output of two NAND gates.

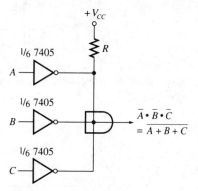

(d) The wire-ANDed output of inverters implements the NOR operation.

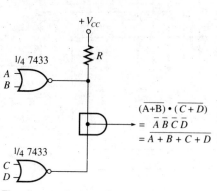

(e) The wire-ANDed output of NOR gates expands the NOR operation.

Figure 10.16
The wired-AND connection.

Three-State Logic and Bus Drivers

A *bus* is any conducting path or set of paths having electrical connections to two or more devices. An example is a ground bus, which may be routed to several devices, chips, or subsystems to provide common ground connections. Figure 10.17(a) shows another

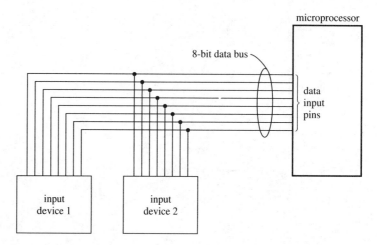

(a) Example of a bus (8-bit) used by two input devices to supply binary data to a common destination. The devices are said to *share* the bus.

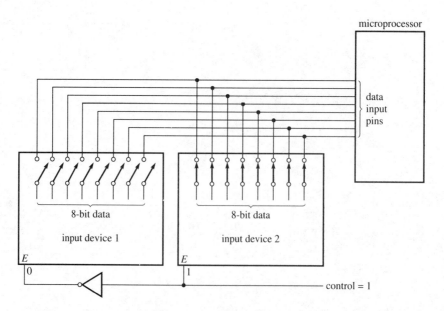

(b) Input device 1 is effectively disconnected from the bus while input device 2 is connected to it. The open switches represent high-impedance states. When its *E* (enable) input is high, a device has *control* of the bus.

Figure 10.17
Three-state devices sharing a bus.

example: an 8-conductor *data bus* connected to two input devices and used by each device to transfer 8-bit words to a microprocessor. One obvious problem with such an arrangement is that one device may produce a high output on the same line and at the same time that the other device is producing a low output. This occurrence would be in effect a short circuit between a high (current source) and a low (current sink) that would clearly have disastrous consequences. We conclude that both devices cannot use the bus *at the same time*. One device must be effectively disconnected from the bus while the other one is electrically connected to it. To accommodate this type of application, specially designed TTL devices called *bus drivers* can be put into a *high-impedance* state by a control signal applied to an appropriate pin. When such a device is in its high-impedance state, there is a very large impedance at its output, effectively isolating the device from whatever circuitry the output normally drives. Figure 10.17(b) shows a simplified equivalent diagram in which the high-impedance isolation is represented by open switches. Note that the control signal is connected to inputs labeled E (for "enable"), which in this example are assumed to be active high. In other words, a 1 applied to an E input activates the device and a 0 puts it in its high-impedance state. (Some devices have enabling inputs that are active low.) In the figure, it can be seen that device 1 is enabled when CONTROL is low and device 2 is enabled when CONTROL is high, so one and only one device is active at any given time. Since such devices have three possible output states, high, low, and high-impedance, they are said to be *three-state* devices, or to have *three-level logic*.

A TTL device is designed for three-state operation by including circuitry that can turn off *both* of the transistors in the totem pole output. When both transistors are off, there is no path for external current to flow into or out of the device. Figure 10.18 shows a simplified schematic of one method that can be used to turn off both output transistors. In this example, the circuit performs as a conventional TTL inverter when the enable input is low and is forced into a high-impedance state when the enable input is high. Thus, the enable input is active low. (Note that we could also label the control input "DISABLE" and say it is active high.) The figure shows the current flow and transistor states when the input labeled $\overline{\text{ENABLE}}$ is V_{CC} (high). Q_5 is off because its base-to-emitter is

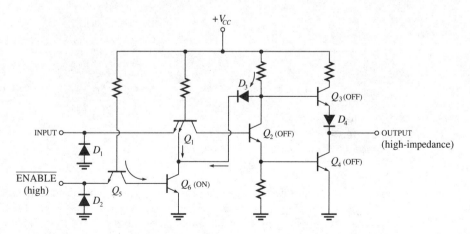

Figure 10.18
An example of a TTL inverter modified for three-state operation. The output is high impedance when $\overline{ENABLE}$ is high, the case illustrated here.

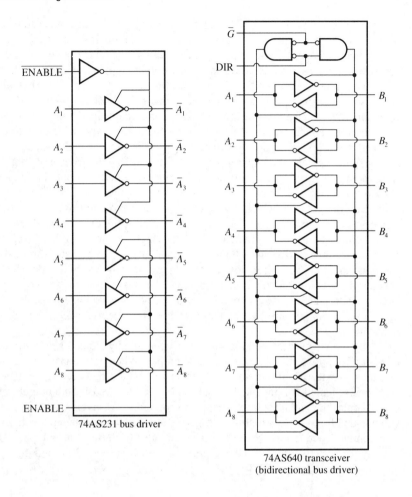

Figure 10.19
A typical three-state bus driver and transceiver.

reverse-biased. Current flows into the base of Q_6, turning it on, thus making one emitter of Q_1 low. As in normal operation, a low input to Q_1 turns off Q_2 and output transistor Q_4. The low produced by Q_6 is also applied to the base of Q_3 via diode D_3. The base voltage of Q_3 is therefore $V_{D3} + V_{CE(sat)} \approx 0.8$ V, which is insufficient to turn Q_3 on. Thus, output transistor Q_3 is also off, and the device is in its high-impedance state. When

$\overline{\text{ENABLE}}$ is low, Q_5 is on and Q_6 is off. Q_1 and the rest of the circuit then behave like a 2-input NAND gate, one of whose inputs is high: $\overline{A \cdot 1} = \overline{A}$.

A typical TTL bus driver contains eight inverters, or eight noninverting drivers, all of whose outputs can be put into a high-impedance state by a single control input. Many have more than one enable input and/or built-in gates that can be used to perform logic operations on the enable input(s). Many versions have separate enable inputs that can be used to enable groups of drivers, as, for example, two enable inputs that control two sets of four drivers each. Bus *transceivers* are two-directional bus drivers: They can be used to pass data in one direction while the other direction is high impedance and vice versa. They are also known as *bidirectional* bus drivers. Figure 10.19 shows logic diagrams for a typical bus driver and a typical bus transceiver.* Note how the three-state input control to each inverter is represented schematically. The 74AS231 has an active-low enable that controls four inverters and an active-high enable that controls the other four inverters. The truth table for the 74AS640 shows how DIR controls the direction in which data is transferred.

10-2   NMOS LOGIC

NMOS technology was introduced in Chapter 9 with a discussion of NMOS inverters having NMOS loads. Recall that NMOS logic is used in large and very large scale integrated circuits for *dedicated* applications, such as microprocessors and computer memories. Thus, input and output terminals of individual logic gates are not externally accessible, and general-purpose logic circuits are not manufactured using NMOS devices.

NMOS logic gates are simple variations of the NMOS inverter. Figure 10.20(a) shows a two-input NMOS NOR gate. Q_1 is the resistor-connected NMOS transistor that serves as a load, and Q_2 and Q_3 are switching transistors controlled by the inputs. Clearly the circuit is equivalent to a resistor and two switches connected in parallel, as shown in Figure 10.20(b). When either or both inputs are high (greater than the positive threshold voltages), the corresponding transistors are on—i.e, the corresponding switches are closed—making the output low. If both inputs are low (less than the threshold voltages), both transistors are off—i.e, both switches are open—and the output is high. The behavior described is that of a positive-logic NOR gate, as shown in the truth tables of Figure 10.20(c).

Figure 10.21(a) shows an NMOS NAND gate. Q_1 is the resistor-connected NMOS transistor, and Q_2 and Q_3 are the switching transistors. Figure 10.21(b) shows the equivalent switching circuit, consisting of a resistor and two switches in series. If either A or B or both are low, the corresponding transistors are off—i.e, the corresponding switches are open—and the output is high. If A and B are both high, the corresponding transistors are on—i.e, the corresponding switches are closed—and the output is low. The behavior described is that of a NAND gate, as shown by the truth tables in Figure 10.21(c).

PMOS gates are similar to NMOS gates, except that negative voltages (equal to the negative supply voltage) are used to turn PMOS transistors on. Positive-logic levels are thus $0 = -V_{DD}$ and $1 = 0$ V. PMOS logic is not used in new designs, since NMOS logic is so much faster.

---

*Section 10-7 gives ANSI/IEEE standard symbols (Figure 10.46).

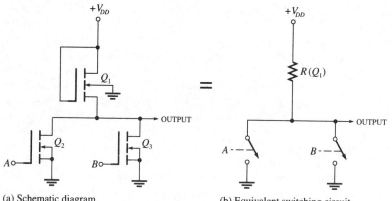

(a) Schematic diagram.

(b) Equivalent switching circuit.

A	B	OUTPUT
LO	LO	III
LO	HI	LO
HI	LO	LO
HI	HI	LO

⇒

A	B	OUTPUT = $\overline{A + B}$
0	0	1
0	1	0
1	0	0
1	1	0

(c) Truth tables.

Figure 10.20
The NMOS NOR gate.

Figure 10.21
The NMOS NAND gate.

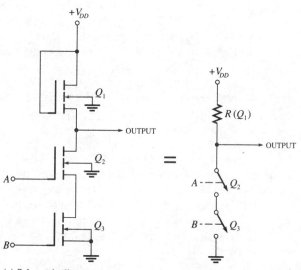

(a) Schematic diagram.

(b) Equivalent switching circuit.

A	B	OUTPUT
LO	LO	HI
LO	HI	HI
HI	LO	HI
HI	HI	LO

⇒

A	B	OUTPUT = $\overline{AB}$
0	0	1
0	1	1
1	0	1
1	1	0

(c) Truth tables.

10-3   CMOS LOGIC

Overview

CMOS technology was introduced in Chapter 9 with a discussion of the CMOS inverter. Recall that CMOS circuits contain both NMOS and PMOS transistors to speed the switching of capacitive loads. Other advantages include extremely small power consumption and the capability of being operated at high voltages, resulting in improved noise immunity. CMOS technology has been used to construct small-, medium-, and large-scale integrated circuits for a wide variety of applications, from general-purpose logic to microprocessors.

An early line of general-purpose CMOS devices has been widely marketed by RCA as its CD4000-series, also available in the functionally equivalent 14000-series from Motorola. The first series of CMOS devices that were the functional equivalents (pin-for-pin replacements) of the TTL 7400-series carried a C designation, as in the 74C00 quad 2-input CMOS NAND gate. The M in MOS stands for "metal", and the devices in all these series have metal gates. A later series of TTL-equivalent CMOS devices have silicon gates and carry an HC designation (for high-speed CMOS), as in the 74HC04 hex inverter. Although the HC-series devices are generally faster and less prone to *latch-up* (which we discuss later), they cannot be operated over as wide a voltage range as the metal-gate devices. One variation of the 74HC00-series is specially designed for interfacing with TTL circuits and carries an HCT designation, as in the 74HCT04 hex inverter. Availability of HCT devices varies with manufacturer, as does the specific TTL subfamilies with which they are compatible. Figure 10.22 is a broad comparison of several performance characteristics of CMOS and TTL devices.

A new series of advanced CMOS devices that are functionally equivalent to the 7400-series but not pin-for-pin replacements is designated by 74AC11000-series numbers. Note that the last *three* digits replace the last two digits of the 7400-series numbers. For example, the 7402 quad 2-input NOR gate becomes a 74AC11002 in the advanced CMOS series. Among other enhancements, the 74AC devices have improved

	Propagation delay	Power per gate	Speed-power product
Silicon-gate CMOS (at 100 kHz)	8 ns	0.17 mW	1.4 pJ
Metal-gate CMOS (at 100 kHz)	105 ns	0.1 mW	10.5 pJ
Standard TTL	10 ns	10 mW	100 pJ
STTL	3 ns	19 mW	57 pJ
LSTTL	10 ns	2 mW	20 pJ
ASTTL	1.5 ns	8.5 mW	12.8 pJ
ALSTTL	4 ns	1 mW	4 pJ

Figure 10.22
Comparison of CMOS and TTL performance characteristics.

pin placement to increase noise immunity. A TTL-compatible version is designated by 74ACT11000 numbers. These devices are designed to interface with TTL circuits but are not pin-for-pin replacements.

Buffered and Unbuffered Gates

Some metal-gate CMOS circuits are available in *buffered* and *unbuffered* versions. The gates in buffered circuits have CMOS inverters in series with their outputs to suppress switching transients and to improve the sharpness of the voltage transition at the output. Referring to the schematic of the unbuffered inverter in Figure 9.16(a), we can see that Q_1 and Q_2 will both conduct briefly when a rising input causes V_{GS} of each transistor to pass through its threshold value, creating a transient current "spike" from the power supply. Increasing the voltage gain through a gate by adding inverters in series with it reduces the time required for it to change state but increases the propagation delay through the gate. The high-speed (HC) series devices are buffered but achieve increased speed through the improved silicon-gate technology.

Figure 10.23(a) shows a 2-input unbuffered CMOS NAND gate. Note that Q_1 and Q_2 are parallel-connected PMOS transistors and that Q_3 and Q_4 are series-connected NMOS transistors. Figure 10.23(b) shows the equivalent switching circuit when both inputs are low. Note that the gates of both PMOS transistors are negative *with respect to their sources*, since the sources are connected to $+V_{DD}$. Therefore, Q_1 and Q_2 are both on. Since the gate-to-source voltages of Q_3 and Q_4 are both 0 V, those transistors are off. The output is therefore connected to $+V_{DD}$ (high) through Q_1 and Q_2 and is disconnected from ground. Figure 10.23(c) shows the equivalent switching circuit when $A = 0$ V and $B = +V_{DD}$. In this case, Q_1 is on because $V_{GS1} = -V_{DD}$ and Q_4 is on because $V_{GS4} = +V_{DD}$. Q_2 and Q_3 are off because their gate-to-source voltages are 0 V. Since Q_1 is on and Q_3 is off, the output is connected to $+V_{DD}$ and is disconnected from ground. When $A = +V_{DD}$ and $B = 0$ V (not shown), the situation is similar: The output is connected to $+V_{DD}$ because Q_2 is on and is disconnected from ground because Q_4 is off. Finally, when both outputs are high, as shown in Figure 10.23(d), Q_1 and Q_2 are both off and Q_3 and Q_4 are both on. The output is therefore connected to ground and disconnected from V_{DD}. The truth tables confirm that the circuit performs the NAND operation. A 3-input NAND gate is constructed similarly, with a third parallel PMOS transistor and a third series NMOS transistor.

Figure 10.24(a) shows an unbuffered 2-input CMOS NOR gate. In this circuit, the PMOS transistors are connected in series and the NMOS transistors are in parallel. The equivalent switching circuits shown in the figure demonstrate how this circuit performs the NOR operation. As in the NAND gate, each PMOS transistor is on when its gate-to-source voltage is $-V_{DD}$ and each NMOS transistor is on when its gate is $+V_{DD}$. A 3-input NOR gate is constructed by adding a third PMOS transistor in series and a third NMOS transistor in parallel.

Transmission Gates

A transmission gate is simply a digitally controlled CMOS switch. When the switch is open (off), the impedance between its terminals is very large. It is used to implement special logic functions, as we will see in a forthcoming example. Since the gate can transmit signals in both directions (and is used in analog circuits that way), it is also called a *bilateral* transmission gate.

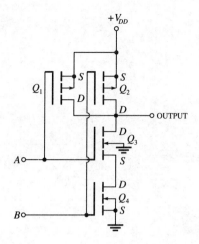

(a) Schematic.

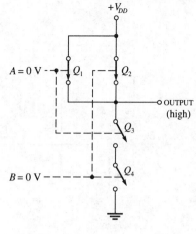

(b) $A = B = 0$ V
$V_{GS1} = V_{GS2} = -V_{DD}$
$V_{GS3} = V_{GS4} = 0$ V

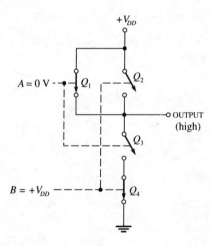

(c) $A = 0$ V, $B = +V_{DD}$
$V_{GS1} = -V_{DD}$
$V_{GS2} = V_{GS3} = 0$ V
$V_{GS4} = +V_{DD}$

(d) $A = B = +V_{DD}$
$V_{GS1} = V_{GS2} = 0$ V
$V_{GS3} = V_{GS4} = +V_{DD}$

A	B	OUTPUT
0 V	0 V	$+V_{DD}$
0 V	$+V_{DD}$	$+V_{DD}$
$+V_{DD}$	0 V	$+V_{DD}$
$+V_{DD}$	$+V_{DD}$	0 V

(e) Voltage truth table.

A	B	OUTPUT $= \overline{AB}$
0	0	1
0	1	1
1	0	1
1	1	0

(f) Positive-logic truth table.

Figure 10.23
The unbuffered CMOS NAND gate.

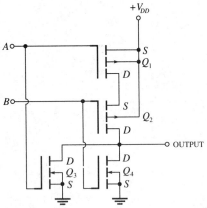

(a) Schematic.

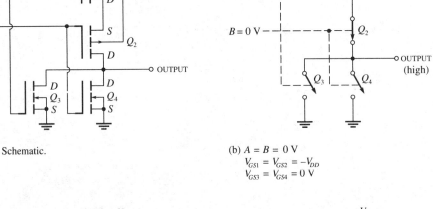

(b) $A = B = 0$ V
$V_{GS1} = V_{GS2} = -V_{DD}$
$V_{GS3} = V_{GS4} = 0$ V

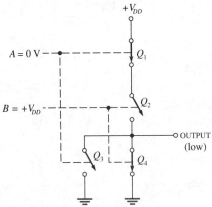

(c) $A = 0$ V, $B = +V_{DD}$
$V_{GS1} = -V_{DD}$
$V_{GS2} = V_{GS3} = 0$ V
$V_{GS4} = +V_{DD}$

(d) $A = B = +V_{DD}$
$V_{GS1} = V_{GS2} = 0$ V
$V_{GS3} = V_{GS4} = +V_{DD}$

A	B	OUTPUT
0 V	0 V	$+V_{DD}$
0 V	$+V_{DD}$	0 V
$+V_{DD}$	0 V	0 V
$+V_{DD}$	$+V_{DD}$	0 V

(e) Voltage truth table.

A	B	OUTPUT $= \overline{A + B}$
0	0	1
0	1	0
1	0	0
1	1	0

(f) Positive-logic truth table.

Figure 10.24
The unbuffered CMOS NOR *gate.*

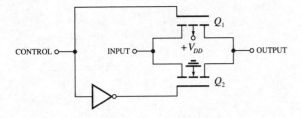

(a) Schematic.

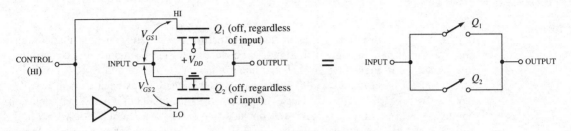

(b) CONTROL high.

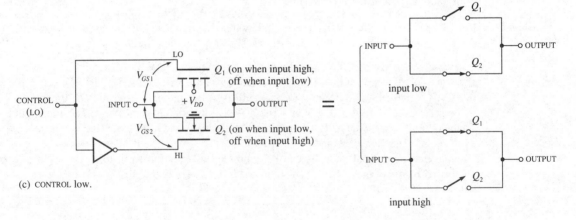

(c) CONTROL low.

Figure 10.25
The CMOS transmission gate. (Continued on p. 346.)

Figure 10.25 shows the schematic diagram of a transmission gate and two symbols used to represent it.* Notice that the NMOS and PMOS transistors are connected in parallel and that control signals are connected to their gates. To analyze the transmission gate let us first suppose that CONTROL is high [Figure 10.25(b)]. Then the gate of PMOS transistor Q_1 is high and the gate of NMOS transistor Q_2 is low. If the (data) input is low, V_{GS1} is positive and V_{GS2} is 0 V, so neither transistor is on. If that input is high, V_{GS1} is

---

*Section 10–7 shows the ANSI/IEEE standard symbol (Figure 10.47).

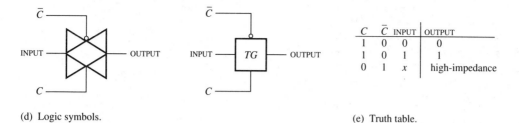

C	$\overline{C}$	INPUT	OUTPUT
1	0	0	0
1	0	1	1
0	1	x	high-impedance

(d) Logic symbols.

(e) Truth table.

Figure 10.25
(Continued)

0 V and V_{GS2} is negative, so again neither transistor is on. We conclude that a high CONTROL signal puts the device in a high-impedance state. Now suppose that CONTROL is low, as shown in Figure 10.25(c). If the input is low, V_{GS1} is 0 V and V_{GS2} is positive. Therefore, Q_1 is off and Q_2 is on. When the input is high, V_{GS1} is negative and V_{GS2} is 0 V, so Q_1 is on and Q_2 is off. Thus, there is always one conducting path from input to output when CONTROL is low, and we conclude that CONTROL is an active-low enabling signal. In the logic symbols shown in Figure 10.25(d), the control input is separated into C and $\overline{C}$. Input C is connected directly to the gate of the N-channel device and $\overline{C}$ to the gate of the P-channel device. Thus, when C is 0, Q_2 is off, $\overline{C}$ is 1, and Q_1 is off, putting both paths in a high-impedance state.

Example 10–2

Find the positive-logic function performed by the circuit whose logic diagram is shown in Figure 10.26(a). (This is one gate in the 74HC86 integrated circuit.)

Solution. We see that the control inputs of transmission gate 1 (TG_1) are $C_1 = \overline{A}$ and $\overline{C}_1 = A$. Therefore, as shown in the truth table in Figure 10.26(b), TG_1 is on when A is 0. TG_2 is off when A is 0 because $\overline{C}_2 = 1$ and $C_2 = 0$. Since TG_1 transmits B, the output is the same as B when A is 0. When A is 1, TG_1 is off and TG_2 is on. Since TG_2 transmits $\overline{B}$, the output is the same as $\overline{B}$ when A is 1. As shown in the truth table, the output is the exclusive-OR function of A and B: $A \oplus B$.

Note that our verbal analysis can be expressed in equation form to confirm the exclusive-OR nature of the circuit: "The output is B if A is 0 or is $\overline{B}$ if A is 1"—i.e, OUTPUT = $\overline{A}B + A\overline{B} = A \oplus B$.

Figure 10.26
Example 10–2.

A	B	C_1	$\overline{C}_1$	TG_1	C_2	$\overline{C}_2$	TG_2	OUTPUT
0	0	1	0	ON	0	1	OFF	$B = 0$
0	1	1	0	ON	0	1	OFF	$B = 1$
1	0	0	1	OFF	1	0	ON	$\overline{B} = 1$
1	1	0	1	OFF	1	0	ON	$\overline{B} = 0$

(b)

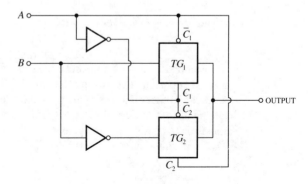

(a)

Open-Drain and High-Impedance Outputs

CMOS logic gates are available with open-drain outputs, similar to their TTL counterparts with open-collector outputs. Figure 10.27(a) shows two inverters at the output of a CMOS gate, used to provide the buffering discussed earlier. As shown in Figure 10.27(b), an open-drain output is obtained by omitting the PMOS transistor in the output inverter. Diode D_1 is connected (internally) to provide protection from electrostatic discharge, which we discuss in more detail later. Open-drain gates can be used with external pull-up resistors to perform wired-AND operations, as described in Section 10–1. Note that the open-drain output stage effectively performs an inversion. Thus, if two of the circuits shown in part (b) of the figure were wire-ANDed, the output would be $\overline{\overline{A}\,\overline{B}} = AB$. Figure 10.27(c) shows two 74HC03 wire-ANDed NAND gates that produce $\overline{AB} \cdot \overline{CD}$. Note that the inverting buffer and the inverting output stage logically cancel each other inside each gate.

CMOS buffers and bus drivers are constructed using a PMOS and an NMOS transistor in an output totem pole, as illustrated in Figure 10.28(a), p. 349. When both transistors are off, the output is in a high-impedance state. The circuit shown is an inverting driver and the logic gates are used to enable it. Observe that the gate of PMOS

Figure 10.27
The CMOS open-drain output.
(Continued on p. 348.)

(Continued on p. 348.)

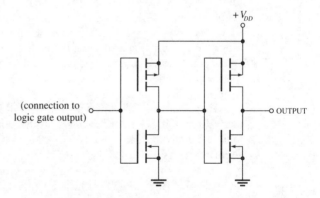

(a) Two series inverters providing output buffering in a CMOS logic gate.

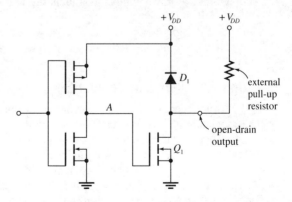

(b) An open-drain output is obtained by omitting the PMOS transistor. D_1 is a protective diode. When A is high, Q_1 is ON and the output is low. When A is low, Q_1 is OFF and the output is high. Thus, the output stage performs inversion.

Figure 10.27
(Continued)

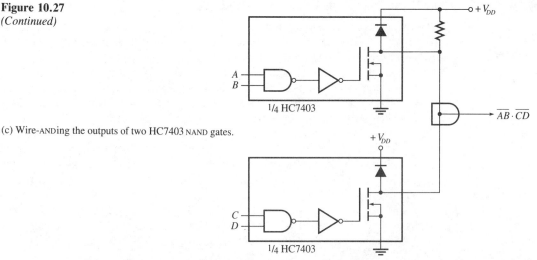

(c) Wire-ANDing the outputs of two HC7403 NAND gates.

transistor Q_1 is $E + I$, where E is the enabling signal and I is the input. The gate of NMOS transistor Q_2 is $\overline{EI}$. As can be seen in the truth table in Figure 10.28(b), E is an active-low enabling signal. When it is low and the input is low, $E + I$ is low, causing Q_1 to be on. $\overline{EI}$ is also low, so Q_2 is off. Therefore, the output is connected through Q_1 to $+V_{DD}$ (high). When E is low and the input is high, $E + I$ and $\overline{EI}$ are both high, turning Q_1 off and Q_2 on. The output is then connected through Q_2 to ground (low). When E is high, $E + I$ is always high and $\overline{EI}$ is always low. Therefore, both transistors are off and the output is in a high-impedance state.

The means used to show active-low control inputs on logic diagrams is confusing and inconsistent throughout the industry. As we know, a small circle ("bubble") is used on logic diagrams to denote inversion. A bar is drawn over the name of a control signal to signify that it is active-low, as, for example, $\overline{\text{ENABLE}}$. Unfortunately, it is a widespread practice to use both the bar and a bubble in a logic diagram showing an active-low control input. Figure 10.29, p. 349, illustrates this practice. Although it is possible to infer that the two inversions cancel each other, meaning that the input is active high, most logic diagrams are drawn without that intention. In this book, wherever possible, we avoid such inversion notation and attempt to show the true nature of digital components using expanded logic diagrams and/or truth tables. When in doubt about the behavior of any digital system having active-low controls, always consult (or construct) a truth table.

Specifications and Standards

The Joint Electron Devices Engineering Council (JEDEC) of the Electronic Industries Association (EIA) has established certain standard specifications for buffered and unbuffered CMOS circuits. These standards are shown in Figure 10.30, pp. 349–350. Buffered devices are identified in the standards as B-types, and unbuffered are identified as UB-types. Manufacturers' component designations have a B or UB suffix to indicate whether they are buffered or unbuffered. For example, the 4011UB is an unbuffered quad 2-input NAND gate and the 14001B is a buffered quad 2-input NOR gate. The supply voltage, V_{DD}, specified by the standards (not shown in Figure 10.30) can range from -0.5 V to $+18$ V. (V_{DD} is referenced to V_{SS}, the voltage to which the substrate of N-channel devices is connected. V_{SS} may be a negative voltage, but for most digital applications it is ground, or 0 V.) Note that the standards depend heavily on the value of V_{DD} as well as on temperature. Not all manufacturers' products conform to all of these standards, due in some cases to special functions or features that selected devices have. On the other hand, the specifications of many basic components exceed the JEDEC standards. The 7400HC-series devices, in particular, do not meet the 18-V specification for the maximum value of V_{DD}, since V_{DD} is limited to 7 V.

Figure 10.31, pp. 351–352, shows a typical set of specifications for a silicon-gate CMOS circuit, in this case a 74HC02 quad 2-input NOR gate. Although we have referred to the supply voltage in CMOS circuits as V_{DD}, note that it is conventional to identify it

Figure 10.28

An inverting CMOS driver with high-imped-ance (three-state) output.

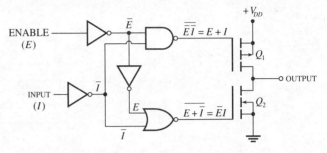

(a) Logic diagram.

Figure 10.29

The conventional means for representing ac-tive-low control inputs. The device is en-abled when the signal labeled ENABLE is low. In most logic diagrams, the bubble at the input is not intended to infer a second inversion.

I	$\bar{I}$	E	$\bar{E}$	$E+I$	$\bar{E}I$	Q_1	Q_2	OUTPUT
0	1	0	1	0	0	ON	OFF	1
1	0	0	1	1	1	OFF	ON	0
X	X	1	0	1	0	OFF	OFF	high-impedance

(b) Truth table.

Figure 10.30

EIA/JEDEC standards for CMOS industry B (buffered) and UB (unbuffered) device specifica-tions. (Mil = military; comm = commercial). (Continued on p. 350.)

ELECTRICAL CHARACTERISTICS

PARAMETER		TEMP RANGE	V_{DD} (Vdc)	CONDITIONS	LIMITS						UNITS
					T_{LOW}*		+25°C		T_{HIGH}*		
					Min	Max	Min	Max	Min	Max	
I_{DD}	Quiescent Device Current	Mil	5			0.25		0.25		7.5	µAdc
			10	$V_{in} = V_{SS}$ or V_{DD}		0.5		0.5		15	
			15			1.0		1.0		30	
	GATES	Comm	5	All valid input combinations		1.0		1.0		7.5	µAdc
			10			2.0		2.0		15	
			15			4.0		4.0		30	
		Mil	5			1.0		1.0		30	µAdc
			10	$V_{IN} = V_{SS}$ or V_{DD}		2.0		2.0		60	
			15			4.0		4.0		120	
	BUFFERS, FLIP-FLOPS	Comm	5	All valid input combinations		4		4.0		30	µAdc
			10			8		8.0		60	
			15			16		16.0		120	
		Mil	5			5		5		150	µAdc
			10	$V_{IN} = V_{SS}$ or V_{DD}		10		10		300	
			15			20		20		600	
	MSI	Comm	5	All valid input combinations		20		20		150	µAdc
			10			40		40		300	
			15			80		80		600	

Figure 10.30
(Continued.)

PARAMETER		TEMP RANGE	V$_{DD}$ (Vdc)	CONDITIONS	LIMITS						UNITS
					T$_{LOW}$*		+25°C		T$_{HIGH}$*		
					Min	Max	Min	Max	Min	Max	
V$_{OL}$	Low-Level Output Voltage	All	5	V$_{IN}$ = V$_{SS}$ or V$_{DD}$		0.05		0.05		0.05	Vdc
			10			0.05		0.05		0.05	
			15	$|I_O|$ < 1µA		0.05		0.05		0.05	
V$_{OH}$	High-Level Output Voltage	All	5	V$_{IN}$ = V$_{SS}$ or V$_{DD}$	4.95		4.95		4.95		Vdc
			10		9.95		9.95		9.95		
			15	$|I_O|$ < 1µA	14.95		14.95		14.95		
V$_{IL}$	Input Low Voltage# B Types	All	5	V$_O$ = 0.5V or 4.5 V		1.5		1.5		1.5	Vdc
			10	V$_O$ = 1.0V or 9.0V		3.0		3.0		3.0	
			15	V$_O$ = 1.5V or 13.5V		4.0		4.0		4.0	
				$|I_O|$ < 1µA							
V$_{IL}$	Input Low Voltage# UB Types	All	5	V$_O$ = 0.5V or 4.5V		1.0		1.0		1.0	Vdc
			10	V$_O$ = 1.0V or 9.0V		2.0		2.0		2.0	
			15	V$_O$ = 1.5V or 13.5V		2.5		2.5		2.5	
				$|I_O|$ < 1µA							
V$_{IH}$	Input High Voltage# B Types	All	5	V$_O$ = 0.5V or 4.5V	3.5		3.5		3.5		Vdc
			10	V$_O$ = 1.0V or 9.0V	7.0		7.0		7.0		
			15	V$_O$ = 1.5V or 13.5V	11.0		11.0		11.0		
				$|I_O|$ < 1µA							
V$_{IH}$	Input High Voltage# UB Types	All	5	V$_O$ = 0.5V or 4.5V	4.0		4.0		4.0		Vdc
			10	V$_O$ = 1.0V or 9.0V	8.0		8.0		8.0		
			15	V$_O$ = 1.5V or 13.5V	12.5		12.5		12.5		
				$|I_O|$ < 1µA							
I$_{OL}$	Output Low (Sink) Current	Mil		V$_O$ = 0.4V,							mAdc
			5	V$_{IN}$ = 0 or 5V	0.64		0.51		0.36		
				V$_O$ = 0.5V,							
			10	V$_{IN}$ = 0 or 10V	1.6		1.3		0.9		
				V$_O$ = 1.5V,							
			15	V$_{IN}$ = 0 or 15V	4.2		3.4		2.4		
		Com		V$_O$ = 0.4V,							mAdc
			5	V$_{IN}$ = 0 or 5V	0.52		0.44		0.36		
				V$_O$ = 0.5V,							
			10	V$_{IN}$ = 0 or 10V	1.3		1.1		0.9		
				V$_O$ = 1.5V,							
			15	V$_{IN}$ = 0 or 15V	3.6		3.0		2.4		
I$_{OH}$	Output High (Source) Current	Mil		V$_O$ = 4.6 V,							mAdc
			5	V$_{IN}$ = 0 or 5V	−0.25		−0.2		−0.14		
				V$_O$ = 9.5V,							
			10	V$_{IN}$ = 0 or 10V	−0.62		−0.5		−0.35		
				V$_O$ = 13.5V,							
			15	V$_{IN}$ = 0 or 15V	−1.8		−1.5		−1.1		
		Com		V$_O$ = 4.6V,							mAdc
			5	V$_{IN}$ = 0 or 5V	−0.2		−0.16		−0.12		
				V$_O$ = 9.5V,							
			10	V$_{IN}$ = 0 or 10V	−0.5		−0.4		−0.3		
				V$_O$ = 13.5V							
			15	V$_{IN}$ = 0 or 15V	−1.4		−1.2		−1.0		
I$_{IN}$	Input Current	Mil	15	V$_{IN}$ = 0 or 15V		±0.1		±0.1		±1.0	µAdc
		Comm	15	V$_{IN}$ = 0 or 15V		±0.3		±0.3		±1.0	µAdc
I$_{OZ}$	3-State Output Leakage Current	Mil	15	V$_{IN}$ = 0 or 15V		±0.4		±0.4		±12	µAdc
		Comm	15	V$_{IN}$ = 0 or 15V		±1.6		±1.6		±12	µAdc
C$_{IN}$	Input Capacitance per unit load	All	—	Any Input				7.5			pF

*T$_{LOW}$ = −55°C for Military temperature range device, −40°C for Commercial temperature range device.
T$_{HIGH}$ = +125°C for Military temperature range device, +85°C for Commercial temperature range device.
#Applies for Worst Case input combinations.

MC54/74HC02

MAXIMUM RATINGS*

Symbol	Parameter	Value	Unit
V_{CC}	DC Supply Voltage (Referenced to GND)	−0.5 to +7.0	V
V_{in}	DC Input Voltage (Referenced to GND)	−1.5 to V_{CC}+1.5	V
V_{out}	DC Output Voltage (Referenced to GND)	−0.5 to V_{CC}+0.5	V
I_{in}	DC Input Current, per Pin	± 20	mA
I_{out}	DC Output Current, per Pin	± 25	mA
I_{CC}	DC Supply Current, V_{CC} and GND Pins	± 50	mA
P_D	Power Dissipation in Still Air, Plastic or Ceramic DIP† SOIC Package†	750 500	mW
T_{stg}	Storage Temperature	−65 to +150	°C
T_L	Lead Temperature, 1 mm from Case for 10 Seconds (Plastic DIP or SOIC Package) (Ceramic DIP)	 260 300	°C

This device contains protection circuitry to guard against damage due to high static voltages or electric fields. However, precautions must be taken to avoid applications of any voltage higher than maximum rated voltages to this high-impedance circuit. For proper operation, V_{in} and V_{out} should be constrained to the range GND ≤ (V_{in} or V_{out}) ≤ V_{CC}.

Unused inputs must always be tied to an appropriate logic voltage level (e.g., either GND or V_{CC}). Unused outputs must be left open.

*Maximum Ratings are those values beyond which damage to the device may occur.
Functional operation should be restricted to the Recommended Operating Conditions.
†Derating — Plastic DIP: − 10 mW/°C from 65° to 125°C
Ceramic DIP: − 10 mW/°C from 100° to 125°C
SOIC Package: − 7 mW/°C from 65° to 125°C

RECOMMENDED OPERATING CONDITIONS

Symbol	Parameter		Min	Max	Unit
V_{CC}	DC Supply Voltage (Referenced to GND)		2.0	6.0	V
V_{in}, V_{out}	DC Input Voltage, Output Voltage (Referenced to GND)		0	V_{CC}	V
T_A	Operating Temperature, All Package Types		− 55	+ 125	°C
t_r, t_f	Input Rise and Fall Time (Figure 1)	V_{CC} = 2.0 V V_{CC} = 4.5 V V_{CC} = 6.0 V	0 0 0	1000 500 400	ns

DC ELECTRICAL CHARACTERISTICS (Voltages Referenced to GND)

Symbol	Parameter	Test Conditions		V_{CC} V	Guaranteed Limit			Unit
					25°C to −55°C	≤85°C	≤125°C	
V_{IH}	Minimum High-Level Input Voltage	V_{out} = 0.1 V or V_{CC} − 0.1 V |I_{out}| ≤20 µA		2.0 4.5 6.0	1.5 3.15 4.2	1.5 3.15 4.2	1.5 3.15 4.2	V
V_{IL}	Maximum Low-Level Input Voltage	V_{out} = 0.1 V or V_{CC} − 0.1 V |I_{out}| ≤20 µA		2.0 4.5 6.0	0.3 0.9 1.2	0.3 0.9 1.2	0.3 0.9 1.2	V
V_{OH}	Minimum High-Level Output Voltage	V_{in} = V_{IH} or V_{IL} |I_{out}| ≤20 µA		2.0 4.5 6.0	1.9 4.4 5.9	1.9 4.4 5.9	1.9 4.4 5.9	V
		V_{in} = V_{IH} or V_{IL}	|I_{out}| ≤4.0 mA |I_{out}| ≤5.2 mA	4.5 6.0	3.98 5.48	3.84 5.34	3.70 5.20	
V_{OL}	Maximum Low-Level Output Voltage	V_{in} = V_{IH} or V_{IL} |I_{out}| ≤20 µA		2.0 4.5 6.0	0.1 0.1 0.1	0.1 0.1 0.1	0.1 0.1 0.1	V
		V_{in} = V_{IH} or V_{IL}	|I_{out}| ≤4.0 mA |I_{out}| ≤5.2 mA	4.5 6.0	0.26 0.26	0.33 0.33	0.40 0.40	
I_{in}	Maximum Input Leakage Current	V_{in} = V_{CC} or GND		6.0	± 0.1	± 1.0	± 1.0	µA
I_{CC}	Maximum Quiescent Supply Current (per Package)	V_{in} = V_{CC} or GND I_{out} = 0 µA		6.0	2	20	40	µA

Figure 10.31

Specifications for the MC54/74HC02 quad 2-input NOR *gate (Courtesy of Motorola, Inc.).*

MC54/74HC02

AC ELECTRICAL CHARACTERISTICS ($C_L = 50$ pF, Input $t_r = t_f = 6$ ns)

Symbol	Parameter	V_{CC} V	Guaranteed Limit			Unit
			25°C to −55°C	≤85°C	≤125°C	
t_{PLH}, t_{PHL}	Maximum Propagation Delay, Input A or B to Output Y (Figures 1 and 2)	2.0	90	115	135	ns
		4.5	18	23	27	
		6.0	15	20	23	
t_{TLH}, t_{THL}	Maximum Output Transition Time, Any Output (Figures 1 and 2)	2.0	75	95	110	ns
		4.5	15	19	22	
		6.0	13	16	19	
C_{in}	Maximum Input Capacitance	—	10	10	10	pF

C_{PD}	Power Dissipation Capacitance (Per Gate) Used to determine the no-load dynamic power consumption: $P_D = C_{PD} V_{CC}^2 f + I_{CC} V_{CC}$	Typical @ 25°C, $V_{CC} = 5.0$ V	
		22	pF

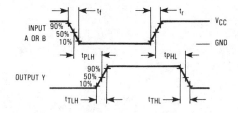

Figure 1. Switching Waveforms

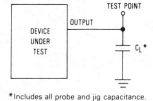

*Includes all probe and jig capacitance.

Figure 2. Test Circuit

EXPANDED LOGIC DIAGRAM
(¼ of the Device)

Figure 10.31
(Continued)

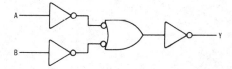

as V_{CC} in CMOS series that are functionally equivalent to TTL. In the portion of the specifications titled DC ELECTRICAL CHARACTERISTICS, we see again that supply voltage and temperature have a strong influence on the values of most of the quantities listed. Note that I_{in} at temperatures less than or equal to 85°C is specified to be no greater than 1 μA, whether the input is low or high. This current is primarily reverse-biased *leakage* current from protective diodes connected at the inputs. Except for C_{PD}, which is discussed presently, the only other specifications that need explanation are t_{TLH} and t_{THL}. These are the low-to-high and high-to-low *transition* times, not to be confused with t_{pLH} and t_{pHL}. As can be seen in the diagram accompanying the specifications, the transition times are simply the rise and fall times of the *output* waveform. Note this important point: Propagation delays are heavily dependent on load capacitance C_L, and the values listed are for the specific case $C_L = 50$ pF.

Power Dissipation and Propagation Delay

The quantity C_{PD} listed in the specifications in Figure 10.31 is the internal *power dissipation capacitance*. For computing power dissipation, its value is the amount of output capacitance *equivalent* to all the internal capacitance of a gate. Thus, it is added to actual load capacitance when computing power dissipation. The total power dissipation is the sum of the *dynamic* dissipation caused by switching the capacitance at a particular frequency and the *static* dissipation, computed as the product of the dc supply voltage and the dc current drawn from the supply. Of course, capacitance itself cannot dissipate power. The dynamic dissipation occurs in the resistive components of a gate when current surges in and out to charge and discharge the capacitance. The higher the frequency at which a gate is switched on and off, the greater the number of current surges in a fixed period of time, so the greater the average power dissipated. The dynamic dissipation can be found from

$$P\text{(dynamic, per gate)} = (C_L + C_{PD})V_{CC}^2 f \tag{10.4}$$

where C_L and C_{PD} are the load and power-dissipation capacitances per gate and f is the switching frequency. The total static dissipation for an integrated circuit is

$$P\text{(static, total)} = V_{CC}I_{CC} \tag{10.5}$$

The total power dissipation for an integrated circuit is the sum of the dynamic dissipations of all of the gates and the total static dissipation of the circuit. This computation is demonstrated in Example 10–3.

As previously indicated, propagation delay is strongly dependent on load capacitance. The total delay is the sum of the delay due to internal capacitance alone and that due to load capacitance. The delay due to internal capacitance alone is called the *intrinsic* propagation delay. The delay due to load capacitance can be approximated as follows:

$$t_p(C_L) \approx 0.5 R_o C_L \text{ seconds} \tag{10.6}$$

where $t_p(C_L)$ is either t_{pLH} or t_{pHL}, R_o is the output resistance of the gate, and C_L is the total load capacitance. R_o depends on the supply voltage, the output buffering (if any), and on whether the output is switching from low to high or from high to low. A typical value for a buffered, HC-series gate at $V_{CC} = +5$ V is $R_o = 150 \ \Omega$. Manufacturers' product literature often contains information on the short-circuit output current, I_{OS}, at a specific value of V_{CC}, which can be used to estimate R_o:

$$Ro \approx \frac{V_{CC}}{I_{OS}} \tag{10.7}$$

The value of I_{OS} may be specified as an output short-circuit *sink* current, which is used in (10.7) to find R_o when the output is switching from high to low or as an output short-circuit *source* current, which is used to find R_o when the output is switching from low to high. In practice, the values of R_o in a well-buffered gate do not depend heavily on whether the output is switching from low to high or from high to low. Recapitulating,

$$t_{pHL} \approx t_{pHL} \text{ (intrinsic)} + 0.5R_oC_L \tag{10.8}$$

$$t_{pLH} \approx t_{pLH} \text{ (intrinsic)} + 0.5R_oC_L \tag{10.9}$$

where the appropriate value of R_o is used in each equation, as just discussed.

Example 10–3

A 74HC02 quad 2-input NOR gate is operated at room temperature with a 4.5-V supply voltage. The total dc current drawn from the supply is 15 μA. Each NOR gate drives another gate whose input capacitance is 6 pF, and each is driven by a 1-MHz square wave. The short-circuit output current (both sink and source) is 17 mA. Find

a. The total power dissipation of the circuit;
b. The intrinsic propagation delays of each gate;
c. The approximate total propagation delays of each gate; and
d. The speed-power product, per gate.

Solution. Room temperature is 25°C, so all values taken from the specifications in Figure 10.31 are those at $T = 25°C$.

a. From the specifications (Figure 10.31), $C_{PD} = 22$ pF per gate. By equation (10.4), the dynamic dissipation in *each* gate is

$$(C_L + C_{PD})V_{CC}^2 f = (6 \text{ pF} + 22 \text{ pF})(4.5 \text{ V})^2(10^6 \text{ Hz}) = 0.567 \text{ mW}$$

Since there are four gates, the total dynamic dissipation is 4(0.567 mW) = 2.268 mW. By equation (10.5), the total static dissipation is

$$V_{CC}I_{CC} = (4.5 \text{ V})(15 \text{ μA}) = 0.0675 \text{ mW}$$

Thus the total dissipation is

$$P_D(\text{total}) = P(\text{total})_{\text{dynamic}} + P(\text{total})_{\text{static}} = 2.268 \text{ mW} + 0.0675 \text{ mW}$$
$$= 2.336 \text{ mW}$$

Note that the dynamic term dominates the total dissipation, due to the relatively high frequency. The total dissipation is very nearly directly proportional to frequency at higher frequencies.

b. From the specifications, the propagation delays t_{pLH} and t_{pHL}, with $C_L = 50$ pF, are no greater than 18 ns. Since t_{pHL} and t_{pLH} are assumed to have the same value ($t_p = 18$ ns), we combine equations (10.8) and (10.9) into the single equation

$$t_p = t_p(\text{instrinsic}) + 0.5R_oC_L$$

From equation (10.7),

$$R_o \approx \frac{V_{CC}}{I_{OS}} = \frac{4.5 \text{ V}}{17 \text{ mA}} = 265 \text{ Ω}$$

Using $t_p = 18$ ns and $C_L = 50$ pF, we find

$$18 \text{ ns} = t_p \text{ (intrinsic)} + 0.5(265 \text{ Ω})(50 \text{ pF})$$
$$t_p \text{ (instrinsic)} = 18 \times 10^{-9} \text{ s} - (0.5)(265 \text{ Ω})(50 \times 10^{-12} \text{ F}) = 11.375 \text{ ns}$$

c. In this case, C_L is the actual load capacitance for each gate (6 pF).

$$t_p = t_p(\text{instrinsic}) + 0.5R_oC_L = 11.275 \text{ ns} + 0.5(265 \ \Omega)(6 \text{ pF})$$
$$= 12.17 \text{ ns}$$

d. Since there are four gates, the power dissipation per gate is the total dissipation divided by 4:

$$P_D(\text{per gate}) = \frac{P_D(\text{total})}{4} = \frac{2.336 \text{ mW}}{4} = 0.584 \text{ mW}$$

The speed-power product is therefore

$$P_D(\text{per gate})t_p = (0.584 \times 10^{-3} \text{ W})(12.17 \times 10^{-9} \text{ s}) = 7.1 \text{ pJ}$$

Drill
10–3

a. Find the total power dissipation in Example 10–3 if the frequency of the square wave is reduced to 1 kHz. Assume all other values remain the same.
b. Find the propagation delay of each gate when the load capacitance on each is 100 pF.

Answer: (a) 68.1 μW; (b) 24.63 ns.

CMOS Hazards

Every MOS device is vulnerable to the buildup of electrical charge on its insulated gate. Recall that the relationship between charge Q and voltage V on a capacitor having capacitance C is

$$V = \frac{Q}{C} \tag{10.10}$$

Since the input capacitance at the gate is usually quite small (a few picofarads), a relatively small amount of charge can create a large voltage. The primary source of charge is ''static'' electricity, usually produced by handling and the motion of various kinds of plastics and textiles. The breakdown voltage of a MOS gate is on the order of 100 V, and it is not unusual for an individual wearing certain kinds of garments or walking on certain floor coverings to generate enough charge to produce hundreds of volts. Circuit boards and cables are themselves charge generators, particularly when they are inserted or removed from connectors. MOS circuits are said to be sensitive to electrostatic discharge (ESD) and must be protected from that hazard. All modern circuits have built-in protective networks at their inputs and outputs, but caution must still be exercised to prevent ESD damage. Figure 10.32 shows a typical (internal) network used to protect

Figure 10.32
A typical network used to protect MOS inputs from ESD damage.

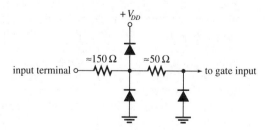

inputs from both positive and negative voltages caused by charge buildup. The resistors in series with the input capacitance of the gate reduce the rate at which a rapidly changing transient can build charge, to give the shorting diodes time to turn on. Advanced CMOS devices (the 74AC11000 series) have more sophisticated ESD circuitry at each input, including several bipolar transistors and a MOS transistor. Input characteristics of CMOS logic gates are in many cases determined primarily by their ESD protective circuitry.

Precautions that should be observed in the handling and design of MOS circuits include the following:

1. Individuals handling MOSFET devices and their workplaces should be grounded. Extensive guidelines on grounding techniques are published in manufacturers' literature.
2. When a chip is out of circuit, it should be inserted in a special conductive foam that effectively shorts all pins together.
3. In system designs, all unused pins should be connected to ground or to the supply voltage.
4. Circuits should have power applied before connecting or disconnecting low-impedance signal sources, such as pulse generators.

CMOS integrated circuits contain *parasitic* PNP and NPN transistors: transistors that exist because of the proximity of P and N materials embedded in the substrate. Their existence is not intentional but is unavoidable. Because of the conducting paths between a pair of such transistors, they form a circuit similar to an SCR (silicon-controlled rectifier), a device that can be triggered into a heavy conduction mode. If the parasitic transistors in a CMOS circuit are triggered into conduction, the circuit is said to *latch-up*, and the heavy current flow that results can destroy the device. Most CMOS circuits contain protective measures to prevent latch-up, but it can still occur if the manufacturers' specified maximum ratings are exceeded. In particular, high-voltage transients, spikes, and ringing at inputs or outputs can trigger latch-up. External diodes should be connected as protection against such transients, especially if the circuit is used in an environment where there are heavy switching loads, such as relays and motor controllers. Power supplies should be well regulated. Current regulators can also be used to limit supply current, should latch-up occur. As always, unused inputs should be grounded or connected to the supply voltage.

Interfacing CMOS and TTL Devices

Specially designed integrated circuits called *level shifters* are available to make devices from different logic families compatible with each other. An example is the 14504B, used to shift levels up or down when interfacing high-voltage (metal-gate) CMOS with the lower-voltage (HC-series) CMOS devices. Since HC-series CMOS devices and TTL devices can be operated with the same value of supply voltage, level shifters are not ordinarily required when interfacing those families. However, attention must be given to values specified for $V_{OH}(min)$ and $V_{IH}(min)$ in the two families or subfamilies of devices to be interfaced. If the lowest possible output voltage produced by one device when the output is high [$V_{OH}(min)$] is less than the smallest input voltage that the other device will recognize as high [$V_{IH}(min)$], then it is clear that we have an unreliable interface. This situation does not generally occur when a CMOS device is driving a TTL load, so CMOS outputs can be connected directly to TTL inputs. The reverse is not necessarily true. For example, $V_{OH}(min)$ for an advanced low-power Schottky (ALS) device is typically 2.7 V, which is less than the 3.5-V specification for $V_{IH}(min)$ in an HCMOS device (when both

Figure 10.33
Use of an external pull-up resistor, R_P, to interface a TTL device to an HCMOS device when $V_{OH}(min)(TTL) < V_{IH}(min)(HCMOS)$.

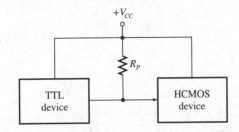

have 5-V supplies). In those situations, an external pull-up resistor can be used at the interface, as shown in Figure 10.33. The pull-up resistance forces the output to rise to $+V_{CC}$ when the output of the TTL device goes high, because the upper transistor in the TTL totem pole is forced off (so the TTL device is effectively in a high-impedance state.) The value of the pull-up resistance depends on load and driver currents, fanout, and the rise-time limitation on the input of an HCMOS device. Design criteria can be found in manufacturers' literature.* A typical value for R_P is 1 kΩ.

Another factor to consider when interfacing devices from different families is the fanout limitation imposed by the maximum current that the driving device can sink. For example, when an HCMOS device drives an ASTTL device, $I_{IL}(max)(ASTTL)$ is 2 mA and $I_{OL}(max)(HCMOS)$ is 4 mA, so the fanout is only 4 mA/2 mA = 2.

Besides using pull-up resistors, another way to interface from TTL devices to HCMOS devices is to use the specially designed HCT-series. Recall that these devices are specifically designed for TTL interfaces. Their inputs are directly compatible with TTL outputs, and their outputs are compatible with both TTL and HCMOS inputs.

10–4   DYNAMIC MOS LOGIC

When power consumption and physical size are prime design considerations, as in digital watches and calculators, dynamic MOS logic is usually the family selected to meet those requirements. Power consumption is minimized by relying on the inherent capacitance of MOS transistors to *store* logic levels—i.e., to remain charged or discharged—and by using clock signals to turn on transistors for very brief intervals of time only. The clock signals turn transistors on to allow the capacitance to recharge or discharge at periodic intervals. Since a transistor is off during most of any given time interval, the average power consumption is quite small. Each transistor used in a dynamic MOS circuit is identical to the other and each can be fabricated in a very small amount of space on a chip. Consequently, large- and very large scale integrations are possible.

The NMOS transmission gate, shown in Figure 10.34, is a fundamental component of dynamic logic circuits. Because the NMOS transistor is completely symmetrical, the drain and source terminals are indistinguishable; that is, current can flow in *either* direction through the device. We therefore label the terminals simply 1 and 2. In dynamic logic applications, there is shunt capacitance at each of these terminals, identified in the figure as C_1 and C_2. Figure 10.34(*b*) shows the circuit when the gate terminal, G, is low. By usual NMOS behavior, the transistor remains off, no matter which or whether both of

---

*See, for example, pages 7-15 through 7-23 in *High-Speed CMOS Logic Data Book*, Texas Instruments, Inc., 1984.

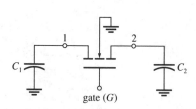

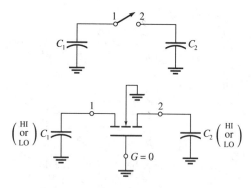

(a) By symmetry, terminals 1 and 2 can serve as either drain or source.

(b) When $G = 0$, the transistor is off, no matter what the levels at C_1 and C_2. Whichever terminal is regarded as the source, V_{GS} is either 0 or negative.

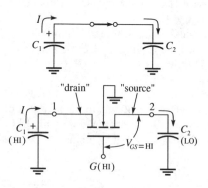

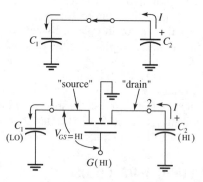

(c) When G and C_1 are both high, C_1 discharges into C_2, transferring the high from 1 to 2.

(d) When G and C_2 are both high, C_2 discharges into C_1, transferring the high from 2 to 1 (or the low from 1 to 2).

Figure 10.34
The NMOS transmission gate (shown with capacitance connected to its terminals).

terminals 1 or 2 are high—i.e, no matter which capacitor(s) are charged. Whether terminal 1 or 2 is regarded as the source, the gate-to-source voltage is either 0 or negative. Figure 10.34(*c*) shows the circuit when G is high, C_1 is charged, and C_2 is discharged. In this case, we can regard terminal 2 as the source and terminal 1 as the drain. Since V_{GS} is high, current flows from terminal 1 to terminal 2 (from drain to source). C_1 discharges into C_2, and, in effect, the high is transferred from terminal 1 to terminal 2. The symmetrical situation is shown in Figure 10.34(*d*), where the high on C_2 is transferred to C_1. In this case, terminal 1 can be regarded as the source and terminal 2 as the drain. If both C_1 and C_2 are low or if both are high, there is no transfer of charge. In summary, we can say that when the gate is high, the level at terminal 2 is ultimately the same as the level that was at terminal 1 and vice versa.

Figure 10.35 shows a dynamic MOS inverter and its timing diagram. The capacitance shown by dashed lines represents inherent device (interelectrode) capacitance. Note that *two* clock signals, ϕ_1 and ϕ_2, are used in the circuit. The two together are called a *two-phase* nonoverlapping clock because ϕ_1 and ϕ_2 are never both high at the same time.

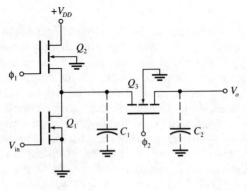

(a) Schematic diagram.

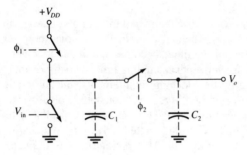

(b) Equivalent switching circuit.

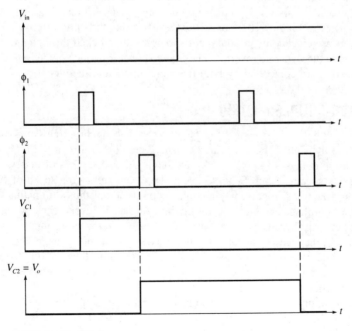

(c) Timing diagram.

Figure 10.35
The dynamic MOS inverter.

Transistors Q_1 and Q_2 act like a conventional MOS inverter [Figure 9.14(c)], with Q_2 serving as a load to switching transistor Q_1. Note, however, that Q_2 serves as an active load only when clock ϕ_1 is high. The rest of the time, Q_2 is off and no current can flow through it. Transistor Q_3 serves as a transmission gate: It transfers charge only when clock ϕ_2 is high. Refer to the timing diagram in Figure 10.35(c) and suppose V_{in} is low. When ϕ_1 goes high, Q_1 remains off because its gate is low, but current flows through Q_2 and charges C_1. When ϕ_1 returns low and while ϕ_2 is still low, C_1 remains charged because there is no path for it to discharge. When ϕ_2 goes high, C_1 discharges into C_2, making V_o high. Thus, the low at V_{in} has created a high at V_o. Suppose now that V_{in} goes high. The next time ϕ_1 goes high, Q_1 turns on, so C_1 cannot charge. It remains discharged until the next ϕ_2 pulse, which allows C_2 to discharge into C_1. Since C_2 discharges, V_o goes low. Thus, the high at V_{IN} has created a low at V_o, and inverter operation is confirmed. The timing diagrams have been idealized for the sake of clarity; we have not shown typical charge and discharge transients.

As is apparent from the previous discussion and the timing diagram, the output of a dynamic logic gate is ''valid'' only when ϕ_2 is high. We say that the gates are *sampled* at the frequency of ϕ_2. A sampled output becomes the input to other gates, whose responses become available only at the next sampling time. The disadvantage of dynamic logic is the complexity added by the clocking requirements, but in practice, successful logic circuits have been constructed with three- and four-phase clocks. The clocks serve another important function in dynamic logic: They allow the capacitance to be recharged at periodic intervals. Without this recharging, called *refreshing*, the charge stored by the capacitance would eventually decay to zero. The minimum clock frequency is therefore determined by the amount of time required for the capacitance to decay significantly. A typical such time is 1 ms, giving a minimum clock frequency of 1 kHz.

Figure 10.36 shows dynamic NOR and NAND gates. Comparing these with the *static* NMOS gates in Figures 10.20 and 10.21, we see that the only difference is that the load transistors are clocked by ϕ_1 and that the outputs are clocked through a transmission gate by ϕ_2, as in the dynamic inverter. The figure shows equivalent switching circuits, having switches that are controlled by both the logic inputs and the clocks.

10–5   EMITTER-COUPLED LOGIC (ECL)

Emitter-coupled logic is named for its use of bipolar junction transistors that are coupled (joined) at their emitters. It has traditionally been the fastest of the logic families, with propagation delays of 1 to 2 ns, although advanced Schottky devices now have comparable speeds. As in Schottky technology, the key to the speed of ECL devices is preventing the transistors from going into deep saturation, thus eliminating storage delays. This is accomplished in the ECL family by using logic levels whose values are so close to each other that a transistor is not driven into saturation when its input switches from low to high. In other words, a transistor is turned on—but not completely on—by a small increase in base voltage. One disadvantage of this approach is that it is difficult to achieve good noise immunity when logic levels are close in value. Also, power consumption is increased when transistors are not saturated. On the other hand, the current drawn from the power supply is more steady than in TTL and CMOS circuits, since ECL gates do not experience the large switching transients found in those types.

As we shall see when we examine ECL circuitry in detail, the logic levels corresponding to 0 and 1 are both negative. Under a positive-logic assumption, typical levels are $0 = -1.7$ V and $1 = -0.9$ V. Since the *magnitude* of the more negative voltage

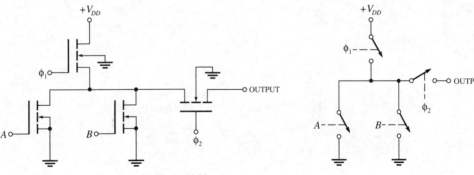

(a) Dynamic NOR gate. Compare with Figure 10.20.

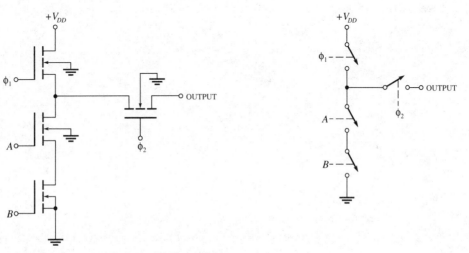

(b) Dynamic NAND gate. Compare with Figure 10.21.

Figure 10.36
Dynamic NOR and NAND gates.

is the greater of the two, some designers prefer to interpret ECL operations using negative logic (0 = −0.9 V, 1 = −1.7 V). We do not use that interpretation in our discussion.

Figure 10.37(*a*) shows a fundamental component of ECL logic gates. Some analysts view this circuit as a *current-switching pair* because the base voltages on Q_1 and Q_2 determine whether current drawn by the negative supply flows through Q_1 or Q_2. However, it is basically a *differential* amplifier. With logic levels as inputs, its operation is best understood by treating the base-emitter junctions of Q_1 and Q_2 as diodes in a diode logic circuit. Note that the base voltage of Q_2 is fixed at −1.3 V, supplied by an internally regulated source. Figure 10.37(*b*) shows the circuit when the input at the base of Q_1 is logical 0: −1.7 V. As can be seen in the equivalent diode circuit, the total voltage tending to forward-bias the base-to-emitter junction of Q_1 is −1.7 − (−5.2 V) = 3.5 V, whereas that for Q_2 is −1.3 V − (−5.2 V) = 3.9 V. Since the base-to-emitter of Q_2 is "most" forward-biased, it is truly forward-biased and is therefore conducting. The base-emitter junctions are specially designed to have forward drops of 0.8 V, so the emitters of both transistors are at −1.3 V − 0.8 V = −2.1 V. Q_2 is on (but not saturated), and Q_1 is off

(a) Schematic diagram.

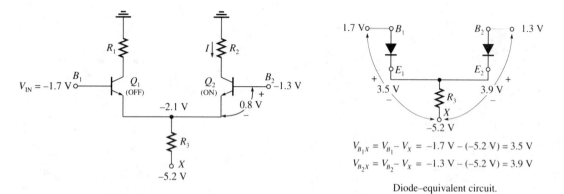

(b) When V_{IN} is low (−1.7 V), Q_1 is off and Q_2 is on, because the base-to-emitter junction of Q_2 is "most" forward-biased.

$V_{B_1X} = V_{B_1} - V_X = -1.7\ \text{V} - (-5.2\ \text{V}) = 3.5\ \text{V}$

$V_{B_2X} = V_{B_2} - V_X = -1.3\ \text{V} - (-5.2\ \text{V}) = 3.9\ \text{V}$

Diode–equivalent circuit.

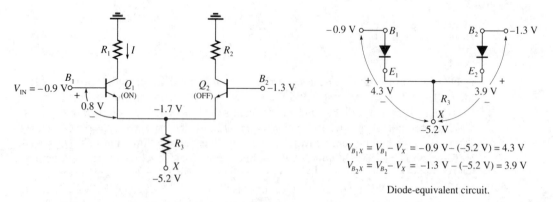

(c) When V_{IN} is high (−0.9 V), Q_1 is on and Q_2 is off, because the base-to-emitter junction of Q_1 is "most" forward-biased.

$V_{B_1X} = V_{B_1} - V_X = -0.9\ \text{V} - (-5.2\ \text{V}) = 4.3\ \text{V}$

$V_{B_2X} = V_{B_2} - V_X = -1.3\ \text{V} - (-5.2\ \text{V}) = 3.9\ \text{V}$

Diode-equivalent circuit.

Figure 10.37
Emitter-coupled transistors at the input of an ECL gate. Note the similarity of the diode-equiva-
lent circuits to a diode OR gate.

because its base-to-emitter voltage is only -1.7 V $- (-2.1$ V$) = 0.4$ V. Figure 10.37(c) shows the circuit when V_{IN} is logical 1: -0.9 V. In this case, Q_1 is on (but not saturated) and Q_2 is off because the total voltage tending to forward-bias the base-to-emitter of Q_1 is 4.3 V and that for Q_2 is 3.9 V. The emitters are at -0.9 V $- 0.8$ V $= -1.7$ V. Notice the similarity of the equivalent diode circuits to a diode-logic OR gate. Resistors R_1 and R_2 in practical ECL gates can be connected externally to a positive V_{CC} supply but are generally grounded for improved noise immunity. The current I_E flowing in R_3 can be found from

$$I_E = \frac{V_{IN} - V_{BE}(\text{on}) + 5.2 \text{ V}}{R_3}$$

$$\approx \frac{V_{IN} + 4.4 \text{ V}}{R_3} \tag{10.11}$$

[Notice that V_{IN} is a negative value in (10.11), about -0.9 V or -1.7 V.] Neglecting the small base current of an on transistor, the collector current I_C in the on transistor is also given by (10.11), since no current flows in the off transistor.

Figure 10.38(a) shows a 2-input ECL NOR/OR gate. Typical of many ECL gates, it has two outputs that are logical complements of each other, a useful feature that can help reduce the chip count of a system. Transistors Q_{1B} and Q_2 are in the same configuration shown in Figure 10.37(a) and discussed in the previous paragraph. Transistor Q_{1A} has been added in parallel with Q_{1B}. Transistors Q_3 and Q_4 are *emitter followers,* whose emitter voltages are the same as their base voltages (less 0.8-V base-to-emitter drops). Figure 10.38(b) shows the circuit when inputs A and B are both low. Treating the base-to-emitter junctions of Q_{1A}, Q_{1B}, and Q_2 as a 3-input diode OR gate, as discussed in the previous paragraph, we find that Q_{1A} and Q_B are both off and Q_2 is on. The value of R_2 is such that current I_2 flowing through Q_2 puts the collector at about -0.9 V. Therefore, the emitter of Q_4 is -0.9 V $- 0.8$ V $= -1.7$ V, and we see that the OR output is low. I_1 is a very small base current flowing into Q_3. The value of R_1 is such that this current puts the collectors of Q_{1A} and Q_{1B} at about -0.1 V, so the emitter of Q_3 is -0.1 V $- 0.8$ V $= -0.9$ V, and we see that the NOR output is high. Figure 10.38(c) shows the circuit when input A is low and input B is high. Using diode logic again, we find Q_{1B} is

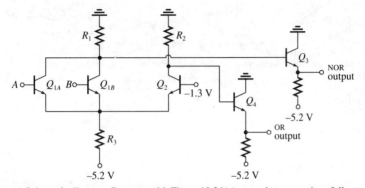

(a) Schematic diagram. Compare with Figure 10.36(a). Q_3 and Q_4 are emitter followers.

Figure 10.38
Two-input ECL OR/NOR gate. (Continued on page 364.)

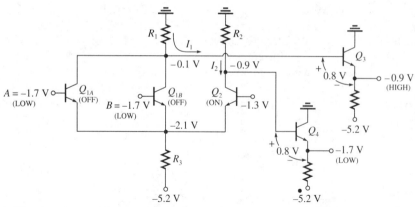

(b) When A and B are both low, Q_2 conducts, the NOR output is high, and the OR output is low.

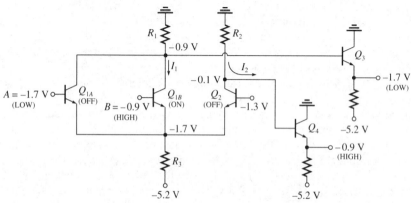

(c) When at least one input is high, the corresponding transistor conducts, the NOR output is low, and the OR output is high.

A	B	OR output	NOR output
−1.7 V	−1.7 V	−1.7 V	−0.9 V
−1.7 V	−0.9 V	−0.9 V	−1.7 V
−0.9 V	−1.7 V	−0.9 V	−1.7V
−0.9 V	−0.9 V	−0.9 V	−1.7 V

Voltage truth table.

A	B	$A + B$	$\overline{A + B}$
0	0	0	1
0	1	1	0
1	0	1	0
1	1	1	0

Positive-logic truth table.

(d) Truth tables confirming that OR and NOR operations are performed by the gate.

(e) Logic symbol.

Figure 10.38
(Continued)

on and Q_{1A} and Q_3 are off. I_1 is now much larger than in part (b), and the collector of Q_{1B} is about −0.9 V, which makes the NOR output −1.7 V, or low. I_2 is the small base current for Q_4, so the collector of Q_2 is about −0.1 V, and the OR output is −0.9 V, or high. By a similar analysis, we find that the NOR output is low and the OR output is high whenever either or both inputs are high. In those cases, the transistors whose inputs are high conduct heavily and Q_2 remains off. The truth tables in Figure 10.38(d) confirm that the gate performs both OR and NOR operations. The circuit can be expanded in an obvious way, by adding additional input transistors in parallel, to construct gates with three or more inputs.

Input and Output Characteristics

ECL NOR gates in modern circuits have internal pull-down resistors connected between each input and the negative supply. These 50-kΩ resistors provide paths for leakage current from the input transistors to prevent the buildup of charge (and voltage) on stray capacitance when inputs are open. Thus, unused inputs in such circuits can be left unconnected. The pull-down resistor effectively holds an unconnected input low. One advantage of the differential input circuitry in ECL gates is that it provides *common-mode rejection*: Power supply noise common to both sides of the differential configuration is effectively canceled (differenced out).

Since an ECL output is produced at an emitter follower, the output impedance is desirably small, typically 7 Ω. As a consequence, ECL gates have very large fanouts and are relatively unaffected by capacitive loads. Some ECL gates are available with multiple outputs, derived from multiple-emitter transistors in the emitter-follower output. For example, one OR/NOR gate may have two OR outputs and two NOR outputs.

ECL Subfamilies and Specifications

Motorola introduced ECL in 1962 with a product line marketed as its MECL I series, followed in 1966 by its MECL II series. These lines are now obsolete, having been replaced by an MECL III series carrying MC1600 numbers, an MECL 10K series carrying MC10000 numbers, and the most recent MECL 10KH series with MC10H000 numbers. ECL subfamilies do not include as wide a range of general-purpose logic gates as do TTL and CMOS families. They do, however, include many complex, special-purpose circuits used in high-speed digital data transmission, arithmetic units, and memories. The most significant enhancements in the 10KH-series devices are (1) a 100% improvement in speed (compared to the 10K-series); (2) replacement of resistor R_3 in Figures 10.37 and 10.38 by a transistor constant-current source, which maintains a constant 4-mA bias current for the emitter-coupled transistors; and (3) an improved internal voltage regulator to supply a constant -1.29 V reference (where the -1.3-V terminals are shown in Figures 10.37 and 10.38). Figure 10.39 compares typical speed-power relations of ECL

		Propagation delay	Power per gate	Speed-power product
	MECL 10KH	1 ns	25 mW	25 pJ
MECL 10K $\{$	10100, 10500 Series	2 ns	25 mW	50 pJ
	10200, 10600 Series	1.5 ns	25 mW	37 pJ
	MECL III	1 ns	60 mW	60 pJ
	ASTTL	1.7 ns	8 mW	13.6 pJ
	ALSTTL	4 ns	1.2 mW	4.8 pJ
	HCMOS	8 ns	0.17 mW (at 100 kHz)	1.4 pJ

Figure 10.39
Comparison of typical speed-power relations of ECL subfamilies and other logic families.

 **MOTOROLA**

MC10H102

QUAD 2-INPUT NOR GATE

The MC10H102 is a quad 2-input NOR gate. The MC10H102 provides one gate with OR/NOR outputs. This MECL 10KH part is a functional/pinout duplication of the standard MECL 10K family part, with 100% improvement in propagation delay, and no increases in power-supply current.

- Propagation Delay, 1.0 ns Typical
- Power Dissipation 25 mW/Gate (same as MECL 10K)
- Improved Noise Margin 150 mV (Over Operating Voltage and Temperature Range)
- Voltage Compensated
- MECL 10K-Compatible

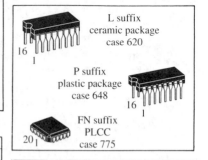

L suffix
ceramic package
case 620

P suffix
plastic package
case 648

FN suffix
PLCC
case 775

MAXIMUM RATINGS

Characteristic	Symbol	Rating	Unit
Power Supply (V_{CC} = 0)	V_{EE}	-8 0 to 0	Vdc
Input Voltage (V_{CC} = 0)	V_I	0 to V_{EE}	Vdc
Output Current — Continuous — Surge	I_{out}	50 100	mA
Operating Temperature Range	T_A	0-75	°C
Storage Temperature Range — Plastic — Ceramic	T_{stg}	-55 to 150 -55 to 165	°C °C

ELECTRICAL CHARACTERISTICS (V_{EE} = -5.2 V ±5%) (See Note)

Characteristic	Symbol	0° Min	0° Max	25° Min	25° Max	75° Min	75° Max	Unit
Power Supply Current	I_E	—	29	—	26	—	29	mA
Input Current High	I_{inH}	—	425	—	265	—	265	μA
Input Current Low	I_{inL}	0.5	—	0.5	—	0.3	—	μA
High Output Voltage	V_{OH}	-1.02	-0.84	-0.98	-0.81	-0.92	-0.735	Vdc
Low Output Voltage	V_{OL}	-1.95	-1.63	-1.95	-1.63	-1.95	-1.60	Vdc
High Input Voltage	V_{IH}	-1.17	-0.84	-1.13	-0.81	-1.07	-0.735	Vdc
Low Input Voltage	V_{IL}	-1.95	-1.48	-1.95	-1.48	-1.95	-1.45	Vdc

AC PARAMETERS

		0° Min	0° Max	25° Min	25° Max	75° Min	75° Max	
Propagation Delay	t_{pd}	0.4	1.25	0.4	1.25	0.4	1.4	ns
Rise Time	t_r	0.5	1.5	0.5	1.6	0.55	1.7	ns
Fall Time	t_f	0.5	1.5	0.5	1.6	0.55	1.7	ns

NOTE:
Each MECL 10KH series circuit has been designed to meet the dc specifications shown in the test table, after thermal equilibrium has been established. The circuit is in a test socket or mounted on a printed circuit board and transverse air flow greater then 500 linear fpm is maintained. Outputs are terminated through a 50-ohm resistor to -2.0 volts.

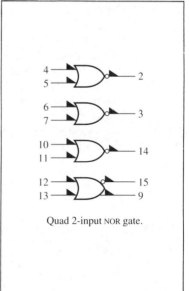

Quad 2-input NOR gate.

Figure 10.40
Manufacturer's specifications for the MC10H102 quad 2-input ECL NOR gate (Courtesy of Motorola, Inc.).

subfamilies and other logic families. We see that ECL is generally faster but has considerably greater power consumption than other families.

Figure 10.40 shows a typical set of manufacturer's specifications for an ECL circuit, in this case the Motorola MC10H102 quad 2-input NOR gate. The special symbols in the logic diagram are used to convey the fact that NOR operations are performed under a positive-logic assumption. Note that one gate has complementary (OR/NOR) outputs.

Example 10–4

Using the specifications from the MC10H102 in Figure 10.40, find

a. the dc noise immunities N_H and N_L;
b. the fanout.

Assume $T = 25°C$ in both cases.

Solution

a. From the specifications,

$$V_{IL}(\text{max}) = -1.48 \text{ V}, \qquad V_{OL}(\text{max}) = -1.63 \text{ V}$$
$$V_{OH}(\text{min}) = -0.98 \text{ V}, \qquad V_{IH}(\text{min}) = -1.13 \text{ V}$$

Therefore,

$$N_L = V_{IL}(\text{max}) - V_{OL}(\text{max}) = -1.48 \text{ V} - (-1.63 \text{ V}) = 0.15 \text{ V}$$
$$N_H = V_{OH}(\text{min}) - V_{IH}(\text{min}) = -0.98 \text{ V} - (-1.13 \text{ V}) = 0.15 \text{ V}$$

b. From the specifications, the maximum continuous output current (identified as I_{out}) is 50 mA. The maximum input current is $I_{inH}(\text{max}) = 265 \text{ }\mu\text{A}$. Therefore, the fanout is

$$\frac{50 \text{ mA}}{265 \text{ }\mu\text{A}} = 188$$

This example demonstrates that fanout limitations are not normally a problem in ECL circuits.

Drill 10–4

a. Find N_H and N_L for the MC10H102 at $T = 75°C$.
b. Find the maximum total power dissipation of the MC10H102 at $T = 25°C$, when $V_{EE} = -5.2 \text{ V}$.

Answers: (a) 0.15 V; 0.15 V; (b) 135.2 mW.

Wired-OR Connections

ECL gates are available with *open-emitter* outputs—i.e., with the resistors in the output emitter followers omitted. In particular, all logic gates in the 10K and 10KH series have open-emitter outputs. Open-emitter outputs can be connected directly together and to an external resistor to perform a *wired-OR* operation. With reference to Figure 10.41(*a*), it is obvious that the wired-OR connection will be high when both outputs are high and will be low when both outputs are low. Figure 10.41(*b*) shows a diode-equivalent circuit of the output transistors when the base of one output transistor is low and the base of the other is high. The circuit behaves like a diode OR gate, so the most forward-biased diode makes the wired-OR connection high (−0.9 V). Figure 10.41(*c*) through (*e*) shows examples of the variety of wired-OR functions that can be implemented with ECL gates.

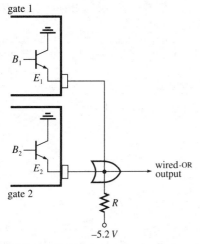

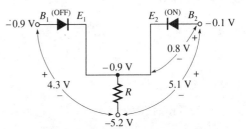

bases		emitters			
B_1	B_2	OUTPUT	B_1	B_2	OUTPUT
−0.9 V	−0.9 V	−1.7 V	0	0	0
−0.9 V	−0.1 V	−0.9 V	0	1	1
−0.1 V	−0.9 V	−0.9 V	1	0	1
−0.1 V	−0.1 V	−0.9 V	1	1	1

(a) Connection of open-emitter outputs to an external resistor to obtain a wired-OR output.

(b) Diode-equivalent circuit of ouput. When one base is low and the other is high, the more forward-biased diode makes the output high. The circuit is equivalent to a diode OR gate.

(c) Wired-OR output of two ECL NOR gates.

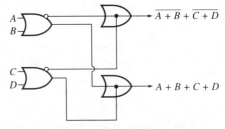

(d) Wired-OR output of two ECL OR/NOR gates.

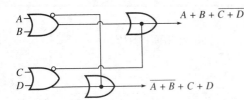

(e) Wired-OR connections of OR outputs with NOR outputs.

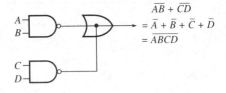

(f) A wired-OR output of two NAND gates expands the NAND operation.

Figure 10.41
The wired-OR connection of open-emitter ECL outputs.

Interfacing ECL Gates

Since ECL logic levels are so different from those of TTL and CMOS circuits, interfacing ECL gates with other families requires more than just a pull-up resistor. Special level-shifting circuits called *translators* are available in the various ECL series to facilitate interfacing ECL with other families. Examples include the MC10H124, which interfaces four TTL or 5-V CMOS outputs to ECL inputs, and the MC10H125, which interfaces four

ECL outputs to TTL or 5-V CMOS inputs. Manufacturers' product literature contains detailed guidelines and design criteria for interfacing ECL circuits with other families.*

10–6   INTEGRATED INJECTION LOGIC (I²L)

Integrated injection logic is the newest of the logic families and is finding widespread use in large-scale integrated circuits. I²L logic gates are constructed using bipolar transistors only; the absence of space-consuming resistors makes it possible to integrate a large number of gates in a single package. Compared to other families, I²L circuits are easily fabricated and economical. Its speed-power product is quite small, on the order of 4 pJ, so it is comparable to advanced low-power Schottky TTL.

To understand the operation of I²L circuitry, it is best first to analyze a simple circuit having the same operating principles. Figure 10.42(a) and (b) shows an inverter that behaves in the same way as an I²L inverter. PNP transistor Q_1 serves as a constant-current source that "injects" current into node X. The direction that the current flows after entering node X depends on the input level. A low input is a current *sink*. When the input is low, the injected current flows into the input, thus diverting current from the base of Q_2. Q_2 is therefore off and the output is high. If the input is high, the injected current flows into the base of Q_2, turning it on and making the output low. Figure 10.42(c) shows an actual I²L inverter. The output transistor has two collectors (sometimes three), making it equivalent to two transistors with parallel bases and emitters [Figure 10.42(d)]. Thus, it produces two equal outputs.

Instead of a collector resistor, the outputs are connected directly to the inputs of other I²L gates. Figure 10.42(e) and (f) shows one I²L inverter driving two others. When the input to Q_1 is high (e), Q_2 is on and its output therefore serves as a current sink to draw current from Q_3 and Q_5. Q_4 and Q_6 are, therefore, both off. When the input to Q_1 is low (f), Q_2 is off and no current is drawn from Q_3 and Q_5. Q_4 and Q_6 are both on and can sink current from other gates. Some analysts interpret I²L logic levels in terms of current flow: Current flowing through a transistor is low and no current is high. Equivalently, an ON transistor capable of sinking current is low and an OFF transistor that prevents current from flowing into it is high.

A typical value for voltage source V in Figure 10.42 is 0.8 V. It should be noted that there is internal resistance (not a resistor) between the base and ground of injector transistor Q_1, which controls the flow of emitter-to-base current. The speed and power dissipation of an I²L gate depends heavily on the value of voltage source V. Increasing V increases dissipation but reduces propagation delay. The speed-power product is very nearly constant.

Figure 10.43 demonstrates the simplicity of I²L gates. The NAND gate in 10.43(a) is simply an inverter with inputs connected directly together at the inverter input. If either or both inputs are low (current sinks), the injected current flows into those inputs and Q_2 remains off (high). If both inputs are high, the injected current turns on Q_2, making the output low. Thus, the NAND operation is performed. Figure 10.43(b) shows that an I²L NOR gate is simply two inverters with their outputs connected together. If either or both inputs are high, the corresponding output transistor is on (low), and the output is a current sink. If both inputs are low, both output transistors are off, so the output is high.

---

*See, for example, *MECL System Design Handbook,* Motorola, Inc., 1983.

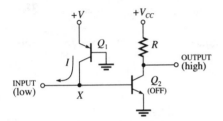

(a) A low input sinks the injected current, thus holding Q_2 off and making the output high.

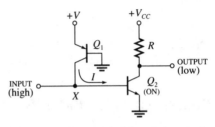

(b) A high input causes the injected current to flow into Q_2, turning it on and making the output low.

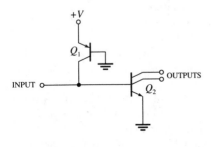

(c) An actual I^2L inverter having two output collectors and no collector resistor.

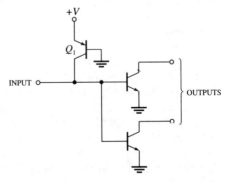

(d) Two-transistor equivalent of the output transistor.

Figure 10.42
The I^2L inverter. Parts (a) and (b) demonstrate I^2L principles in a conventional circuit, whereas (c) is an actual I^2L inverter.

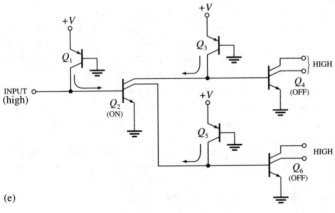

(e)

(f) One I^2L inverter driving two others.

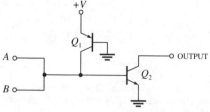

A	B	Q_2	OUTPUT = $\overline{AB}$
0	0	OFF	1
0	1	OFF	1
1	0	OFF	1
1	1	ON	0

0 = current sink (ON transistor)
1 = OFF transistor

(a) I²L NAND gate.

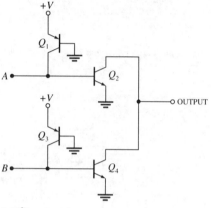

A	B	Q_2	Q_4	OUTPUT = $\overline{A + B}$
0	0	OFF	OFF	1
0	1	OFF	ON	0
1	0	ON	OFF	0
1	1	ON	ON	0

0 = current sink (ON transistor)
1 = OFF transistor

(b) I²L NOR gate.

Figure 10.43
I²L logic gates.

10–7   ANSI/IEEE STANDARD SYMBOLS

Open and Three-State Outputs

Figure 10.44, p. 372, summarizes the qualifying symbols used to represent open-collector and open-drain, open-emitter, and three-state outputs in the ANSI/IEEE standards.

Figure 10.45, p. 372, shows examples of logic symbols for 7400-series integrated circuits having open-collector and open-drain outputs. Note that the symbol qualifying the type of output appears only once on each diagram.

Control Blocks and the EN Input

For circuits having three-state outputs, the enabling input in the ANSI/IEEE standard is designated EN. In integrated circuits containing multiple gates, drivers, or other devices having individual inputs and outputs, the enabling input often enables several devices simultaneously. The ANSI/IEEE standards specify that any such single input simultaneously affecting several devices in a circuit can be shown on a special *common control block* at the top of the circuit's logic symbol. This control block is the only distinctive,

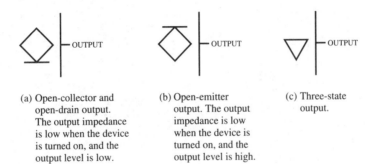

(a) Open-collector and
open-drain output.
The output impedance
is low when the device
is turned on, and the
output level is low.

(b) Open-emitter
output. The output
impedance is low
when the device is
turned on, and the
output level is high.

(c) Three-state
output.

Figure 10.44
ANSI/IEEE symbols used to represent open-collector and open-drain, open-emitter, and three-state outputs.

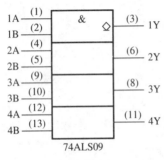

74ALS09

(a) 74ALS09 quad 2-input TTL AND
gate with open-collector outputs.

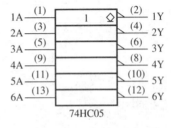

74HC05

(b) 74HC05 hex CMOS inverter with
open-drain outputs.

Figure 10.45
Examples of ANSI/IEEE standard symbols for integrated circuits with open-collector and open-drain outputs.

non-rectangular outline used in the standards. Figure 10.46(*a*), p. 373, shows an example: the logic symbol for the 74AS241 octal buffer/bus driver. In this example, active-low enabling input $1\overline{G}$ controls four buffers and active-high input 2G controls the other four, so two separate control blocks are shown. Note that EN for the active-low input is not written with a negation bar over it. The triangle in the center, pointing to the right, is the ANSI/IEEE symbol for a device, such as a buffer, having more than usual drive capability. It points in the direction of signal flow. Figure 10.46(*b*) shows an example of a circuit in which logic operations are performed on the control inputs to produce EN—in this case, EN = $\overline{G}1\overline{G}2$, meaning that both $\overline{G}1$ and $\overline{G}2$ must be low to enable all eight drivers in the circuit. Note that the outputs are the complements of the inputs.

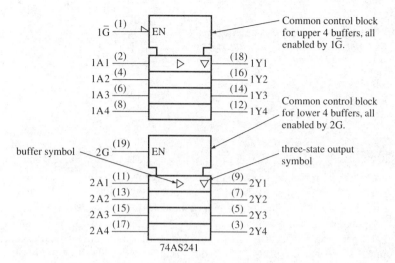

(a) The 74AS241 octal buffer/bus driver.

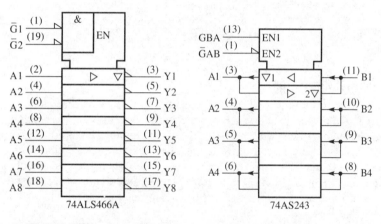

(b) The 74ALS466A octal buffer/bus driver. All buffers are enabled when G1 and G2 are both low.

(c) The 74AS243 quad bidirectional bus driver, or transceiver.

Figure 10.46
Examples of ANSI/IEEE standard symbols for integrated circuits with three-state outputs, showing common control blocks and enabling (EN) inputs.

Dependency Notation

EN is one of eleven types of *dependencies* that have been defined in ANSI/IEEE standards. Dependency notation is a means for identifying signals whose states depend on the states of other signals. In enable (EN) dependency, the notation identifies outputs that are enabled when EN is high. A numerical suffix is appended to EN and the same number is placed at outputs that depend on EN. For example, an output labeled 1 is enabled when EN1 is high and is in a high-impedance state (disabled) when EN1 is low. Figure 10.46(*c*)

Figure 10.47
ANSI/IEEE symbol and schematic diagram of the CMOS transmission gate, illustrating transmission dependency (X).

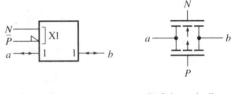

(a) Symbol.   (b) Schematic diagram.

shows an example: the 74AS243 bidirectional bus driver, or transceiver. When GBA is active (high), EN1 is high, so the output labeled 1 is enabled and right-to-left (B-to-A) signal flow is established. When $\overline{GAB}$ is active (low), EN2 is high, output 2 is enabled, and left-to-right (A-to-B) signal flow is established. When GBA is low and $\overline{GAB}$ is high, all outputs are in a high-impedance state. Note the use of arrows to show right-to-left signal flow. No arrows are shown on the left-to-right path because that direction is assumed when not otherwise identified. As in other ANSI/IEEE symbols, the notation in the top rectangles is not repeated in the remaining rectangles but is assumed to apply to all.

EN dependency can also be shown in the symbols for circuits having conventional (two-state) outputs. In these cases, making EN low forces all outputs to their inactive states (high or low, as the case may be). An example is discussed in Chapter 13 [Figure 13.14(*b*)].

Transmission Gates

Figure 10.47 shows the ANSI/IEEE symbol for a CMOS transmission gate, along with a schematic diagram of the gate. This symbol contains an example of *transmission dependency*, represented by the letter X. By definition of transmission dependency, all input/outputs depending on X are bidirectionally connected together when X is high. In the present example, a 1 appears in the two places where signals a and b are connected, so a and b are bidirectionally connected when X1 is high and disconnected when X1 is low. The *grouping* bracket,], shows that X1 is determined by inputs N and $\overline{P}$. When N is high and $\overline{P}$ is low, X1 is high.

10-8   SPICE EXAMPLES

Example 10-5

Use SPICE to obtain a truth table for the TTL NAND gate shown in Figure 10.4(*a*). Logic levels are 1 = +5 V and 0 = 0 V. The output is loaded by 1 kΩ of resistance. Compare the high output voltage computed by SPICE with that predicted by equation (10.2).

Solution.   The gate in its SPICE format and the input data file are shown in Figure 10.48(*a*). Note that the two-emitter input transistor is simulated by two transistors having their collectors joined and their bases joined. Inputs V1 and V2 are stepped through all combinations of 0 V and 5 V to generate a truth table. The output voltage is V(10). The .PRINT statement also requests values for V(8,9), the value of V_{CE} for Q_3, and V(9,10), the value of the drop across the diode, so that values can be substituted in equation (10.2).

The results of the simulation, shown in Figure 10.48(*b*), confirm that the gate performs the NAND operation. V1 is 0 V for the first two values of V2 and is 5 V for the second two values of V2, so the voltage and logic truth tables are as follows on page 376.

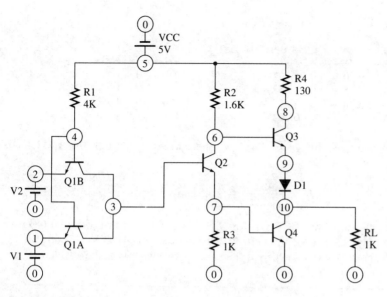

```
EXAMPLE 10.5
V1 1 0
V2 2 0
Q1A 3 4 1 TRANS
Q1B 3 4 2 TRANS
R1 4 5 4K
VCC 5 0 5V
R2 5 6 1.6K
Q2 6 3 7 TRANS
R3 7 0 1K
R4 5 8 130
Q3 8 6 9 TRANS
D1 9 10 DIODE
Q4 10 7 0 TRANS
RL 10 0 1K
·MODEL TRANS NPN
·MODEL DIODE D
.DC V2 0 5 5 V1 0 5 5
.PRINT DC V(10) V(8,9) V(9,10)
.END
```

(a)

EXAMPLE 10.5

****   DC TRANSFER CURVES   TEMPERATURE =   27.000 DEG C

****

V2	V(10)	V(8,9)	V(9,10)
0.000E+00	3.452E+00	4.163E-01	6.871E-01
5.000E+00	3.452E+00	4.163E-01	6.871E-01
0.000E+00	3.452E+00	4.163E-01	6.871E-01
5.000E+00	1.785E-02	4.622E+00	3.601E-01

(b)

Figure 10.48
Example 10–5.

V1	V2	V(9)	A	B	$\overline{AB}$
0 V	0 V	3.452 V	0	0	1
0 V	5 V	3.452 V	0	1	1
5 V	0 V	3.452 V	1	0	1
5 V	5 V	0.1785 V	1	1	0

Since the high output voltage is 3.452 V, the current I_L through the 1-kΩ load resistance is

$$I_L = \frac{3.452 \text{ V}}{1 \text{ k}\Omega} = 3.452 \text{ mA}$$

The computed results show that VCE(sat) for Q_3 is 0.4163 V and that VD = 0.6871 V when the output is high. From equation (10.2), VHI = 5 V − 0.4163 V − 0.6871 V − (3.542 mA)(130 Ω) = 3.44 V, which is in close agreement with the results of the simulation.

Example 10–6

Use SPICE to determine the propagation delay, t_{pLH}, of the NMOS NOR gate in Figure 10.20(a) when the output is loaded by 10 pF of capacitance. The value of β for Q_2 and Q_3 is 0.25×10^{-3} A/V² and that for Q_1 is 0.025×10^{-3} A/V². All transistors have $V_T = 2$ V. The supply voltage is 10 V.

Solution. Figure 10.49(a) shows the gate in its SPICE format and the input data file. To determine t_{pLH}, we must specify input voltages that will cause the output to switch from low to high. For that purpose, input A is connected to ground (0 V) and input B is switched from 8 V to 0 V (1 to 0). Thus, the output of the NOR gate switches from

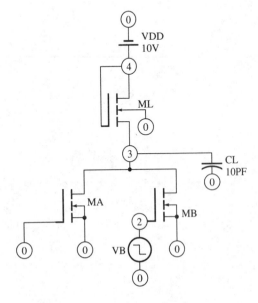

```
EXAMPLE 10.6
VB 2 0 PULSE(8 0 0 1PS 1PS)
MA 3 0 0 0 MOSFET1
MB 3 2 0 0 MOSFET1
ML 4 4 3 0 MOSFET2
VDD 4 0 10V
CL 3 0 10PF
.MODEL MOSFET1 NMOS KP = 0.25E-3 VTO = 2
.MODEL MOSFET2 NMOS KP = 0.025E-3 VTO = 2
.TRAN 10NS 200NS
.PLOT TRAN V(3)
.END
```

(a)

Figure 10.49
Example 10–6.

EXAMPLE 10.6

****   TRANSIENT ANALYSIS   TEMPERATURE =   27.000 DEG C

****

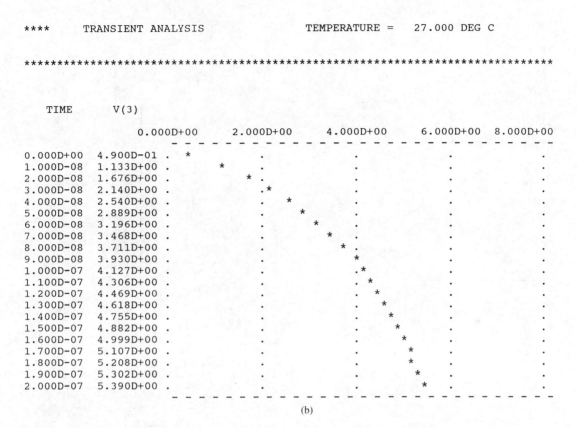

```
       TIME        V(3)

                          0.000D+00    2.000D+00    4.000D+00    6.000D+00  8.000D+00
                          - - - - - - - - - - - - - - - - - - - - - - - - - - - - -
   0.000D+00   4.900D-01 .    *              .            .            .            .
   1.000D-08   1.133D+00 .          *        .            .            .            .
   2.000D-08   1.676D+00 .               *  .            .            .            .
   3.000D-08   2.140D+00 .                 .*             .            .            .
   4.000D-08   2.540D+00 .                 .    *         .            .            .
   5.000D-08   2.889D+00 .                 .        *     .            .            .
   6.000D-08   3.196D+00 .                 .           *  .            .            .
   7.000D-08   3.468D+00 .                 .             * .           .            .
   8.000D-08   3.711D+00 .                 .               * .         .            .
   9.000D-08   3.930D+00 .                 .                 *         .            .
   1.000D-07   4.127D+00 .                 .                  .*       .            .
   1.100D-07   4.306D+00 .                 .                  . *      .            .
   1.200D-07   4.469D+00 .                 .                  .    *   .            .
   1.300D-07   4.618D+00 .                 .                  .     *  .            .
   1.400D-07   4.755D+00 .                 .                  .       * .           .
   1.500D-07   4.882D+00 .                 .                  .         * .         .
   1.600D-07   4.999D+00 .                 .                  .           * .       .
   1.700D-07   5.107D+00 .                 .                  .            * .      .
   1.800D-07   5.208D+00 .                 .                  .            * .      .
   1.900D-07   5.302D+00 .                 .                  .             * .     .
   2.000D-07   5.390D+00 .                 .                  .              * .    .
                          - - - - - - - - - - - - - - - - - - - - - - - - - - - - -
```

(b)

Figure 10.49
(Continued)

$\overline{0 + 1} = 0$ to $\overline{0 + 0} = 1$. Note that the first voltage specified in the PULSE voltage statement is 8 V and that the second is 0 V. Thus, input B is switched from logical 1 to logical 0 at the beginning of the simulation. The pulsewidth specification (PW) in a PULSE statement always refers to the duration of the *second* voltage, so in this example the pulse is momentarily 8 V and then remains at 0 V throughout the duration of the transient (since PW defaults to TSTOP).

The results of the simulation are shown in Figure 10.49(*b*). When SPICE obtained an "Initial Transient Solution" (not shown), the initial value of the output was computed to be $V_{LO} = 0.49$ V, so the initial output voltage in the plot (at $t = 0$) is seen to be 0.49 V. Although the duration of the transient computation is not long enough to observe the final value of the output, it is known to be $V_{HI} = V_{DD} - V_T = 10$ V $- 2$ V $= 8$ V (See Example 9–10). Thus, the voltage at the 50% point is $0.5(V_{HI} - V_{LO}) = 0.5(8$ V $- 0.49$ V$) = 3.755$ V. Since the input pulse is for all practical purposes ideal (the PULSE statement specifies t_r and t_f to be 1 ps), the value of t_{pLH} is the time required for the output to fall from 0.49 V to 3.755 V. The output plot shows that the output reaches a voltage near that value (3.711 V) at $t = 80$ ns, so $t_{pLH} \approx 80$ ns.

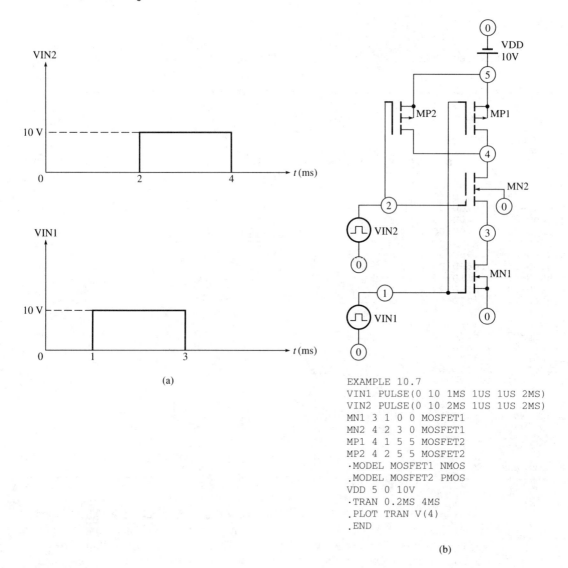

EXAMPLE 10.7
VIN1 PULSE(0 10 1MS 1US 1US 2MS)
VIN2 PULSE(0 10 2MS 1US 1US 2MS)
MN1 3 1 0 0 MOSFET1
MN2 4 2 3 0 MOSFET1
MP1 4 1 5 5 MOSFET2
MP2 4 2 5 5 MOSFET2
·MODEL MOSFET1 NMOS
.MODEL MOSFET2 PMOS
VDD 5 0 10V
·TRAN 0.2MS 4MS
.PLOT TRAN V(4)
.END

(b)

Figure 10.50
Example 10–7.

Example 10–7   The inputs to the CMOS NAND gate in Figure 10.23(*a*) are the pulses shown in Figure 10.50(*a*). Use SPICE to obtain a plot of the output of the gate.

Solution.   The gate in its SPICE format and the input data file are shown in Figure 10.50(*b*). Note that the PULSE statements defining VIN1 and VIN2 specify delays of 1 ms and 2 ms, respectively.

The plot of the output at node 3 is shown in Figure 10.50(*c*). This plot confirms that the gate performs the NAND function, since the output is low only during the interval between 2 ms and 3 ms, the only interval during which both inputs are high.

EXAMPLE 10.7

```
****      TRANSIENT ANALYSIS                 TEMPERATURE =    27.000 DEG C

***********************************************************************


      TIME       V(4)

                   -5.000D+00      0.000D+00      5.000D+00      1.000D+01   1.500D+01
                   - - - - - - - - - - - - - - - - - - - - - - - - - - - - - - - - -
0.000D+00  1.000D+01 .                .              .              *              .
2.000D-04  1.000D+01 .                .              .              *              .
4.000D-04  1.000D+01 .                .              .              *              .
6.000D-04  1.000D+01 .                .              .              *              .
8.000D-04  1.000D+01 .                .              .              *              .
1.000D-03  1.000D+01 .                .              .              *              .
1.200D-03  1.000D+01 .                .              .              *              .
1.400D-03  1.000D+01 .                .              .              *              .
1.600D-03  1.000D+01 .                .              .              *              .
1.800D-03  1.000D+01 .                .              .              *              .
2.000D-03  1.000D+01 .                .              .              *              .
2.200D-03  2.773D-07 .                *              .              .              .
2.400D-03  2.773D-07 .                *              .              .              .
2.600D-03  2.773D-07 .                *              .              .              .
2.800D-03  2.773D-07 .                *              .              .              .
3.000D-03  2.773D-07 .                *              .              .              .
3.200D-03  1.000D+01 .                .              .              *              .
3.400D-03  1.000D+01 .                .              .              *              .
3.600D-03  1.000D+01 .                .              .              *              .
3.800D-03  1.000D+01 .                .              .              *              .
4.000D-03  1.000D+01 .                .              .              *              .
                   - - - - - - - - - - - - - - - - - - - - - - - - - - - - - - - - -
```

(c)

Figure 10.50
(Continued)

Example 10–8

The ECL gate in Figure 10.38(*a*) has the following component values: R1 = 200 Ω, R2 = 50 Ω, RE = 777 Ω, R3 = 1 kΩ, and R4 = 1 kΩ. Use SPICE to confirm the dc voltages shown in Figure 10.38(*c*) when input A is low (−1.7 V) and input B is high (−0.9 V).

Solution.   The gate in its SPICE format and the input data file are shown in Figure 10.51. Note that the .MODEL statement specifies that the forward voltage drop across the base-to-emitter junctions of the transistors is 0.8 V (VJE = 0.8), in accordance with ECL technology. The following table compares the voltages computed by SPICE with the voltages shown in Figure 10.38(*c*):

	V(5)	V(4)	V(7)	V(9)	V(8)
Figure 10.38(*c*)	−0.9 V	−1.7 V	−0.1 V	−1.7 V	−0.9 V
SPICE	−0.8956 V	−1.713 V	−0.1075 V	−1.702 V	−0.9015 V

We see there is excellent agreement between corresponding voltage values.

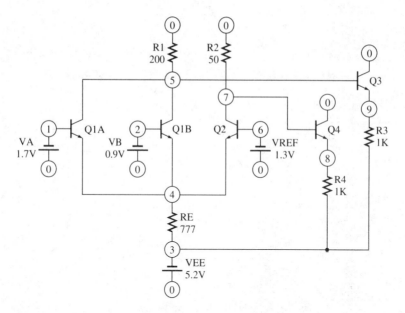

```
EXAMPLE 10.8
VA 0 1 1.7V
VB 0 2 0.9V
Q1A 5 1 4 TRANS
Q1B 5 2 4 TRANS
RE 4 3 777
VEE 0 3 5.2V
R1 5 0 200
R2 7 0 50
Q2 7 6 4 TRANS
VREF 0 6 1.3
Q4 8 7 0 TRANS
R4 8 3 1K
Q3 0 5 9 TRANS
R3 9 3 1K
.MODEL TRANS NPN VJE = 0.8
.DC VA 1.7 1.7 1
.PRINT DC V(5) V(4) V(7) V(9) V(8)
.END
```

Figure 10.51
Example 10–8

EXERCISES

Section 10–1

10.1 The TTL NAND gate in Figure 10.4(*a*) supplies 2 mA to an external load when the output is high. Assuming V_{CC} = +5 V, find the output voltage when inputs have the given values:

(a) A = 0 V and B = +5 V
(b) $A = B = +5$ V

10.2 The TTL NAND gate in Figure 10.4(*a*) has a 1-kΩ load connected to its output. Assuming V_{CC} = +5 V, find the output voltage when inputs have the given values:

(a) $A = +5$ V and $B = 0$ V

(b) $A = B = +5$ V

10.3 Manufacturer's specifications for a certain logic gate state that the maximum values of I_{IL} and I_{IH} are 40 μA and −1.2 mA, respectively. How much current must a totem pole be able to (a) sink and (b) supply when the totem pole is connected to 5 inputs of the specified gate?

10.4 Manufacturer's specifications for a certain logic gate state that I_{IH} for each input is no greater than 50 μA, under specified operating conditions. The output of a 74L00 NAND gate [Figure 10.8(b)] is connected to 10 inputs of the specified gate. Assuming the specified operating conditions and that $V_{CC} = +5$ V, what is the minimum output voltage of the 74L00 when $A = B = 0$ V?

10.5 Under the test conditions for which the specifications in Figure 10.14 are given,

(a) Find the maximum current flow out of an input terminal of a 74AS00 NAND gate when the input is low;

(b) Find the maximum current flow into an input terminal of a 74ALS00A NAND gate when the input is high;

(c) Find the maximum current that the output of a 74AS00 NAND gate can sink;

(d) Find the minimum propagation delay of a 74ALS00A NAND gate when its output switches from low to high.

10.6 Using the specifications given in Figure 10.14, and assuming the test conditions specified, compute

(a) The fanout of a 74AS00 NAND gate;

(b) The noise immunities N_L and N_H of a 74AS00 NAND gate (assume $V_{CC} = +4.5$ V).

10.7 Figure 10.52 shows manufacturer's specifications for the 74AS02 quad 2-input NOR gate. Assuming the test conditions specified, find

(a) The fanout of a NOR gate;

(b) The noise immunities N_L and N_H (assume $V_{CC} = +4.5$ V);

(c) The maximum power supplied by a +5.5-V power supply to the 74AS02 circuit when the inputs to all NOR gates are 0 V. (*Hint:* Power = VI watts.)

10.8 Refer to the specifications for the 74AS02 given in Figure 10.52. Under the test conditions specified, and assuming that $V_{CC} = +4.5$ V, find

(a) The absolute maximum input voltage to any gate;

(b) The maximum output voltage of a gate when both of its inputs are high;

(c) The minimum output voltage of a gate when both of its inputs are low;

(d) The minimum input voltage that will be interpreted by a gate as a high input.

10.9 Assuming each of the gates shown in Figure 10.53 has an open-collector output, write and simplify the Boolean algebra expression represented by each OUTPUT signal.

10.10 A certain TTL system has the following unused gates:

1. Two inverters in a 7405 hex inverter with open-collector outputs;
2. One gate in a 7403 quad 2-input NAND gate with open-collector outputs;
3. One gate in a 7433 quad 2-input NOR gate with open-collector outputs.

Find two ways that unused gates can be wire-ANDed to implement $\overline{A}\,\overline{B}C + \overline{A}\,B\overline{D}$. Draw the logic diagrams.

10.11 Each of two three-state bus drivers is enabled by an active-low control signal. Driver 1 is to have control of the bus if the decimal value of the binary number A_1A_0 is less than 3. Otherwise, driver 2 is to have control. Draw a logic diagram showing how the drivers can be controlled.

10.12 Using two 74AS231 bus drivers, design a system that transmits the 1's complement of 8-bit words to a single data bus under the following conditions: Driver 1 transmits when the binary number A_1A_0 has a decimal value less than 1, driver 2 transmits when A_1A_0 equals 1 or 2, and both drivers are in a high-impedance state when A_1A_0 equals 3. Draw the logic diagram.

Section 10–2

10.13 Draw a schematic diagram of a 3-input NMOS NOR gate. Also draw the equivalent

ALS and AS CIRCUITS
SN54AS02, SN74AS02
QUADRUPLE 2-INPUT POSITIVE-NOR GATES

---

absolute maximum ratings over operating free-air temperature range (unless otherwise noted)

Supply voltage, V_{CC} . . . . 7 V
Input voltage . . . . 7 V
Operating free-air temperature range: SN54AS02 . . . . −55°C to 125°C
　　　　　　　　　　　　　　　　　　　　 SN74AS02 . . . . 0°C to 70°C
Storage temperature range . . . . −65°C to 150°C

recommended operating conditions

		SN54AS02			SN74AS02			UNIT
		MIN	NOM	MAX	MIN	NOM	MAX	
V_{CC}	Supply voltage	4.5	5	5.5	4.5	5	5.5	V
V_{IH}	High-level input voltage	2			2			V
V_{IL}	Low-level input voltage			0.8			0.8	V
I_{OH}	High-level output current			−2			−2	mA
I_{OL}	Low-level output current			20			20	mA
T_A	Operating free-air temperature	−55		125	0		70	°C

electrical characteristics over recommended operating-free-air temperature range (unless otherwise noted)

PARAMETER	TEST CONDITIONS		SN54AS02			SN74AS02			UNIT
			MIN	TYP[†]	MAX	MIN	TYP[†]	MAX	
V_{IK}	$V_{CC} = 4.5$ V,	$I_I = -18$ mA			−1.2			−1.2	V
V_{OH}	$V_{CC} = 4.5$ V to 5.5 V,	$I_{OH} = -2$ mA	$V_{CC}-2$			$V_{CC}-2$			V
V_{OL}	$V_{CC} = 4.5$ V,	$I_{OL} = 20$ mA		0.35	0.5		0.35	0.5	V
I_I	$V_{CC} = 5.5$ V,	$V_I = 7$ V			0.1			0.1	mA
I_{IH}	$V_{CC} = 5.5$ V,	$V_I = 2.7$ V			20			20	μA
I_{IL}	$V_{CC} = 5.5$ V,	$V_I = 0.4$ V			−0.5			−0.5	mA
I_O[‡]	$V_{CC} = 5.5$ V,	$V_O = 2.25$ V	−30		−112	−30		−112	mA
I_{CCH}	$V_{CC} = 5.5$ V,	$V_I = 0$ V		3.7	5.9		3.7	5.9	mA
I_{CCL}	$V_{CC} = 5.5$ V,	$V_I = 4.5$ V		12.5	20.1		12.5	20.1	mA

[†] All typical values are at $V_{CC} = 5$ V, $T_A = 25$°C.
[‡] The output conditions have been chosen to produce a current that closely approximates one half of the true short-circuit output current, I_{OS}.

switching characteristics

PARAMETER	FROM (INPUT)	TO (OUTPUT)	V_{CC} = 4.5 V to 5.5 V, C_L = 50 pF, R_L = 500 Ω, T_A = MIN to MAX				UNIT
			SN54AS02		SN74AS02		
			MIN	MAX	MIN	MAX	
t_{PLH}	A or B	Y	1	5	1	4.5	ns
t_{PHL}	A or B	Y	1	5	1	4.5	

Figure 10.52
Exercises 10.7 and 10.8 (Courtesy of Texas Instruments, Inc.).

Figure 10.53
Exercise 10.9.

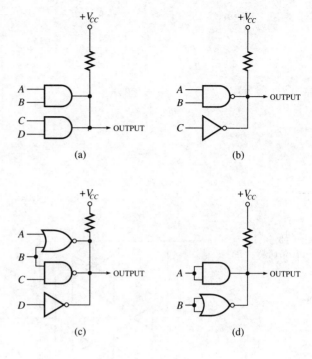

(a)

(b)

(c)

(d)

Figure 10.54
Exercise 10.16.

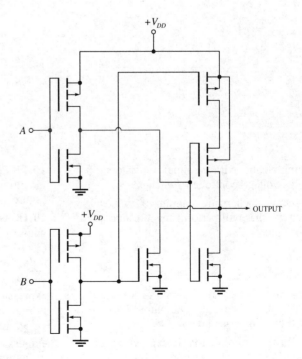

Figure 10.55
Exercise 10.17.

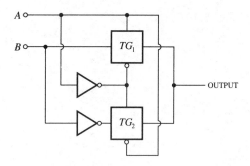

Figure 10.56
Exercise 10.18.

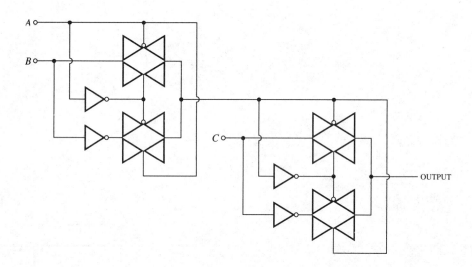

switching circuit, with the input transistors shown as voltage-controlled switches.

10.14 Draw a schematic diagram of an NMOS logic circuit that will perform the AND operation on two inputs.

Section 10–3

10.15 Draw a schematic diagram of a buffered 3-input CMOS NAND gate. The output is buffered by two series inverters.

10.16 Write a simplified logic expression for the output of the circuit in Fig. 10.54 (383).

10.17 Find the logic operation performed by the circuit whose logic diagram is shown in Figure 10.55.

10.18 Write the truth table for the circuit whose logic diagram is shown in Figure 10.56. (This circuit produces the *sum bit* in a *full-adder*, to be discussed in a later chapter.)

10.19 Write a simplified Boolean algebra expression for the output of the circuit shown in Figure 10.57.

10.20 Write a simplified Boolean algebra expression for the output of the circuit shown in Figure 10.58.

Figure 10.57
Exercise 10.19.

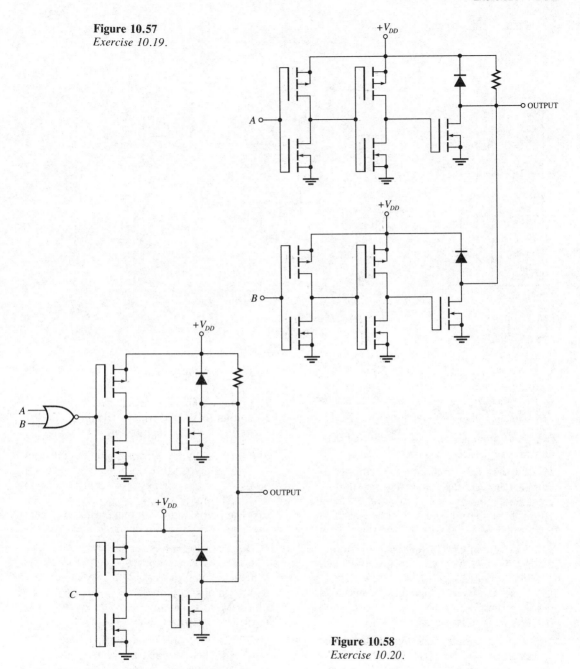

Figure 10.58
Exercise 10.20.

10.21 Figure 10.59 shows a logic diagram of a 74HC126 quad bus driver driving a two-line bus, B_1B_2. Note that the amplifiers are non-inverting and that the enabling inputs are active high. The driver is to be used to perform the following functions:

1. If A_1 and A_2 are both high, A_3 and A_4 are to be transferred to B_1 and B_2 (subject to other conditions listed below).
2. If A_3 and A_4 are both high, A_1 and A_2 are to be transferred to B_1 and B_2, (subject to other conditions listed below).

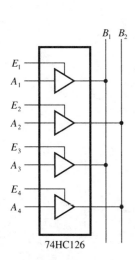

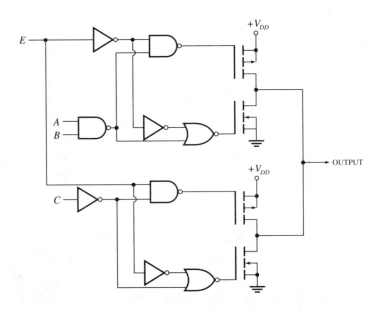

Figure 10.59
Exercise 10.21.

Figure 10.60
Exercise 10.22.

3. If A_1, A_2, A_3, and A_4 are all high, the bus is to be isolated.

4. If neither A_1 and A_2 nor A_3 and A_4 are both high, the bus is to be isolated.

Draw a logic diagram showing how the data transfer can be accomplished.

10.22 Construct a truth table showing the output state resulting from every possible combination of inputs A, B, C, and E in Figure 10.60. Describe in words what the output represents and write a logic expression for it.

10.23 A certain CMOS logic gate conforms to the EIA/JEDEC specifications given in Figure 10.30. Assuming $V_{DD} = +10$ V and $T = 25°C$, find the noise immunities N_L and N_H if
(a) It is a buffered gate;
(b) It is an unbuffered gate.

10.24 Under the test conditions given in the specifications for the 74HC02 (Figure 10.31), find the noise immunities N_L and N_H when $V_{CC} = 4.5$ V at $T = 25°C$. Assume $|I_{out}| \le 20$ μA.

10.25 A CMOS circuit contains two logic gates, each of which has the following characteristics:

1. Intrinsic propagation delay $= t_p = t_{pHL} = t_{pLH} = 25$ ns

2. Power dissipation capacitance $= 34$ pF

3. Output resistance when switching from low to high or from high to low $= 300$ Ω

The total dc current drawn by the circuit from its 5-V supply is 10 μA. If each gate drives a 50-pF capacitive load at a switching frequency of 40 kHz, find
(a) The propagation delay in each gate;
(b) The total power dissipation in each gate.

10.26 The short-circuit output current of a CMOS logic gate with a 5-V supply is 20 mA. When the gate drives a 42-pF capacitive load, its total propagation delay is $t_p = t_{pHL} = t_{pLH} = 27$ ns. What is the approximate value of its intrinsic propagation delay?

10.27 List two precautions that should be observed when building and testing an experimental circuit using a MOS device in a laboratory setting.

10.28 (a) How much static charge accumulated on the gate of a MOS transistor having gate capacitance 5 pF will cause the gate voltage to reach its breakdown value of 100 V?

(b) Explain the origin and nature of latch-up in a MOS integrated circuit.

10.29 Draw block diagrams showing any interface circuitry that would be necessary when

(a) A 74AS00 circuit drives a 74HC02 circuit;

(b) A 74HC02 circuit drives a 74AS00 circuit.

Refer to the specifications given in Figures 10.14 and 10.31. Assume both circuits are operated at 25°C with a 4.5-V supply and that $|I_{out}|$ for the 74HC02 is less than 20 μA.

10.30 A certain logic gate can sink 10 mA when its output is low and can drive (source) 1.2 mA when its output is high. It is to be used to drive logic gates from another family having $I_{IL}(max) = -100$ μA and $I_{IH}(max) = 50$ μA. Find the fanout.

Section 10–4

10.31 Draw a schematic diagram of a 3-input dynamic NMOS NOR gate whose output is connected to a dynamic NMOS inverter. Also draw an equivalent switching circuit showing voltage-controlled switches.

10.32 Find the logic function performed by the dynamic NMOS circuit whose schematic is shown in Figure 10.61.

Section 10–5

10.33 Assuming the base-to-emitter voltage drop of an on transistor in Figure 10.62 is 0.8 V and that $V_{IN} = -0.8$ V, find

(a) The emitter-to-ground voltage, V_E;

(b) The emitter current, I_E;

(c) The collector-to-ground voltage, V_1.

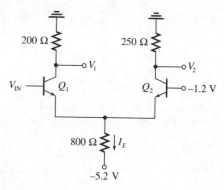

Figure 10.62
Exercises 10.33 and 10.34.

10.34 Assuming the base-to-emitter voltage drop of an on transistor in Figure 10.62 is 0.8 V and that $V_{IN} = -1.6$ V, find

(a) The emitter-to-ground voltage, V_E;

(b) The emitter current, I_E;

(c) The collector-to-ground voltage, V_2.

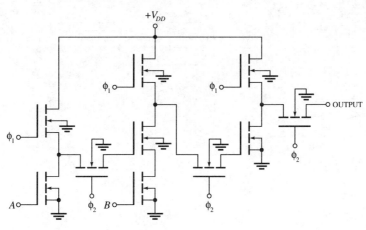

Figure 10.61
Exercise 10.32.

Figure 10.63
Exercise 10.35.

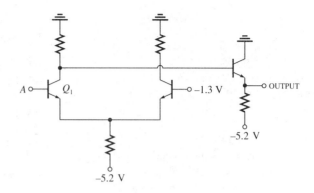

10.35 Logic levels for the circuit shown in Figure 10.63 are -1.7 V $= 1$ and -0.9 V $= 0$. The base-to-emitter drop of an on transistor is 0.8 V. When Q_1 is on, its collector is at -0.1 V. Find the logic operation performed by the circuit.

10.36 Assuming the same logic levels and circuit voltages as shown in Figure 10.38, determine the logic functions performed by the circuit shown in Figure 10.64 at OUT$_1$ and OUT$_2$.

10.37 Assuming maximum values of the parameters given in the specifications in Figure

10.40, find the speed-power product of an MC10H102 NOR gate at $T = 25°C$.

10.38 Repeat Exercise 10.37 with $T = 75°C$.

10.39 Write simplified Boolean algebra expressions for each output in Figure 10.65. Assume each output gate is of the open-emitter ECL type.

10.40 Draw a logic diagram showing how the MC10H102 integrated circuit could be connected to produce logic outputs ABC and $\overline{ABC}$.

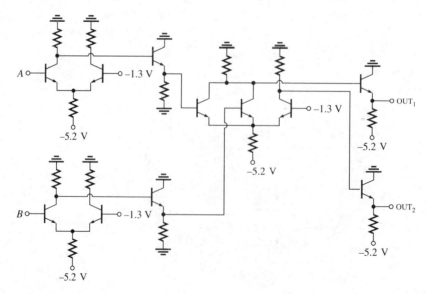

Figure 10.64
Exercise 10.36.

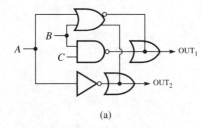
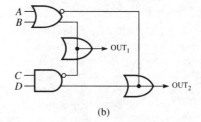

(a)

(b)

Figure 10.65
Exercise 10.39.

Figure 10.66
Exercise 10.41

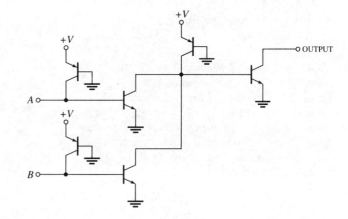

Figure 10.67
Exercise 10.42.

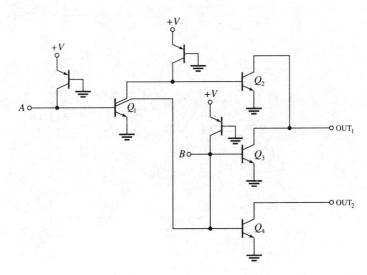

Section 10–6

10.41 Find the logic operation performed by
the circuit in Figure 10.66.

10.42 Find the logic operations performed by
the circuit in Figure 10.67. (*Hint:* Construct a
truth table to determine which transistors are on
and which are off for every input combination.)

CHALLENGING EXERCISES

10.43 A certain logic gate has a propagation delay of 12 ns with no capacitive load. When load capacitance is added, the propagation delay increases in direct proportion to the capacitance added. The total delay is 25 ns with a load of 50 pF. The power consumed by the gate is 4 mW. It is to be used to drive several other gates, each representing a load capacitance of 10 pF. If the speed-power product cannot exceed 95 pJ, what is the maximum number of gates it can drive?

10.44 Write a simplified Boolean algebra expression for the output of the circuit shown in Figure 10.68.

10.45 A 3-input CMOS NAND gate and a 3-input CMOS NOR gate are to be used to implement the logic shown in the truth table in Figure 10.69. If each gate has an open-drain output, draw a logic diagram showing how the gates should be connected to perform the logic. Be certain to show connections to all inputs, including unused inputs.

A	B	C	D	OUTPUT
0	0	0	0	1
0	0	0	1	0
0	0	1	0	1
0	0	1	1	0
0	1	0	0	1
0	1	0	1	0
0	1	1	0	1
0	1	1	1	0
1	0	0	0	1
1	0	0	1	0
1	0	1	0	1
1	0	1	1	0
1	1	0	0	1
1	1	0	1	0
1	1	1	0	0
1	1	1	1	0

Figure 10.69
Exercise 10.45.

10.46 Each gate in a quad 2-input CMOS NOR gate has intrinsic propagation delay $t_{pLH} =$

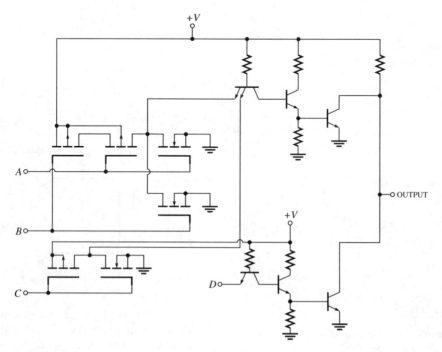

Figure 10.68
Exercise 10.44.

Figure 10.70
Exercise 10.47.

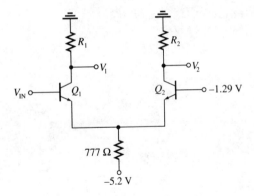

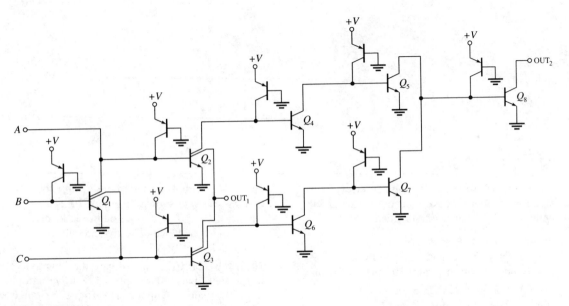

Figure 10.71
Exercise 10.48.

t_{pHL} = 20 ns. The integrated circuit draws a total dc current of 25 μA from a 5-V supply when all the gates are operated at a switching frequency of 50 kHz. Each gate has output resistance 250 Ω and has a total propagation delay (with loads connected) of 38 ns. When operated under the conditions described, it is found that the integrated circuit generates 3.582 J of heat energy in a 1-h period. What is the power dissipation capacitance, C_{PD}, of each gate, assuming it is the same for all gates?

10.47 When V_{IN} in Figure 10.70 is -1.69 V, V_2 is -0.98 V. When V_{IN} is -0.89 V, V_1 is -0.98 V. Assuming the base-to-emitter drop of an on transistor is 8.0 V, find R_1 and R_2. (Neglect base currents.)

10.48 Write simplified logic expressions for OUT$_1$ and OUT$_2$ in the circuit shown in Figure 10.71 (*Hint:* Construct a truth table to determine which transistors are on and which are off for every input combination.)

Figure 10.72
Exercise 10.50.

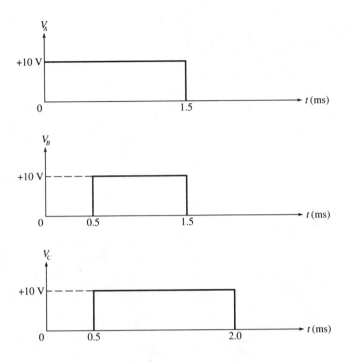

SPICE EXERCISES

Note: In each of the following exercises, assume all devices have their default parameter values, unless otherwise specified.

10.49 Use SPICE to determine approximate values for $V_{IL}(\text{max})$ and $V_{IH}(\text{min})$ in the TTL NAND gate shown in Figure 10.4(a).

10.50 A 3-input NMOS NAND gate has the inputs shown in Figure 10.72. The three switching transistors have $\beta = 0.25 \times 10^{-3}$ A/V^2 and $V_T = 2$ V. The NMOS load transistor has $\beta = 0.025 \times 10^{-3}$ A/V^2 and $V_T = 2$ V. The supply voltage is 10 V. Use SPICE to obtain a plot of the output waveform. Interpret the results.

10.51 Use SPICE to determine t_{pHL} for the CMOS NOR gate shown in Figure 10.24(a) when $V_{DD} = 10$ V and the output is loaded by 10 pF of capacitance. The logic levels are 0 = 0 V and 1 = +10 V.

10.52 Use SPICE to investigate the sensitivity of the ECL gate in Example 10–8 to variations in the V_{EE} supply voltage. Determine the range of the output high voltage and of the output low voltage when V_{EE} drifts ± 1 V. Assuming that the specifications for the MC10H102 at $T = 25°C$ apply, determine if the variation in output logic levels is such that an output low or high would not be properly interpreted at the input to an MC10H102 gate.

Part III

STORAGE DEVICES AND SEQUENTIAL LOGIC

Chapter 11

FLIP-FLOPS

INTRODUCTION

In our study of monostable multivibrators, we noted that those devices have *one* stable state. A flip-flop, known more formally as a *bistable* multivibrator, has *two* stable states. It is a device that can be triggered so that its output *remains* high (one stable state) or triggered again so that its output remains low (the other stable state). As such, the flip-flop serves as a *storage* device, a very important function in digital systems. When its output is high, it is storing a 1, and when its output is low, it is storing a 0. A set of flip-flops called a *storage register* stores the individual bits of a binary number, each flip-flop storing the state of one bit in the number. To change the number being stored, we simply change the states of the flip-flops so that they correspond to the bits of the new number.

Flip-flops also serve as the fundamental components of another important digital circuit: the digital *counter*. As its name implies, this device is used to keep track of events that occur in a digital system, such as the number of pulses that occur in a given period of time. It is also used as a *frequency divider*, as we shall see in later discussions.

Flip-flops are constructed in a wide variety of types, each type having the special features or characteristics necessary for particular applications. In this chapter, we discuss the various types and introduce some of their important applications. Before the advent of integrated circuits, discrete flip-flops were constructed using the classic Eccles-Jordan circuit. Virtually all modern flip-flops are constructed by interconnecting logic gates in integrated circuits. Since the flip-flop is such an important ingredient of so many kinds of digital systems, it is found in small, medium, large, and very large scale integrated circuits and in all of the logic families discussed in Chapter 10.

11-2   THE RS LATCH

Figure 11.1(*a*) shows the logic symbol for the simplest type of flip-flop, called an *RS latch*. Note that it has two outputs, labeled Q and $\overline{Q}$. The state of the latch corresponds to the level of Q (high or low, 1 or 0), and $\overline{Q}$ is, of course, the complement of that state. Although both outputs are not used in all applications, the availability of both makes the flip-flop more versatile. The figure shows that the latch also has two inputs: set (*S*) and

Figure 11.1
The RS latch.

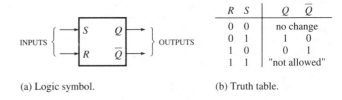

R	S	Q	$\overline{Q}$
0	0	no change	
0	1	1	0
1	0	0	1
1	1	"not allowed"	

(a) Logic symbol. (b) Truth table.

reset *(R)*. The name of the latch, RS, or reset-set, is derived from the names of its inputs, as is the case in other flip-flops we will study. When the set input is made high, Q becomes 1 (and $\overline{Q}$ 0). When the reset input is made high, Q becomes 0 (and $\overline{Q}$ 1). This behavior is shown in the truth table in Figure 11.1(b). If both inputs are low, there is *no change* in the state of the latch. Thus, the latch remains in whatever state it was last in; i.e., it continues to store that state until it is changed by a new set or reset input. If both inputs are made high, we have a contradictory situation, since we cannot set and reset the latch simultaneously. Depending on the design of the device, both outputs in that situation may be low, both may be high, or the output state may be unpredictable. This condition is described variously as *not allowed, unpredictable, invalid,* or *indeterminate*. (There are, however, applications where it is convenient to be able to make both outputs high or both low, despite the resulting contradiction of Q and $\overline{Q}$.)

There is a great deal of duplicate terminology associated with flip-flops. For example, reset is sometimes referred to as *clear* (since we "clear out" the 1 by resetting it to 0). Therefore, the RS latch is also called an SC latch (also an SR latch). It is called a latch because it "latches onto" a 1 or 0 (changes state) immediately upon receiving a set or reset input. As we shall see, there can be a significant time delay in changing the state of other, more complex types of flip-flops. In fact, the term *flip-flop* is now widely used to refer *exclusively* to those more sophisticated devices. Set and reset are used as verbs as well as adjectives: For example, we set the flip-flop, and we say that it is in the set state. We also say that a flip-flop is storing a 1, or is in the one-state, when it is set, and that it is storing a 0, or is in the zero-state, when it is reset.

The NOR-Gate Latch

Figure 11.2 shows how an RS latch can be constructed using two NOR gates. Note that the NOR gates are *cross-coupled:* one of the inputs to each gate is the output of the other gate. As a reference aid for analyzing this circuit, the figure shows the truth table for a 2-input NOR gate. Let us assume that the latch is initially set, so $Q = 1$ and $\overline{Q} = 0$. We can verify that it remains set when $R = S = 0$ by observing that the inputs to NOR gate 2 are then 0 and 1, which holds $\overline{Q}$ at 0. Since $\overline{Q}$ is 0, the inputs to NOR gate 1 are both 0, holding Q

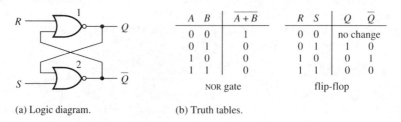

A	B	$\overline{A + B}$
0	0	1
0	1	0
1	0	0
1	1	0

NOR gate

R	S	Q	$\overline{Q}$
0	0	no change	
0	1	1	0
1	0	0	1
1	1	0	0

flip-flop

(a) Logic diagram. (b) Truth tables.

Figure 11.2
Using cross-connected NOR gates to construct an RS latch.

at 1. We see that $Q = 1$ forces $\overline{Q}$ to be 0 and that $\overline{Q} = 0$ forces Q to be 1. Thus, the latch must remain in that set state indefinitely, as long as R and S are both 0. With R still 0, if we now make $S = 1$, *nothing changes:* The output of gate 2 $(\overline{Q})$ remains low when both of its inputs are made high. We conclude that making the set input high when the latch is already set has no effect on it. Suppose that we now make $R = 1$, with $S = 0$. The high input to gate 1 causes its output (Q) to change from 1 to 0. That 0 is applied to gate 2, which now has two low inputs. Therefore, its output $(\overline{Q})$ changes from 0 to 1. That 1 is now an input to gate 1, so Q is held at 0. We have reset the latch. An analysis similar to the foregoing confirms that the latch will remain reset when R and S are both made low. It is also easy to verify that making $S = 1$, with $R = 0$, will once again set the latch. If R and S are *both* 1, both gates have at least one high input, so the output of both gates is 0. As mentioned previously, this combination is not normally encountered.

Careful contemplation of the truth table for the latch reveals the important insight that it is necessary only to *pulse* a set or reset input to change the state of the latch. For example, if the latch is initially reset, a pulse applied to its set input is the same as making S momentarily 1, followed by 0. The 1 sets the latch, after which R and S are once again 0, the no-change condition. Since a pulse must remain high long enough for both NOR gates to change states, the minimum pulsewidth is the sum of the propagation delays through the gates. One gate must change from low to high and the other gate from high to low, so

$$PW_{(\text{min})} = t_{p\text{LH}} + t_{p\text{HL}} \tag{11.1}$$

To reinforce understanding of latch dynamics, Figure 11.3 shows a timing diagram that illustrates how an RS latch responds to arbitrarily selected waveforms at its R and S

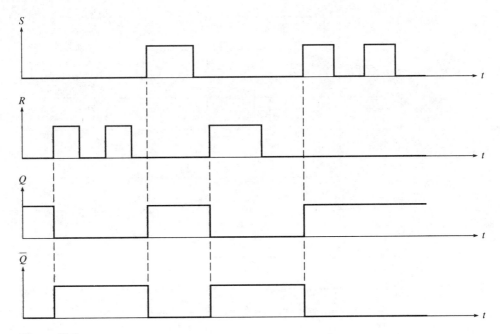

Figure 11.3
Timing diagram illustrating latch response to waveforms applied to its S and R inputs.

inputs. We assume the latch is initially set, so Q is 1 until the R input goes high and resets it. When the R input returns low, $R = 0$ and $S = 0$, so there is no change. Another pulse at the R input has no effect, since the latch is already reset. When the set input goes high, we have $S = 1$ and $R = 0$, so the latch sets and Q becomes 1. As an instructive exercise, verify the response of the latch shown in the rest of the timing diagram.

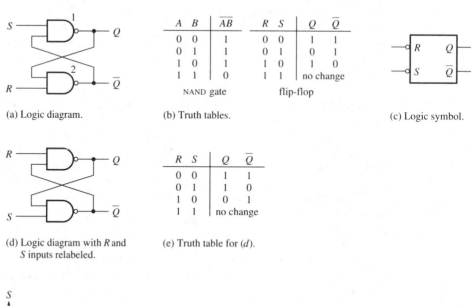

A	B	$\overline{AB}$
0	0	1
0	1	1
1	0	1
1	1	0

NAND gate

R	S	Q	$\overline{Q}$
0	0	1	1
0	1	0	1
1	0	1	0
1	1	no change	

flip-flop

(a) Logic diagram. (b) Truth tables. (c) Logic symbol.

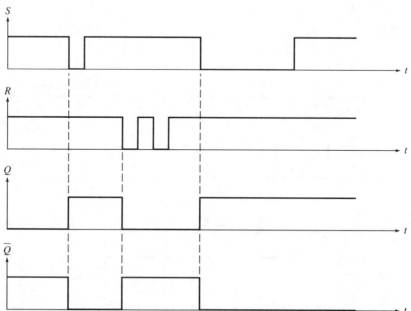

R	S	Q	$\overline{Q}$
0	0	1	1
0	1	1	0
1	0	0	1
1	1	no change	

(d) Logic diagram with R and S inputs relabeled. (e) Truth table for (d).

(f) Example of the response of the latch in (a) to waveforms at its S and R inputs.

Figure 11.4
The NAND-gate RS latch.

The NAND-Gate Latch

Figure 11.4(a) shows how two NAND gates can be cross-coupled to construct an RS latch. The truth table shows that $R = S = 1$ is the no-change condition in this case. For example, if the latch is initially set, then $\overline{Q} = 0$ and $S = 1$ will hold the output of gate 1 high, whereas $Q = 1$ and $R = 1$ will hold the output of gate 2 low. If R then goes low, the output of NAND gate 2 goes high. Since the inputs to NAND gate 1 are then both high, Q goes low. Thus, the latch has reset, and we see that the reset input is *active-low*. A similar analysis reveals that the set input is active-low. The input combination $R = S = 0$ is not normally allowed, since that makes $Q = \overline{Q} = 1$. Figure 11.4(c) shows the logic symbol for the latch having active-low inputs. Note that we *could* interchange the set and reset labels on the inputs, as shown in Figure 11.4(d), producing the truth table in 11.4(e). The inputs can now be regarded as active-high, but that somewhat contradicts the fact that $R = S = 1$ is still the no-change condition. Note that interchanging the R and S labels in Figure 11.4(a) has the same effect as leaving them alone and interchanging the Q and $\overline{Q}$ labels. The point of these observations is that there is nothing sacred about how inputs and outputs are labeled. Any confusion arising from ambiguity of labeling should always be resolved by consulting (or constructing) a truth table. Figure 11.4(f) is a timing diagram illustrating the response of the latch in 11.4(a) to arbitrary waveforms at its R and S inputs.

11-3   GATED (CLOCKED) RS LATCHES

The RS latches we have discussed so far respond immediately to pulses applied to their set or reset inputs (except for small propagation delays). In many digital systems, events such as these must be *synchronized,* so that they all occur at the same time, under the control of a clock. In these synchronous systems, RS latches change state only when a clock pulse is applied to them. Thus, the basic design we have studied must be modified so that the latch cannot change state, regardless of its R and S inputs, unless a clock pulse is present. The modified circuit is called a *gated, clocked,* or *synchronous* RS latch. Understand that this type of latch does not *necessarily* change state when it is clocked; the clock pulse merely allows it to change state if the R and S inputs prescribe a change. Conversely, when no clock pulse is present, changes in the R and S inputs have no effect on the latch. Since the basic RS designs we studied earlier are not synchronized by a clock, they are said to be *asynchronous*.

Figure 11.5(a) shows how a pair of AND gates is added to the basic NOR-gate latch to convert it to a gated RS latch. When the clock is low, the output of both AND gates is 0. Therefore, the NOR gate inputs (that were R and S in the original design) are both 0, which is the no-change condition. Thus, the latch cannot change state when the clock is low, regardless of how the R and S inputs to the AND gates change. When the clock is high, the output of each AND gate is the same as its R or S input. In other words, the clock *gates* the R and S inputs to the cross-coupled NOR gates, which then respond like the NOR-gate latch. The response occurs when the clock goes high—i.e., on the leading edge of a clock pulse. However, the device will continue to respond to changes in R and S as long as the clock remains high. For that reason, it is said to be *level*-triggered (rather than edge-triggered, which we discuss later.) Figure 11.5(b) shows the logic symbol for the gated RS latch. Figure 11.5(c) shows the truth table, which for clocked flip-flops is often called a *state table*. Q_n is called the *present* state, the state of the flip-flop before arrival of the next clock pulse. Q_{n+1} is the *next* state, the state that results from the clocking action. Figure 11.5(d) shows a more compact way of presenting the same information as

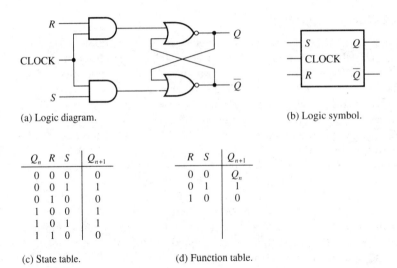

(a) Logic diagram.

(b) Logic symbol.

(c) State table.

Q_n	R	S	Q_{n+1}
0	0	0	0
0	0	1	1
0	1	0	0
1	0	0	1
1	0	1	1
1	1	0	0

(d) Function table.

R	S	Q_{n+1}
0	0	Q_n
0	1	1
1	0	0

Figure 11.5
The gated RS NOR-gate latch

the state table. Called a *function table*, it shows that Q_{n+1} is the same as Q_n when $R = S = 0$ (no change) and that Q_{n+1} is 1 or 0 otherwise, depending on R and S.

Set-Up Time and Other Specifications

Although the gated latch is unaffected by changes in R and S while the clock is low, the R and S inputs must be constant (static) for a certain minimum time before a clock pulse is applied. This minimum time is called the *set-up time*, t_{su}, and its value is given in manufacturers' specifications. Some flip-flops also require a certain minimum *hold time*, t_h, which is the minimum time that the inputs must remain constant after the clock pulse occurs. Both t_{su} and t_h are illustrated in Figure 11.6. Another important specification is the maximum clocking frequency. The frequency of clock pulses is limited by the propagation delay through the logic gates from which the latch is constructed. The frequency cannot be so great that a new clock pulse arrives before the device has had time to respond to the previous clock pulse.

Figure 11.6
R or S must not change for at least t_{su} seconds before the clock edge occurs nor for at least t_h seconds after the clock edge occurs.

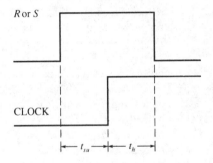

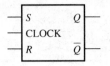

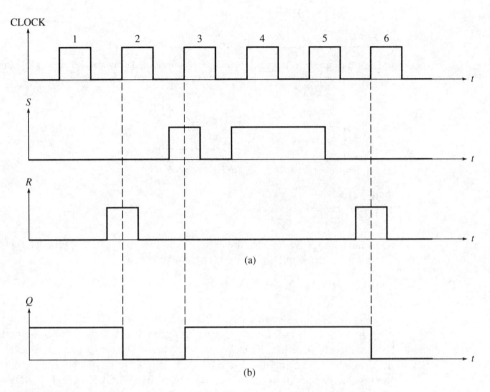

Figure 11.7
Example 11-1.

Example 11-1

Complete the timing diagram in Figure 11.7(a) to show the response of the gated RS latch to the R, S, and CLOCK waveforms.

Solution. The response of the latch is shown in part (b) of the figure. When the leading edge of clock pulse 1 occurs, R and S are both 0, so the latch remains set. When the leading edge of clock pulse 2 occurs, S is 0 and R is 1, so the latch resets at that time. As an instructive exercise, verify the rest of the timing diagram. Note carefully that the latch can change state only when the clock is high and that changes in R and S when the clock is low have no immediate effect on the state.

Although R and S in this example did not change while the clock was high, it is important to remember that any such changes in a level-triggered device will have an immediate effect on the output state. Exercise 11.10 is an instructive example.

Drill 11-1

a. If the S and R waveforms in Figure 11.7(b) are interchanged, at which clock pulses will the latch then change state?

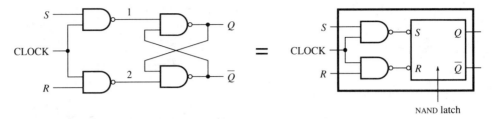

(a) Logic diagram and its equivalent. The input NAND gates make the S and R inputs active-high.

R	S	Q_{n+1}
0	0	Q_n
0	1	1
1	0	0

(b) Function table.

(c) Logic symbol.

Figure 11.8
The gated RS NAND-gate latch.

b. If the S and R waveforms are as shown in Figure 11.7(b) but the latch is initially reset, at which clock pulses will it then change state?

Answer: (a) 3 and 6; (b) 3 and 6.

Figure 11.8(a) shows a gated RS NAND-gate latch. In this circuit, the S and R inputs are gated by NAND gates, so the inputs (labeled 1 and 2) to the cross-coupled NAND latch are effectively inverted. This inversion converts the NAND latch to one that responds to active-high R and S inputs. When CLOCK is low, inputs 1 and 2 are both high, which is the no-change condition for the NAND latch. When CLOCK is high, inputs 1 and 2 are the complements of S and R. For example, if S is 0 and R is 1, then input 1 is 1 and input 2 is 0, the combination that resets the NAND latch. The function table in Figure 11.8(b) confirms that the latch responds to active-high inputs.

Asynchronous (Direct) Set and Reset

In many applications where clocked flip-flops are used, it is necessary to be able to set or reset a flip-flop whether a clock pulse is present or not. For that reason, many clocked flip-flops are equipped with *asynchronous,* or *direct,* set and reset inputs that can be used to control their states irrespective of the clock. Figure 11.9(a) shows how direct inputs can be added to a clocked NAND latch. Since these inputs are connected directly to the cross-coupled NAND gates, they are active-low. If either one is made low, the output of the corresponding NAND gate immediately becomes high. Note, however, that if both are low or if the clock is high and we attempt to reset and direct set the latch simultaneously, we obtain the usual contradictory outputs: $Q = \overline{Q} = 1$. Such is not the case in more sophisticated designs we study later, where the direct inputs have priority over all others, irrespective of the clock. Direct inputs are often labeled *preset* and *clear* to distinguish them from the synchronous set and reset inputs. The initial state of a flip-flop is unpredictable when power is first applied, so preset and clear inputs are useful for *initializing* states, prior to normal system operation.

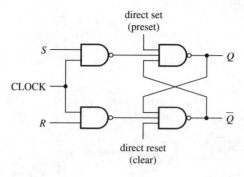

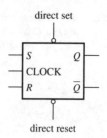

(a) Logic diagram.

(b) Logic symbol. Note that the direct inputs are active-low.

Figure 11.9
Clocked NAND-gate latch with direct set and reset.

11–4   THE TRANSPARENT D LATCH

In many applications, it is not necessary to have separate R and S inputs to a latch. If the input combinations $R = S = 0$ and $R = S = 1$ are never needed, then R and S are always complements of each other. Thus, as shown in Figure 11.10(a), we can construct a latch with a single input *(S)* and obtain the R input by inverting S. The single input is labeled D (for data) and the device is called a D latch. When $D = 1$, we have $S = 1$ and $R = 0$, causing the latch to set when clocked. When $D = 0$, we have $S = 0$ and $R = 1$, causing the latch to reset when clocked. When the clock is high, the output follows the input, so the latch is said to be *transparent*. When the clock goes low, Q remains in the last state it was in when the clock was high.

Practical D Latches and Applications

A very common application for D-type latches is temporary storage of binary data that is to be transferred to another device at a specified time. The latch is typically operated in its transparent mode (clock high) until such time as a transfer is required. At that time, the

Figure 11.10
The transparent D latch.

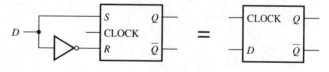

(a) Logic diagram.

(b) Logic symbol.

Q_n	D	Q_{n+1}
0	0	0
0	1	1
1	0	0
1	1	1

(c) State table.

D	Q_{n+1}
0	0
1	1

(d) Function table.

Figure 11.11
A practical transparent D latch. Note that the output is latched (fixed) when ENABLE is low and that Q follows D when ENABLE is high.

A	B	$\overline{AB}$
0	0	1
0	1	1
1	0	1
1	1	0

NAND truth table
(for reference)

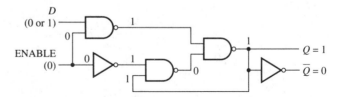

(a) Logic states when ENABLE is low and Q is 1. Q remains 1, regardless of changes in D.

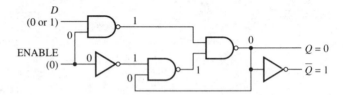

(b) Logic states when ENABLE is low and Q is 0. Q remains 0, regardless of changes in D.

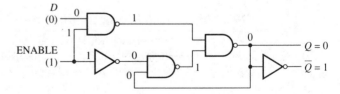

(c) Logic states when ENABLE is high and D is 0.

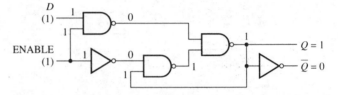

(d) Logic states when ENABLE is high and D is 1.

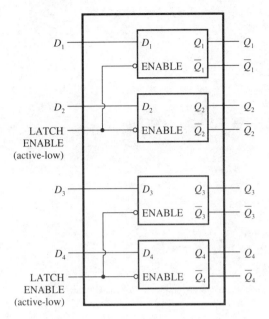

(a) Logic diagram.

Figure 11.12
The 7475 dual 2-bit transparent D latch.

clock goes low, and the output data remain stable, being insensitive to further changes in D. Because of this type of application, the clock input to a D latch is usually designated ENABLE (not to be confused with the enable signal used in three-state circuits). Since the data is effectively latched, or held constant, at the output when the clock goes low, the ENABLE signal is regarded as active-low. In other words, enabling the latch means fixing its output, so it no longer responds to variations at the D input.

Although a D latch can be constructed as shown in Figure 11.10(a) using a cross-coupled RS latch, practical D latches are more often constructed in a functionally equivalent configuration such as that shown in Figure 11.11. Figure 11.11(a) shows the circuit when ENABLE is low and Q is 1. It is clear that Q remains 1, irrespective of changes in D. (Check the logic levels shown in the figure when $D = 0$ and when $D = 1$.) Similarly, Figure 11.11(b) shows that when ENABLE is low and Q is 0, Q is unaffected by changes in D. Figure 11.11(c) and (d) shows that Q follows D when ENABLE is high. [Change D to 1 in (c) and change D to 0 in (d) to verify transparency.]

Figure 11.12(a) shows a logic diagram* of a typical integrated circuit containing transparent D-type latches, in this case the 7475-series dual ''2-bit'' latch. The (four) latches are arranged so that each pair has a common enable signal, so each pair is called a 2-bit latch. (The circuit is also called a 4-bit latch.) Note that the LATCH ENABLE input is shown as active-low. The Q and $\overline{Q}$ outputs of each latch are available at separate

---

*ANSI/IEEE standard symbols for integrated-circuit flip-flops are shown in Section 11–8.

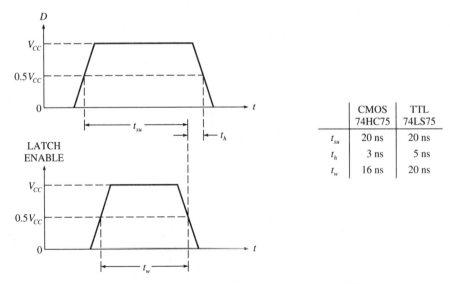

	CMOS 74HC75	TTL 74LS75
t_{su}	20 ns	20 ns
t_h	3 ns	5 ns
t_w	16 ns	20 ns

(b) Typical specifications for set-up and hold times, and for minimum width of the ENABLE pulse, in the 7475.

Figure 11.12
(Continued)

pins of the 16-pin circuit. Some integrated circuits are constructed with a single clock input common to all devices and with $\overline{Q}$-outputs omitted, so that more devices can be integrated into one circuit without requiring an excessive number of pins. The 7475 is available in TTL and CMOS versions. Figure 11.12(b) shows typical specifications for the set-up and hold times for two versions. Although a maximum clocking frequency is not specified, note that there is a minimum specified pulsewidth, t_w, for the ENABLE input.

11–5   MASTER-SLAVE FLIP-FLOPS

Clock Skew and the Time Race

In many synchronous systems, flip-flop outputs are connected directly, or through intervening logic circuitry, to the inputs of other flip-flops. The propagation delay of a flip-flop and/or the delays of intervening gates make it difficult to predict precisely when the changing state of one flip-flop will be experienced at the input of another. Furthermore, the clock signal, which is applied simultaneously to all flip-flops in a synchronous system, may undergo varying degrees of delay due to wiring between components. This delay is called *clock skew*. If the clock skew is minimal, a flip-flop may be clocked before it receives its new input (derived from the output of another clocked flip-flop). See Figure 11.13. On the other hand, if the clock pulse is delayed significantly, the inputs to a flip-flop may have changed before the clock pulse arrives. In these and similar situations, we have what amounts to a *race* between two competing signals that are attempting to accomplish opposite effects. The "winner" in such a race depends on largely unpredictable propagation delays, delays that can vary from device to device and that can change with environmental conditions. It is clear that reliable system operation is not possible when flip-flop responses depend on the outcome of a race.

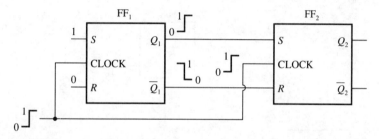

Figure 11.13
An example of a timing race. If the clock input to FF_2 goes high before FF_1 changes state, FF_2 will reset. If the clock input to FF_2 goes high after FF_1 changes state, FF_2 will set. Small differences in propagation delays can affect which event occurs first, making the actual response unpredictable.

The Master-Slave (Pulse-Triggered) Flip-Flop

The master-slave flip-flop was developed to make synchronous operation of flip-flops more predictable. This improvement is achieved by introducing a known time delay (equal to the width of one clock pulse) between the time that a flip-flop responds to a clock pulse at its input and the time that that response appears at its *output*. Figure 11.14(*a*) shows how a master-slave (M/S) RS flip-flop can be constructed using two RS latches. The latch to which external RS connections are made is called the *master*. The latch providing outputs Q and $\overline{Q}$ is called the *slave*. When the rising edge of a clock pulse appears at the master, it responds as prescribed by its *S* and *R* inputs. Note that the same

Figure 11.14
The master-slave RS flip-flop.

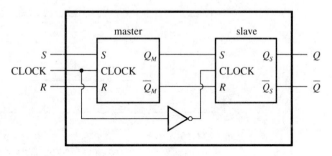

(a) Construction of an RS master-slave flip-flop using two clocked RS latches.

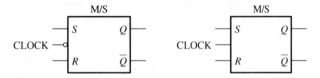

(b) Logic symbol for a master-slave RS flip-flop. The state clocked into the master appears at the output of the flip-flop on the trailing edge of the clock pulse.

(c) Logic symbol for a master-slave RS flip-flop whose output appears on the leading edge of the clock pulse.

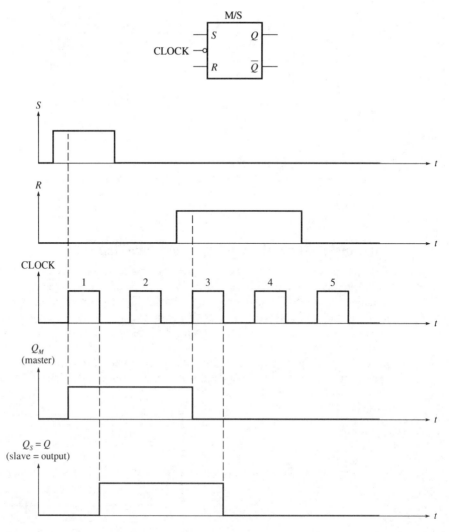

Figure 11.15
Example of a timing diagram for a master-slave RS flip-flop.

clock pulse is inverted and applied to the slave. Thus, the slave experiences a *falling* clock edge and does nothing. However, when the clock pulse goes low, a rising edge then appears at the slave, causing it to respond: Since its S and R inputs are the Q and $\overline{Q}$ outputs of the master, the slave *follows* the master. We say that a 1 or 0 is ''clocked into'' the master on the rising edge of the clock pulse and is clocked into the slave on the falling edge. Thus, the output of the master-slave flip-flop is delayed by the width of one clock pulse. Figure 11.14(b) shows the logic symbol for the master-slave RS flip-flop. Note the inversion bubble at the clock input. The inversion symbol means that the state clocked into the master appears at the output (the slave) on the *trailing* edge of the clock, as we have described. Figure 11.14(c) shows the logic symbol for a clocked flip-flop whose output appears at the leading edge of the clock. (In this device, data is clocked into the

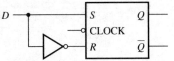

(a) Construction of a D-type master-slave
flip-flop using an RS master-slave.

(b) Logic symbol.

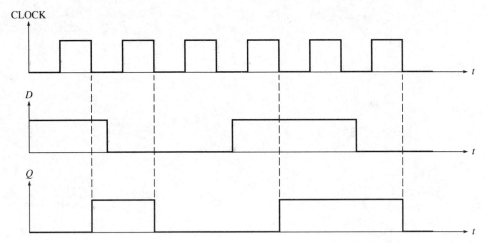

(c) Timing diagram.

Figure 11.16
The D-type master-slave flip-flop.

master on the trailing edge of the clock and into the slave on the leading edge.) A master-slave flip-flop is also called a *pulse-triggered* flip-flop because the length of time required for its output to change state equals the width of one (clock) pulse.

Figure 11.15 shows a timing diagram illustrating the behavior of a master-slave RS flip-flop. We assume that the master and slave are both initially reset. When the leading edge of clock pulse 1 arrives, S is 1 and R is 0. Therefore, a 1 is clocked into the master ($Q_M = 1$). On the trailing edge of clock pulse 1, the slave follows the master and is itself set, making the output ($Q_S = Q$) equal to 1. During clock pulse 2, $S = R = 0$, so there is no change in either the master or the slave. As an instructive exercise, verify the rest of the timing diagram.

Figure 11.16(a) shows how a D-type master-slave flip-flop can be constructed from an RS master-slave. As in the D latch, the output follows D, except in this case the state of D does not appear at the output until the trailing edge of the clock occurs. The timing diagram in Figure 11.16(c) illustrates this fact.

11−6   THE JK FLIP-FLOP

Recall that the RS flip-flop has one combination of inputs that is disallowed: Either $R = S = 1$ or $R = S = 0$, depending on design. The flip-flop would be more versatile if every input combination were available to accomplish some useful function. Such is the

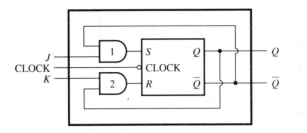

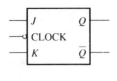

(a) Construction of a JK flip-flop from a master-slave RS flip-flop.　　　(b) Logic symbol.

$$S_{n+1} = J_{n+1}\overline{Q}_n \qquad R_{n+1} = K_{n+1}Q_n$$

	J_{n+1}	K_{n+1}	Q_n	$\overline{Q}_n$	S_{n+1}	R_{n+1}	Q_{n+1}	$\overline{Q}_{n+1}$
no change	0	0	0	1	0	0	0	1
	0	0	1	0	0	0	1	0
reset	0	1	0	1	0	0	0	1
	0	1	1	0	0	1	0	1
set	1	0	0	1	1	0	1	0
	1	0	1	0	0	0	1	0
toggle	1	1	0	1	1	0	1	0
	1	1	1	0	0	1	0	1

(c) State table.

	J_{n+1}	K_{n+1}	Q_{n+1}
no change	0	0	Q_n
reset	0	1	0
set	1	0	1
toggle	1	1	$\overline{Q}_n$

(d) Function table.

Figure 11.17
The JK flip-flop.

case with the JK flip-flop. In this device, J and K serve the same role as R and S in resetting and setting the flip-flop, but the combination $J = K = 1$ can be used additionally to *change the state* of the flip-flop, regardless of its previous, or initial, state. This action is called *toggling* the flip-flop: If it is initially reset, a new clock pulse sets it, and if it is initially set, a new clock pulse resets it.

Figure 11.17(a) shows that a master-slave JK flip-flop can be constructed from a master-slave RS flip-flop by connecting outputs back to the input side. If the flip-flop is initially reset ($\overline{Q} = 1$), then AND gate 1 is enabled and the state of J is transmitted to S. If the flip-flop is initially set ($Q = 1$), then AND gate 2 is enabled and the state of K is transmitted to R. Thus, $S = J\overline{Q}$ and $R = KQ$. More precisely, since the flip-flop is a master-slave whose output does not appear until the end of the present clock pulse, $S_{n+1} = J_{n+1}\overline{Q}_n$ and $R_{n+1} = K_{n+1}Q_n$. In other words, the set and reset inputs (S_{n+1} and R_{n+1}) when the next clock pulse arrives depend on the states of J and K when that pulse arrives (J_{n+1} and K_{n+1}) as well as on the *present* states Q_n and $\overline{Q}_n$. If $J_{n+1} = K_{n+1} = 1$, then $S_{n+1} = \overline{Q}_n$ and $R_{n+1} = Q_n$, meaning that the flip-flop will set on the next clock pulse if it was initially reset ($\overline{Q}_n = 1$) and will reset if it was initially set ($Q_n = 1$). If $J = K = 0$, then the output of both AND gates is 0, so $R = S = 0$, the no-change condition. The state and function tables in Figure 11.17(c) and (d) shows that J and K play the same roles as S and R in the remaining input combinations. (As a memory aid, think of K as standing for "KLEAR.")

Figure 11.18 shows a common application of a JK flip-flop. Since J and K are both held high, ($J = K = 1$ at all times), the flip-flop toggles on the trailing edge of every clock

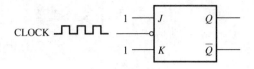

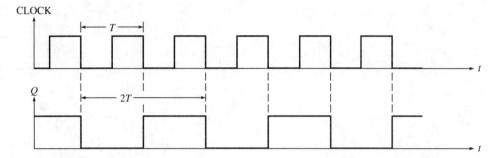

Figure 11.18
Use of a JK flip-flop in frequency division. Since J = K = 1, the flip-flop toggles on every clock pulse, making the frequency of Q one-half that of the clock.

T_{n+1}	Q_{n+1}
0	Q_n
1	$\overline{Q_n}$

(a) Conversion of a JK flip-flop to a T flip-flop.   (b) Function table.

Figure 11.19
The T (toggle) flip-flop.

pulse. Consequently, the frequency of the output waveform is one-half that of the clock, as shown in the timing diagram. This application is called *frequency division*.

T Flip-Flops

A T flip-flop has a single control input, labeled *T* for toggle. When *T* is high, the flip-flop toggles on every new clock pulse. When *T* is low, the flip-flop remains in whatever state it was last in. Although T flip-flops are not widely available as commercial items, it is easy to convert a JK flip-flop to the functional equivalent of a T flip-flop. As shown in Figure 11.19, we simply connect *J* and *K* together and label the common connection *T*. Thus, when *T* = 1, we have *J* = *K* = 1 and the flip-flop toggles. When *T* = 0, we have *J* = *K* = 0 and there is no change.

11–7   EDGE-TRIGGERED FLIP-FLOPS

The practical difficulty with the clocked flip-flops we have discussed so far is that in most applications the inputs *(R, S, D, J,* or *K)* must remain *constant* while the clock pulse is high. Any change in the inputs while the clock is high affects the state of the flip-flop. Recall that this type of flip-flop is said to be *level-triggered*. Changes in the outputs of

Figure 11.20
Creating a narrow clocking pulse for use in an edge-triggered flip-flop.

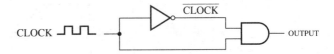

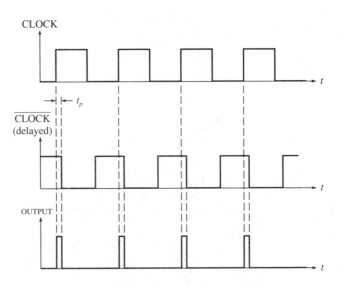

these flip-flops are observed on the leading or trailing edge of a clock pulse only *provided* that the inputs stay constant while the clock is high. They are not, therefore, truly edge-triggered devices. Since they remain sensitive to input changes for relatively long time intervals, they are susceptible to noise for undesirably long times.

Most modern flip-flops are edge-triggered, the JK variety being the most popular. Edge-triggering is achieved by performing a logical operation between the clock pulse and an inverted and slightly delayed version of itself. The delayed version is obtained by passing the clock signal through an inverting gate having a prescribed propagation delay. Figure 11.20 shows one way to accomplish the desired result. The inverter introduces the propagation delay and the output of the AND gate is a very narrow pulse. With this type of clock pulse, the control inputs to a flip-flop need to be held constant only for a very brief time.

A master-slave flip-flop whose master is edge-triggered is said to have *data lockout* because changes in the data inputs *(D, R, S, J, or K)* while the clock is high have no effect. Edge-triggering is also called *dynamic* triggering, since changes in state occur only at the changing edge of the clock. A triangle is drawn at the clock input on the flip-flop symbol to signify dynamic triggering. Figure 11.21 shows examples of symbols for leading- and trailing-edge-triggered flip-flops. Note that both a bubble and a triangle are used to denote a trailing-edge-triggered flip-flop.

Edge-Triggered JK Flip-Flops

In a practical edge-triggered flip-flop, a narrow clock pulse may not be a separately identifiable signal because the edge-triggering principle previously described can be embodied into the flip-flop circuitry in a way that effectively conceals the presence of a

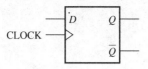

(a) Leading-edge-triggered D flip-flop.

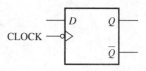

(b) Trailing-edge-triggered D flip-flop.

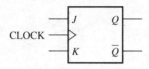

(c) Leading-edge-triggered JK flip-flop.

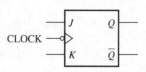

(d) Trailing-edge-triggered JK flip-flop.

Figure 11.21
Examples of symbols for edge-triggered flip-flops.

pulse. Figure 11.22 shows a case in point. This configuration is widely used to construct a trailing-edge-triggered JK flip-flop. NAND gates 1 and 2 are used to invert the clock pulse and to introduce the necessary propagation delay. (Gate 1 performs that function when $J = \overline{Q} = 1$ and gate 2 performs it when $K = Q = 1$, the only two conditions under which the flip-flop must change state.) The delaying function is not evident or relevant in the *static* conditions shown in this figure but is discussed shortly. Figure 11.22(*a*) shows the logic levels in the circuit when the clock is low and Q is 1. Since the outputs of NAND gates 1 and 2 remain 0 when the clock is 0, J and K can change (0 or 1) without affecting the state of the flip-flop. It is also easy to verify that J and K have no effect when the clock is low and Q is 0. In each case, cross-coupling has the usual effect of making Q and $\overline{Q}$ hold each other in opposite states. Figure 11.22(*b*) demonstrates the characteristic that truly distinguishes this flip-flop from the level-triggered types. The figure shows the logic states when the clock is high and $Q = 1$. Changes in K do not affect NAND gate 2 because one of its inputs is 0. Changes in J can make the output of NAND gate 1 equal to 0 or 1, but the figure shows that these changes do not affect the output of NOR gate 8 $(\overline{Q})$, which remains 0. A similar analysis confirms that changes in J and K do not affect the state of the flip-flop when the clock is high and $Q = 0$. In summary, we see that changes in J and K have no effect on the flip-flop when the clock is a constant low *or* when it is a constant high.

Figure 11.23 demonstrates how the flip-flop triggers on the trailing edge of a clock pulse. We assume that it is initially reset, with $J = 1$ and $K = 0$, the conditions that call for changing Q from 0 to 1 when the trailing edge occurs. Part (*a*) of the figure shows the logic levels just before the clock goes low. Recall that NAND gates 1 and 2 are designed to have considerable propagation delay in comparison to other gates in the circuit. Thus, when the clock goes low, we see in the timing diagram that the rising output of NAND gate 2 is delayed. The interval labeled t_p is where all the switching action takes place. The output of AND gate 3 goes low without delay because it is driven directly by the clock. However, the output of AND gate 4 is delayed because it is driven by the delayed output from NAND gate 2. As illustrated in the timing diagram, the inputs to NOR gate 5 are then both low, which causes its output to go high. Thus, Q goes high, making the output of AND

A	B	C	$\overline{ABC}$
0	0	0	1
0	0	1	1
0	1	0	1
0	1	1	1
1	0	0	1
1	0	1	1
1	1	0	1
1	1	1	0

A	B	$\overline{A + B}$
0	0	1
0	1	0
1	0	0
1	1	0

Reference truth tables.

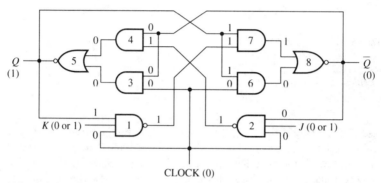

(a) Logic states when the clock is low and $Q = 1$. J and K have no effect because the outputs of NAND gates 1 and 2 remain high.

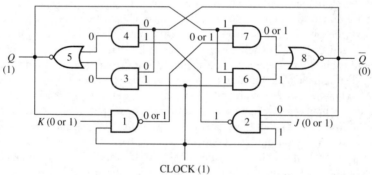

(b) Logic states when the clock is high and $Q = 1$. Changes in J and K do not affect the state of the flip-flop.

Figure 11.22
A trailing-edge-triggered JK flip-flop. Logic states are shown under static conditions (clock not changing).

gate 7 go high, which in turn causes the output of NOR gate 8 to go low. That low coupled back through gates 4 and 5 ensures that the output of gate 5 *(Q)* remains high, so the flip-flop sets. At no time in this sequence is the output of AND gate 4 allowed to go high: $\overline{Q}$ switches one input to gate 4 low *before* the other input, from NAND gate 2, has time to switch high. Thus, the two inputs to NOR gate 5 both stay at 0, causing Q to stay high. The key ideas in the sequence are that Q switches high momentarily (during t_p) because of the

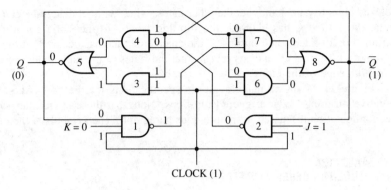

(a) Static logic levels just before the clock goes low.

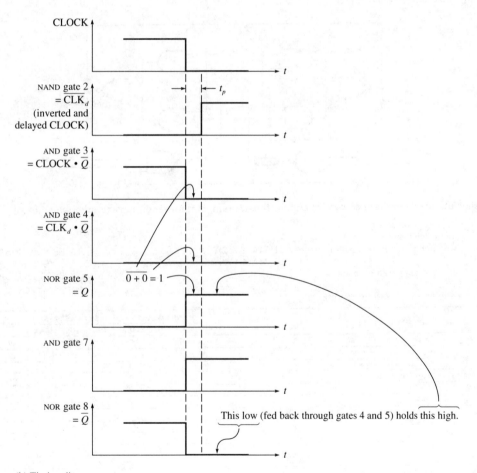

(b) Timing diagram.

Figure 11.23
Trailing-edge-triggering of the JK flip-flop.

delayed input to gate 4 and that the other gates switch without delay to lock Q into its new state. After time t_p has elapsed, the flip-flop reverts to the static conditions described earlier. For all intents and purposes, the change of state has occurred at the trailing edge of the clock. A similar analysis reveals that the states of the logic gates are unaffected by a *leading* edge of the clock.

Figure 11.24 shows manufacturer's specifications for the 54/74ALS112A dual JK negative- (trailing-) edge-triggered flip-flops. Note that these devices are constructed in the configuration discussed in the previous paragraph. The symbols at the output of each

SN54ALS112A, SN74ALS112A
DUAL J-K NEGATIVE-EDGE-TRIGGERED FLIP-FLOPS
WITH CLEAR AND PRESET

logic diagram (positive logic)

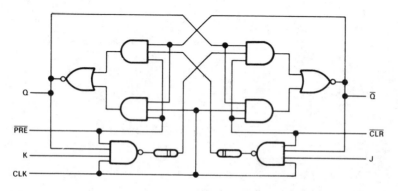

absolute maximum ratings over operating free-air temperature range (unless otherwise noted)

Supply voltage, V_{CC} . . . . 7 V
Input voltage . . . . 7 V
Operating free-air temperature range: SN54ALS112A . . . . $-55\,°C$ to $125\,°C$
                                   SN74ALS112A . . . . $0\,°C$ to $70\,°C$
Storage temperature range . . . . $-65\,°C$ to $150\,°C$

recommended operating conditions

			SN54ALS112A			SN74ALS112A			UNIT
			MIN	NOM	MAX	MIN	NOM	MAX	
V_{CC}	Supply voltage		4.5	5	5.5	4.5	5	5.5	V
V_{IH}	High-level input voltage		2			2			V
V_{IL}	Low-level input voltage				0.7			0.8	V
I_{OH}	High-level output current				−0.4			−0.4	mA
I_{OL}	Low-level output current				4			8	mA
f_{clock}	Clock frequency		0		25	0		30	MHz
t_w	Pulse duration	$\overline{PRE}$ or $\overline{CLR}$ low	15			10			ns
		CLK high	20			16.5			
		CLK low	20			16.5			
t_{su}	Setup time before CLK↓	Data	25			22			ns
		$\overline{PRE}$ or $\overline{CLR}$ inactive	22			20			
t_h	Hold time, data after CLK↓		0			0			ns
T_A	Operating free-air temperature		−55		125	0		70	°C

Figure 11.24
Specifications (Courtesy of Texas Instruments, Inc.).

electrical characteristice over recommended operating free-air temperature range (unless otherwise noted)

PARAMETER		TEST CONDITIONS		SN54ALS112A			SN74ALS112A			UNIT
				MIN	TYP†	MAX	MIN	TYP†	MAX	
V_{IK}		$V_{CC} = 4.5$ V,	$I_I = -18$ mA			-1.5			-1.5	V
V_{OH}		$V_{CC} = 4.5$ V to 5.5 V,	$I_{OH} = -0.4$ mA	$V_{CC}-2$			$V_{CC}-2$			V
V_{OL}		$V_{CC} = 4.5$ V,	$I_{OL} = 4$ mA		0.25	0.4		0.25	0.4	V
		$V_{CC} = 4.5$ V	$I_{OL} = 8$ mA					0.35	0.5	
I_I	J, K, or CLK	$V_{CC} = 5.5$ V,	$V_I = 7$ V			0.1			0.1	mA
	$\overline{PRE}$ or $\overline{CLR}$					0.2			0.2	
I_{IH}	J, K, or CLK	$V_{CC} = 5.5$ V,	$V_I = 2.7$ V			20			20	μA
	$\overline{PRE}$ or $\overline{CLR}$					40			40	
I_{IL}	J, K, or CLK	$V_{CC} = 5.5$ V,	$V_I = 0.4$ V			-0.2			-0.2	mA
	$\overline{PRE}$ or $\overline{CLR}$					-0.4			-0.4	
I_O‡		$V_{CC} = 5.5$ V,	$V_O = 2.25$ V	-30		-112	-30		-112	mA
I_{CC}		$V_{CC} = 5.5$ V,	See Note 1		2.5	4.5		2.5	4.5	mA

† All typical values are at $V_{CC} = 5$ V, $T_A = 25\,^{\circ}$C.
‡ The output conditions have been chosen to produce a current that closely approximates one half of the true short-circuit output current, I_{OS}.
NOTE 1: I_{CC} is measured with J, K, CLK, and $\overline{PRE}$ grounded, then with J, K, CLK, and $\overline{CLR}$ grounded.

switching characteristics (see Note 2)

PARAMETER	FROM (INPUT)	TO (OUTPUT)	$V_{CC} = 4.5$ V to 5.5 V, $C_L = 50$ pF, $R_L = 500\ \Omega$, T_A = MIN to MAX				UNIT
			SN54ALS112A		SN74ALS112A		
			MIN	MAX	MIN	MAX	
f_{max}			25		30		MHz
t_{PLH}	$\overline{PRE}$ or $\overline{CLR}$	Q or $\overline{Q}$	3	26	3	15	ns
t_{PHL}			4	23	4	18	
t_{PLH}	CLK	Q or $\overline{Q}$	3	23	3	15	ns
t_{PHL}			5	24	5	19	

NOTE 2: Load circuit and voltage waveforms are shown in Section 1.

Figure 11.24
(Continued)

NAND gate represent the time delays necessary to achieve edge triggering. Also note that direct, active-low preset and clear inputs are provided. When $\overline{PRE}$ is made low, there will always be at least one low input to the Q NOR gate, making Q high, irrespective of other inputs (including the clock). Similarly, when $\overline{CLR}$ is made low, $\overline{Q}$ is held high. Although the combination $\overline{PRE} = \overline{CLR} = 0$ makes Q and $\overline{Q}$ both high, this combination is not normally used, and the manufacturer does not guarantee specifications for it. The specifications show that the 74ALS112A has a maximum clock frequency of 30 MHz, a set-up time of 22 ns, and a hold time of zero. Comparable specifications for the 74HC112 CMOS version are 20 MHz, 20 ns, and 3 ns. Note that flip-flop specifications include values for the same kinds of parameters that we studied in connection with logic gates, including I_{IH}, t_{pLH}, and so on. These quantities are defined in the same way they are for logic gates, but notice in many cases there is a variation in value, depending on which input (J, K, $\overline{PRE}$, $\overline{CLR}$, or CLOCK) is specified. For example, in the 74ALS112A, propagation delays (t_{pLH} and t_{pHL}) between the input and output of the flip-flop are specified for the cases where $\overline{PRE}$ and $\overline{CLR}$ are considered as inputs as well as for the case where the clock is considered the input. Also, the input currents I_H and I_L depend on which input

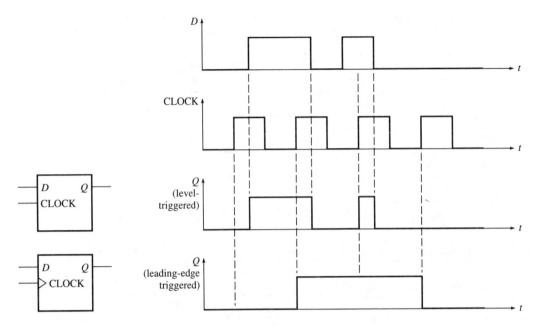

Figure 11.25
Comparison of responses of level-triggered and edge-triggered D flip-flops.

is specified. The values can be used to compute fanouts in the same way they are for logic gates, except the fanout depends on which kind of input the output drives. See Exercises 11.25 and 11.26.

Edge-Triggered D Flip-Flops

Figure 11.25 compares the response of a level-triggered D flip-flop (such as the 7475 in Figure 11.12) to the response of a leading-edge-triggered D flip-flop, such as the 74174 (to be discussed presently). Note that the level-triggered device responds to changes in D whenever the clock is high, whereas the output of the edge-triggered device is latched to whatever state *D* has at the leading edge of the clock.

Figure 11.26 shows manufacturer's specifications for the 74HC174 CMOS hex edge-triggered D flip-flop. In contrast to the 7475 discussed earlier, note that all devices have a common clock, there is no direct set input, and that the $\overline{Q}$ outputs are not externally accessible. These measures permit the fabrication of six flip-flops in a single 16-pin circuit. The logic diagram shows that the individual flip-flops are trailing-edge-triggered, but since the clock is inverted internally, the effect is to make them leading-edge-triggered. The non-inverting amplifiers shown at each *Q* output are the output buffers commonly found in CMOS circuits.

Example 11–2

Assume the 74HC174 is to be used with a 4.5-V supply at $T = 25°C$.

a. What is the minimum permissible width of a clock pulse if the clock has a 50% duty cycle?

b. If *D* changes from low to high, for what minimum amount of time must it remain high before the arrival of the leading edge of a clock pulse, assuming *Q* is to be high after the clock pulse?

MC54/74HC174

Symbol	Parameter	Value	Unit
V_{CC}	DC Supply Voltage (Referenced to GND)	−0.5 to +7.0	V
V_{in}	DC Input Voltage (Referenced to GND)	−1.5 to V_{CC}+1.5	V
V_{out}	DC Output Voltage (Referenced to GND)	−0.5 to V_{CC}+0.5	V
I_{in}	DC Input Current, per Pin	±20	mA
I_{out}	DC Output Current, per Pin	±25	mA
I_{CC}	DC Supply Current, V_{CC} and GND Pins	±50	mA
P_D	Power Dissipation in Still Air, Plastic or Ceramic DIP†	750	mW
	SOIC Package†	500	
T_{stg}	Storage Temperature	−65 to +150.	°C
T_L	Lead Temperature, 1 mm from Case for 10 Seconds		°C
	(Plastic DIP or SOIC Package)	260	
	(Ceramic DIP)	300	

*Maximum Ratings are those values beyond which damage to the device may occur.
  Functional operation should be restricted to the Recommended Operating Conditions.
†Derating — Plastic DIP: −10 mW/°C from 65° to 125°C
  Ceramic DIP: −10 mW/°C from 100° to 125°C
  SOIC Package: −7 mW/°C from 65° to 125°C
For high frequency or heavy load considerations, see Chapter 4 subject listing on page 4-2.

This device contains protection circuitry to guard against damage due to high static voltages or electric fields. However, precautions must be taken to avoid applications of any voltage higher than maximum rated voltages to this high-impedance circuit. For proper operation, V_{in} and V_{out} should be constrained to the range GND ≤ (V_{in} or V_{out}) ≤V_{CC}.

Unused inputs must always be tied to an appropriate logic voltage level (e.g., either GND or V_{CC}). Unused outputs must be left open.

EXPANDED LOGIC DIAGRAM

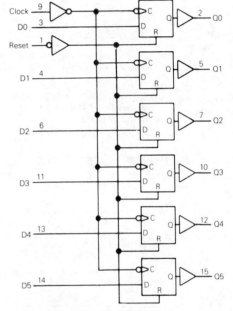

Figure 11.26 *(on pages 419 and 420)*
Specifications (Courtesy of Motorola, Inc.).

DC ELECTRICAL CHARACTERISTICS (Voltages Referenced to GND)

Symbol	Parameter	Test Conditions		V_{CC} V	Guaranteed Limit			Unit
					25°C to −55°C	≤85°C	≤125°C	
V_{IH}	Minimum High-Level Input Voltage	V_{out}=0.1 V or V_{CC}−0.1 V $|I_{out}|$≤20 μA		2.0 4.5 6.0	1.5 3.15 4.2	1.5 3.15 4.2	1.5 3.15 4.2	V
V_{IL}	Maximum Low-Level Input Voltage	V_{out}=0.1 V or V_{CC}−0.1 V $|I_{out}|$≤20 μA		2.0 4.5 6.0	0.3 0.9 1.2	0.3 0.9 1.2	0.3 0.9 1.2	V
V_{OH}	Minimum High-Level Output Voltage	V_{in}=V_{IH} or V_{IL} $|I_{out}|$≤20 μA		2.0 4.5 6.0	1.9 4.4 5.9	1.9 4.4 5.9	1.9 4.4 5.9	V
		V_{in}=V_{IH} or V_{IL}	$|I_{out}|$≤4.0 mA $|I_{out}|$≤5.2 mA	4.5 6.0	3.98 5.48	3.84 5.34	3.70 5.20	
V_{OL}	Maximum Low-Level Output Voltage	V_{in}=V_{IH} or V_{IL} $|I_{out}|$≤20 μA		2.0 4.5 6.0	0.1 0.1 0.1	0.1 0.1 0.1	0.1 0.1 0.1	V
		V_{in}=V_{IH} or V_{IL}	$|I_{out}|$≤4.0 mA $|I_{out}|$≤5.2 mA	4.5 6.0	0.26 0.26	0.33 0.33	0.40 0.40	
I_{in}	Maximum Input Leakage Current	V_{in}=V_{CC} or GND		6.0	±0.1	±1.0	±1.0	μA
I_{CC}	Maximum Quiescent Supply Current (per Package)	V_{in}=V_{CC} or GND I_{out}=0 μA		6.0	8	80	160	μA

NOTE: Information on typical parametric values can be found in Chapter 4 subject listing on page 4-2.

AC ELECTRICAL CHARACTERISTICS ($C_L = 50$ pF, Input $t_r = t_f = 6$ ns)

Symbol	Parameter	V_{CC} V	Guaranteed Limit			Unit
			25°C to −55°C	≤85°C	≤125°C	
f_{max}	Maximum Clock Frequency (50% Duty Cycle) (Figures 1 and 4)	2.0 4.5 6.0	6.0 30 35	4.8 24 28	4.0 20 24	MHz
t_{PLH}, t_{PHL}	Maximum Propagation Delay, Clock to Q (Figures 1 and 4)	2.0 4.5 6.0	165 33 28	205 41 35	250 50 43	ns
t_{PHL}	Maximum Propagation Delay, Reset to Q (Figures 2 and 4)	2.0 4.5 6.0	165 33 28	205 41 35	250 50 43	ns
t_{TLH}, t_{THL}	Maximum Output Transition Time, Any Output (Figures 1 and 4)	2.0 4.5 6.0	75 15 13	95 19 16	110 22 19	ns
C_{in}	Maximum Input Capacitance	−	10	10	10	pF

NOTES:
1. For propagation delays with loads other than 50 pF, see Chapter 4 subject listing on page 4-2.
2. Information on typical parametric values can be found in Chapter 4.

C_{PD}	Power Dissipation Capacitance (Per Flip-Flop) Used to determine the no-load dynamic power consumption: $P_D = C_{PD} V_{CC}^2 f + I_{CC} V_{CC}$ For load considerations, see Chapter 4 subject listing on page 4-2.	Typical @ 25°C, $V_{CC} = 5.0$ V	
		35	pF

TIMING REQUIREMENTS (Input $t_r = t_f = 6$ ns)

Symbol	Parameter	V_{CC} V	Guaranteed Limit			Unit
			25°C to −55°C	≤85°C	≤125°C	
t_{su}	Minimum Setup Time, Data to Clock (Figure 3)	2.0 4.5 6.0	100 20 17	125 25 21	150 30 26	ns
t_h	Minimum Hold Time, Clock to Data (Figure 3)	2.0 4.5 6.0	5 5 5	5 5 5	5 5 5	ns
t_{rec}	Minimum Recovery Time, Reset Inactive to Clock (Figure 2)	2.0 4.5 6.0	5 5 5	5 5 5	5 5 5	ns
t_w	Minimum Pulse Width, Clock (Figure 1)	2.0 4.5 6.0	80 16 14	100 20 17	120 24 20	ns
t_w	Minimum Pulse Width, Reset (Figure 2)	2.0 4.5 6.0	80 16 14	100 20 17	120 24 20	ns
t_r, t_f	Maximum Input Rise and Fall Times (Figure 1)	2.0 4.5 6.0	1000 500 400	1000 500 400	1000 500 400	ns

NOTE: Information on typical parametric values can be found in Chapter 4 subject listing on page 4-2.

RECOMMENDED OPERATING CONDITIONS

Symbol	Parameter		Min	Max	Unit
V_{CC}	DC Supply Voltage (Referenced to GND)		2.0	6.0	V
V_{in}, V_{out}	DC Input Voltage, Output Voltage (Referenced to GND)		0	V_{CC}	V
T_A	Operating Temperature, All Package Types		−55	+125	°C
t_r, t_f	Input Rise and Fall Time (Figure 1)	$V_{CC} = 2.0$ V $V_{CC} = 4.5$ V $V_{CC} = 6.0$ V	0 0 0	1000 500 400	ns

Figure 11.26
(Continued)

c. After arrival of the leading edge of the clock pulse in *(b)*, for what minimum time must *D* remain high?

Solution

a. The specifications in Figure 11.26 show that the minimum pulse width of the clock, t_w, is 16 ns. However, if the clock has a 50% duty cycle, the specifications show that the maximum clock frequency is 30 MHz. Therefore, the minimum width of a pulse in a clock having 50% duty cycle is

$$t_w \text{ (min)} \frac{1}{2} T_{CLOCK} = \frac{1}{2}\left(\frac{1}{f_{max}}\right) = \frac{0.5}{30 \times 10^6 \text{ Hz}} = 16.16 \text{ ns}$$

b. From the specifications, the set-up time is $t_{su} = 20$ ns.
c. From the specifications, the hold time is $t_h = 5$ ns.

Drill
11–2

If all six flip-flops in the 74HC174 are operated at their maximum permissible frequency, with $V_{CC} = 6$ V, what is the maximum power dissipation of the circuit? Assume that there is no load capacitance, that C_{PD} is independent of V_{CC}, and that $T = 25°C$. *Answer:* 264.648 mW.

Edge-Triggered Flip-Flops with Three-State Outputs

D-type flip-flops are available with three-state outputs, making them useful for the kinds of bus-driving applications that were discussed in Chapter 10. Figure 11.27(*a*) shows the logic diagram of the 74374 octal edge-triggered D flip-flop with three-state outputs. Data inputs are clocked into the flip-flops on the leading edge of the clock. When output $\overline{\text{ENABLE}}$ is high, the outputs are in a high-impedance state. Three-state control is asynchronous and is independent of the clocking of data. In other words, if the outputs are in a high-impedance state, new data can still be clocked into the flip-flops on the leading edges of the clock. The state table is shown in Figure 11.27(*b*). Figure 11.27(*c*) compares specifications of the AS, ALS, F, and HCMOS versions of the circuit. Some three-state flip-flop circuits have fewer flip-flops than the 74374 but more complex control circuitry. For example, the 74173 is a quad edge-triggered D flip-flop with two inputs for three-state control and two inputs that enable the clocking of input data as well as a separate clock input.

11–8   ANSI/IEEE STANDARD SYMBOLS

Control (C) Dependency

The ANSI/IEEE qualifying symbols for flip-flop inputs are the same as those we have previously used: R, S, D, J, K, and T. The clock input is designated C, but, in general, C stands for *control* and is another example of dependency notation, as first discussed in connection with EN dependency in Chapter 10. A numerical suffix with C refers to inputs that have the same numerical *prefix* and that depend on C. For example, C1 could be the clock input that causes input data at 1D to be clocked to the output of a D flip-flop. As before, a triangle drawn inside the outline at the C input means flip-flop control is dynamic—i.e., that it is edge-triggered. The ANSI/IEEE inversion triangle outside the outline at a dynamic C input means that the flip-flop is trailing-edge-triggered.

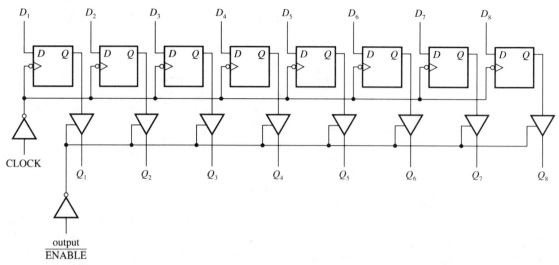

(a) Logic diagram.

output ENABLE	CLOCK	D	Q
0	↑	1	1
0	↑	0	0
0	↓	X	no change
0	0	X	no change
0	1	X	no change
1	X	X	high-impedance

(b) State table, each flip-flop.

	AS	ALS	F	HC
t_{su}	2 ns	10 ns	2 ns	5 ns
t_h	2 ns	0	2 ns	20 ns
f_{CLOCK}	125 MHz	35 MHz	100 MHz	30 MHz

(c) Comparison of specifications.

Figure 11.27
The 74374 octal edge-triggered D flip-flop with three-state outputs.

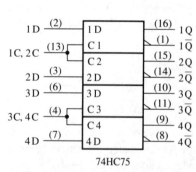

74HC75

(a) The 74HC75 quad D latch. Note the control
(C) dependency: 1D depends on C1, 2D
depends on C2, etc.

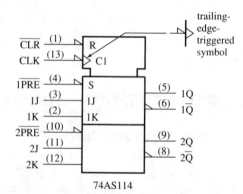

74AS114

(b) The 74AS114 dual trailing-edge-triggered
JK flip-flop. Since C1 and R are in the control
block, they apply to both flip-flops. 1J and
1K depend on C1, but R and S do not.

Figure 11.28
Examples of ANSI/IEEE standard symbols for latches and flip-flops.

Figure 11.29
An example of the logic symbol for a JK master-slave (pulse-triggered) flip-flop.

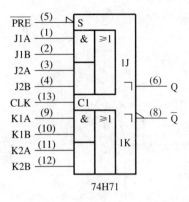

74H71

Figure 11.28 shows examples of ANSI/IEEE standard symbols for a quad D latch and a dual, trailing-edge-triggered JK flip-flop. Note the use of a control block in the symbol for the latter. Recall that inputs to the control block affect all devices in the circuit. In this case, we see that both flip-flops have the same clock and both can be cleared directly by the same $\overline{\text{CLR}}$ input to R. However, each flip-flop has a separate direct-set input, $\overline{\text{PRE}}$, connected to *S*. Inputs 1J and 1K depend on C1, but S and R do not. Thus, the flip-flop can be directly set and reset irrespective of the clock (asynchronously).

Figure 11.29 shows an example of the logic symbol for a JK master-slave (pulse-triggered) flip-flop. As can be seen in the figure, logic operations are performed on the inputs to generate J and K:

$$1J = (J1A \cdot J1B) + (J2A \cdot J2B)$$

$$1K = (K1A \cdot K1B) + (K2A \cdot K2B)$$

The ⌐ symbol is the ANSI/IEEE standard symbol for representing a *delayed* output. The symbol appears at the Q and $\overline{\text{Q}}$ outputs in this example because the flip-flop is the master-slave type and the outputs are delayed by one clock pulse. Note that this flip-flop does not have data lockout, since the clock input is not shown to be dynamic.

11–9   SPICE EXAMPLES

Example 11–3

Model an RS latch using cross-connected NOR gates in a SPICE simulation. Test the model by driving it with input pulses that reset and set it.

Solution.   In this example, we are not as interested in the semiconductor devices used to construct the latch as we are in developing the simplest possible model of its *functional* equivalent. Accordingly, we elect to use NMOS transistors with resistive loads as a very simple way to model NOR gates (two transistors and one resistor per gate). Figure 11.30(a) shows the NMOS NOR gate we will use, configured as a *subcircuit* (SUBCKT) for the model. Part *(b)* of the figure shows how the NOR-gate subcircuits are cross-connected in the main circuit to create the latch. Note how the main-circuit nodes are listed in each of the subcircuit calls (X1 and X2) so that they connect to the subcircuit nodes, as shown in the figure. The figure also shows the pulses we will use to reset and set the latch. V1 starts at $+10$ V, whereas V2 is 0 V ($R = 1$, $S = 0$), so the latch is initially reset. In the interval between 1 ms and 1.5 ms, V1 and V2 are both 0 V ($R = 0$, $S = 0$), so there is no change. At $t = 1.5$ ms, V2 goes high ($R = 0$, $S = 1$), so the latch will set.

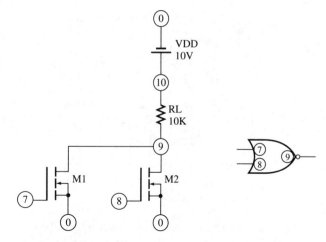

(a) Subcircuit NOR.

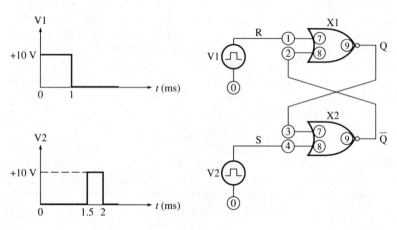

```
EXAMPLE 11.3
.SUBCKT NOR 7 8 9
M1 9 7 0 0 MOS
M2 9 8 0 0 MOS
RL 9 10 10K
VDD 10 0 10V
.MODEL MOS NMOS KP = 0.25E - 3 VTO = 2
.ENDS NOR
X1 1 2 3 NOR
X2 3 4 2 NOR
V1 1 0 PULSE(10 0 1MS 1US 1US)
V2 4 0 PULSE(0 10 1.5MS 1US 1US)
.TRAN 0.1MS 2MS
.PLOT TRAN V(3)
.PLOT TRAN V(2)
.END
```

(b)

Figure 11.30
Example 11–3.

EXAMPLE 11.3

****   TRANSIENT ANALYSIS   TEMPERATURE =   27.000 DEG C

****

```
    TIME        V(3)

                  -5.000D+00      0.000D+00      5.000D+00      1.000D+01  1.500D+01
                  - - - - - - - - - - - - - - - - - - - - - - - - - - - - - - - - - -
0.000D+00   2.476D-01 .            .*             .              .              .
1.000D-04   2.476D-01 .            .*             .              .              .
2.000D-04   2.476D-01 .            .*             .              .              .
3.000D-04   2.476D-01 .            .*             .              .              .
4.000D-04   2.476D-01 .            .*             .              .              .
5.000D-04   2.476D-01 .            .*             .              .              .
6.000D-04   2.476D-01 .            .*             .              .              .
7.000D-04   2.476D-01 .            .*             .              .              .
8.000D-04   2.476D-01 .            .*             .              .              .
9.000D-04   2.476D-01 .            .*             .              .              .
1.000D-03   2.476D-01 .            .*             .              .              .
1.100D-03   4.905D-01 .            .*             .              .              .
1.200D-03   4.905D-01 .            .*             .              .              .
1.300D-03   4.905D-01 .            .*             .              .              .
1.400D-03   4.905D-01 .            .*             .              .              .
1.500D-03   4.905D-01 .            .*             .              .              .
1.600D-03   1.000D+01 .            .              .              *              .
1.700D-03   1.000D+01 .            .              .              *              .
1.800D-03   1.000D+01 .            .              .              *              .
1.900D-03   1.000D+01 .            .              .              *              .
2.000D-03   1.000D+01 .            .              .              *              .
                  - - - - - - - - - - - - - - - - - - - - - - - - - - - - - - - - - -

    TIME        V(2)

                  -5.000D+00      0.000D+00      5.000D+00      1.000D+01  1.500D+01
                  - - - - - - - - - - - - - - - - - - - - - - - - - - - - - - - - - -
0.000D+00   1.000D+01 .            .              .              *              .
1.000D-04   1.000D+01 .            .              .              *              .
2.000D-04   1.000D+01 .            .              .              *              .
3.000D-04   1.000D+01 .            .              .              *              .
4.000D-04   1.000D+01 .            .              .              *              .
5.000D-04   1.000D+01 .            .              .              *              .
6.000D-04   1.000D+01 .            .              .              *              .
7.000D-04   1.000D+01 .            .              .              *              .
8.000D-04   1.000D+01 .            .              .              *              .
9.000D-04   1.000D+01 .            .              .              *              .
1.000D-03   1.000D+01 .            .              .              *              .
1.100D-03   1.000D+01 .            .              .              *              .
1.200D-03   1.000D+01 .            .              .              *              .
1.300D-03   1.000D+01 .            .              .              *              .
1.400D-03   1.000D+01 .            .              .              *              .
1.500D-03   1.000D+01 .            .              .              *              .
1.600D-03   2.476D-01 .            .*             .              .              .
1.700D-03   2.476D-01 .            .*             .              .              .
1.800D-03   2.476D-01 .            .*             .              .              .
1.900D-03   2.476D-01 .            .*             .              .              .
2.000D-03   2.476D-01 .            .*             .              .              .
                  - - - - - - - - - - - - - - - - - - - - - - - - - - - - - - - - - -
```

(c)

Figure 11.30
(Continued)

Voltage Truth Table   *Logic Truth Table*

V1 (R)	V2 (S)	V(3) (Q)	V(2) $\overline{Q}$	R	S	Q	$\overline{Q}$
0 V	0 V	0.4905 V	10 V	0	0	0	1
0 V	10 V	10 V	0.2476 V	0	1	1	0
10 V	0 V	0.2476 V	10 V	1	0	0	1
10 V	10 V	0.4905 V	0.4904 V	1	1	0	0

(d)

Figure 11.30
(Continued)

Note how the V1 pulse is described in the .PULSE statement. Since the pulsewidth (PW) refers to the duration of 0 V (not 10 V), we must specify a 1-ms delay in order to maintain V1 high for 1 ms.

Figure 11.30(c) shows the results of the simulation. Output V(3), representing Q, is low until $t = 1.5$ ms, at which time it switches high. Output V(2), representing $\overline{Q}$, is the complement of Q. These results confirm the validity of the model. Further confirmation can be obtained by a .DC analysis that produces a truth table. Replacing the inputs by dc sources and the .TRAN and .PLOT statements by .DC V2 0 10 10 V1 0 10 10 and .PRINT DC V(3) V(2), we obtain the voltage truth table shown in Figure 11.30(d). (A .NODESET statement is used to make the latch initially reset: .NODESET V(2) = 10.) The corresponding logic truth table, also shown in the figure, is that of a NOR-gate latch.

Example 11–4

Construct a SPICE model of a gated (clocked) RS latch.

Solution. We use the model of the RS latch developed in Example 11–3 and add to it a pair of AND gates, as shown in Figure 11.31. In the continued interest of simplicity, we will use diode AND gates, as shown in Figure 11.31(a). Since two AND gates are required, we define the AND gate in a new subcircuit named AND. Figure 11.31(b) shows how the nodes in the main circuit are connected to the subcircuit nodes. Also shown is the input data file and the waveforms we will use to test the model. Note that a .NODESET statement is used to make the latch initially set [Q = V(3) = 10 V]. When the clock (VCLK) first goes high, the reset (VR) and set (VS) inputs are both low, so the latch should remain set. At $t = 0.4$ ms, VR goes high while the clock is still high, so the latch should reset. VS is high between $t = 0.8$ ms and 1.0 ms, but since the clock is low, the latch should remain reset. Finally at $t = 1.4$ ms, VS goes high while the clock is high, so the latch should set. The SPICE plots in Figure 11.31(c) show the results just described and confirm the validity of the model.

Example 11–5

Construct a SPICE model of an RS master-slave flip-flop.

Solution. Continuing to expand the models developed in the previous examples, we can use the gated RS latch of Example 11–4 to serve as the master and as the slave. Toward that end, the entire latch is now defined to be a subcircuit named LATCH. See Figure 11.32. Note that the LATCH subcircuit itself contains the previously defined subcircuits

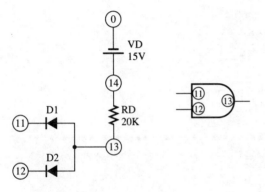

(a) Subcircuit AND.

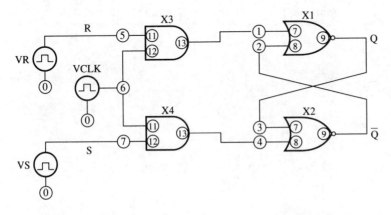

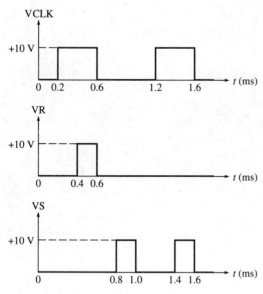

(b)

```
EXAMPLE 11.4
.SUBCKT NOR 7 8 9
M1 9 7 0 0 MOS
M2 9 8 0 0 MOS
.MODEL MOS NMOS KP = 0.25E - 3 VTO = 2
RL 9 10 10K
VDD 10 0 10V
.ENDS NOR
X1 1 2 3 NOR
X2 3 4 2 NOR
.SUBCKT AND 11 12 13
D1 13 11 DIODE
D2 13 12 DIODE
RD 13 14 20K
VD 14 0 15V
.MODEL DIODE D
.ENDS AND
X3 5 6 1 AND
X4 6 7 4 AND
.NODESET V(3) = 10
VR 5 0 PULSE(0 10 0.4MS 1US 1US 0.2MS)
VCLK 6 0 PULSE(0 10 0.2MS 1US 1US 0.4MS 1MS)
VS 7 0 PULSE(0 10 0.8MS 1US 1US 0.2MS 0.6MS)
.TRAN 0.1MS 1.6MS
.PLOT TRAN V(3)
.PLOT TRAN V(2)
.END
```

Figure 11.31
Example 11–4.

EXAMPLE 11.4

```
****      TRANSIENT ANALYSIS              TEMPERATURE =    27.000 DEG C


**********************************************************************************


      TIME      V(3)

                     -5.000D+00      0.000D+00      5.000D+00      1.000D+01   1.500D+01
                  - - - - - - - - - - - - - - - - - - - - - - - - - - - - - - - - -
 0.000D+00   1.000D+01 .                .              .              *          .
 1.000D-04   1.000D+01 .                .              .              *          .
 2.000D-04   1.000D+01 .                .              .              *          .
 3.000D-04   1.000D+01 .                .              .              *          .
 4.000D-04   1.000D+01 .                .              .              *          .
 5.000D-04   2.387D-01 .               .*              .              .          .
 6.000D-04   2.387D-01 .               .*              .              .          .
 7.000D-04   4.905D-01 .               .*              .              .          .
 8.000D-04   4.905D-01 .               .*              .              .          .
 9.000D-04   4.905D-01 .               .*              .              .          .
 1.000D-03   4.905D-01 .               .*              .              .          .
 1.100D-03   4.905D-01 .               .*              .              .          .
 1.200D-03   4.905D-01 .               .*              .              .          .
 1.300D-03   4.905D-01 .               .*              .              .          .
 1.400D-03   4.905D-01 .               .*              .              .          .
 1.500D-03   1.000D+01 .                .              .              *          .
 1.600D-03   1.000D+01 .                .              .              *          .
                  - - - - - - - - - - - - - - - - - - - - - - - - - - - - - - - - -


      TIME      V(2)

                     -5.000D+00      0.000D+00      5.000D+00      1.000D+01   1.500D+01
                  - - - - - - - - - - - - - - - - - - - - - - - - - - - - - - - - -
 0.000D+00   4.905D-01 .               .*              .              .          .
 1.000D-04   4.905D-01 .               .*              .              .          .
 2.000D-04   4.905D-01 .               .*              .              .          .
 3.000D-04   4.905D-01 .               .*              .              .          .
 4.000D-04   4.905D-01 .               .*              .              .          .
 5.000D-04   1.000D+01 .                .              .              *          .
 6.000D-04   1.000D+01 .                .              .              *          .
 7.000D-04   1.000D+01 .                .              .              *          .
 8.000D-04   1.000D+01 .                .              .              *          .
 9.000D-04   1.000D+01 .                .              .              *          .
 1.000D-03   1.000D+01 .                .              .              *          .
 1.100D-03   1.000D+01 .                .              .              *          .
 1.200D-03   1.000D+01 .                .              .              *          .
 1.300D-03   1.000D+01 .                .              .              *          .
 1.400D-03   1.000D+01 .                .              .              *          .
 1.500D-03   2.387D-01 .               .*              .              .          .
 1.600D-03   2.387D-01 .               .*              .              .          .
                  - - - - - - - - - - - - - - - - - - - - - - - - - - - - - - - - -
```

(c)

Figure 11.31
(Continued)

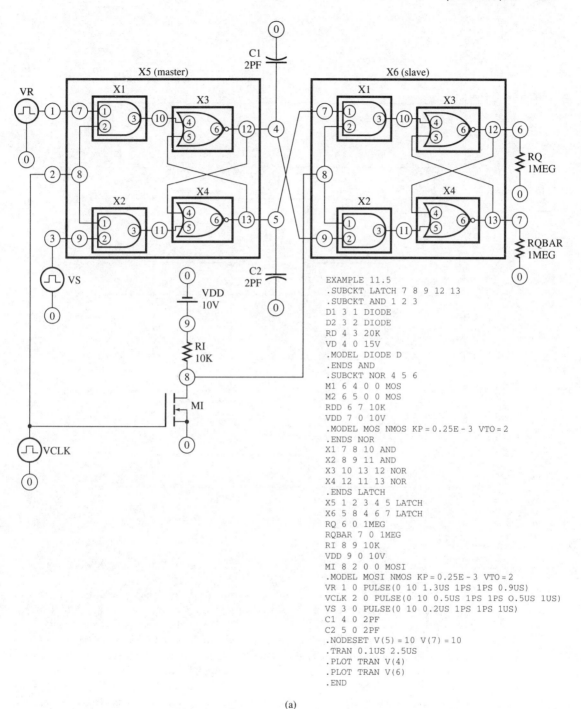

```
EXAMPLE 11.5
.SUBCKT LATCH 7 8 9 12 13
.SUBCKT AND 1 2 3
D1 3 1 DIODE
D2 3 2 DIODE
RD 4 3 20K
VD 4 0 15V
.MODEL DIODE D
.ENDS AND
.SUBCKT NOR 4 5 6
M1 6 4 0 0 MOS
M2 6 5 0 0 MOS
RDD 6 7 10K
VDD 7 0 10V
.MODEL MOS NMOS KP = 0.25E - 3 VTO = 2
.ENDS NOR
X1 7 8 10 AND
X2 8 9 11 AND
X3 10 13 12 NOR
X4 12 11 13 NOR
.ENDS LATCH
X5 1 2 3 4 5 LATCH
X6 5 8 4 6 7 LATCH
RQ 6 0 1MEG
RQBAR 7 0 1MEG
RI 8 9 10K
VDD 9 0 10V
MI 8 2 0 0 MOSI
.MODEL MOSI NMOS KP = 0.25E - 3 VTO = 2
VR 1 0 PULSE(0 10 1.3US 1PS 1PS 0.9US)
VCLK 2 0 PULSE(0 10 0.5US 1PS 1PS O.5US 1US)
VS 3 0 PULSE(0 10 0.2US 1PS 1PS 1US)
C1 4 0 2PF
C2 5 0 2PF
.NODESET V(5) = 10 V(7) = 10
.TRAN 0.1US 2.5US
.PLOT TRAN V(4)
.PLOT TRAN V(6)
.END
```

(a)

Figure 11.32
Example 11–5. (Continued on pages 430–432.)

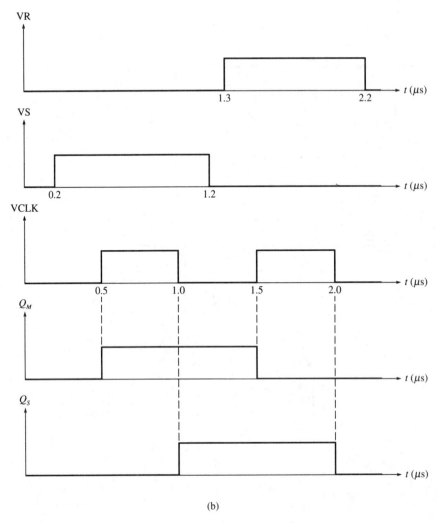

Figure 11.32
(Continued)

AND and NOR, which are said to be *nested* within LATCH. Thus, there are four sets of independent nodes: those within the nested subcircuits AND and NOR (1, 2, 3 and 4, 5, 6), those of the latch subcircuit (7, 8, 9, 10, 11, 12, 13), and those of the main circuit (1, 2, 3, 4, 5, 6, 7, 8, 9). Remember that node 0 (ground) is the only node common to all subcircuits and the main circuit. As shown in the figure, the clock signal to the slave is inverted by an NMOS inverter in the main circuit. One-megohm resistors RQ and RQBAR are connected to the Q and $\overline{Q}$ outputs of the slave, since SPICE requires at least two connections to every node. Note that 2-pF capacitors are connected to the Q and $\overline{Q}$ outputs of the master to simulate the propagation delay needed to eliminate a race problem between the inverted clock and the inputs to the slave. The waveforms used to test the model are shown in Figure 11.32*(b)*, along with the waveforms that should result

EXAMPLE 11.5


```
****      TRANSIENT ANALYSIS              TEMPERATURE =    27.000 DEG C


**************************************************************************


    TIME       V(4)

                     0.000D+00     5.000D+00     1.000D+01     1.500D+01  2.000D+01
                    - - - - - - - - - - - - - - - - - - - - - - - - - - - -
  0.000D+00   7.547D-01 . *            .             .             .             .
  1.000D-07   7.547D-01 . *            .             .             .             .
  2.000D-07   7.547D-01 . *            .             .             .             .
  3.000D-07   7.547D-01 . *            .             .             .             .
  4.000D-07   7.547D-01 . *            .             .             .             .
  5.000D-07   7.547D-01 . *            .             .             .             .
  6.000D-07   9.948D+00 .              .             *             .             .
  7.000D-07   1.000D+01 .              .             *             .             .
  8.000D-07   1.000D+01 .              .             *             .             .
  9.000D-07   1.000D+01 .              .             *             .             .
  1.000D-06   1.000D+01 .              .             *             .             .
  1.100D-06   1.088D+01 .              .             . *           .             .
  1.200D-06   1.088D+01 .              .             . *           .             .
  1.300D-06   1.088D+01 .              .             . *           .             .
  1.400D-06   1.088D+01 .              .             . *           .             .
  1.500D-06   1.088D+01 .              .             . *           .             .
  1.600D-06   4.106D-01 .*             .             .             .             .
  1.700D-06   4.092D-01 .*             .             .             .             .
  1.800D-06   4.093D-01 .*             .             .             .             .
  1.900D-06   4.093D-01 .*             .             .             .             .
  2.000D-06   4.093D-01 .*             .             .             .             .
  2.100D-06   7.548D-01 . *            .             .             .             .
  2.200D-06   7.547D-01 . *            .             .             .             .
  2.300D-06   7.547D-01 . *            .             .             .             .
  2.400D-06   7.547D-01 . *            .             .             .             .
  2.500D-06   7.547D-01 . *            .             .             .             .
                    - - - - - - - - - - - - - - - - - - - - - - - - - - - -
                                        (c)
```

Figure 11.32
(Continued)

at the outputs of the master (Q_M) and slave (Q_S). The SPICE plots shown in Figure 11.32(c) confirm these results.

Example 11-6

Construct a SPICE model of a master-slave JK flip-flop.

Solution.   To expand the M/S flip-flop of Example 11-5 to a JK flip-flop, we add an AND gate at each input to the master so that the Q output of the slave can be ANDed with a new K input and the $\overline{Q}$ output of the slave can be ANDed with a new J input. See Figure 11.33(a). The new AND gates are implemented by 2-input diode gates. Although it would be possible to expand the existing X1 and X2 AND gates in the master to 3-input gates, we could not then use them as subcircuits in the slave latch. Note that the 1-MΩ resistors at

EXAMPLE 11.5

```
****     TRANSIENT ANALYSIS               TEMPERATURE =     27.000 DEG C


******************************************************************************


       TIME       V(6)

                          -5.000D+00   0.000D+00   5.000D+00   1.000D+01  1.500D+01
                          - - - - - - - - - - - - - - - - - - - - - - - - - - - - -
  0.000D+00   2.279D-01 .              .*            .             .            .
  1.000D-07   2.279D-01 .              .*            .             .            .
  2.000D-07   2.279D-01 .              .*            .             .            .
  3.000D-07   2.279D-01 .              .*            .             .            .
  4.000D-07   2.279D-01 .              .*            .             .            .
  5.000D-07   2.279D-01 .              .*            .             .            .
  6.000D-07   4.965D-01 .              .*            .             .            .
  7.000D-07   4.965D-01 .              .*            .             .            .
  8.000D-07   4.965D-01 .              .*            .             .            .
  9.000D-07   4.965D-01 .              .*            .             .            .
  1.000D-06   4.965D-01 .              .*            .             .            .
  1.100D-06   9.901D+00 .              .             .             *            .
  1.200D-06   9.901D+00 .              .             .             *            .
  1.300D-06   9.901D+00 .              .             .             *            .
  1.400D-06   9.901D+00 .              .             .             *            .
  1.500D-06   9.901D+00 .              .             .             *            .
  1.600D-06   9.901D+00 .              .             .             *            .
  1.700D-06   9.901D+00 .              .             .             *            .
  1.800D-06   9.901D+00 .              .             .             *            .
  1.900D-06   9.901D+00 .              .             .             *            .
  2.000D-06   9.901D+00 .              .             .             *            .
  2.100D-06   2.279D-01 .              .*            .             .            .
  2.200D-06   2.279D-01 .              .*            .             .            .
  2.300D-06   2.279D-01 .              .*            .             .            .
  2.400D-06   2.279D-01 .              .*            .             .            .
  2.500D-06   2.279D-01 .              .*            .             .            .
                          - - - - - - - - - - - - - - - - - - - - - - - - - - - - -
```

Figure 11.32   (c) (c continues from 431)
(Continued)

the outputs of the slave have been deleted, as they are no longer required, and that 2-pF capacitors have been added to simulate propagation delay through the slave. Figure 11.33(b) shows the waveforms we use to test the model and the expected results at the output of the master (Q_M) and at the output of the slave (Q_S). Since the flip-flop is initially reset, Q_M should set on the leading edge of the first clock pulse (where $J = 1$ and $K = 0$), reset on the leading edge of the second clock pulse (where $J = 0$ and $K = 1$), and set again on the leading edge of the third clock pulse, since both J and K are high at that time. The SPICE plots shown in Figure 11.33(c) confirm these results.

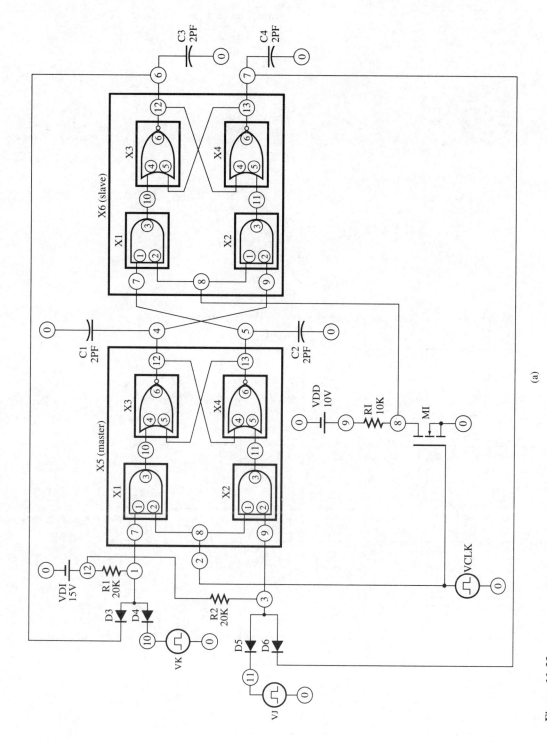

Figure 11.33
Example 11−6. (Continues on pages 434−436.)

Figure 11.33 *(Continued)*

(a)

```
EXAMPLE 11.6                                    VDD 9 0 10V
.SUBCKT LATCH 7 8 9 12 13                       MI 8 2 0 0 MOSI
.SUBCKT AND 1 2 3                               .MODEL MOSI NMOS KP = 0.25E-3 VTO = 2
D1 3 1 DIODE                                    VK 10 0 PULSE(0 10 1.4US 1PS 1PS 0.7US 1US)
D2 3 2 DIODE                                    VCLK 2 0 PULSE(0 10 0.5US 1PS 1PS 0.5US 1US)
RD 4 3 20K                                      VJ 11 0 PULSE(0 10 0.4US 1PS 1PS 0.7US 2US)
VD 4 0 15V                                      C1 4 0 2PF
.MODEL DIODE D                                  C2 5 0 2PF
.ENDS AND                                       D3 1 6 DIODE
.SUBCKT NOR 4 5 6                               D4 1 10 DIODE
M1 6 4 0 0 MOS                                  D5 3 11 DIODE
M2 6 5 0 0 MOS                                  D6 3 7 DIODE
RDD 6 7 10K                                     .MODEL DIODE D
VDD 7 0 10V                                     R1 1 12 20K
.MODEL MOS NMOS KP = 0.25E-3 VTO = 2            R2 3 12 20K
.ENDS NOR                                       VDI 12 0 15V
X1 7 8 10 AND                                   C3 6 0 2PF
X2 8 9 11 AND                                   C4 7 0 2PF
X3 10 13 12 NOR                                 .NODESET V(5) = 10 V(7) = 10
X4 12 11 13 NOR                                 .TRAN 0.1US 3.2US
.ENDS LATCH                                     .PLOT TRAN V(4)
X5 1 2 3 4 5 LATCH                              .PLOT TRAN V(6)
X6 5 8 4 6 7 LATCH                              .END
RI 8 9 10K
```

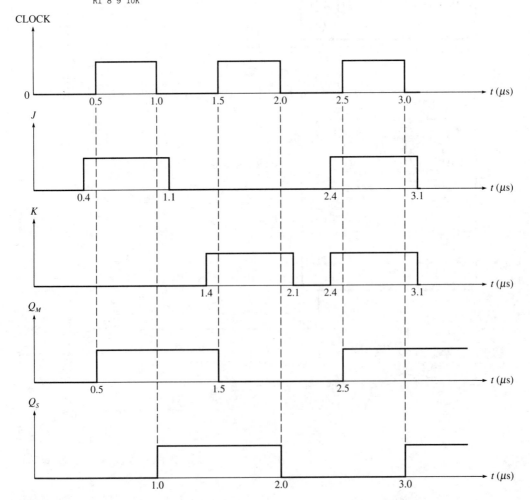

(b)

434

EXAMPLE 11.6

```
****      TRANSIENT ANALYSIS                TEMPERATURE =    27.000 DEG C

*************************************************************************

     TIME        V(4)

                    0.000D+00    5.000D+00    1.000D+01    1.500D+01  2.000D+01
                    - - - - - - - - - - - - - - - - - - - - - - - - - - - -
  0.000D+00   7.547D-01 .  *              .              .              .         .
  1.000D-07   7.547D-01 .  *              .              .              .         .
  2.000D-07   7.547D-01 .  *              .              .              .         .
  3.000D-07   7.547D-01 .  *              .              .              .         .
  4.000D-07   7.547D-01 .  *              .              .              .         .
  5.000D-07   7.547D-01 .  *              .              .              .         .
  6.000D-07   9.951D+00 .                 .              *              .         .
  7.000D-07   9.999D+00 .                 .              *              .         .
  8.000D-07   1.000D+01 .                 .              *              .         .
  9.000D-07   1.000D+01 .                 .              *              .         .
  1.000D-06   1.000D+01 .                 .              *              .         .
  1.100D-06   1.088D+01 .                 .              . *            .         .
  1.200D-06   1.088D+01 .                 .              . *            .         .
  1.300D-06   1.088D+01 .                 .              . *            .         .
  1.400D-06   1.088D+01 .                 .              . *            .         .
  1.500D-06   1.088D+01 .                 .              . *            .         .
  1.600D-06   4.100D-01 .*                .              .              .         .
  1.700D-06   4.088D-01 .*                .              .              .         .
  1.800D-06   4.088D-01 .*                .              .              .         .
  1.900D-06   4.088D-01 .*                .              .              .         .
  2.000D-06   4.088D-01 .*                .              .              .         .
  2.100D-06   7.547D-01 .  *              .              .              .         .
  2.200D-06   7.547D-01 .  *              .              .              .         .
  2.300D-06   7.547D-01 .  *              .              .              .         .
  2.400D-06   7.547D-01 .  *              .              .              .         .
  2.500D-06   7.547D-01 .  *              .              .              .         .
  2.600D-06   9.677D+00 .                 .             *.              .         .
  2.700D-06   9.734D+00 .                 .             *.              .         .
  2.800D-06   9.734D+00 .                 .             *.              .         .
  2.900D-06   9.734D+00 .                 .             *.              .         .
  3.000D-06   9.734D+00 .                 .             *.              .         .
  3.100D-06   1.088D+01 .                 .              . *            .         .
  3.200D-06   1.088D+01 .                 .              . *            .         .
                    - - - - - - - - - - - - - - - - - - - - - - - - - - - -
                                          (c)
```

Figure 11.33
(Continued)

EXAMPLE 11.6

```
****     TRANSIENT ANALYSIS              TEMPERATURE =    27.000 DEG C


*****************************************************************************


     TIME        V(6)
                    0.000D+00      5.000D+00      1.000D+01      1.500D+01  2.000D+01
              - - - - - - - - - - - - - - - - - - - - - - - - - - - - - - - - - -
0.000D+00   2.267D-01 .*              .              .              .           .
1.000D-07   2.267D-01 .*              .              .              .           .
2.000D-07   2.267D-01 .*              .              .              .           .
3.000D-07   2.267D-01 .*              .              .              .           .
4.000D-07   2.267D-01 .*              .              .              .           .
5.000D-07   2.256D-01 .*              .              .              .           .
6.000D-07   4.854D-01 .*              .              .              .           .
7.000D-07   4.854D-01 .*              .              .              .           .
8.000D-07   4.854D-01 .*              .              .              .           .
9.000D-07   4.854D-01 .*              .              .              .           .
1.000D-06   4.854D-01 .*              .              .              .           .
1.100D-06   9.959D+00 .               .              *              .           .
1.200D-06   1.000D+01 .               .              *              .           .
1.300D-06   1.000D+01 .               .              *              .           .
1.400D-06   1.000D+01 .               .              *              .           .
1.500D-06   1.008D+01 .               .              *              .           .
1.600D-06   1.008D+01 .               .              *              .           .
1.700D-06   1.008D+01 .               .              *              .           .
1.800D-06   1.008D+01 .               .              *              .           .
1.900D-06   1.008D+01 .               .              *              .           .
2.000D-06   1.008D+01 .               .              *              .           .
2.100D-06   3.894D-01 .*              .              .              .           .
2.200D-06   2.267D-01 .*              .              .              .           .
2.300D-06   2.267D-01 .*              .              .              .           .
2.400D-06   2.267D-01 .*              .              .              .           .
2.500D-06   3.873D-01 .*              .              .              .           .
2.600D-06   1.157D+00 .        *      .              .              .           .
2.700D-06   1.157D+00 .        *      .              .              .           .
2.800D-06   1.157D+00 .        *      .              .              .           .
2.900D-06   1.157D+00 .        *      .              .              .           .
3.000D-06   1.157D+00 .        *      .              .              .           .
3.100D-06   1.008D+01 .               .              *              .           .
3.200D-06   1.000D+01 .               .              *              .           .
              - - - - - - - - - - - - - - - - - - - - - - - - - - - - - - - - - -
```

(c)

Figure 11.33
(Continued)

EXERCISES

Section 11–2

11.1 Construct a truth table for the circuit shown in Figure 11.34.

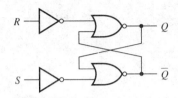

Figure 11.34
Exercise 11.1.

11.2 Construct a truth table for the circuit shown in Figure 11.35.

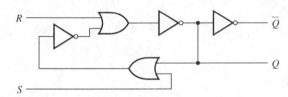

Figure 11.35
Exercise 11.2.

11.3 Complete the timing diagram in Figure 11.36, p. 438, to show the outputs of the NOR-gate latch when the inputs are the waveforms shown. The latch is initially reset.

11.4 Complete the timing diagram in Figure 11.37, p. 438, to show the outputs of the NOR-gate latch when the inputs are the waveforms shown. The latch is initially set.

11.5 Construct a truth table for the circuit shown in Figure 11.38.

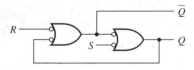

Figure 11.38
Exercise 11.5

11.6 Construct a truth table for the circuit shown in Figure 11.39.

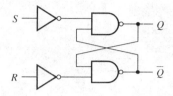

Figure 11.39
Exercise 11.6.

Exercises continue on p. 439.

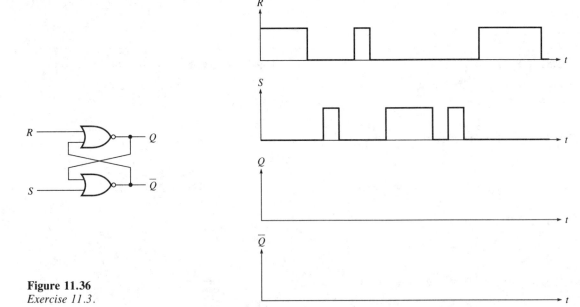

Figure 11.36
Exercise 11.3.

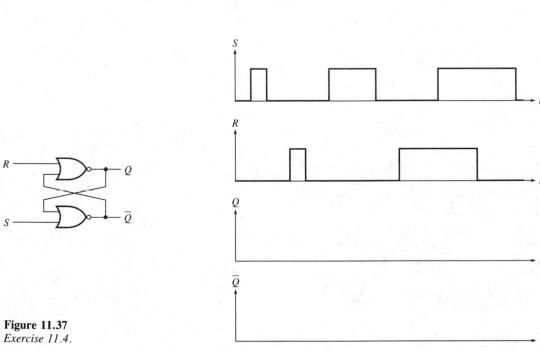

Figure 11.37
Exercise 11.4.

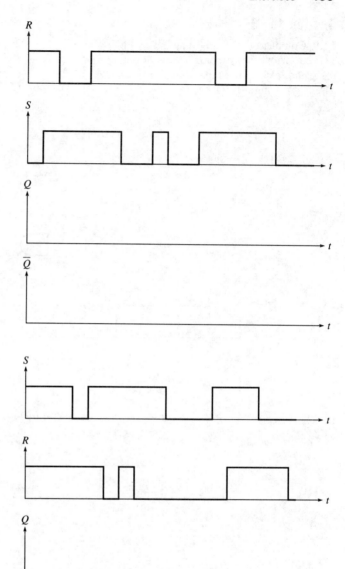

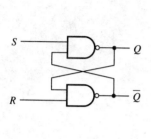

Figure 11.40
Exercise 11.7.

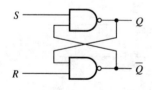

Figure 11.41
Exercise 11.8.

11.7 Complete the timing diagram in Figure 11.40 to show the outputs of the NAND-gate latch when the inputs are the waveforms shown. The latch is initially set.

11.8 Complete the timing diagram in Figure 11.41 to show the outputs of the NAND-gate latch when the inputs are the waveforms shown. The flip-flop is initially reset.

Section 11–3

11.9 Complete the timing diagram in Figure 11.42 to show the response of the Q output of the gated latch to the input waveforms shown. The latch is initially set.

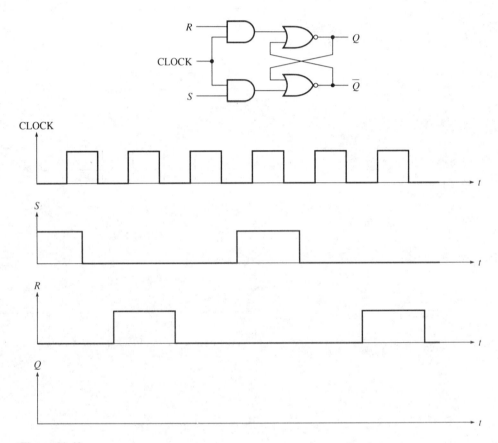

Figure 11.42
Exercise 11.9.

11.10 Complete the timing diagram in Figure 11.43 to show the response of the Q output of the gated latch to the input waveforms shown. The latch is initially reset.

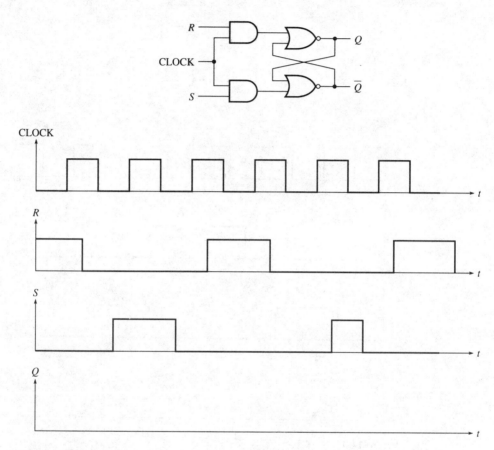

Figure 11.43
Exercise 11.10.

11.11 Complete the timing diagram shown in Figure 11.44 to show the response of the Q output of the gated latch to the input waveforms shown. The latch is initially set.

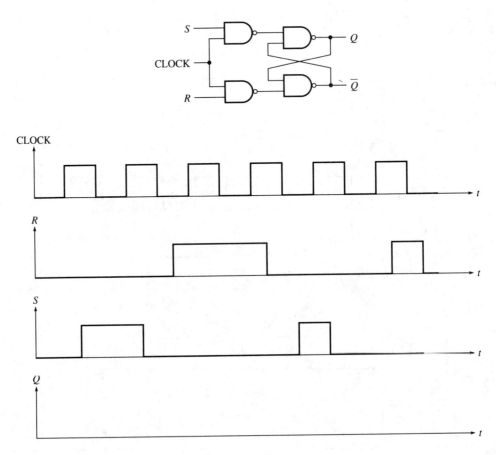

Figure 11.44
Exercise 11.11.

11.12 Complete the timing diagram shown in Figure 11.45 to show the response of the Q output of the gated latch to the input waveforms shown. The latch is initially reset.

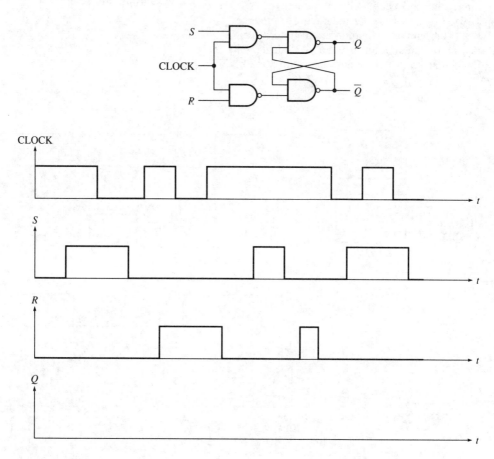

Figure 11.45
Exercise 11.12.

Section 11–4

11.13 Complete the timing diagram shown in Figure 11.46 to show the Q_1 and Q_2 outputs of the transparent D latches. Both latches are initially reset.

11.14 When the input to a circuit is a 4-bit binary number whose decimal value reaches $13)_{10}$, it is necessary for the circuit to produce that 4-bit number as output and to hold it for 1 ms. Design the circuit, using a 7475 dual 2-bit D latch and a 74121 monostable multivibrator.

Section 11–5

11.15 Figure 11.47 shows two master-slave RS flip-flops having a common clock input. Construct a timing diagram showing the responses of the master and slave latches in *each* flip-flop to two clock pulses. Assume all latches are initially reset.

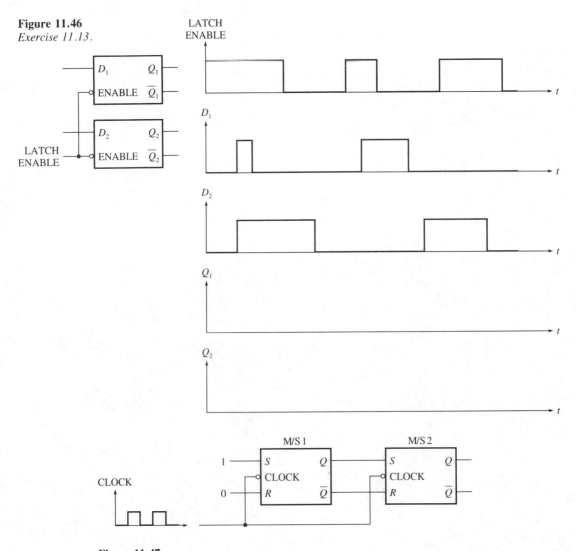

Figure 11.46
Exercise 11.13.

Figure 11.47
Exercise 11.15.

11.16 Complete the timing diagram in Figure 11.48 to show the responses of the master and slave latches in the D-type master-slave flip-flop to the input shown. Note the absence of an inversion bubble at the clock input, meaning the slave is loaded on the leading edge of the clock. Assume the master and slave are both initially reset.

11.17 Construct a timing diagram showing the outputs of gates 1 through 8 when a single clock pulse is applied to the circuit shown in Figure 11.49. Assume the outputs of gates 3 and 7 are initially 0 and the outputs of gates 4 and 8 are initially 1. What type of circuit is represented, and how should inputs 1 and 2 and outputs 1 and 2 be labeled?

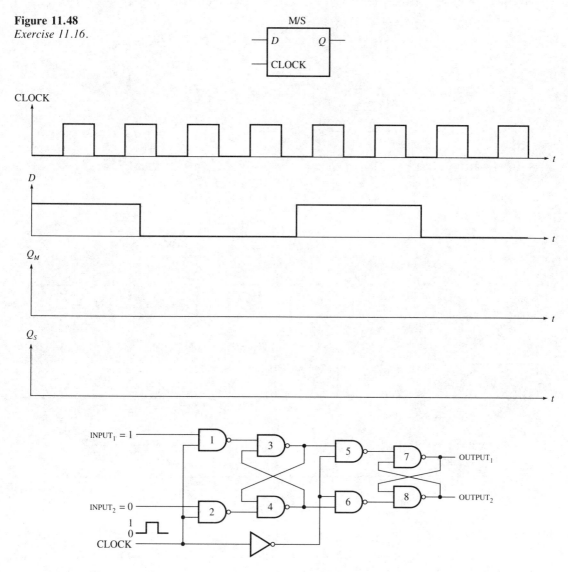

Figure 11.48
Exercise 11.16.

Figure 11.49
Exercise 11.17.

11.18 Repeat Exercise 11.17 for the circuit shown in Figure 11.50, except assume that the outputs of gates 3 and 7 are initially 1 and that the outputs of gates 4 and 8 are initially 0.

Section 11–6

11.19 Complete the timing diagram in Figure 11.51 to show the response of the JK master-slave flip-flop to the input waveforms shown. The flip-flop is initially reset.

11.20 Construct a function table for the circuit whose logic diagram is shown in Figure 11.52.

11.21 Draw a logic diagram and a timing diagram showing how the waveform in Figure 11.53 could be used with T flip-flops to produce a 25-kHz square wave.

11.22 Complete the timing diagram in Figure 11.54 to show the responses (Q_1 and Q_2) of the circuit to the inputs shown. Assume both flip-flops are initially reset.

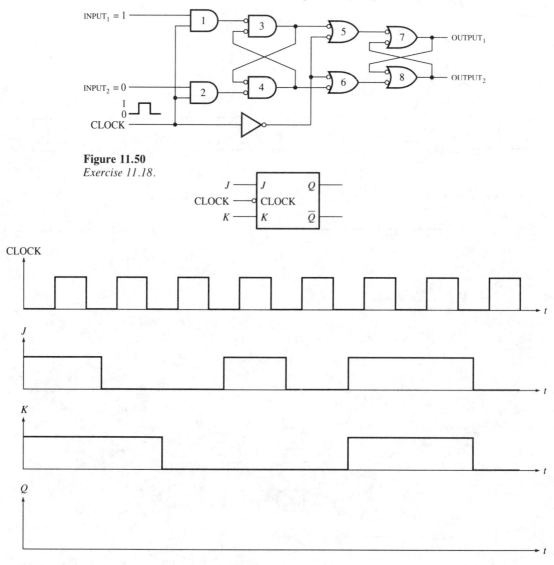

Figure 11.50
Exercise 11.18.

Figure 11.51
Exercise 11.19.

Figure 11.52
Exercise 11.20.

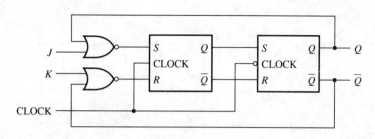

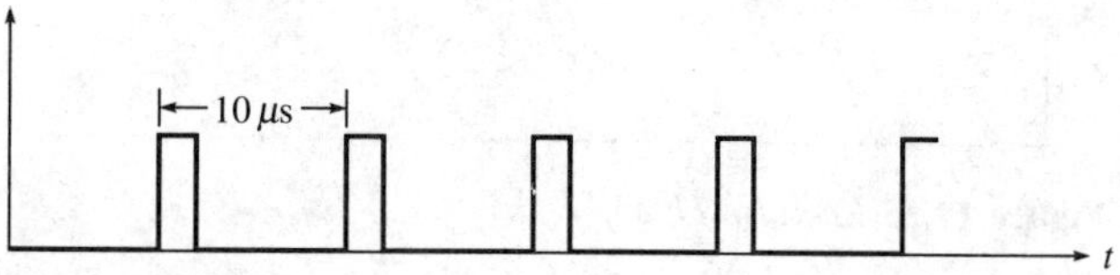

Figure 11.53
Exercise 11.21.

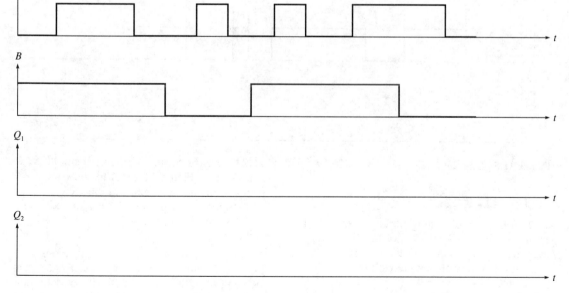

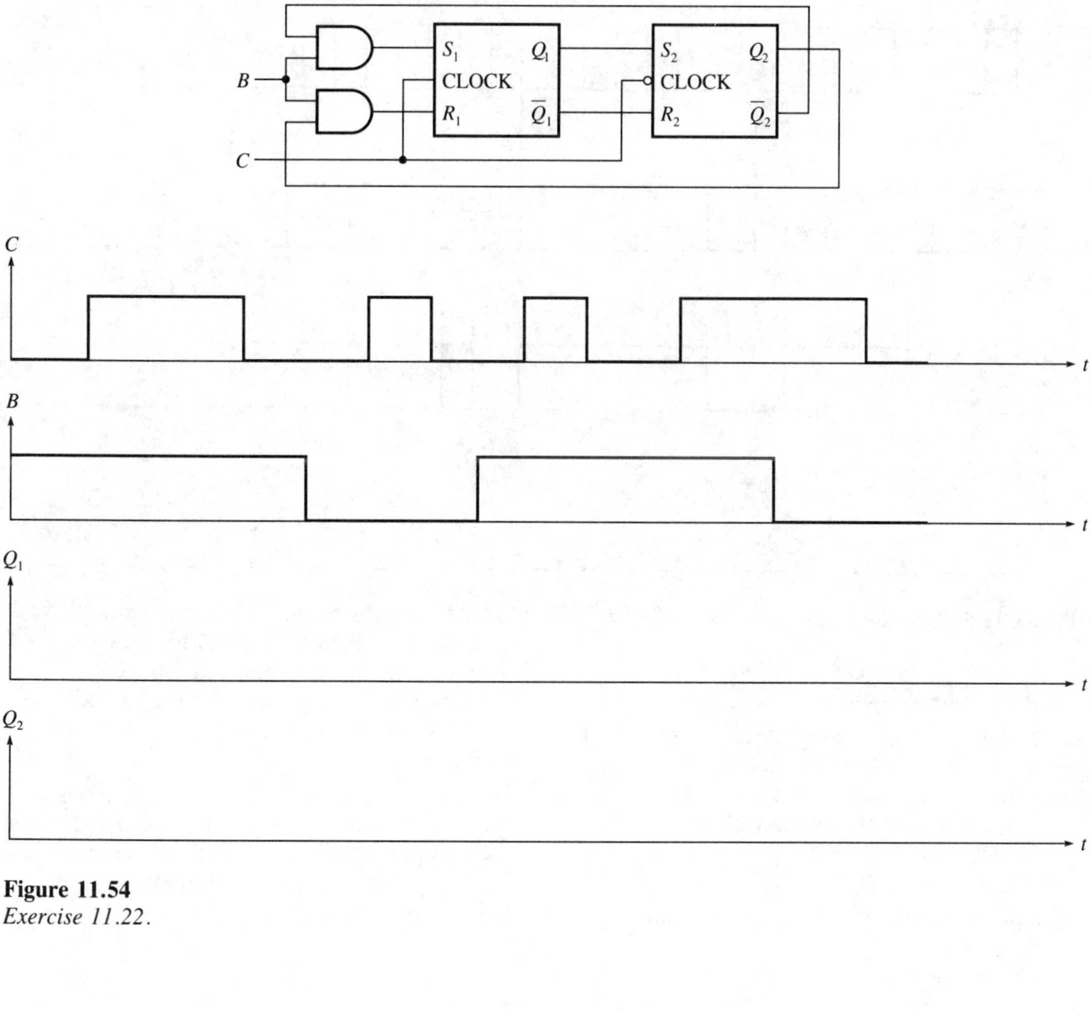

Figure 11.54
Exercise 11.22.

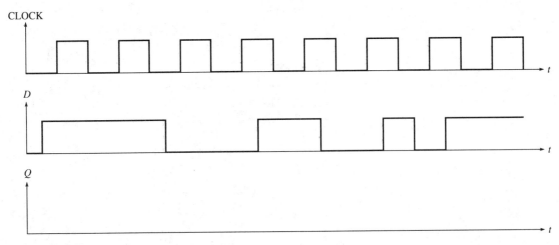

Figure 11.55 *Exercise 11.27.*

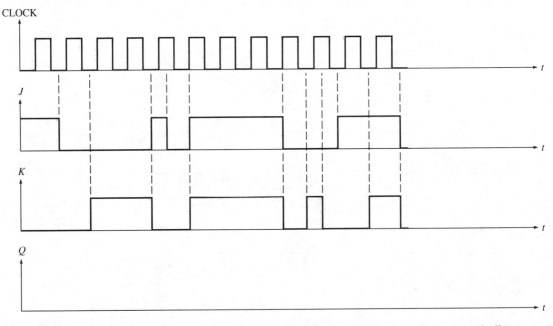

Figure 11.56 *Exercise 11.28.*

Section 11–7

11.23 Make a table listing the logic state (0 or 1) of the output of each gate of the JK edge-triggered flip-flop shown in Figure 11.22 for each of the following static conditions:
(a) CLOCK = 0, Q = 0
(b) CLOCK = 1, Q = 0

Assume that each of J and K can be either 0 or 1.

11.24 Construct a timing diagram similar to that given in Figure 11.23(*b*) to show the response of the same gates in the JK edge-triggered flip-flop to the *leading* edge of a clock pulse. Assume J = 1, K = 0, Q = 0, and $\overline{Q}$ = 1.

11.25 With reference to the specifications given in Figure 11.24 for the 74ALS112A dual JK negative-edge-triggered flip-flops, determine each of the following, assuming the test conditions specified:

(a) The maximum time between the 50% point
   of the trailing edge of the clock and the
   50% point of the Q output when it changes
   from low to high.
(b) The minimum time between the 50% point
   of the $\overline{PRE}$ input when it changes from
   high to low and the 50% point of the Q
   output when it changes from high to low.
(c) The maximum number of J or K inputs to
   a 74ALS112A that the Q output of a
   74ALS112A flip-flop can drive.

11.26 With reference to the specifications
given in Figures 11.24 and 11.26, determine
each of the following, assuming the test
conditions specified:
(a) The maximum time between the 50% point
   of the leading edge of a clock input to a
   74HC174 flip-flop when $D = 0$ and $Q = 1$
   and the 50% point of the resulting Q output
   of the flip-flop, assuming $V_{CC} = 4.5$ V and
   $T = 25°C$.
(b) The maximum time between the 50% point
   of the trailing edge of the clock input to a
   74ALS112A when $Q = 1$, $J = 1$, and
   $K = 1$ and the 50% point of the resulting
   Q output.
(c) The maximum number of $\overline{CLR}$ inputs to a
   74ALS112A that the Q output of a
   74ALS112A can drive.

11.27 Complete the timing diagram in Figure
11.55 to show the response of a 74174 flip-flop
to the inputs shown. Assume the flip-flop is
initially reset.

11.28 Complete the timing diagram in Figure
11.56 to show the response of a 74112 flip-flop
to the inputs shown. Assume the flip-flop is
initially reset.

Section 11–8

11.29 With reference to the circuit whose logic
symbol is shown in Figure 11.57, answer the
following questions:
(a) What type of circuit does the symbol
   represent?
(b) Under what circumstances are the outputs
   in a high-impedance state?

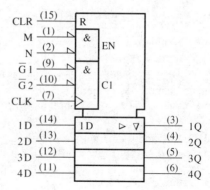

Figure 11.57
Exercise 11.29.

(c) Under what circumstances is input data
   latched?
(d) Does clearing depend on the clock input?
(e) What is the significance of the
   right-pointing triangle in the symbol?

11.30 With reference to the circuit whose logic
symbol is shown in Figure 11.58, answer the
following questions:
(a) What type of circuit does the symbol
   represent?
(b) Describe what action is necessary to make
   the outputs at pins 8 and 13 both low,
   irrespective of the clock and of the J and K
   inputs.
(c) Describe what action is necessary to make
   the outputs at pins 9 and 12 both low,
   irrespective of the clock and of the J and K
   inputs.
(d) If $1Q = 1$, $1J = 0$, and $1K = 1$, when
   (with respect to the clock) does a change
   appear in the 1Q output?
(e) Does the circuit have data lockout?

Figure 11.58 *Exercise 11.30.*

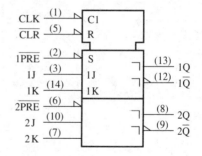

CHALLENGING EXERCISES

11.31 What function does the circuit in Figure 11.59 perform? Construct a truth table.

11.32 Draw a schematic diagram of an NMOS RS latch. Label R, S, Q, and $\overline{Q}$ on the schematic and construct the truth table for the circuit.

11.33 Draw a logic diagram showing how a JK master-slave flip-flop can be constructed using *only* NOR gates. The slave output should appear on the trailing edge of the clock.

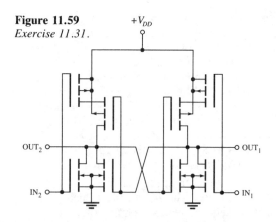

Figure 11.59
Exercise 11.31.

SPICE EXERCISES

11.34 Construct a SPICE model of an RS latch using cross-connected NMOS NAND gates. Define the NAND gate in a subcircuit. Test your model by obtaining a truth table showing Q and $\overline{Q}$ for every combination of R and S.

11.35 Construct a SPICE model of a gated (clocked) RS latch using cross-connected NMOS NAND gates. Test your model by driving it with the inputs shown in Figure 11.60. Use a .NODESET statement to make the latch initially reset.

11.36 Construct a SPICE model of the transparent D latch shown in Figure 11.11 using NMOS NAND gates only. Simulate

propagation delay by connecting a 2-pF capacitor between the Q-output and ground. Test your model by driving it with the inputs shown in Figure 11.61. The values for TR and TF in the PULSE statements defining D and ENABLE should be 0.1 μs.

11.37 Construct a SPICE model of an RS latch using cross-connected CMOS NOR gates. Test your model by obtaining a truth table showing Q and $\overline{Q}$ for every combination of R and S. Use a .NODESET statement to make the latch initially set. What do you notice about the voltage levels obtained for Q and $\overline{Q}$ in comparison to those obtained when using NMOS models?

Figure 11.60
Exercise 11.35.

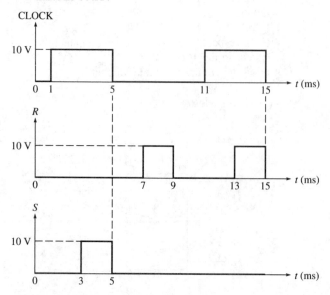

Figure 11.61
Exercise 11.36.

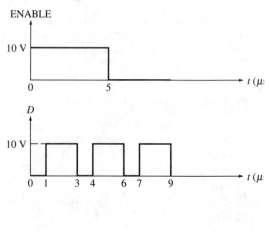

Chapter 12

SHIFT REGISTERS

12–1   SERIAL AND PARALLEL DATA

Multi-bit data is said to be in *parallel* form when all the bits are available (accessible) *simultaneously*. For example, the 8-bit data at the output of the 74374 in Figure 11.27(a) is in parallel form, since all 8 outputs are present and equally accessible at any given time. Parallel *data transfer* is the simultaneous transmission of all bits from one device to another. For example, if the 8 outputs of a 74374 were clocked into the 8 inputs of another 74374, we would say that the 8-bit word had been transferred in parallel from the first 74374 to the second. Data is in *serial* form when the bits appear sequentially (one after the other, in time) at a single terminal. Serial data must be transmitted under the synchronization of a clock, since the sequence of 1s and 0s at a serial terminal change with time. The clock provides the means to specify the time at which each new bit is sampled. Figure 12.1 shows one example of how the binary number 10110 could be generated in serial form. Note that each new bit appears and remains (at the single serial terminal) during the time that each successive clock pulse is high. In this example, the sequence begins with the least significant bit (0) of the binary number. Serial data is also generated in other formats and with different synchronization schemes, but the key idea is that only one bit is available at any given time.

A register is simply a set of flip-flops used to store binary data. *Loading* a register means setting or resetting the individual flip-flops so that their states correspond to the bits of the data to be stored. Parallel loading (parallel input) occurs when the data is transferred to the register in parallel form, meaning that all flip-flops are triggered into their new states at the same time. The register output is in parallel form when the data it stores is available in parallel. Clearly, parallel input requires that the set and/or reset controls of every flip-flop be accessible, and parallel output requires that the Q output of every flip-flop be accessible.

A register is said to have a serial input when it can be loaded by data in serial form. Similarly, serial output means the data can be transferred out of the register in serial form. Registers are classified according to the methods that can be used to transfer their input and output data. Examples include serial-in, serial-out and serial-in, parallel-out registers.

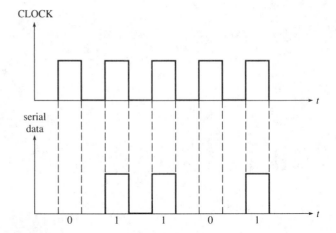

Figure 12.1
An example of serial data. The bits of the binary number 10110 occur in a time sequence synchronized by the clock, beginning with the least significant bit.

Some registers have inputs and/or outputs that can be operated in either the serial or the parallel mode.

12–2 SERIAL-IN, SERIAL-OUT SHIFT REGISTERS

When serial data is transferred into a register, each new bit is clocked into the first flip-flop in the register. The bit that was previously stored by that flip-flop is transferred to the second flip-flop. Similarly, the bit that was stored by the second flip-flop is transferred to the third, and so on. Figure 12.2 illustrates this process in a 4-bit register consisting of

after CLOCK pulse	serial input	Q_1	Q_2	Q_3	Q_4	
0	1	0	0	0	0	(initial contents)
1	0	1	0	0	0	
2	0	0	1	0	0	
3	1	0	0	1	0	
4		1	0	0	1	

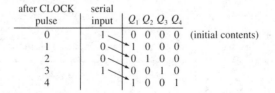

(a) Initial states (0000).

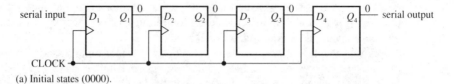

(b) After the first clock pulse (1000).

Figure 12.2
A 4-bit serial-in, serial-out shift register loaded with the serial input 1001. See also the timing diagram in Figure 12.3. (Continued on p. 453.)

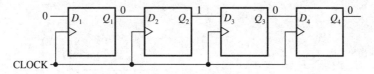

(c) After the second clock pulse (0100).

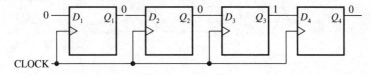

(d) After the third clock pulse (0010).

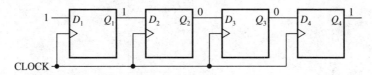

(e) After the fourth clock pulse (1001).

Figure 12.2
(Continued)

four clocked D flip-flops. D_1 is the serial input terminal and the Q output of each flip-flop is the D input to the next. Figure 12.3 is a timing diagram showing how the register is loaded with the serial data 1001. Notice that the bits are effectively *shifted* from one flip-flop to the next on the leading edge of each clock pulse. Hence, this type of register is called a *shift register*. In the example, we assume that all flip-flops are initially reset, although the initial states have no effect on the loading of new data. The initial states represent old data, whose bits are effectively shifted out of the register in serial form at Q_4 as the new bits are shifted in. Since 4 bits must be shifted into the register, the time required to load it equals the total time between four clock pulses. If the clock continues beyond that time, the bits are serially shifted out at Q_4. We see that this register is of the serial-in, serial-out type.

Notice that the shift register is a *synchronous* device: All flip-flops are clocked simultaneously. As such, there is a potential for the problems discussed earlier in connection with a timing race. Examination of the timing diagram in Figure 12.3 confirms that possibility. For example, at the leading edge of the second clock pulse, Q_1 goes low but must remain high long enough to set Q_2. It is also clear that level-triggered flip-flops cannot be used in this application. For example, if Q_2 were level-triggered, we can see from the timing diagram that it would prematurely set during the time that the first clock pulse is high because its input ($D_2 = Q_1$) is high during that time. The flip-flops must be edge-triggered or master-slave types. Figure 12.4 shows specifications for the 74LS91 8-bit shift register, which uses master-slave RS flip-flops, Notice that the register has two inputs, labeled A and B. If serial input is connected to A, it will be loaded into the register only if B is high and vice versa. This gating allows one of the inputs to serve as a control:

Figure 12.3
Timing diagram showing the loading of the serial input 1001 in the register shown in Figure 12.2. Notice that the register is fully loaded ($Q_1Q_2Q_3Q_4 = 1001$) after the fourth clock pulse.

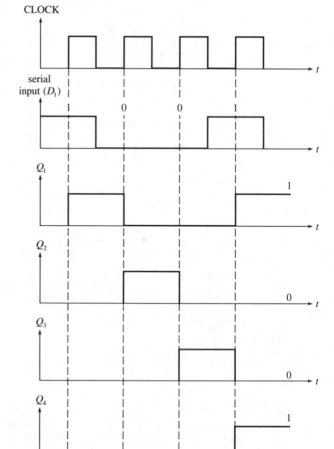

When it is high, serial input on the other line is enabled and when it is low, serial input is inhibited.

Example 12–1

What is the maximum amount of time that could be required to load the 74LS91 if it were operated at its maximum clock frequency?

Solution.   The specifications show that the maximum clock frequency, f_{MAX}, has a *minimum* value of 10 MHz. Since we cannot be certain that an arbitrarily selected device will operate reliably at a frequency greater than 10 MHz, we must use that value in our computations. The period of the clock is then

$$T = \frac{1}{10 \times 10^6 \text{ Hz}} = 0.1 \text{ μs}$$

Since an 8-bit serial register is fully loaded after 8 clock pulses, the maximum time required is $8T = 0.8$ μs.

SN54LS91
SN74LS91

8-BIT SHIFT REGISTERS

LOW POWER SCHOTTKY

The SN54LS/74LS91 is an 8-Bit Serial-In/Serial Out Shift Register. This device features eight R-S master-slave flip-flops, input gating and a clock driver. By gating single-rail data and input control thru inputs A, B, and an internal inverter, complementary inputs to the first bit of the shift register are formed. An inverting clock driver provides the drive for the internal common clock line. The clock pulse inverter driver causes this circuitry to shift information one-bit on the positive edge of the input clock pulse.

FUNCTION TABLE

INPUTS AT t_n		OUTPUTS AT t_{n+8}	
A	B	Q_H	$\bar{Q}_H$
H	H	H	L
L	X	L	H
X	L	L	H

H = HIGH, L = LOW
X = Irrelevant
t_n = Reference bit time
t_{n+8} = Bit time after 8
    LOW to High Clock
    transition

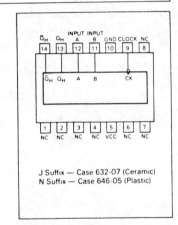

J Suffix — Case 632-07 (Ceramic)
N Suffix — Case 646-05 (Plastic)

FUNCTIONAL BLOCK DIAGRAM

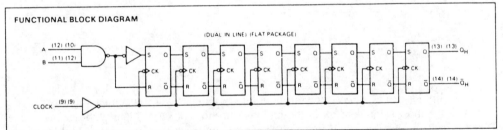

GUARANTEED OPERATING RANGES

SYMBOL	PARAMETER		MIN	TYP	MAX	UNIT
V_{CC}	Supply Voltage	54	4.5	5.0	5.5	V
		74	4.75	5.0	5.25	
T_A	Operating Ambient Temperature Range	54	−55	25	125	°C
		74	0	25	70	
I_{OH}	Output Current — High	54 , 74			−0.4	mA
I_{OL}	Output Current — Low	54			4.0	mA
		74			8.0	

Figure 12.4
Specifications for the 74LS91 8-bit serial-in, serial-out shift register (Courtesy of Motorola, Inc.).

DC CHARACTERISTICS OVER OPERATING TEMPERATURE RANGE (unless otherwise specified)

SYMBOL	PARAMETER		LIMITS			UNITS	TEST CONDITIONS	
			MIN	TYP	MAX			
V_{IH}	Input HIGH Voltage		2.0			V	Guaranteed Input HIGH Voltage for All Inputs	
V_{IL}	Input LOW Voltage	54			0.7	V	Guaranteed Input LOW Voltage for All Inputs	
		74			0.8			
V_{IK}	Input Clamp Diode Voltage			−0.65	−1.5	V	V_{CC} = MIN, I_{IN} = −18 mA	
V_{OH}	Output HIGH Voltage	54	2.5	3.5		V	V_{CC} = MIN, I_{OH} = MAX, V_{IN} = V_{IH} or V_{IL} per Truth Table	
		74	2.7	3.5		V		
V_{OL}	Output LOW Voltage	54,74		0.25	0.4	V	I_{OL} = 4.0 mA	V_{CC} = V_{CC} MIN, V_{IN} = V_{IL} or V_{IH} per Truth Table
		74		0.35	0.5	V	I_{OL} = 8.0 mA	
I_{IH}	Input HIGH Current				20	μA	V_{CC} = MAX, V_{IN} = 2.7 V	
					0.1	mA	V_{CC} = MAX, V_{IN} = 7.0 V	
I_{IL}	Input LOW Current				−0.4	mA	V_{CC} = MAX, V_{IN} = 0.4 V	
I_{OS}	Short Circuit Current		−20		−100	mA	V_{CC} = MAX	
I_{CC}	Power Supply Current				20	mA	V_{CC} = MAX	

AC CHARACTERISTICS: T_A = 25°C, V_{CC} = 5.0 V

SYMBOL	PARAMETER	LIMITS			UNITS	TEST CONDITIONS
		MIN	TYP	MAX		
f_{MAX}	Maximum Clock Frequency	10	18		MHz	
t_{PLH}	Propagation Delay LOW to HIGH		24	40	ns	V_{CC} = 5.0 V C_L = 15 pF
t_{PHL}	Propagation Delay HIGH to LOW		27	40		

AC SETUP REQUIREMENTS: T_A = 25°C, V_{CC} = 5.0 V

SYMBOL	PARAMETER	LIMITS			UNITS	TEST CONDITIONS
		MIN	TYP	MAX		
t_W	Clock Pulse Width Low	25			ns	
t_S	Setup Time	25			ns	V_{CC} = 5.0 V
t_h	Hold Time	0			ns	

Figure 12.4
(Continued)

Drill 12–1
The flip-flops in a 74LS91 are storing a binary number whose decimal value is $213)_{10}$, with the most significant bit in the leftmost flip-flop. A serial input with decimal value $83)_{10}$ is connected to the input, with the least significant bit occurring first. What is the decimal value of the number stored in the register after the fifth clock pulse occurs? *Answer:* $158)_{10}$.

12–3 SERIAL-IN, PARALLEL-OUT SHIFT REGISTERS

A serial-in, parallel-out register can be constructed by simply providing externally accessible connections to the Q output of each flip-flop in a shift register. The register is loaded serially in the same manner already discussed, and, when loaded, the register's output is available in parallel form. Figure 12.5 shows an example: the 74164, 8-bit shift register. Note that this register also has an active-low, asynchronous clear input that can be used to reset every flip-flop in the register. The serial inputs are gated in the same way

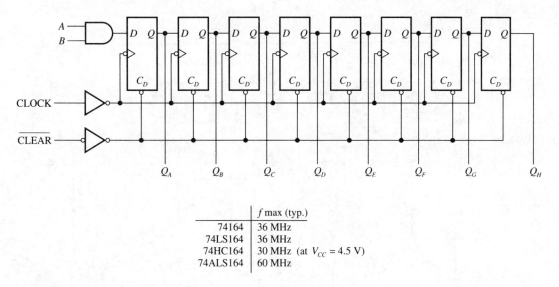

	f max (typ.)
74164	36 MHz
74LS164	36 MHz
74HC164	30 MHz (at $V_{CC} = 4.5$ V)
74ALS164	60 MHz

Figure 12.5
The 74164-series 8-bit serial-in, parallel-out shift register.

as in the 7491. Note that serial output can also be obtained from this register, by taking the output at Q_H.

It is clear that parallel data transfer is much faster than serial data transfer, since the latter requires a number of clock pulses equal to the number bits to be transferred. In a synchronous system, data transfer in parallel form requires only one clock pulse, regardless of the number of bits. For example, if the 74164 shift register is operated at 10 MHz, the time required for output is $1/(10 \times 10^6) = 0.1$ μs, but the time required for input is $8(0.1$ μs$) = 0.8$ μs.

12–4   PARALLEL-IN, SERIAL-OUT SHIFT REGISTERS

To load a shift register in parallel, it is necessary that at least one control input (set or reset) from each flip-flop be externally accessible. When loading data, the set and reset functions are always complements of each other, so most parallel-in registers use only a single control input for each flip-flop and use an internal inversion to generate the other. The 74165, whose logic diagram is shown in Figure 12.6, is an example of a parallel-load register that uses *direct* (asynchronous) inputs for loading. Note that a serial input (SER) is also provided, so the register can be loaded in either serial or parallel form. The input labeled shift/$\overline{\text{load}}$ serves two functions: When it is high, bits are shifted on the leading edge of the clock, and new data can be loaded at the SER input. When shift/$\overline{\text{load}}$ is low, parallel data can be loaded into the register via inputs A through H. Since these inputs are asynchronous, parallel loading is independent of the clock and of any data at the SER input. Notice how the NAND gates provide the logic necessary to implement the shift/$\overline{\text{load}}$ function and to invert the input data for the direct reset on each flip-flop. When shift/$\overline{\text{load}}$ is low, we see that a 1 at a data input produces a 0 at the active-low direct set and a 1 at the active-low reset. Note also that the clock input to the register is logically

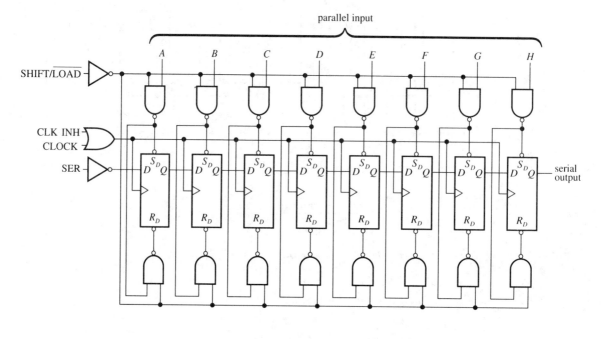

	f max (typ.)
74165	26 MHz
74LS165A	35 MHz
74HC165	30 MHz (at $V_{CC} = 4.5$ V)
74ALS165	25 MHz

Figure 12–6
The 74165-series 8-bit, parallel-in, serial-out shift register.

ORed with a clock inhibit (CLK INH) function. When this inhibit is high, the output of the OR gate is a constant high, so no clocking occurs.

12–5 PARALLEL-IN, PARALLEL-OUT SHIFT REGISTERS

Figure 12.7 shows the logic diagram of 74195 4-bit parallel-in, parallel-out shift register. Note the shift/$\overline{\text{load}}$ control, which performs the same function as that in the 74165. The 74195 also has two serial inputs labeled J and $\overline{K}$. These can be used to control the state of the first (A) flip-flop in the same way that J and K control a clocked JK flip-flop, except that K is active-low:

J	$\overline{K}$	CLOCK	Q_A (shift/$\overline{\text{load}}$ = 1)	
0	0	↑	0	(reset)
0	1	↑	Q_A	(no change)
1	0	↑	$\overline{Q}_A$	(toggle)
1	1	↑	1	(set)

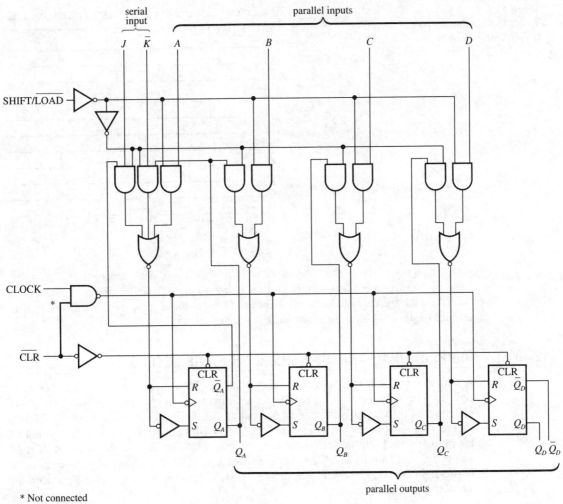

Figure 12.7
The 74195 4-bit, parallel-in, parallel-out shift register.

	f max (typ.)
74195	39 MHz
74S195	105 MHz
74LS195	39 MHz (at V_{CC} = 4.5 V)
74HC195	30 MHz
74AS195	70 MHz

* Not connected
in some versions.

With this arrangement, conventional serial input can be obtained by connecting the input simultaneously to J and $\overline{K}$. (The truth table shows that Q_A resets when J and $\overline{K}$ are both 0 and that Q_A sets when J and $\overline{K}$ are both 1.) The AND/OR/INVERT (AOI) logic at the parallel inputs is used to implement the shift/load function and to supply inverted inputs to the

Figure 12.8
*Typical timing diagram for
the 74195 shift register.*

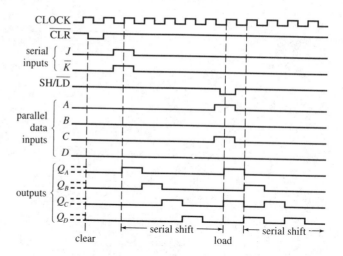

active-high resets. Note that the output of each flip-flop is not connected directly to an input of the succeeding flip-flop, but is gated through the AOI logic. The AOI gates in the first *(A)* stage are also used to implement the $J\overline{K}$ input logic. Figure 12.8 shows a typical timing diagram, illustrating clear, shift, and load sequences.

12–6   BIDIRECTIONAL AND UNIVERSAL SHIFT REGISTERS

A *bidirectional* shift register is one whose bits can be shifted from left to right or from right to left. A *universal* shift register is a bidirectional register whose input can be in either serial or parallel form and whose output can be in either serial or parallel form. Figure 12.9 shows the logic diagram of the 74194 4-bit universal shift register. Note that the output of each flip-flop is routed through AND/OR/INVERT logic to the stage on its right and to the stage on its left. The mode control inputs $S0$ and $S1$ are used to enable the left-to-right connections when it is desired to shift right and the right-to-left connections when it is desired to shift left:

S1	S0	Clock	Action
0	0	x	no change
0	1	↑	shift right
1	0	↑	shift left
1	1	↑	parallel load

The truth table shows that no shifting occurs when $S0$ and $S1$ are both low or when they are both high. When $S0 = S1 = 0$, there is no change in the contents of the register, and when $S0 = S1 = 1$, the parallel input data at *A*, *B*, *C*, and *D* are loaded into the register on the rising edge of a clock pulse. The combination $S0 = S1 = 0$ is said to *inhibit* the loading of serial or parallel data, since the register contents cannot change under that condition. The register has an asynchronous active-low clear that can be used to reset all flip-flops irrespective of the clock and any serial or parallel inputs. Figure 12.10 shows a typical timing diagram.

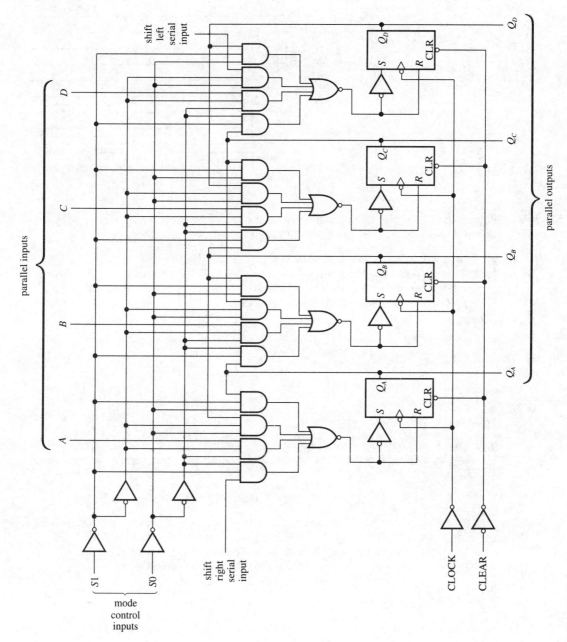

Figure 12.9
The 74194 4-bit universal shift register.

Figure 12.10
Typical timing diagram for the 74194 universal shift register.

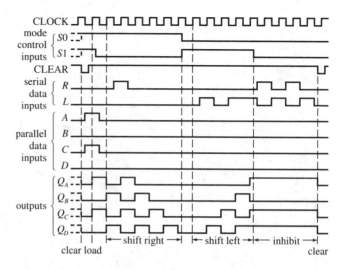

Figure 12.11(*a*) shows the logic circuitry connected to the set and reset (*S* and *R*) inputs of flip-flop *B* in the 74194 universal shift register.

a. Write the Boolean algebra expression for *S*.

b. Use the expression to determine the values of *S* and *R* for each of the following combinations of the other variables. In each case, interpret the action that occurs in the register on the leading edge of the next clock pulse:

Q_A	Q_B	Q_C	S0	S1	B
0	0	1	0	0	0
1	0	1	0	1	1
1	1	0	1	0	0
0	1	1	1	1	1

Solution

a. Figure 12.12 shows an expanded logic diagram of the AOI logic, with the inputs to each AND gate labeled in accordance with the connections shown in part (*a*) of the figure. We see that

$$S = S0 \cdot \overline{S1} \cdot Q_A + S0 \cdot S1 \cdot B + \overline{S0} \cdot S1 \cdot Q_C + \overline{S0} \cdot \overline{S1} \cdot Q_B$$

Note that Q_A is the state of the stage to the left of flip-flop *B*, Q_C is the state of the stage to its right, and *B* is the state of the parallel input to flip-flop *B*. Therefore, the expression is simply a logical restatement of the truth table given earlier to describe the action of the register. Note that *R* is always the complement of *S*.

b.

Q_A	Q_B	Q_C	S0	S1	B	S	R	Action when clocked (↑)
0	0	1	0	0	0	0	1	Q_B is loaded back in (no change).
1	0	1	0	1	0	1	0	Q_B follows Q_C (left shift).
0	1	0	1	0	1	0	1	Q_B follows Q_A (right shift).
0	0	1	1	1	1	1	0	Q_B follows *B* (parallel load).

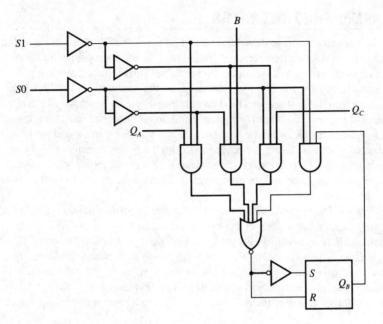

Figure 12.11
Example 12–2.

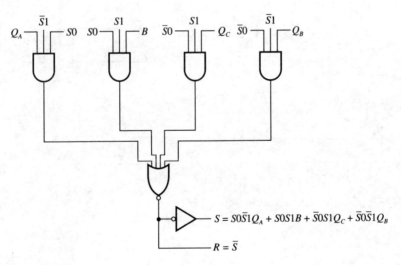

$$S = S0\bar{S}1Q_A + S0S1B + \bar{S}0S1Q_C + \bar{S}0\bar{S}1Q_B$$

$$R = \bar{S}$$

Figure 12.12
Example 12–2.

Drill
12–2

What is the value of S at flip-flop A when $Q_A = 0$, $Q_B = 1$, $S0 = 0$, $S1 = 1$, $A = 0$, and the shift-right serial input to the register is high?
Answer: 1.

12-7   DYNAMIC SHIFT REGISTERS

All the registers we have discussed up to this point use flip-flops to store data, so each bit of the stored data can be retained indefinitely in an appropriate flip-flop. In that sense, such registers are said to be *static*. In a *dynamic* shift register, storage is accomplished by continually shifting the bits from one stage to the next and *recirculating* the output of the last stage into the first stage. The data continually circulates through the register under the control of a clock. To obtain output, a serial output terminal must be accessed at a *specific* clock pulse, for otherwise the sequence of bits will not correspond to the data stored. For example, if a 64-bit word is circulating through a 64-bit register, serial output must begin at multiples of 64 clock pulses. To store new data in such a register, the recirculation path between the last stage and the first stage is interrupted, and the new data is serially loaded into the first stage.

Since each stage of a dynamic shift register needs to retain a bit only for a time equal to one clock period, it is not necessary that each stage be a flip-flop. In particular, the dynamic MOS technology discussed in Section 10-4 is used to construct dynamic shift registers in which each stage consists simply of a dynamic inverter. The clock pulses cause bits to be transferred from one inverter stage to the next by transferring charge stored on the inherent capacitance of the MOS devices, as described in Chapter 10. Recall that this design *requires* the use of a clock having a certain minimum frequency to ensure

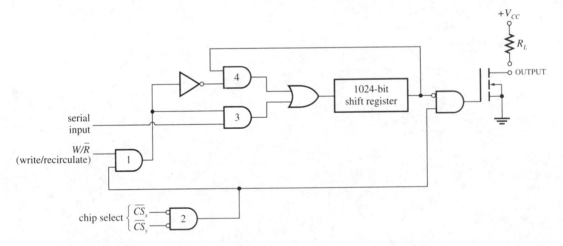

(a) One-half of the 2401 dual 1024-bit dynamic NMOS shift register.

W/$\overline{R}$	$\overline{CS}_x$	$\overline{CS}_y$	Function
1	0	0	write
0	X	X	recirculate
X	1	X	recirculate
X	X	1	recirculate
X	0	0	read

(b) Truth table.

Figure 12.13
The 2401 dynamic NMOS shift register.

that the capacitance does not fully discharge between "refresh" cycles. The principal advantages of dynamic MOS registers are their small power consumption and their simplicity, which permits a very large number of stages to be fabricated in a single integrated circuit. Their disadvantage is that all data transfer must be in serial form, which is much slower than parallel data transfer.

Dynamic MOS registers are widely used as *memory* devices in digital systems that operate on serial data. Because of their small power consumption and the inherent slowness of serial systems, they are used in applications where power consumption and physical size are more important considerations than speed, such as in pocket calculators. In the context of memory applications, loading a register is called *writing* into it and taking output from it is called *reading*. Figure 12.13 shows a logic diagram and truth table for the 2401 dual 1024-bit dynamic NMOS shift register. Note that it has a write/recirculate *(W/$\overline{R}$)* control that is used to govern whether new serial data is written (loaded) into the register or whether the existing data is recirculated (stored) by the register. When $W/\overline{R}$ is high, AND gate 3 is enabled and serial input is transferred through it to the register. When $W/\overline{R}$ is low, AND gate 4 is enabled, and the recirculation path from output to input is completed. Serial data appears at the output regardless of the state of $W/\overline{R}$. The circuit also has two active-low *chip-select* inputs, labeled $\overline{CS}_x$ and $\overline{CS}_y$. These are similar to the enable inputs we have discussed previously: Both $\overline{CS}_x$ and $\overline{CS}_y$ must be low in order to read or write data. Note, however, that recirculation is independent of $\overline{CS}_x$ and $\overline{CS}_y$ and that recirculation is independent of $W/\overline{R}$ if at least one of the chip-selects is high. This example of a dynamic shift register demonstrates why parallel data transfer is not practical in many such devices. (How many pins would the chip be required to have if parallel output were available?)

12-8 APPLICATIONS OF SHIFT REGISTERS

Time Delays

In many digital systems, it is necessary to delay the transfer of data until such time as operations on other data have been completed or to synchronize the arrival of data at a subsystem where it is to be processed with other data. A shift register can be used to delay the arrival of serial data by a specific number of clock pulses, since the number of stages corresponds to the number of clock pulses required to shift each bit completely through the register. The total time delay can be controlled by adjusting the clock frequency and by prescribing the number of stages in the register. In practice, it is often the case that the clock frequency is fixed and the total delay can be adjusted only by controlling the number of stages through which the data is shifted. By using a serial-in, parallel-out register and taking the serial output at any one of the intermediate stages, we have the flexibility to delay the output by any number of clock pulses equal to or less than the number of stages in the register. The next example demonstrates such a case.

Example 12-3

A 16-bit serial data word is to be delivered to one subsystem 4 μs after it arrives and to another subsystem 10 μs after it arrives at the first subsystem. Show how two 74164 shift registers operating with a 1-MHz clock can be used to obtain the prescribed delays.

Solution. As shown in Figure 12.5, the 74164 is an 8-bit serial-in, parallel-out shift register. When it is operated at 1 MHz, the time delay introduced by each stage is, therefore, 1 μs. Thus, the 16-bit data must be shifted through 4 stages to obtain an output delayed by 4 μs and through an additional 10 stages to obtain another output delayed by

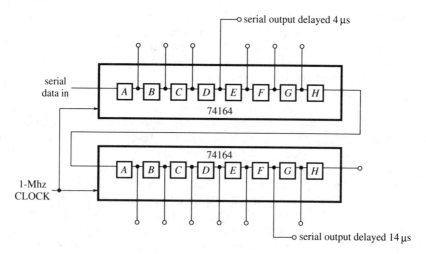

Figure 12.14
Example 12–3.

a total of 14 μs. As shown in Figure 12.14, we can *cascade* the two 8-bit registers by connecting the output of one to the input of the other. The parallel outputs of the two registers can then be selected to provide any delay between 1 clock period and 16 clock periods. The figure shows that the 4-μs delay is obtained by taking serial output from the fourth stage of the first register (Q_D), and the 14-μs delay is obtained by taking serial output from the sixth stage of the second register (Q_F). Notice that the number of bits in the serial data word (16 in this case) is irrelevant.

Drill 12–3
If the registers in Figure 12.14 are clocked at 2.5 MHz, which outputs should be used to obtain delays of 2.4 μs and 5.2 μs in a serial data word?
Answer: Q_F of the first register and Q_E of the second register.

Serial/Parallel Data Conversion

We have seen that parallel data transfer is faster than its serial counterpart. Similarly, the processing of data, such as that which occurs in the arithmetic operations performed by microprocessors and digital computers, is much faster when all data bits are available and can be operated on simultaneously. For that reason, digital systems in which speed is an important consideration are designed to operate on data in parallel form. However, when data is transmitted to or from such a system, particularly when it must be transmitted over long distances, it is impractical to provide all the parallel lines that are necessary to transmit data in parallel. It is more economical to transmit serial data, since only one line is required. Consequently, a means must be provided for converting serial data to parallel form, so that serial input can be processed by a parallel system, and for converting parallel data to serial form, so that parallel output can be transmitted serially. Shift registers are widely used for those purposes.

A serial-in, parallel-out register is used to perform serial-to-parallel conversion, and a parallel-in, serial-out register is used to perform parallel-to-serial conversion. A universal shift register can be used to perform both functions. Since serial data must be

Figure 12.15

A UART contains the registers and synchronizing circuitry necessary to receive data in serial form and to convert and transmit it in parallel form and vice versa.

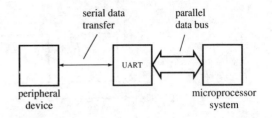

sampled at specific time intervals to obtain the individual bits in proper sequence, format conversion involves certain synchronization problems. For example, special bits must be added to serial data to mark the beginning and end of a word, and these bits must be "stripped off" when the word is converted to parallel form. A specially designed integrated circuit called a *universal asynchronous receiver/transmitter* (UART) contains all the registers and synchronizing circuitry necessary to receive data in serial form and transmit it in parallel and vice versa. It is widely used to interface microprocessor systems with peripheral devices that transmit and receive data in serial form, as illustrated in Figure 12.15. The two-headed arrows show that data transmission can occur in both directions.

In some systems, serial data are transmitted or received with the least significant bit (LSB) occurring first, and in other systems or subsystems the most significant bit (MSB) occurs first. To make these systems compatible, the order of serial data in one system must be reversed. A bidirectional shift register can be used for that purpose. The serial data are first loaded into the register by shifting from left to right. Once the register is fully loaded, the data are removed by shifting them out from right to left, thus reversing the order of the bits.

Ring Counters

A ring counter is constructed by connecting the output of the last stage of a shift register to the input of the first stage, as illustrated in Figure 12.16(a). In its most common application, the register is loaded so that exactly one stage contains a 1, and all other stages contain 0s. Thus, as the register is continually clocked, the 1 continually circulates through it, resulting in a timing diagram like that shown in Figure 12.16(b). Note that one and only one output is high at any given time. We see that the outputs can be used as a sequence of *synchronizing pulses*, as discussed in Section 8–2 and illustrated in Figure 8.9, using monostable multivibrators. The device is called a *ring counter* because each output can be regarded as providing a count of the number of clock pulses that have occurred. For example, when the output of the second stage is high, we know that two clock pulses have occurred, and when the output of the third stage goes high, we know that three clock pulses have occurred. In that sense, it is a *decimal* counter. Of course, the count *resets* after a number of clock pulses equal to the number of stages. Counters are discussed in more detail in a later chapter. The ring counter can also be used as a *divide-by-n* device, where *n* is the number of stages. Note that the frequency of each output in the 4-bit counter shown in Figure 12.16 is one-fourth the frequency of the clock.

For the ring counter to operate as shown in Figure 12.16, it is necessary to load it with a single 1, which means a separate initializing operation involving additional logic and timing circuitry is required. Figure 12.17 shows a 4-bit *self-starting* ring counter. When power is first applied, the initial states of the flip-flops are unpredictable. No matter what those initial states, the circuit shown will eventually (after a maximum of four clock

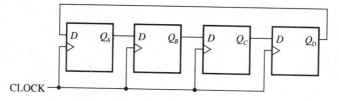

(a) A ring counter is constructed by connecting the output of the last stage to the input of the first stage.

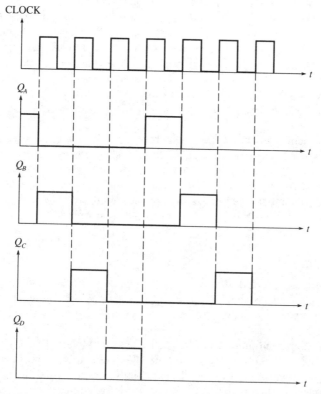

(b) Timing diagram when the counter is initially loaded with 1000. The 1 circulates around the register producing the sequence 1000, 0100, 0010, 0001, 1000, etc.

Figure 12.16
The ring counter.

pulses) circulate a single 1. This fact is demonstrated in part *(b)* of the figure, in which we assume all states are initially 1. After four clock pulses, we see that a single 1 continues to circulate. If the initial states are any of those shown after the first through fourth clock pulses, the ultimate result is seen to be the same. It is easy to verify that any other initial states will also produce the same result, since the output of the NOR gate must eventually become 1.

Figure 12.17
A self-starting ring counter.

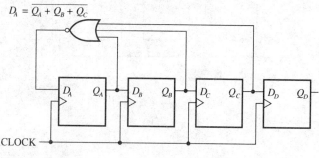

$$D_A = \overline{Q_A + Q_B + Q_C}$$

(a) Logic diagram.

	Q_A	Q_B	Q_C	Q_D	D_A
initial states	1	1	1	1	0
after pulse 1	0	1	1	1	0
after pulse 2	0	0	1	1	0
after pulse 3	0	0	0	1	1
after pulse 4	1	0	0	0	0
after pulse 5	0	1	0	0	0
after pulse 6	0	0	1	0	0
after pulse 7	0	0	0	1	1
after pulse 8	1	0	0	0	0

(b) Output sequence when the initial contents of the register are 1111. No matter what the initial contents, a single 1 will eventually circulate.

Johnson Counters

Figure 12.18(a) shows a 4-bit *Johnson counter*, which is identical to a ring counter, except that the *complement* of the output of the last stage is connected to the input of the first stage. For that reason, it is also called a *twisted-ring counter*. When the counter is initially loaded with all 0s, the count sequence and timing diagram that result are shown in Figure 12.18(b). Notice that the frequency of each output is one-eighth that of the clock. The Johnson counter is a divide-by-$2n$ device, where n is the number of stages. Figure 12.18(c) shows a self-starting Johnson counter.

12–9 ANSI/IEEE STANDARD SYMBOLS

The qualifying symbol that identifies a shift register is SRGn, where n is the number of bits. For example, SRG4 and SRG8 identify 4- and 8-bit shift registers, respectively. Figure 12.19 shows the symbol for an 8-bit serial-in, parallel-out shift register, the 74HC164. The arrow indicates that shifting occurs from left to right—i.e., from Q_A toward Q_H. Since $\overline{CLR}$ and CLK are inputs to the control block, we know that all eight states are clocked simultaneously and can be cleared simultaneously. Clearing is independent of the clock (asynchronous), because if it were synchronous, a clock dependency would be shown by using the symbol 1R instead of R at the $\overline{CLR}$ input. Inputs A and B are ANDed to create the clock-dependent data input 1D. Note that D here does not refer to the D stage. In ANSI/IEEE notation, the letter D is always used to represent input data to a storage element, in this case the A stage of the register. Since the CLK input at C1 is shown to be dynamic and non-inverting, we know that bits are shifted synchronously on the leading edge of the clock.

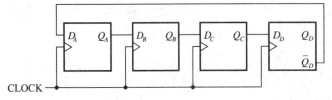

(a) The complement of the output of the last stage ($\bar{Q}_D$) is the input to the first stage.

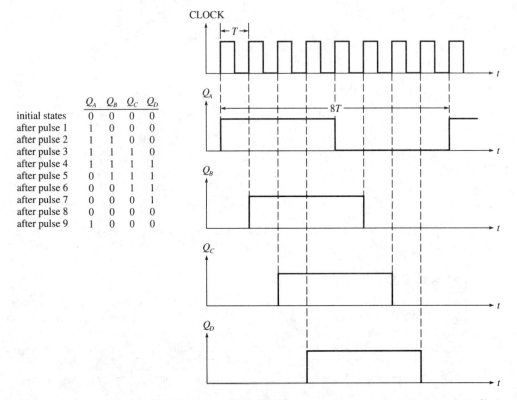

	Q_A	Q_B	Q_C	Q_D
initial states	0	0	0	0
after pulse 1	1	0	0	0
after pulse 2	1	1	0	0
after pulse 3	1	1	1	0
after pulse 4	1	1	1	1
after pulse 5	0	1	1	1
after pulse 6	0	0	1	1
after pulse 7	0	0	0	1
after pulse 8	0	0	0	0
after pulse 9	1	0	0	0

(b) When the initial contents of the 4-bit Johnson counter are 0000, the output of each stage has a frequency that is $1/8$ the clock frequency.

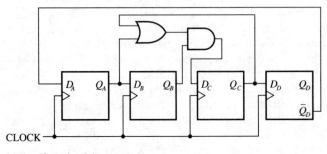

(c) A self-starting Johnson counter.

Figure 12.18
The Johnson counter.

470

Figure 12.19
ANSI/IEEE logic symbol for the 74HC164 8-bit serial-in, parallel-out shift register.

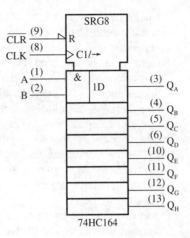

74HC164

Mode Dependency

Shift registers, like other digital devices, are often designed so that they can be operated in one of several *modes,* under the control of one or more binary inputs. We have seen, for example, that the S0 and S1 inputs to the 74194 universal shift register determine whether it is shifted right, shifted left, or loaded in parallel. Mode (M) dependency notation is the ANSI/IEEE means for showing inputs and outputs that are affected by the operating mode of the device. If a device has 4 operating modes, the symbol M_3^0 is used to mean that the 4 modes are 0, 1, 2 and 3, corresponding to binary control inputs having states 00, 01, 10, and 11. Figure 12.20(*a*) shows this mode dependency in the 74194 universal shift register. Note that M_3^0 applies to control inputs S0 and S1, as indicated by the brace, }, called a *binary grouping* symbol. (Modes can also be enumerated within the outline, as, for example: M1 load or M2 shift.) The clock input to the 74194 is at *C4*,

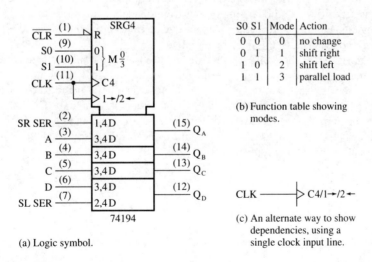

(a) Logic symbol.

S0 S1	Mode	Action
0 0	0	no change
0 1	1	shift right
1 0	2	shift left
1 1	3	parallel load

(b) Function table showing modes.

CLK ———▷ C4/1→/2←

(c) An alternate way to show dependencies, using a single clock input line.

Figure 12.20
ANSI/IEEE logic symbol for the 74194 universal shift register, illustrating mode (M) dependency notation.

where the numeral 4 is used to show C dependency, since numerals 0, 1, 2, and 3 already been used for mode dependency. Note that each stage has both mode and dependency. The shift-right serial input (SR SER) has dependency 1,4D, meaning input data is accepted at that point when in mode 1 and that the data (D) clock-dependent (4)—i.e., loaded synchronously. Parallel inputs A, B, C, and D all have dependency 3,4D, since parallel loading occurs in mode 3 and data is clock-dependent (synchronous). Similarly, shift-left serial input data (SL SER) has mode dependency (2) and clock dependency (4), as indicated by 2,4D. Note that $\overline{CLR}$ is asynchronous, for otherwise the symbol 4R would be used instead of R. The clock input is further qualified by 1→/2←, meaning that right shifting occurs in mode 1 and left shifting in mode 2. Although the CLK input is shown connected to the control block in two places, this does *not* mean that the physical circuit has two pins for clock connections. The two CLK lines are used in the symbol to separate the C- and M-dependencies. It is permissible to place multiple dependency notations on one line, as shown in Figure 12.20(c). Note that C4 appears first, followed by the mode dependencies, and that all are separated by slashes (called *solidi*, the plural of *solidus*).

12–10   SPICE EXAMPLES

Example 12–4

Construct a SPICE model of a 3-bit shift register using clocked *RS* flip-flops. Test the model by shifting 101 into it.

Solution. Figure 12.21(b), pages 473–74, shows how we can incorporate the model of the clocked RS latch developed in Example 11.4 to model a 3-bit shift register. Here, we define the entire model of the latch to be a subcircuit named LATCH. For the sake of simplicity (to avoid the use of JK flip-flops), we will simulate edge-triggered RS flip-flops by specifying very narrow clock pulses. Note that inverters are inserted between each stage to provide buffering (isolation) and to provide the interstage delays necessary for synchronous operation. The inverter is defined in a subcircuit named BUFFER, as shown in Figure 12.21(a). The threshold voltage of the NMOS transistor in BUFFER ($V_T =$ VT0) is set to 5 V, so a changing output from one stage must rise to a higher level than in previous models before the change is experienced in the succeeding stage. This behavior contributes to the interstage delay.

Figure 12.21(c) shows the waveforms we use to test the model. Since the serial input to the register is at the R_1 and S_1 terminals, we can shift in 101 by making $R_1 S_1 = 01$ at the first clock pulse, 10 at the second clock pulse, and 01 at the third clock pulse. Notice that the clock pulses are very narrow (0.01 μs) in comparison to the R_1 and S_1 pulses. The timing diagram shows the expected waveforms at Q_1, Q_2, and Q_3, and the SPICE plots in Figure 12.21(d), pages 474–76, confirm the results. Note that the contents of the register after three clock pulses are $Q_1 Q_2 Q_3 = 101$.

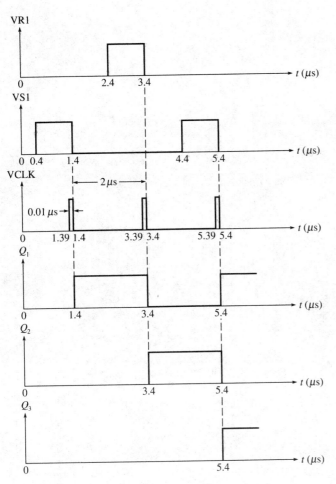

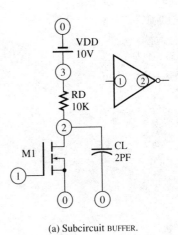

(a) Subcircuit BUFFER.

Figure 12.21
Example 12−4.
(Continued on pages
474−476.)

All pulse amplitudes 10 V.

(c)

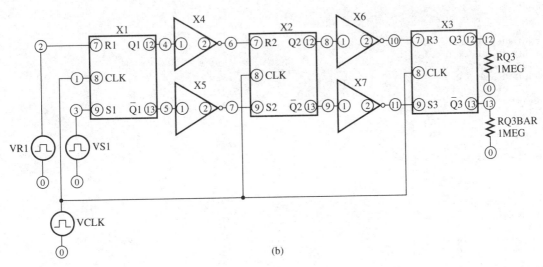

(b)

(Continued on p. 474.)

Figure 12.21 *(Continued)*

(b) EXAMPLE 12.4
```
.SUBCKT LATCH 7 8 9 12 13                      .MODEL MOS NMOS KP = 0.25E-3 VTO = 5
.SUBCKT AND 1 2 3                              RD 2 3 10K
D1 3 1 DIODE                                   CL 2 0 2PF
D2 3 2 DIODE                                   VDD 3 0 10V
RD 4 3 20K                                     .ENDS BUFFER
VD 4 0 15V                                     X1 2 1 3 4 5 LATCH
.MODEL DIODE D                                 X2 6 1 7 8 9 LATCH
.ENDS AND                                      X3 10 1 11 12 13 LATCH
.SUBCKT NOR 4 5 6                              X4 4 6 BUFFER
M1 6 4 0 0 MOS                                 X5 5 7 BUFFER
M2 6 5 0 0 MOS                                 X6 8 10 BUFFER
RDD 6 7 10K                                    X7 9 11 BUFFER
VDD 7 0 10V                                    RQ 12 0 1MEG
.MODEL MOS NMOS KP = 0.25E-3 VTO = 2           RQ BAR 13 0 1MEG
.ENDS NOR                                      .NODESET V(5) = 10 V(9) = 10 V(13) = 10
X1 7 8 10 AND                                  VCLK 1 0 PULSE(0 10 1.39US 1PS 1FS 0.01US 2US)
X2 8 9 11 AND                                  VR1 2 0 PULSE(0 10 2.4US 1PS 1PS 1US 10US)
X3 10 13 12 NOR                                VS1 3 0 PULSE(0 10 0.4US 1PS 1PS 1US 4US)
X4 12 11 13 NOR                                .TRAN 0.2US 6US
C1 12 0 2PF                                    .PLOT TRAN V(4)
C2 13 0 2PF                                    .PLOT TRAN V(8)
.ENDS LATCH                                    .PLOT TRAN V(12)
.SUBCKT BUFFER 1 2                             .END
M1 2 1 0 0 MOS
```

(d) ****   TRANSIENT ANALYSIS   TEMPERATURE =   27.000 DEG C

```
*******************************************************************************
     TIME       V(4)

                       0.000D+00     5.000D+00     1.000D+01     1.500D+01  2.000D+01
                       -  -  -  -  -  -  -  -  -  -  -  -  -  -  -  -  -  -  -  -  -  -
 0.000D+00   4.905D-01 .*              .             .             .              .
 2.000D-07   4.905D-01 .*              .             .             .              .
 4.000D-07   4.905D-01 .*              .             .             .              .
 6.000D-07   4.905D-01 .*              .             .             .              .
 8.000D-07   4.905D-01 .*              .             .             .              .
 1.000D-06   4.905D-01 .*              .             .             .              .
 1.200D-06   4.905D-01 .*              .             .             .              .
 1.400D-06   3.906D+00 .               *  .          .             .              .
 1.600D-06   1.000D+01 .               .             *             .              .
 1.800D-06   1.000D+01 .               .             *             .              .
 2.000D-06   1.000D+01 .               .             *             .              .
 2.200D-06   1.000D+01 .               .             *             .              .
 2.400D-06   1.000D+01 .               .             *             .              .
 2.600D-06   1.000D+01 .               .             *             .              .
 2.800D-06   1.000D+01 .               .             *             .              .
 3.000D-06   1.000D+01 .               .             *             .              .
 3.200D-06   1.000D+01 .               .             *             .              .
 3.400D-06   3.904D-01 .*              .             .             .              .
 3.600D-06   4.905D-01 .*              .             .             .              .
 3.800D-06   4.905D-01 .*              .             .             .              .
 4.000D-06   4.905D-01 .*              .             .             .              .
 4.200D-06   4.905D-01 .*              .             .             .              .
 4.400D-06   4.905D-01 .*              .             .             .              .
 4.600D-06   4.905D-01 .*              .             .             .              .
 4.800D-06   4.905D-01 .*              .             .             .              .
 5.000D-06   4.905D-01 .*              .             .             .              .
 5.200D-06   4.905D-01 .*              .             .             .              .
 5.400D-06   3.904D+00 .               *  .          .             .              .
 5.600D-06   1.000D+01 .               .             *             .              .
 5.800D-06   1.000D+01 .               .             *             .              .
 6.000D-06   1.000D+01 .               .             *             .              .
                       -  -  -  -  -  -  -  -  -  -  -  -  -  -  -  -  -  -  -  -  -  -
```

Figure 12.21 *(d, continued)*

```
    TIME        V(8)

                      0.000D+00    5.000D+00    1.000D+01    1.500D+01  2.000D+01
                      - - - - - - - - - - - - - - - - - - - - - - - - - -
0.000D+00   4.905D-01 .*            .            .            .           .
2.000D-07   4.905D-01 .*            .            .            .           .
4.000D-07   4.905D-01 .*            .            .            .           .
6.000D-07   4.905D-01 .*            .            .            .           .
8.000D-07   4.905D-01 .*            .            .            .           .
1.000D-06   4.905D-01 .*            .            .            .           .
1.200D-06   4.905D-01 .*            .            .            .           .
1.400D-06   3.149D-01 .*            .            .            .           .
1.600D-06   4.905D-01 .*            .            .            .           .
1.800D-06   4.905D-01 .*            .            .            .           .
2.000D-06   4.905D-01 .*            .            .            .           .
2.200D-06   4.905D-01 .*            .            .            .           .
2.400D-06   4.905D-01 .*            .            .            .           .
2.600D-06   4.905D-01 .*            .            .            .           .
2.800D-06   4.905D-01 .*            .            .            .           .
3.000D-06   4.905D-01 .*            .            .            .           .
3.200D-06   4.905D-01 .*            .            .            .           .
3.400D-06   1.088D+00 .   *         .            .            .           .
3.600D-06   1.000D+01 .             .            *            .           .
3.800D-06   1.000D+01 .             .            *            .           .
4.000D-06   1.000D+01 .             .            *            .           .
4.200D-06   1.000D+01 .             .            *            .           .
4.400D-06   1.000D+01 .             .            *            .           .
4.600D-06   1.000D+01 .             .            *            .           .
4.800D-06   1.000D+01 .             .            *            .           .
5.000D-06   1.000D+01 .             .            *            .           .
5.200D-06   1.000D+01 .             .            *            .           .
5.400D-06   4.553D-01 .*            .            .            .           .
5.600D-06   4.904D-01 .*            .            .            .           .
5.800D-06   4.905D-01 .*            .            .            .           .
6.000D-06   4.905D-01 .*            .            .            .           .
                      - - - - - - - - - - - - - - - - - - - - - - - - - -
```

Figure 12.21 *(d, continued)*

```
         TIME       V(12)

                    -5.000D+00      0.000D+00      5.000D+00      1.000D+01  1.500D+01
                    - - - - - - - - - - - - - - - - - - - - - - - - - - - - - - - - -
     0.000D+00   4.964D-01 .               .*             .              .           .
     2.000D-07   4.965D-01 .               .*             .              .           .
     4.000D-07   4.965D-01 .               .*             .              .           .
     6.000D-07   4.965D-01 .               .*             .              .           .
     8.000D-07   4.965D-01 .               .*             .              .           .
     1.000D-06   4.965D-01 .               .*             .              .           .
     1.200D-06   4.965D-01 .               .*             .              .           .
     1.400D-06   2.418D-01 .               .*             .              .           .
     1.600D-06   4.965D-01 .               .*             .              .           .
     1.800D-06   4.965D-01 .               .*             .              .           .
     2.000D-06   4.965D-01 .               .*             .              .           .
     2.200D-06   4.965D-01 .               .*             .              .           .
     2.400D-06   4.965D-01 .               .*             .              .           .
     2.600D-06   4.965D-01 .               .*             .              .           .
     2.800D-06   4.965D-01 .               .*             .              .           .
     3.000D-06   4.965D-01 .               .*             .              .           .
     3.200D-06   4.965D-01 .               .*             .              .           .
     3.400D-06   3.171D-01 .               .*             .              .           .
     3.600D-06   4.965D-01 .               .*             .              .           .
     3.800D-06   4.965D-01 .               .*             .              .           .
     4.000D-06   4.965D-01 .               .*             .              .           .
     4.200D-06   4.965D-01 .               .*             .              .           .
     4.400D-06   4.965D-01 .               .*             .              .           .
     4.600D-06   4.965D-01 .               .*             .              .           .
     4.800D-06   4.965D-01 .               .*             .              .           .
     5.000D-06   4.965D-01 .               .*             .              .           .
     5.200D-06   4.965D-01 .               .*             .              .           .
     5.400D-06   1.089D+00 .               .    *         .              .           .
     5.600D-06   9.901D+00 .               .             .              *            .
     5.800D-06   9.901D+00 .               .             .              *            .
     6.000D-06   9.901D+00 .               .             .              *            .
                    - - - - - - - - - - - - - - - - - - - - - - - - - - - - - - - - -
```

Figure 12.22
Exercise 12.3.

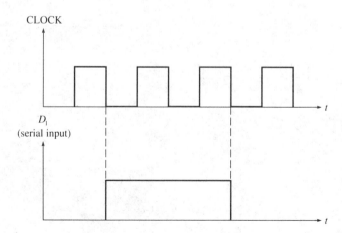

EXERCISES

Section 12–2

12.1 An 8-bit serial-in, serial-out shift register is storing a binary number whose decimal value is 154. The least significant bit of the number is stored in the rightmost flip-flop. The register is to be loaded with a serial number whose decimal value is 167 by shifting the bits into the leftmost flip-flop, least significant bit first. Construct a table showing the contents of the shift register in binary form after each of eight successive clock pulses.

12.2 An 8-bit serial-in, serial-out shift register having no direct reset must be cleared (all flip-flops reset) before commencing another operation. Describe how the register could be cleared and determine how long it would take, if the clock frequency is 1.25 MHz.

12.3 Construct a timing diagram (similar to that shown in Figure 12.3) to show the state of each flip-flop in the register in Figure 12.2(*a*) when the input is as shown in Figure 12.22, p. 476. Assume the initial contents of the register are 0000. What are the binary contents of the register after the fourth clock pulse?

12.4 Construct a timing diagram (similar to that shown in Figure 12.3) to show the state of each flip-flop in the register in Figure 12.2(*a*) when the input is as shown in Figure 12.23. Assume the initial contents of the register are 0000. What are the binary contents of the register after the eighth clock pulse?

Section 12–3

12.5 Serial data is loaded into a 74164 shift register and then transmitted in parallel form for storage in a computer memory whose capacity is 2^{16} 8-bit words. After each 8-bit word is shifted into the register, the next clock pulse is used to transfer the word in parallel form to memory, and the loading of a new serial word into the register commences on the succeeding clock pulse. If the register is clocked at 2 MHz, how long does it take to load memory to its full capacity?

12.6 If the computer memory in Exercise 12.5 must be completely loaded in no less than 100

Figure 12.23
Exercise 12.4.

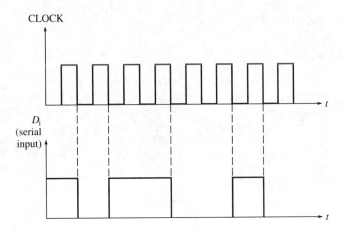

ms, what is the minimum clock frequency at which the system can be operated?

Section 12–4

12.7 A digital *counting circuit* is used to count the number of vehicles passing through an intersection. It generates the binary number *ABCDEFGH* in parallel form, which increases by 1 each time a vehicle passes: 00000000, 00000001, 00000010, 00000011, ... Notice that this sequence is *asynchronous*—i.e., independent of any clock. When 200 vehicles have passed, the binary number $11001000 = 200)_{10}$ is to be transmitted in serial form to a traffic controller, under the synchronization of a separately available clock signal. Design a circuit using a 74165 shift register that will perform the required function. Assume that the register must not be clocked until after the required count is reached and that other circuitry is used to "freeze" the count *ABCDEFGH* once it has been reached. The serial output of the system must remain 0 until the count is reached.

12.8 Modify the design in Exercise 12.7 so that the serial output is toggled each time the count changes. Once the count reaches $200)_{10}$, it is to be serially transmitted as before. The design should also provide an OUT signal that goes high to signify that the serial output is valid after the count is reached.

Section 12–5

12.9 The binary number *ABCD* is loaded in parallel form into a 74195 shift register. If the parallel output does not equal 1111, parallel loading is to continue. If the output equals 1111, it should be changed to 0111 at the leading edge of the next clock pulse. Design the circuit. What happens when the *ABCD* input changes thereafter?

12.10 Modify the design in Exercise 12.9 so that the output remains 1111 on the leading edges of all future clock pulses, once it reaches 1111, irrespective of any future changes in *ABCD*.

Section 12–6

12.11 A 74194 universal shift register is to be used to perform the following sequence of events:

1. Load the 4-bit word $A_1B_1C_1D_1$ in parallel form; A_1 is the least significant bit.
2. Convert $A_1B_1C_1D_1$ to serial form, with least significant bit occurring first.
3. Delay for two clock periods.
4. Load $A_2B_2C_2D_2$ in parallel form. D_2 is the least significant bit.
5. Convert $A_2B_2C_2D_2$ to serial form with least significant bit occurring first.

Assuming each parallel load operation requires one clock period, construct a timing diagram showing the clock and the *S0* and *S1* waveforms that must be generated to drive the register. Indicate the (groups of) clock pulses during which each of the 5 required operations are performed.

12.12 Figure 12.24 shows three stages in a shift register that is to be designed for bidirectional operation using a single directional control signal, DIR. When DIR = 0, right shifting is to be enabled, and when DIR = 1, left shifting is to be enabled. Assuming no parallel input is necessary, design the logic circuitry required for stage *B*. Draw a logic diagram.

Figure 12.24
Exercise 12.12.

Section 12–7

12.13 One-half of a 2401 dynamic shift register is clocked at 500 kHz. What is the *maximum* length of time that could be required to obtain access to (begin shifting out) an 8-bit word stored in the register?

12.14 Draw a schematic diagram of two stages of a dynamic NMOS shift register.

Section 12-8

12.15 Figure 12.25 shows a system used to determine whether there is noise in either one of two data transmission paths. One serial data word is transmitted simultaneously via satellite and via a telephone line to a receiving station. If the output of the coincidence (exclusive-NOR) gate ever goes low, there is a discrepancy between bits transmitted by the two paths, and it can be presumed that noise has affected one of the transmissions. The serial data is generated at a rate of 64×10^3 bits/s and a synchronized 64-kHz clock signal is available at the receiver. The time of transmission via satellite is 0.2720 s and the time of transmission via telephone line is 0.2560 s. Specify the delay circuitry that will ensure that corresponding bits arrive at the coincidence gate at the same time. Draw a block diagram.

12.16 A 32-bit shift register with parallel output is to be used to delay serial data by 7.2 μs.

(a) At what frequency should it be clocked if the delayed output is to be taken from the 18th stage?

(b) Using the clock frequency in *(a)*, at what stage should the output be taken if the total delay is to be 12.4 μs?

12.17 The register shown in Figure 12.26 is initially loaded with $Q_A Q_B Q_C Q_D = 1011$. Construct a timing diagram showing the output of each stage for a total of 8 clock pulses. What is the frequency of each output, as a fraction of the clock frequency?

12.18 Repeat Exercise 12.17 when $\overline{Q_D}$ instead of Q_D is the D input to the first stage. How many times must the register be clocked before

Figure 12.25
Exercise 12.15.

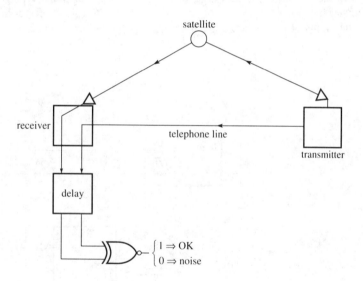

Figure 12.26
Exercise 12.17.

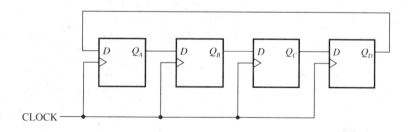

the contents are the same as the initial contents (1011)?

12.19 If the initial contents of the self-starting ring counter in Figure 12.17(a) are $Q_A Q_B Q_C Q_D = 1100$, how many clock pulses are required before the contents are 1000?

12.20 If the initial contents of the self-starting Johnson counter in Figure 12.18(c) are $Q_A Q_B Q_C Q_D = 1010$, how many clock pulses are required before the contents are 0011?

Section 12–9

12.21 With reference to Figure 12.27, answer the following questions.
(a) What type of register is represented (number of bits and serial or parallel inputs and outputs)?
(b) Is clearing synchronous or asynchronous?
(c) How many operating modes does the register have?
(d) If the inputs at pins 2 and 3 are both held

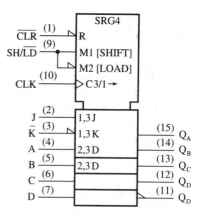

Figure 12.27
Exercise 12.21.

low and pin 9 is high, explain what happens as the register is clocked.

12.22 What type of circuit is shown in Figure 12.28? Describe the functions it can perform with reference to switch settings a and b.

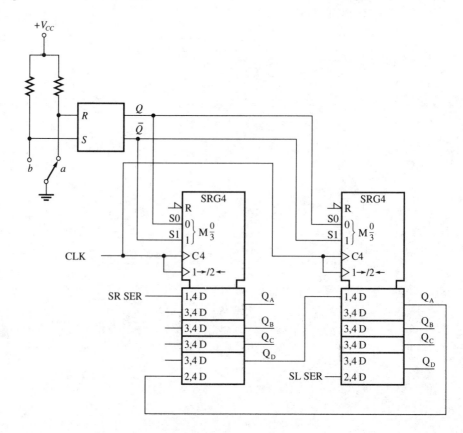

Figure 12.28
Exercise 12.22.

CHALLENGING EXERCISES

12.23 Complete the timing diagram in Figure 12.29 to show the Q_A, Q_B, Q_C, and Q_D outputs of a 74194 shift register in response to the inputs shown. All outputs are initially 0.

12.24 Design a 4-bit shift-register counter that will produce the following sequence:

$$0\ 0\ 0\ 0$$
$$0\ 1\ 0\ 1$$
$$1\ 1\ 1\ 1$$
$$1\ 0\ 1\ 0$$
$$0\ 0\ 0\ 0$$
$$\vdots$$
etc.

Assume the initial contents of the register will always be 0000. What is the frequency of the Q output of each flip-flop when the clock frequency is 100 kHz?

12.25 Design a 4-output logic circuit, all of whose outputs are zero until the self-starting ring counter in Figure 12.17(a) begins circulating a single 1. After that time, the outputs of the logic circuit should be the same as the outputs of the counter.

12.26 Design a self-starting 4-bit shift-register counter that will eventually circulate a single 0, irrespective of the initial contents of the counter. Write the sequence of contents produced by your design when the initial contents are 0000.

12.27 Show how one additional output can be obtained from the 4-bit ring counter shown in Figure 12.16(a) without adding another flip-flop. In other words, design a 5-output ring counter that circulates a single 1 using only four flip-flops. Your design should be self-starting.

SPICE EXERCISES

12.28 Construct a SPICE model of a 2-bit ring counter, clocked at 500 kHz. Load the counter with a single 1 in the least significant stage and obtain plots of the output of each stage over a period of 6 µs.

12.29 Construct a SPICE model of a 3-bit Johnson counter, clocked at 500 kHz. Preclear all stages and then obtain plots of the output of each stage over a period of time sufficient to cycle the contents from 000 and back to 000. Then preload the counter with 101 and obtain plots to demonstrate that it is not self-starting.

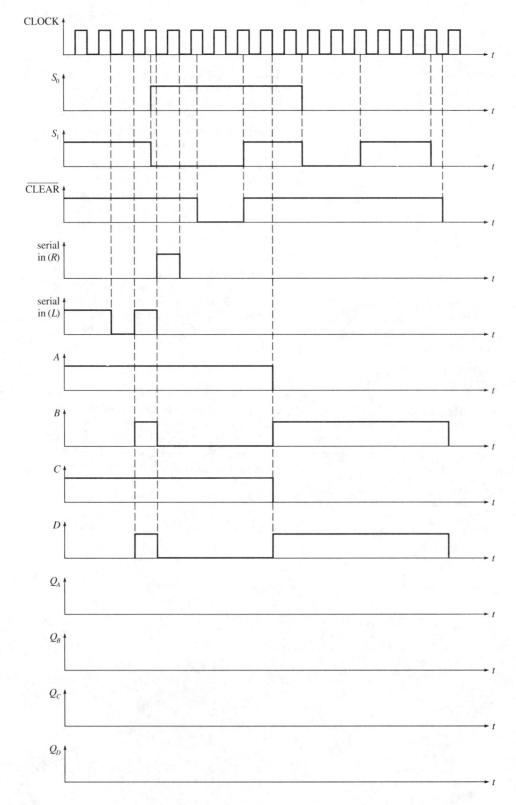

Figure 12.29 *Exercise 12.23.*

Chapter 13

BINARY CODES AND CODING CIRCUITS

INTRODUCTION

We have seen that digital data is represented, stored, and transmitted as groups of binary digits (bits). Previous discussions have probably left the impression that such data always represents numerical values derived from the binary number system. In fact, digital data is more often *encoded,* whereby groups of bits, called *bit patterns,* are used to represent both numbers and letters of the alphabet as well as many special characters and control functions. In these cases, the equivalent decimal ''value'' of a sequence of bits may or may not bear any relation to the ''value'' of the data it represents. In some systems, ''pure'' binary numbers—those whose decimal values are based solely on the weights of the bit positions—are found only where arithmetic operations must be performed on numerical data. Even in those systems, the numerical data may be stored and/or transmitted using binary codes rather than binary numbers.

Binary codes can be classified as numeric or alphanumeric. Numeric codes are used to represent numbers. Often they have special characteristics that facilitate mathematical operations, such as a sign bit to indicate algebraic sign. Indeed, the 1's- and 2's-complement methods of representing signed numbers, as studied in Chapter 2, are examples of numeric codes. Alphanumeric codes are used to represent *characters:* alphabetic letters and numerals. In these codes, a *numeral* is treated simply as another symbol rather than as a number or numerical value. Alphanumeric codes are widely used to transmit data, for example, from a keyboard to a computer. Numerals transmitted in an alphanumeric code are then converted to a numeric code so that mathematical operations can be performed. The numeric results can then be converted back to alphanumeric form and transmitted to an output display, such as a video terminal. Data can be stored in computer memory in both numeric and alphanumeric form.

13–2 **8-4-2-1 BCD**

Binary-coded decimal (BCD) is a numeric code in which *each* digit of a *decimal* number is represented by a separate group of bits. The most common BCD code is 8-4-2-1 BCD,

Table 13.1

Decimal digit	8-4-2-1 BCD code
0	0000
1	0001
2	0010
3	0011
4	0100
5	0101
6	0110
7	0111
8	1000
9	1001

in which each decimal digit is represented by a 4-bit binary number. It is called 8-4-2-1 BCD because the usual weights associated with a 4-bit binary number are determined by the positions of the bits *within each group*. Table 13.1 shows the 4-bit code groups used to represent a single decimal digit. 8-4-2-1 BCD is so widely used that it is common practice to refer to it simply as BCD, with the understanding that 8-4-2-1 BCD is meant.

Following are two examples showing how decimal numbers are encoded in 8-4-2-1 BCD:

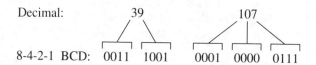

Decimal: 39   107

8-4-2-1 BCD: 0011   1001   0001   0000   0111

Note that each binary-coded decimal digit is a weighted code, but multiple-digit coding is not. For example, if the sequence of bits representing 39 in BCD, 00111001, were treated as a pure, weighted binary number, it would have value $57)_{10}$. Conversely, $39)_{10}$ expressed as a pure binary number is $100111)_2$.

As a means for representing numbers, 8-4-2-1 BCD is less *efficient* than pure binary, in the sense that it requires more bits. For example, 8 bits are necessary in 8-4-2-1 BCD to represent the 100 decimal numbers from 0 through 99. However, in pure binary only 7 bits are required to represent numbers up to 99, and 8 bits can represent the $2^8 = 256$ numbers from 0 through 255. The reason for this difference is the fact—important to remember—that the six bit patterns 1010, 1011, 1100, 1101, 1110, and 1111 are *never used* in 8-4-2-1 BCD. Those six patterns have decimal values 10 through 15, which are not single decimal digits. They are, therefore, said to be *not allowed*, or invalid, and any operation on BCD numbers that produces one of those patterns must be corrected.

The advantage of a BCD code is that it is easy to convert between it and decimal. The principal disadvantage of 8-4-2-1 BCD, besides its low efficiency, is that arithmetic operations are more complex than they are in pure binary. This fact is demonstrated in the next paragraph.

8-4-2-1 BCD Arithmetic

To add two BCD numbers, we simply add the binary numbers corresponding to each decimal digit. Following is an example:

$$
\begin{array}{r}
24 = 0010\ 0100 \\
+\ 13 = 0001\ 0011 \\
\hline
37 = 0011\ 0111 \\
\end{array}
$$

$$
\underbrace{3}\ \underbrace{7}
$$

A difficulty arises when the sum of two decimal digits is greater than 9, for in that case one of the disallowed bit patterns is produced. For example,

$$
\begin{array}{r}
25 = 0010\quad 0101 \\
+\ 38 = 0011\quad 1000 \\
\hline
63 \neq 0101\quad 1101 \\
\end{array}
$$

It also possible that the sum of two 4-bit numbers will produce a 5-bit number. For example,

$$
\begin{array}{r}
8 = \quad 1000 \\
+\ 9 = \quad 1001 \\
\hline
17 \quad 10001 \\
\end{array}
$$

5 bits, not allowed

We recognize that in both of these cases decimal arithmetic requires the propagation of a carry to the next-higher decimal digit. In 8-4-2-1 BCD arithmetic, we must also propagate that carry; in addition, we must *correct* the bit pattern. The correction is accomplished by adding $6)_{10} = 0110)_2$ to the affected bits. Thus, the rule for 8-4-2-1 BCD addition is as follows.

> If the sum of two BCD digits is greater than 1001, add 0110 and propagate a carry to the next most significant sum digit.

The carry is added to the sum bits of the next most significant digit of the sum. The carry bit itself is always produced automatically: If the sum bits are a 5-bit number, the fifth (highest-order) bit is the carry; otherwise, this bit results when the sum correction (adding 0110) is performed. In either case, sum correction is required. Note that adding the carry to the next-higher-order digit may itself produce a sum greater than 9, requiring that the same correction be repeated.

Example 13–1

Perform each of the following decimal additions in 8-4-2-1 BCD.

a. $\begin{array}{r} 25 \\ +\ 17 \end{array}$ b. $\begin{array}{r} 38 \\ +\ 48 \end{array}$ c. $\begin{array}{r} 172 \\ +\ 429 \end{array}$

d. $\begin{array}{r} 268 \\ +\ 245 \end{array}$ e. $\begin{array}{r} 579 \\ +\ 299 \end{array}$ f. $\begin{array}{r} 76 \\ +\ 52 \end{array}$

Solution

a.

25	0010	0101
+ 17	+ 0001	0111
42	0011	1100

$1100 > 9)_{10}$

add $6)_{10}$

	+ 0110
0011	①0010
+ 1 ←	
0100	0010

propagate carry

corrected sum

b.

38	0011	1000
+ 48	+ 0100	1000
86	0111	①0000

5-bit sum

+ 1 ← 0110	
1000	0110

propagate carry and add $6)_{10}$

corrected sum

c.

172	0001	0111	0010
+ 429	+ 0100	0010	1001
601	0101	1001	1011

$1011 > 9)_{10}$

add $6)_{10}$

		+ 0110
0101	1001	①0001
	+ 1 ←	

propagate carry

0101	1010	0001
	+ 0110	

$1010 > 9)_{10}$

add $6)_{10}$

0101	①0000	0001
+ 1 ←		
0110	0000	0001

propagate carry

corrected sum

d.

268	0010	0110	1000
+ 245	+ 0010	0100	0101
513	0100	1010	1101

1010 and $1101 > 9)_{10}$

add $6)_{10}$ to 1101

		+ 0110
0100	1010	①0011
	+ 1 ←	

propagate carry

0100	1011	0011
	+ 0110	

$1011 > 9)_{10}$

add $6)_{10}$

0100	①0001	0011
+ 1 ←		
0101	0001	0011

propagate carry

corrected sum

e.

579	0101	0111	1001
+ 299	+ 0010	1001	1001
878	0111	①0000	①0010

two 5-bit sums

+ 1 ←	+ 1 ←	
1000	0001	0010

propagate carries

	+ 0110	+ 0110
1000	0111	1000

add $6)_{10}$ to both digits

corrected sum

f.

76	0111	0110
+ 52	+ 0101	0010
128	1100	1000

$1100 > 9)_{10}$

add $6)_{10}$

	+ 0110	
1	0010	1000

corrected sum (with overflow carry)

Note in *(f)* that the sum of the two 2-digit decimal numbers is a 3-digit decimal number (128). Consequently, the BCD addition produces a carry bit out of the most significant sum, indicating an overflow.

Drill 13–1

How many times do sum corrections have to be made when adding $6267)_{10}$ and $1735)_{10}$ in 8-4-2-1 BCD?

Answer: 3

Subtraction in 8-4-2-1 BCD is performed by subtracting the binary numbers corresponding to each decimal digit. In practice, subtraction is performed using the complement methods discussed in Chapter 2. Since we are subtracting decimal digits, we must form the 9's (or 10's) complement of the decimal subtrahend and encode that number in 8-4-2-1 BCD. The resulting BCD numbers are then added following the rules already discussed. Remember how carries are treated in complementary arithmetic. Following is an example:

$$
\begin{array}{cc}
78 & 78 \\
-\ 25 & +\ 74 \leftarrow \text{9's complement of 25} \\
\hline
53 & \textcircled{1}\ 52 \\
& \llcorner\!\!\rightarrow +\ 1 \quad \text{end-around carry} \\
& \hline
& 53
\end{array}
$$

$$
\begin{array}{llll}
78)_{10} = & 0111 & 1000 & \\
74)_{10} = & 0111 & 0100 & \\
\hline
& 1110 & 1100 & 1100 \text{ and } 1110 > 9)_{10} \\
& & +\ 0110 & \text{add } 6)_{10} \text{ to } 1100 \\
\hline
& 1110 & \textcircled{1}\ 0010 & \\
& +\ 1\ \llcorner & & \text{propagate carry} \\
\hline
& 1111 & 0010 & 1111 > 9)_{10} \\
& +\ 0110 & & \text{add } 6)_{10} \\
\hline
& \textcircled{1}\ 0101 & 0010 & \\
& \llcorner\!\!\longrightarrow & +\ 1 & \text{end-around carry} \\
\hline
53)_{10} = & 0101 & 0011 & \text{corrected difference}
\end{array}
$$

Note that the binary complement of a bit pattern does *not* correspond to the decimal complement of the digit it represents. For example, the 9's complement of 3 is 6, and the bit pattern for 3 is 0011. However, the 1's complement of 0011 is 1100, which equals 12, not 6. Note also that the end-around carry is added to the least significant group of sum bits, *not* the least significant bit of the leftmost group.

13–3   EXCESS-THREE (XS-3) CODE

As can be seen in the foregoing example, 8-4-2-1 BCD arithmetic has the additional complication that the 9's or 10's complement of a number must be computed before a subtraction can be performed. The process would be much simpler if the 1's (or 2's) complement of a bit pattern produced a pattern corresponding to the 9's (or 10's) complement of the decimal digit it represents. A code with that property is said to be *self-complementing*. Excess-three (XS-3) is an example of a self-complementing BCD code. Its name is derived from the fact that each bit pattern has a decimal value that is

Table 13.2

Decimal digit	XS-3 code
0	0011
1	0100
2	0101
3	0110
4	0111
5	1000
6	1001
7	1010
8	1011
9	1100

greater by 3 than the decimal digit it actually represents. For example, XS-3 code for the decimal digit 5 is 1000, which has decimal value 8. Table 13.2 shows the entire code.

Note that the disallowed bit patterns in XS-3 code are those whose decimal values are 0, 1, 2, 13, 14, and 15—i.e., 0000, 0001, 0010, 1101, 1110, and 1111. Also note the self-complementing characteristic of the code. For example, the 9's complement of 3 is 6 and the 1's complement of 0110 is 1001, which is the bit pattern for 6. Also, the 10's complement of 3 is 7, and the 2's complement of 0110 is 1010, which is the bit pattern for 7. Excess-three is *not* a weighted code: There is no way to assign weights to the bit positions in such a way that sums of the weights consistently equal the decimal values of numbers in the code.

Example 13–2

Write the XS-3 code for $492)_{10}$ and for its 9's complement.

Solution. With reference to Table 13.2,

$$492)_{10} = 0111\ 1100\ 0101$$

The 9's complement of $492)_{10}$ is $507)_{10}$:

$$507)_{10} = 1000\ 0011\ 1010$$

Note that this bit pattern is the 1's complement of the bit pattern for $492)_{10}$.

Drill 13–2

1001 1010 1100 1010 in XS-3 code represents the 10's complement of what decimal number?
Answer: $3706)_{10}$.

Excess-Three Arithmetic

Since each XS-3 code group has a value that is three greater than the decimal digit it represents, the sum of two XS-3 code groups will have a value that is six greater than the sum of the two decimal digits they represent. Thus, each XS-3 sum must be corrected by *subtracting* $3)_{10}$, or $0011)_2$. If a carry is generated when adding two XS-3 code groups, it is an indication that an invalid code group has been created by the addition, and in that case the sum must be corrected by *adding* $3)_{10} = 0011)_2$. The carry is propagated to the next more significant code group of the sum, which is then corrected in the same way (by adding 3 if a new carry is generated and by subtracting 3 if not).

Example
13–3

Perform each of the following decimal additions in XS-3 code.

a.	4	**b.**	17	**c.**	108
	+ 3		+ 29		+ 296

Solution

a.

```
    4        0111  = 4)₁₀
  + 3      + 0110  = 3)₁₀
    7        1101    no carry
           - 0011    subtract 3)₁₀
7)₁₀  =      1010    corrected sum, in XS-3
```

b.

```
     17      0100   1010  = 17)₁₀
   + 29    + 0101   1100  = 29)₁₀
     46      1001①0110       carry generated
               + 1↵          propagate carry
             1010   0110
                  + 0011    add 3)₁₀ to correct 0110
             1010   1001
           - 0011            subtract 3)₁₀ to correct 1010
46)₁₀  =     0111   1001    corrected sum, in XS-3
```

c.

```
    108      0100   0011   1011  = 108)₁₀
  + 296      0101   1100   1001  = 296)₁₀
    404      1001   1111 ①0100      carry generated
                         + 1↵        propagate carry
             1001 ① 0000   0100      carry generated
                + 1↵                 propagate carry
             1010   0000   0100      no carry generated
                      + 0011 + 0011   add 3)₁₀ to correct 0000 and
             1010   0011   0111      0100
           - 0011                    subtract 3)₁₀ to correct 1010
404)₁₀  =    0111   0011   0111      corrected sum, in XS-3
```

Drill
13–3

How many times is 3)₁₀ added and how many times is it subtracted when correcting the XS-3 addition of 2865)₁₀ and 4079)₁₀?
Answer: Added twice and subtracted twice.

As already noted, XS-3 code is self-complementing, so subtraction using the method of complement addition is more direct than in 8-4-2-1 BCD. After forming the complement of the subtrahend, we add the individual code groups and correct the resulting sums, following the rules already given. Remember that 9's-complement subtraction requires an end-around carry to the least significant sum digit. In 10's-complement subtraction, a carry generated in the most significant sum digit is discarded. The 10's complement of a decimal number in XS-3 code is found by adding 1 to the least significant bit of the 9's complement.

Example
13-4

Perform the subtraction 735)$_{10}$ − 249)$_{10}$ in XS-3 code using the 9's-complement method.

Solution.   The subtrahend, 249)$_{10}$, in XS-3 code and its complements are:

Decimal	9's complement	XS-3 (249)			1's complement (750)		
249	750	0101	0111	1100	1010	1000	0011

735)$_{10}$ in XS-3 is 1010 0110 1000.

$$
\begin{array}{rr}
735 & 735 \\
-\ 249 & +\ 750 \\
\hline
486 & ①\ 485 \\
& ↳ +\ 1 \\
\hline
& 486
\end{array}
$$

	1010	0110	1000	= 735)$_{10}$
+	1010	1000	0011	= 750)$_{10}$
①	0100	1110	1011	
			→ + 1	end-around carry
	0100	1110	1100	
+ 0011				correct 0100 by adding 3)$_{10}$
	0111	1110	1100	correct 1110 and 1100 by subtracting 3)$_{10}$
	− 0011	− 0011		
486)$_{10}$ =	0111	1011	1001	corrected difference in XS-3

Drill
13-4

Repeat Example 13-4 using the 10's-complement method.

13-4   OTHER 4-BIT BCD CODES

2-4-2-1 Code

The 2-4-2-1 BCD is another self-complementing code that, unlike XS-3, has the additional feature that its 4-bit code groups are weighted. In this case, the weights are 2-4-2-1, meaning that two bit positions have the same weight (2). It is sometimes referred to as 2*-4-2-1 code, where the asterisk simply distinguishes one position with weight 2 from the other. Since two positions have the same weight, there are two possible bit patterns that could be used to represent some decimal digits, but only one of those patterns is actually assigned. Table 13.3 shows the code assignments.

Note the self-complementing characteristic of the code. An easy way to remember this code is to ignore the leftmost column (make it 0) and count in binary from 0 through 4 in the remaining bit positions. Digits 5 through 9 are then simply the 1's complements of the code groups for 4 through 0 (the 9's complements of 5 through 9).

There are numerous other weighted 4-bit BCD codes, each developed to have certain properties useful for special applications. Table 13.4 shows several, identified by the weights assigned to their bit positions. The $74\overline{2}\overline{1}$ code is somewhat different than the others, in that a 1 occurring in either of the two rightmost positions means that the weight of that position is *subtracted*, rather than added, to determine the decimal value.

Table 13.3

Decimal digit	2-4-2-1 code
0	0000
1	0001
2	0010
3	0011
4	0100
5	1011
6	1100
7	1101
8	1110
9	1111

Table 13.4
Weighted 4-bit BCD codes.

Decimal	3321	4221	5211	5311	5421	6311	7421	742$\overline{1}$
0	0000	0000	0000	0000	0000	0000	0000	0000
1	0001	0001	0001	0001	0001	0001	0001	0111
2	0010	0010	0011	0011	0010	0011	0010	0110
3	0011	0011	0101	0100	0011	0100	0011	0101
4	0101	1000	0111	0101	0100	0101	0100	0100
5	1010	0111	1000	1000	1000	0111	0101	1010
6	1100	1100	1001	1001	1001	1000	0110	1001
7	1101	1101	1011	1011	1010	1001	1000	1000
8	1110	1110	1101	1100	1011	1011	1001	1111
9	1111	1111	1111	1101	1100	1100	1010	1110

13–5 ERROR-DETECTING CODES

Parity

When binary data is transmitted and processed, it is—as are all electrical signals—susceptible to noise that can alter or distort its content. 1s may be effectively changed to 0s and 0s to 1s, as discussed in Chapter 6. Although a numeric BCD code must have a minimum of 4 bits to encode the 10 decimal digits, one or more additional bits are often added as an aid in detecting errors caused by noise. The most common of these is a *parity* bit that signifies whether the total number of 1s in a code group is odd or even. In an odd-parity system, the parity bit is made 0 or 1 as necessary to make the total number of 1s odd (counting the parity bit itself). In an even-parity system, the parity bit is chosen to make the total number of 1s even. Table 13.5 shows how parity bits would be added to 8-4-2-1 BCD code groups in both systems. When digital data is received, a *parity checking circuit* generates an error signal if the total number of 1s is odd in an even-parity system or if it is even in an odd-parity system. Note that this parity check always detects a *single* error (one bit changed from 0 to 1 or from 1 to 0) but may not detect two or more errors. Odd parity is used more often than even parity because even parity does not detect a situation where all 0s are created due to a short circuit or other fault condition.

Table 13.5
Odd and even parity in 8-4-2-1 BCD.

Decimal	8-4-2-1 BCD $ABCD$	Parity bit	
		Odd parity	Even parity
0	0000	1	0
1	0001	0	1
2	0010	0	1
3	0011	1	0
4	0100	0	1
5	0101	1	0
6	0110	1	0
7	0111	0	1
8	1000	0	1
9	1001	1	0

Example 13-5

Design a parity circuit that will assign a parity bit, P, to the 8-4-2-1 BCD code group $ABCD$ in an odd-parity system.

Solution.   With reference to the odd-parity column in Table 13.5, we see that

$$P = \overline{A}\,\overline{B}\,\overline{C}\,\overline{D} + \overline{A}\,\overline{B}CD + \overline{A}BC\overline{D} + \overline{A}BC\overline{D} + A\overline{B}\,\overline{C}D$$

Since the input to the circuit is an 8-4-2-1 BCD code group, we know that the (invalid) combinations having decimal values 10 through 15 will never occur. Therefore, those combinations can be treated as *don't cares:*

$$d = \Sigma\ 10,\ 11,\ 12,\ 13,\ 14,\ 15$$
$$= \Sigma\ A\overline{B}C\overline{D},\ A\overline{B}CD,\ AB\overline{C}\,\overline{D},\ AB\overline{C}D,\ ABC\overline{D},\ ABCD$$

Figure 13.1(a) shows the Karnaugh map for P. We see that the simplified expression is

$$P = AD + B\overline{C}\,\overline{D} + \overline{B}CD + BC\overline{D} + \overline{A}\,\overline{B}\,\overline{C}\,\overline{D}$$

Figure 13.1(b) shows the logic diagram. Figure 13.1(c) shows an alternate method for producing the parity bit, using coincidence (exclusive-NOR) gates: $P = A \odot B \odot C \odot D$. There is an exercise at the end of this chapter to show that, for decimal digits 0 through 9, this expression is equivalent to the simplified expression. Although the coincidence-gate expression does not take advantage of *don't-care* conditions, it does have the advantage that it can be implemented using three identical gates.

Drill 13-5

Find the simplified expression for the even parity bit assigned to an 8-4-2-1 BCD code group, $ABCD$.
Answer: $P = BCD + \overline{B}\,\overline{C}\,\overline{D} + \overline{A}\,\overline{B}C\overline{D} + A\overline{C}\,\overline{D}$, (or) $P = BCD + \overline{B}\,\overline{C}\,\overline{D} + \overline{A}\,\overline{B}C\overline{D} + A\overline{B}\,\overline{D}$.

Note that the parity circuit described in Example 13-5 can also be used as a parity-checking circuit for incoming data. The P output of the circuit is compared with the

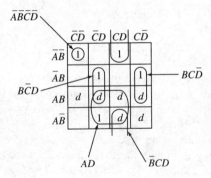

$$P = AD + B\bar{C}D + BC\bar{D} + \bar{B}CD + \bar{A}\bar{B}\bar{C}\bar{D}$$

(a)

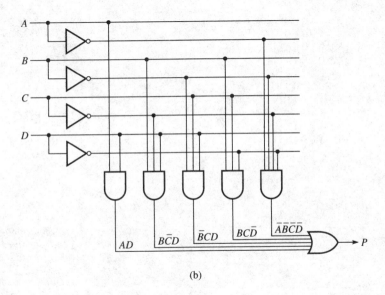

(b)

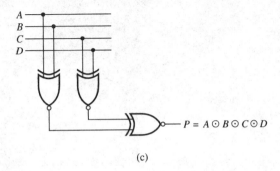

(c)

Figure 13.1
Example 13–5.

actual parity bit accompanying each code group. If these bits are not the same, there is a parity error.

Block Parity

When several binary words are transmitted or stored in succession, the resulting collection of bits can be regarded as a *block* of data, having rows and columns. For example, eight 6-bit words in succession form an 8 × 6 block. Parity bits can then be assigned to both rows and columns, as illustrated in Figure 13.2(*a*). This scheme makes it possible to *correct* any single error occurring in a data word and to *detect* any two errors in a word. For example, as shown in Figure 13.2(*b*), suppose the second bit in the third data word is in error, having been changed from a 0 to a 1. Then the number of 1s in the second column is even, and the number of 1s in the third row is even. Therefore, the (odd) parity bits in the second column and third row both detect the error, and we know that the second

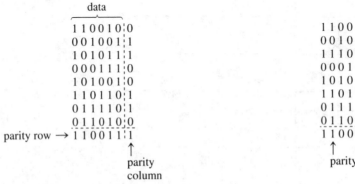

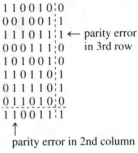

(a) *Adding (odd) parity bits to the rows and columns of a 6 × 8 data block.*

(b) *Parity errors in the second column and third row mean that the second bit in the third word is in error, so it can be corrected.*

(c) *Parity errors in two columns mean that two errors have occurred in one data word.*

Figure 13.2
Block parity examples.

Table 13.6
Five-bit BCD codes.

Decimal	63210	Two-out-of-five	Shift-counter	51111
0	00110	00011	00000	00000
1	00011	00101	00001	00001
2	00101	00110	00011	00011
3	01001	01001	00111	00111
4	01010	01010	01111	01111
5	01100	01100	11111	10000
6	10001	10001	11110	11000
7	10010	10010	11100	11100
8	10100	10100	11000	11110
9	11000	11000	10000	11111

bit in the third word must be changed from a 1 to a 0. Figure 13.2(c) shows how two errors in one data word are changed, transforming 000111 to 000100. Since the total number of 1s is still odd, the parity bit in the fourth row does not detect the error. However, the parity bits in the fifth and sixth columns do detect the error. We are not able to correct the error in this case because there is no information revealing the row where the errors occurred. The parity row is often called a *parity word,* and the block parity technique, also called word parity, is widely used for data stored on magnetic tape.

Five-Bit Codes

Table 13.6 shows some 5-bit BCD codes that have useful characteristics, including ease of error detection. 63210 code is a weighted code (except for decimal digit 0) that has the useful error-detecting feature that there are exactly two 1s in every code group. It is used for storing data on magnetic tape. Two-out-of-five code is unweighted and also has exactly two 1s in each code group. It is used in the telephone and communications industries. Shift-counter code, also called Johnson code, has the bit patterns produced by a 5-bit Johnson counter, as discussed in Chapter 12 (see Figure 12.18). 51111 code is similar to Johnson code but is weighted.

Biquinary Code

Biquinary code is the weighted 7-bit BCD code shown in Table 13.7. Note that each code group can be regarded as consisting of a 2-bit subgroup and a 5-bit subgroup, and each of these contains a single 1. Thus, it has the error-checking feature that each code group must contain exactly two 1s, and each subgroup must contain exactly one 1. The weights of the bit positions are 50 43210. Since there are two positions with weight 0, it is possible to encode decimal 0 with a group containing 1s, unlike other weighted codes.

Ring-Counter Code

The ring counter was discussed in Section 12–8 (Figure 12.16). Recall that the output of one and only one flip-flop is high at any given time. Therefore, a 10-bit ring counter will produce a sequence of 10-bit groups having the property that each group has a single 1. Ring-counter code, shown in Table 13.8, is the code obtained by assigning a decimal digit to each of those ten patterns. It is a weighted code (9876543210) because each bit position

Table 13.7

Decimal digit	Biquinary 50 43210
0	01 00001
1	01 00010
2	01 00100
3	01 01000
4	01 10000
5	10 00001
6	10 00010
7	10 00100
8	10 01000
9	10 10000

Table 13.8

Decimal digit	Ring-counter code 9876543210
0	0000000001
1	0000000010
2	0000000100
3	0000001000
4	0000010000
5	0000100000
6	0001000000
7	0010000000
8	0100000000
9	1000000000

has a weight equal to 1 of the 10 decimal digits. Although the code is inefficient (1024 numbers could be encoded in pure binary with 10 bits), it has excellent error-detecting properties and is simple to implement.

13–6 UNIT-DISTANCE CODES

A unit-distance code is one having the property that the bit patterns for two *consecutive* numbers differ in one bit position only. Note that this is certainly not a characteristic of binary numbers, nor of any of the codes we have discussed thus far. For example, the binary representations of the two consecutive numbers $3)_{10}$ and $4)_{10}$ are 011 and 100, which differ in all three bit positions.

Unit-distance codes are used in instrumentation and data-acquisition systems where linear or angular displacement is measured. In these systems, *transducers* are used to sense displacements and to produce coded numerical values proportional to angular or linear position. Figure 13.3 shows an example, in this case an optical *shaft encoder* used to measure angular displacement (rotation). The shaded segments prevent the passage of light through the disk, and the unshaded segments admit light. Light-sensitive devices aligned behind the disk produce a binary 1 when they receive light and a binary 0 when

Figure 13.3
An optical shaft encoder.

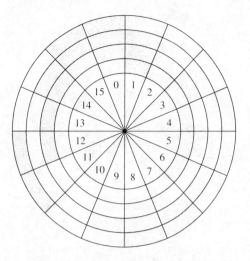

(a) The pattern of shaded and unshaded segments forms a unit-distance code.

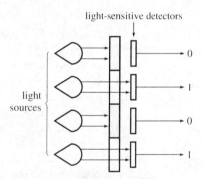

light-sensitive detectors

light sources

(b) End view, when the shaft has rotated to position 6 (0101).

shaded. The pattern of shaded and unshaded segments is such that the binary numbers produced when the disk rotates form a unit-distance code (the Gray code, to be discussed shortly).

The reason that a unit-distance code is used in these applications is that it reduces errors caused by small misalignments of the transducers. To illustrate, suppose that the code on the shaft encoder in Figure 13.3 were pure binary instead of a unit-distance code and that the shaft had rotated to a position *between* $7)_{10} = 0111)_2$ and $8)_{10} = 1000)_2$. Since all 4 bits must change when rotating from $7)_{10}$ to $8)_{10}$, it is possible for the shaft to be in an in-between position, where the most significant bit changes (from 0 to 1) before the remaining 3 bits change (from 1s to 0s). Thus, the binary number detected would be $1111)_2 = 15)_{10}$, which is greatly in error. On the other hand, since only one bit changes in a unit-distance code, the position detected by transducers using such a code would never differ by more than one adjacent position from the true position.

Table 13.9

Decimal	Gray code
0	0000
1	0001
2	0011
3	0010
4	0110
5	0111
6	0101
7	0100
8	1100
9	1101
10	1111
11	1110
12	1010
13	1011
14	1001
15	1000

Gray Code

The most popular unit-distance code is Gray code. The bit patterns assigned to decimal numbers 0 through 15 are shown in Table 13.9, which shows that any two adjacent code groups differ in one bit position only. Note that Gray code is *not* a BCD code. Table 13.9 shows only the first 16 Gray-coded numbers, but additional bits can be used to expand the table to an arbitrarily large number.

Gray code is an example of a *reflected* code. Notice that the two least significant bits for $4)_{10}$ through $7)_{10}$ are the mirror images of those for $0)_{10}$ through $3)_{10}$. Similarly, the three least significant bits for $8)_{10}$ through $15)_{10}$ are the mirror images of those for $0)_{10}$ through $7)_{10}$. In general, the n least significant bits for 2^n through $2^{n+1} - 1$ are the mirror images of those for 0 through $2^n - 1$. This observation gives us a method for counting in Gray code. However, it is often easier to determine the code pattern assigned to an arbitrary number or to decode an arbitrary Gray-code pattern, using one of the conversion methods discussed in later paragraphs.

Another property of Gray code is that the Gray-coded number corresponding to the decimal number $2^n - 1$, for any n, differs from Gray-coded 0 in one bit position only. For example, for $n = 2$, 3, and 4, we see that $3)_{10} = 0010)_{\text{Gray}}$, $7)_{10} = 0100)_{\text{Gray}}$, and $15)_{10} = 1000)_{\text{Gray}}$ each differ from 0000 in one bit position only. This property places the Gray code for the largest n-bit binary number at unit distance from 0. Thus, error minimization is maintained when the rotary shaft encoder is midway between its minimum and maximum positions, which are physically adjacent to each other on the disk. Referring to Figure 13.3(*a*), we see that those adjacent positions in this example are $0)_{10}$ and $15)_{10}$.

Gray-to-Binary Conversion

The procedure used to convert a Gray-code number to binary is as follows:

1. The most significant bit of the binary number is the same as the most significant bit of the Gray-code number. Write it down.

2. To obtain the next binary digit, perform an exclusive-OR operation between the bit just written down and the next Gray-code bit. Write down that result.

3. Repeat Step 2 until all Gray-code bits have been exclusive-ORed with binary digits. The sequence of bits that have been written down is the binary equivalent of the Gray-code number.

Example 13-6

Find the binary equivalent of the Gray-code number 1101011.

Solution

$$\text{Gray-code:} \quad 1 \quad 1 \quad 0 \quad 1 \quad 0 \quad 1 \quad 1$$

$$\text{binary:} \quad 1 \quad 0 \quad 0 \quad 1 \quad 1 \quad 0 \quad 1$$

Thus, the binary equivalent is $1001101)_2$, or $77)_{10}$.

Drill 13-6

Find the decimal equivalent of the Gray-code number 01011001.
Answer: 110.

Binary-to-Gray Conversion

As an aid to understanding the procedure for converting a binary number to its Gray-code equivalent, let the binary number be represented by

$$B_1 B_2 B_3 \cdots B_n$$

and its Gray-code equivalent by

$$G_1 G_2 G_3 \cdots G_n$$

Note that B_1 and G_1 are the most significant bits of the two numbers. The Gray-code bits are obtained from the binary bits as follows:

$$G_1 = B_1$$
$$G_2 = B_1 \oplus B_2$$
$$G_3 = B_2 \oplus B_3$$
$$\vdots$$
$$G_n = B_{n-1} \oplus B_n$$

Example 13-7

Find the Gray-code equivalent of $1011001)_2$.

Solution

$$\text{binary:} \quad 1 \oplus 0 \oplus 1 \oplus 1 \oplus 0 \oplus 0 \oplus 1$$

$$\text{Gray:} \quad 1 \quad 1 \quad 1 \quad 0 \quad 1 \quad 0 \quad 1$$

Thus, the Gray-code equivalent is 1110101.

Drill 13-7

Find the Gray-code equivalent of $83)_{10}$.
Answer: 1111010.

Table·13.10

Decimal digit	Excess-3 Gray code
0	0010
1	0110
2	0111
3	0101
4	0100
5	1100
6	1101
7	1111
8	1110
9	1010

Another way to write down the Gray-code bits in a binary-to-Gray conversion is to exclusive-OR the bits of the binary number with those of the binary number shifted right one position. The least significant bit of the shifted number is discarded, and the most significant bit of the Gray-code number is the same as the most significant bit of the original binary number. To illustrate, we find the Gray-code equivalent of the binary number given in Example 13–7:

$$
\begin{array}{ll}
1011001 & \text{binary number} \\
\oplus \ 101100\rlap{/}1 & \text{binary number shifted right} \\
\hline
1110101 & \text{discard}
\end{array}
$$

Gray-code number:

Excess-Three Gray Code

Gray code can be used for binary-coded decimals by encoding each decimal digit using the Gray-code bit patterns for 0 through 9. However, Gray-coded 9 (1101) is not at unit distance from 0. Excess-three Gray code, in which each decimal digit is encoded with the Gray-code pattern of the decimal digit that is greater by 3, does have unit distance between the patterns for 0 and 9. The code is shown in Table 13.10.

13–7   ALPHANUMERIC CODES

Alphanumeric codes are used to encode the characters of the alphabet in addition to the decimal digits. They are used primarily for transmitting data between a computer and its input/output (I/O) devices, such as printers, keyboards, and video display terminals. Because the number of bits used in most alphanumeric codes is much greater than that required to encode the 10 decimal digits and the 26 alphabetic characters, the codes can include bit patterns for a wide range of other symbols and functions. Mathematical symbols such as $+$, $-$, $=$, $>$, and $<$ can be included. Also, special *control characters*, such as carriage return, are included. Two early 5-bit alphanumeric codes were the Baudot code, used in punched paper tape, and the Hollerith code, used with punched cards ("IBM" cards). The most popular modern codes are ASCII and EBCDIC, which are described in the paragraphs that follow.

ASCII Code

American Standard Code for Information Interchange (ASCII), referred to as "AS-KEE," is a widely used 7-bit alphanumeric code. Since the number of different bit patterns that can be created with 7 bits is $2^7 = 128$, ASCII can be used to encode both the lowercase and uppercase characters of the alphabet (52 symbols), the 10 decimal digits, 33 control characters, and 33 special symbols.

Table 13.11 shows the ASCII code groups. For ease of presentation, these are listed with the three most significant bits of each group along the top row and the four least

Table 13.11
ASCII code.

	Most significant bits							
	000	001	010	011	100	101	110	111
0000	NUL	DEL	Space	0	@	P	'	p
0001	SOH	DC1	!	1	A	Q	a	q
0010	STX	DC2	"	2	B	R	b	r
0011	ETX	DC3	#	3	C	S	c	s
0100	EOT	DC4	$	4	D	T	d	t
0101	END	NAK	%	5	E	U	e	u
0110	ACK	SYN	&	6	F	V	f	v
0111	BEL	ETB	'	7	G	W	g	w
1000	BS	CAN	(	8	H	X	h	x
1001	VT	ESC	)	9	I	Y	i	y
1010	LF	SUB	*	:	J	Z	j	z
1011	VT	ESC	+	;	K	[	k	{
1100	FF	FS	,	<	L	\	l	|
1101	CR	GS	–	=	M	]	m	}
1110	SO	RS	.	>	N	∧	n	~
1111	SI	US	/	?	O	_	o	DLE

Least significant bits (row labels at left)

ACK	Acknowledge	EOT	End of transmission	NUL	Null
BEL	Bell	ESC	Escape	RS	Record separator
BS	Backspace	ETB	End of transmission block	SI	Shift in
CAN	Cancel	ETX	End text	SO	Shift out
CR	Carriage return	FF	Form feed	SOH	Start of heading
DC1–DC4	Direct control	FS	Form separator	STX	Start text
DEL	Delete idle	GS	Group separator	SUB	Substitute
DLE	Data link escape	HT	Horizontal tab	SYN	Synchronous idle
EM	End of medium	LF	Line feed	US	Unit separator
ENQ	Enquiry	NAK	Negative acknowledge	VT	Vertical tab

significant bits along the left column. For example, the ASCII code for the decimal digit 9 is 011 1001 and that for A is 100 0001. Notice that the four least significant bits for the decimal digits are the same as 8-4-2-1 BCD code for those digits. Thus, for the decimal digits ASCII code is easy to convert to numeric BCD: Simply delete the first three bits.

EBCDIC Code

Extended Binary-Coded Decimal Interchange Code (EBCDIC, or "eb-see-dick") is an 8-bit alphanumeric code. Since 256 (2^8) different bit patterns can be formed with 8 bits, EBCDIC code can be used to encode all the symbols and control characters found in ASCII as well as many others. In fact, many of the bit patterns are unassigned. There are 50 control characters assigned to the 64 patterns found from 00000000 through 00111111, leaving 14 unassigned. Table 13.12 shows just the alphanumeric codes in EBCDIC, which are in the range from 01000000 through 11111111 (192 bit patterns). Only 89 of

Table 13.12
*EBCDIC code.**

Least significant bits	Most significant bits											
	0100	0101	0110	0111	1000	1001	1010	1011	1100	1101	1110	1111
0000	space	&	—									0
0001			/		a	j			A	J		1
0010					b	k	s		B	K	S	2
0011					c	l	t		C	L	T	3
0100					d	m	u		D	M	U	4
0101					e	n	v		E	N	V	5
0110					f	o	w		F	O	W	6
0111					g	p	x		G	P	X	7
1000					h	q	y		H	Q	Y	8
1001					i	r	z		I	R	Z	9
1010	¢	!		:								
1011	.	$	,	#								
1100	<	*	%	@								
1101	(	)	—	'								
1110	+	;	>	=								
1111	|	¬	?	"								

*0000 0000 through 0011 1111 not shown. There are 50 control characters in this range.

these are actually assigned. Note that the ten decimal digits are all encoded with 1111 as the four most significant bits and with 8-4-2-1 code as the four least significant bits. Thus, the decimal digits can be converted to 8-4-2-1 BCD by simply deleting the first four bits of the EBCDIC code groups.

13-8   ENCODERS

When we say that the input to a digital device is a decimal number, we mean that the device actually has 10 binary inputs, one for each of the 10 decimal digits. Normally, only one of those 10 inputs will be high at any given time: the one corresponding to whichever decimal digit is, in fact, the decimal input at that time. Similarly, a device whose input is one of the 26 alphabetic characters actually has 26 binary inputs. An *encoder* is a device whose inputs are decimal digits and/or alphabetic characters and whose outputs are the coded representations of those inputs. For example, a decimal-to-8-4-2-1-BCD encoder has 10 binary inputs and 4 binary outputs, the outputs being the 4 bits of 8-4-2-1 BCD code groups.

Keyboard Encoders

Figure 13.4 shows a typical *keyboard encoder* consisting of a *diode matrix*, used to encode the 10 decimal digits in 8-4-2-1 BCD. RS flip-flops are used to store the BCD output. When a key corresponding to one of the decimal digits is depressed, a positive voltage forward-biases selected diodes connected to the set and reset inputs of the flip-flops. The diodes are arranged so that each flip-flop sets or resets as necessary to produce the 4-bit code corresponding to the decimal digit. For example, when the 5-key is depressed, the diodes connected to the set-inputs of Q_4 and Q_1 are forward-biased, as are those connected to the reset-inputs of Q_8 and Q_2. Thus, the output is 0101. Note that the diode configuration at each set and reset input is essentially a diode OR gate. Diode matrix encoders are found on printed-circuit boards in many devices having a keyboard as the means of data entry.

Figure 13.5 shows how four OR gates can be used to encode decimal inputs into 8-4-2-1 BCD. The OR-gate outputs produce the 8-4-2-1 BCD code groups $B_8B_4B_2B_1$. The table of 8-4-2-1 codes is effectively four truth tables implemented by the gates. As discussed earlier, each decimal input is actually a binary signal whose value is 1 when the decimal digit it represents is the selected input and whose value is 0 otherwise. Notice that there is no explicit input for decimal digit 0; the BCD output is 0000 when decimal inputs 1 through 9 are all zero.

Priority Encoders

The encoders we have discussed thus far operate correctly *provided* that one and only one decimal input is high at any given time. In some practical systems, two or more decimal inputs may inadvertently become high at the same time. For example, a person operating a keyboard might depress one key before releasing another. What is the output of a conventional encoder when more than one input is high at the same time? Suppose, for example, that D_3 and D_4 were both high in Figure 13.5. We see that the output would be $0111 = 7)_{10}$, which is neither 3 nor 4. A *priority encoder* responds to just one input among those that may be simultaneously high, in accordance with some priority system. The most common priority system is based on the relative magnitudes of the inputs:

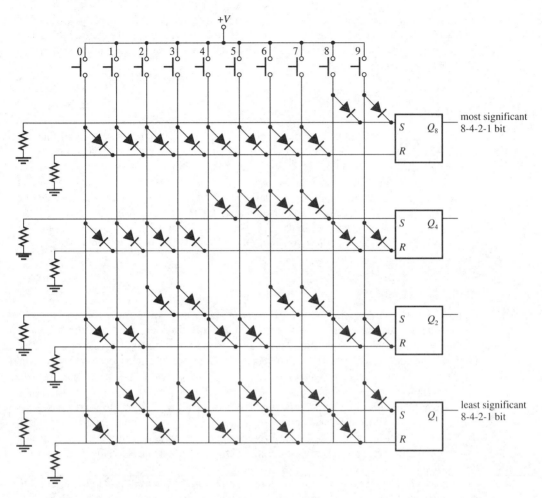

Figure 13.4
A keyboard encoder employing a diode matrix to convert decimal inputs to 8-4-2-1 BCD outputs. When a decimal key is depressed, forward-biased diodes cause appropriate flip-flops to be set or reset.

whichever decimal input is largest is the one that is encoded. Thus, in the foregoing example, a priority encoder would encode decimal 4 if both 3 and 4 were simultaneously high.

In some practical applications, priority encoders may have several inputs that are routinely high at the same time, and the principal function of the encoder in those cases is to select the input with highest priority. This function is called *arbitration*. A common example is found in computer systems where there are numerous input devices, several of which may attempt to supply data to the computer at the same time. A priority encoder is used to enable the input device having the highest priority among those competing for access to the computer at the same time.

For a priority encoder that assigns the highest priority to the input with the largest magnitude, the logic that must be implemented can be deduced from the code table in Figure 13.5. Referring to that table, we see that the least significant output bit, B_1, is high

Figure 13.5
Decimal-to-8-4-2-1-BCD encoder.

decimal	8-4-2-1 BCD output			
input	B_8	B_4	B_2	B_1
D_0	0	0	0	0
D_1	0	0	0	1
D_2	0	0	1	0
D_3	0	0	1	1
D_4	0	1	0	0
D_5	0	1	0	1
D_6	0	1	1	0
D_7	0	1	1	1
D_8	1	0	0	0
D_9	1	0	0	1

$$B_8 = D_8 + D_9 \quad B_4 = D_4 + D_5 + D_6 + D_7 \quad B_2 = D_2 + D_3 + D_6 + D_7$$
$$B_1 = D_1 + D_3 + D_5 + D_7 + D_9$$

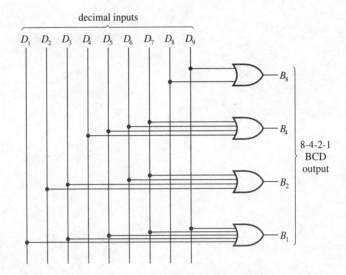

only when the decimal input is an odd number (1, 3, 5, 7, or 9). If the decimal input is $1)_{10}$, then B_1 should be high, provided that decimal inputs 2, 4, 6, and 8 are low. (If any of the latter were high, then B_1 should be low, since 2, 4, 6, and 8 have higher priorities.) Following this line of reasoning, we can therefore fully characterize B_1 under magnitude priority by stating that B_1 should be high if:

D_1 is high and D_2, D_4, D_6, and D_8 are low

OR D_3 is high and D_4, D_6, and D_8 are low

OR D_5 is high and D_6 and D_8 are low

OR D_7 is high and D_8 is low

OR D_9 is high

Thus, $B_1 = D_1\overline{D_2}\overline{D_4}\overline{D_6}\overline{D_8} + D_3\overline{D_4}\overline{D_6}\overline{D_8} + D_5\overline{D_6}\overline{D_8} + D_7\overline{D_8} + D_9$. Clearly, this logic can be implemented by four AND gates and a 5-input OR gate. The logic for B_2, B_4, and B_8 can be similarly derived from the code table (Exercise 13.45).

Figure 13.6
The 74147 priority encoder.

decimal inputs									BCD output			
D_1	D_2	D_3	D_4	D_5	D_6	D_7	D_8	D_9	B_8	B_4	B_2	B_1
1	1	1	1	1	1	1	1	1	1	1	1	1
x	x	x	x	x	x	x	x	0	0	1	1	0
x	x	x	x	x	x	x	0	1	0	1	1	1
x	x	x	x	x	x	0	1	1	1	0	0	0
x	x	x	x	x	0	1	1	1	1	0	0	1
x	x	x	x	0	1	1	1	1	1	0	1	0
x	x	x	0	1	1	1	1	1	1	0	1	1
x	x	0	1	1	1	1	1	1	1	1	0	0
x	0	1	1	1	1	1	1	1	1	1	0	1
0	1	1	1	1	1	1	1	1	1	1	1	0

$x = don't\ care$

(a) Truth table.

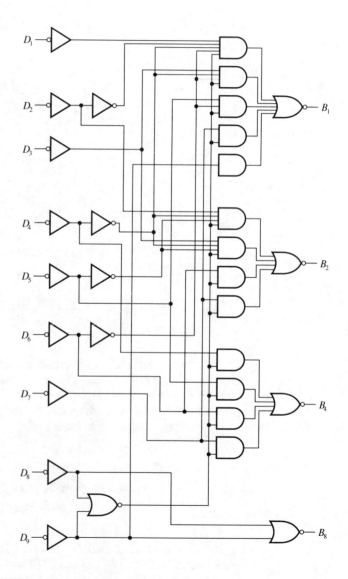

(b) Logic diagram.

Figure 13.6 shows the logic diagram and truth table of the 74147 priority encoder. In this device, the inputs *and* outputs are *active-low*. For example, if D_8 is low (active) and D_9 is high (inactive), the output is 0111 (the 1's complement of $1000 = 8)_{10}$). The truth table clearly shows that the magnitudes of the decimal inputs determine their priorities: If any decimal input is active, it is encoded provided all higher-value inputs are inactive and regardless of the states of all lower-value inputs.

13–9 DECODERS

A "decimal output" from a digital device, like a decimal input, is actually 10 binary signals, only 1 of which is normally high, or active, at any given time. The input to a decimal *decoder* is a binary code group and the output is the decimal digit it represents — i.e., one active signal line — in the sense just described. Some decoders have (three) binary inputs and *octal* outputs: eight output lines, one and only one of which is active at any given time. An 8-4-2-1 BCD-to-decimal decoder is also called a *4-line-to-10-line* decoder, or simply a 4-to-10 decoder, since its input is a 4-bit code group and its output is one of the 10 decimal digits. Since 16 numbers can be represented by 4 bits, 4-to-16 decoders are also available. These devices are properly referred to as *binary*-to-decimal decoders, since 8-4-2-1 BCD does not include decimal digits 10 through 15. The logic necessary to implement a 4-to-10 decoder is easily deduced from the code table, as shown in Figure 13.7.

Example 13–8

Write the logic expression for each of the following:

a. The zero-output (D_0) of an XS-3-to-decimal decoder.
b. The eight output (D_8) of a 2-4-2-1-BCD-to-decimal decoder.
c. The E output of an ASCII-to-alphabetic character decoder.

Solution

a. $0)_{10}$ in XS-3 code is 0011. Letting the input bits be *ABCD,* we have $D_0 = \overline{A}\overline{B}CD$.
b. Since $8)_{10}$ in 2-4-2-1 is 1110,

$$D_8 = B_2 B_4 B_2 \overline{B}_1$$

c. From Table 13.11, E in ASCII code is 1000101. Letting the seven input bits be $A_1 A_2 A_3 A_4 A_5 A_6 A_7$, we have

$$E = A_1 \overline{A}_2 \overline{A}_3 \overline{A}_4 A_5 \overline{A}_6 A_7$$

Drill 13–8

Write the logic expression for the *V*-output of an EBCDIC-to-alphabetic character decoder.
Answer: $A_1 A_2 A_3 \overline{A}_4 \overline{A}_5 A_6 \overline{A}_7 A_8$.

In practical applications, decoders are often used as a means of enabling a selected one of several devices in a computer system. For example, the computer may generate a *device number* in binary code, such as number 6 (0110), indicating that, for instance, a printer designated number 6 is to be activated. The decoder receives 0110 as input and activates the D_6 line, which enables the printer. Decoder outputs are often active-low because the enable input to many devices is active-low. Figure 13.8(*a*) and (*b*) shows the logic diagram and function table for the 74138 3-line-to-8-line decoder, which has active-low outputs and is itself enabled by $G_1 \overline{G}_{2A} \overline{G}_{2B}$.

decimal digit	8-4-2-1 BCD code				logic expression
	B_8	B_4	B_2	B_1	
0	0	0	0	0	$D_0 = \bar{B}_8\,\bar{B}_4\,\bar{B}_2\,\bar{B}_1$
1	0	0	0	1	$D_1 = \bar{B}_8\,\bar{B}_4\,\bar{B}_2\,B_1$
2	0	0	1	0	$D_2 = \bar{B}_8\,\bar{B}_4\,B_2\,\bar{B}_1$
3	0	0	1	1	$D_3 = \bar{B}_8\,\bar{B}_4\,B_2\,B_1$
4	0	1	0	0	$D_4 = \bar{B}_8\,B_4\,\bar{B}_2\,\bar{B}_1$
5	0	1	0	1	$D_5 = \bar{B}_8\,B_4\,\bar{B}_2\,B_1$
6	0	1	1	0	$D_6 = \bar{B}_8\,B_4\,B_2\,\bar{B}_1$
7	0	1	1	1	$D_7 = \bar{B}_8\,B_4\,B_2\,B_1$
8	1	0	0	0	$D_8 = B_8\,\bar{B}_4\,\bar{B}_2\,\bar{B}_1$
9	1	0	0	1	$D_9 = B_8\,\bar{B}_4\,\bar{B}_2\,B_1$

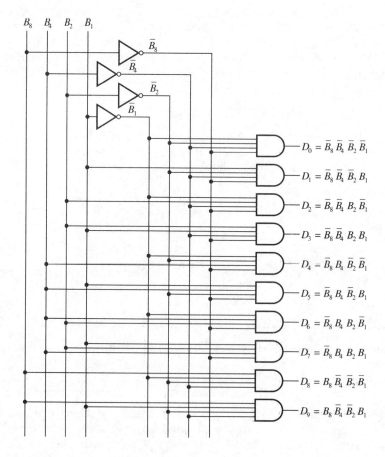

Figure 13.7
An 8-4-2-1 BCD-to-decimal decoder (4-line-to-10-line decoder).

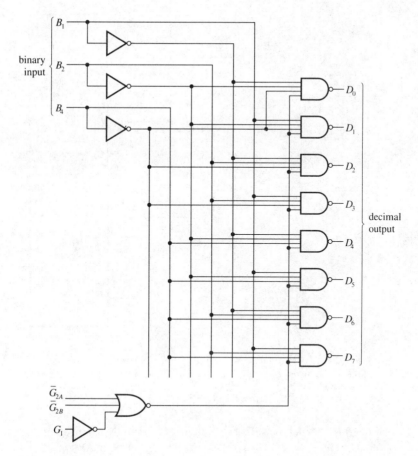

binary input B_1, B_2, B_4

decimal output

$\overline{G}_{2A}$
$\overline{G}_{2B}$
G_1

(a) Logic diagram.

enable inputs			binary input			decimal outputs							
G_1	$\overline{G}_{2A}$	$\overline{G}_{2B}$	B_4	B_2	B_1	D_0	D_1	D_2	D_3	D_4	D_5	D_6	D_7
x	1	x	x	x	x	1	1	1	1	1	1	1	1
x	x	1	x	x	x	1	1	1	1	1	1	1	1
0	x	x	x	x	x	1	1	1	1	1	1	1	1
1	0	0	0	0	0	0	1	1	1	1	1	1	1
1	0	0	0	0	1	1	0	1	1	1	1	1	1
1	0	0	0	1	0	1	1	0	1	1	1	1	1
1	0	0	0	1	1	1	1	1	0	1	1	1	1
1	0	0	1	0	0	1	1	1	1	0	1	1	1
1	0	0	1	0	1	1	1	1	1	1	0	1	1
1	0	0	1	1	0	1	1	1	1	1	1	0	1
1	0	0	1	1	1	1	1	1	1	1	1	1	0

$x = don't\ care$

(b) Function table.

Figure 13.8
The 74138 3-line-to-8-line decoder.

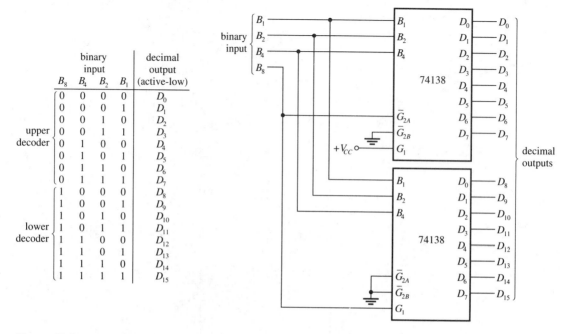

	binary input				decimal output
	B_8	B_4	B_2	B_1	(active-low)
	0	0	0	0	D_0
	0	0	0	1	D_1
	0	0	1	0	D_2
upper	0	0	1	1	D_3
decoder	0	1	0	0	D_4
	0	1	0	1	D_5
	0	1	1	0	D_6
	0	1	1	1	D_7
	1	0	0	0	D_8
	1	0	0	1	D_9
	1	0	1	0	D_{10}
lower	1	0	1	1	D_{11}
decoder	1	1	0	0	D_{12}
	1	1	0	1	D_{13}
	1	1	1	0	D_{14}
	1	1	1	1	D_{15}

Figure 13.9

Connecting two 74138 3-to-8 decoders to obtain a 4-to-16 decoder. Note that the upper decoder (for D_0 through D_7) is enabled only when B_8 is low and that the lower decoder (for D_8 through D_{15}) is enabled only when B_8 is high.

Binary-to-decimal decoding is often *expanded* by connecting the binary inputs through appropriate logic gates to the enabling circuitry of two or more decoders. Many decoders, such as the 74138, have built-in logic at their enabling inputs to facilitate expansion. Figure 13.9 shows how two 74138 decoders can be connected to function as a 4-line-to-16-line decoder. Note that the most significant input bit (B_8) is connected to $\overline{G}_{2A}$ on the upper decoder (for D_0 through D_7) and to G_1 on the lower decoder (for D_8 through D_{15}). Thus, when B_8 is low, the upper decoder is enabled and D_8 through D_{15} are all high (inactive). The code table shown in the figure confirms that B_8 is low for all decimal outputs from D_0 through D_7. When B_8 is high, the lower decoder is enabled and D_0 through D_7 are all inactive.

13–10   BCD-TO-SEVEN-SEGMENT DECODERS

Figure 13.10(*a*) shows a *seven-segment display*, consisting of seven light-emitting segments. The segments are designated by letters *a* through *g*, as shown. By illuminating various combinations of segments, as shown in 13.10(*b*), the numerals 0 through 9 can be displayed. Seven-segment displays are commonly constructed with light-emitting diodes (LEDs) and with liquid-crystal displays (LCDs). LEDs generally provide greater illumination levels but require much greater power than LCDs. Figure 13.10(*c*) and (*d*) shows two types of LED displays: the common-anode and the common-cathode types. In the common-anode type, a low voltage applied to an LED cathode allows current to flow

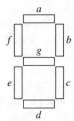

(a) A seven-segment display, showing the letters used to designate the segments.

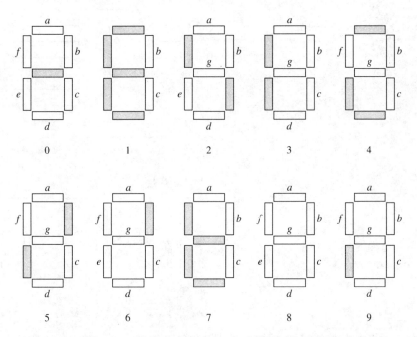

(b) By causing different combinations of the segments to illuminate (shown unshaded), the numerals 0 through 9 can be displayed.

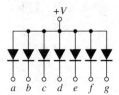

(c) A common-anode LED display. When the cathode of an LED is low, current flows through that LED and it emits light.

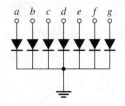

(d) A common-cathode LED display. When the anode of an LED is high, current flows through that LED and it emits light.

Figure 13.10
The seven-segment display.

through the diode, which causes it to emit light. In the common-cathode type, a high voltage applied to an LED anode causes the current flow and light emission.

An 8-4-2-1 BCD-to-seven-segment decoder is a logic circuit whose input is a BCD number and whose output drives a seven-segment display. Thus, the input consists of 4 bits, $B_8B_4B_2B_1$, and the output consists of 7 lines to drive segments a through g. See Figure 13.11(a). The decoder must activate only those segments that display the numeral corresponding to the BCD input. Figure 13.11(b) shows a function table for such a decoder, based on the illuminated segments shown in Figure 13.10(b). Since a 1 (high) on any output line means that that output is activated, we are assuming the display is of

Figure 13.11
8-4-2-1 BCD-to-seven-segment decoder.

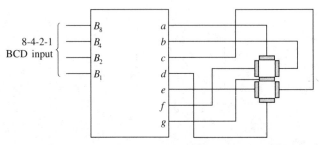

(a) An 8-4-2-1 BCD-to-seven-segment decoder has 4 inputs and 7 outputs.

decimal digit	8-4-2-1 BCD B_8	B_4	B_2	B_1	a	b	c	d	e	f	g
0	0	0	0	0	1	1	1	1	1	1	0
1	0	0	0	1	0	1	1	0	0	0	0
2	0	0	1	0	1	1	0	1	1	0	1
3	0	0	1	1	1	1	1	1	0	0	1
4	0	1	0	0	0	1	1	0	0	1	1
5	0	1	0	1	1	0	1	1	0	1	1
6	0	1	1	0	1	0	1	1	1	1	1
7	0	1	1	1	1	1	1	0	0	0	0
8	1	0	0	0	1	1	1	1	1	1	1
9	1	0	0	1	1	1	1	1	0	1	1

(b) Function table.

$$a = \bar{B}_8\bar{B}_4\bar{B}_2\bar{B}_1 + \bar{B}_8\bar{B}_4B_2\bar{B}_1 + \bar{B}_8\bar{B}_4B_2B_1 + \bar{B}_8B_4\bar{B}_2B_1 + \bar{B}_8B_4B_2\bar{B}_1$$
$$+ \bar{B}_8B_4B_2B_1 + B_8\bar{B}_4\bar{B}_2\bar{B}_1 + B_8\bar{B}_4\bar{B}_2B_1$$
$$= \Sigma\ 0, 2, 3, 5, 6, 7, 8, 9 \quad d = \Sigma\ 10, 11, 12, 13, 14, 15$$

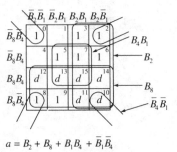

$$a = B_2 + B_8 + B_1B_4 + \bar{B}_1\bar{B}_4$$

(c) Derivation of simplified expression for driving segment a.

the common-cathode type. Figure 13.11*(c)* shows a Karnaugh map used to simplify the logic expression for driving segment a. Note that we use *cell numbers* in the map, as discussed in Section 5.5 (Figure 5.10). This application is ideally suited to the cell-number convention, since each BCD input represents a numerical value. As usual, the *don't cares* are 10 through 15. Derivation of simplified expressions for driving segments *b* through *g* is left as an exercise at the end of this chapter.

Many practical BCD-to-seven-segment decoders incorporate other features useful to display applications. Since light-emitting diodes require considerable power, decoders often contain output *drivers,* capable of supplying that power. In a typical open-collector decoder/driver (the 7446), each of the seven-output segment lines can sink 40 mA and dissipate 320 mW. Some decoders offer variations in the type and style of output display provided. For example, some display numerals 6 and 9 without the "tails" shown in Figure 13.10*(b)* (i.e., as b and q). Many also provide special displays when the BCD input has values 10 through 14—for example, ⊏ (10), ⊐ (11), ⊔ (12), ⊑ (13), and ⊨ (14). Practical decoder designs for circuits such as these often do not utilize the simplest Boolean expressions [Figure 13.11*(c)*] for the segment drivers.

Another useful feature found in many decoder/drivers is *ripple blanking*. In multi-digit displays, it is often desirable to suppress (blank out) leading and/or trailing zeros. For example, in a 6-decimal-digit display, it may be desired to display 004.200 simply as 4.2. In the 7446 through 7449 series and in the 9317 series TTL decoder/drivers, each decoder is equipped with an active-low ripple-blanking input ($\overline{\text{RBI}}$) and an active-low ripple-blanking output ($\overline{\text{RBO}}$). These can be interconnected in multi-digit display to suppress leading and/or trailing zeros.

13–11 CODE CONVERTERS

Code converters are logic circuits whose inputs are bit patterns representing numbers (or characters) in one code and whose outputs are the corresponding representations in a different code. For example, an 8-4-2-1-BCD-to-XS-3 code converter has four BCD input lines, $B_8B_4B_2B_1$, and four XS-3 output lines, *ABCD*. When the input is 0000, for instance, the output should be 0011, and so forth. To design a code converter, we use a code table, treating it as a truth table, to express *each* output as a Boolean algebra function of *all* inputs. In the example just cited, we would treat an 8-4-2-1-BCD-to-XS-3 code table as four truth tables to derive expressions for *A, B, C,* and *D*. Each of these expressions would, in general, contain all four input variables, B_8, B_4, B_2, and B_1. For example, the simplified expression for *A* is $A = B_8 + B_4B_2 + B_4B_1$. Thus, the code converter is actually four logic circuits, one for each output.

The logic expressions derived for the code converter can be simplified using the usual techniques, including Karnaugh maps and *don't cares*. When the input is 8-4-2-1 BCD, it is convenient to use the cell-numbering scheme described in Section 5.5 for Karnaugh maps. If the input is an unweighted code, cell numbering can still be used, but the cell numbers must correspond to the input combinations as if they *were* 8-4-2-1 weighted code. For example, the XS-3 number $\overline{A}B\overline{C}D = 0101$, which represents $2)_{10}$, is assigned cell number 5, *not* cell number 2. Be careful to determine which input combinations, if any, will never occur and can be treated as *don't cares*. Of course, it is *input* bit patterns that determine *don't cares,* not output patterns. The next example illustrates the procedure for designing a code converter. The designs for other converters are left as exercises at the end of the chapter.

***Example
13–9***

Design an 8-4-2-1-BCD-to-XS-3 code converter.

Solution. The 8-4-2-1-BCD-to-XS-3 code table is shown in Figure 13.12(a). Using the cell-numbering scheme and treating the code table as four truth tables, we see that $A = \Sigma\ 5, 6, 7, 8, 9$; $B = \Sigma\ 1, 2, 3, 4, 9$; $C = \Sigma\ 0, 3, 4, 7, 8$; and $D = \Sigma\ 0, 2, 4, 6, 8$. Since the input is 8-4-2-1 BCD, the *don't care* conditions in every case are $d = \Sigma\ 10, 11, 12, 13, 14, 15$. The Karnaugh maps for obtaining simplified expressions are shown in Figure 13.12(b), and the complete logic circuit is shown in Figure 13.12(c).

decimal	8-4-2-1 BCD input				excess-3 output			
	B_8	B_4	B_2	B_1	A	B	C	D
0	0	0	0	0	0	0	1	1
1	0	0	0	1	0	1	0	0
2	0	0	1	0	0	1	0	1
3	0	0	1	1	0	1	1	0
4	0	1	0	0	0	1	1	1
5	0	1	0	1	1	0	0	0
6	0	1	1	0	1	0	0	1
7	0	1	1	1	1	0	1	0
8	1	0	0	0	1	0	1	1
9	1	0	0	1	1	1	0	0

$A = \Sigma\ 5, 6, 7, 8, 9$
$B = \Sigma\ 1, 2, 3, 4, 9$
$C = \Sigma\ 0, 3, 4, 7, 8$
$D = \Sigma\ 0, 2, 4, 6, 8$
$d = \Sigma\ 10, 11, 12, 13, 14, 15$

(a)

$A = B_8 + B_4 B_2 + B_4 B_1$

$B = B_8 B_1 + B_8 B_2 + B_4 \bar{B_2} \bar{B_1}$

$C = B_2 B_1 + \bar{B_2} \bar{B_1}$

$D = \bar{B_1}$

(b)

Figure 13.12
(Example 13–9) Design of an 8-4-2-1-BCD-to-XS-3 code converter.

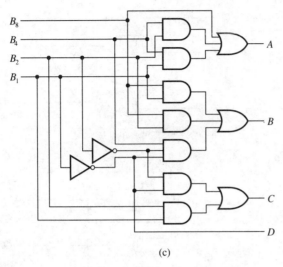

(c)

Figure 13.12
(Continued)

Drill
13–9

Write the cell numbers occupied by 1s in the Karnaugh maps for an 8-4-2-1-BCD-to-2-4-2-1 code *(ABCD)* converter.
Answer: $A = \Sigma$ 5, 6, 7, 8, 9; $B = \Sigma$ 4, 6, 7, 8, 9; $C = \Sigma$ 2, 3, 5, 8, 9; $D = \Sigma$ 1, 3, 5, 7, 9.

Binary-to-Gray and Gray-to-Binary Converters

The same procedure used to design the code converter of the previous example can be used to design binary-to-Gray or Gray-to-binary code converters. Of course, there are no *don't care* conditions, and the number of input and output lines is as many as necessary to accommodate a given word size. However, by referring to the methods described in Section 13.6 for converting between binary and Gray, it is easy to deduce the logic circuitry needed to perform those conversions. Figure 13.13*(a)* shows a 4-bit binary-to-Gray code converter, whose design consists simply of three exclusive-OR gates. This design follows immediately from the procedure used to convert binary numbers to Gray-coded numbers: $G_1 = B_1$, $G_2 = B_1 \oplus B_2$, $G_3 = B_2 \oplus B_3$, $G_4 = B_3 \oplus B_4$, where G_1 and B_1 are the most significant bits (MSB) and G_4 and B_4 are the least significant bits (LSB). The converter can be expanded in an obvious way to accommodate any number of bits. Figure 13.13*(b)* shows a Gray-to-binary code converter. Again, the design follows immediately from the procedure used to convert Gray-coded numbers to binary numbers: $B_1 = G_1$, $B_2 = B_1 \oplus G_2$, $B_3 = B_2 \oplus G_3$, $B_4 = B_3 \oplus G_4$. It is left as an exercise to show that these designs are identical to those that would be obtained using the Karnaugh-map method.

Multi-Digit BCD/Binary Code Converters

It is obvious that a code converter is not necessary when converting between binary and 8-4-2-1 BCD, provided that the range of decimal numbers is restricted to 0 through 9—

Figure 13.13
The design of binary-to-Gray and Gray-to-binary code converters can be deduced from the procedures used to convert between binary and Gray-coded numbers.

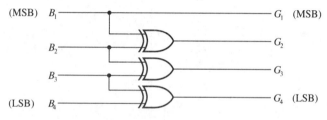

(a) A 4-bit binary-to-Gray code converter.

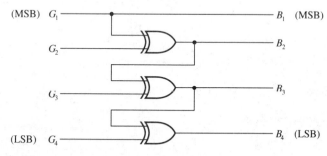

(b) A 4-bit Gray-to-binary code converter.

that is, to a *single* decimal digit. However, since the representations of multi-digit decimal numbers in 8-4-2-1 BCD and binary are quite different, code converters are required in those cases. For example, if the input to an 8-4-2-1-BCD-to-binary code converter is decimal 25 (0010 0101), the output must be $11001)_2$.

Integrated circuits are available to convert between multi-digit BCD and binary. The 74484 converts 2-decimal digit BCD (8-bit) inputs to binary outputs. The 74485 converts binary inputs to 2-decimal digit (8-bit) BCD outputs. These circuits can be cascaded to perform conversions between larger numbers.

Binary/BCD conversions are most often encountered in connection with computer applications. Numerical data transmitted in BCD form from input devices must be converted to binary so that arithmetic operations can be performed on it. The binary results of arithmetic operations must be converted to BCD for transmission to output devices. Therefore, conversions are often accomplished using major components of the computer system itself rather than special converter circuits. For example, every computer system is designed so it can retrieve data from a *memory* component. Thus, conversions can be accomplished by accessing a *read-only memory* (ROM, to be described in a later chapter) containing a *look-up table*. This table contains the binary equivalent of every BCD number in the range of numbers to be converted and/or the BCD equivalent of every binary number in the range. In some systems, conversions are accomplished by the computer itself through execution of a specially designed program. This is called a *software* conversion, as opposed to the hardware conversion performed by logic circuits.

13–12   ANSI/IEEE STANDARD SYMBOLS

The qualifying symbol for a decoder, encoder, or code converter is X/Y, where X represents the input code and Y represents the output code. The letters X and Y can be

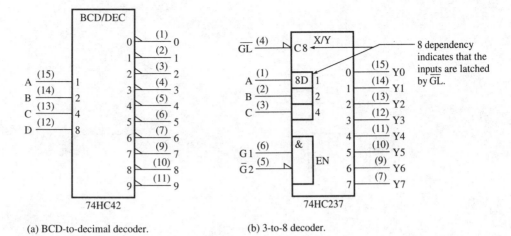

(a) BCD-to-decimal decoder.

(b) 3-to-8 decoder.

Figure 13.14
Examples of ANSI/IEEE logic symbols for decoders.

used in the symbol, or one or both of them can be replaced by characters that identify the codes they represent, such as BIN, DEC, etc. The ANSI/IEEE standards permit input and output lines to be identified in several different ways. If the input code is binary, the lines can be identified by the weights of the bit positions they represent: 1, 2, 4, ... Alternatively, the binary grouping symbol, }, can be used, as is discussed shortly in connection with a new type of dependency notation. In cases where the input or output code is apparent from the way in which the lines are identified, the letters X or Y in the qualifying symbol need not be replaced by characters representing the code. Output lines can be numbered consecutively inside the outline and identified outside the outline by characters that refer to a function table defining the code.

Figure 13.14 shows examples of the logic symbols for two integrated-circuit decoders: the 74HC42 BCD-to-decimal decoder and the 74HC237 3-to-8 decoder. Note that the BCD/DEC decoder has active-low outputs. The 3-to-8 decoder has latched inputs, as evident from the C8 dependency (8D). The EN dependency in this example means that all outputs are inactive (low) when G1 is low or $\overline{G2}$ is high. Note that EN = G1 · $\overline{G2}$ and that the three-state symbol does not appear at the output lines.

AND Dependency

Another of the eleven types of dependencies is called AND dependency and is symbolized by the letter G. As we shall see, it is often used in the logic symbols of decoders to show that the outputs depend on the AND function of the binary inputs. Figure 13.15(*a*) and (*b*) shows examples of AND dependency notation and logic diagrams to which they are equivalent. In (*a*), note that a numeral (1 in this example) is used to associate input *b* with inputs *a* and *c*. The bar over the 1 at input *c* means that *b* is ANDed with the complement of *c*. Part (*b*) is an example showing how an output (*b*) affecting an input (*a*) is symbolized with G-dependency. Figure 13.15(*c*) shows how G-dependency can be used with a binary grouping symbol applied to the input lines of a decoder or code converter. In this case, the input lines are numbered consecutively, beginning with 0, rather than being identified by their weights.

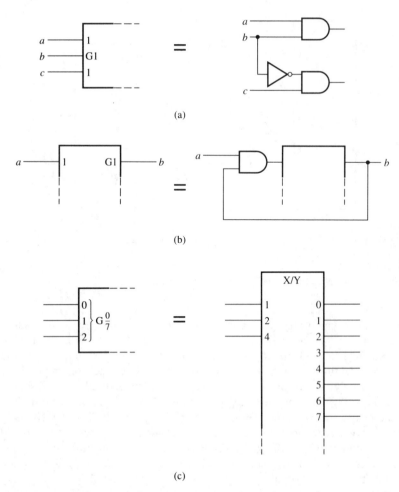

(a)

(b)

(c)

Figure 13.15
Examples of AND (G) dependency notation.

13–13   SPICE EXAMPLES

***Example
13–10***

Construct a SPICE model of a parity checking circuit for 3-bit binary data in an even-parity system.

Solution.   The correct even-parity bit for a 3-bit number, *ABC*, can be generated by the operation $A \oplus B \oplus C$. To demonstrate the validity of the SPICE model that we will develop, we compare $A \oplus B \oplus C$ with $P = 0$ for every combination of *A*, *B*, and *C*. For those combinations for which *P* should equal 1, an active-low error signal will be generated. This comparison can be performed by an exclusive-NOR (coincidence) gate, since its output will be 0 whenever its inputs, $A \oplus B \oplus C$ and *P*, do not coincide. To implement the exclusive-OR operations required in the model, we use NOR gates and inverters to form

$$\overline{\overline{A + B} + \overline{\overline{A} + \overline{B}}} = (A + B)(\overline{A} + \overline{B}) = A \oplus B$$

Figure 13.16(*a*) shows NMOS subcircuits for the NOR and inversion (INV) operations and shows how they are nested within an exclusive-OR (XOR) subcircuit. Figure 13.16(*b*) shows the model of the parity checking circuit, utilizing the XOR subcircuits. Also shown is the input data file. Note that an inverter is connected to the output of the X3 exclusive-OR gate

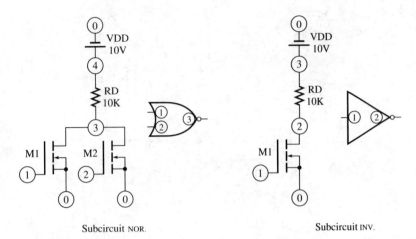

Subcircuit NOR.   Subcircuit INV.

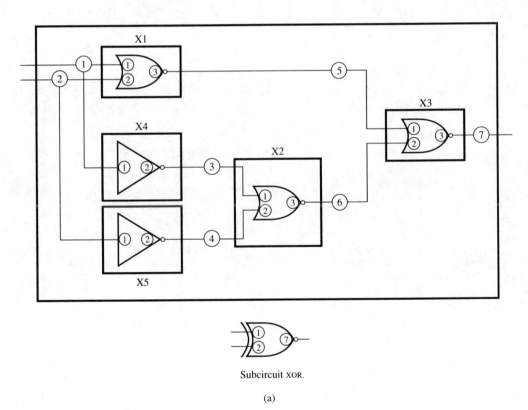

Subcircuit XOR.

(a)

Figure 13.16
Example 13–10. (Continued on pages 520–521.)

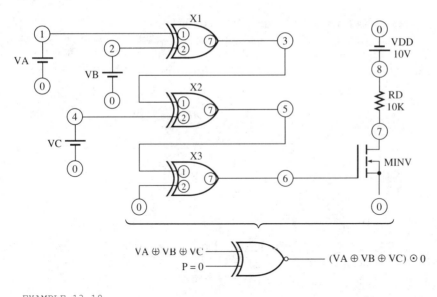

$$VA \oplus VB \oplus VC$$
$$P = 0$$
$$(VA \oplus VB \oplus VC) \odot 0$$

```
EXAMPLE 13.10
.SUBCKT EXOR 1 2 7
.SUBCKT NOR 1 2 3
M1 3 1 0 0 MOS
M2 3 2 0 0 MOS
.MODEL MOS NMOS KP = 0.25E - 3 VTO = 2
RDD 3 4 10K
VDD 4 0 10V
.ENDS NOR
.SUBCKT INV 1 2
M1 2 1 0 0 MOS
.MODEL MOS NMOS KP = 0.25E - 3 VTO = 2
RD 2 3 10K
VDD 3 0 10V
.ENDS INV
X1 1 2 5 NOR
X2 3 4 6 NOR
X3 5 6 7 NOR
X4 1 3 INV
X5 2 4 INV
.ENDS EXOR
X1 1 2 3 EXOR
X2 3 4 5 EXOR
X3 5 0 6 EXOR
MINV 7 6 0 0 MOS
.MODEL MOS NMOS KP = 0.25E - 3 VTO = 2
RD 7 8 10K
VDD 8 0 10V
VA 1 0
VB 2 0
VC 4 0
.DC VC 0 10 10 VB 0 10 10
.PRINT DC V(1) V(2) V(4) V(7)
.END
```

(b)

Figure 13.16
(Continued)

EXAMPLE 13.10

```
****      DC TRANSFER CURVES                  TEMPERATURE =    27.000 DEG C


***************************************************************************


   VC            V(1)         V(2)          V(4)         V(7)


 0.000E+00      0.000E+00    0.000E+00    0.000E+00    1.000E+01
 1.000E+01      0.000E+00    0.000E+00    1.000E+01    4.904E-01
 0.000E+00      0.000E+00    1.000E+01    0.000E+00    4.905E-01
 1.000E+01      0.000E+00    1.000E+01    1.000E+01    1.000E+01


   VC            V(1)         V(2)          V(4)         V(7)


 0.000E+00      1.000E+01    0.000E+00    0.000E+00    4.905E-01
 1.000E+01      1.000E+01    0.000E+00    1.000E+01    1.000E+01
 0.000E+00      1.000E+01    1.000E+01    0.000E+00    1.000E+01
 1.000E+01      1.000E+01    1.000E+01    1.000E+01    4.904E-01
```

V1	V2	V4	correct	circuit	output
A	B	C	P	P	V(7)
0	0	0	0	0	1
0	0	1	1	0	0 ←
0	1	0	1	0	0 ←
0	1	1	0	0	1
1	0	0	1	0	0 ← — errors
1	0	1	0	0	1
1	1	0	0	0	1
1	1	1	1	0	0 ←

(c)

Figure 13.16
(Continued)

to form a coincidence gate, which is used to compare $P = 0$ with $A \oplus B \oplus C$. The simulation is performed with VA = 0 V while VB and VC are stepped through all combinations of 0 V and 5 V and is repeated with VA = 5 V (not shown). Thus, all combinations of *ABC*, from 000 through 111, are generated. The results are shown in Figure 13.16(c). Also shown is the corresponding logic table, which demonstrates that the model correctly detects the four cases where the parity bit should have been 1.

Example 13–11   Construct a SPICE model of a 3-bit binary-to-Gray code converter using TTL open-collector NAND gates and wired (dotted) AND logic.

Solution.   Referring to Figure 13.13(a), we see that a 3-bit binary-to-Gray code converter requires the use of 2 exclusive-OR gates. As shown in Figure 13.17(a), p. 523, we can implement the exclusive-OR operation using two inverters and two open-collector

NAND gates whose outputs are wire-ANDed, since $A \oplus B = (A + B) \cdot \overline{AB} = (\overline{\overline{A}\overline{B}}) \cdot (\overline{AB})$. Figure 13.17(b) shows NAND and inverter (INV) subcircuits that we will nest within an exclusive-OR (XOR) subcircuit. Figure 13.17(c) shows how the nodes of the nested subcircuits are connected to form the XOR subcircuit. Note that a pull-up resistor (RUP) is connected to the output at node 5 to implement wire ANDing. Figure 13.17(d) shows how the XOR subcircuits are connected to form the code converter. The 1-MΩ resistors R1 and R2 are used to ensure that there are at least two connections to nodes 5 and 6, as required by SPICE. The input data file is shown in Figure 13.17(e). The simulation is performed with the most significant bit (B_1 = V1) set to 0 while V2 and V3 are stepped through 0 V and 5 V and then repeated with V1 set to 5 V (not shown). Thus, results are obtained for all eight combinations of the binary inputs, from 000 through 111. The results are shown in Figure 13.17(f). Also shown is the corresponding logic table, from which we see that binary-to-Gray code conversion is achieved.

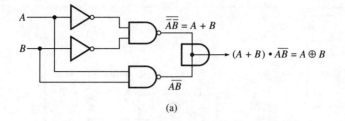

(a)

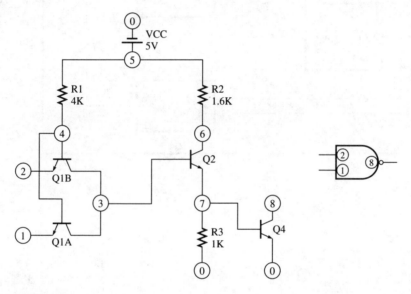

Subcircuit NAND

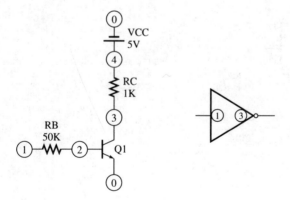

Subcircuit INV

(b)

Figure 13.17
Example 13–11. (Continued on pages 523 and 524.)

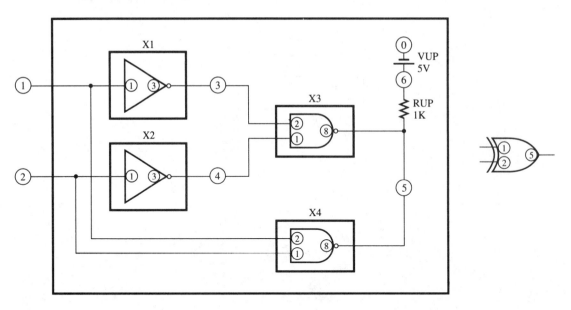

Subcircuit XOR

(c)

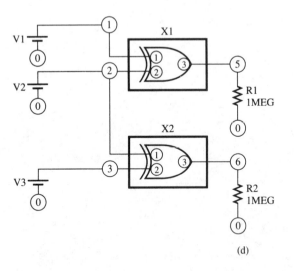

(d)

```
EXAMPLE 13.11
.SUBCKT XOR 1 2 5          X1 1 3 INV
.SUBCKT NAND 1 2 8         X2 2 4 INV
Q1A 3 4 1 TRANS           X3 4 3 5 NAND
Q1B 3 4 2 TRANS           X4 2 1 5 NAND
R1 4 5 4K                 RUP 5 6 1K
VCC 5 0 5V                VUP 6 0 5V
R2 5 6 1.6K               .ENDS XOR
Q2 6 3 7 TRANS            X1 1 2 5 XOR
R3 7 0 1K                 X2 2 3 6 XOR
Q4 8 7 0 TRANS            R1 5 0 1MEG
.MODEL TRANS NPN          R2 6 0 1MEG
.ENDS NAND                V3 3 0
.SUBCKT INV 1 3           V2 2 0
Q1 3 2 0 TRANS            V1 1 0
RB 1 2 50K                .DC V3 0 5 5 V2 0 5 5
RC 3 4 1K                 .PRINT DC V(1) V(2) V(3)
VCC 4 0 5V                .PRINT DC V(1) V(5) V(6)
.MODEL TRANS NPN          .END
.ENDS INV
```

(e)

Figure 13.17
(Continued)

EXAMPLE 13.11

****

V3	V(1)	V(2)	V(3)
0.000E+00	0.000E+00	0.000E+00	0.000E+00
5.000E+00	0.000E+00	0.000E+00	5.000E+00
0.000E+00	0.000E+00	5.000E+00	0.000E+00
5.000E+00	0.000E+00	5.000E+00	5.000E+00

V3	V(1)	V(5)	V(6)
0.000E+00	0.000E+00	3.378E-02	3.378E-02
5.000E+00	0.000E+00	3.378E-02	4.995E+00
0.000E+00	0.000E+00	4.995E+00	4.995E+00
5.000E+00	0.000E+00	4.995E+00	3.378E-02

V3	V(1)	V(2)	V(3)
0.000E+00	5.000E+00	0.000E+00	0.000E+00
5.000E+00	5.000E+00	0.000E+00	5.000E+00
0.000E+00	5.000E+00	5.000E+00	0.000E+00
5.000E+00	5.000E+00	5.000E+00	5.000E+00

V3	V(1)	V(5)	V(6)
0.000E+00	5.000E+00	4.995E+00	3.378E-02
5.000E+00	5.000E+00	4.995E+00	4.995E+00
0.000E+00	5.000E+00	3.378E-02	4.995E+00
5.000E+00	5.000E+00	3.378E-02	3.378E-02

| B_1 | B_2 | B_3 | G_1 | G_2 | G_3 |
(V_1)	(V_2)	(V_3)	(V_1)	(V_5)	(V_6)
0	0	0	0	0	0
0	0	1	0	0	1
0	1	0	0	1	1
0	1	1	0	1	0
1	0	0	1	1	0
1	0	1	1	1	1
1	1	0	1	0	1
1	1	1	1	0	0

(f)

Figure 13.17
(Continued)

EXERCISES

Section 13–2

13.1 Encode each of the following decimal numbers in 8-4-2-1 BCD.
(a) 25
(b) 640
(c) 1853

13.2 Determine the decimal value of each of the following 8-4-2-1 BCD numbers.
(a) 0001 0011
(b) 1000 0000 0000 0111
(c) 0000 1001 1001

13.3 **(a)** How many bits are required to represent decimal numbers from 0 through 99,999 in 8-4-2-1 BCD?

(b) How many numbers could be represented in pure binary using the same number of bits as required in *(a)*?

13.4 A computer system has a word size that permits it to store binary numbers having a decimal value up to 65,535. What is the largest decimal number that can be represented by one word if an 8-4-2-1 BCD format is used instead?

13.5 Perform each of the following decimal additions in 8-4-2-1 BCD.
(a) $\begin{array}{r} 38 \\ +21 \\ \hline \end{array}$
(b) $\begin{array}{r} 17 \\ +65 \\ \hline \end{array}$
(c) $\begin{array}{r} 49 \\ +49 \\ \hline \end{array}$
(d) 3 + 897

13.6 Perform each of the following decimal additions in 8-4-2-1 BCD.
(a) $\begin{array}{r} 105 \\ +234 \\ \hline \end{array}$
(b) $\begin{array}{r} 188 \\ +288 \\ \hline \end{array}$
(c) 4996 + 8

13.7 Perform each of the following decimal subtractions in 8-4-2-1 BCD using the 9's-complement method.
(a) $\begin{array}{r} 82 \\ -59 \\ \hline \end{array}$
(b) $\begin{array}{r} 697 \\ -103 \\ \hline \end{array}$
(c) $\begin{array}{r} 258 \\ -\ 16 \\ \hline \end{array}$

13.8 Perform each of the following subtractions in 8-4-2-1 BCD using the 10's-complement method.
(a) $\begin{array}{r} 75 \\ -58 \\ \hline \end{array}$
(b) $\begin{array}{r} 968 \\ -420 \\ \hline \end{array}$
(c) 352 − 16

Section 13–3

13.9
(a) Find the decimal equivalent of each of the following XS-3 BCD numbers:
   (i) 1011 0011 1000
   (ii) 1010 0101 0111 1001
(b) Each of the following XS-3 coded numbers represents the 9's complement of what decimal number?
   (i) 1100 1001 0011
   (ii) 0100 0110 1000 0111

13.10 Express each of the following numbers in XS-3 BCD code. Also write the XS-3 code for the 9's complement of each number.
(a) $725)_{10}$
(b) $9381)_{10}$
(c) $01101101)_2$
(d) $146)_8$
(e) $2AF)_{16}$

13.11 Perform each of the following additions in XS-3 code.

(a)   5
        +4

(b)   34
       +27

(c)   607
       +195

13.12 Perform each of the following additions in XS-3 code.

(a)   15
       +31

(b)   792
       +118

(c) 306 + 45

13.13 Perform each of the following subtractions using the 9's-complement method in XS-3 code.

(a)   86
       −34

(b)   509
       −172

(c)   876
       −498

13.14 Perform each of the following subtractions using the 10's-complement method in XS-3 code.

(a)   9
        −2

(b)   35
       −17

(c) 468 − 21

Section 13–4

13.15 Express each of the following numbers in 2-4-2-1 code. Also write the 2-4-2-1 code for the 9's complement of each number.
(a) $263)_{10}$
(b) $1001)_{10}$
(c) $AB)_{16}$
(d) $101101)_2$

13.16
(a) Find the decimal equivalent of each of the following 2-4-2-1 coded numbers:
  (i) 0100 1110 0000
  (ii) 1101 1011 0001 0100

(b) Each of the following 2-4-2-1 coded numbers represents the 9's complement of what decimal number?
  (i) 1100 0011 1111
  (ii) 0000 1101 0001 1011

13.17 Which, if any, of the 4-bit BCD codes shown in Table 13.4 are self-complementing?

13.18 Find the decimal equivalent of each of the following $742\overline{1}$-coded numbers:
(a) 0101 1001 0111
(b) 1110 0100 0000 1111

Section 13–5

13.19 Show that implementation of the odd parity bit for 8-4-2-1 BCD code using coincidence gates, $P = A \odot B \odot C \odot D$, is equivalent to the simplified expression for P given in Example 13–5. [*Hint:* The coincidence function is associative: $(A \odot B) \odot C = A \odot B \odot C$.]

13.20 Show that the even parity bit for 8-4-2-1 BCD code can be implemented by $P = A \oplus B \oplus C \oplus D$. [*Hint:* The exclusive-OR function is associative: $(A \oplus B) \oplus C = A \oplus B \oplus C$.]

13.21 Design a logic circuit that generates an even-parity bit for 2-4-2-1 BCD code. Draw the logic diagram.

13.22 Design a logic circuit that generates an odd-parity bit for 3-3-2-1 BCD code. Draw the logic diagram.

13.23 The last row in the accompanying data block is an odd-parity row and the rightmost column is an odd-parity column. If there is a correctible error in the data, locate and correct it.

                    0110111
                    1010001
                    0111000
                    1101011
                    0101100
                    1110110
                    0010111
                    1010111
                    1010010

13.24 Suppose that two errors occurred in one column of data in a data block having a parity row and a parity column. How would this occurrence be revealed? Could the errors be corrected?

13.25 Serial data in 63210 code is clocked into a binary counting circuit (a counter). The counter produces a binary number in parallel form representing the total number of 1s that have occurred in all the data that have been clocked in, up to any given time. After each code group representing a decimal digit is clocked into the counter, the new contents of the counter are checked to determine if there is a single error in the bits of that code group. Describe how the check could be implemented.

13.26 The word size in a certain computer system is such that it can represent 2,097,152 numbers in pure binary. What is the largest decimal number that can be represented by one word if the format is changed to biquinary BCD code?

Section 13–6

13.27 Using Table 13.9 and the reflecting property of Gray code, write the Gray-code bit patterns for decimal numbers 16 through 31. (Note that $16)_{10}$ is 11000 in Gray code; it would be helpful to rewrite Table 13.9 in 5-bit code groups.)

13.28 Describe how to obtain the Gray-code pattern for the decimal number 2^n, where n is any integer.

13.29 Convert the following Gray-coded numbers to binary.
(a) 10100
(b) 011110
(c) 10011011

13.30 Convert the following Gray-coded numbers to decimal.
(a) 10101
(b) 011011
(c) 11101001

13.31 Convert the following binary numbers to Gray code.
(a) 11101
(b) 0100110
(c) 10110101

13.32 Convert the following decimal numbers to Gray code.
(a) 22
(b) 63
(c) 135

13.33 Convert the following decimal numbers to binary-coded decimals in XS-3 Gray code.
(a) 25
(b) 104
(c) 9637

13.34 Convert the following XS-3 Gray-coded BCD numbers to decimal.
(a) 0100 1110
(b) 0010 0111 1010

Section 13–7

13.35 Decode the following ASCII message:

1010010   1101111   1110101   1110100
1100101   0100000   0110001   0110000

13.36 Encode the following phrase in ASCII:

AND gate 2

13.37 Encode the following message in EBCDIC:

Call 29

13.38 Decode the following EBCDIC message:

11100010   10000101   10010101   10000100
01000000   01011011   11110101

13.39 Decode the following message, which is written in EBCDIC, using hexadecimal notation:

C2   96   A7   40   F1   F3   F8   F4

13.40 Encode the following message in EBCDIC using hexadecimal notation:

Apt. 3D

Section 13–8

13.41 Draw a schematic diagram showing a diode matrix for a keyboard encoder that can be used to encode the decimal digits 0 through 9 in 2-4-2-1 code.

13.42 Draw a schematic diagram showing a diode matrix for a keyboard encoder that can be used to encode the decimal digits 0 through 9 in XS-3 BCD code.

13.43 Draw a logic diagram for a decimal-to-XS-3 encoder.

13.44 Draw a logic diagram for a decimal (0 to 9)-to-Gray code encoder. (Although most such encoders are opto-mechanical devices such as shown in Figure 13.3, your design should use logic gates.)

13.45 Draw a logic diagram for a decimal-to-8-4-2-1-BCD priority encoder. Inputs and outputs should be active-high.

13.46 Draw a logic diagram for a decimal-to-XS-3-BCD priority encoder. Inputs and outputs should be active-low.

Section 13–9

13.47 Draw a logic diagram for an XS-3-to-decimal decoder. Inputs and outputs should be active-high.

13.48 Draw a logic diagram for a 2-4-2-1-to-decimal decoder. Inputs should be active-high and outputs, active-low.

13.49 Figure 13.18 shows the 74154 4-line-to-16-line decoder, which is enabled when both $\overline{G}_1$ and $\overline{G}_2$ are low. Draw a logic diagram showing how two of these decoders could be connected to function as a 5-line-to-32-line decoder.

13.50 The 74154 4-to-16 decoder in Figure 13.18 has active-high inputs and active-low outputs. In a certain application, the 74154 is to be used as an 8-4-2-1 BCD-to-decimal decoder. If the binary input is invalid (10 through 15), all outputs are to be inactive. Draw a logic diagram (including external logic gates) showing how the 74154 could be used in this way.

Section 13–10

13.51 Derive the simplest possible expressions for driving segments b through g in an 8-4-2-1-BCD-to-seven-segment decoder for decimal digits 0 through 9. Outputs should be active-high.

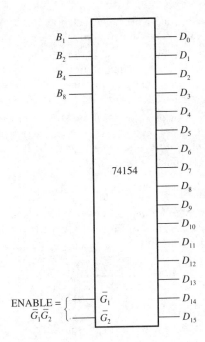

Figure 13.18
Exercises 13.49 and 13.50.

13.52 Derive the simplest possible expressions for driving segments a through g in an 8-4-2-1-BCD-to-seven-segment decoder that can be used for a common-anode seven-segment LED display of decimal digits 0 through 9. Decimal 6 should be displayed as ᗷ and decimal 9 as ꝯ.

Section 13–11

13.53 Design an 8-4-2-1-BCD-to-2-4-2-1-BCD code converter. Use the simplest possible logic expressions and draw the logic diagram.

13.54 Find the simplest possible logic expressions for a 2-4-2-1-BCD-to-51111-BCD code converter.

13.55 Show that the circuit for the 4-bit binary-to-Gray code converter in Figure 13.13(*a*) can be obtained by using a code table and Karnaugh maps to simplify the logic expressions for G_1 through G_4.

13.56 Show that the circuit for bits B_1 through B_3 in the 4-bit Gray-to-binary code converter in Figure 13.13(*b*) can be obtained by using a code table and Karnaugh maps to simplify the logic expressions for B_1 through B_3.

13.57 The input to a BCD-to-binary code converter is a 3-decimal digit 8-4-2-1 BCD number. How many output lines must the converter have?

13.58 A binary-to-8-4-2-1-BCD code converter has 12 binary input lines. How many output lines must it have in order to convert the full range of input values?

Section 13–12

13.59 With reference to the circuit whose logic symbol is shown in Figure 13.19, answer the following questions:

(a) What type of circuit does the symbol represent?

(b) Under what circumstances are all outputs high?

(c) When the circuit is enabled and the input is $ABCD = 0110$, which output is low?

13.60 With reference to the circuit whose logic symbol is shown in Figure 13.20, answer the following questions:

(a) What type of circuit does the symbol represent?

(b) Under what circumstances are all outputs low?

(c) What is the significance of the label 20D at pin 2?

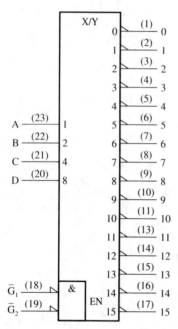

Figure 13.19
Exercise 13.59.

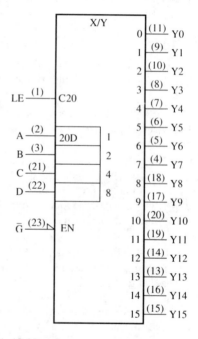

Figure 13.20
Exercise 13.60.

CHALLENGING EXERCISES

13.61 Invent a weighted, self-complementing, 4-bit, BCD code.

13.62 Invent a 5-bit unit-distance BCD code. The code group for $9)_{10}$ should be at unit distance from $0)_{10}$.

13.63 Design a decimal-to-8-4-2-1-BCD priority encoder based on a priority system in which the decimal number having the smallest magnitude has the greatest priority. Inputs and outputs should be active-high. Draw the logic diagram.

13.64 Design a 5-line-to-24-line decoder using 3 74138 3-to-8 decoders and no external logic gates.

13.65 Design a two-out-of-five-to-8-4-2-1 BCD code converter. Draw a logic diagram that implements the simplest possible logic expressions.

SPICE EXERCISES

13.66 Construct a SPICE model that generates the output $A \odot B \odot C$. Use exclusive-OR gates constructed from NMOS NAND gates and inverters defined by subcircuits. Test the model by driving it with every binary combination from 000 through 111. Recalling that $A \oplus B \oplus C$ generates an even-parity bit for 3-bit numbers (Example 13–10) and that $A \odot B \odot C \odot D$ generates an odd-parity bit for BCD numbers, use your results to form a conclusion about whether or not $A \odot B \odot C$ generates an odd-parity bit for 3-bit numbers. How are the logic functions $A \oplus B \oplus C$ and $A \odot B \odot C$ related?

13.67 Construct a SPICE model of a 3-bit binary-to-Gray Code converter using CMOS transmission gates to implement exclusive-OR operations (see Figure 10.26). Use subcircuits and test your model by driving it with all combinations of the binary inputs from 000 through 111.

13.68 Construct a SPICE model of a 2-bit binary-to-decimal decoder (2-to-4) using diode AND gates and CMOS inverters defined by subcircuits. Obtain plots of the four decoder outputs when the binary inputs and those shown in Figure 13.21.

Figure 13.21
Exercise 13.66.

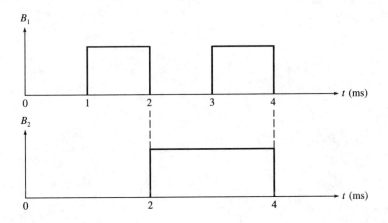

Chapter 14

COUNTERS

14–1   INTRODUCTION

A digital counter is a set of flip-flops whose states change in response to pulses applied at the input to the counter. The flip-flops are interconnected so that their combined state at any time is the binary equivalent of the total number of pulses that have occurred up to that time. Thus, as its name implies, a counter is used to count pulses. For example, when five pulses have occurred at the input to a counter consisting of three flip-flops, the contents of the counter are 101, and after the sixth pulse occurs, the contents are 110. The input pulses may recur at some fixed frequency, as, for example, when the input is a clock signal, or they may occur at unpredictable intervals, signifying the occurrence of "events" in a larger system. An example of the latter is a counter used to count automobiles passing through an intersection, each vehicle generating one input pulse when it passes over a pressure plate embedded in the pavement.

All counters are binary, in the sense that the contents of individual flip-flops represent 1s and 0s. However, we hear of binary counters, decimal counters, BCD counters, and so on, names that refer to how the contents are interpreted in the various binary codes studied in Chapter 13. The code determines the count at which the counter must *reset* itself—i.e., the maximum number to which it counts before it (automatically) starts counting over again from its initial contents, usually zero. For example, a binary counter consisting of five flip-flops can count up to $11111 = 31)_{10}$ ($2^5 - 1$) and resets to 00000 on the occurrence of the 32nd pulse. On the other hand, a *decimal* (BCD) counter, consisting of four flip-flops, must reset to 0000 after its count reaches 1001—i.e., $9)_{10}$.

The contents of counters are decoded using logic circuitry similar to that described for decoders in Chapter 13. Thus, for example, a BCD-to-decimal decoder would be used to decode a BCD counter, making one out of 10 possible outputs lines high, depending on the 4-bit input supplied from the counter.

Counters are widely used in digital systems because they can perform many important functions other than just keeping track of the number of pulses that have occurred. For example, we will see that they are often used as *frequency dividers* to obtain waveforms with frequencies that are a specific fraction of the clock frequency. They are also used to perform *timing* functions, as in digital watches and in the creation of time delays, as well as to produce non-sequential binary counts, such as 001, 010, 100, 001, ...

14–2   BINARY RIPPLE COUNTERS

The simplest type of counter, the easiest to design, and the one requiring the least amount of hardware is the binary ripple counter. It consists of toggle (T) flip-flops, or other types connected to behave as T flip-flops, interconnected so that the Q-output of one stage is the toggling input to the next stage. Recall that a toggle flip-flop changes state on every leading (or trailing) edge that occurs at its input. Figure 14.1(a) shows a 3-bit ripple counter consisting of three trailing-edge-triggered T flip-flops. The flip-flop outputs are designated Q_1, Q_2, and Q_4, with subscripts corresponding to the weights of the bit positions in the binary count. The timing diagram in Figure 14.1(b) shows how the flip-flops change state in response to a clock signal applied to the least significant stage (Q_1). As can be seen in the count table, Q_1 must change state at every new count, and that action occurs in the circuit because Q_1 is toggled directly by trailing edges of the clock. Q_2 toggles on the trailing edge of Q_1, so it changes state after every two clock pulses. Q_4 toggles on the trailing edge of Q_2, so it changes state after every four clock pulses. Note that the binary contents of the counter after any given number of clock pulses (trailing edges) have occurred equals that total number. Also note that the counter resets to 000 after the eighth clock pulse, the one occurring after the contents have reached their maximum value of $111 = 7)_{10}$.

The timing diagram in Figure 14.1(b) does not show how the propagation delays in the flip-flops affect the output waveforms. Since each flip-flop is toggled by the changing state of the preceding flip-flop, it is apparent that no flip-flop can change state until after the propagation delay of the flip-flop that precedes it. This delay accumulates as we proceed through additional stages. The detrimental effects of that delay on the operation of a ripple counter are discussed in a later section. For now, it is sufficient to understand that this kind of operation is responsible for the name *ripple* counter: A given flip-flop cannot change state until all preceding state changes have occurred; that is, a change of state in the first stage must effectively ripple through the succeeding stages. Ripple counters are also called *serial*, or *series*, counters because of the sequential nature of these state changes. They are also called *asynchronous* counters because the clock does not directly control the time at which every stage changes state.

The number of stages in a ripple counter can be extended in an obvious way, by simply cascading additional T flip-flops. The largest count that an *n*-stage binary counter can reach is $2^n - 1$, which occurs when every flip-flop is set. The next pulse causes every flip-flop to reset, and the counter repeats its count sequence as succeeding pulses occur.

The timing diagram in Figure 14.1(b) is expanded in Figure 14.2 to show the waveforms resulting from the application of 16 clock pulses. We see that the period of the waveform at Q_1 is twice the period of the clock. Therefore, the waveform at Q_1 has one-half the frequency of the clock. Similarly, the waveform at Q_2 has one-half the frequency of the waveform at Q_1, or one-fourth the frequency of the clock. Finally, the waveform at Q_4 has one-eighth the frequency of the clock. In general, the frequency f_n at the output of the *n*th stage of a binary ripple counter is

$$f_n = \frac{f_c}{2^n}$$

(14.1)

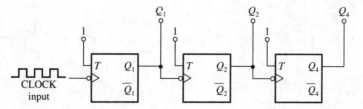

(a) Logic diagram.

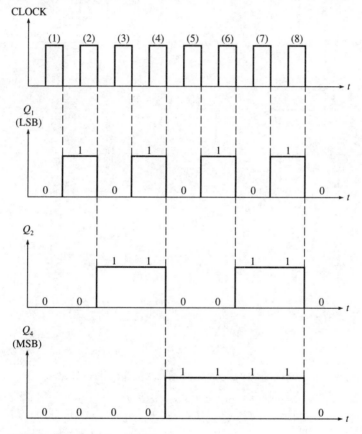

(b) Timing diagram and count table.

Figure 14.1
A three-stage binary ripple counter. Note that the binary contents at any time correspond to the total number of pulses (trailing edges) that have occurred up to that time.

where f_c is the frequency of the clock signal driving the least significant stage. Counters are often used in this way for the express purpose of performing frequency division. A 3-bit counter is called a divide-by-8 counter, and an n-bit counter is a divide-by-2^n counter.

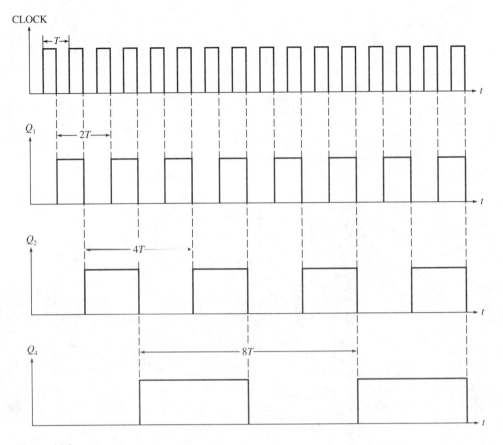

Figure 14.2
Expanded timing diagram for the 3-bit ripple counter, showing period and frequency relations of the waveforms. Note that $f_4 = f_c/8$.

Example 14–1

A binary ripple counter must be capable of counting up to $16{,}383)_{10}$.

a. How many flip-flops are required?
b. If the clock frequency is 8.192 MHz, what is the frequency at the output of the most significant stage?

Solution

a. $2^n - 1 = 16{,}383$
   $2^n \quad = 16{,}384$
   $n \quad = \log_2(16{,}384) = 14$

b. From equation (14.1),

$$f_{14} = \frac{f_c}{2^{14}} = \frac{8.192 \text{ MHz}}{16{,}384} = 500 \text{ Hz}$$

Drill
14–1

At what stage of the counter in Example 14–1 is the frequency of the waveform equal to 8 kHz?
Answer: Tenth.

Down Counters

The counters we have discussed so far are called *up counters* because they count *up:* Their binary counts increase as the number of pulses increases. A *down counter* counts down: Its binary contents decrease as the number of pulses increases. For example, a typical sequence of counts in a three-stage down counter is 111, 110, 101, 100, 011, 010, 001, 000, 111, ... Note that this type of counter resets itself from 000 to 111. One use of down counters is in applications where it is necessary to display the amount of time remaining until the execution of some event, as, for example, in microwave ovens, clothes dryers, racing clocks, or launch clocks.

When the binary contents of a counter are increased by 1, we say that the counter is *incremented*. When the contents are decreased by 1, we say that the counter is *decremented*. Thus, each clock pulse increments an up counter and each clock pulse decrements a down counter.

The count sequence of a down counter can be realized by simply using the complemented ($\overline{Q}$) outputs of the flip-flops in an up counter as the output waveforms. This fact is demonstrated in Figure 14.3, where the down counter outputs are relabeled Q'_1, Q'_2, and Q'_4. A down counter can also be realized by using the Q-outputs of cascaded, *leading-edge*-triggered T flip-flops. Verifications of this method and others are exercises at the end of the chapter.

14–3   MODULUS COUNTERS

The number of input pulses that causes a counter to reset to its initial count is called the *modulus* of the counter. Thus, the modulus equals the total number of distinct *states* (counts), including zero, that a counter can store. For example, the three-stage counter in Figure 14.1 has modulus 8 because it resets to 000 after the eighth clock pulse, and it has the eight states 000 through 111. It is said to be a mod-8 counter. A binary counter with n stages is a mod-2^n counter. Note that the largest count a mod-N counter can achieve is $N - 1$; that is, a mod-N counter never reaches the binary number equal to its modulus.

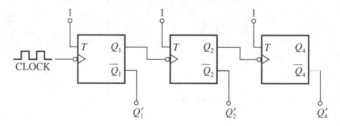

(a) Logic diagram.

Figure 14.3
Using the complemented outputs of an up counter to obtain a down counter. (Continued on 538.)

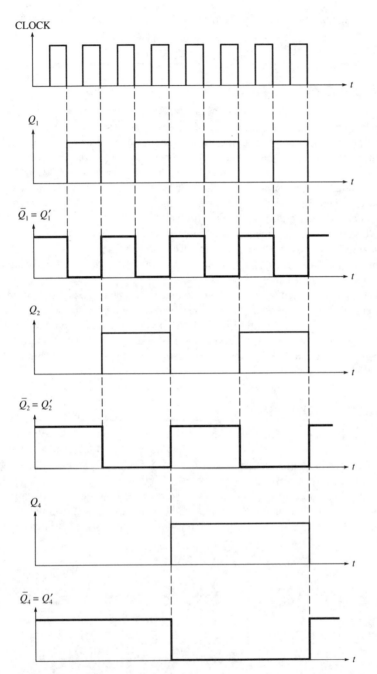

(b) Timing diagram and count table.

Figure 14.3
(Continued)

   The output from the most significant stage of a mod-N counter has a frequency equal to f_c/N, where f_c is the clock frequency. Thus, a mod-N counter is a divide-by-N counter. In some applications, it is necessary to use a counter whose modulus is a number that cannot be expressed as 2^n. For example, if we wish to divide the clock frequency by 5, we must use a mod-5 counter. As another example, a mod-60 counter is needed for real-time clock displays, since the count of seconds must reset every minute and the count of minutes must reset every hour. If the modulus of a counter cannot be expressed in the form $N = 2^n$, the output waveforms may not be symmetrical square waves (with 50% duty cycle) like those shown in Figures 14.1 and 14.2. However, that is not a critical consideration when the waveforms are used to drive edge-triggered devices.

   Figure 14.4 illustrates one method that can be used to construct a mod-N counter for any N. In this example, a mod-5 counter, the direct-clear inputs on the flip-flop are used to clear the count to 000 whenever it reaches $101 = 5)_{10}$. The input to the NAND gate is $Q_1\overline{Q}_2Q_4$, so when $Q_1Q_2Q_4 = 101$, the low output from the NAND gate clears the flip-flops. The timing diagram shows the asymmetrical waveform that results at Q_4 and confirms that it has the clock frequency divided by 5. Note the narrow spike, or "glitch," appearing in Q_1 when the counter resets. This occurrence is due to the fact that the count must momentarily reach 101 before it is cleared. In many applications, the narrow spike is not important, either because it is of such short duration (typically a few nanoseconds) or because the Q_1 waveform is not used. For example, the counter may be used solely to divide frequency by 5, so only the Q_4 waveform is needed. As another example, if the outputs drive indicating lamps, the pulse may be too narrow to create a visible response in a lamp. However, for some N, a mod-N counter of this design may create a spike in the divide-by-N (most significant) output.

Example 14–2   Design a decade (mod-10) counter using the method demonstrated in Figure 14.4 for a mod-5 counter. By constructing timing diagrams, determine which outputs will contain glitches when the counter resets.

Solution.   The number of flip-flops required to construct a mod-N counter equals the smallest n for which $N \leqslant 2^n$. For a mod-10 counter, four stages are required, since $10 \leqslant 2^4$ and $10 > 2^3$. Figure 14.5(a) shows the logic diagram. The counter resets when it reaches a count of $10)_{10} = 1010)_2$—i.e., when $Q_8\overline{Q}_4Q_2\overline{Q}_1$ makes the output of the NAND gate go low.

   The timing diagram in Figure 14.5(b) shows that a glitch appears in the Q_2 waveform because the count must momentarily reach 1010. Note that glitches will always occur in any output that must momentarily change from 0 to 1 to 0 when the count of a mod-N

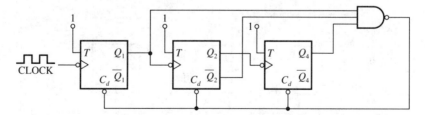

(a) Logic diagram. When the count reaches 101, the output of the NAND gate goes low and resets all flip-flops.

Figure 14.4
A mod-5 ripple counter. (Continued on p. 540.)

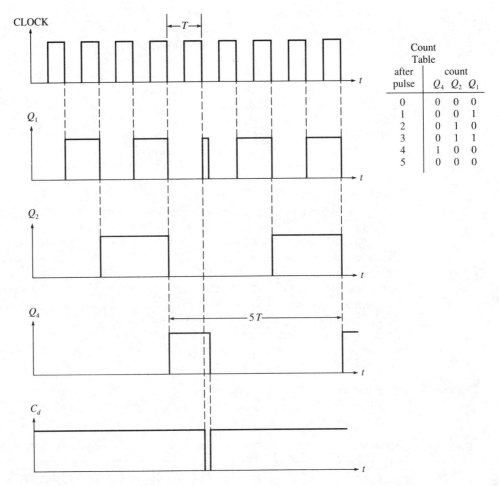

Count
Table

after pulse	count		
	Q_4	Q_2	Q_1
0	0	0	0
1	0	0	1
2	0	1	0
3	0	1	1
4	1	0	0
5	0	0	0

(b) Timing diagram and count table. Note that the frequency of Q_4 is the clock frequency divided by 5.

Figure 14.4
(Continued)

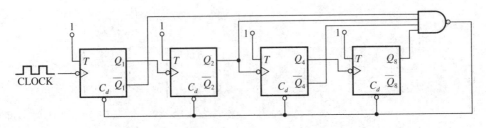

(a) Logic diagram.

Figure 14.5
(Example 14–2) A mod-10 (decade) ripple counter. (Continued on p. 541.)

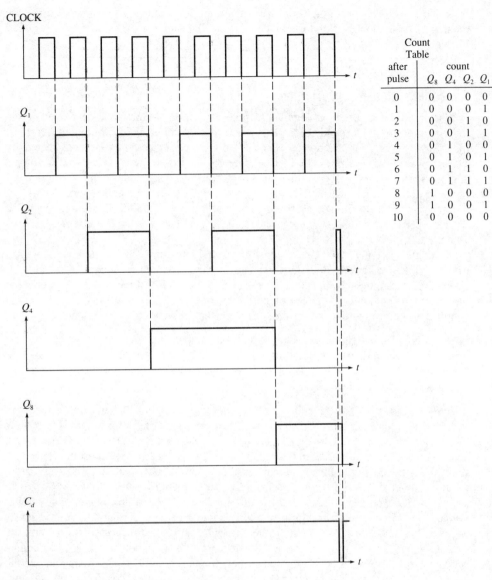

CLOCK

Q_1

Q_2

Q_4

Q_8

C_d

after pulse	count			
	Q_8	Q_4	Q_2	Q_1
0	0	0	0	0
1	0	0	0	1
2	0	0	1	0
3	0	0	1	1
4	0	1	0	0
5	0	1	0	1
6	0	1	1	0
7	0	1	1	1
8	1	0	0	0
9	1	0	0	1
10	0	0	0	0

Count Table

(b) Timing diagram and count table.

Figure 14.5
(Continued)

counter changes from $N - 1$ to N to 0. In the decade counter, this sequence is 1001, 1010, 0000, and we see that Q_2 must change from 0 to 1 to 0:

$$
\begin{array}{rccccc}
 & Q_8 & Q_4 & Q_2 & Q_1 \\
N - 1 = \;\; 9 = & 1 & 0 & 0 & 1 \\
N = 10 = & 1 & 0 & 1 & 0 \\
0 = \;\; 0 = & 0 & 0 & 0 & 0
\end{array}
$$

In which waveforms will glitches occur when a mod-N counter is constructed using the method of Example 14–2 and the value of N is (a) 7 and (b) 12? *Answers:* (a) Q_1; (b) Q_4.

Cascading Mod-*N* Counters

When the output of the most significant stage of a mod-N counter is connected to the toggling input of the least significant stage of a mod-M counter, the combination forms a mod-MN counter. For example, a decade (mod-10) counter could be constructed by cascading a mod-5 counter and a mod-2 counter. (A mod-2 counter is a single flip-flop.) A mod-MN counter constructed this way will divide the clock frequency by MN regardless of the order in which the mod-M and mod-N counters are connected (mod-M driving mod-N or vice versa). However, the duty cycle of the most significant output may depend on the order in which the counters are cascaded. This fact is demonstrated in Exercises 14.15 and 14.16.

Modulus counters are often constructed by cascading lower-modulus counters because of the availability of certain standard-modulus counters in integrated circuits. Of course, more than two counters can be cascaded; the modulus of the overall counter is the product of the moduli of the individual counters. Again, the duty cycle of the most significant output may depend on the order in which the counters are cascaded.

Example
14–3

A real-time 12-hour digital clock must display seconds, minutes, and hours. A 60-Hz clock signal derived from the ac power lines is available, as are trailing-edge-triggered mod-6 and mod-10 counters. Draw a block diagram showing how the counters and displays should be connected.

Solution. To drive the seconds' display, we need a signal that generates 1 pulse/s (1 Hz). This signal can be obtained using a divide-by-60 counter whose input is the 60-Hz clock. See Figure 14.6. The 1-Hz signal drives a cascaded mod-10 and mod-6 counter. The mod-10 counter counts from 0 through 9 and resets every 10 s, so its count represents the least significant digit of the total number of seconds that have elapsed. The 4-bit output of this counter drives a BCD-to-7-segment decoder, which in turn drives a 7-segment display. After every 10 s, the mod-10 counter increments the mod-6 counter. This counter contains the most significant digit of the total number of seconds that has elapsed, up to 59. Note that it drives a 7-segment display that will register decimal digits from 0 through 5. After every 60 s (1 min), the mod-10 and mod-6 counters both reset, so the decimal seconds display becomes 00. At the same time, another mod-60 counter is incremented to count the total number of minutes. Its construction is identical to that of the seconds counter and will cause the total number of minutes, up to 59, to be displayed.

Finally, the hours display must be capable of registering the 12 counts from $00)_{10}$ through $11)_{10}$. (Note that the largest number displayed is 11 59 59: 11 hours, 59 minutes and 59 seconds; the next input pulse resets the entire clock to 00 00 00.) Although a mod-12 counter must be used to count the total number of hours, the outputs of a single mod-12 counter would be difficult to decode for driving the two BCD hour displays. Instead, we can use a mod-10 counter driving a standard BCD-to-7-segment decoder for the least significant digit of the total number of hours. The most significant digit is either 0 or 1. In practical circuits, a single flip-flop is used to drive the display for the most significant digit. Although the combination of a mod-10 and mod-2 counter forms a mod-20 counter, a *feedback* arrangement is used, whereby the mod-10 counter is reset to

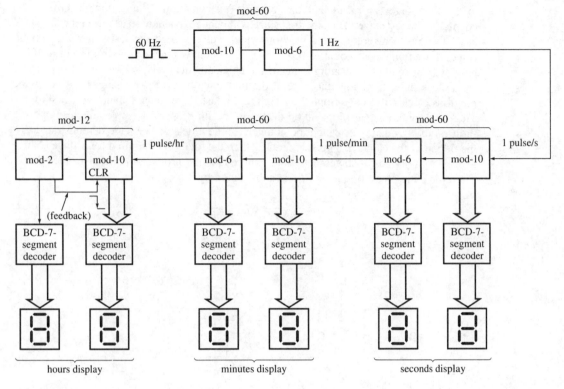

Figure 14.6
A 12-hour real-time clock. (Example 14–3)

0000 by the flip-flop each time its state changes from 1 to 0. Thus, the count changes from $1\ 1001 = 11)_{10}$ to $0\ 0000$, and the mod-20 counter is converted to a mod-12 counter.

Drill *14–3*	What is the frequency of the output of the mod-2 counter in Figure 14.6? *Answer:* 23.1481 Hz.

14–4   PRACTICAL CONSIDERATIONS IN RIPPLE COUNTER APPLICATIONS

The Effects of Propagation Delay

Recall from Section 14–2 that no flip-flop in a ripple counter can change state until all state changes in preceding flip-flops have occurred. That is not to say that *every* flip-flop must change state after *every* new clock pulse, for that occurs only when the count changes from a number such as 0111 to 1000.

However, in cases such as those, where every flip-flop must change state, it is easy to visualize how propagation delays could adversely affect the operation of the counter: The most significant stage cannot change state until every preceding stage has changed state—i.e., until the cumulative sum of all the propagation delays through the preceding stages has elapsed. Suppose the count is 0111 and a new clock pulse occurs at the input to the first stage *before* all the state changes just mentioned have rippled through the counter to the last stage. Then the first stage will change to a new state before the last stage has had time to respond to the previous clock pulse. When the last stage does finally respond, we will have a count of 1001. Thus, the count will have gone from 0111 to 1001, having skipped 1000. Recapitulating, if the clock frequency is so high that it is possible for a clock pulse to change the state of the first stage before state changes caused by the previous clock pulse have rippled through to the last stage, then a count will be skipped. Thus, it is obvious that propagation delays in the flip-flops of a ripple counter impose a limit on the frequency at which the counter can be clocked. We have demonstrated that the period of the clock, T_c, must be greater than the total time required for a stage change to ripple through all n stages: $T_c > nt_p$, where t_p is the propagation delay in each stage. Therefore, the clock frequency, f_c, is constrained by

$$\frac{1}{T_c} = f_c < \frac{1}{nt_p} \tag{14.2}$$

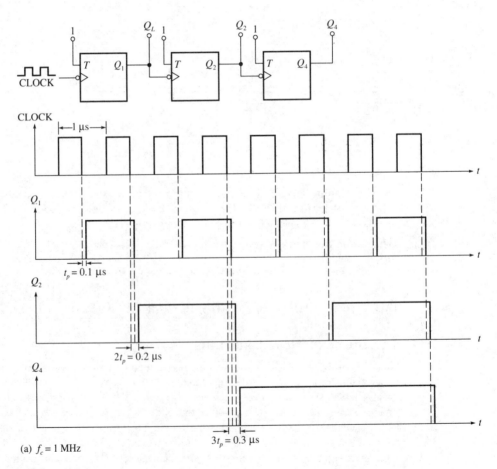

(a) $f_c = 1$ MHz

Figure 14.7
Example 14–4.

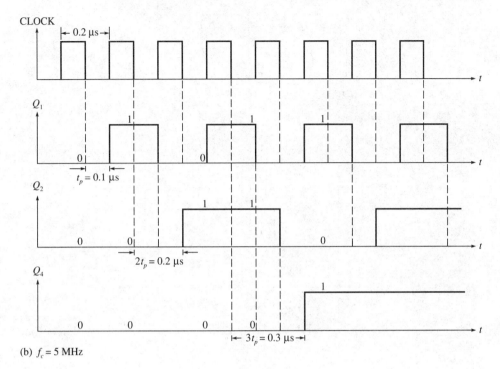

(b) $f_c = 5$ MHz

Figure 14.7
(Continued)

Example 14-4

Each trailing-edge-triggered flip-flop in a 3-bit ripple counter has propagation delay $t_{pHL} = t_{pLH} = t_p = 100$ ns. Sketch the waveforms at the output of each stage when the clock frequency is (a) 1 MHz and (b) 5 MHz.

Solution

a. Figure 14.7(*a*) shows the waveforms when $f_c = 1$ MHz ($T_c = 1$ μs). Note how the propagation delay accumulates, the total delay at the last stage being $3t_p = 300$ ns $= 0.3$ μs. Since $T_c > 3t_p$ in this case, no counts are skipped.

b. Figure 14.7(*b*) shows the waveforms when $f_c = 5$ MHz ($T_c = 0.2$ μs). In this case, $T_c < 3t_p$—i.e., $f_c > 1/(3t_p)$—and we see that the count 100 is never reached. (Note the appearance of combinations of flip-flop states *between* the normal combinations, giving rise to an invalid count sequence. This effect is discussed in the next section.)

Drill 14-4

For what minimum value of propagation delay in each flip-flop will a 10-bit ripple counter skip a count when it is clocked at 10 MHz?
Answer: 10 ns.

Decoding

Figure 14.8(*a*) shows a binary-to-decimal decoder used with a 2-bit ripple counter. The four AND gates produce high outputs in succession, one for each of the four successive counts 0, 1, 2, and 3, corresponding to states 00 ($\overline{Q}_2\overline{Q}_1$), 01 ($\overline{Q}_2Q_1$), 10 ($Q_2\overline{Q}_1$), and 11

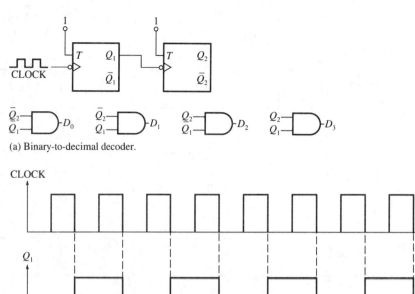

(a) Binary-to-decimal decoder.

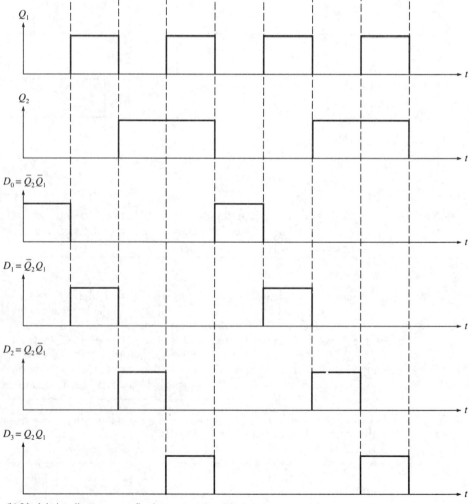

(b) Ideal timing diagram, not reflecting propagation delays.

Figure 14.8
Decoding a 2-bit ripple counter.

(Q_2Q_1). The waveforms in Figure 14.8(*b*) show that one and only one decoder output is high at any given time. This is yet another way to produce timing pulses for synchronizing computer operations. Recall that we have seen the same kinds of waveforms produced by one-shots and by ring counters. The waveforms in Figure 14.8(*b*) are *ideal*, in the sense that they ignore the effects of propagation delays.

Figure 14.9 shows the actual decoder waveforms that result when the propagation delays of a 2-bit ripple counter are taken into account. Propagation delays of the AND gates are neglected. We see that glitches occur in the D_0 and D_2 waveforms due to intermittent combinations of states that activate the corresponding AND gates. The width of these glitches equals the width of the propagation delay, t_p. Figure 14.7(*b*) in Example 14–4 shows a more extreme case, where propagation delays are a significant portion of a clock period. Note that the sequence of counts, including the intermittent combinations between legitimate counts, is actually 000, 001, 000, 010, 011, 010, 000, 101, 100, 110, 111. As discussed earlier, the practical consequences of the glitches depend on the application. If the decoder outputs drive indicator lamps, the glitches may be too narrow to create any response. On the other hand, if the decoder outputs drive edge-triggered devices, the glitches may be responsible for unsatisfactory performance. In those cases, the outputs can be "deglitched" by ANDing them with the clock. However, this remedy reduces the width of a deglitched output to the width of a clock pulse, which causes gaps to appear between decoder outputs.

Integrated-Circuit Ripple Counters

Because of the inherent simplicity of ripple counters, they are available in integrated circuits with a relatively large number of stages in a single package. Many also have additional features that enhance their versatility and make it possible to use one circuit for a variety of applications. An example is the 7493-series TTL counter, which can be used as a mod-2, mod-8, mod-10, and mod-12, or mod-16 counter. The logic diagram of the 7493 is shown in Figure 14.10. (In this and subsequent integrated-circuit diagrams, flip-flops are assumed to toggle on clock edges.) Note that three flip-flops, Q_B, Q_C, and Q_D, are connected in cascade and driven by clock input CLK*B*, and a fourth flip-flop, Q_A, has a separate clock input, CLK*A*. Thus, Q_A can be used as a mod-2 counter, while Q_B, Q_C, and Q_D are used as a mod-8 counter. By connecting output Q_A to input CLK*B* or output Q_D to input CLK*A*, the combination can be used as a mod-16 counter. All flip-flops can be cleared by making inputs R_{01} and R_{02} both high. This feature can be used to configure the circuit as a mod-10 or mod-12 counter, as is demonstrated in the next example.

Example 14–5

Show how the counter in Figure 14.10 can be configured to operate as (a) a mod-10 counter and (b) a mod-12 counter.

Solution

a. Figure 14.11(*a*) shows the external connections necessary to operate the circuit as a mod-10 counter. Note that output Q_A is connected to input CLK*B*, so Q_A is the least significant stage and Q_D is the most significant stage. When a count of $10)_{10}$ is reached, $Q_DQ_CQ_BQ_A = 1010$—i.e., $Q_D = 1$ and $Q_B = 1$. This count is the only one between 0 and 10 for which both Q_D and Q_B are 1. Since Q_B and Q_D are connected to R_{01} and R_{02}, the counter resets immediately on reaching 1010 and therefore serves as a mod-10 counter.

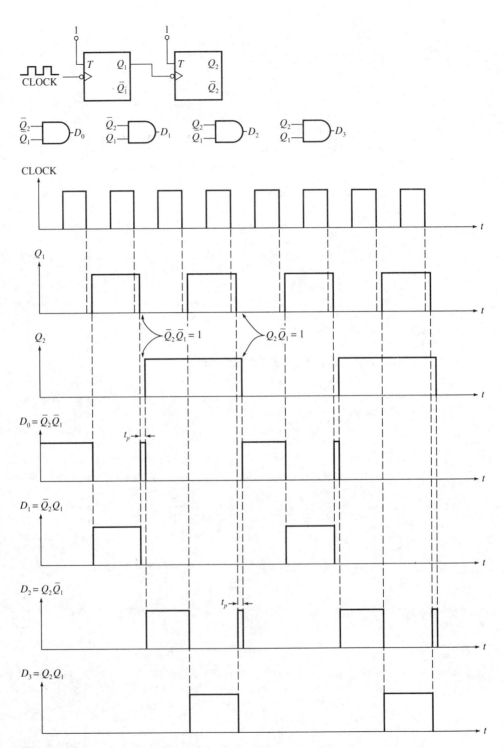

Figure 14.9
Glitches appear in the practical decoder waveforms due to propagation delays of the flip-flops.

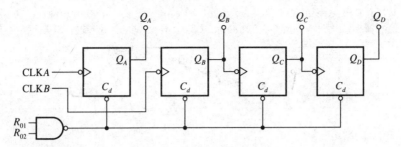

Figure 14.10
Logic diagram of the 7493-series TTL counter.

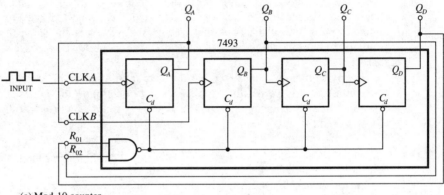

(a) Mod-10 counter.

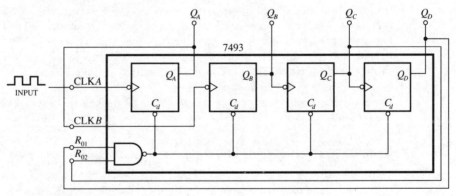

(b) Mod-12 counter.

Figure 14.11
Example 14–5.

b. Figure 14.11(*b*) shows the connections necessary for a mod-12 counter. Once again, Q_A is the least significant stage. When a count of 12)$_{10}$ is reached, $Q_D Q_C Q_B Q_A = 1100$. This is the only count between 0 and 12 for which Q_C and Q_D are both 1. Since Q_C and Q_D are connected to R_{01} and R_{02}, the counter resets immediately on reaching 1100 and therefore serves as a mod-12 counter.

Drill
14–5 What is the frequency of the (negative-going) pulses at the output of the NAND gate in Figure 14.11(*b*) when the frequency of the input at CLK*A* is 690 kHz? *Answer:* 57.5 kHz.

14–5   SYNCHRONOUS BINARY COUNTERS

Introduction

A synchronous binary counter has the same sequence of count states as a ripple (asynchronous) counter. However, if a flip-flop changes state when a synchronous counter is incremented, the change occurs on the leading or trailing edge of a *clock* pulse rather than on the leading or trailing edge of the preceding flip-flop's output. Thus, as is characteristic of all synchronous devices, the clock is connected to every flip-flop in the counter, and all those flip-flops that change state do so at the same time. (Recall the similarity of this operation to that of another synchronous device we have studied: the shift register.)

In a synchronous counter, the flip-flops are most often of the *JK* type. In general, the (*present*) states of all the flip-flops are used to determine the *next* state of any given flip-flop, through logic circuitry connected to the *J* and *K* inputs of that flip-flop. In other words, when the next clock pulse occurs, a flip-flop changes state only if the previous state of the counter dictates that it should. To illustrate, when the present state is $Q_4Q_2Q_1 = 100$, the logic circuitry permits Q_1 to change state on the next clock pulse so that the new state will be 101.

As we have indicated, logic circuitry is connected to every *J* and *K* input on every flip-flop of a synchronous counter, and this circuitry controls the state changes at every new count through the entire sequence of the counter's states. It is apparent, then, that synchronous counters are more complex than asynchronous, or ripple, counters. Designing the logic circuitry for some synchronous counters can be a very complex task. Indeed, the general design procedure, which we study in a later section, requires simplification of *n*-variable Boolean algebra expressions for each and every *J* and *K* input of all *n* flip-flops in an *n*-stage counter. However, the logic required for a *binary* synchronous counter is very easy to deduce from the sequence of count states, as we shall see presently.

Synchronous Binary Up Counters

Figure 14.12(*a*) shows the sequence of states of a 3-bit binary counter, from which we can deduce the logic required to control the flip-flops in a 3-bit synchronous counter (Figure 14.12(*b*)). Note that Q_1 changes state at every new count. Thus, we can simply hold J_1 and K_1 high, as shown in Figure 14.12(*b*), causing Q_1 to toggle on every clock pulse. We also note that Q_2 changes state only when Q_1 is 1. That is, the present state of Q_2 should be changed on the next clock pulse if and only if the present state of Q_1 is 1. This logic is implemented by connecting Q_1 to the J_2 and K_2 inputs of Q_2. If Q_1 is high, $J_2 = K_2 = 1$, and Q_2 toggles on the next clock pulse. If Q_1 is low, $J_2 = K_2 = 0$, and there is no change in Q_2 on the next clock pulse. Finally, we see that Q_4 changes state only when Q_1 and Q_2 are *both* 1. That is, the present state of Q_4 should be changed on the next clock pulse if and only if the present states of Q_1 and Q_2 are both 1. This logic is implemented by connecting J_4 and K_4 to the logical AND of Q_1 and Q_2. The timing diagram in Figure

	Q_4	Q_2	Q_1
0	0	0	0
1	0	0	1
2	0	1	0
3	0	1	1
4	1	0	0
5	1	0	1
6	1	1	0
7	1	1	1
8	0	0	0

(a) The sequence of counts shows that Q_1 changes on every clock pulse, Q_2 changes when $Q_1 = 1$, and Q_4 changes when $Q_2Q_1 = 1$.

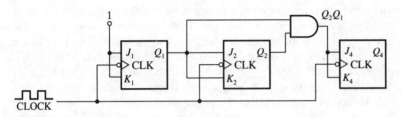

(b) Logic diagram.

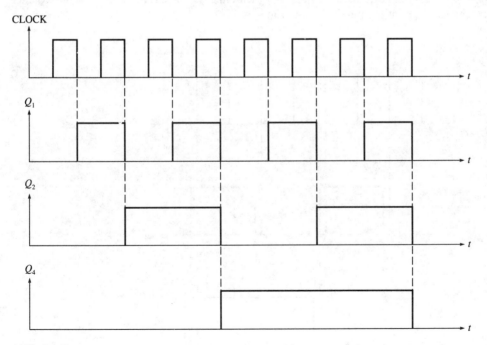

(c) Timing diagram.

Figure 14.12
A 3-bit synchronous binary up counter.

551

14.12(c) shows the waveforms that result when the JK flip-flops are trailing-edge-triggered.

By constructing tables showing the count sequences, it is easy to see how the 3-bit synchronous counter can be expanded to 4 bits or more. In a 4-bit counter, we find $J_8 = K_8 = Q_1Q_2Q_4$, and in a 5-bit counter, $J_{16} = K_{16} = Q_1Q_2Q_4Q_8$. In general, $J_{2^n} = K_{2^n} = Q_1Q_2Q_4 \cdots Q_{2^{n-1}}$.

Propagation Delays in Synchronous Counters

The timing diagram in Figure 14.12(c) appears to be identical to that for the 3-bit asynchronous (ripple) counter shown in Figure 14.1(b). However, propagation delays are not shown in either figure, and when those delays are taken into account we will see that there is a significant difference. Figure 14.13 shows a comparison of the actual waveforms of 2-bit synchronous and asynchronous counters, assuming that the propagation delay in the flip-flops of each is 20 ns. We should note that a synchronous counter with edge-triggered flip-flops *relies* on the propagation delays for proper operation. We can see that such is the case by referring to Figures 14.12(b) and 14.13(b): Q_1 must remain high for a short time (the setup time) *after* the occurrence of the clock edge to ensure that J_2 and K_2 remain high long enough for Q_2 to toggle. The propagation delay provides that time. Returning to our comparison, note that the propagation delay in the ripple counter accumulates to 40 ns, as previously described, while the propagation delay in the synchronous counter is a *constant* 20 ns. Thus, the flip-flops in the synchronous counter truly change state at the same time (in synchronism), whereas those in the ripple counter change states at progressively later times.

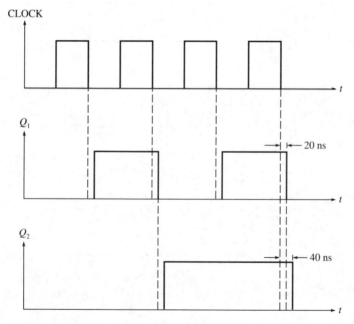

(a) Asynchronous (ripple) counter.

Figure 14.13
Comparison of waveforms of 2-bit synchronous and asynchronous counters. (Continued, 553.)

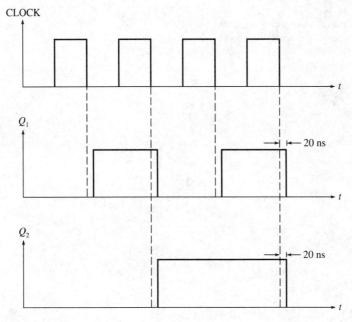

(b) Synchronous counter.

Figure 14.13
(Continued)

As in ripple counters, propagation delays can cause glitches when the outputs of synchronous counters are decoded. However, in the case of synchronous counters, it is small *differences* between the propagation delays of the flip-flops and differences between the values of t_{pHL} and t_{pLH} that are responsible for the glitches, rather than the delays themselves. These differences prevent the flip-flop outputs from changing in perfect synchronism. The creation of glitches due to this lack of perfect synchronism is illustrated in Figure 14.14. Here, the difference between t_{pHL} *in* Q_1 and t_{pLH} in Q_2 creates a glitch in the D_0 decoder output. Some integrated circuit decoders are equipped with ENABLE inputs that can be used to deglitch the decoder outputs. The ENABLE input is *strobed* (pulsed) by an external signal that enables the decoder at regular time intervals which exclude the times where glitches occur.

Synchronous Binary Down Counters

Figure 14.15 shows the sequence of count states, the logic diagram, and the timing diagram of a 3-bit synchronous binary down counter. We see that Q_1 changes state on every clock pulse, Q_2 changes state if and only if Q_1 is 0, and Q_4 changes state if and only if both Q_1 and Q_2 are 0. Thus, $J_2 = K_2 = \overline{Q}_1$ and $J_4 = K_4 = \overline{Q}_1\overline{Q}_2$. In the general, *n*-bit down counter, $J_{2^n} = K_{2^n} = \overline{Q}_1\overline{Q}_2\overline{Q}_4 \cdots Q_{2^{n-1}}$. The timing diagram in Figure 14.15 includes propagation delays, and it can be seen that these are necessary for proper operation of the edge-triggered flip-flops. Exercises 14.31 and 14.32 demonstrate another way to construct a synchronous binary down counter.

Figure 14.14
Illustration of how lack of true synchronism in the outputs of a synchronous counter can create glitches in decoder outputs.

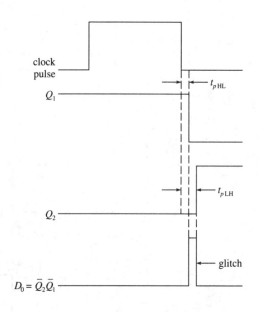

Synchronous Up/Down Counters

An up/down counter is one that can be made to count up or to count down, depending on the state of an externally applied input control. We have seen that the nth stage of an up counter has $J_n = K_n = Q_1 Q_2 \cdots Q_{2^{n-1}}$ and the nth stage of a down counter has

after pulse	count Q_4 Q_2 Q_1			decimal count
0	1	1	1	7
1	1	1	0	6
2	1	0	1	5
3	1	0	0	4
4	0	1	1	3
5	0	1	0	2
6	0	0	1	1
7	0	0	0	0
8	1	1	1	7

(a)

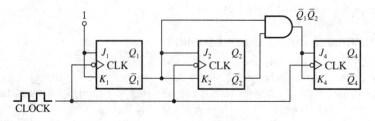

(b) Logic diagram.

Figure 14.15
A 3-bit synchronous down counter. (Continued on p. 555.)

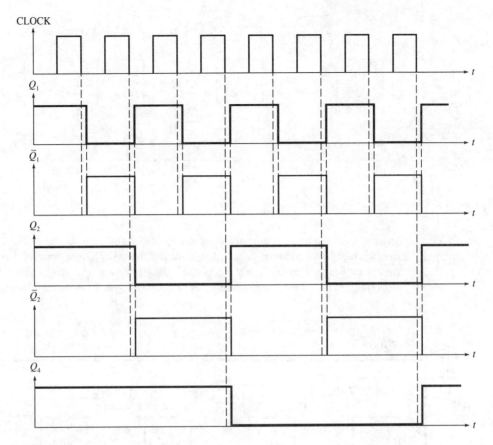

(c) Timing diagram.

Figure 14.15
A 3-bit synchronous down counter (continued).

$J_n = K_n = \overline{Q}_1\overline{Q}_2 \cdots \overline{Q}_{2^{n-1}}$. Figure 14.16 shows a 4-bit up/down counter whose operation is controlled by the input labeled UP/$\overline{\text{DOWN}}$. Up counting occurs when UP/$\overline{\text{DOWN}}$ is high and down counting occurs when UP/$\overline{\text{DOWN}}$ is low. The logic that must be implemented is obvious: If UP/$\overline{\text{DOWN}}$ is high, make $J_n = K_n = Q_1Q_2 \cdots Q_{2^{n-1}}$, or if UP/$\overline{\text{DOWN}}$ is low, make $J_n = K_n = \overline{Q}_1\overline{Q}_2 \cdots \overline{Q}_{2^{n-1}}$. Thus, $J_n = K_n = (\text{UP/}\overline{\text{DOWN}}) \cdot Q_1Q_2 \cdots Q_{2^{n-1}} + \overline{(\text{UP/}\overline{\text{DOWN}})} \cdot \overline{Q}_1\overline{Q}_2 \cdots \overline{Q}_{2^{n-1}}$. We see that the gates between each stage route either $Q_1Q_2 \cdots$ or $\overline{Q}_1\overline{Q}_2 \cdots$ to the J and K inputs of the next stage, depending on the state of UP/$\overline{\text{DOWN}}$. The logic circuitry is sometimes called *steering* logic because of this routing function it performs.

Cascading Synchronous Counters

As with ripple counters, mod-*N* and mod-*M* synchronous counters can be cascaded to create a mod-*MN* counter—i.e., a divide-by-*MN* counter. In the case of synchronous binary counters, we must be certain that the logic controlling the *J* and *K* inputs is carried

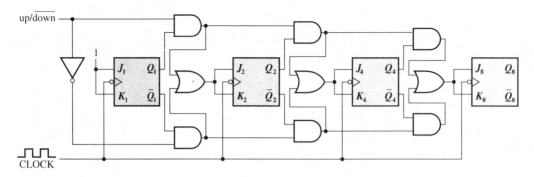

Figure 14.16
A 4-bit synchronous up/down counter.

forward from the less significant stages to the more significant stages. For example, if we are cascading two mod-16 counters, we must carry $Q_1Q_2Q_4Q_8$ forward from the least significant four stages to J_{16} and K_{16} of the most significant four stages. This is illustrated in Figure 14.17. Many integrated-circuit versions of synchronous counters provide both

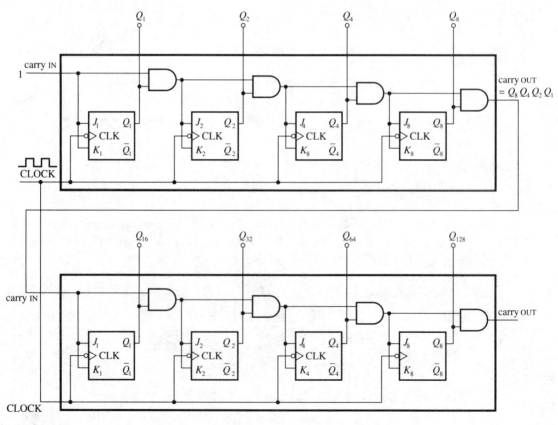

Figure 14.17
Cascading two mod-16 synchronous binary counters to obtain a mod-256 binary counter.

a *carry-out* and a *carry-in* connection for facilitating expansion. As shown in Figure 14.17, the carry out of the counter with the least significant stages is connected to the carry in of the counter with the most significant stages.

Note that the carry in of the least significant counter in Figure 14.17 is held high. This is necessary to make the least significant bit, Q_1, toggle on every clock pulse. However, there is another way to view the function of carry in. If it is made low, Q_1 never toggles, meaning that all counting ceases. From this perspective, carry in serves as an ENABLE input: When it is high, counting is enabled, and when it is low, counting is inhibited. Thus, we can regard the more significant counter in Figure 14.7 as one that is enabled (by its carry in) for the duration of one clock pulse only, each time the less significant counter cycles through its complete sequence. Adopting that viewpoint for the carry-in function, integrated circuit manufacturers usually designate it as an enabling input to a counter. We will see examples in Section 14–7 when we discuss integrated-circuit versions of synchronous counters.

14–6   SYNCHRONOUS COUNTER DESIGN

The purpose of this section is to introduce a systematic procedure that can be used to design a synchronous counter that will count in *any* prescribed sequence. It is not even necessary that the counts in a sequence be consistently "up" or "down" from each other, in terms of their binary values. For example, we can design a counter to produce the Gray-code sequence 000, 001, 011, 010, 110, 111, 101, 100, 000, . . . , which is an up sequence of Gray-coded numbers but is neither up nor down in terms of binary numbers. The only restriction we place on a count sequence is that every count can appear at most *once* before the sequence repeats itself. For example, we cannot use the procedure to design a sequence such as 00, 01, 00, 10, 00, 01, 00, 10, 00, 01, . . . , since 00 appears twice before the sequence repeats.

The logic circuitry used to control the ways in which flip-flops change states in response to clock pulses, in order to achieve a desired sequence of states, is an example of *sequential* logic. By contrast, logic circuitry that merely provides high or low outputs based on the combination of inputs supplied to it, without any reference to the sequence in which those outputs occur, is called *combinational* logic. Our purpose now is to develop the principles of sequential logic design.

State Diagrams

In the context of synchronous counters, the combined states of the flip-flops at any given time, i.e., the count, is usually called the state of the counter. For example, we would say that the state of a 4-bit up counter is 0101 after it has counted five input pulses. A *state diagram,* also called a transition diagram, is a graphical means for depicting the sequence of states through which a counter progresses. Each state is shown inside a circle, and arrows join the circles to show the sequence. To illustrate, Figure 14.18 shows state diagrams for 3-bit binary up and down counters.

In some *n*-stage counters, the sequence of states may not include every one of the 2^n possible states that can be realized with *n* flip-flops. In those cases, the initial state, when power is first applied, may not be one of the states in the sequence. A *self-starting* counter is one that will eventually enter its proper sequence of states regardless of its initial state. The state diagram for a self-starting counter shows every possible state, including those not in the desired sequence, and the progression of states leading from those not in the

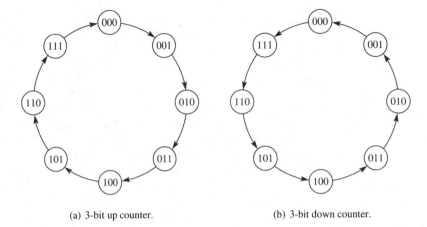

(a) 3-bit up counter.   (b) 3-bit down counter.

Figure 14.18
Examples of state diagrams.

Figure 14.19
Example of a state diagram for a self-starting counter. The diagram shows how a counter sequence is entered from states outside the sequence.

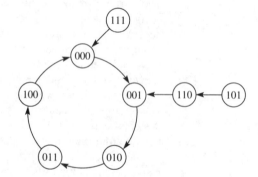

sequence to one in the sequence. Figure 14.19 shows an example. Here, the counter sequence is 000, 001, 010, 011, 100, 000, . . . and the diagram shows how that sequence is entered from states 101, 110, and 111.

The arrows linking states in a state diagram are frequently labeled to show the condition(s) under which a transition from one state to another takes place. Figure 14.20 shows two examples. Figure 14.20(*a*) shows the state diagram for a 2-bit binary counter that will not count unless an enable input, *E*, is high. The arrows circling back on each state and labeled $E = 0$ mean that the counter remains in whatever state it was last in when *E* was made 0. When $E = 1$, the count sequence is enabled. Figure 14.20(*b*) shows the state diagram of a 2-bit up/down counter that counts up when $U/\overline{D} = 1$ and counts down when $U/\overline{D} = 0$. In this case, it is understood that the labels on the arrows refer to the state of $U/\overline{D}$.

State Tables

State diagrams are used to construct *state tables,* which were first discussed in connection with flip-flops in Chapter 11. The state tables in turn are used to derive the logic necessary to control the *J* and *K* inputs of each flip-flop, as necessary to achieve a desired state sequence. Figure 14.21 shows an example of a state table derived from a state diagram,

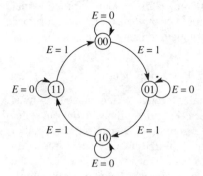

(a) Binary counter that is enabled when $E = 1$.

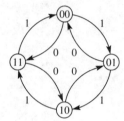

(b) Binary up/down counter. The labels on the arrows refer to the state of $U/\overline{D}$. Up counting occurs when $U/\overline{D} = 1$ and down counting occurs when $U/\overline{D} = 0$.

Figure 14.20
Examples of state diagrams for counters whose transitions from state to state are conditional on other control signals.

(a) State diagram.

present state $Q_4 \ Q_2 \ Q_1$	next state $Q_4 \ Q_2 \ Q_1$
0 0 0	0 0 1
0 0 1	0 1 0
0 1 0	0 1 1
0 1 1	1 0 0
1 0 0	1 0 1
1 0 1	1 1 0
1 1 0	1 1 1
1 1 1	0 0 0

(b) State table.

Figure 14.21
Example of a state table derived from a state diagram.

in this case for a 3-bit binary counter. The table lists every possible state of the counter (the present states) and shows the next state for each—i.e., the new state the counter should enter on the occurrence of the next clock pulse. The state table provides a convenient means for determining how each flip-flop should respond to a clock pulse, as a function of the (present) state of the counter. Thus, we can regard the state table as a kind of truth table for sequential logic design. As an illustration, the state table in Figure 14.21 tells us that if $Q_1 = 0$ and $Q_2 = 0$ and $Q_4 = 0$, then Q_1 should set on the next clock pulse, and Q_2 and Q_4 should not change state.

Design Procedure

There are at least two approaches we can take to designing the logic for a synchronous counter, based on the types of flip-flops used. First, we can use T-type flip-flops and develop logic that will make each flip-flop either toggle or not toggle on the occurrence of each new clock pulse. Note that this is the approach used to control the flip-flops in the binary counters discussed in the last section: Tying the J and K inputs together is the same as having a single T input. Secondly, we can develop logic to control the *individual J* and *K* inputs of each flip-flop. In either case, for each flip-flop we form a logic expression that contains every present state for which the flip-flop should change state on the occurrence of the next clock pulse. The next example illustrates the method using T-type flip-flops (or JK flip-flops operated as T-type flip-flops).

Example 14-6

Design a 3-bit synchronous Gray-code counter using T-type flip-flops.

Solution. Figure 14.22(a) shows the sequence of states through which the counter must progress. When designing with T-type flip-flops, it is easier to determine the present states for which each flip-flop must change state by simply examining the state sequence rather than by constructing a state table. This is illustrated in the redrawn sequence shown in Figure 14.22(b). Each arrow represents a state change. Note that it is important to list the reset state 000 following 100. T_1, T_2, and T_3 are the toggle inputs for the three flip-flops. Each is a sum-of-products logic expression representing the sum of all present states for which the corresponding flip-flop must toggle. For example, we see that $T_2 = \bar{Q}_3\bar{Q}_2Q_1 +$

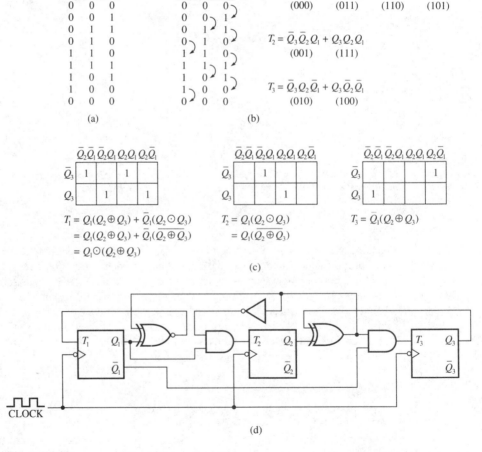

Figure 14.22
Design of a synchronous 3-bit Gray-code counter using T flip-flops (Example 14-6).

$Q_3 Q_2 Q_1$, meaning that when the present state of the counter is either 001 or 111, Q_2 must be toggled on the next clock pulse.

The next step in the design procedure is to simplify the toggling expressions as far as possible. Figure 14.22(c) shows Karnaugh maps, which reveal that no simplifications are possible in this case. However, the expressions can be rewritten in terms of exclusive-OR and coincidence functions, as shown. Figure 14.22(d) shows the logic diagram of the counter, wherein each T input is implemented according to the toggling expressions.

Example 14–7

Design a synchronous decade (BCD) counter using T-type flip-flops.

Solution. Figure 14.23(a) shows the state sequence with arrows indicating state changes in each flip-flop. To facilitate simplification of the toggling expressions using Karnaugh maps, we use decimal cell numbers to represent present states for which toggling is required. For example, we see that $T_4 = \Sigma\, 3, 7$, meaning that when the present state of the counter is either 3 (0011) or 7 (0111), Q_4 must be toggled on the next clock pulse.

Since Q_1 toggles on every clock pulse, it is obvious that $T_1 = 1$. Figure 14.23(b) shows the Karnaugh maps used to simplify T_2, T_4, and T_8. Note that we can make states 10, 11, 12, 13, 14, and 15 *don't cares*, since these do not occur in the BCD sequence. Figure 14.23(d) shows the logic diagrams of the counter, with each T input implemented according to the simplified expressions.

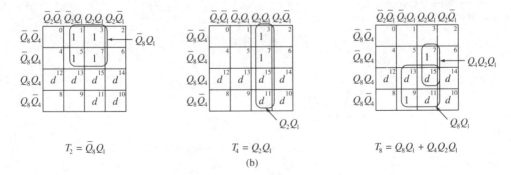

(a)

(b)

Figure 14.23
Design of a synchronous 4-bit BCD counter using T flip-flops. (Example 14–7)

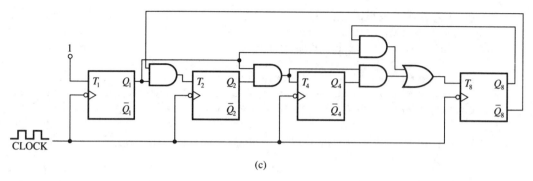

(c)

Figure 14.23
(Continued)

The BCD counter designed in Example 14–7 is self-starting. We can verify that fact by determining the sequence of states through which the counter progresses from each of the *don't care* states 10, 11, 12, 13, 14, and 15. For example, if $Q_8Q_4Q_2Q_1 = 1010$, then $T_1 = 1, T_2 = \overline{Q}_8Q_1 = 0, T_4 = Q_2Q_1 = 0$ and $T_8 = Q_8Q_1 + Q_4Q_2Q_1 = 0$. Therefore, the next state is 1011. This combination makes $T_1 = 1, T_2 = 0, T_4 = 1, T_8 = 1$, so the next state is 0110, which is in the counter sequence. All subsequent states remain in the sequence. Note that we can quickly determine the states of T_2, T_4, and T_8 for any counter state by examining the Karnaugh maps in Figure 14.22(b). The states of T_2, T_4, and T_8 for any *don't care* state can also be determined from the maps: The state of each is the same as the state we assigned to its *don't care* position in each map. For example, when the state of the counter is 1101 (13), $T_8 = 1$ because we let $d = 1$ in cell 13 of the map for T_8. Figure 14.24 shows the complete state diagram for the synchronous BCD counter, using decimal numbers for the states rather than binary numbers.

Designing with JK Flip-Flops

If we design with JK flip-flops and wish to control each J and K input separately, it is helpful to refer to a summary of the states that J and K must have in order to accomplish

Figure 14.24
State diagram for the synchronous BCD counter in Figure 14.23(c), demonstrating that it is self-starting.

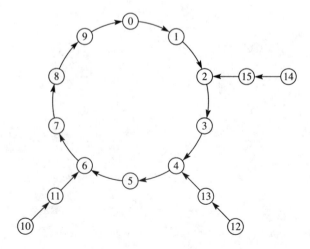

Figure 14.25

State table showing the necessary present states of J and K to achieve all possible flip-flop actions when a clock pulse occurs.

(Present) Q_n	(Next) Q_{n+1}	(Present) J_n	K_n
0	0	0	x
0	1	1	x
1	0	x	1
1	1	x	0

$x = don't\ care$

desired actions in a flip-flop. Figure 14.25 shows such a summary, in the form of a state table for a JK flip-flop. To illustrate, if the present state of Q_n is 0 and its next state is to be 0, we must make the present state of J equal to 0, and we don't care what the present state of K is. (The flip-flop stays reset if $J = 0$ and $K = 1$ or if $J = 0$ and $K = 0$.) The use of this table is demonstrated in the next example.

Example 14–8

Design a synchronous decade counter using JK flip-flops with separate logic circuitry for each J and K input.

Solution.   Figure 14.26(*a*) shows the state table. The present states of all J and K inputs necessary to achieve the next states of the flip-flops are listed, in accordance with the state table in Figure 14.25. For example, when the present state of the counter is 0000, we see that we must have $J_1 = 1$, $K_1 = x$, $J_2 = 0$, $K_2 = x$, $J_4 = 0$, $K_4 = x$, $J_8 = 0$, and $K_8 = x$ in order that the next state of the counter be 0001.

Figure 14.26(*b*) shows the Karnaugh maps used to simplify the expressions for each J and K input. Treating each J and K column of the state table as a separate truth table, we plot the 1s using decimal cell notation. The *don't care* conditions for the J and K inputs, indicated in the state table as x's, are plotted on the maps as x's. The *don't cares* corresponding to the disallowed states 10, 11, 12, 13, 14, and 15 are plotted as d's. However, both types of *don't cares* are the same as far as map simplification is concerned. Figure 14.26(*c*) shows the logic diagram with the implementation of the simplified expressions. We note that this design is somewhat simpler to implement than that of

present decimal	Q_8 Q_4 Q_2 Q_1	next Q_8 Q_4 Q_2 Q_1	J_8 K_8	J_A K_A	J_2 K_2	J_1 K_1
0	0 0 0 0	0 0 0 1	0 x	0 x	0 x	1 x
1	0 0 0 1	0 0 1 0	0 x	0 x	1 x	x 1
2	0 0 1 0	0 0 1 1	0 x	0 x	x 0	1 x
3	0 0 1 1	0 1 0 0	0 x	1 x	x 1	x 1
4	0 1 0 0	0 1 0 1	0 x	x 0	0 x	1 x
5	0 1 0 1	0 1 1 0	0 x	x 0	1 x	x 1
6	0 1 1 0	0 1 1 1	0 x	x 0	x 0	1 x
7	0 1 1 1	1 0 0 0	1 x	x 1	x 1	x 1
8	1 0 0 0	1 0 0 1	x 0	0 x	0 x	1 x
9	1 0 0 1	0 0 0 0	x 1	0 x	0 x	x 1

(a) $d = \Sigma$ 10, 11, 12, 13, 14, 15 $x = don't\ care$

Figure 14.26
Design of a synchronous decade counter using JK flip-flops (Example 14–8).

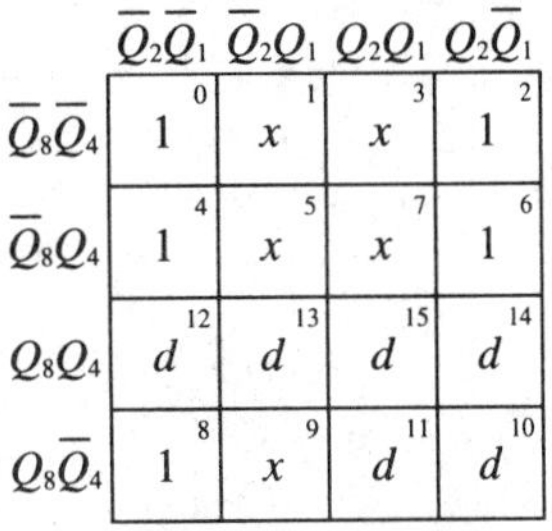

$J_1 = \Sigma\, 0, 2, 4, 6, 8$
$x = \Sigma\, 1, 3, 5, 7, 9$
$J_1 = 1$

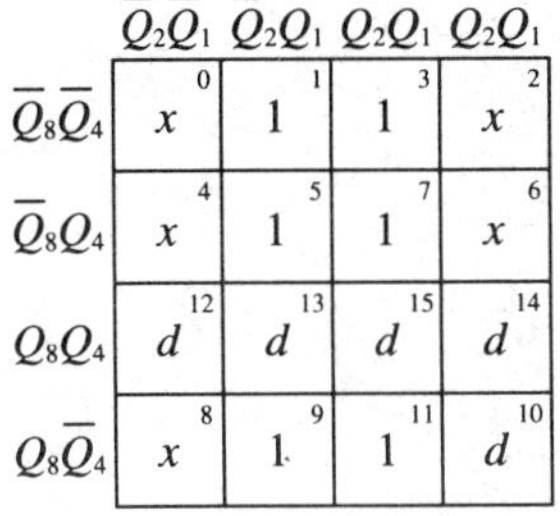

$K_1 = \Sigma\, 1, 3, 5, 7, 9$
$x = \Sigma\, 0, 2, 4, 6, 8$
$K_1 = 1$

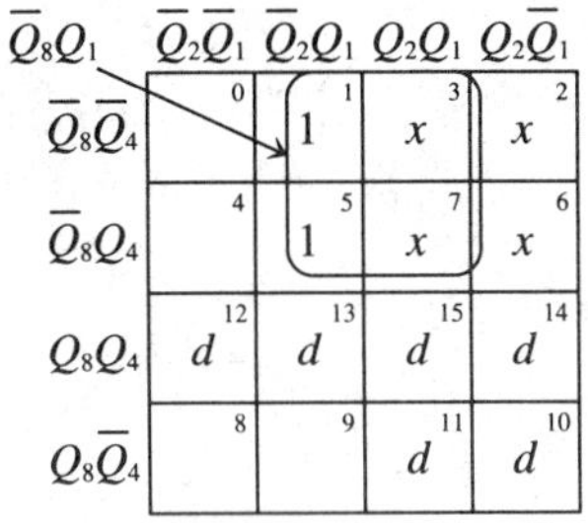

$J_2 = \Sigma\, 1, 5$
$x = \Sigma\, 2, 3, 6, 7$
$J_2 = \overline{Q}_8 Q_1$

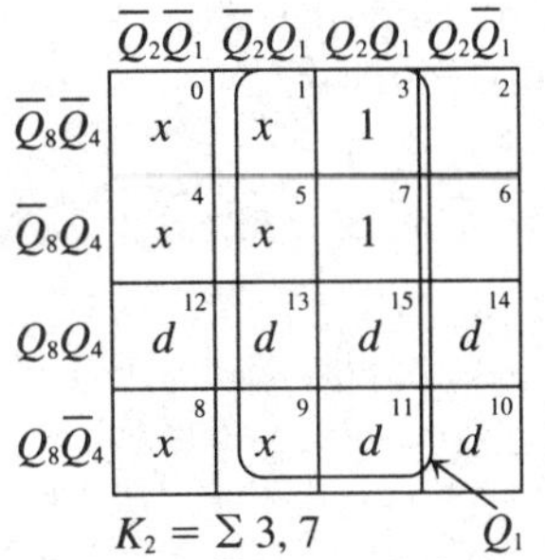

$K_2 = \Sigma\, 3, 7$
$x = \Sigma\, 0, 1, 4, 5, 8, 9$
$K_2 = Q_1$

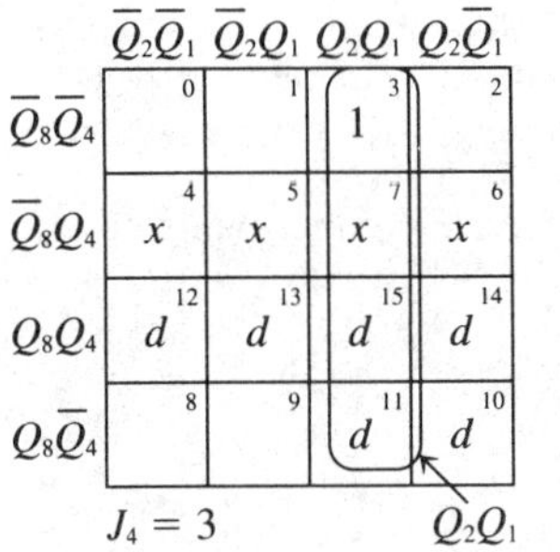

$J_4 = 3$
$x = \Sigma\, 4, 5, 6, 7$
$J_4 = Q_2 Q_1$

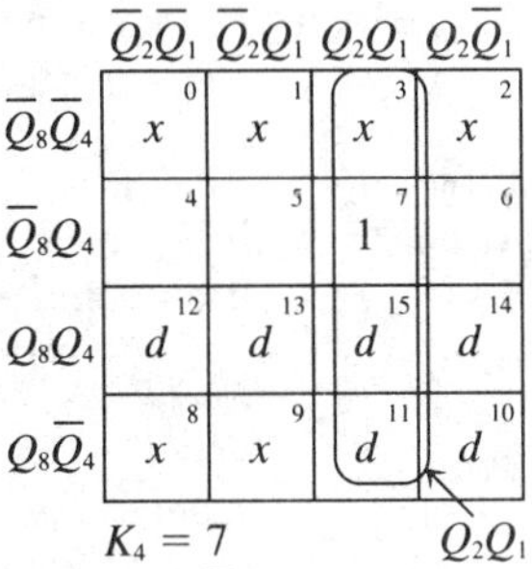

$K_4 = 7$
$x = \Sigma\, 0, 1, 2, 3, 8, 9$
$K_4 = Q_2 Q_1$

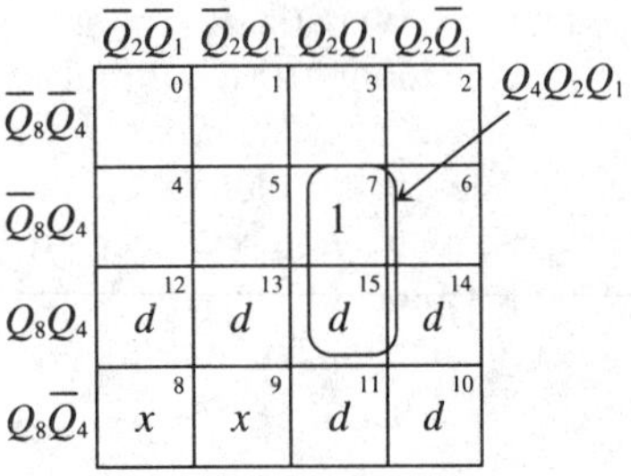

$J_8 = 7$
$x = \Sigma\, 8, 9$
$J_8 = Q_4 Q_2 Q_1$

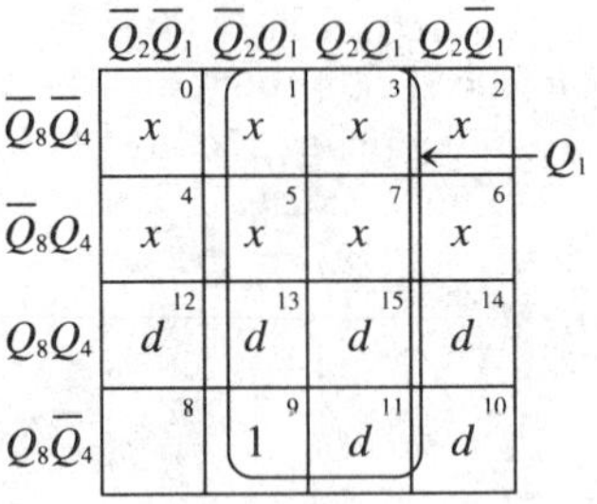

$K_8 = 9$
$x = \Sigma\, 1, 2, 3, 4, 5, 6, 7$
$K_8 = Q_1$

(b)

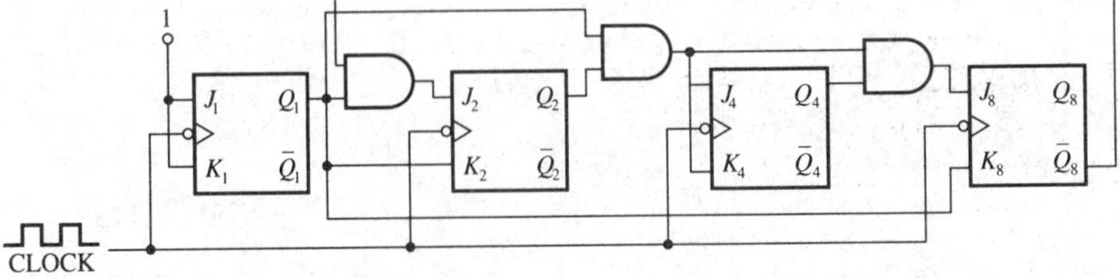

Figure 14.26
(Continued)

564

Example 14–7: Three 2-input AND gates are required, as opposed to four 2-input AND gates and an OR gate in the T flip-flop design.

As is the case in ripple counters, synchronous counters having a modulus other than 2^n may produce asymmetrical waveforms—i.e., flip-flop outputs with other than 50% duty cycles. The next example is another example of synchronous counter design using JK flip-flops and demonstrates lack of symmetry in the outputs.

Example 14–9

a. Design a mod-5 synchronous counter using JK flip-flops with separate logic circuitry for each J and K input.
b. Construct a timing diagram and determine the duty cycle of the output of the most significant stage.
c. Construct a state diagram to determine if the counter is self-starting.

Solution

a. Figure 14.27(a) shows the state table and 14.27(b) shows the Karnaugh maps. Note that the *don't cares* corresponding to disallowed states are 5, 6, and 7 in this case. Figure 14.27(b) shows the logic diagram.
b. Figure 14.28(a) shows the timing diagram. This diagram should be studied carefully to verify how each waveform ensures that the proper flip-flops change states to

present	next	present (required)
decimal	Q_4 Q_2 Q_1	Q_4 Q_2 Q_1	J_4 K_4	J_2 K_2	J_1 K_1
0	0 0 0	0 0 1	0 x	0 x	1 x
1	0 0 1	0 1 0	0 x	1 x	x 1
2	0 1 0	0 1 1	0 x	x 0	1 x
3	0 1 1	1 0 0	1 x	x 1	x 1
4	1 0 0	0 0 0	x 1	0 x	0 x

(a) $d = \Sigma\, 5, 6, 7$ $x =$ don't care

$J_1 = \Sigma\, 0, 2$ $x = \Sigma\, 1, 3$
$J_1 = \overline{Q_4}$

$K_1 = \Sigma\, 1, 3$ $x = \Sigma\, 0, 2, 4$
$K_1 = 1$

$J_2 = \Sigma\, 1$ $x = \Sigma\, 2, 3$
$J_2 = Q_1$

$K_2 = \Sigma\, 3$ $x = \Sigma\, 0, 1, 4$
$K_2 = Q_1$

$J_4 = \Sigma\, 3$ $x = \Sigma\, 4$
$J_4 = Q_2 Q_1$

$K_4 = \Sigma\, 4$ $x = \Sigma\, 0, 1, 2, 3$
$K_4 = 1$

(b)

Figure 14.27
Design of a synchronous mod-5 counter using JK flip-flops (Example 14–9).

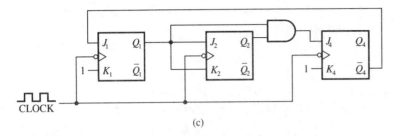

(c)

Figure 14.27
(Continued)

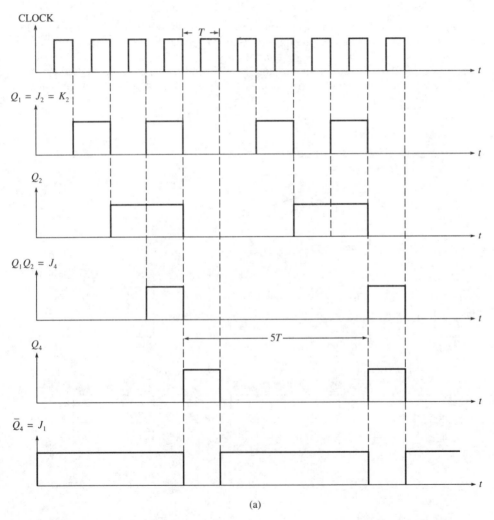

(a)

Figure 14.28
Timing diagram and state diagram for the mod-5 counter in Figure 14.27 (Example 14–9).

| | (initial) present states | | | | | | | next state | | |
decimal	Q_4 Q_2 Q_1	J_4	K_4	J_2	K_2	J_1	K_1	Q_4 Q_2 Q_1	decimal
5	1 0 1	0	1	1	1	0	1	0 1 0	2
6	1 1 0	0	1	0	0	0	1	0 1 0	2
7	1 1 1	1	1	1	1	0	1	0 0 0	0

(b)

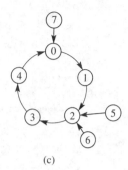

(c)

Figure 14.28
(Continued)

produce the desired sequence. When constructing a timing diagram for a synchronous counter, it is important to show the waveforms that control the J and K inputs. For example, in this design it is necessary to show $Q_1 Q_2$, which controls J_4, and $\overline{Q}_4$, which controls J_1. We see that the duty cycle of the most significant output is $T/5T = 0.2$. Note also that the frequency of that output is one-fifth the clock frequency, as expected in a mod-5 counter.

c. Figure 14.28(b) is a table from which we can determine the counter's sequence when its initial present state is 5, 6, or 7. Given each present state, we can determine the state of each J and K input from the logic diagram. Thus, we can determine each next state. We see that present states 5 and 6 both lead to state 2 and present state 7 leads to state 0. Thus, the counter is self-starting, as shown in the state diagram in Figure 14.28(c).

The Biquinary Counter

The decade counters in Examples 14–7 and 14–8 also produce asymmetrical outputs at their most significant stages. If a divide-by-10 output having a 50% duty cycle is required, the *biquinary* counter shown in Figure 14.29(a) can be used. Note that this counter consists simply of cascaded mod-2 and mod-5 synchronous counters: hence the name bi(2)-qui(5)-nary. The timing diagram in Figure 14.29(b) confirms that the most significant output has a 50% duty cycle. (The waveforms for Q_1, Q_2, and Q_4 are the same as those shown in Figure 14.28(a) for the mod-5 counter.) Note that the count sequence in this case is *not* BCD. The state diagram in Figure 14.29(c) shows the actual sequence of (decimal) states and shows that the counter is self-starting.

Hybrid Counters

Another way to obtain a symmetrical divide-by-N output is to use a *hybrid* counter: a synchronous counter whose output drives the clock input of another counter. For example, when N is any number divisible by 2, we can obtain a symmetrical divide-by-N waveform using the method illustrated in Figure 14.30. Here, a synchronous counter with modulus $N/2$ drives the clock input of a mod-2 counter (a single flip-flop). The output of the mod-2 counter has frequency f_c/N, since the overall modulus of the hybrid counter is $(N/2)(2) = N$.

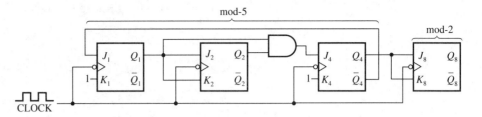

(a) Logic diagram.

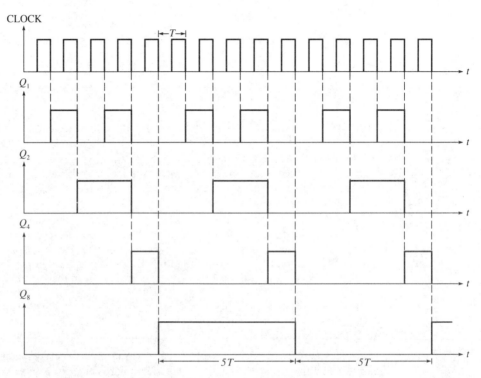

(b) Timing diagram. Q_8 has $1/10$ the frequency of the clock and has a 50% duty cycle.

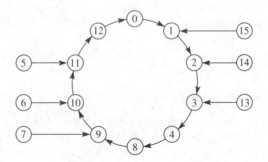

(c) State diagram. Note that the normal count sequence is not BCD.

Figure 14.29
The biquinary counter, used to obtain a symmetrical, divide-by-10 output.

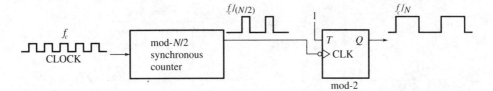

Figure 14.30
Using a hybrid counter to obtain a symmetrical divide-by-N waveform when N is a number divisible by 2.

14-7 INTEGRATED-CIRCUIT SYNCHRONOUS COUNTERS

Programmable Counters

Most synchronous counters available in integrated circuits can be *preloaded* with a binary number in parallel form (as with parallel-in shift registers) prior to initiation of counting. This preloading capability makes it possible to begin a count sequence from zero or any other number. Such counters are said to be *programmable*. In a counter with *asynchronous loading,* the initial state is loaded using direct set and direct clear inputs on the flip-flops. Thus, loading occurs irrespective of the clock. In a counter with *synchronous loading,* the initial state is loaded on the occurrence of a clock edge, using the *J* and *K* inputs of the flip-flops. Synchronous loading generally requires the use of rather elaborate logic circuitry connected to the *J* and *K* inputs, since those inputs are also needed for control of the normal counting sequence. In either case, the circuit will have a separate load control input, which must be made active to achieve loading. Some circuits can be loaded either synchronously or asynchronously and have separate control inputs, such as SLOAD and ALOAD, to enable one type of loading or the other.

Look-Ahead Carry

We have seen that the *J* and *K* inputs in a synchronous binary up counter are controlled by $J_n = K_n = Q_1 Q_2 \cdots Q_{2^{n-1}}$. Figure 14.31(a) shows the implementation of this logic in a 4-bit counter, using cascaded AND gates as we have in the past. Recall that the logic $Q_1 Q_2 \cdots$ is called a *carry* that is brought forward to each stage. Note that the carry in Figure 14.31(a) must *ripple* through successive AND gates; i.e., the output of the rightmost AND gate is not valid until the outputs of all preceding AND gates are valid. Since the propagation delays of the AND gates accumulate, the output of the rightmost AND gate is not valid until after the sum of the propagation delays of all the AND gates has elapsed. This cumulative delay limits the counting speed of a synchronous counter. To increase counting speed, many integrated-circuit versions of synchronous counters use the *look-ahead* carry illustrated in Figure 14.31(b). Notice that the *J* and *K* control logic is the same as that in the ripple-carry circuit, but the accumulation of propagation delays is eliminated. The total delay at each stage in this case equals the propagation delay of just one AND gate.

The 74160- through 74163-Series Counters

The 74160 through 74163 series are examples of 4-bit integrated-circuit synchronous counters having synchronous loading and look-ahead carry. The Schottky versions are

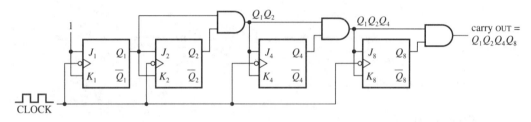

(a) Sychronous counter with ripple carry. Counting speed is limited by the accumulation of propagation delays through the AND gates.

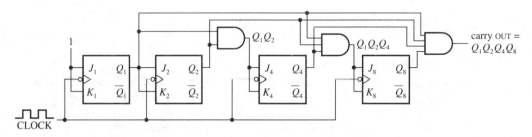

(b) Synchronous counter with look-ahead carry. The carry input to each stage is delayed by the propagation delay of just one AND gate.

Figure 14.31
Comparison of ripple carry and look-ahead carry in synchronous counters.

constructed with D flip-flops instead of JK or T-type flip-flops, and the logic diagrams are somewhat too complex to reproduce here. Figure 14.32 shows a pin diagram applicable to all circuits. The 74160 and 74162 are decade counters and the 74161 and 74163 are binary counters. Figure 14.32 gives a summary of the functions of the inputs and outputs. Specifications for devices in the series give the following as the minimum values of the maximum clocking frequency:

74160 through 74163:   25 MHz
74LS160A through 74LS163A: 25 MHz
74ALS160B through 74ALS163B: 40 MHz
74AS160 through 74AS163:   75 MHz

Figure 14.33 shows a typical timing diagram for the 74163-series binary counter with synchronous clear. Assuming that the initial state of the counter is 1111, the sequence of events shown in the timing diagram is as follows:

1. $\overline{\text{CLR}}$ goes low, making $Q_D Q_C Q_B Q_A = 0000$ on the leading edge of the next clock pulse.
2. $\overline{\text{LOAD}}$ goes low, and DCBA = 1100 = $12)_{10}$ is loaded into $Q_D Q_C Q_B Q_A$ on the leading edge of the next clock pulse.
3. Enable inputs ENP and ENT go high and up counting commences. The count goes to 1101 = $13)_{10}$ on the leading edge of the clock pulse.
4. Counting continues to 1111 = $15)_{10}$, at which time RCO goes high. The counter resets to 0000 and continues upcounting.

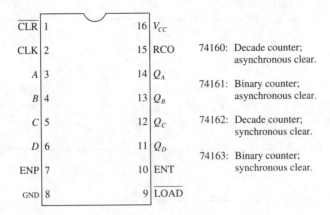

74160:	Decade counter; asynchronous clear.	
74161:	Binary counter; asynchronous clear.	
74162:	Decade counter; synchronous clear.	
74163:	Binary counter; synchronous clear.	

$\overline{\text{CLR}}$ Active-low clear. When made low, the count clears immediately to 0000 in the 74160 and 74161. In the 74162 and 74163, the count clears to 0000 on the leading edge of the next clock pulse.

CLK   Clock input. Flip-flops are triggered on the leading edge of the clock.

A, B, C, D   Input data for preloading. A is the least significant bit and D is the most significant bit. Data is loaded synchronously when $\overline{\text{LOAD}}$ is low.

ENP, ENT   Active-high enable inputs. When either input is low, counting is inhibited. The counter retains the last state it was in.

Q_A, Q_B, Q_C, Q_D Flip-flop outputs. Q_A is least significant and Q_D is most significant.

RCO   Ripple carry output, enabled by ENT. In binary counters, RCO = $(Q_A Q_B Q_C Q_D) \cdot$ ENT. In decade counters, RCO = $(Q_A \overline{Q_B}\, \overline{Q_C}\, Q_D) \cdot$ ENT.

$\overline{\text{LOAD}}$ Synchronous input loading control. When $\overline{\text{LOAD}}$ is made low, the data at ABCD is loaded into Q_A, Q_B, Q_C, Q_D on the rising edge of the next clock pulse (irrespective of the states of enable inputs ENP and ENT).

Figure 14.32
The 74160- through 74163-series 4-bit synchronous counters.

5. After the count of 0010 = 2)$_{10}$ is reached, enable input ENP goes low, so counting ceases. The count 0010 is retained in $Q_D Q_C Q_B Q_A$.

Figure 14.34 shows two ways to cascade 74160- through 74163-series counters using the ripple-carry outputs and the enable inputs. In cascaded binary counters, each 4-bit stage generates an RCO pulse when its count reaches 1111, and the RCO pulse enables the next higher stage so its count can be incremented by 1. In cascaded decimal counters, the same action occurs when the count reaches 1001 = 9)$_{10}$. Thus, the contents of n cascaded 4-bit decimal counters represent an n-decimal-digit BCD number. In Figure 14.34(a), called a ripple-mode carry circuit, every ENP enabling input is held high for normal counting, and the RCO output of one stage activates the ENT enable of the next higher stage. Since the next higher stage is enabled only for the duration of a single clock pulse, its count is incremented only by 1. In Figure 14.34(b), called a look-ahead carry mode, ENP of the first stage and ENT of the second stage are held high, and RCO from the first stage activates ENP of the other stages. At the same time, RCO of each stage

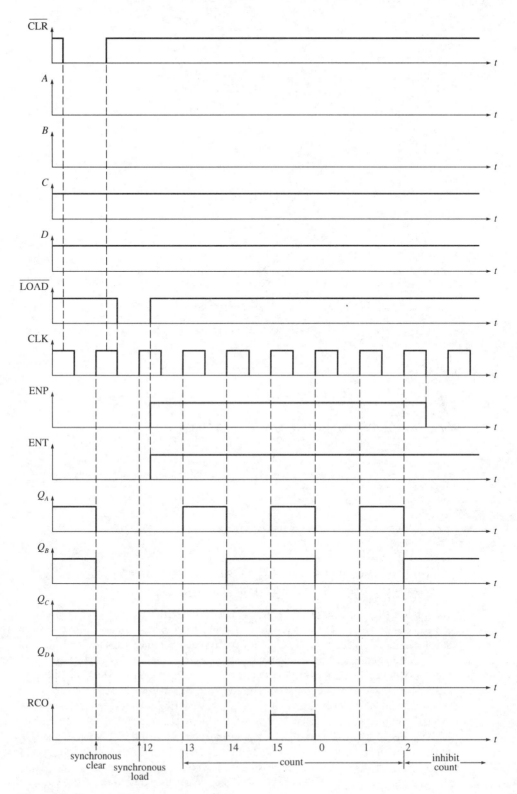

Figure 14.33
Typical timing diagram for the 74163A-series counter.

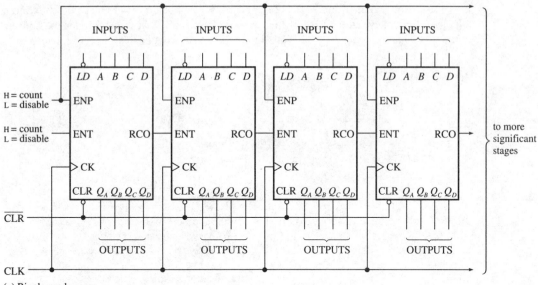

(a) Ripple-mode carry.

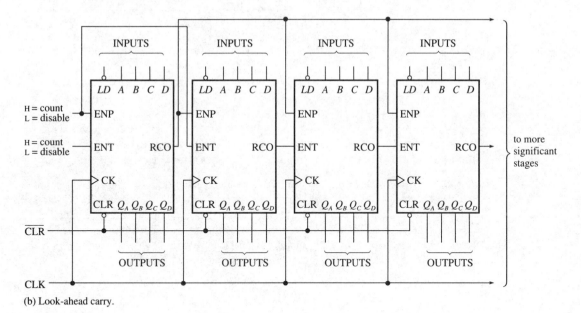

(b) Look-ahead carry.

Figure 14.34
Cascading 74160- through 74163-series counters.

beyond the first activates ENT of the next higher stage. The carry mode, "ripple" or "look-ahead," refers in this case to how the carry is propagated between 4-bit stages. We already know that carry look-ahead is implemented within each stage—i.e., between the bits of each 4-bit stage.

Cascading stages of the 74160- through 74163-series counters reduces the maximum clocking frequency. In the case of the ripple-mode circuit, that reduction depends on the number of stages cascaded. The maximum clocking frequencies can be calculated as follows:

Ripple-mode carry:

$$f_{\max} = \frac{1}{t_{pLH} \text{ (CLK to RCO)} + (n - 2)t_{pLH} \text{ (ENT to RCO)} + t_{su} \text{ (ENT)}} \quad \textbf{(14.3)}$$

where n is the number of cascaded 4-bit stages.
Look-ahead carry:

$$f_{\max} = \frac{1}{t_{pLH} \text{ (CLK to RCO)} + t_{su} \text{ (ENP)}} \quad \textbf{(14.4)}$$

Manufacturers' specifications for the maximum values of the quantities appearing in equations (14.3) and (14.4) are as follows:

Series	Specification of maximum value			
	t_{pLH} **(CLK to RCO)**	t_{pLH} **(ENT to RCO)**	t_{su} **(ENT)**	t_{su} **(ENP)**
74160–74163	35 ns	16 ns	20 ns	20 ns
74LS160A–74LS163B	35 ns	14 ns	20 ns	20 ns
74ALS160B–74ALS163B	20 ns	13 ns	*15 ns	*15 ns
74AS160–74AS163	12.5 ns	9 ns	*8 ns	*8 ns

*Minimum values

Example 14–10   A BCD counter capable of counting up to 999 and having an asynchronous clear is to be constructed using low-power Schottky technology. The counter is to be clocked so that the output of the most significant flip-flop has frequency 16 kHz. Design the circuit.

Solution.   To count to 999, three decade counters are required. The decade counter with asynchronous clear in the LS series is the 74LS160A. Since three decade (mod-10) counters divide the clock frequency by 10^3, the counter must be clocked at $10^3 \times 16$ kHz $= 16$ MHz in order to produce 16 kHz at the most significant output. To determine whether ripple or look-ahead cascading is required, we use equations (14.3) and (14.4) to calculate the maximum clocking frequencies of each type.

Ripple-mode carry:

$$f_{\max} = \frac{1}{35 \text{ ns} + (3 - 2)14 \text{ ns} + 20 \text{ ns}} = \frac{1}{69 \times 10^{-9} \text{ s}} = 14.5 \text{ MHz}$$

Look-ahead carry:

$$f_{\max} = \frac{1}{35 \text{ ns} + 20 \text{ ns}} = \frac{1}{55 \times 10^{-9} \text{ s}} = 18.2 \text{ MHz}$$

We see that look-ahead cascading is necessary to accommodate a 16-MHz clocking frequency, and the 74LS160A circuits must be connected as shown in Figure 14.34(*b*).

Drill
14–10
What is the maximum permissible set up time for the ENT input of each 74ALS161B in a ripple-cascaded binary counter whose maximum count is 1,048,576, if the counter must be clocked at 10.1 MHz?
Answer: 40 ns.

The 74190/74191-Series Up/Down Counters

The 74190 is a decade up/down counter and the 74191 is a 4-bit binary up/down counter. Figure 14.35(*a*) shows a pin diagram applicable to both counters. Both can be preloaded asynchronously when the control input $\overline{\text{LOAD}}$ is made low. Neither has a synchronous or asynchronous clear. Up/down counting is controlled by a D/$\overline{\text{U}}$ input that causes up counting when high and down counting when low. $\overline{\text{RCO}}$ is an active-low ripple carry output that goes low for one clock pulse when the 74190 counts to 9 or 0 and when the 74191 counts to 15 or 0. MAX/MIN produces a high output for one clock *period* when the counter counts *up* to 15 or counts *down* to 0. $\overline{\text{CTEN}}$ is an active-low counter enable input. $\overline{\text{RCO}}$, MAX/MIN, and $\overline{\text{CTEN}}$ are provided to facilitate cascading, as discussed in Section 14–5 (''Cascading Synchronous Counters''). The counters are available in standard, LS, and ALS technologies. Figure 14.35(*b*) shows a typical timing diagram for a 74190 decade counter. Note that flip-flops change state on the leading edges of clock pulses.

The 74ALS560A/74ALS561A Counters

The 74ALS560A/74ALS561A 4-bit synchronous counters are similar to the 74160-through 74163-series but have considerably more versatility. The 74ALS560A is a decade counter and the 74ALS561A is a binary counter. Both have internal look-ahead carry, both are programmable, and both have enable and carry-out functions to facilitate cascading. In addition, the 74ALS560A and 74ALS561A have three-state outputs. The outputs can be put into a high-impedance state by making control input $\overline{G}$ high. (Counting is not inhibited when the outputs are in a high-impedance state.) Furthermore, these counters can be cleared either synchronously or asynchronously (using $\overline{\text{ACLR}}$ and $\overline{\text{SCLR}}$ inputs), and can be loaded either synchronously or asynchronously (using $\overline{\text{ALOAD}}$ and $\overline{\text{SLOAD}}$ inputs). Figure 14.36(*a*) shows the pin diagram for both counters. ENT and ENP perform the same enabling functions as in the 74160–74163-series; both must be high to enable counting. The clocked-carry output (CCO) is similar to the ripple-carry output (RCO) but does not have the glitches commonly associated with RCO. However, because of timing considerations, the manufacturer recommends that RCO be used for cascading in high-frequency applications. Figure 14.36(*b*) shows a typical timing diagram for the 74ALS561A binary counter. Manufacturers' specifications state that the minimum value of the maximum counting frequency is 20 MHz for the 74ALS560A and 30 MHz for the 74ALS561A.

14–8   ANSI/IEEE STANDARD SYMBOLS

Asynchronous Counters

The qualifying symbol for an asynchronous (ripple) counter is RCTR*n*, where *n* is the number of bits. Figure 14.37 shows an example: the 74HC4024 7-bit binary counter. The

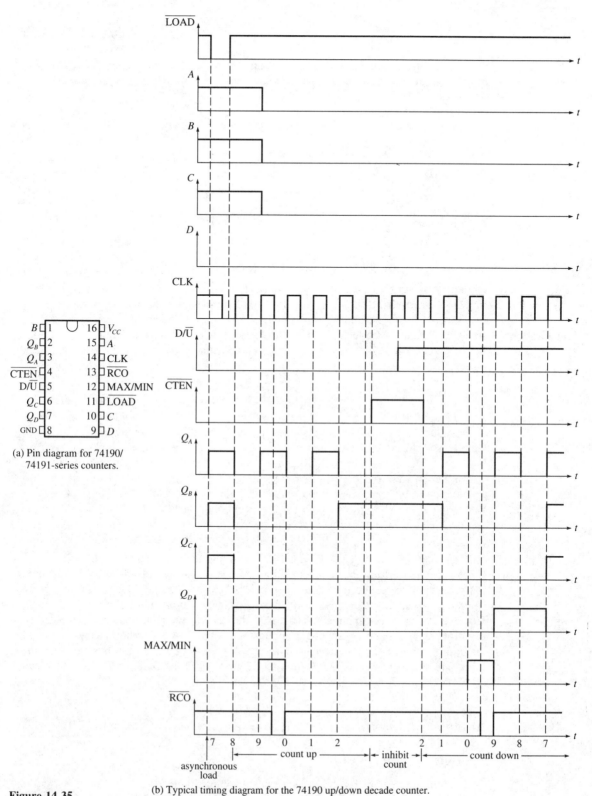

(a) Pin diagram for 74190/
74191-series counters.

(b) Typical timing diagram for the 74190 up/down decade counter.

Figure 14.35
Pin diagram and timing diagram for 74190/74191-series up/down counters.

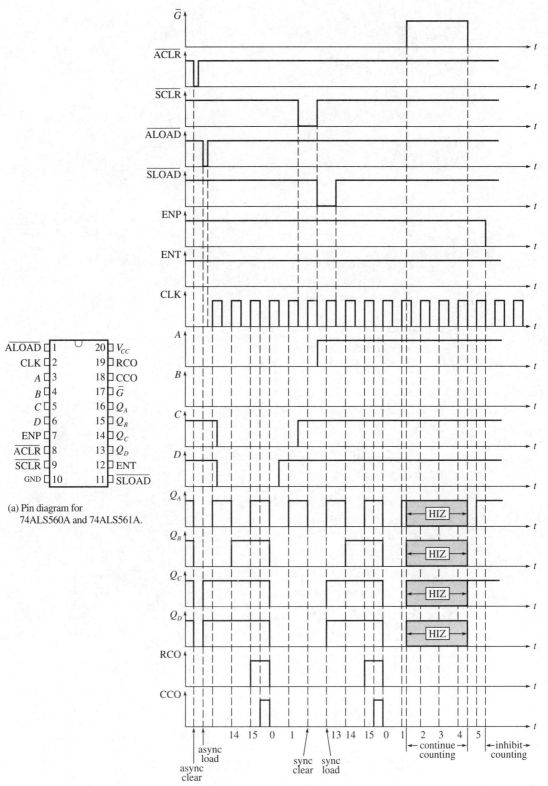

(a) Pin diagram for 74ALS560A and 74ALS561A.

Figure 14.36
(b) Typical timing diagram for the 74ALS561A binary counter.
Pin and timing diagrams for the 74ALS560A and 74ALS561A counters.

Figure 14.37

Logic symbol for a 7-bit binary ripple counter.

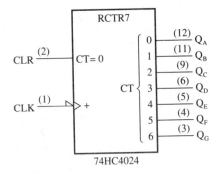

+ symbol at the clock input means it is an up counter; a − symbol is used for a down counter. Note the use of the binary grouping symbol applied to the seven outputs, numbered 0 through 6. As in past examples, this means that the output assumes all $2^7 = 512$ states from $0)_{10}$ through $511)_{10}$. The notation CT = 0 means that the counter output becomes $0)_{10} = 0000000)_2$ when the CLR input is active (high). Note that clearing is asynchronous, since there is no clock dependency shown at the CLR input.

Synchronous Counters

The qualifying symbol for a synchronous binary counter is CTR*n*, where *n* is the number of bits. The symbol CTR DIV *n* is also used for both synchronous and asynchronous counters to identify a divide-by-*n* counter. Note that CTR*n* is equivalent to CTR DIV 2^n. Figure 14.38 shows the logic symbol for the 74AS162 synchronous decade counter. In this case, we see that clearing is synchronous: The $\overline{\text{CLR}}$ input is connected to 5CT = 0, indicating clock dependency, since the CLK input is connected to C5. The LOAD input creates two mode dependencies, M1 and M2, where, as we shall see, mode 1 is the parallel-load mode ($\overline{\text{LOAD}}$ low) and mode 2 is the up-count mode ($\overline{\text{LOAD}}$ high). Note the AND dependency notation for inputs ENT (G3) and ENP (G4). As discussed in Section 13–12, we know that other symbols having the prefix 3 or 4 are dependent through the AND function on these inputs. Thus, the numeral 3 in the notation 3CT = 9 at the RCO output indicates that RCO will be active (high) when the count is $9)_{10}$ *and* ENT is active (high). The notation at the CLK input, C5/2,3,4 +, means that up counting (+) occurs

Figure 14.38

Logic symbol for a synchronous decade counter.

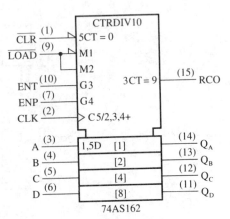

Figure 14.39
Logic symbol for a 4-bit binary up/down counter.

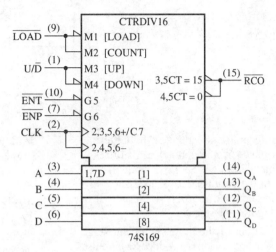

in mode 2, provided ENT *and* ENP are both high. The notation 1,5D at the input to the A stage means that data (D) is loaded synchronously (5) when in mode 1. Since this notation is understood to apply to the inputs of all four stages, we know that parallel, synchronous loading occurs in mode 1 ($\overline{\text{LOAD}}$ low).

Up/Down Counters

Figure 14.39 shows the logic symbol for the 74S169 4-bit, binary up/down counter. Note that four mode dependencies are defined:

$$M1 = \text{load}, \quad \text{when } \overline{\text{LOAD}} \text{ is low}$$
$$M2 = \text{count}, \quad \text{when } \overline{\text{LOAD}} \text{ is high}$$
$$M3 = \text{up}, \quad \text{when } U/\overline{D} \text{ is high}$$
$$M4 = \text{down}, \quad \text{when } U/\overline{D} \text{ is low}$$

The notation at the upper CLK input, 2,3,5,6+/C7, means that up counting (+) occurs in modes 2 and 3 provided $\overline{\text{ENT}}$ and $\overline{\text{ENP}}$ (5, 6) are active (low). The notation at the lower CLK input, 2,4,5,6−, means that down counting (−) occurs in modes 2 and 4 provided $\overline{\text{ENT}}$ and $\overline{\text{ENP}}$ are active. The notation at the upper ripple-carry output ($\overline{\text{RCO}}$), 3,5CT = 15, means that $\overline{\text{RCO}}$ is active (low) in mode 3 when the count is $15)_{10}$ *and* $\overline{\text{ENT}}$ is active. The notation 4,5CT = 0 means that $\overline{\text{RCO}}$ is active in mode 4 when the count is $0)_{10}$ *and* $\overline{\text{ENT}}$ is active. Thus, $\overline{\text{RCO}}$ is active when an up count reaches 15 and when a down count reaches 0. The notation 1,7D applying to the inputs of the four stages means that in mode 1 (load), parallel, synchronous (7) loading of data (D) occurs.

EXERCISES

Section 14–2

14.1 Draw a logic diagram of a binary ripple counter that can count up to $15)_{10}$. Construct a timing diagram showing the output at each stage over an interval of time during which 16 input pulses occur. Construct a count table.

14.2 Draw a logic diagram of a binary ripple counter whose output at its most significant stage has frequency 2 kHz when the input clock frequency is 64 kHz. What is the decimal equivalent of the largest number this counter can store?

Figure 14.40
Exercise 14.5.

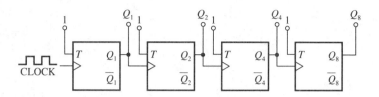

14.3 The decimal equivalent of the largest number a binary ripple counter can store is 4095. What should be the clock frequency at the input to the counter if it is necessary to obtain a frequency of 550 Hz at the output of the most significant stage?

14.4 The output of the most significant stage of a binary ripple counter produces 875 pulses for every 896,000 clock pulses at the input to the counter. The clock frequency is 128 kHz.
(a) How many flip-flops does the counter have?
(b) What is the period of the output at the most significant stage?

14.5 Draw a timing diagram showing the full range of outputs at each stage of the counter shown in Figure 14.40. (Assume all flip-flops are initially set.) Construct a count table. Is the counter an up counter or a down counter?

14.6 Repeat Exercise 14.5 for the counter shown in Figure 14.41. (Assume all flip-flops are initially reset.)

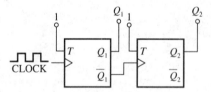

Figure 14.41
Exercise 14.6.

14.7 Repeat Exercise 14.5 for the counter shown in Figure 14.42. The outputs are Q_1', Q_2', Q_4'. (Assume all flip-flops are initially set.)

14.8 Repeat Exercise 14.5 for the counter shown in Figure 14.43. (Assume all flip-flops are initially set.)

Section 14–3

14.9 Draw a logic diagram for a mod-7 ripple counter. Construct a timing diagram showing the output of each stage over an interval of time in which eight input clock pulses occur. Be certain to show any glitches that appear in any outputs.

14.10 Determine the modulus of the counter shown in Figure 14.44. Draw a timing diagram that shows any glitches that may occur in any output(s). Construct a count table.

14.11 Draw a timing diagram showing the output waveform at each stage of the counter shown in Figure 14.45. Construct a count table. Assume all flip-flops are initially reset. What is the frequency of the waveform at Q_4 when the frequency of the clock is 750 kHz?

14.12 Design a ripple counter that counts in the sequence 111, 110, 101, 100, 011, 000, 111, ... Draw a timing diagram to confirm that your design produces the required sequence.

14.13 The clock signal in a certain digital system has period 250 ns. It is necessary to

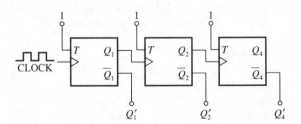

Figure 14.42
Exercise 14.7.

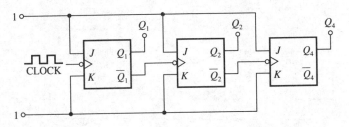

Figure 14.43
Exercise 14.8.

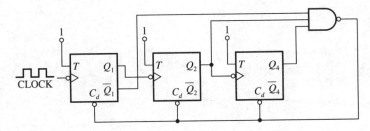

Figure 14.44
Exercise 14.10.

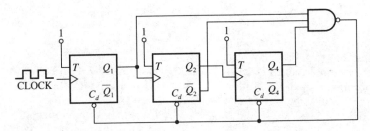

Figure 14.45
Exercise 14.11.

produce two synchronizing signals having frequencies 50 kHz and 1.25 kHz using standard mod-5, mod-8, and mod-10 counters. Draw a block diagram showing how the standard-modulus counters could be cascaded to produce the required frequencies.

14.14 The clock signal in a digital system has frequency 2.4 MHz. Draw a block diagram showing how mod-5, mod-8, mod-10, and mod-12 counters could be connected to obtain signals having frequencies 25 kHz and 4 kHz. Use the fewest possible number of counters.

14.15 Draw a timing diagram showing the waveforms of Q_A, Q_B, Q_C, and Q_D in Figure

14.46. If the frequency of the clock is 1 MHz, what is the frequency of the waveform at Q_D? What is the duty cycle of the waveform at Q_D?

14.16 Draw a timing diagram showing the waveforms of Q_A, Q_B, Q_C, and Q_D in Figure 14.47. If the frequency of the clock is 1 MHz, what is the frequency of the waveform at Q_D? What is the duty cycle of the waveform at Q_D?

Section 14–4

14.17 The propagation delay of each flip-flop in an eight-stage ripple counter is 20 ns. What is the limiting clock frequency of the counter if

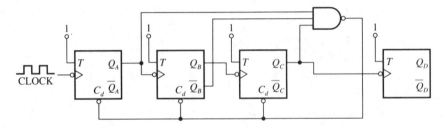

Figure 14.46
Exercise 14.15.

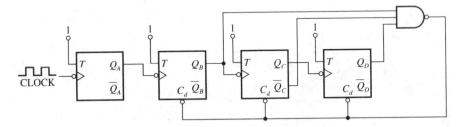

Figure 14.47
Exercise 14.16.

it must successfully register every count in its sequence?

14.18 A ripple counter is to be constructed using flip-flops whose propagation delays are 35 ns. If the counter is to be clocked at 2 MHz, what is the maximum modulus of the counter if it must successfully register every count in its sequence?

14.19 The period of the clock driving a ripple counter is 0.25 μs. The propagation delay of each flip-flop is 30 ns. What is the value of the maximum binary number the counter can reach if it must register every count in its sequence?

14.20 A mod-65,536 ripple counter is driven by a clock having period 0.25 μs. What is the maximum propagation delay that each flip-flop in the counter can have if the counter must register every count in its sequence?

14.21 Draw a logic diagram of a 3-bit ripple counter driving a binary-to-decimal decoder. Construct a timing diagram showing the outputs of the flip-flops in the counter and the outputs of the decoder, ignoring all propagation delays.

14.22 Repeat Exercise 14.21 when the clock frequency is 10 MHz and the propagation delay

in each flip-flop is 20 ns. Show all glitches that occur in the decoder outputs and indicate their widths.

14.23
(a) Draw a logic diagram of a mod-5 ripple counter driving a binary-to-decimal decoder.
(b) Construct a timing diagram showing all glitches that appear in the decoder outputs.
(c) Modify the diagram in *(a)* to show how the decoder outputs having glitches can be deglitched.

14.24
(a) Draw a logic diagram of a mod-3 ripple counter using *leading-edge*-triggered flip-flops and driving a binary-to-decimal decoder.
(b) Construct a timing diagram showing all glitches that appear in the decoder outputs.
(c) Modify the diagram in *(a)* to show how the decoder outputs having glitches can be deglitched.
(d) Construct a timing diagram showing the outputs of the deglitched decoder.

14.25 A 7493 counter is to be used to produce a 7.5-kHz output at its most significant stage

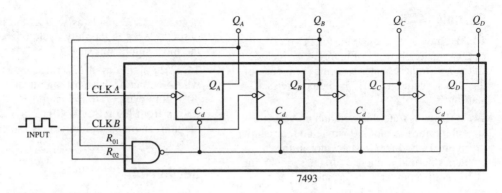

Figure 14.48
Exercise 14.26.

when the clock frequency at its input is 45 kHz. Draw a logic diagram showing how the 7493 should be connected to meet this requirement. Label the input and then identify the least significant and most significant outputs.

14.26 What is the modulus of the 7493 counter connected as shown in Figure 14.48?

Section 14–5

14.27 Draw a logic diagram and a timing diagram for a 4-bit synchronous binary up-counter using trailing-edge-triggered JK flip-flops. What is the modulus of this counter?

14.28 Draw a logic diagram showing how a 4-bit synchronous binary up counter could be constructed using trailing-edge-triggered T-type flip-flops.

14.29 The trailing-edge-triggered flip-flops in a 3-bit synchronous up counter each have $t_{pLH} = 10$ ns and $t_{pHL} = 20$ ns. Draw a timing diagram showing the glitches that appear in the outputs of a binary-to-decimal decoder driven by the counter. Indicate the widths of the glitches.

14.30 Repeat Exercise 14.29 when each flip-flop has $t_{pLH} = 20$ ns and $t_{pHL} = 10$ ns.

14.31 Replace the flip-flops in Figure 14.12(b) with leading-edge-triggered JK flip-flops. Then construct a timing diagram for the circuit. What type of counter does the revised circuit represent?

14.32 Draw a timing diagram for the circuit shown in Figure 14.49. What type of counter does this circuit represent?

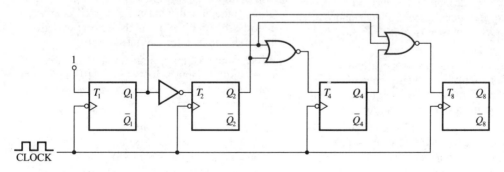

Figure 14.49
Exercise 14.32.

Section 14–6

14.33 Draw a state diagram for the mod-5 ripple counter shown in Figure 14.4. Include states not in the counter's normal sequence of the states, including 101. Is the counter self-starting?

14.34 Draw a state diagram and construct a state table for the mod-10 ripple counter shown in Figure 14.5. Include states not in the counter's normal sequence of states, including 1010. Is the counter self-starting?

14.35 Design a mod-5 synchronous counter using T-type flip-flops.

14.36 Design a mod-13 synchronous counter using T-type flip-flops. Draw a state diagram and determine if your design is self-starting.

14.37 A synchronous counter is required to count up from $0)_{10}$ to $10)_{10}$ through the even numbers and then down to 1 through the odd numbers—i.e., in the sequence $0,2,4,6,8,10,9,7,5,3,1,0, \ldots$ Design the counter using T-type flip-flops. Draw a state diagram and determine if your design is self-starting.

14.38 Design a synchronous counter using T-type flip-flops that counts in the sequence $7, 6, 5, 4, 3, 7, 6, 5, 4, 3, \ldots$ Construct a state diagram and determine if your design is self-starting. Construct a timing diagram showing the states of the T-inputs as well as the flip-flop outputs. By what factor is the clock frequency divided at the most significant output?

14.39 To verify the design of the 3-bit synchronous up counter in Figure 14.12(b), use the procedure for designing synchronous counters with individually controlled J and K inputs to derive expressions for J_1, K_1, J_2, K_2, J_4 and K_4.

14.40 Design a mod-6 synchronous counter using flip-flops with individually controlled J and K inputs. Draw a state diagram to determine if your design is self-starting.

14.41 Design a mod-12 synchronous counter using flip-flops with individually controlled J and K inputs. Draw a state diagram to determine if your design is self-starting. Construct a timing diagram showing the levels of the J and K inputs as well as the flip-flop outputs. What is the duty cycle of the output of the most significant stage?

14.42 Design a mod-12 synchronous counter whose output at the most significant stage has a 50% duty cycle. (*Hint:* Incorporate the mod-6 design from Exercise 14.40). Draw a timing diagram and a state diagram. Is the counter self-starting?

Section 14–7

14.43 The timing diagram in Figure 14.50 shows inputs to the 74AS161 counter. Complete the timing diagram to show outputs Q_A, Q_B, Q_C, Q_D, and RCO. Assume that the initial output states are $Q_A = 0, Q_B = 1, Q_C = 1$, and $Q_D = 0$.

14.44 The timing diagram in Figure 14.51 shows inputs to the 74ALS162B counter. Complete the timing diagram to show outputs Q_A, Q_B, Q_C, Q_D, and RCO. Assume that the initial output states are $Q_D = 1, Q_C = 0, Q_B = 1$, and $Q_A = 0$.

14.45 A BCD counter is to be constructed by cascading 74AS162 counters in the configuration shown in Figure 14.34(b).
(a) How many 74AS162 circuits must be used if the counter must be capable of counting the total number of seconds in one 24-h day?
(b) What is the maximum permissible setup time for the ENP input of each circuit if the counter is to be clocked at 40 MHz?

14.46 A binary counter is to be constructed by cascading 74LS161 counters in the configuration shown in Figure 14.34(a). The clock frequency driving the cascaded counters is to be 10 MHz. What is the largest overall modulus the counter can have?

14.47 Draw a timing diagram showing the clock and all input signals to a 74191 up/down counter necessary to obtain the following sequence of events:

1. Load the counter with count $2)_{10}$.
2. Beginning with the next clock pulse after loading, count down to $13)_{10}$.
3. Inhibit counting for the next two clock periods.

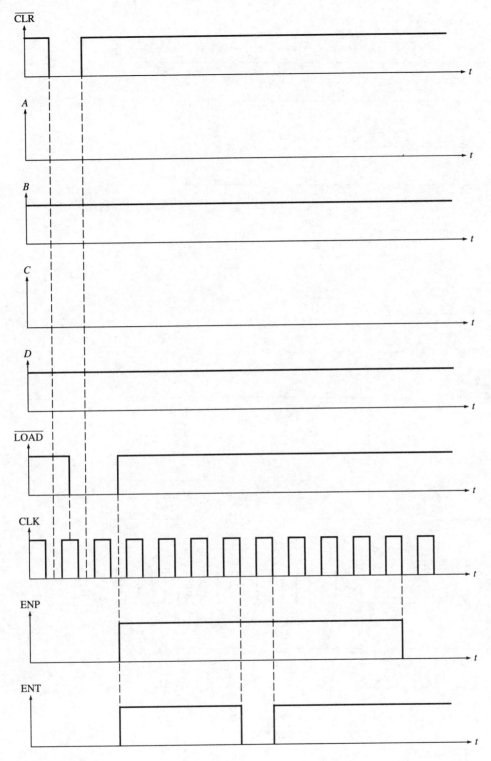

Figure 14.50
Exercise 14.43.

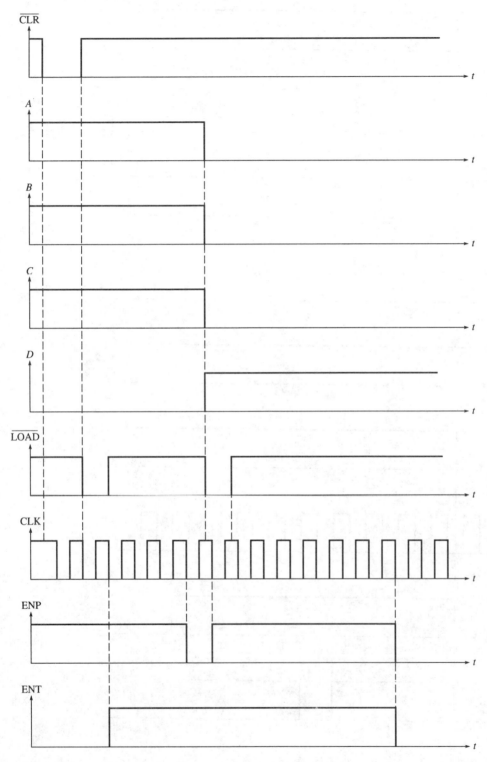

Figure 14.51
Exercise 14.44.

586

4. Count up to $1)_{10}$.

5. Inhibit counting.

Your timing diagram should also show outputs Q_A, Q_B, Q_C, Q_D, MAX/MIN, and RCO.

14.48 Draw a wiring diagram showing how two 4-bit binary counters (*A* and *B*) should be connected to perform the following functions:

1. The 4-bit outputs of both counters drive a common bus.

2. Both counters are clocked simultaneously by the same clock signal.

3. Counter A drives the bus with counts 0 through 7 while the outputs of counter B are in a high-impedance state.

4. Counter B drives the bus with counts 8 through 15 while the outputs of counter A are in a high-impedance state.

5. Closing one externally connected switch clears both counters asynchronously.

6. Closing another externally connected switch loads both counters asynchronously with $15)_{10}$.

Your wiring diagram should show all necessary connections to pins on the pin diagrams of both counters and any external logic circuitry necessary. Assume that V_{CC} is the logic level for 1.

Section 14–8

14.49 With reference to the (dual) counter whose logic symbol is shown in Figure 14.52, answer the following questions.

(a) Each counter is of what type (number of bits, synchronous, or asynchronous)?

(b) Do state changes occur on the leading or trailing edge of the clock?

(c) Are the counters programmable?

(d) Are the counters cleared synchronously or asynchronously?

(e) What pins should be connected together to create a divide-by-256 counter?

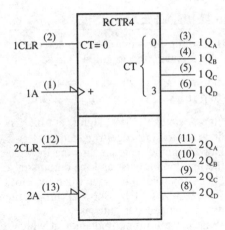

Figure 14.52
Exercise 14.49.

14.50 With reference to the counter whose logic diagram is shown in Figure 14.53, answer the following questions.

(a) Is clearing synchronous or asynchronous?

(b) Is loading synchronous or asynchronous?

(c) Under what circumstances is output $\overline{CO}$ low?

(d) Under what circumstances is output $\overline{BO}$ low?

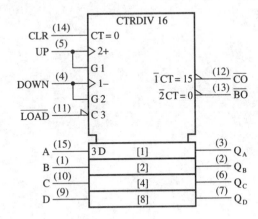

Figure 14.53
Exercise 14.50.

CHALLENGING EXERCISES

14.51 The flip-flops in a 2-bit ripple counter have the following propagation delays:

Flip-flop A: $t_{pLH} = 20$ ns, $t_{pHL} = 40$ ns
Flip-flop B: $t_{pLH} = 60$ ns, $t_{pHL} = 30$ ns

Flip-flop A stores the least significant bit. The flip-flops are trailing-edge-triggered.

(a) Draw a timing diagram showing the clock, the flip-flop outputs, and the outputs of a binary-to-decimal decoder when the clock frequency is 1 MHz. Show all propagation delays and decoder glitches.

(b) What is the limiting frequency at which the counter can be clocked without skipping any counts?

14.52 Draw a logic diagram showing how a 3-bit synchronous up counter could be constructed using D flip-flops.

14.53 Design a synchronous mod-23 up counter using T-type flip-flops.

14.54 Draw a wiring diagram showing how a 74163 counter could be connected with four external switches to obtain a divide-by-N output for any N from 2 through 16, depending on the switch settings. Make a table showing the switch settings required for each N. Using your design, determine which output (Q_A, Q_B, Q_C, or Q_D) provides the divide-by-2 signal and which provides the divide-by-5 signal. Also determine the duty cycle of each of those signals.

Part IV

DATA PROCESSING, MANIPULATION, AND STORAGE

Chapter 15

ARITHMETIC CIRCUITS, COMPARATORS, AND MULTIPLEXERS

15–1   HALF- AND FULL-ADDERS

The Half-Adder

Figure 15.1(a) is a table showing the sum (S) and carry (C) bits that result when two 1-bit numbers, A and B, are added. Every possible combination of the values of the two numbers is listed. (Seę also Section 2–4). Our objective is to construct a circuit that will produce the correct sum and carry bits in response to every possible combination of the two numbers. Toward that end, we can treat the addition table as two truth tables, one for S and one for C. As shown in the figure, it is clear that

$$S = A\bar{B} + \bar{A}B = A \oplus B \qquad\qquad (15.1)$$

and

$$C = AB \qquad\qquad (15.2)$$

Figure 15.1(b) shows a logic diagram that implements the expressions for S and C. For reasons that will become apparent shortly, this circuit is called a *half-adder*.

The Full-Adder

When we add two 2-bit binary numbers, we add the two least significant bits of each number and *propagate any carry that is generated into the addition of the two most significant bits*. For example, when we add 01 to 01, we know that the carry generated by adding the two 1s must be added to the two 0s in the most significant column:

$$
\begin{array}{r}
\text{carry:} \quad \mathbf{1} \hookleftarrow \\
0 \ \vdots \ 1 \\
+\ 0 \ \vdots \ 1 \\
\hline
1 \ \vdash 0
\end{array}
$$

We see that *three* bits must be added in the most significant column: the two most significant bits of the binary numbers plus the carry bit generated in the least significant column. The half-adder we discussed in the previous paragraph cannot be used for this

Figure 15.1
The half-adder.

A	B	sum S	carry C
0	0	0	0
0	1	1	0
1	0	1	0
1	1	0	1

$S = A\bar{B} + \bar{A}B = A \oplus B$
$C = AB$

(a) Truth table.

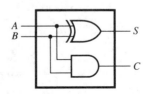

(b) Logic diagram.

addition because it has only two inputs. To add three input bits, we need a *full-adder*. A full-adder adds 2 bits to a carry bit, called the *carry in,* that is generated in the next less significant stage. The outputs of the full-adder are a sum bit and a *carry-out* bit. If we designate the input bits by A, B, and C_{in}, then the full-adder produces the sum: $S = A$ plus B plus C_{in} and the carry out, C_{out} resulting from that addition. For example, if $A = 1$, $B = 0$, and the carry from the next less significant stage is 1, the full-adder produces $S = 0$ and $C_{out} = 1$:

$$
\begin{array}{r}
1 = A \\
0 = B \\
+\ 1 = C_{in} \\
\hline
1\ 0 \\
\end{array}
$$
$$ C_{out} \overset{\nearrow}{\quad} \overset{\nwarrow}{\quad} S $$

Figure 15.2(*a*) is the addition table for a full-adder, in which the sum and carry-out bits are shown for every possible combination of the three inputs, A, B, and C_{in}. As we did for the half-adder, we can treat the addition table for the full-adder as truth tables for the logic functions that generate outputs S and C_{out}. Figure 15.2(*b*) shows Karnaugh maps and logic diagrams based on this interpretation. We see that S cannot be simplified:

$$ S = \bar{A}\bar{B}C_{in} + \bar{A}B\bar{C}_{in} + A\bar{B}\bar{C}_{in} + ABC_{in} \tag{15.3} $$

Equation (15.3) simply states that $S = 1$ if one and only one input is 1 or if all three inputs are 1. The expression for C_{out} simplifies to a statement that C_{out} is 1 if any two inputs are 1:

$$ C_{out} = AB + AC_{in} + BC_{in} \tag{15.4} $$

It is an exercise at the end of this chapter to show that the logic expression for S (equation (15.3)) is equivalent to

$$ S = A \oplus B \oplus C_{in} \tag{15.5} $$

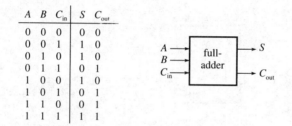

A	B	C_{in}	S	C_{out}
0	0	0	0	0
0	0	1	1	0
0	1	0	1	0
0	1	1	0	1
1	0	0	1	0
1	0	1	0	1
1	1	0	0	1
1	1	1	1	1

(a) Addition table.

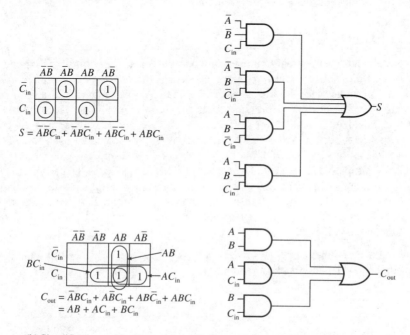

$S = \overline{A}\overline{B}C_{in} + \overline{A}B\overline{C}_{in} + A\overline{B}\overline{C}_{in} + ABC_{in}$

$C_{out} = \overline{A}BC_{in} + A\overline{B}C_{in} + AB\overline{C}_{in} + ABC_{in}$
$= AB + AC_{in} + BC_{in}$

(b) Simplification and implementation of logic expressions for S and C_{out}.

Figure 15.2
The full-adder.

It is another exercise to show that the logic expression for C_{out} (equation (15.4)) is equivalent to

$$C_{out} = (A \oplus B)C_{in} + AB \qquad (15.6)$$

Figure 15.3 shows how two half-adders (and an OR gate) can be used to generate these expressions for S and C_{out}. Since a full-adder can be constructed from two half-adders, the origin of the name half-adder is now apparent. The disadvantage of constructing a full-adder from half-adders is that the bits must propagate through several gates in succession, which makes the total propagation delay greater than that of the circuits in Figure 15.2.

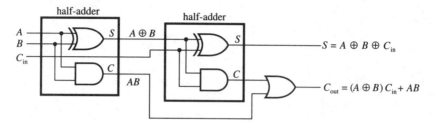

Figure 15.3
Using two half-adders to construct a full-adder.

15–2   PARALLEL ADDERS

Recall that when a binary number is in parallel form, every bit of the number is available (accessible) simultaneously. A parallel adder is used to add two numbers in parallel form and to produce the sum bits as parallel output. A block diagram of a 4-bit parallel adder capable of adding two 4-bit numbers is shown in Figure 15.4. The 4-bit binary numbers that are added in parallel are designated $A_3A_2A_1A_0$ and $B_3B_2B_1B_0$, and the corresponding output sum bits are $S_3S_2S_1S_0$. With the exception of the adder used to add the two least significant bits (A_0 and B_0), one full-adder is required for each pair of corresponding bits in the numbers to be added. The carry out from each stage is the carry in to the next more significant stage. In practical parallel adders, the least significant stage is also a full-adder to facilitate cascading, as is discussed shortly. In Figure 15.4, we use a full-adder in the least significant stage and connect its carry in to logical 0. Thus, it performs the same function as a half-adder.

Example 15–1

Find the sum and carry-out bits at each stage of the 4-bit adder in Figure 15.4 when the numbers 1110 and 0111 are added.

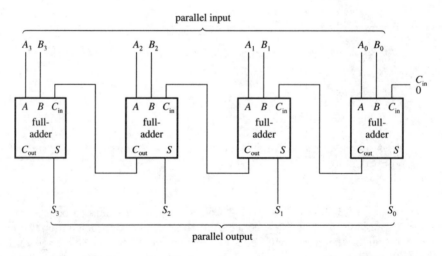

Figure 15.4
A 4-bit full-adder that adds binary numbers $A_3A_2A_1A_0$ and $B_3B_2B_1B_0$ in parallel form. Note that the carry out from each stage is the carry in to the next most significant stage.

Solution.   Letting $A_3A_2A_1A_0 = 1110$ and $B_3B_2B_1B_0 = 0111$, we have

$$\begin{bmatrix} A_0 & B_0 & C_{\text{in}} & S_0 & C_{\text{out}} \\ 0 & + & 1 & + & 0 & = & 1 & 0 \end{bmatrix} \quad \text{least significant stage}$$

$$\begin{bmatrix} A_1 & B_1 & C_{\text{in}} & S_1 & C_{\text{out}} \\ 1 & + & 1 & + & 0 & = & 0 & 1 \end{bmatrix}$$

$$\begin{bmatrix} A_2 & B_2 & C_{\text{in}} & S_2 & C_{\text{out}} \\ 1 & + & 1 & + & 1 & = & 1 & 1 \end{bmatrix}$$

$$\begin{bmatrix} A_3 & B_3 & C_{\text{in}} & S_3 & C_{\text{out}} \\ 1 & + & 0 & + & 1 & = & 0 & 1 \end{bmatrix} \quad \text{most significant stage}$$

We see that $S_3S_2S_1S_0 = 0101$. Since the carry out from the most significant stage is 1, we have an *overflow;* i.e., the sum (10101) must be expressed in 5 bits.

Drill 15–1   If $A_3A_2A_1A_0 = 1001$ in Example 15–1, how many different numbers represented by $B_3B_2B_1B_0$ will cause an overflow to be generated?
Answer: 9.

Cascading Parallel Adders

Figure 15.5 shows how two 4-bit adders can be cascaded to construct an 8-bit parallel adder. One 4-bit adder is used to add the four least significant bits of the input (the low-order adder), and the other (the high-order adder) adds the four most significant bits. Note that the carry out from the low-order adder is the carry in to the least significant stage of the high-order adder. Additional adders can be cascaded in an obvious way to create parallel adders for any number of bits.

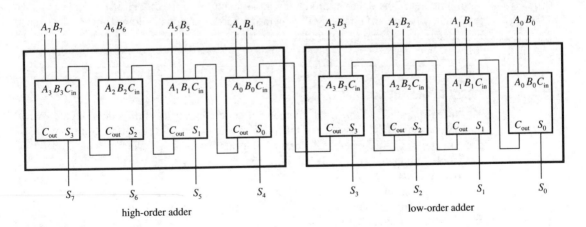

Figure 15.5
Cascading two 4-bit parallel adders to create an 8-bit parallel adder.

15-3 RIPPLE CARRY AND CARRY LOOK-AHEAD

Ripple Carry

As usual, propagation delays in the logic circuitry of a full-adder create a delay between the time that its inputs appear and the time that its S and C outputs are valid. In other words, a certain time must elapse before the S and C outputs can change in response to a change in the inputs. Referring to the 4-bit parallel adder in Figure 15.4, we can see that the S_1 output is not valid until after the delay of the first *two* full-adders, since the carry out from the least significant adder is delayed there. Similarly, the S_2 output is not valid until after the cumulative delay of three adders. At each stage, the sum bit is not valid until after the carry bits in *all* preceding stages are valid. In effect, carry bits must propagate, or *ripple*, through all stages before the most significant sum bit is valid. Thus, the total sum (the parallel output) is not valid until after the cumulative delay of all the adders. A parallel adder of this design, where the carry out of each full-adder is the carry in to the next more significant adder, is called a *ripple-carry* adder. Clearly, the greater the number of bits that a ripple-carry adder must add, the greater the time required for it to perform a valid addition.

Example 15-2

The time required to generate each carry-out bit in the parallel adder shown in Figure 15.5 is 24 ns. Once all inputs to an adder are valid, there is a delay of 32 ns until the output sum bit is valid. What is the maximum number of additions per second that it can perform?

Solution. Since the carry must ripple through 7 stages to reach the most significant stage, that delay is 7×24 ns = 168 ns. Another 32 ns is required for the most significant sum bit to be valid, so the total delay is 168 ns + 32 ns = 200 ns. Since one addition can be performed each 200 ns, the number of additions per second is

$$\frac{1 \text{ addition}}{200 \times 10^{-9} \text{ s}} = 5 \times 10^6 \text{ additions/s}$$

Drill 15-2

Assuming the 32-ns delay in producing a valid sum bit in the adder of Example 15-2 is the same, what maximum delay in generating carry-out bits is allowed if the adder must be capable of performing 10^7 additions/s?
Answer: 9.71 ns.

Carry Look-Ahead

To speed parallel addition, a method of interconnecting full-adders called carry *look-ahead* is used. The essence of this method is to examine all input bits simultaneously and to determine and generate the carry-input bits for *all* stages simultaneously. How is it possible to know in advance what the carry input to any stage must be? We can answer this question by studying the logic that creates carry inputs. Let C_0 be the carry input to the least significant stage, C_1 be the carry input to the next more significant stage, and so forth. Since the carry out from the least significant stage is C_1, the carry in to the next stage, we can use equation (15.4) to write

$$\begin{aligned} C_1 &= A_0B_0 + A_0C_0 + B_0C_0 \qquad &\textbf{(15.7)} \\ &= A_0B_0 + (A_0 + B_0)C_0 \qquad &\textbf{(15.8)} \end{aligned}$$

Similarly,

$$C_2 = A_1B_1 + A_1C_1 + B_1C_1 \qquad \text{(15.9)}$$
$$= A_1B_1 + (A_1 + B_1)C_1 \qquad \text{(15.10)}$$

The carry input to the nth full-adder, C_n, is

$$C_n = A_{n-1}B_{n-1} + (A_{n-1} + B_{n-1})C_{n-1} \qquad \text{(15.11)}$$

Equation (15.8) tells us that there are basically two ways that C_1 can equal 1. First, if the two data inputs A_0 and B_0 are both 1, then C_1 will be 1. This is called a *carry generate:* the input data itself generates a carry. We write the carry-generate term as

$$G_0 = A_0B_0 \qquad \text{(15.12)}$$

The second way that C_1 can be 1 is if C_0 is already 1 and if either A_0 or B_0 is 1. In this case, since C_0 is already 1, A_0 and B_0 merely serve to propagate that carry to the next higher stage, so we call the term $(A_0 + B_0)$ a *carry propagate, P_0:*

$$P_0 = A_0 + B_0 \qquad \text{(15.13)}$$

Similarly, a carry can be generated or propagated at stage 1:

$$G_1 = A_1B_1$$
$$P_1 = A_1 + B_1 \qquad \text{(15.14)}$$

For the nth stage,

$$G_n = A_nB_n \qquad \text{(15.15)}$$
$$P_n = A_n + B_n \qquad \text{(15.16)}$$

The equations for C_1, C_2, C_3, . . . , C_n can now be written in terms of the carry generates and carry propagates at each preceding stage:

$$C_1 = G_0 + P_0C_0 \qquad \text{(15.17)}$$
$$C_2 = G_1 + P_1C_1 \qquad \text{(15.18)}$$
$$C_3 = G_2 + P_2C_2 \qquad \text{(15.19)}$$
$$\vdots$$
$$C_n = G_{n-1} + P_{n-1}C_{n-1} \qquad \text{(15.20)}$$

Substituting (15.17) into (15.18),

$$C_2 = G_1 + P_1(G_0 + P_0C_0)$$
$$= G_1 + P_1G_0 + P_1P_0C_0 \qquad \text{(15.21)}$$

Substituting (15.21) into (15.19),

$$C_3 = G_2 + P_2(G_1 + P_1G_0 + P_1P_0)C_0$$
$$= G_2 + P_2G_1 + P_2P_1G_0 + P_2P_1P_0C_0 \qquad \text{(15.22)}$$

Continuing in this manner, we find

$$C_n = G_{n-1} + P_{n-1}G_{n-2} + P_{n-1}P_{n-2}G_{n-3} + \cdots$$
$$+ P_{n-1}P_{n-2} \cdots P_1G_0 + P_{n-1}P_{n-2} \cdots P_0C_0 \qquad \text{(15.23)}$$

Figure 15.6

A full-adder that produces carry-generate (G) and carry-propagate (P) outputs for use in carry look-ahead.

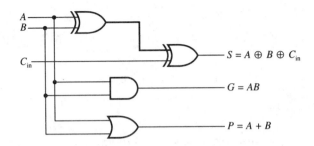

The important point to notice about this set of equations [(15.21) through (15.23)] is that each equation contains one and only one carry term, namely, C_0. Note that each carry-generate and carry-propagate term is a function only of the parallel input data. Thus, we can produce the carry input to any stage by logical operations on the bits of the parallel inputs (and C_0), all of which are available simultaneously.

Figure 15.6 shows a full-adder designed to produce a sum bit, a carry-generate output, and a carry-propagate output. Of course, it is no longer necessary to produce a carry out, since the carry-generate and carry-propagate outputs are used to implement carry look-ahead.

Figure 15.7 shows a 4-bit parallel adder incorporating carry look-ahead. It is apparent that the penalty paid to achieve the greater speed this method affords is the greater complexity of the logic. In particular, note that the *n*th stage requires *n* AND gates, one of

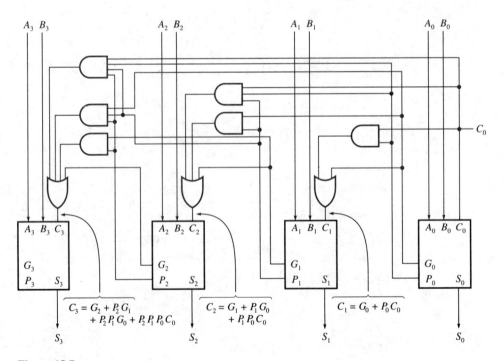

Figure 15.7

A 4-bit parallel adder incorporating carry look-ahead. Each full adder is of the type shown in Figure 15.6.

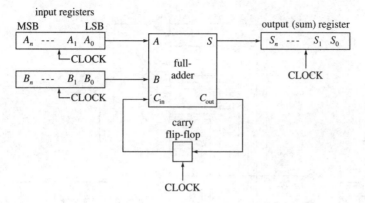

Figure 15.8
A serial adder.

which has n inputs, and an n-input OR gate. A look-ahead adder capable of adding large binary numbers is substantially more complex than its ripple-carry counterpart.

15–4   SERIAL ADDERS

A serial adder is used to add serial binary numbers. As shown in Figure 15.8, only one full-adder is required for this application. The serial input bits appear in synchronism, first A_0 and B_0 (the least significant bits), then—one clock pulse later—A_1 and B_1, then A_2 and B_2, and so forth. The carry bit generated in any one of these additions must be saved and added to the next-higher-order pair of input bits. The purpose of the flip-flop is to store the carry bit (C_{out}) for the duration of one clock pulse and then present it as C_{in} when the next pair of input bits is added. As can be seen in the figure, the output sum bits are shifted into the output register as the input bits are shifted out of the input registers. In practice, the output bits are often shifted into one of the input registers behind the data being shifted out. This register, which contains one of the binary numbers to be added before the addition commences and contains the sum after the addition is completed, is called an *accumulator*. It is obvious that serial adders are slower than parallel adders, since they require one clock pulse per pair of bits added. Serial adders are used where circuit minimization is more important than speed, as in pocket calculators.

15–5   INTEGRATED-CIRCUIT ADDERS

Parallel adders are available as integrated circuits in several technologies, including TTL and CMOS. Figure 15.9 shows the logic diagram for the 74283-series 4-bit parallel adder. Note that both carry in (C_0) and carry out (C_3) are available to implement cascading. The design incorporates carry look-ahead, and manufacturers' specifications state that each carry term is typically generated in 10 ns for the 74LS283 and in 7.5 ns for the 74S283.

15–6   THE ARITHMETIC/LOGIC UNIT (ALU)

Although a digital computer can be used to perform extremely complex mathematical computations, the internal hardware dedicated to arithmetic computations is relatively

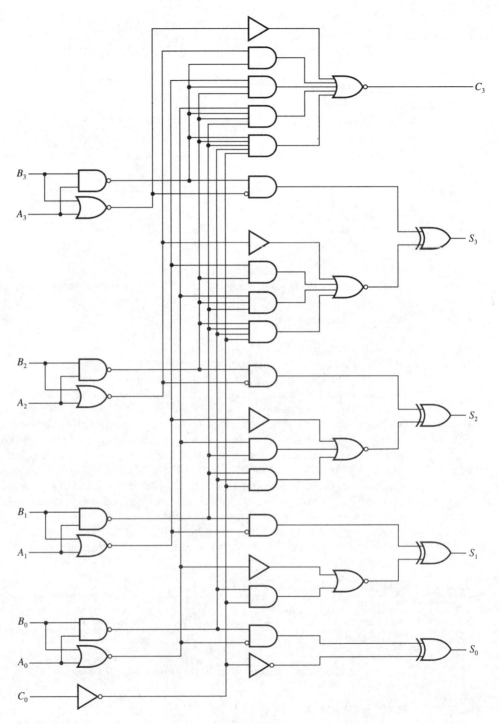

Figure 15.9
Logic diagram for the 74283 4-bit parallel adder with carry look-ahead.

simple. The key to the computational power of a computer is its ability to perform a very large number of simple arithmetic and logic operations at a very high speed. In many cases, the types and sequences of the calculations that a computer performs when engaged in a complex computation seem inefficient by human standards because of the large number of repetitions of very simple operations. However, the speed at which these operations can be performed makes it possible for the computer to produce a solution much faster than a human being using more efficient methods.

The circuitry used to perform the fundamental logic and arithmetic operations upon which more complex operations are based is contained in the *arithmetic/logic unit* (ALU). In many computers, the ALU is a custom-designed unit that provides no more or no less than a certain minimum set of basic operations necessary to support the tasks for which the computer is designed. General-purpose ALUs with capabilities that may or may not be necessary for a particular application are also available. We discuss one such unit presently. No matter what the design, an ALU will almost certainly contain an adder and circuitry that can be used to perform bit-by-bit logic operations (AND, OR, and so on) on binary words.

An example of a general-purpose, integrated-circuit ALU is the 74181 series, whose pin diagram is shown in Figure 15.10. The data inputs to this unit are two 4-bit words, designated $A_3A_2A_1A_0$ and $B_3B_2B_1B_0$. Control inputs S_3, S_2, S_1, and S_0 are used to specify the type of arithmetic or logic operation that is to be performed on the input data. The results of the selected operation appear at outputs F_3, F_2, F_1, F_0. Mode control input M is used to specify whether the operation to be performed is arithmetic ($M = 0$) or logical ($M = 1$).

The 74181, which is available in standard TTL, Schottky (S), low-power Schottky (LS), advanced-Schottky (74AS181A), and CMOS technologies, can perform 16 different arithmetic operations and 16 different logic operations. Logic operations are performed on a bit-by-bit basis. For example, if input word A is 1011 and input word B is 1101, then A AND $B = 1001$. In addition to the standard logic operations (AND, OR, NAND, and so forth), the 74181 can perform a variety of combinations of logic operations on the data, such as $\overline{A} + B$ and $A\overline{B}$. Since these can be regarded as Boolean algebra *functions* of the data, the 74181 is also called an ALU/function generator. Input $\overline{C}_n$ in Figure 15.10 is the carry in for arithmetic operations. Note that it is active-low, so a carry is added to a

Figure 15.10
The 74181 ALU.

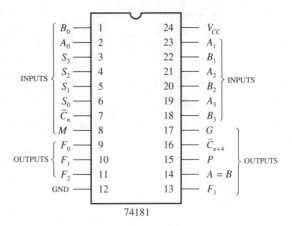

74181

Table 15.1
Arithmetic and logic functions performed by the 74181 ALU.

S_3	S_2	S_1	S_0	$M = 1$ Logic functions	$M = 0$ Arithmetic operations $\overline{C}_n = 1$ (no carry)	$M = 0$ Arithmetic operations $\overline{C}_n = 0$ (with carry)
0	0	0	0	$F = \overline{A}$	$F = A$	$F = A$ plus 1
0	0	0	1	$F = \overline{A + B}$	$F = A + B$	$F = (A + \overline{B})$ plus 1
0	0	1	0	$F = \overline{A}B$	$F = A + \overline{B}$	$F = (A + \overline{B})$ plus 1
0	0	1	1	$F = 0$	$F = $ minus 1 (2's complement)	$F = 0$
0	1	0	0	$F = \overline{AB}$	$F = A$ plus $A\overline{B}$	$F = A$ plus $A\overline{B}$ plus 1
0	1	0	1	$F = \overline{B}$	$F = (A + B)$ plus $A\overline{B}$	$F = (A + B)$ plus $A\overline{B}$ plus 1
0	1	1	0	$F = A \oplus B$	$F = A$ minus B minus 1	$F = A$ minus B
0	1	1	1	$F = A\overline{B}$	$F = A\overline{B}$ minus 1	$F = A\overline{B}$
1	0	0	0	$F = \overline{A} + B$	$F = A$ plus AB	$F = A$ plus AB plus 1
1	0	0	1	$F = A \odot B$	$F = A$ plus B	$F = A$ plus B plus 1
1	0	1	0	$F = B$	$F = (A + \overline{B})$ plus AB	$F = (A + \overline{B})$ plus AB plus 1
1	0	1	1	$F = AB$	$F = AB$ minus 1	$F = AB$
1	1	0	0	$F = 1$	$F = A$ plus A	$F = A$ plus A plus 1
1	1	0	1	$F = A + \overline{B}$	$F = (A + B)$ plus A	$F = (A + B)$ plus A plus 1
1	1	1	0	$F = A + B$	$F = (A + \overline{B})$ plus A	$F = (A + \overline{B})$ plus A plus 1
1	1	1	1	$F = A$	$F = A$ minus 1	$F = A$

sum when $\overline{C}_n = 0$. Table 15.1 summarizes the arithmetic and logic operations performed by the 74181. The next example illustrates the use and interpretation of the table.

Example 15–3

The data inputs to a 74181 are $A = 0101$ and $B = 0011$. The control inputs are $S_3S_2S_1S_0 = 1001$. If $\overline{C}_n = 1$, find $F = F_3F_2F_1F_0$ when (a) $M = 1$ and (b) $M = 0$.

Solution

a. When $M = 1$, a bit-by-bit logical operation is performed on the data. From Table 15.1, when $S_3S_2S_1S_0 = 1001$ we see that the operation is $A \odot B$, the exclusive NOR, or coincidence, function.

$$F_3 = A_3 \odot B_3 = 0 \odot 0 = 1$$
$$F_2 = A_2 \odot B_2 = 1 \odot 0 = 0$$
$$F_1 = A_1 \odot B_1 = 0 \odot 1 = 0$$
$$F_0 = A_0 \odot B_0 = 1 \odot 1 = 1$$

b. When $M = 0$, an arithmetic operation is performed on the data. From Table 15.1, we see that the operation is A *plus* B (addition). Since $\overline{C}_n = 1$, no carry is added to the sum:

$$
\begin{array}{rcccc}
A = & 0 & 1 & 0 & 1 \\
+B = & 0 & 0 & 1 & 1 \\
\hline
F = & 1 & 0 & 0 & 0 \\
 & F_3 & F_2 & F_1 & F_0
\end{array}
$$

Drill
15–3

Repeat Example 15–3 when $\overline{C}_n = 0$.
Answer: (a) $F = 1000$; (b) $F = 1001$.

The input data and the outputs can be treated as active-low. Under that interpretation, the logic and arithmetic operations performed by the 74181 are different from those shown in Table 15.1. For example, the coincidence function, $A \odot B$, becomes the exclusive -OR function, $A \oplus B$, when the data are regarded as active-low. Manufacturer's product literature can be consulted to determine all the functions performed by the 74181 for active-low data.

The 74181 can also be used as a *comparator* to provide outputs that signify whether input word A is equal to, greater than, or less than word B. With the control inputs set to select subtraction and with $\overline{C}_n = 1$, the output labeled $A = B$ in Figure 15.10 (pin 14) goes high when A and B have equal magnitude. The output labeled $\overline{C}_{n+4}$ (pin 16) is an active-low carry out used for cascading but can also be used in conjunction with different values of $\overline{C}_n$ to signify which of inputs A or B has greater magnitude. The 74181 features internal carry look-ahead that enables one addition to be performed in the AS series in 5 ns. The $\overline{C}_{n+4}$ carry out can be connected directly to the $\overline{C}_n$ carry in of another 74181 to create an 8-bit ALU that can perform an addition in 10 ns. Note, however, that this type of cascading amounts to *ripple carry* between units having internal carry look-ahead. Carry look-ahead *generators* are available in separate integrated circuits and can be used to implement carry look-ahead *between* cascaded ALUs, thus speeding operations on large data words. The carry-propagate *(P)* and carry-generate *(G)* outputs of the 74181 (pins 15 and 17) are used with a carry look-ahead generator for that purpose.

15–7 COMPARATORS

A comparator is a logic circuit used to compare the magnitudes of two binary numbers. Depending on design, it may simply provide an output that is active (goes high, for example) when the two numbers are equal, or it may additionally provide outputs that signify which of the numbers is larger when equality does not hold.

The basic building block of a comparator is the coincidence (exclusive-NOR) gate. As we know, $A \odot B$ is logical 1 if and only if bits A and B *coincide* (both 0 or both 1). Two binary numbers are equal if and only if *all* their corresponding bits coincide. For example, $A_2A_1A_0 = B_2B_1B_0$ if and only if $A_2 = B_2$, $A_1 = B_1$, and $A_0 = B_0$. Thus, equality holds when A_2 coincides with B_2 AND A_1 coincides with B_1 AND A_0 coincides with B_0. Implementation of this logic, EQUALITY $= (A_2 \odot B_2) \cdot (A_1 \odot B_1) \cdot (A_0 \odot B_0)$, is straightforward and is shown in Figure 15.11. The circuit can be expanded in an obvious way to accommodate binary numbers with any number of bits.

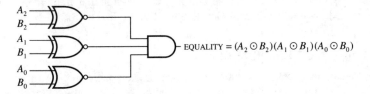

Figure 15.11
A comparator used to determine equality of two 3-bit binary numbers: $A_2A_1A_0$ and $B_2B_1B_0$. The output is 1 if and only if all corresponding bits of the numbers coincide.

A comparator capable of signifying which of two binary numbers has greater magnitude is somewhat more complex. Consider a procedure that we could use to determine which of the numbers $A = A_3A_2A_1A_0$ and $B = B_3B_2B_1B_0$ is greater:

1. Compare the *most* significant bits, A_3 and B_3, of the number. If A_3 is 1 and B_3 is 0, then A is greater than B, *regardless of how the remaining bits compare*. Similarly, if A_3 is 0 and B_3 is 1, then B is greater than A. If (and only if) A_3 and B_3 coincide, go to Step 2.
2. If A_2 is 1 and B_2 is 0, then A is greater than B, regardless of how the remaining bits compare. If A_2 is 0 and B_2 is 1, then B is greater than A. If (and only if) A_2 and B_2 coincide, go to Step 3.
3. Continue in this fashion, checking each pair of less significant bits, until finding a pair that do not coincide. The bit that is 1 belongs to the larger number. If all pairs of bits coincide, then, of course, the numbers are equal.

To illustrate this procedure, let us compare $A = 1001$ with $B = 1010$. A_3 and B_3 coincide (both are 1), so we check A_2 and B_2. A_2 and B_2 coincide (both are 0), so we check A_1 and B_1. Since $B_1 = 1$ and $A_1 = 0$, we conclude that B is greater than A and *do not check any further pairs of bits*.

When implementing the logic described in the previous paragraph, we must be certain that a pair of bits is compared only if *all* preceding pairs of bits coincide. Let us summarize the logic we could use to determine if A is greater than B ($A > B$) in a set of statements, as follows:

(1) If $A_3 = 1$ and $B_3 = 0$, then $A > B$.
OR (2) If A_3 and B_3 coincide and if $A_2 = 1$ and $B_2 = 0$, then $A > B$.
OR (3) If A_3 and B_3 coincide, and if A_2 and B_2 coincide, and if $A_1 = 1$ and $B_1 = 0$, then $A > B$.
OR (4) If A_3 and B_3 coincide, and if A_2 and B_2 coincide, and if A_1 and B_1 coincide, and if $A_0 = 1$ and $B_0 = 0$, then $A > B$.

From these statements, we see that the logic expression for $A > B$ can be written as:

$$\begin{aligned} A > B = A_3\overline{B}_3 &+ (A_3 \odot B_3)A_2\overline{B}_2 \\ &+ (A_3 \odot B_3)(A_2 \odot B_2)A_1\overline{B}_1 \\ &+ (A_3 \odot B_3)(A_2 \odot B_2)(A_1 \odot B_1)A_0\overline{B}_0 \end{aligned} \tag{15.24}$$

In a completely parallel fashion, we obtain the logic expression for $B > A$:

$$\begin{aligned} B > A = \overline{A}_3B_3 &+ (A_3 \odot B_3)\overline{A}_2B_2 \\ &+ (A_3 \odot B_3)(A_2 \odot B_2)\overline{A}_1B_1 \\ &+ (A_3 \odot B_3)(A_2 \odot B_2)(A_1 \odot B_1)\overline{A}_0B_0 \end{aligned} \tag{15.25}$$

Figure 15.12 shows a logic diagram of a comparator that implements the logic we have described. Note that it provides three active-high outputs: $A > B$, $A < B$, and $A = B$.

Integrated-Circuit Comparators

Comparators of the design we have described are available in integrated-circuit form. Figure 15.13(a) shows the pin diagram for a typical example, the 7485-series 4-bit magnitude comparator. The time required to perform a comparison is 11 ns for the Schottky (S) version and 24 ns for the LS version. Note that pins 2, 3, and 4 are designated $(A < B)_{in}$, $(A = B)_{in}$, and $(A > B)_{in}$. These are used for cascading. Figure 15.13(b) shows

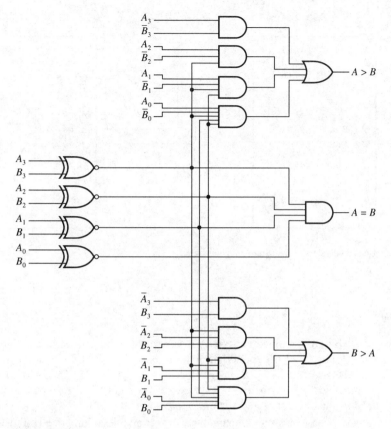

Figure 15.12
A 4-bit magnitude comparator used to determine which of inputs $A = A_3A_2A_1A_0$ and $B = B_3B_2B_1B_0$ is larger or if they are equal.

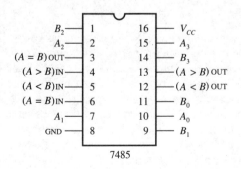

(a) Pin diagram.

Figure 15.13
The 7485 4-bit comparator. (Continued on p. 606.)

605

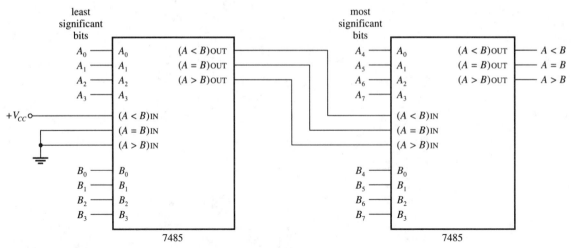

(b) Cascading two 7485 4-bit comparators to create an 8-bit comparator.

Figure 15.13
(Continued)

how two 4-bit comparators are cascaded to perform 8-bit comparisons. The $(A < B)_{out}$, $(A = B)_{out}$, and $(A > B)_{out}$ outputs from the comparator used for the least significant 4 bits are connected to the $(A < B)_{in}$, $(A = B)_{in}$, and $(A > B)_{in}$ inputs of the high-order comparator. Notice how the $(A < B)_{in}$, $(A = B)_{in}$, and $(A > B)_{in}$ inputs to the low-order comparator must be connected for this application.

A wide variety of 8-bit comparators is also available in integrated circuits. Variations in design and capabilities include open-collector outputs, latched outputs, and active-low outputs. Some comparators, such as the 74518 series, provide equality comparison $(A = B)$ only and are referred to as *identity* comparators. The 74ALS526 is a 16-bit identity comparator that can be preprogrammed to compare a 16-bit input for equality with a fixed 16-bit word.

15-8   MULTIPLEXERS

Reduced to simplest terms, multiplexing means *sharing*. A common example occurs when several peripheral devices share a single transmission line, or bus, to communicate with a computer. To accomplish this sharing, each device *in succession* is allocated a brief time to send or receive data. At any given time, one and only one device is using the line, but each device receives its turn. This is an example of *time* multiplexing, since each device is given specific time intervals to use the line. (In *frequency* multiplexing, several devices share a common line by transmitting at different frequencies.) In large mainframe computer installations, numerous users are time-multiplexed to the computer in such rapid succession that all appear to be using the computer simultaneously.

A digital data multiplexer (MUX) is a logic circuit having several data inputs and a single output. A set of *data-select* inputs is used to control which of the data inputs is routed to the single output. A multiplexer is also called a *data selector* because of this

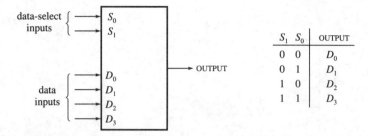

Figure 15.14
A multiplexer capable of selecting one of four data inputs (a 4-to-1 multiplexer).

ability to select which data input is connected to the output. The logic symbol in Figure 15.14 can be used to illustrate the concept. The 2-bit binary number at the data-select inputs, S_1 and S_0, specifies which of the four data inputs is to be routed to the output. If $S_1S_0 = 00$, then input D_0 is connected to the output. If $S_1S_0 = 01$, then input D_1 is connected to the output, and so forth. Of course, the number of data-select inputs in a particular design determines the number of data inputs that can be selected. With three data-select inputs, $2^3 = 8$ different data inputs can be selected. That type of device is called an 8-to-1 multiplexer; in general, multiplexers are said to be 2^n-to-1, where n is the number of data-select inputs. If we were to drive S_1S_0 in Figure 15.14 with the output of a 2-bit binary counter, so that S_1S_0 continually progressed through the states 00, 01, 10, 11, 00 . . . , then we would be continually selecting the data inputs in turn and thus would accomplish the time-multiplexing described previously.

It is easy to conceive and implement the logic necessary to construct a multiplexer. As shown in Figure 15.15, we simply decode the data-select inputs and use them to enable the appropriate data inputs. The circuit can be expanded in an obvious way to construct multiplexers with 8, 16, . . . , 2^n data inputs.

Example 15-4

Complete the timing diagram in Figure 15.16(*a*) to show the output of the multiplexer of Figure 15.15 when its inputs are as shown.

Solution. Note that S_1S_0 is sequenced through a 2-bit binary count: 00, 01, 10, 11, 00, ... Thus, data inputs D_0, D_1, D_2, and D_3 are connected to the output in sequence. When $S_1S_0 = 00$, data input D_0 is high, so the output is high during that time. When $S_1S_0 = 01$, data input D_1 is low and the output is low. Figure 15.16(*b*) shows the waveform that results at the output and indicates the data input that is selected during each time interval.

15-9 INTEGRATED-CIRCUIT MULTIPLEXERS

Multiplexers are available in integrated circuits in a wide variety of designs, including 2-to-1, 4-to-1, 8-to-1, and 16-to-1. Some packages contain more than one multiplexer, as, for example, the 74157 quad 2-to-1 multiplexer (four 2-to-1 multiplexers having the same data-select inputs) and the 74153 dual 4-to-1 multiplexer. Some designs have three-state outputs and others have open-collector outputs. Most have enable inputs to facilitate

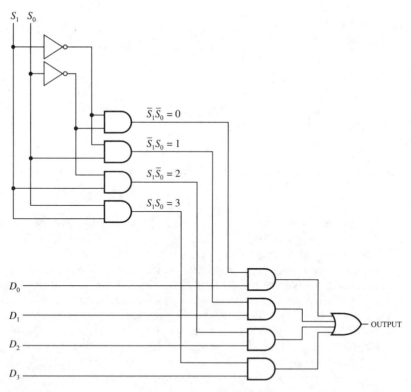

Figure 15.15
Logic diagram for the 4-to-1 multiplexer shown in Figure 15.14.

cascading. Figure 15.17(a) shows a functional block diagram of the 74151A 3-to-8 multiplexer, which has an active-low enable ($\overline{EN}$) input. Note that both the selected data and its complement are available as outputs. Figure 15.17(b) shows how two of these devices can be cascaded to create a 4-to-16 multiplexer. The most significant data-select input (S_3) controls the $\overline{EN}$ inputs of the two multiplexers. It is connected directly to $\overline{EN}$ on the low-order multiplexer (whose data inputs are the least significant 8 bits of the data) and it is inverted and connected to the $\overline{EN}$ input of the high-order multiplexer. Thus, when $S_3 = 0$, only the low-order data, selected by $0S_2S_1S_0$, are routed to the output, and when $S_3 = 1$, the high-order data selected by $1S_2S_1S_0$ are routed to the output.

15–10   APPLICATIONS OF MULTIPLEXERS

Generation of Logic Functions

A multiplexer can be used in place of logic gates to implement a logic expression. As we shall see, it can be connected so that it duplicates the logic of any truth table; i.e., it can generate any Boolean algebra function of a set of input variables. In such applications, the multiplexer can be viewed as a function generator because we can easily set or change the logic function it implements. One advantage of using a multiplexer in place of logic gates is that a single integrated circuit can perform a function that might otherwise require

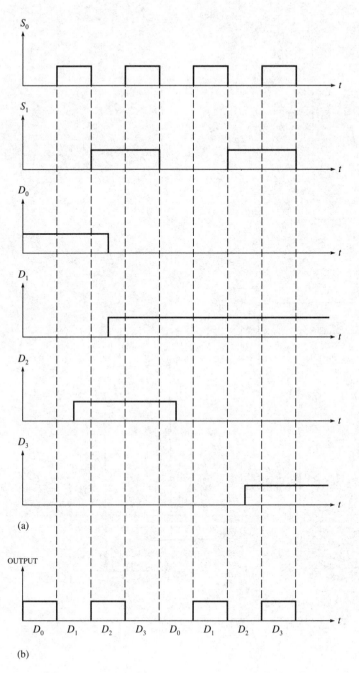

(a)

(b)

Figure 15.16
Example 15–4.

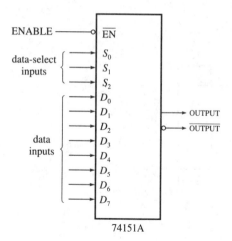

(a) Logic diagram.

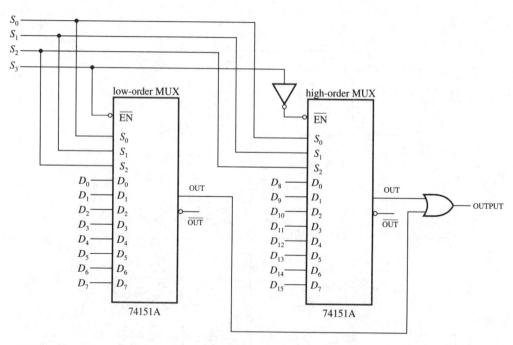

(b) Cascading two 74151A 8-to-1 multiplexers to create a 16-to-1 multiplexer.

Figure 15.17
The 74151A multiplexer.

numerous integrated circuits. Another advantage is that it is very easy to change the logic function implemented if redesign of a system becomes necessary.

The first step in the design of a function generator using a multiplexer is to construct a truth table for the function to be implemented. We then connect logical 1 to each data input of the multiplexer corresponding to each combination of input variables for which

Figure 15.18
Example 15–5.

S_2 S_1 S_0 A B C	AB	$\bar{B}C$	$\bar{A}BC$	$F = AB + \bar{B}C + \bar{A}BC$
0 0 0	0	0	0	0
0 0 1	0	1	0	1
0 1 0	0	0	0	0
0 1 1	0	0	1	1
1 0 0	0	0	0	0
1 0 1	0	1	0	1
1 1 0	1	0	0	1
1 1 1	1	0	0	1

(a)

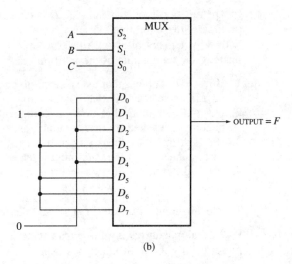

(b)

the truth table shows the function to equal 1. Logical 0 is connected to the remaining data inputs. The variables themselves are connected to the data-select inputs of the multiplexer. For example, suppose the truth table specifies that the function F equals 1 for input combination 101. Then, when $S_2S_1S_0 = 101$, data input 5, which is connected to logical 1, will be selected and will route a 1 to the output of the multiplexer. The next example illustrates the procedure.

Example 15–5

Use a multiplexer to implement the logic function $F = AB + \bar{B}C + \bar{A}BC$.

Solution.   The truth table for F is shown in Figure 15.18(a). Since there are three input variables, we use a multiplexer with 3 data-select inputs (an 8-to-1 MUX). The truth table shows that we will use data-select inputs S_2, S_1, and S_0 for input variables A, B, and C, respectively. Since $F = 1$ when $ABC = 001, 011, 101, 110,$ and 111, we connect logical 1 to data inputs 1, 3, 5, 6, and 7. As shown in Figure 15.18(b), logical 0 is connected to the other data inputs (0, 2, and 4). When the data-select inputs are any of the combinations for which $F = 1$, the output will be 1, and when the data-select inputs are any combination for which $F = 0$, the output will be 0. Thus, the multiplexer behaves in exactly the same way that a set of logic gates implementing the function F would behave.

Drill
15–5

If a design change in Example 15–5 dictates that F must equal $A \oplus B \oplus C$, what data inputs should be connected to logical 1?
Answer: D_1, D_2, D_4, D_7.

In the previous example, we saw that a multiplexer having three data-select inputs can be used to implement a Boolean-algebra function of three variables. With a slight variation in design, it can also be used to implement any function of four variables. In fact, a multiplexer with n data-select inputs can implement any function of $n + 1$ variables. The key to this design is to use the most significant input variable and its complement to drive certain of the data inputs. To illustrate, suppose we wish to implement a 4-variable logic function using a multiplexer with three data-select inputs. Let the input variables be $ABCD$, so A is the most significant bit. As before, a truth table for the function F of A, B, C, and D is constructed. In the truth table, we note that BCD progresses *twice* through the sequence 000, 001, . . . , 111: once with $A = 0$ and again with $A = 1$. The following rules are used to determine the connections that should be made to the data inputs of the multiplexer:

1. If $F = 0$ both times that the same combination of BCD occurs, connect logical 0 to the data input selected by that combination.
2. If $F = 1$ both times that the same combination of BCD occurs, connect logical 1 to the data input selected by that combination.
3. If F is different for the two occurrences of a combination of BCD and if $F = A$ in each case, connect A to the data input selected by that combination.
4. If F is different for the two occurrences of a combination of BCD and if $F = \overline{A}$ in each case, connect $\overline{A}$ to the data input selected by that combination.

The next example illustrates how these rules are used to implement a 4-variable function.

Example
15–6

Use a multiplexer having three data-select inputs to implement the logic for function F prescribed by the truth table shown in Figure 15.19(a).

Solution.   Since $F = 1$ both times that $BCD = 000$ appears in the table, data input D_0 is connected to logical 1. Since $F = 0$ both times that $BCD = 001$ appears in the table, data input D_1 is connected to logical 0. Now, $F = 1$ the first time that $BCD = 010$ appears, but $F = 0$ the second time that $BCD = 010$ appears. However, $F = \overline{A}$ in both cases, so $\overline{A}$ is connected to data input D_2. Also, $F = 0$ the first time that $BCD = 011$ appears and $F = 1$ the second time that $BCD = 011$ appears. Since $F = A$ in both cases, A is connected to data input D_3. Continuing our analysis of the truth table, we find the proper connection for each of the remaining data inputs, as shown in the figure. Figure 15.19(b) shows the resulting logic diagram.

Drill
15–6

To what data inputs would A and $\overline{A}$ be connected in Example 15–6 if the logic function were changed to $F = \Sigma\ 0, 1, 2, 3, 4, 10, 11, 14, 15$?
Answer: A, D_6 and D_7; $\overline{A}$, D_0, D_1, and D_4.

Multiplexing Seven-Segment Displays

Optical output displays, such as seven-segment displays, typically consume considerable power. In applications where power consumption is a major concern, such as pocket calculators, and where several displays must be illuminated simultaneously, multiplexers

Figure 15.19
Example 15–6.

A	S_2 B	S_1 C	S_0 D	F	Value of F:	For both occurrences of:
0	0	0	0	1	$F = 1$	$BCD = 000$
0	0	0	1	0	$F = 0$	$BCD = 001$
0	0	1	0	1	$F = \overline{A}$	$BCD = 010$
0	0	1	1	0	$F = A$	$BCD = 011$
0	1	0	0	0	$F = 0$	$BCD = 100$
0	1	0	1	1	$F = 1$	$BCD = 101$
0	1	1	0	1	$F = 1$	$BCD = 110$
0	1	1	1	0	$F = A$	$BCD = 111$
1	0	0	0	1	$F = 1$	$BCD = 000$
1	0	0	1	0	$F = 0$	$BCD = 001$
1	0	1	0	0	$F = \overline{A}$	$BCD = 010$
1	0	1	1	1	$F = A$	$BCD = 011$
1	1	0	0	0	$F = 0$	$BCD = 100$
1	1	0	1	1	$F = 1$	$BCD = 101$
1	1	1	0	1	$F = 1$	$BCD = 110$
1	1	1	1	1	$F = A$	$BCD = 111$

(a)

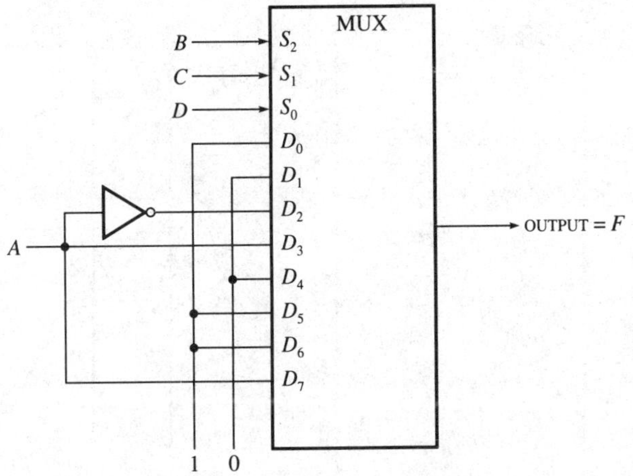

(b)

are used to reduce power consumption. Instead of illuminating all displays simultaneously, a multiplexer selects each display in turn. If the rate at which the displays are illuminated is great enough (usually around 30 times per second), the human eye will not be able to detect any flicker, and it will appear that all displays are illuminated simultaneously. The power consumption will be no greater than if a single display were illuminated continuously.

Figure 15.20 shows how a 74157 multiplexer can be used to multiplex two seven-segment displays. The 74157 contains four 1-to-2 multiplexers, all of which have the same data-select input. See the logic diagram in part (a) of the figure and note that when S_0 is 0, all four D_0 inputs are selected; when S_0 is 1, all four D_1 inputs are selected.

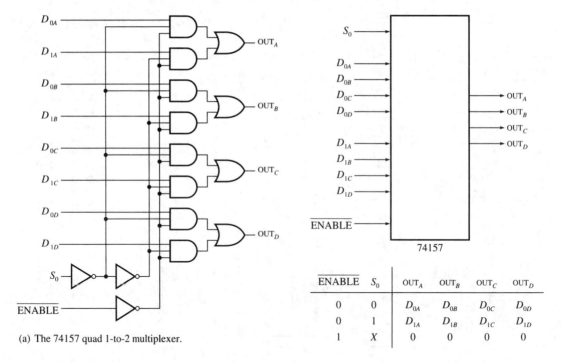

(a) The 74157 quad 1-to-2 multiplexer.

ENABLE	S_0	OUT_A	OUT_B	OUT_C	OUT_D
0	0	D_{0A}	D_{0B}	D_{0C}	D_{0D}
0	1	D_{1A}	D_{1B}	D_{1C}	D_{1D}
1	X	0	0	0	0

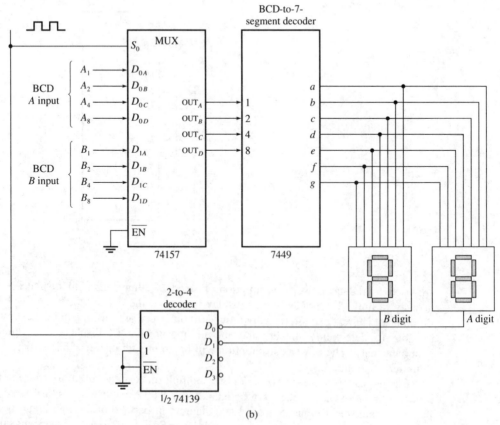

(b)

Figure 15.20
Multiplexing seven-segment displays.

Thus, the output of the circuit is four parallel bits that have been selected from either the four D_0 inputs or from the four D_1 inputs. As shown in part *(b)* of the figure, the two 8-4-2-1 BCD numbers whose values are to be displayed, $A = A_8A_4A_2A_1$ and $B = B_8B_4B_2B_1$, are connected to the data inputs of the multiplexer. The A bits are connected to the D_0 inputs and the B bits are connected to the D_1 inputs. A square wave drives the data-select input. When the square wave is low, the D_0 inputs are selected, so the A bits are routed to the output. When the square wave is high, the D_1 inputs are selected and the B bits are routed to the output. Note that the parallel output of the multiplexer is connected to the input of a 7449 BCD-to-7-segment-display decoder. Recall that this device provides the seven outputs needed to drive the segments that must be illuminated when its BCD input corresponds to any of the decimal digits from 0 through 9. We see that the multiplexer alternately supplies the decoder with $A_8A_4A_2A_1$ and $B_8B_4B_2B_1$ at the frequency of the square wave, so the decoder alternately activates the segments for the decimal digits corresponding to $A_8A_4A_2A_1$ and $B_8B_4B_2B_1$. The outputs of the 7-segment decoder are connected in parallel to two common-cathode seven-segment displays. However, we want only one display, A or B, depending on which BCD input has been selected, to be illuminated. Recall that a common-cathode display is activated when its common line is made low. The 74139 2-to-4 decoder is used to select the seven-segment display that is to be activated. Since input 1 of the decoder is connected to logical 0 and the square wave is connected to input 0, active-low decoder output D_0 is low when the square wave is low (input = 00) and output D_1 is low when the square wave is high (input = 01). These active-low outputs enable the seven-segment displays in the proper sequence.

15–11 DEMULTIPLEXERS

A demultiplexer (DMUX) routes a single data input to one of several outputs. Thus, whereas a multiplexer is an *N*-to-1 device, a demultiplexer is a 1-to-*N* (or 2^n) device. A demultiplexer can be thought of as a *distributor,* since it distributes the same data to different destinations. In fact, the distributor in an automobile ignition system is a good conceptual example of a demultiplexer: It distributes a single high-voltage input to the spark plugs, each in succession. As illustrated in Figure 15.21*(a)*, a digital demultiplexer has a set of input lines used to select the output to which data is to be delivered. The input data is a single bit, D_{in}. In Chapter 13, we discussed a very similar device: the binary-to-decimal decoder. Recall that the input to a decoder is an *n*-bit binary number and the (decimal) output consists of 2^n lines. The decimal output line that is activated is the one whose value corresponds to the value of the binary input. Figure 15.21*(b)* shows how we can use a 3-to-8 decoder as a 1-to-8 demultiplexer. The 3-bit binary input is used for the select lines, $S_2S_1S_0$, that select the output to which data is routed. The input data, D_{in}, is inverted and connected to the active-low $\overline{\text{ENABLE}}$ input of the decoder. We assume the decoder outputs are active-high. Thus, when D_{in} is 1, the decoder is enabled and the output selected by $S_2S_1S_0$ is active (1), whereas all others are low. When D_{in} is 0, the decoder is disabled and the selected output is inactive (0), as are all others. In effect, we route D_{in} to the selected output.

15–12 ANSI/IEEE STANDARD SYMBOLS

Adders and Arithmetic/Logic Units

The qualifying symbol for an adder is the Greek letter sigma: Σ. Figure 15.22*(a)* shows the logic symbol for the 74AS283 4-bit adder. As in past examples, binary grouping

Figure 15.21

Examples of demultiplexers.

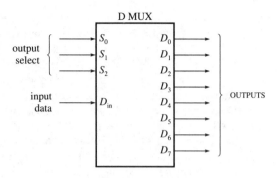

(a) A 1-to-8 demultiplexer.

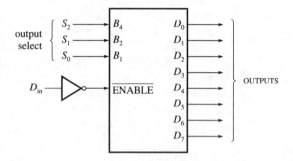

(b) Using a 3-to-8 decoder as a 1-to-8 demultiplexer.

symbols are used to show that all possible combinations of the 4-bit inputs, P and Q, produce all combinations of the output sum, Σ. The input labeled CI is carry in, and the output labeled CO is carry out. Figure 15.22(b) shows the logic symbol for the 74S381 arithmetic/logic unit, for which the qualifying symbol is ALU. Mode dependency is created by inputs S0, S1, and S2, as shown in the function table. Outputs P and G are active-low carry-propagate (CP) and carry-generate (CG) outputs, respectively, used for cascading with look-ahead carry. As indicated by the mode dependencies shown on the symbol, (1/2/3), these outputs are active only when the ALU is in one of the three arithmetic modes: B minus A, A minus B, or A plus B. We see that the input labeled Cn can serve as an active-low *borrow* input (BI) in subtraction modes 1 and 2 or as an active-high carry input (CI) in the addition mode (3). The mode dependencies shown in this symbol are an example of *factoring* in dependency notation: (1/2)BI is the same as 1BI/2BI, and (1/2/3)CP is the same as 1CP/2CP/3CP.

Comparators

The qualifying symbol for a comparator is COMP. As shown in the symbol for the 74HC85 4-bit comparator in Figure 15.23, the standard symbols for comparator outputs are the same as those we have used in previous discussions: P < Q, P = Q, and P > Q. The inputs labeled <, =, and > are used for cascading, as also discussed previously.

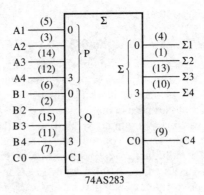

(a) Logic symbol for a 4-bit adder.

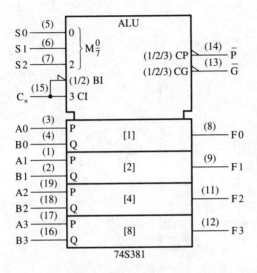

S2	S1	S0	mode	operation
0	0	0	0	CLEAR
0	0	1	1	B minus A
0	1	0	2	A minus B
0	1	1	3	A plus B
1	0	0	4	A ⊕ B
1	0	1	5	A + B
1	1	0	6	AB
1	1	1	7	preset

Function table.

(b) Logic symbol for a 4-bit arithmetic/logic unit.

Figure 15.22
ANSI/IEEE standard logic symbols.

Figure 15.23
Logic symbol for a 4-bit comparator.

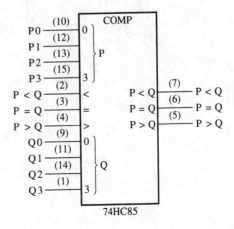

Multiplexers and Demultiplexers

The qualifying symbol for a multiplexer is MUX. Figure 15.24(a) shows the logic symbol for the 74S251 8-to-1 multiplexer with three-state output and bus-driving capability. Note that the qualifying symbol includes the right-pointing triangle that signifies bus-driving capability. The output and its complement have the three-state symbols and are enabled by input $\overline{G}$. The binary grouping symbol shows that the output depends on the eight possible binary combinations of the data-select inputs, A, B, and C. The qualifying symbol for a demultiplexer is DMUX. Figure 15.24(b) shows the 74HC38 1-to-8 demultiplexer. The data input in this case is $G1 \cdot \overline{G}2A \cdot \overline{G}2B$, so a single active-high input (G1) could be used by connecting $\overline{G}2A$ and $\overline{G}2B$ both low, or either active-low input could be used with G1 connected high. Recall that a demultiplexer can be regarded as a decoder (Figure 15.21(b)). Figure 15.24(b) shows an equivalent logic symbol for the 74HC38 when its function is interpreted as a 3-to-8 (BIN/OCT) decoder. In this

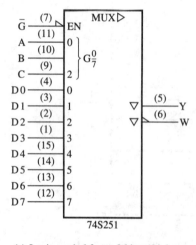

(a) Logic symbol for an 8-bit multiplexer.

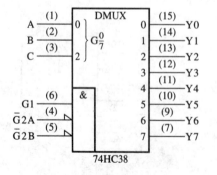

(b) Logic symbol for a 1-to-8 demultiplexer.

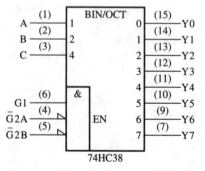

(c) Alternate symbol for the 1-to-8 demultiplexer in (b), when used as a 3-to-8 (binary-to-octal) decoder.

Figure 15.24
ANSI/IEEE standard logic symbols.

interpretation, the inputs G1, G2A, and G2B create an enable (EN) dependency rather than serving as data inputs. Any combination that makes $G1 \cdot \overline{G2A} \cdot \overline{G2B} = 0$ causes all outputs to be inactive (low).

EXERCISES

Section 15–1

15.1 Show that the logic expression $(A \oplus B)C_{in} + AB$ generates the carry-out bit for a full-adder.

15.2 Show that the logic expression $A \oplus B \oplus C_{in}$ generates the sum bit for a full-adder.

Section 15–2

15.3 Draw a logic diagram showing how four exclusive-OR gates, four AND gates, and two OR gates can be connected to construct a 2-bit parallel adder.

15.4 Determine the sum and carry-out bits at each stage of the 4-bit adder in Figure 15.4 when the numbers 0110 and 1010 are added.

15.5 One of the binary numbers to be added by an 8-bit parallel adder is 11001010. How many values of the other number to be added would cause output overflows?

15.6 An adder circuit is to be constructed for a computer by cascading three 8-bit parallel adders. If the most significant sum bit is to be reserved as a sign bit, what is the largest possible decimal sum that the adder can produce without overflow?

Section 15–3

15.7 The time required to generate the carry-out bit in each adder of a 16-bit ripple-carry

adder is 19 ns. Once all the inputs to each adder are valid, there is a 35-ns delay before the output sum bit is valid. What is the maximum number of additions per second that the adder is capable of performing?

15.8 The inputs to an 8-bit parallel adder with carry look-ahead are $A_7A_6A_5A_4A_3A_2A_1A_0$ and $B_7B_6B_5B_4B_3B_2B_1B_0$. Write the logic expression for C_5 in terms of carry propagates, carry generates, and C_0.

15.9 Show that the circuit in Figure 15.25 can be used as a full-adder for a parallel adder employing carry look-ahead.

15.10 In each full-adder shown in Figure 15.7, there is a propagation delay of 4 ns before the carry-generate and carry-propagate outputs are valid. The delay in each external logic gate is 3 ns. Once all inputs to an adder are valid, there is a delay of 6 ns before the output sum bit is valid. What is the maximum number of additions per second that the adder can perform?

Section 15–4

15.11 A serial adder is clocked at 12.5 MHz. What is the largest (decimal) sum it can generate without overflow in 640 ns?

15.12 The initial contents of the registers of a 4-bit serial adder are shown in Figure 15.26. Note that the sum bits are shifted into the

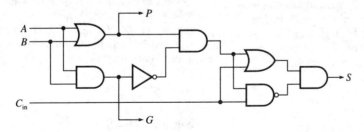

Figure 15.25
Exercise 15.9.

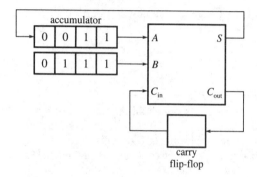

Figure 15.26
Exercise 15.12.

accumulator when addition commences. List the contents of the accumulator after each of the first four clock pulses.

Section 15–5

15.13 By deriving a logic expression, show that the S_0 output in the logic diagram for the 74283 adder (Figure 15.9) is the sum bit when bits A_0, B_0, and C_0 are added.

15.14 By deriving a logic expression, show that the S_1 output in the logic diagram for the 74283 adder is the sum bit when bits A_1, B_1, and a carry look-ahead bit are added.

Section 15–6

15.15 The control inputs to a 74181 ALU are $S_3S_2S_1S_0 = 0110$. The data inputs are $A = 1010$ and $B = 0110$. If $\bar{C}_n = 0$, find $F = F_3F_2F_1F_0$ for each value of M.
(a) $M = 1$
(b) $M = 0$

15.16 In one application of a 74181 ALU, it is desired to make outputs $F_3F_2F_1F_0 = 0000$ whenever $M = 1$ and the data inputs A and B are equal. What should be the states of the control inputs, $S_3S_2S_1S_0$, to accomplish that result? (*Hint:* The output of what logic gate is 0 whenever both inputs to it are equal?)

Section 15–7

15.17 If the coincidence gates in the comparator shown in Figure 15.11 were replaced by exclusive-OR gates, what would a high output from the comparator signify?

15.18 How could the comparator in Figure 15.12 be redesigned so that the four AND gates and the OR gate used to generate the $B > A$ output are eliminated? Draw a logic diagram. (*Hint:* If A does not equal B and if A is not greater than B, then B must be greater than A.)

15.19 The outputs of a 4-bit binary up counter are connected to the A inputs of a 7485 magnitude comparator ($Q_8 = A_3$, $Q_4 = A_2$, $Q_2 = A_1$, and $Q_1 = A_0$). The B input to the comparator is $B_3B_2B_1B_0 = 1000$. Construct a timing diagram showing 15 clock pulses, $A_3A_2A_1A_0$, and the three outputs of the comparator.

15.20 Draw a logic diagram showing how two 7485 comparators could be used to generate a 1-ms wide pulse 100 ms after an 8-bit binary up counter clocked at 1 kHz begins counting from 00000000.

Section 15–8

15.21 Complete the timing diagram in Figure 15.27 to show the output of the multiplexer when the inputs are as shown.

15.22 Construct a timing diagram showing the output of the multiplexer in Figure 15.28 for the duration of eight clock pulses.

15.23 The multiplexer in Figure 15.29 is used to select serial data inputs D_0, D_1, D_2, and D_3 in succession. The serial data is clocked into the selected inputs at a rate of 5×10^5 bits/s. Each input must be selected long enough to allow one 8-bit serial word to be transmitted. What should be the frequency of the clock shown in the figure?

15.24 Draw a logic diagram showing how two 4-to-1 multiplexers could be used to construct a system to accomplish the following task: Data-select inputs S_0 and S_1 are used to select data inputs D_0, D_1, D_2, and D_3. If the data on the selected input is 0, then another data input, D_4, should be routed to the output of the system. Otherwise, D_5 should be routed to the output of the system.

Section 15–9

15.25 Using a 74151A multiplexer, design a circuit that accomplishes the following:

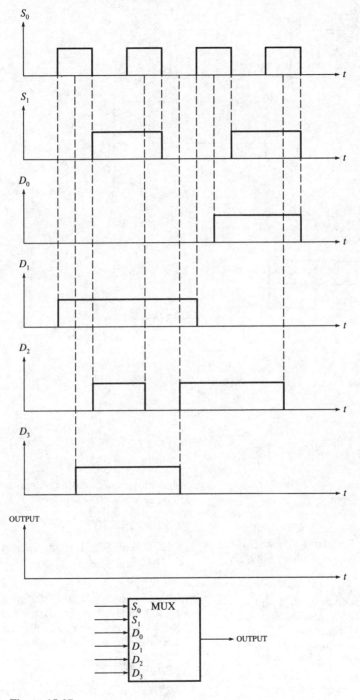

Figure 15.27
Exercise 15.21.

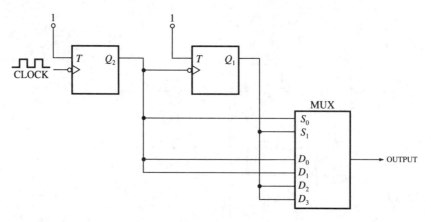

Figure 15.28
Exercise 15.22.

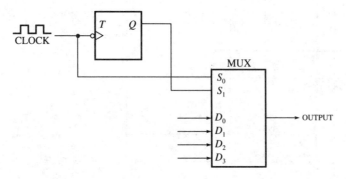

Figure 15.29
Exercise 15.23.

1. If data inputs D_0 and D_7 are identical, no data inputs are to be connected to the output.
2. So long as D_0 and D_7 are different, data inputs D_0 through D_7 are to be connected continually, in succession, to the output.

Draw a logic diagram of your design.

15.26 Using 74151A multiplexers, design a circuit that connects even-numbered data inputs $D_0, D_2, \ldots, D_{14}$, in succession, to the output, then connects odd-numbered data inputs $D_1, D_3, \ldots, D_{15}$, in succession, to the output, and continually repeats the entire sequence. Draw a logic diagram of your design. (*Hint:* S_0 must change state after every eight clock pulses, and the low-order, high-

order multiplexers must be alternately enabled after every four clock pulses.)

Section 15–10

15.27 Use an 8-to-1 multiplexer to generate the logic function $F = AC + BC + \overline{A}\,\overline{B}\,\overline{C}$. Draw a logic diagram.

15.28 Use an 8-to-1 multiplexer to implement a logic function that equals 1 if $B_8B_4B_2B_1$ is an XS-3 BCD code group and that equals 0 otherwise. Draw a logic diagram.

Section 15–11

15.29 Complete the timing diagram to show the outputs of the demultiplexer in Figure 15.30.

Figure 15.30
Exercise 15.29.

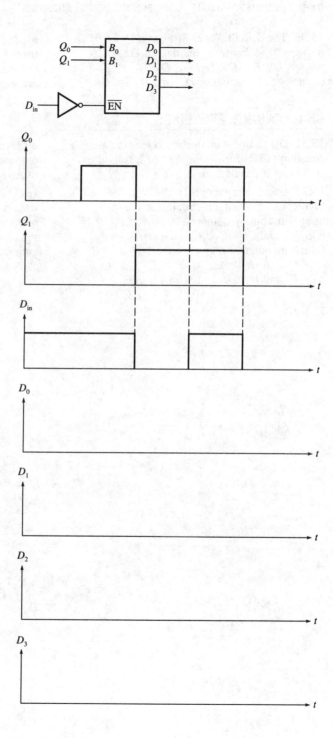

15.30 The demultiplexer in Figure 15.21(b) is to be used to distribute the data at D_{in} to outputs $D_0, D_1, D_2, D_3, D_4, D_0, D_1, \ldots$, in succession. Outputs D_5, D_6, and D_7 are not used. D_{in} is to appear at each selected output for 0.5 ms. Draw a logic diagram showing all components and interconnections necessary to accomplish this task.

CHALLENGING EXERCISES

15.31 Using Boolean algebra to manipulate equation (15.3) for the sum bit of a full-adder, show that it equals $A \oplus B \oplus C_{in}$.

15.32 Using comparators of the type shown in Figure 15.12, design a circuit that compares three 3-bit binary numbers, $A = A_2A_1A_0$, $B = B_2B_1B_0$, and $C = C_2C_1C_0$, and provides the following outputs: $A = B = C$, $A > B > C$, $A > C > B$, $B > A > C$, $B > C > A$, $C > A > B$, and $C > B > A$. Draw a logic diagram. Will at least one of these seven outputs be high for every possible combination of A, B, and C? Explain.

15.33 Use two 8-to-1 multiplexers and whatever other external logic gates are necessary to design an adder that adds four 1-bit inputs: A_3, A_2, A_1, and A_0. Note that the outputs of the adder will be a sum bit and two carry bits. Draw a logic diagram.

Chapter 16

MEMORIES

THE ROLE OF MEMORY IN A COMPUTER SYSTEM

Program and Data Memory

A computer *program* is a set of instructions that a computer executes to achieve a desired result. As mentioned in Chapter 1, one of the significant advances in computer technology was the realization that computer programs could be stored in *memory*. Virtually all modern computers are of the *stored-program* type, where programs are stored as a set of *machine-language* instructions—i.e., binary codes—in memory. This, then, is one of the principal roles of computer memory: to provide a convenient and simple means for storing, retrieving, and/or altering computer programs. When a computer is given a command to execute a program, it retrieves the machine-language instructions, in sequence, from its memory and performs the tasks specified by those instructions.

Computer memory is also used for *data* storage. In this context, data are numerical values or alphanumeric information that a program uses during its computations or that it produces as a result of its computations. An example is the set of account balances of the clients of a bank. The computer periodically executes a program that directs it to retrieve that data, compute new balances, and store the results back in memory. Memory used for that purpose is called *data memory,* as opposed to program memory. By far the largest portion of memory in general-purpose computer systems is devoted to data storage. In small, special-purpose computer systems, such as a microprocessor designed to control a traffic signal, there may be little or no data memory. Whether memory is used for program or data storage, it always consists, at its most basic level, of a means for storing groups of 1s and 0s—i.e., binary words.

Main and Peripheral Memory

Because of the variety of ways that memories are used in computer systems, a number of different types have evolved, each suited to particular applications. For example, a general-purpose computer system (one that can be readily programmed and reprogrammed to perform a wide variety of tasks), has *main* memory and *peripheral* memory. Main memory is an integral part of the system hardware and is very *fast*, in the sense that information can be stored and retrieved from it very quickly. Peripheral memory, also called auxiliary memory, is typically "add-on" memory with very large storage capacity but much slower than main memory. It often serves as data memory for storing very large quantities of data, a role called *mass storage*. In a typical system, users store programs and data in peripheral memory, the computer retrieves a program and its associated data, stores it temporarily in main memory while it executes the program, and then stores the results back in peripheral memory. In modern computers, main memory is typically constructed from semiconductor devices in integrated circuits, whereas peripheral memory consists of magnetic disks or tapes. In older computers, main memory was constructed from tiny electromagnets, called magnetic *cores*, and the term *core memory*, or simply *core*, was used synonomously with main memory.

16-2   MEMORY TYPES AND TERMINOLOGY

Memory Organization

As we have mentioned, all memory, regardless of its type or use, consists of a means for storing *binary* information, or bits. A *word* is the fundamental group of bits used to represent one entity of information, such as one numerical value. *Word size*, the number of bits in a word, varies widely among computer systems and may range from 4 to 32 or more bits. A group of 8 bits is called a *byte*, so word size can also be expressed as a certain number of bytes. For example, a 16-bit word is 2 bytes. A group of 4 bits is sometimes called a *nibble*.

A memory *location* is a set of devices capable of storing one word. For example, each location in a memory used in an 8-bit microcomputer (one that uses 8-bit words) might consist of 8 latches. Each device, such as a latch, in each memory location stores one bit of a word. Each such device is called a *cell*. The capacity, or *size*, of a memory is the total number of bits (sometimes bytes or words) that it can store. For reasons that will become apparent shortly, memory size is often a power of 2. For convenience, the size of such a memory is frequently expressed as a multiple of $2^{10} = 1024$, which is abbreviated by K. For example, a memory of size 2^{11}, which equals 2048, is said to be 2K. A memory of size 2^{16} (65,536) has size 64K.

Figure 16.1 illustrates how memory organization can be visualized, in this case for a small memory capable of storing ten 8-bit words. Each memory location is shown as a row containing 8 cells. Note that each location is assigned a specific *address*, from 00 through 09. We say, for example, that "the contents of memory location (or address) 08 are 01011100." Although the 8-bit word stored at a particular address might represent any quantity (it could be one instruction in a program), it is conventional to refer to the bits as *data* bits D_0 through D_7. Hereafter, we use the word data with that understanding. Although the figure is a useful way to envision memory, the physical arrangement is not necessarily like that shown. For example, one integrated circuit might be used to store all the D_0 bits in a memory, a different integrated circuit for all the D_1 bits, and so forth.

Reading and Writing

The process of storing data in memory is called *writing* in memory. Retrieving data from memory is called *reading* memory. Figure 16.2 illustrates how reading and writing into

Figure 16.1
Organization of a memory containing ten 8-bit words. Each row represents a memory location with a specific address (arbitrary bits shown in cells).

address	D_7	D_6	D_5	D_4	D_3	D_2	D_1	D_0
00	0	1	1	0	1	1	0	1
01	1	0	0	1	0	0	1	1
02	0	1	0	1	1	1	1	0
03	1	1	0	0	1	0	1	0
04	0	0	1	0	0	1	1	1
05	1	1	0	1	1	0	0	1
06	0	1	0	1	1	1	0	1
07	1	1	1	0	1	1	1	0
08	0	1	0	1	1	1	0	0
09	1	1	0	1	1	0	1	1

memory is accomplished in an 8-bit microcomputer system. The microprocessor serves as the *central processing unit* (CPU) for the computer. It contains an arithmetic/logic unit (ALU) and numerous registers and logic circuitry that it uses to perform read and write operations as well as to execute programs. Note the control signals labeled *read* and *write*. The CPU activates these when a read or write operation is to be performed. The wide two-headed arrow represents an 8-bit data bus consisting of 8 lines on which data bits D_0 through D_7 are transmitted. It is called a *bidirectional* data bus because words can be transmitted from the CPU to memory (when writing) or from memory to the CPU (when reading). The unidirectional address bus is the set of lines over which the CPU transmits the address bits corresponding to the memory address to be read or written into. In the example shown, the address bus is a 16-bit bus (A_0 through A_{15}), meaning that the CPU could access (read or write into) up to $2^{16} = 65,536$ different memory addresses. Following is a typical sequence of events, during which a byte is read from one memory location and written into another:

1. The CPU activates the read control and transmits 16-bit address $000A)_{16}$ to memory, via the address bus.
2. As a result of Step 1, the 8-bit word stored in address $000A)_{16}$, say $45)_{16}$, is placed on the data bus and transmitted to the CPU.

Figure 16.2
Reading and writing into memory in a microcomputer system is accomplished via data and address buses.

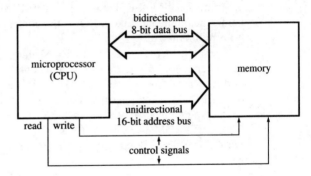

3. The CPU activates the write control and transmits the 16-bit address $000B)_{16}$ to memory. It also transmits the 8-bit data word $45)_{16}$ to memory via the data bus.
4. As a result of Step 3, the data word $45)_{16}$ is stored at address $000B)_{16}$. (The original contents of that address are lost.)

RAMs, ROMs, and PROMs

The type of memory we have discussed is called *read/write* memory because it can be accessed for both of those kinds of operations. *Read-only memory* (ROM) cannot be written into. It is used for permanent storage of programs or data in *dedicated* applications. For example, the program executed by a microprocessor controlling a microwave oven would be stored in ROM because there is never any reason to alter the program by writing new instructions. Another example is a *look-up table* containing the binary equivalents of BCD numbers. Again, it is necessary only to read memory when using such a table.

The data stored in most read/write memories constructed from semiconductor devices will be lost if power is removed. Such memory is said to be *volatile*. On the other hand, ROM memory is non-volatile, the principal feature that justifies its existence. It would clearly be very inconvenient if it were necessary to restore the program controlling a microwave oven every time the power was turned off. RAM memory is *random-access memory:* memory whose addresses can be accessed directly and immediately. By contrast, to access a memory location on magnetic tape, it is necessary to wind or unwind the tape and search through a series of addresses before reaching the one desired. It is conventional to use the term RAM to mean read/write memory—that is, in contrast to ROM. However, most ROM is random access in the sense we have described. From a system designer's viewpoint, there are three types of read-only memory available in integrated circuits. The term ROM by itself generally refers to an integrated-circuit memory whose contents are permanently stored by the manufacturer at the time the circuit is fabricated, a so-called factory-programmed ROM. (Storing the contents of a ROM is called *programming* the ROM, although the contents are not necessarily a computer program.) The manufacturing process for integrated circuits involves the use of a *mask,* which is like a template, that controls the structure of the circuit. For that reason, a factory-programmed ROM is often called a mask programmable ROM, or simply a mask ROM. Once a mask ROM has been manufactured, its contents cannot be altered. Manufacturers provide certain standard ROMs, such as those containing look-up tables, and many will custom-build a ROM to a user's specifications.

A second type of ROM available from manufacturers is called a *programmable read-only memory* (PROM). A PROM is basically a "blank" ROM that can be programmed by a user having special PROM programming apparatus. It is said to be field-programmable. Like a ROM, the contents of a PROM cannot be altered once it has been programmed.

Finally, there is a type of PROM that can be field-programmed and subsequently reprogrammed, if desired. That is, the contents can be "erased" and new contents stored in their place. This type of PROM is called an *erasable* PROM (EPROM). The contents of an EPROM are still permanent, in the sense that they are non-volatile and can be erased and reprogrammed only through the use of special equipment. Actually, there are two types of EPROMs—those that can be erased by exposing them to ultraviolet light and those that can be erased by subjecting them to certain electrical voltages. The latter type is called an *electrically erasable* PROM, or EEPROM. It is also called an electrically alterable PROM (EAPROM), an electrically erasable ROM (EEROM), or an electrically alterable ROM (EAROM).

Memories Are Made of This

Recall from our study of logic families that the two principal types of transistors used in the manufacture of digital integrated circuits are bipolar junction transistors (BJTs) and metal-oxide-semiconductor field-effect transistors (MOSFETs). Integrated-circuit memories are available in both of these technologies. However, not every type of memory we have discussed is available in bipolar technology. Because MOSFETs are more easily manufactured than BJTs and occupy much less space on an integrated-circuit chip, that technology has become dominant in the production of large-capacity integrated-circuit memories. However, BJT memory is considerably faster than MOSFET memory, and the BJT types are used in applications where high speed is more important than large storage capacity. Memory circuits constructed with I^2L technology are also available. These feature higher speeds than MOSFET memories and greater capacity per circuit than BJT memories.

In Chapter 10, we discussed *dynamic* logic, in which logic levels are preserved as charge (or absence of charge) on capacitances. Dynamic memory is memory constructed with that technology. Its principal advantages are that a very large number of memory cells can be fabricated in a single circuit and that it has very small power consumption. Recall, however, that such circuitry must be periodically *refreshed* to replenish the stored charges. This requirement adds somewhat to the complexity of systems that incorporate dynamic logic. Dynamic memory is available only in MOSFET circuits. To distinguish between dynamic memory and memory that utilizes conventional storage devices such as latches, the latter type is said to be *static* memory. Static memory is available in both BJT and MOSFET technology. To distinguish between static RAM and dynamic RAM, the terms SRAM and DRAM are sometimes used. Figure 16.3 summarizes the technologies used in the manufacture of integrated-circuit ROMS and read/write memories.

16–3   ROM CELLS AND CIRCUITS

Cells and Cell Arrays

Conceptually, the structure of a ROM is quite simple. As illustrated in Figure 16.4, we can view it as consisting of an array of row lines and column lines. A cell is located at each position where a row line crosses a column line. Words are stored along the rows, so each of the 6 words (rows) in the example shown has 4 bits (columns). At cell locations where a 1 is stored, a voltage-controlled switch is connected between the row line and the column line at that location. (As we know, diodes, BJTs and MOSFETs can all be regarded as voltage-controlled switches, and all are used that way in the construction of ROMs.) At cell locations where a 0 is stored, there is no connection between the row line and the column line. To read a memory location (row), we apply a voltage to that row line. The voltage closes all switches that are connected to the row, so a high voltage, V, appears on each column line connected to a cell storing a 1. No voltage appears on the column lines connected to cells where 0s are stored. Thus, the word stored in the row appears as output on the column lines. For example, in the figure we see that reading

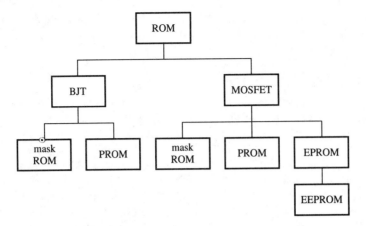

(a) Read-only memories.

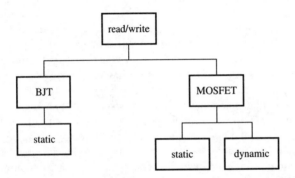

(b) Read/write memories (also called RAM memories).

Figure 16.3
Technologies used in the fabrication of integrated-circuit memories.

location 3 by applying a voltage to row 3 causes V to appear on columns 0 and 1 and 0 to appear on columns 2 and 3, so the word read is 1100. It is apparent that the row numbers are just the addresses of the memory locations, and the column numbers are the bit positions of the data stored. It is also apparent that the presence or absence of switches determines whether 1s or 0s are stored, so the removal of power has no effect on the contents of the memory.

The voltage-controlled switches in Figure 16.4 can be diodes, BJTs, or MOSFETs, and they can be connected to rows and columns in a variety of ways. In some practical designs, the closing of a switch connects 0 V to a column line, so the presence of a switch represents storing a 0. Figure 16.5, p. 632, shows some of the ways that semiconductor devices are used as voltage-controlled switches in ROM circuits. Parts *(c)* and *(e)* of the figure are examples in which the presence of a device represents storing a 0. In every case, pull-up or pull-down resistors are shown connected to the column lines. Note that these are required because, in their absence, an unactivated column would be floating (open

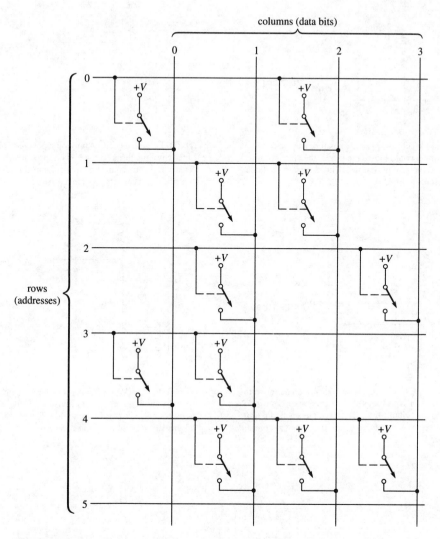

Figure 16.4
*Representation of a ROM as an array of row-lines and column-lines that cross at cell locations.
A 1 is stored at each cell where a voltage-controlled switch is shown. Activating a row-line
causes the switches in that row to close and connects +V to column-lines.*

circuit) instead of at 0 V or high. In part *(e)*, the function of the pull-up resistors is
performed by NMOS transistors serving as resistors, the usual case in MOSFET circuits.

***Example
16–1***

With reference to the MOSFET ROM shown in Figure 16.6, p. 633, determine

a. The word size;
b. The number of words stored;
c. The binary contents of the memory. *(Solution follows on pages 633–634.)*

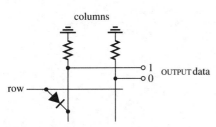

(a) The diode cell. Applying a positive voltage to the row-line forward biases the diode; thus the voltage appears on the column-line. The absence of a diode represents a stored 0.

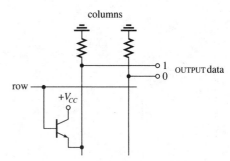

(b) The BJT (bipolar) cell. Applying a positive voltage to the row-line turns on the NPN transistor, which causes V_{CC} to be connected to a column-line. The absence of a transistor represents a stored 0.

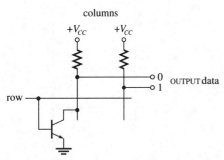

(c) Another BJT cell. Applying a positive voltage to the row-line turns on the NPN transistor and connects 0 V to the column-line. Thus, the presence of a transistor represents a stored 0, and the absence represents a stored 1.

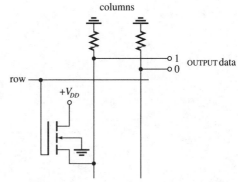

(d) A MOSFET cell. Applying a positive voltage to the row-line turns on the NMOS transistor, causing V_{DD} to be connected to the column line. The absence of a transistor represents a stored 0.

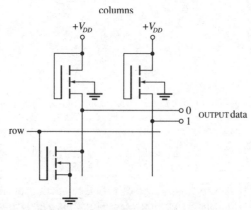

(e) Another MOSFET cell. When a positive voltage is applied to the row-line, the NMOS transistor turns on and connects 0 V to the column-line. The absence of a transistor represents a stored 1. Note the use of NMOS transistors connected as resistors.

Figure 16.5
Examples of diode, BJT, and MOSFET cells in a ROM.

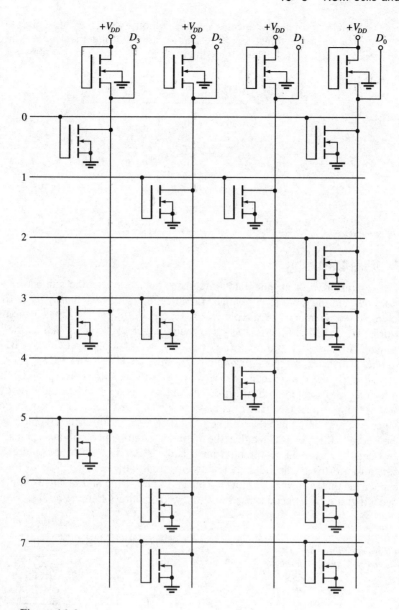

Figure 16.6
Example 16–1.

Solution

a. The word size is the number of column-lines, 4, since each bit of a word is read on one column.

b. The number of words stored is the number of rows, 8.

c. The NMOS transistors connect 0 V to the columns when they are activated. Therefore, the presence of a transistor represents a stored 0 and the absence of a transistor represents a stored 1. Thus, the contents are

Address	D_3	D_2	D_1	D_0
0	0	1	1	0
1	1	0	0	1
2	1	1	1	0
3	0	0	1	0
4	1	1	0	1
5	0	1	1	1
6	1	0	1	0
7	1	0	1	0

Address Decoding

The *organization* of an integrated-circuit memory refers to the number of words it can store *(N)* and the word size *(W)*. The organization is expressed as N by W ($N \times W$). For example, the memory in Example 16–1 (Figure 16.6) is 8×4, since it stores eight 4-bit words. Actual MOSFET ROMs have much larger capacities, such as the TMS47256, which is organized as 32K $\times$ 8 (32,768 $\times$ 8). As mentioned previously, BJT ROMs are smaller but faster, a typical example being the 256 $\times$ 4 74187 ROM.

As we have seen, to read a ROM we must activate a line corresponding to the address where a word is stored. In practice, the address is supplied to the circuit as a binary number. Input pins designated as address bits A_0, A_1, . . . are used for that purpose. Since one and only one address line must be activated for each possible combination of the address bits, it is clear that the memory circuit must contain a binary-to-decimal decoder. For example, a circuit with 8 address bits has $2^8 = 256$ addresses and must contain an 8-to-256 decoder. It is now apparent why integrated-circuit memories have capacities that are various powers of 2. Figure 16.7 shows an example of address decoding: a 3-to-8 decoder used to activate the address lines of an 8×8 memory.

Figure 16.7
Illustration of address decoding. The 3 address bits are decoded to activate one of the 8 possible address lines.

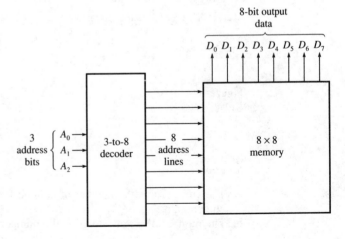

To simplify decoding, many practical ROMs, especially those with large capacities, use two address decoders instead of one. In these designs, additional row or column lines are connected to the cells in such a way that *both* a column line and a row line must be activated to read a particular cell at a row-column intersection. One address decoder activates the row line in which the cell is located and another decoder activates the column line in which it is located. Figure 16.8 shows how a MOSFET memory with 5 address bits could use a 2-to-4 decoder to activate rows and a 3-to-8 decoder to activate columns. The row decoder is often called the X decoder and the column decoder is called the Y decoder. Note that the outputs of the Y decoder are active-low. When a 5-bit address is supplied to the circuit, bits A_4 and A_3 are decoded to select a row line, and the gates of all the MOSFETs in that row go high. Address bits A_2, A_1, and A_0 are decoded by the Y decoder, so the source terminals of every MOSFET in the selected column go low. Only the MOSFET at the intersection of the selected row and column has a positive voltage on its gate and a grounded source, so only that MOSFET turns on. Since the drain is connected to the data output line and to a resistor-connected MOSFET, the output data bit is low (0). The absence of a MOSFET at a row-column intersection represents a stored 1, since the data output remains high (V_{DD}) when such a row-column intersection is selected. In this example, there is only one data bit stored at each address, so the memory is organized as 32×1. The next example illustrates how the basic circuit can be expanded so that 2 (or more) bits can be stored at each address.

Example 16-2

With reference to the ROM memory shown in Figure 16.9, determine

a. The organization;
b. The binary contents of the memory.

Solution

a. Note that each output from the X decoder is connected to two rows, one for output data bit D_0 and one for output data bit D_1. Since there are three address bits in all, the memory is organized as 8×2.
b. As in Figure 16.8, the presence of a MOSFET represents a stored 0 and the absence represents a stored 1. Thus, the binary contents are

Address			Contents	
A_2	A_1	A_0	D_1	D_0
0	0	0	1	0
0	0	1	1	1
0	1	0	1	0
0	1	1	0	1
1	0	0	0	1
1	0	1	0	0
1	1	0	1	1
1	1	1	1	0

Access Time

The access time, t_a, of a ROM is the time required to obtain valid output when reading it. It can be specified as the interval between the time that the input address bits are valid

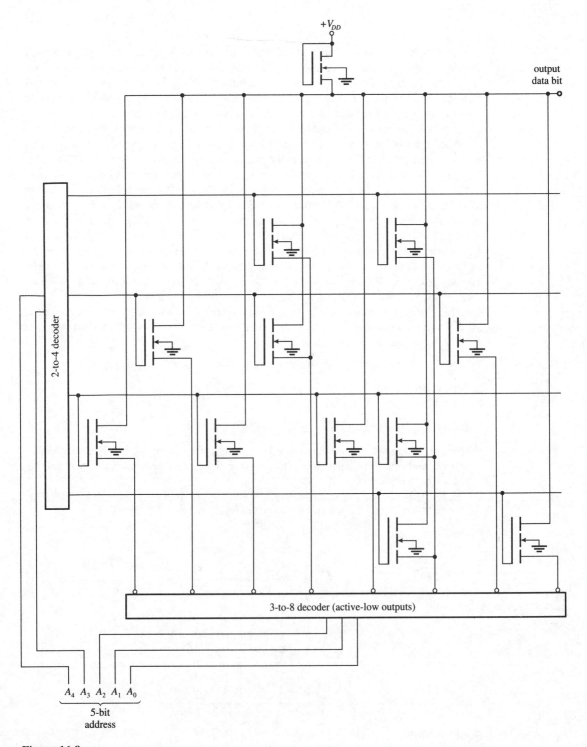

Figure 16.8
Example of address decoding using two decoders. An NMOS transistor in a selected cell has a positive gate voltage and a grounded source, so it connects 0 V to the output. The absence of a transistor represents a stored 1.

636

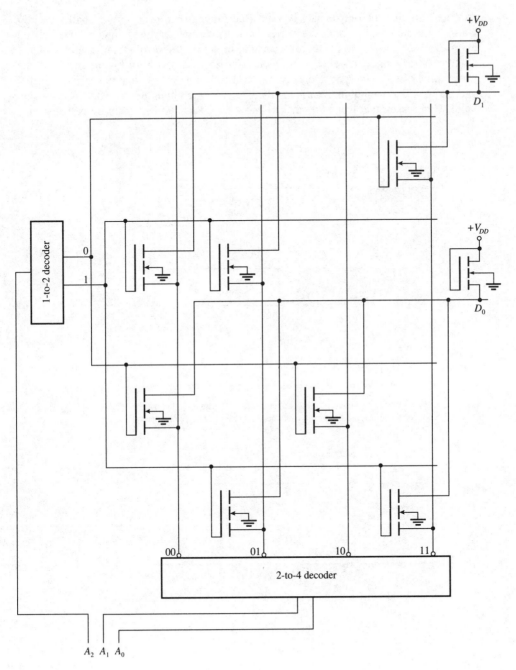

Figure 16.9
Example 16–2.

and the time that the output data is valid (see Figure 16.10*(a)*). As we shall presently learn, most integrated-circuit memories have active-low enabling inputs, often called chip-select *(CS)* inputs, to facilitate expansion. In Figure 16.10*(a)*, t_a is measured with the assumption that the chip-select input is already low. As shown in Figure 16.10*(b)*, the access time can also be specified as the interval between the time that the chip is selected and the time that the output data is valid under the assumption that the address is already valid. One reason that it is important to know the access time of memory is that it may

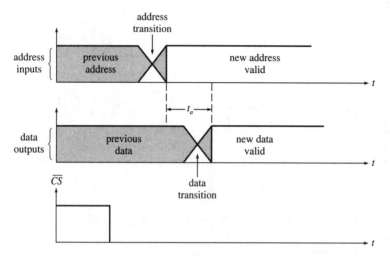

(a) Access time, t_a, defined as the time required for new output data to be valid after a new address is valid, assuming the chip-select input, $\overline{CS}$, is already active (low).

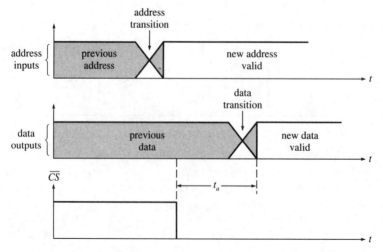

(b) Access time, t_a, defined as the time required for new output data to be valid after the chip-select input ($\overline{CS}$) becomes active (low), assuming the address inputs are already valid.

Figure 16.10
Access time for reading a ROM.

Figure 16.11
The 7488 bipolar 32 × 8 ROM.

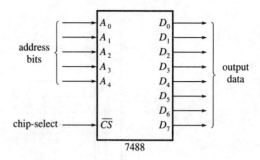

be the limiting factor in the speed at which a computer system can operate. Recall that the CPU repeatedly accesses memory to read and execute program instructions. The frequency at which memory is read must not be so great that a new address appears at its input before sufficient time, t_a, has elapsed for the data stored at the previous address to be valid. The access time for bipolar ROMs is on the order of 30 ns, whereas MOSFET ROMs have access times in the range from 200 to 400 ns.

Examples of Integrated-Circuit ROMs

Figure 16.11 shows the logic symbol* for a typical bipolar ROM, the 7488, which has a capacity of 256 bits. With 5-bit addressing and an 8-data-bit output, we see that it is organized as 32 × 8. This ROM could be used to store a small number of data bytes or a short program in an 8-bit microprocessor application where high speed is important. A memory of greater capacity could be constructed by using two or more of these ROMs, as is described in a later discussion of memory expansion. Another example of a bipolar ROM is the 74187, which has larger capacity (256 × 4) than the 7488. However, at least two of these must be used in any system with 8-bit word size.

Figure 16.12 shows a block diagram of a large-capacity MOSFET ROM, the 8K × 8 M36000H. This ROM has a number of features that are typical of modern MOSFET ROMS, including latched address inputs and three-state outputs. The ROM is enabled on the negative-going *edge* of the chip enable $(\overline{CE})$ input, at which time the address bits are latched. With this feature, a new address can be supplied to the ROM in anticipation of the next read operation and before the output data from the current read is valid. The outputs are in a high-impedance state when $\overline{CE}$ is high (inactive), so the outputs can share a data bus with other devices or other memories. Other features found in modern high-capacity ROMs are reduced power consumption when the ROM is not enabled and user-specified chip-select inputs. Two or more such inputs can be factory programmed to be active-low or active-high.

Use of a ROM as a Look-Up Table

ROMs are frequently used as *look-up tables* for code conversions and mathematical functions. An example of the latter is a ROM that stores values of the trigonometric sine function. The ''addresses'' in such an application are actually binary-coded values of angles, and the outputs are the values of the sines of the angles. Figure 16.13(a) shows a logic diagram for a preprogrammed code-converter ROM, the 74184 BCD-to-binary

---

*ANSI/IEEE standard logic symbols for memory circuits are shown in Section 16–13.

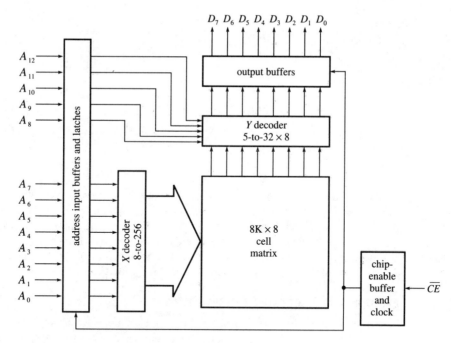

Figure 16.12
Block diagram of the M36000H 8K × 8 NMOS ROM.

converter. Notice that the BCD input appears on 5 bits. Since the least significant bit of the binary output is always the same as the least significant bit of the BCD input, we can convert 6 bits of BCD input, i.e., 1½ BCD digits.* The most significant BCD digit (MSD) must appear on 2 bits, so the largest decimal value that can be converted is 39. Outputs Y_6, Y_7, and Y_8 are not used in BCD-to-binary conversion. As shown in Figures 16.13(*b*) and *(c)*, the 74184 can also be used to produce the 9's or 10's complement of a 4-bit BCD input. Y_6, Y_7, and Y_8 are used as outputs for those applications. The BCD input is applied to DCBA, and 9's-complement conversion occurs when $E = 0$, 10's-complement when $E = 1$. A companion ROM, the 74185, is available for binary-to-BCD code conversion. Both types can be cascaded to perform conversions of multidigit numbers.

16-4   PROMs, EPROMs, AND EEPROMs

Fusible-Link PROMs

Mask-programmed ROMs are not economically feasible unless produced in large quantities. Manufacturing costs are generally too high to justify fabrication of one or a small number of units. On the other hand, the field programmable ROM, or PROM, is ideally suited for development work and small production quantities. Manufacturers produce these blank ROMs in quantities large enough to keep the cost of a single PROM

---

*As is discussed in Chapter 17, the term ½ BCD digit often refers to a *single* bit rather than two.

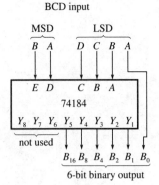

(a) BCD-to-binary decoder.

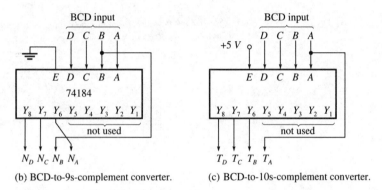

(b) BCD-to-9s-complement converter. (c) BCD-to-10s-complement converter.

Figure 16.13
The 74184 ROM BCD-to-binary code converter and complement generator.

at a reasonable level. PROMs are often used for testing and refining new system designs because they can be replaced with relatively little effort and cost. Once all requirements have been firmly established, mask-type ROMs can be ordered for large production quantities.

PROMs are manufactured with *fusible links* that can be selectively "blown" by a user to create open circuits in a cell array. Figure 16.14(a) shows an array of bipolar cells having fusible links in series with the emitters of all transistors. Since every collector is connected to V_{CC}, 1s are stored at every memory location. Using a special apparatus called a *PROM programmer*, a user causes a large current to flow through the transistors at cells where a 0 is to be stored. The current burns open the fuses in those cells, thus disconnecting the emitters. Figure 16.14(b) shows an example. Once a PROM has been programmed, its contents cannot be changed (except to store additional 0s) because the burned-open links cannot be restored.

Fusible links are constructed using a metal, such as nichrome or titanium-tungsten, or using polycrystalline silicon. PROM programmers subject selected cells to current pulses whose amplitudes and durations are closely controlled, as necessary to burn open the type of links used. A typical commercially available programmer is fully automatic, in the sense that the data to be stored in the PROM can first be loaded into the programmer, which then automatically sequences through the memory addresses of the PROM and supplies the necessary current pulses at the correct cell locations.

Figure 16.14

Bipolar PROM array before and after pro-
gramming.

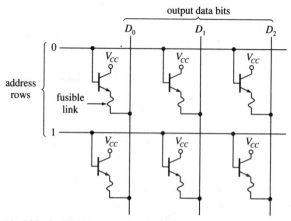

(a) A bipolar PROM array before programming. All fusible links are intact, so a 1 is stored in every cell.

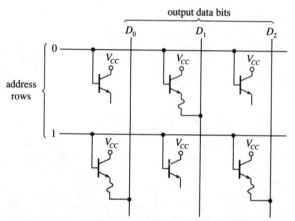

(b) Example of the PROM array after programming. The fusible links have been permanently burned open to store 0s at selected locations. The word 010 is stored at address 0 and 101 at address 1.

Bipolar PROMs are available in a wide variety of organizations, including 32×8 (74S188/288), 256×4 (74S287/387), 512×4 (74S570/71), 256×8 (74LS471), 512×8 (74S472/73), 1024×4 (74S72/73), 1024×8 (77S180 series), 2048×4 (77S184 series), 2048×8 (77S190 series), 4096×4 (77S195 series), and 4096×8 (77S321 series). All have enabling/chip-select inputs, and different versions are available with open-collector or three-state outputs to facilitate expansion. Many are manufactured with 0s stored in all cells instead of 1s. Most MOSFET PROMs are erasable (EPROMs), but some large-capacity MOSFET PROMs are available. Examples include the $128K \times 8$ TMX27PC010 and the $64K \times 16$ TMX27PC210.

EPROMs

Like the fusible-link PROM, the erasable PROM (EPROM) is useful for developmental and experimental work because its contents can be programmed with relative ease and at

Figure 16.15

Floating-gate NMOS transistor used in EPROM cells.

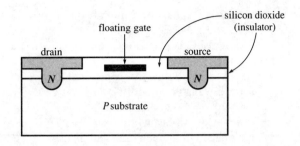

small cost. The principal feature of the EPROM is that its contents can be erased and reprogrammed, so the same device can be used repeatedly. The term *ultraviolet-* (UV-) *erasable* PROM is often used synonomously with EPROM, since that is the mechanism used to erase it.

EPROM cells are constructed from MOSFET transistors whose gates have a special design. Figure 16.15 is a cross-sectional view of one such NMOS transistor. As shown, the gate is actually a region that is completely surrounded by silicon dioxide insulating material. There are no electrical connections to this gate region, and since it is completely isolated, it is said to be ''floating'' (in the electrical sense). If a high drain-to-source voltage is applied to the transistor, electrons are injected into the floating gate, where they must remain, trapped, indefinitely. The effect on an NMOS transistor is the same as applying a negative gate-to-source voltage: The transistor is off. Thus, cells in an EPROM are programmed by selectively applying high voltages to charge floating gates, rendering selected transistors permanently off. The transistors remain off (until the EPROM is reprogrammed) because there is no path through which the trapped charge in the gate regions can escape. The transistors whose gates are not charged remain on because conduction between drain and source is not inhibited. EPROM programmers similar to those described for PROMs are available for automatically generating the pulses necessary to charge selected gates.

To erase an EPROM, the cells must be exposed to relatively intense ultraviolet radiation. The radiation imparts enough energy to the trapped electrons to allow them to escape through the insulating layers, and the transistors are all once again in the on state. As shown in Figure 16.16, an EPROM has a transparent quartz window to permit exposure of the cells to UV radiation when erasing. Commercially available EPROM

Figure 16.16

An EPROM has a transparent window to permit exposure of the cells to ultraviolet radiation when erasing.

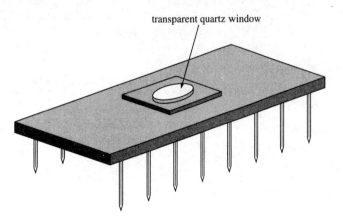

erasers are equipped with ultraviolet lamps and timers to facilitate the process. Complete erasure usually takes less than one-half hour. Once an EPROM has been programmed, an opaque label is placed over the window to prevent stray radiation from causing gradual erasure.

A popular line of commercially available EPROMs is the 27XX and 27CXX series, available in both NMOS and CMOS (27CXX) versions. The XX designates the capacity in K bits. For example, the 2716 has a capacity of 16K bits. All devices in the series store 8-bit words, so the 2716 is organized as 2K × 8. As with most integrated-circuit memories, all devices in the series have three-state outputs to facilitate bus sharing and memory expansion. These EPROMs are manufactured with a 1 stored in every cell location. One input, designated V_{pp}, is used for programming. When V_{pp} is raised to +25 V, charge is injected into the floating gate of a selected cell, resulting in the storage of a 0.

EEPROMs

The electrically erasable PROM (EEPROM) represents the most recent development in the evolution of ROM technology. Unlike the UV-erasable PROM, the EEPROM does not necessarily have to be removed from a circuit and exposed to a different environment to be erased. Many EEPROMS can be both erased and programmed with modest power requirements, so it is possible to integrate erasing and reprogramming circuitry into the system utilizing the memory. Although these capabilities bring the EEPROM a step closer to becoming nonvolatile read/write memory, the time required to "write" (erase and reprogram) is still in the millisecond range, far slower than conventional read/write memory. Another advantage of the EEPROM over the EPROM is that it is possible to erase and restore a single byte in an array.

The structure of an NMOS transistor used in an EEPROM cell is shown in Figure 16.17. Like the UV-erasable EPROM, it has a floating gate that is electrically isolated by silicon dioxide. A second gate, called the control gate, is used for erasing. The cell is programmed in the same way as an EPROM cell, by injecting electrons into the floating gate. When a positive voltage is applied to the control gate, the electrons escape through the silicon dioxide by virtue of a semiconductor phenomenon known as *tunneling* (the principle of the tunnel diode). Thus, selected transistors in an EEPROM can be both turned on and turned off by the application of electrical voltages.

A typical EEPROM, such as the 2816 (16K × 8), can be operated in a number of different modes. The 2816 has three active-low control inputs: $\overline{CE}$ (chip enable), $\overline{OE}$ (output enable), and $\overline{WE}$ (write enable), that determine the operation performed. Table 16.1 summarizes the operations.

Figure 16.17

Structure of an NMOS transistor used in an EEPROM cell

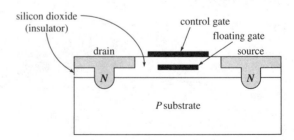

Table 16.1
Operating modes of the 2816 EEPROM.

$\overline{CE}$	$\overline{OE}$	$\overline{WE}$	Mode	Operation
1	x	x	Standby	Outputs in high-Z state; low power consumption
0	0	1	READ	Data output
0	1	0	BYTE WRITE	Data input
0	1	1	R/W INHIBIT	Outputs in high-Z state
0	1	HV	BYTE ERASE	All bits in an addressed byte go to 1
0	HV	HV	CHIP ERASE	All bits in all addresses go to 1

x = don't care
HV = high voltage (21 V)

Note that a single byte or the entire contents of the memory can be erased. Although the CHIP ERASE mode requires about 10 ms to erase all contents (every bit becomes 1), this time is far shorter than the 30 min or so required to erase a UV-erasable EPROM.

16–5   STATIC READ/WRITE (RAM) MEMORY

As previously mentioned, it is customary to refer to read/write random-access memory as RAM to contrast it with ROM. Despite the fact that ROM is also random-access memory, we will hereafter observe that convention. Static RAMs are constructed in bipolar (TTL) and MOSFET technologies.

Bipolar RAMs

As an aid in understanding the operation of a bipolar RAM cell, consider first the two cross-connected transistors shown in Figure 16.18(a). Note that the collector of each is connected to the base of the other. This arrangement forms a simple latch, in which the presence or absence of current flow represents logic levels. In the figure, Q_2 is assumed to be off, so current flows into the base of Q_1, holding it on. Since Q_1 is on, no current flows into the base of Q_2, so it remains off. We see that the fact that one transistor is on ensures that the other transistor is off and vice versa. Therefore, the states remain unchanged until and unless an external signal is used to reverse them. If power is removed, the contents of the latch are lost, which accounts for the fact that the static RAM cell is *volatile*. We assume that the latch is storing a 1 when Q_2 is off and is storing a 0 when Q_2 is on.

Figure 16.18(b) shows the basic bipolar RAM cell, in which the cross-connected transistors forming a latch are now the multiple-emitter transistors used in TTL circuitry. The multiple emitters provide three *parallel* paths for current flow through each transistor. Note that two sets of emitters are used to address the cell (X-address, Y-address), using the same row-column scheme we discussed in connection with cell arrays in ROMs. To address a cell, both its X-address line and its Y-address line are made high. In that case,

Figure 16.18
The bipolar RAM cell.

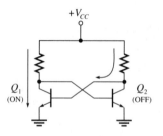

(a) Two cross-connected transistors forming a latch.

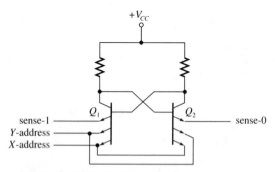

(b) The bipolar (TTL) RAM cell constructed from multiemitter transistors cross-connected to form a latch.

no current can flow through either of the emitters connected to address lines. If the latch is storing a 0, Q_2 is on, and current flows into the sense-0 line. Q_1 is off, so no current flows into the sense-1 line. If the latch is storing a 1, Q_1 is on, current flows into the sense-1 line, and no current flows into the sense-0 line. We have just described how a cell is read. Note that when the cell is *not* addressed, either or both of the address lines will be low, so current will flow into an address line instead of into a sense line. To write into a cell, we must again address it by making both address lines high. The sense lines are then used to change the state of the latch (if a change is necessary). Suppose Q_2 is on and Q_1 is off (storing a 0). If we now make sense-0 a 1, we shut off current flow through Q_2 (all its emitters are high), turning it off. At the same time, we make sense-1 a 0, providing a path for current flow through Q_1 and allowing it to turn on. Thus, the state of the latch reverses, as we have written a 1 into it. If sense-0 is made 0 and sense-1 is made 1, we write a 0 into the cell.

Figure 16.19 shows a 4 × 4 array of bipolar RAM cells. Note that the sense-0 line is connected to every cell, as is the sense-1 line. When reading, the addressed cell is the only one that can produce current in one of the sense lines, since all other cells have at least one address line low. Also note that the same sense lines are driven by inputs W_0 and W_1. These are used to write a 0 or a 1 in an addressed cell, as previously described. The array shown in the figure represents a 16 × 1 RAM memory, since a single bit is read or written for each address selected. A memory having N-bit words would have separate sense and write lines for each bit.

Practical bipolar RAMs have additional buffering and logic circuitry, such as address decoders, not shown in Figure 16.19. A typical integrated circuit has one set of data pins

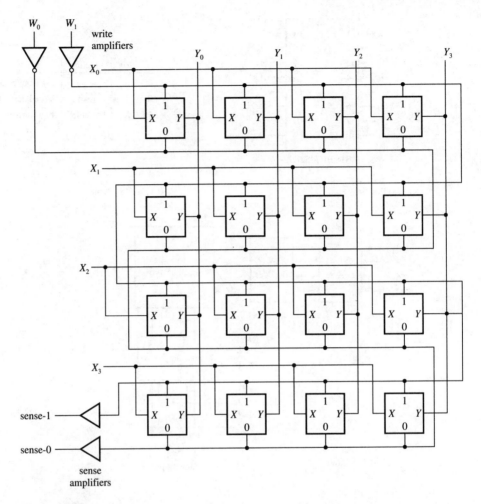

Figure 16.19
A 4 × 4 array of bipolar RAM cells forming a 16 × 1 memory.

used to read data out and another set to write data in, a set of address pins, a chip-select input ($\overline{CE}$), and read/write control input(s). An example is the 74S189B/74S289B-series 64-bit RAMs, which are organized 16 × 4. Figure 16.20 shows the pin diagram, block diagram, and function table. The 74S189B has three-state outputs and the 74S289B has open-collector outputs to facilitate expansion. When the active-low chip-select is high (inactive) the data outputs of the 74S189B are in a high-impedance state and those of the 74S289B are off. Note that the outputs, $\overline{Q_1}$, $\overline{Q_2}$, $\overline{Q_3}$, and $\overline{Q_4}$, are active-low. Reading the contents of an address by making $\overline{CE} = 0$ and $R/\overline{W} = 1$ causes the complement of the data stored at the address to appear at the output. Therefore, when writing data in (with $\overline{CE} = 0$ and $R/\overline{W} = 0$), the inputs at D_1, D_2, D_3, and D_4 should be the complements of the bits it is desired to store. Note that the outputs are in a high-impedance or off state during write operations. Typical access time for this Schottky-clamped series is 25 ns.

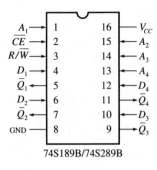

(a) Pin diagram.

$\overline{CE}$	$R/\overline{W}$	function	OUTPUTS
0	0	write	high-impedance
0	1	read	stored data
1	X	disable	high-impedance

(b) Function table.

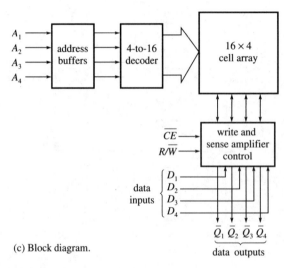

(c) Block diagram.

Figure 16.20
The 74S189B/74S289B 16 × 4 TTL RAM.

MOSFET RAMs

As is the case with read-only memory, MOSFET RAMs are generally slower than bipolar RAMs but have the significant advantage that they can be manufactured with much greater capacity in a single integrated circuit. Also, they consume much less power. Current MOSFET technology has produced faster circuits, and most modern RAM memory is of the MOSFET type. MOSFET RAMs are constructed using NMOS and CMOS technology.

Figure 16.21 shows the basic NMOS cell, consisting of cross-connected transistors Q_1 and Q_2. Q_3 and Q_4 serve as resistor-connected transistors. The arrangement forms a bistable latch similar to that of the bipolar cell. When Q_1 is on, its drain voltage is low, making the gate of Q_2 low and holding Q_2 off. Since Q_2 is off, its drain voltage is high, which holds Q_2 on. We assume those states correspond to storing a 1, and the opposite set of states corresponds to storing a 0. Transistors Q_5 and Q_6 serve as voltage-controlled switches (transmission gates) that permit reading or writing into the cell. When the

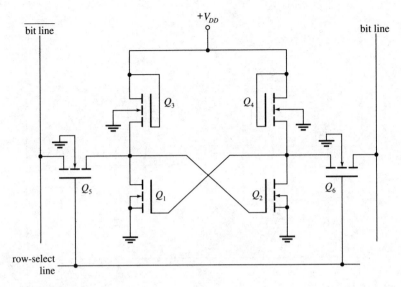

Figure 16.21
The static NMOS RAM cell.

row-select line is high, Q_5 and Q_6 are turned on and the high or low voltage at each of Q_1 and Q_2 appears on the bit lines. In an array of such cells, the row-line is connected to all cells in the row. When the row-line is made high, the bits appear on the several bit lines, so all the cells in the row are read. Since a row of cells can be considered to store a multi-bit word, the row-line is sometimes called a word-line. It is apparent that a static MOSFET cell, like a static bipolar cell, is volatile and has non-destructive readout. To write into a cell, we again make the row-line high and turn on Q_5 and Q_6. Suppose that Q_2 is off and Q_1 is on (storing a 1). To write a 0 into the cell, we make bit line = 0, which connects a low to the gate of Q_1 and turns it off. Also, we make $\overline{\text{bit line}}$ = 1, which connects a high to the gate of Q_2 and turns it on. Thus, the latch changes to the 0-state and we have written a 0 into the cell. Making bit line = 1 and $\overline{\text{bit line}}$ = 0 causes a 1 to be written into the selected cell.

Figure 16.22 shows the basic cell used in CMOS RAMs. Like the NMOS cell, it is a bistable latch whose state can be read or altered via transistors Q_5 and Q_6. Transistors Q_3 and Q_4 are PMOS devices. If the gates of Q_4 and Q_2 are high (positive), then NMOS transistor Q_2 is on and PMOS transistor Q_4 is off. Since Q_2 is on, its drain is low, and so therefore are the gates of Q_3 and Q_1. Then NMOS transistor Q_1 is off and PMOS transistor Q_3 is on, making their drains high. This high holds the other pair of transistors in the states we originally assumed (Q_2 on, Q_4 off), so the latch remains locked in one stable state. As in the NMOS latch, a change of state as accomplished by voltages applied to the bit lines when the row-select line is high.

Because they are more widely used than bipolar RAMs, MOSFET RAMs are available in a great variety of sizes and organizations. For example, in 1989 the Texas Instruments line of CMOS RAMs ranged from 16K in three different organizations (1 bit, 4 bit, and 8 bit) to 288K organized as 32K × 9.

An example of a static MOSFET RAM is the 2K × 8 4016, produced by several manufacturers. A pin diagram and function table are shown in Figure 16.23. Unlike the bipolar RAM discussed earlier, the 4016 has a single set of data pins (D_1–D_8) that can be

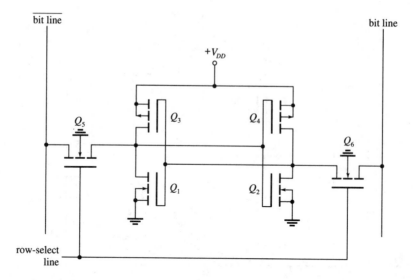

Figure 16.22
The static CMOS RAM cell.

used for both reading data out and writing data in. This feature, which is shared by many modern RAMs, makes it more compatible with the bidirectional data bus found in microcomputer systems (see Figure 16.2). As usual, an active-low chip-select input is available to facilitate bus sharing and memory expansion.

ECL RAMs

RAMs constructed using ECL technology have the highest speeds among those we have considered. The access time for bipolar TTL RAMs is on the order of 25–50 ns; for

A_7	1	24	V_{CC}
A_6	2	23	A_8
A_5	3	22	A_9
A_4	4	21	$\overline{W}$
A_3	5	20	$\overline{G}$
A_2	6	19	A_{10}
A_1	7	18	$\overline{CS}$
A_0	8	17	D_8
D_1	9	16	D_7
D_2	10	15	D_6
D_3	11	14	D_5
V_{SS}	12	13	D_4

4016

(a) Pin diagram.

$\overline{W}$	$\overline{CS}$	$\overline{G}$	function	$D_1 - D_8$
0	0	X	write	data inputs
1	0	0	read	data outputs
X	1	X	device disabled	high-impedance
1	0	1	output disabled	high-impedance

(b) Function table.

Figure 16.23
The 4016 2K × 8 static MOSFET RAM.

MOSFET RAMs it is on the order of 200–400 ns (although some new CMOS designs have speeds approaching those of bipolar RAMs). The access time for ECL RAMs is on the order of 5–10 ns. They consume considerable power and are not available in large sizes, so they are used where speed is the most important consideration. Examples include *cache* memory, where data from a slower memory can be stored for quick access by a CPU, and *scratch-pad* memory, used for storage of intermediate results of computations. In these applications, data must be stored and retrieved very rapidly in order not to delay high-speed computations performed by an ALU.

ECL RAMs typically have open-emitter outputs to facilitate expansion and wire-ORing. They also have separate pins for input and output data, a feature that can be used to reduce delays between read and write operations: New data to be written can be applied while old data is still being read. As an illustration of the varieties available, the National Semiconductor line of ECL RAMs ranges from 256 × 1 to 1K × 4 and 4K × 1. Note that these sizes, while small compared to MOSFET RAMs, are considerably greater than those of TTL RAMs.

16–6 DYNAMIC RAMS

Dynamic RAMs are constructed only in MOSFET technology. Considerable effort has been expended in recent years to develop dynamic RAMs with smaller cell sizes and greater densities, to the extent that dynamic RAMs now represent the highest-capacity memories available.

Early dynamic RAMs had four NMOS transistors per cell and relied on inherent gate capacitance of the transistors for storage of charge. The 4-transistor cell was complex because transistors were needed to buffer and sense the tiny charges stored on the capacitance. Later designs relocated the buffering and sensing circuitry so that it could be used for multiple cells in rows and columns of a cell matrix. This tactic reduced the complexity of a single cell and allowed more cells to be fabricated in one circuit. Figure 16.24 shows the improved 3-transistor cell and the most current design: a single-transistor cell.

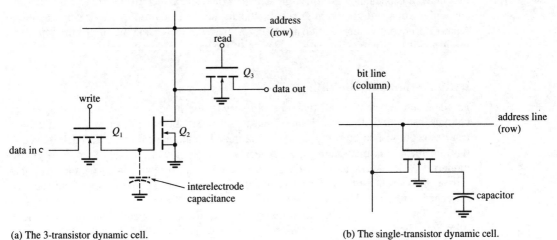

(a) The 3-transistor dynamic cell.

(b) The single-transistor dynamic cell.

Figure 16.24
Dynamic memory cells.

The principle of the 3-transistor cell is very much like that of a dynamic inverter, as discussed in Section 10–4. Transistors Q_1 and Q_3 are transmission gates used to gain access to the gate of transistor Q_2 (for writing) or to its drain (for reading). The drain of Q_2 is connected to another NMOS transistor, not shown, that is common to all cells in a row. Suppose the capacitance is charged, representing a stored 1. If the READ line is made high, Q_2 turns on because its gate is positive. Consequently, a low, or logic 0, is read out. If the capacitor is initially discharged, Q_2 remains off and a 1 is read. We see that the readout is inverting. To write, the write-line is made high, and the capacitance is either charged (to write a 1) or discharged (to write a 0) through Q_1. In addition to the fact that it uses three transistors and inverts the data, a disadvantage of this design is that charge on the capacitance is dissipated when the cell is read. In other words, it has a destructive readout, and the data must be written back in each time the cell is read.

The one-transistor cell in Figure 16.24(b) is the ultimate in simplicity. The transistor serves as a transmission gate controlled by the address line. To read, the address line is made high, turning on the transmission gate, and the capacitor voltage appears on the bit line. To write, the address line is again made high, and the voltage on the bit line charges or discharges the capacitor through the transmission gate. Readout is destructive, so every read operation must be followed by a write operation. We should note that the capacitor is a separate component of the cell and the space it occupies is at least as great as the space required for the transistor. (This despite the fact that the capacitance is less than 0.1 pF.) Therefore, the 1-transistor cell is not truly one-third the size of the 3-transistor cell. Figure 16.25 shows an array of 1-transistor cells. Note that one sense amplifier is required for each column of cells sharing a common bit line. The transistors along the top row, Q_0, Q_1, . . . , act as switches to connect a selected bit line to the data line. When a particular row and a particular column address line are both made high, the only data transferred to the data line (if reading) is the bit stored in the cell at the intersection of the selected row and column.

Although the 1-transistor cell is quite simple, the fact that many of them share the same sense amplifier adds to the complexity of that circuitry. The capacitance of a long bit line connected to many cells adversely affects the ability to sense charges. The bit line may have a capacitance 20 times greater than the capacitance of a cell, and it creates a capacitive voltage divider that greatly attenuates the voltage produced by a cell when it is read. Consequently, there is a design trade-off between increasing the size of the capacitor in each cell and increasing the complexity of the sensing circuitry.

Dynamic Memory Control

Another complication of all dynamic circuitry is the need to *refresh*—i.e., to restore charges that would otherwise drain off the capacitances. In dynamic memories, refreshing can be accomplished by reading, since data must be automatically written back into cells that are read. The maximum time between refresh cycles is typically 2, 4, or 8 ms in modern RAMs. Although it would be possible to read every cell in succession and thus refresh the entire memory during each refresh cycle, most dynamic memories can be refreshed one entire row at a time. In *burst*-mode refreshing, every row of cells is refreshed in succession, with all normal memory operations suspended until all rows have been refreshed. Alternatively, row refreshing can be interspersed with other memory operations. In either case, refresh control circuitry is necessary to synchronize refresh cycles and to ensure that every row is refreshed within the specified time. *Dynamic memory controllers* are available for that purpose. Some dynamic RAMs have built-in

column address lines

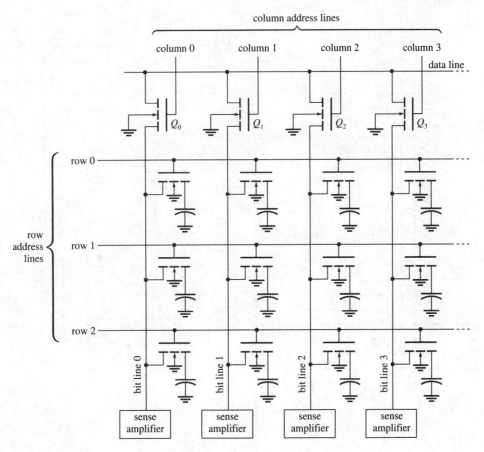

Figure 16.25
An array of 1-transistor cells in a dynamic memory.

refresh control circuitry and are said to be *pseudodynamic* (or *semidynamic*), because a user is not required to provide external hardware to accomplish the refresh task.

To refine some of the concepts we have introduced, let us examine a specific dynamic RAM, the TMS4C1624, 1M × 1 (1,048,576 × 1), manufactured by Texas Instruments and available from numerous other sources. The architecture (component arrangement) and control requirements for this RAM are typical of most modern dynamic memories. Figure 16.26 shows a functional block diagram. Twenty address lines are required to address $10^{20} = 1,048,576$ cells. However, to conserve pins, only 10 address pins, $A_0, A_1,$ . . . , A_9 are used. Notice that the 10 address lines are connected simultaneously to a set of 10 row-address latches and a set of 10 column-address latches. To access a particular cell, the user first supplies the 10-bit row address, which is latched when the external control signal $\overline{RAS}$ (row-address strobe) goes low. The 10-bit column address is supplied next, and it is latched when the control signal $\overline{CAS}$ (column-address strobe) goes low. Thus, a 20-bit address is acquired in a two-step process. The process is called *address multiplexing*, since address pins A_0–A_9 are multiplexed (shared) for both row and column addresses.

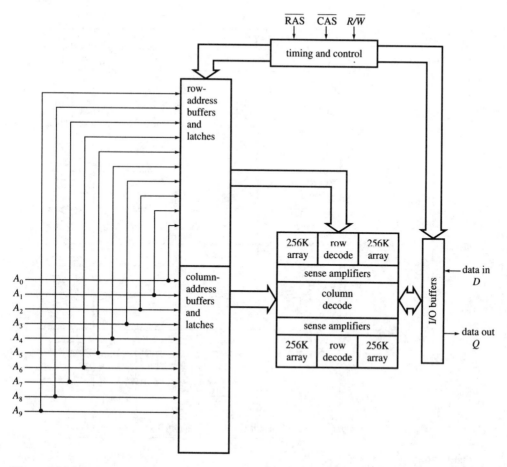

Figure 16.26
Functional block diagram of a 1M × 1 dynamic memory with address multiplexing.

Figure 16.27 shows a timing diagram for a memory read operation, called a *read cycle*. The control signal R/$\overline{\text{W}}$ is made high for memory read and low for memory write. Note the exceptionally large number of parameters listed under "timing requirements" and shown in the timing diagram. Minimum values for these parameters are given in manufacturer's specifications, which attest to the complexity of dynamic memory operations. Similar diagrams and specifications are available for write cycles.

Refreshing the TMS4C1024 can be accomplished in several ways. When $\overline{\text{RAS}}$ goes low, all cells in the row address contained in the row-address latches are refreshed. Thus, any normal read or write cycle will refresh a row of cells. A burst-mode refresh could be accomplished by cycling address inputs through all 1024 row addresses and pulsing $\overline{\text{RAS}}$ after each new address is applied. Since the column address is irrelevant, $\overline{\text{CAS}}$ can be left inactive (high) during the entire process. Refreshing with $\overline{\text{CAS}}$ high is called $\overline{\text{RAS}}$-only refresh, and it has the advantage that the inactive $\overline{\text{CAS}}$ signal disables the output buffers (the output is in a high-impedance state), which conserves power. Another way to refresh is called $\overline{\text{CAS}}$-before-$\overline{\text{RAS}}$. If $\overline{\text{CAS}}$ is held low while $\overline{\text{RAS}}$ is pulsed, rows are refreshed

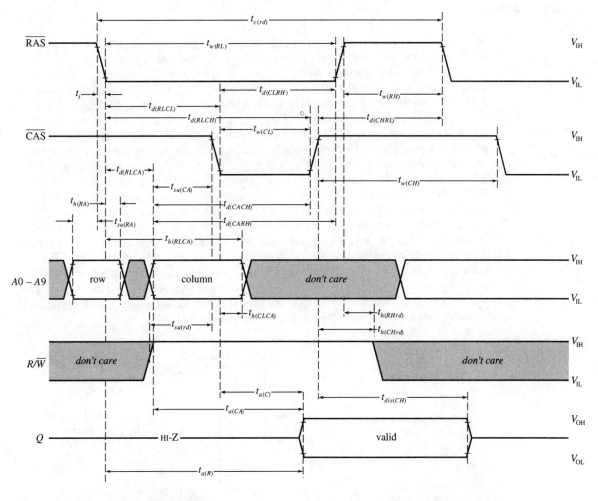

Switching characteristics

$t_{a(C)}$	access time from $\overline{CAS}$ low
$t_{a(CA)}$	access time from column address
$t_{a(R)}$	access time from $\overline{RAS}$ low
$t_{dis(CH)}$	OUTPUT disable time after $\overline{CAS}$ high

Timing requirements

$t_{c(rd)}$	read cycle time
$t_{w(CH)}$	pulse duration, $\overline{CAS}$ high
$t_{w(CL)}$	pulse duration, $\overline{CAS}$ low
$t_{w(RH)}$	pulse duration, $\overline{RAS}$ high (precharge)
$t_{w(RL)}$	pulse duration, $\overline{RAS}$ low
$t_{su(CA)}$	column-address set-up time before $\overline{CAS}$ low
$t_{su(RA)}$	row-address set-up time before $\overline{RAS}$ low
$t_{su(rd)}$	read set-up time before $\overline{CAS}$ low
$t_{h(RA)}$	row-address hold time after $\overline{RAS}$ low
$t_{h(RLCA)}$	column-address hold time after $\overline{RAS}$ low
$t_{h(CHrd)}$	read hold time after $\overline{CAS}$ high
$t_{h(RHrd)}$	read hold time after $\overline{RAS}$ high
$t_{d(RLCH)}$	delay time, $\overline{RAS}$ low to $\overline{CAS}$ high
$t_{d(CHRL)}$	delay time, $\overline{CAS}$ high to $\overline{RAS}$ low
$t_{d(CLRH)}$	delay time, $\overline{CAS}$ low to $\overline{RAS}$ high
$t_{d(RLCL)}$	delay time, $\overline{RAS}$ low to $\overline{CAS}$ low
$t_{d(RLCA)}$	delay time, $\overline{RAS}$ low to column address
$t_{d(CARH)}$	delay time, column address to $\overline{RAS}$ high
$t_{d(CACH)}$	delay time, column address to CAS high
t_t	transition time

Figure 16.27
Read cycle of the TMS4C1024 dynamic RAM.

655

in a sequence of addresses generated internally. The contents of the address latches are ignored, and the user is relieved of the task of cycling the address inputs.

It is clear that incorporating dynamic memory into a system creates some complications with respect to timing, addresses, and refreshing. Nevertheless, dynamic RAMs are widely used because they are available in such large sizes. Eight-megabit ($1M \times 8$) devices are now being manufactured. To reduce the complexity of the circuitry required to support dynamic RAMs, a large variety of dynamic memory controllers have been developed. Figure 16.28 shows a functional block diagram of a typical controller, one that could be used with a $1M \times 1$ dynamic RAM. Notice that it accepts a 20-bit address input and delivers a 10-bit address output. The multiplexer alternately connects A_0–A_9 (the row address) to the output and A_{10}–A_{19} (the column address) to the output under the control of the row/column ($R/\overline{C}$) enable input. Thus, the controller provides the address multiplexing function required by the dynamic RAM for normal read/write operations. When the refresh control input is made active, the contents of the 10-bit counter are delivered to the output. Note that refresh overrides $R/\overline{C}$ enable. For a burst-type row refresh, the counter (driven by an external clock input labeled COUNT) would be cycled through its 1024 states and the zero-detect output could be used to signify completion of the refresh. Of course, the user still has the responsibility of providing correctly timed and synchronized addresses and control signals.

Figure 16.28
A dynamic memory controller for a 1M × 1 memory.

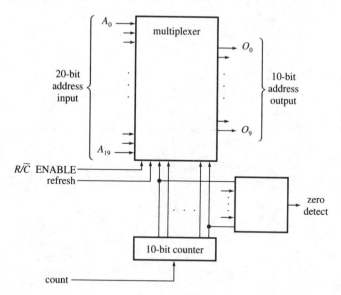

Function Table

refresh	$R/\overline{C}$ ENABLE	function
0	1	$A_0 - A_9$ (row address) delivered to output
0	0	$A_{10} - A_{19}$ (column address) delivered to output
1	x *(don't care)*	counter contents delivered to output

Example
16–3

A 64K $\times$ 1 dynamic RAM with address multiplexing is to be refreshed using $\overline{RAS}$-only burst-mode refreshing. The memory must be refreshed at least every 4 ms and the entire memory must be refreshed in no longer than 100 μs.

a. How many outputs should a dynamic controller used for this memory circuit have?
b. At what frequency should the counter in the controller be clocked?
c. At what frequency should the $\overline{RAS}$ input to the memory circuit be pulsed?
d. At what frequency should the REFRESH input to the controller be driven? What should be the duty cycle of that input?
e. If the access time for the memory is 200 ns, how many read cycles can occur between refreshes?

Solution

a. Since a 64K memory has 16-bit addresses ($2^{16} = 65,536$), each row address and each column address has 8 bits. Since address multiplexing is used, the controller should have 8 outputs.
b. The counter must count through $2^8 = 256$ states to generate all row addresses in 100 μs. Therefore, it should be clocked at $256/(100 \ \mu s) = 2.56$ MHz.
c. The $\overline{RAS}$ input must go low once for each row address generated by the counter, so it should also be pulsed at 2.56 MHz.
d. Refreshing must occur no less frequently than every 4 ms, so the frequency of the REFRESH input should be no less than $1/(4 \ ms) = 250$ Hz. The refresh input must remain active for the 100-μs refresh time, so its duty cycle should be $(100 \ \mu s)/(4 \ ms) = 0.025$.
e. Memory can be read when the REFRESH input is inactive, which occurs for $4 \ ms - 100 \ \mu s = 3.9$ ms between every refresh cycle. Therefore, memory can be read $(3.9 \ ms)/(200 \ ns) = 19,500$ times between refresh cycles.

Drill
16–3

To provide a margin of safety, it was decided to burst-mode refresh the memory in Example 16–3 every 2.4 ms. How many read cycles can then occur between refresh cycles?
Answer: 11,500.

16–7 MEMORY EXPANSION

The two principal reasons for expanding memory are to increase the word size or increase the capacity. In a *bit-organized* memory, each integrated circuit stores 1 bit of each word. For example, in a bit-organized 64K $\times$ 8 memory, the least significant bit of every word is stored in one 64K $\times$ 1 circuit, the next least significant bit in another 64K $\times$ 1 circuit, and so forth. A memory in which every bit of a word is stored in each circuit is said to be *word-organized*. For example, a word-organized 64K $\times$ 8 memory might consist of eight 8K $\times$ 8 circuits. In each of the examples cited, memory expansion is required: In the bit-organized memory, we must interconnect 8 64K $\times$ 1 circuits, and in the word-organized memory, we must interconnect eight 8K $\times$ 8 circuits. Some organizations require an expansion in word size as well as an expansion in capacity. For example, if we wished to construct a 128K $\times$ 8 memory using 16K $\times$ 4 circuits, we would need two of the latter for each 16K of 8-bit words, or a total of $2(128K/16K) = 16$ circuits.

The examples of memory expansion that follow were chosen to conserve space, so memory sizes are relatively small. Although there are older systems with small-capacity

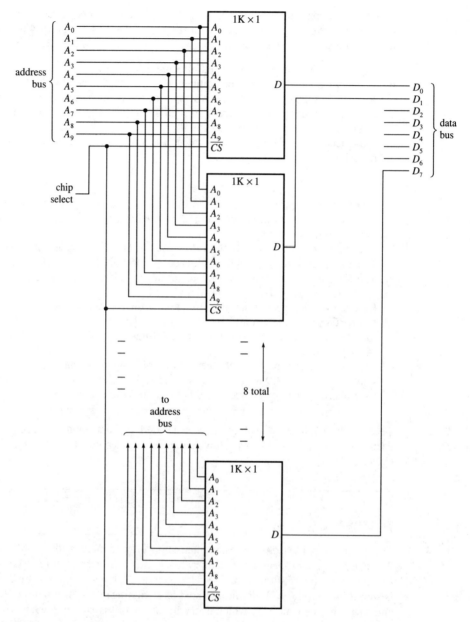

Figure 16.29
Construction of a 1 K × 8 memory from 1K × 1 circuits.

circuits still in use, in most cases new memories of the sizes illustrated would be contained in a single integrated circuit.

Figure 16.29 shows how eight 1K × 1 circuits are interconnected to form a 1K × 8 memory. Notice that the address inputs are simply connected in parallel. Thus, when an

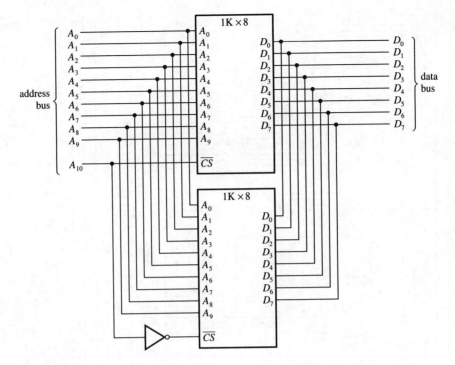

Figure 16.30
Interconnection of two 1K × 8 memories to form a 2K × 8 memory. The upper circuit is selected when $A_{10} = 0$, so it contains the least significant 1K of addresses.

address appears on the address bus during a read operation, each cell having that address in each circuit is read. The outputs of the 8 cells that are read appear as $D_0, D_1, \ldots ,$ D_7 on the data bus. The method illustrated here for increasing word size applies to all of the memory types we have discussed, including ROMs, PROMs, and static and dynamic RAMs. If the memory circuits are dynamic, then the $\overline{RAS}$ and $\overline{CAS}$ inputs are also paralleled.

Figure 16.30 shows an example of increasing memory capacity by interconnecting two 1K × 8 circuits to form a 2K × 8 memory. Notice that the data lines are in parallel in this case. Address inputs A_0 through A_9 are also in parallel, but A_{10} and $\overline{A}_{10}$ are connected to active-low chip-select inputs. When a $\overline{CS}$ input is inactive (high), the data outputs (or input/outputs) are in a high-impedance state. Thus, when $A_{10} = 0$, the upper circuit is selected and the lower one is effectively disconnected from the data bus. The upper circuit therefore contains addresses 00000000000 through 01111111111. When $A_{10} = 1$, the lower circuit is selected and the upper one is disconnected, so the lower circuit has addresses 10000000000 through 11111111111.

Figure 16.31 shows how 4 1K × 8 circuits are interconnected to form a 4K × 8 memory. Note that a 2-to-4 decoder with active-low outputs is used to select one of the four memory circuits. When $A_{11}A_{10} = 00$, the least significant 1K addresses are selected; when $A_{11}A_{10} = 01$, the next least significant 1K addresses are selected, and so forth. As before, the data bus is shared by all circuits, but only one can be active at any given time.

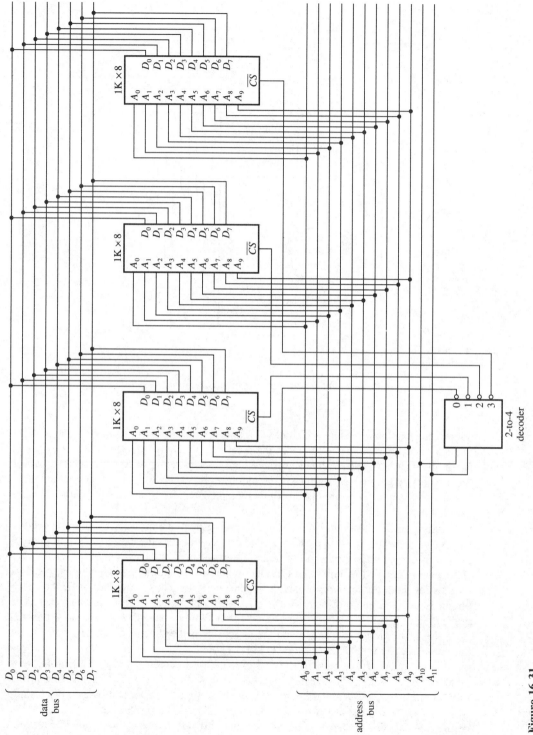

Figure 16.31
Construction of a 4K × 8 memory from 1K × 8 circuits. The 2-to-4 decoder ensures that no more than one set of data lines is active at any given time.

16–8   PROGRAMMABLE LOGIC ARRAYS

A programmable logic array (PLA) is an integrated circuit that can be programmed (in the same way a PROM is programmed) to implement specific logic functions. The principal advantage of such a circuit is that in many applications it can replace numerous other circuits and thereby reduce the chip count and cost of a system. As a simple example, to implement the function $F = A\overline{B}C + D\overline{E} + FG$ would require 3 AND gates, 2 inverters, and an OR gate, which might require 3 or 4 separate chips to implement in 7400-series logic. A single PLA programmed to implement the function could represent a significant cost savings. Programmable array logic, identified as PAL® (a registered trademark of Monolithic Memories, Inc.), is a particular family of PLA devices that are widely used and available from a number of manufacturers. It is the subject of our ensuing discussion.

PAL circuits consist of a set of AND gates whose inputs can be programmed and whose outputs are connected to an OR gate. Some also allow output inversion to be programmed. Thus, like AND-OR and AND-OR-INVERT logic, they implement a sum-of-products logic function. Figure 16.32(a) shows a small-scale example of the basic structure. The fuse symbols represent fusible links that can be burned open using equipment similar to a PROM programmer. Note that every input variable and its complement can be left connected or disconnected from every AND gate. We say that the AND gates are programmable. Figure 16.32(b) shows how the circuit is programmed to implement $F = A\overline{B}C + \overline{A}BC$. Note this important point: All input variables and their complements are left connected to the unused AND gate, whose output is therefore $A\overline{A}B\overline{B}C\overline{C} = 0$. The 0 has no effect on the output of the OR gate. On the other hand, if all inputs to the unused AND gate are burned open, the output of the AND gate "floats" high (logic 1), and the output of the OR gate in that case is permanently 1. Actual PAL circuits have several groups of AND gates, each group providing inputs to separate OR gates.

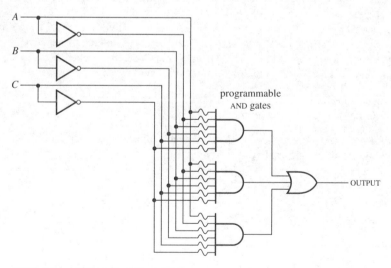

(a) The basic structure of a PAL circuit.

Figure 16.32
Structure and example of a programmed PAL circuit. (Continued on p. 662.)

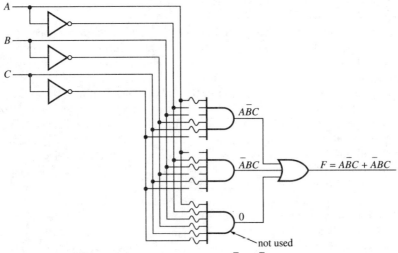

(b) The circuit in (a) programmed to implement $F = A\overline{B}C + \overline{A}BC$. Note that all inputs to the unused AND gate are left connected, so its output is 0.

Figure 16.32
(Continued)

Figure 16.33(a) shows a conventional means for abbreviating PLA connection diagrams. Note that the AND gate is drawn with a single input line, whereas, in reality, it has 3 inputs. An X denotes a connection through an intact fusible link, and a dot represents a permanent connection. The absence of any symbol represents an open (no connection), by virtue of a burned-open link. In the example shown, input A is connected to the gate through a fusible link, input B is disconnected, and input C is permanently connected. Therefore, the output of the gate is AC. Figure 16.33(b) shows an example of how the PAL structure is represented using the abbreviated connections. In this example, the circuit is unprogrammed because all fusible links are intact. Note that the 3-input OR gates are also drawn with a single input line.

Example 16–4   Using connection abbreviations, redraw the circuit in Figure 16.33(b) to show how it would be programmed to implement $F_1 = \overline{A}BC + A\overline{B}C + A\overline{C}$ and $F_2 = \overline{A}\,\overline{B}\,\overline{C} + BC$.

Solution.   See Figure 16.34. Note that the unused AND gate has all links intact. The diagram is sometimes called a *fuse map*.

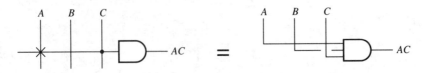

(a) Conventional method for abbreviating connections in PLA diagrams.

Figure 16.33
Simplified method for showing connections in PLA circuits. (Continued on p. 663.)

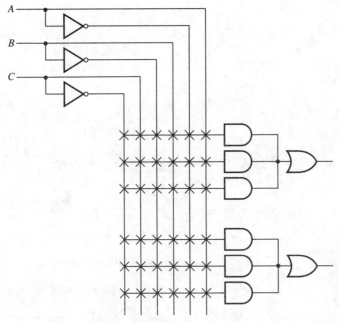

(b) The basic PAL structure in an abbreviated connection diagram. All fusible links are intact.

Figure 16.33
(Continued)

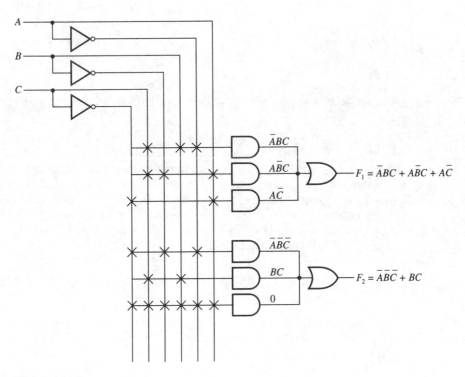

$$F_1 = \bar{A}BC + A\bar{B}C + A\bar{C}$$

$$F_2 = \bar{A}\bar{B}\bar{C} + BC$$

Figure 16.34
Example 16–4.

663

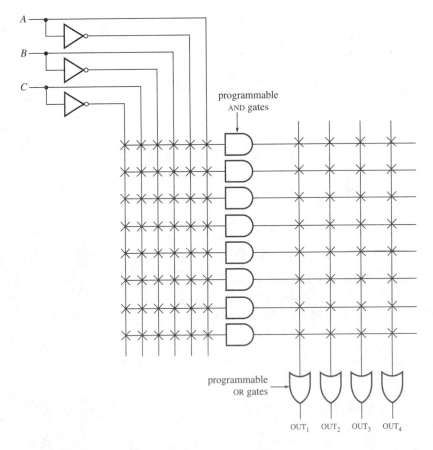

Figure 16.35
Structure of an (unprogrammed) FPLA circuit. All AND and OR gates are programmable.

FPLA

FPLA (field-programmable logic array) represents another type of programmable logic with a slightly different architecture. (The PAL family is also field-programmable; do not be confused by the presence or absence of an *F* in these acronyms.) In FPLA circuits, the output of every AND gate has a programmable connection to the input of every OR gate. Thus, all AND and OR gates are programmable. Figure 16.35 demonstrates the FPLA structure, with every fusible link intact.

Example 16–5

Show how the FPLA circuit in Figure 16.35 would be programmed to implement the sum and carry outputs of a full-adder.

Solution. Recall that the sum output of a full-adder is $S = \overline{A}\,\overline{B}C_{in} + \overline{A}B\overline{C}_{in} + A\overline{B}\,\overline{C}_{in} + ABC_{in}$ and the carry output can be simplified to $C_{out} = AB + AC_{in} + BC_{in}$. To implement these expressions we need a 4-input OR gate and a 3-input OR gate. Since the inputs to the OR gates of the FPLA can be programmed, we can implement the expressions as shown in Figure 16.36.

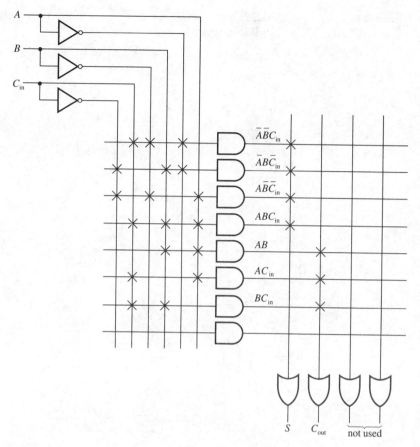

Figure 16.36
Example 16–5.

The PROM as a PLA

A programmable ROM can be viewed as a type of programmable logic array and can be used for that purpose. The address inputs to the PROM serve as logic variable inputs and the data output as the node where the output of a logic function is realized. For example, stating that a 1 is stored at address 0110 is the same as stating that the logic function being implemented equals 1 when the input combination is $\overline{A}BC\overline{D}$. In both cases, the output will be 1 when the input is 0110. When we regard a PROM as a PLA, we realize that the AND gates are not programmable. In effect, an AND gate is already in place for every possible combination of the inputs, corresponding to every possible address of the PROM. Therefore, to program a PROM as a PLA, we must have a truth table that specifies the value of the function being implemented for every possible combination of the inputs. For each combination where $F = 1$, we leave the output of the corresponding AND gate connected to the output OR gate. For each combination where $F = 0$, we burn open the connection to the OR gate. We see that a PROM is a PLA with fixed AND gates and a programmable OR gate. An $M \times N$ PROM can be regarded as a PLA having N programmable OR gates, capable of implementing N different logic functions of M

variables. As demonstrated in the next example, a PROM is ideally suited for implementing a logic function directly from a truth table.

Example 16–6

Show how an 8×1 PROM should be programmed to implement the logic function whose truth table is given in Figure 16.37(a).

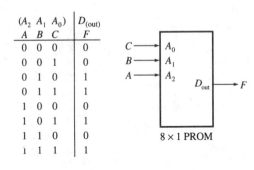

(a)

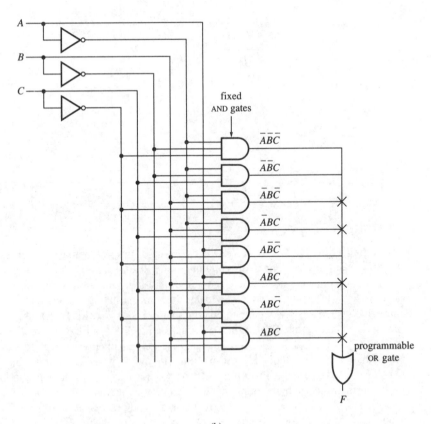

(b)

Figure 16.37
Example 16–6.

Solution.   Figure 16.37*(b)* shows the programmed PROM in the simplified connection format of a PLA. A logical 1 or 0 is stored at every address combination corresponding to a combination of the input variables for which the function equals 1 or 0.

Practical PLA Circuits

Integrated circuits in the PAL family are available with many more inputs (up to 20) than those we have used in the examples. Many have features that enhance their versatility, such as programmable output inversion. This is accomplished by connecting the output through an exclusive-OR gate whose other input is a programmable 0 or 1. Since $X \oplus 1 = \overline{X}$ and $X \oplus 0 = X$, we see that inversion or no inversion is determined by the state of the programmed input to the exclusive-OR gate. Other features and options available include latched and 3-state outputs and fuse-programmable input latches.

16—9   MAGNETIC DISK MEMORY

A magnetic disk is a flat, circular plate that has been coated with a magnetic material. Binary data is stored on a disk by magnetizing tiny regions of the surface, and data is read from the disk by sensing that magnetization. We will discuss the technical details presently. Magnetic disk memories are used primarily for peripheral or auxiliary memory and for mass storage because access times are slower than those of semiconductor memories. They are less costly, per bit of storage capacity, than are semiconductor memories, and they have become the dominant type of peripheral memory in computer systems in all sizes. The disks themselves range in size from a 2-in.-diameter ''floppy'' disk used with microcomputer systems to a 14-in.-diameter hard disk used in larger systems. Floppy disks are made of plastic, are inexpensive, and are readily portable. Hard disks are made of aluminum, have much greater storage capacity than ''floppies,'' and are used in complex, more expensive systems where higher speeds and large capacities are required.

Read and Write Mechanisms

Whether a disk is small or large, hard or floppy, the basic principle of writing and reading this type of memory is the same. Figure 16.38*(a)* shows a *write head* used to magnetize spots on the surface of the disk. By the principle of electromagnetism, a magnetic field is produced in the iron-core frame when an electrical current flows through the winding. The magnetic field crosses through the air gap at the surface of the disk and magnetizes a tiny spot there. The magnetized spot is then like a small magnet that is itself producing a magnetic field, and that magnetism remains when the current through the winding is removed. Thus, storage is non-volatile. The *direction* (polarity) of the magnetic field produced by the spot determines whether a 0 or a 1 has been stored. As shown in Figure 16.38*(b),* the direction of the field is reversed by reversing the direction of the current through the winding. To write a sequence of 0s and 1s, the disk is rotated and the winding is driven by a sequence of current pulses, each pulse having one direction or the other. Thus, 0s and 1s are stored in a sequence around a circular path on the disk.

    To read data on a disk, a *read head* similar in construction to the write head is used. See Figures 16.39*(a)* and *(b).* As the disk rotates, the magnetic field generated by each passing spot is coupled into the core and induces a voltage in the winding. The polarity

Figure 16.38
The write head.

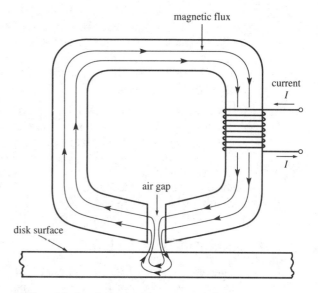

(a) Current in the winding produces a magnetic field that magnetizes
a spot on the disk.

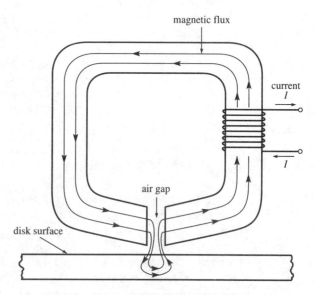

(b) Reversing the direction of the current in the winding reverses the
direction of the magnetic field and the polarity of the magnetized spot.

of the induced voltage depends on the direction of the magnetic field produced by each
spot, so a stored 1 produces an output voltage pulse of one polarity and a stored 0
produces a pulse of opposite polarity. As shown in Figure 16.39(c), the read and write
heads can be combined in a single unit.

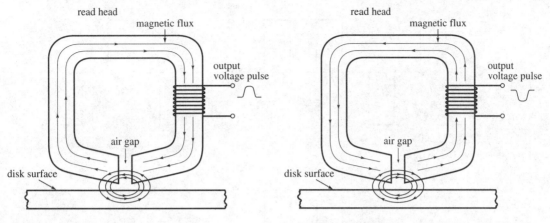

(a) The magnetized spot induces a voltage in the winding when the disk passes under the read head.

(b) A spot magnetized with opposite polarity induces an output voltage of opposite polarity.

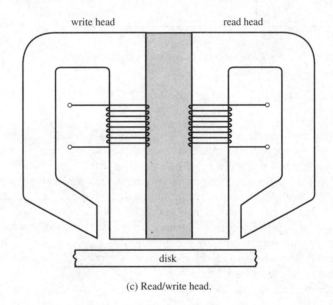

(c) Read/write head.

Figure 16.39

Recording Formats

Our description of bit storage as a sequence of discrete dots having one magnetic polarity or the other implied that the regions between the dots were unmagnetized. That may or may not be the case, depending on the format used to record data. In *return-to-zero* (RZ) recording, there is, in fact, zero magnetization between every bit. Figure 16.40(*a*) illustrates the RZ format. As is conventional, we regard one magnetic polarity as positive (that corresponding to logical 1) and the other as negative (logical 0). The important point

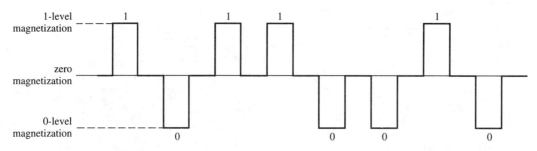

(a) The return-to-zero (RZ) format.

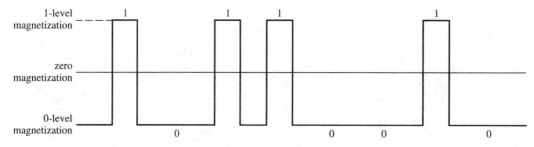

(b) Return-to-bias, in which the bias level is the magnetization corresponding to logical 0.

Figure 16.40
Return-to-zero and return-to-bias formats for recording data.

to note is that magnetization returns to zero between every bit, including adjacent 1s and adjacent 0s. A generalization of RZ is called return-to-reference (or bias), wherein the level to which magnetization returns between bits can be any level between 1 and 0. Figure 16.40(*b*) shows an example in which magnetization returns to the polarity representing logical 0 between every bit. Although there is no "return" involved between adjacent 0s, this format has the advantage that the total *change* in magnetization between a 0 and a 1 is large, making it easier to detect such cases. In this format, as well as others we discuss, reading and writing are synchronized by a clock signal, usually recorded on the disk itself. A clock is necessary to ascertain the precise time interval, called the *bit time*, during which successive bits occur.

In the non-return-to-zero (NRZ) format, there is no change in magnetization between adjacent 0s or between adjacent 1s. This format is illustrated in Figure 16.41. The advantage of NRZ is that no space or time is required for inserting zero magnetization (or bias-level magnetization) between adjacent bits, making greater storage density possible. Obviously, a synchronizing clock must be used to enable the reading of data at bit times when this format is used.

Another recording format is called phase (or biphase) recording, in which 1s and 0s are distinguished by their magnetization levels in the first and second *half* of each bit time. Figure 16.42(*a*) shows an example. Note that a 1 is represented by a positive polarity in the first half of a bit time, followed by a negative polarity in the second half. A 0 is represented by a negative polarity in the first half. Thus, a 1 is detected by a 1-to-0 transition in the middle of a bit time and a 0 by a 0-to-1 transition in the middle of a bit time. In a variation of this format, called the *Manchester* format, a 1-to-0 transition at the

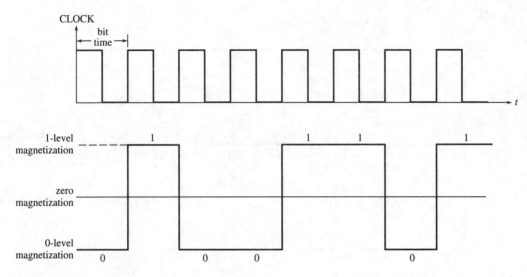

Figure 16.41
An example of data (01001101) recorded in a non-return-to-zero (NRZ) format.

beginning of a bit time represents a 0 and no transition represents a 1. This format is illustrated in Figure 16.42(b).

Whatever method is used to record data, the process is "transparent" to the computer utilizing disk memory. That is, the computer is not required to perform any tasks related to interpreting or converting pulses or magnetization levels into a form suitable for its use. These are the tasks of the *disk controller,* an integral part of disk memory that contains

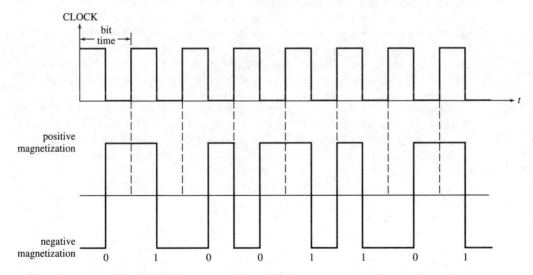

(a) An example of data stored in a biphase format.

Figure 16.42
The biphase and Manchester formats. (Continued on p. 672.)

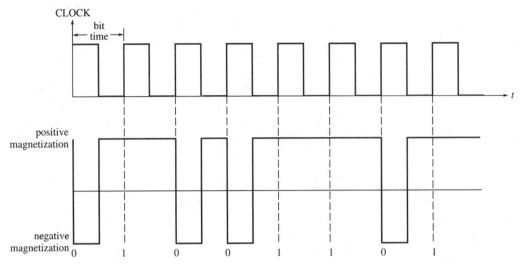

(b) An example of data stored in a Manchester format.

Figure 16.42
(Continued)

all the electronics necessary for data conversion and for controlling the position of the read/write head.

Floppy Disks

Floppy disks (or diskettes) are available in 2-in., 3½-in., 5¼-in., and 8-in. sizes. They are permanently enclosed in a square jacket, as shown in Figure 16.43. A small index hole in the disk is used for referencing the beginning point of stored data. When the disk is installed in a *disk drive,* the disk spins during read or write operations, *inside* the jacket. A window in the jacket exposes the surface of the disk so that it can contact the read/write head. Unlike larger, hard-disk systems, the read/write head does physically contact the disk. A *write-protect* notch in the jacket can be covered with a piece of tape to protect stored data—i.e., to prevent writing new data to the disk.

Data on a disk is stored in concentric circles, called *tracks,* as illustrated in Figure 16.44(a). The disk is further divided into *sectors,* as shown in Figure 16.44(b). Each track and each sector is numbered, so the portion of a particular track within a particular sector defines one block of data having a specific track and sector address. The address is stored at the beginning of each such block, so the read/write head, as it moves across the spinning disk, can identify and access any sector on any track. Typical access times are 200–500 ms.

The index hole (Figure 16.43) marks the beginning of sector 0 on track 0. A *soft-sectored* disk is one for which all remaining sectors must be defined and identified by the disk controller and computer system before the disk can be used. This procedure is called *formatting* the disk. The disk is said to be soft-sectored because formatting is performed under the control of a program (software) and because the formatting can be changed at will. A *hard-sectored* disk has all sectors defined at the time of manufacture. It typically has index holes marking the beginning of each sector. The number of tracks and the number of sectors on a floppy disk vary widely with the system where it is used,

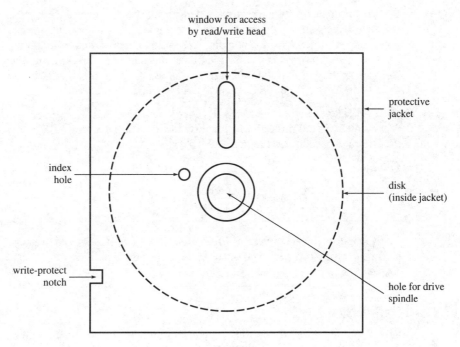

Figure 16.43
The floppy disk.

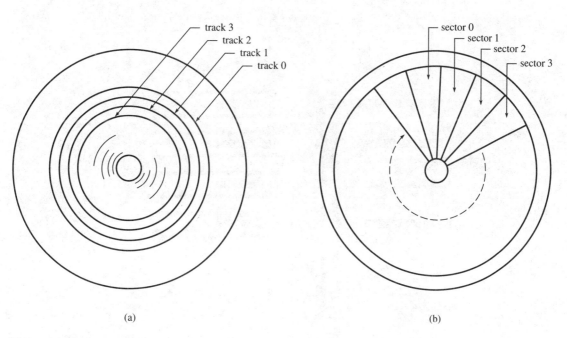

(a)

(b)

Figure 16.44
Tracks and sectors on a disk.

the size of the disk, and the recording density for which it is designed. Disks are available in *single density* and *double density* and for single-sided and double-sided recording. Each sector on each track typically contains 128 or 256 bytes of data in addition to bits reserved for addresses and various other identification and error-checking functions. The total storage capacity of floppy disks currently in use ranges from about 78 kilobytes to more than 1 megabyte, for double-density, double-sided, 8-in. disks.

Example
16–7

A single-sided, double-density, 8-in. floppy disk has tracks numbered 0 through 76 and sectors numbered 0 through 25. Each sector in each track stores 256 bytes. What is the total storage capacity of the disk?

Solution

$$(77 \text{ tracks})\left(26\frac{\text{sectors}}{\text{track}}\right)\left(256\frac{\text{bytes}}{\text{sector}}\right) = 512,512 \text{ bytes}$$

Drill
16–7

The total storage capacity of a floppy disk having 80 tracks and storing 128 bytes per sector is 163,840 bytes. How many sectors does the disk have?
Answer: 16

Hard Disks

Like floppy disks, hard disks are organized in tracks and sectors but have many more of them. A single 14-in. hard disk may have as many as 500 tracks and store 35 or more megabytes. In a large installation, a single disk drive may be used for four or more hard disks, so the total capacity of such a system is much greater than that of a floppy disk. Hard disks are permanently attached to a spindle and multiple disks are stacked, as illustrated in Figure 16.45. Note that multiple read/write heads may also be used, thus

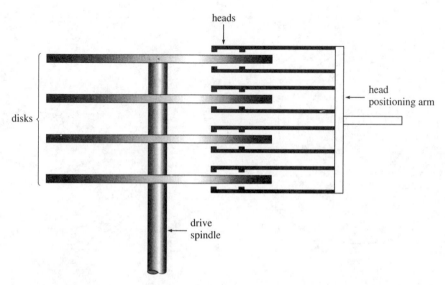

Figure 16.45
A hard-disk unit consisting of four stacked disks.

reducing access times. A hard disk spins about 10 times faster than a floppy disk (about 3600 rpm versus 360 rpm), which further reduces access time. The access time in a hard disk system is on the order of 20 ms versus 200 ms or more in a floppy disk system. Another difference between hard and soft disk systems is that the heads do not physically contact the surface of a hard disk. Instead, they "float" on a cushion of air generated by the spinning disk. They are called *flying heads,* and their clearance above the surface of the disk is so small that smoke or dust particles pose a serious threat of contamination. For that reason, the disks are sealed in protective cases and cannot ordinarily be handled by a user.

16–10   MAGNETIC BUBBLE MEMORY

According to the *domain theory of magnetism,** certain materials (magnetic materials) are composed of tiny regions called domains, each of which behaves like a small magnet. Depending on the material and its previous history, the magnetic fields produced by these domains may all be in the same direction, or in nearly the same direction, or in opposite directions, or in random directions. The directions of the fields can be altered by subjecting the material to an *externally* applied magnetic field.

Structure

Magnetic bubble memory is constructed from a magnetic material (yttrium-iron garnet) that has the property that small cylindrically shaped domains called bubbles can be created by subjecting the material to a strong, externally applied magnetic field. Furthermore, the bubbles can be moved under the influence of another external field. The presence of a magnetic bubble represents a stored 1, and its absence represents a stored 0.

Figure 16.46 shows a cross-sectional view of the structure of a bubble memory. The permanent magnet generates the strong magnetic field necessary to create magnetic bubbles at the surface of the magnetic garnet layer. Notice that the field is perpendicular to that layer and that the magnetic polarity of the bubbles is opposite that of the external field. Thus, the bubbles can be viewed as tiny magnets floating in a field of opposite polarity. The field coils are used to generate a magnetic field parallel to the magnetic garnet layer for the purpose of moving the bubbles. The propagating elements provide magnetic paths over which the bubbles move. These are made of a highly permeable (easily magnetized) nickel-iron alloy called *permalloy.*

The magnetic bubbles are moved across the propagating elements when the magnetic field produced by the field coils *rotates* in a plane parallel to the surface. This is illustrated in Figure 16.47(*a*). A rotating field is one whose polarity is constantly changing, and it is created by constantly changing the magnitude and direction of the current in the field coils. The figure shows the current waveforms, which, when applied to the X and Y field coils in Figure 16.46, create the rotating field. As shown in Figure 16.47(*b*), there are two basic shapes used for propagating elements: the T-bar and the chevron. Figure 16.47(*c*) illustrates the movement of magnetic bubbles across chevron elements under the influence of the rotating field.

---

*See, for example, Theodore F. Bogart, *Electric Circuits,* Macmillan, 1988, p. 340.

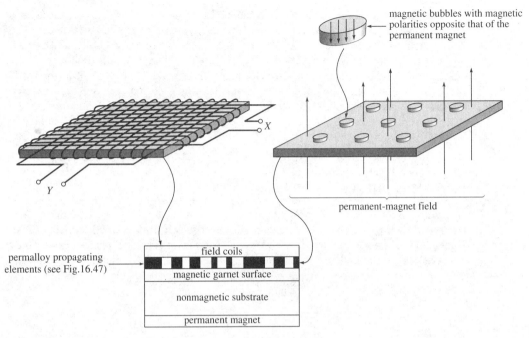

permanent-magnet field

permalloy propagating
elements (see Fig.16.47)

field coils

magnetic garnet surface

nonmagnetic substrate

permanent magnet

Figure 16.46
Structure of a magnetic bubble memory.

Major and Minor Loop Architecture

Data is stored in bubble memory by continually circulating magnetic bubbles, representing 1s (or absence of bubbles, representing 0s), around *loops* of propagating elements. As shown in Figure 16.48, the bits circulate around a number of *minor* loops, each of which can transfer bits to a *major* loop. A typical memory has 256 minor loops. The 256 bits occupying the same positions in the 256 loops constitute one *page* of stored data. Thus, data storage can be regarded as parallel (across a page) with pages circulating in series. When a read operation is performed, an entire page is transferred, by the transfer gates, to the major loop. The bits then circulate in series around the major loop and are detected serially.

Figure 16.49 shows additional components of a bubble memory. The *replicator/annihilator* is a loop of wire that "stretches" a bubble until it splits into two bubbles when a low current is passed through the wire. This replicating action occurs when a bit is read: One of the bubbles is transferred to the detector and its twin rejoins the major loop to continue circulating. Thus, readout is nondestructive. The bubble *generator* is a loop of wire that creates a new bubble when a pulse of current is passed through it. When performing a write operation, existing bubbles in the major loop are annihilated instead of replicated by the replicator/annihilator, and the bubble generator creates new bubbles wherever a 1 is to be stored. The new page of data in the major loop is then transferred to the minor loops for storage. Magnetic bubble memory is nonvolatile, since bubbles are not destroyed when power is removed. However, circulation ceases when power is removed, so some means must be incorporated to retain knowledge of the location of data when circulation stops.

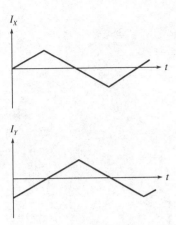

Current applied to field windings to create the rotating field.

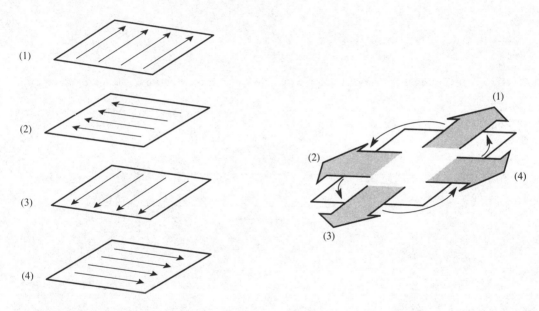

(a) The rotating magnetic field. (Although only four directions are shown, the field continuously changes through every direction in between.)

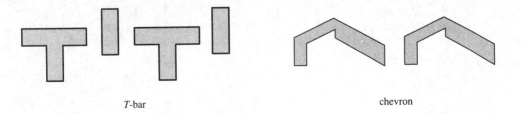

T-bar

chevron

(b) Patterns used for propagating elements.

Figure 16.47
A rotating magnetic field is used to move magnetic bubbles across propagating elements.
(Continued on p. 678.)

677

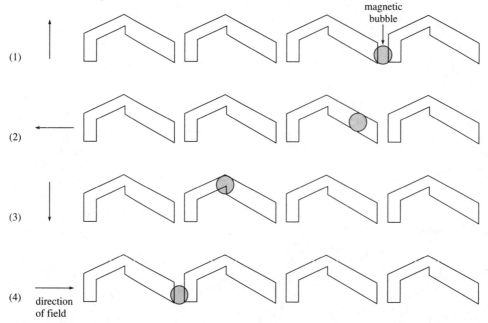

(1)

(2)

(3)

(4) direction
of field

(c) Movement of a magnetic bubble (from right to left) along propagating elements under the influence of the rotating magnetic field.

Figure 16.47
(Continued)

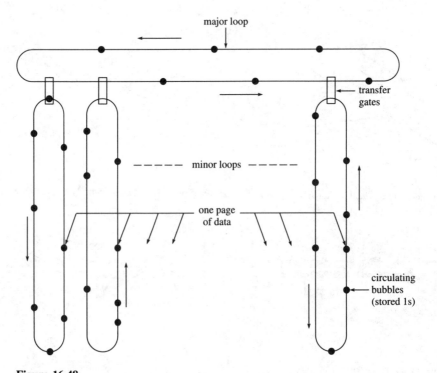

Figure 16.48
Magnetic bubbles circulate around minor loops and are transferred to and from a major loop, one page at a time, for the purpose of reading and writing.

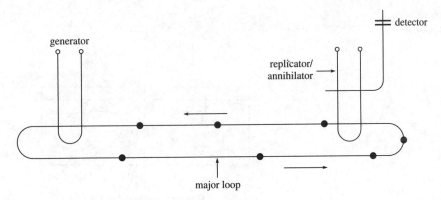

Figure 16.49
Components used for reading and writing in magnetic bubble memory.

The access time of magnetic bubble memory is on the order of a few milliseconds, so it is much faster than floppy-disk memory but not fast enough for main memory. It has a large capacity, on the order of a megabit, which makes it comparable to a floppy disk in that respect. A principal advantage of bubble memory is that, unlike disk memory, it has no moving parts and no mechanisms that wear or require periodic maintenance and alignment. It does require some separate support circuitry, such as clock generators, field coil drivers, and amplifiers. The cost of bubble memory is still somewhat greater than that of floppy-disk memory, but its speed and the advantages it enjoys are likely to make it a viable contender for replacing floppy disks when volume production creates reduced costs.

16–11   MAGNETIC TAPE

The primary role of magnetic tape is mass storage and *backup:* the duplication and preservation of data stored in other media. Backup is important in many systems because critical records, computer programs, and/or scientific data are susceptible to loss through power failures, mechanical malfunctions, or human error. Magnetic tape storage is nonvolatile and has immense capacity at relatively low cost. A single magnetic tape used in a large computer system may be 2400 ft long and contain hundreds of megabits. The tape is wound on large reels that must be unwound to read or write at a specific location. Consequently, depending on that location, access time can be quite long, on the order of a minute or more. The technology of magnetic tape storage and the read/write mechanism is quite similar to that of magnetic disks. Adjacent regions on the surface of the tape are magnetized with one polarity or the other, to represent 1s and 0s, and read/write heads are used to sense or alter the magnetization. The recording formats discussed in connection with magnetic disk memory can be used for magnetic tape storage, the NRZ format being the most common. Tapes used in large systems typically have nine parallel *tracks* and a read/write head for each track. Seven of the tracks are used for storing data in ASCII code and the other two for parity and timing. Storage density on a single track can range from 200 to 1600 bits per inch.

Audiocassette tape is sometimes used for auxiliary storage in small personal computer systems, although floppy disks have largely displaced tape in that role. A *frequency modulation* (FM) format is used for recording on cassette tape, whereby 0s and 1s are

Figure 16.50
Structure of a single MOS device in a CCD memory.

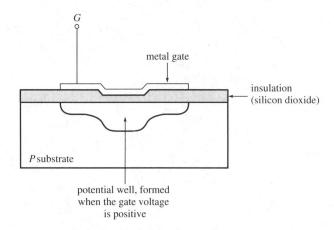

represented by pulse trains having different frequencies. In *Kansas City standard* FM, a 0 is represented by 4 cycles of a 1200-Hz signal and a 1 by 8 cycles of a 2400-Hz signal.

16–12 CHARGE-COUPLED DEVICES

Charge-coupled device (CCD) memory is a type of dynamic memory in which packets of charge are continuously transferred from one MOS device to another. The structure of a single MOS device is quite simple, as shown in Figure 16.50. When a high voltage is applied to the metal gate, holes are repelled from a region beneath the gate in the *P*-type substrate. This region, called a *potential well,* is then capable of accepting a packet of negative charge (electrons). Data in the form of charge is transferred from one device to an adjacent one by clocking their gates.

CCD memory is inherently serial. Practical memories are constructed in the form of shift registers, each shift register being a line of charge-coupled devices. By controlling the timing of the clock signals applied to the shift registers, data can be accessed one bit at a time from a single register or several bits at a time from multiple registers. The principal advantage of CCD memory is that its simple cell structure makes it possible to construct large-capacity memories at low cost. On the other hand, like other dynamic memories, it must be periodically refreshed, and it must be driven by rather complex, multiphase clock signals. Since data are stored serially, average access times are long in comparison to semiconductor RAM memory, on the order of 100 μs.

16–13 ANSI/IEEE STANDARD SYMBOLS

Address (A) Dependency

The qualifying symbol for an integrated-circuit memory is any one of the acronyms we have already used (ROM, RAM, PROM, and so on), followed by numbers representing the organization (words × bits), such as 512 × 8. Address (A) dependency is used in memory circuits to show that data inputs and/or outputs depend on the address inputs. Figure 16.51 shows the logic symbol for the 74HC219 16 × 4 static RAM. Data inputs are D1 through D4 and data outputs are G1 through G4. Note that address dependency is

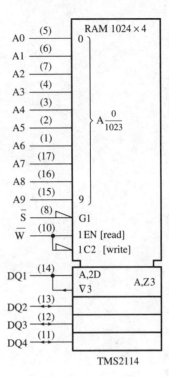

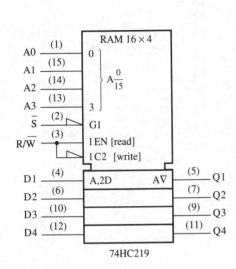

Figure 16.51
Logic symbol for a 16 × 4 static RAM, illustrating address (A) dependency.

Figure 16.52
Logic symbol for a 1K × 4 static RAM, illustrating interconnection (Z) dependency.

indicated by the A at the inputs and outputs and by the binary grouping symbol labeled A%₁₅ at the address inputs. The outputs are seen to be three-state and are enabled when R/W is high AND $\overline{S}$ is low. The AND (G) dependency here is inferred by the G1 label at the $\overline{S}$ input and the prefix 1 in the label 1EN at the R/W input. The 2D label at the data input means data is written when control C2 is active. We see that C2 is active when R/W is low and $\overline{S}$ is low, where the AND dependency is once again inferred by the G1 label at $\overline{S}$ and the 1 prefix in the label 1C2. In summary, if the chip is selected ($\overline{S}$ low), data is read out when R/W is high, and data is written in when R/W is low. If $\overline{S}$ is inactive (high), the outputs are in a high-impedance state.

Interconnection (Z) Dependency

The memory circuit just discussed (Figure 16.51) has separate data inputs and outputs. As usual, data inputs are shown on the left and outputs on the right. In memory circuits having common data-in/data-out lines, an *interconnection* (Z) dependency is used to indicate that ''outputs'' on the right are actually interconnected to inputs on the left. As in other dependency notation, an input or output with the label Z*n* means that it is interconnected to any other input or output labeled *n*. Figure 16.52 shows an example, the TMS2114 1K × 4 static RAM. Note the output labeled A,Z3, meaning that it is interconnected on the left at the point labeled ∇3. The label A,2D means that the data input (D) is address-dependent (A) and C2-dependent. The labels G1, 1EN, and 1C2 have

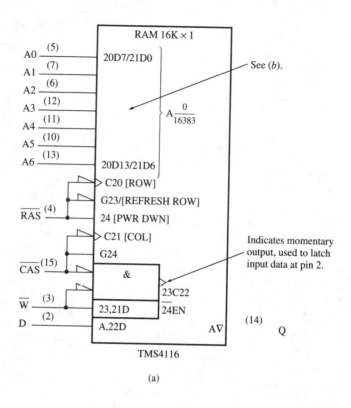

(a)

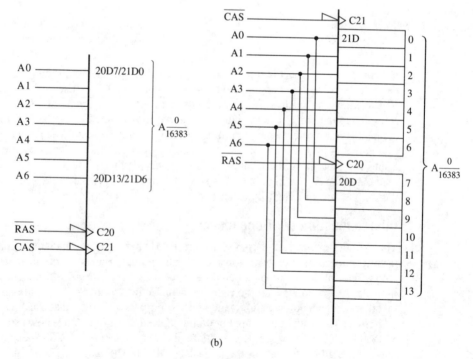

(b)

Figure 16.53
Logic symbol for a 16K × 1 dynamic RAM.

the same meaning as in the previous example (Figure 16.51). As usual, the labels in the upper rectangle are understood to apply to all those below it.

Dynamic Memories

The logic symbol for a dynamic memory is somewhat more complex than those discussed earlier because of the need to show multiplexed address inputs and refresh operations. Figure 16.53(a) shows the symbol for the TMS4116 16K × 1 dynamic RAM. Note the control dependencies C20 created by $\overline{RAS}$ and C21 created by $\overline{CAS}$. The address inputs are labeled 20D7/21D0 through 20D13/21D6. Address registers are not shown explicitly here but are understood to exist. Figure 16.53(b) shows an equivalent representation of the address inputs. The address labels mean that data inputs D7 through D13 are latched by $\overline{RAS}$ (C20) and address inputs D0 through D16 are latched by $\overline{CAS}$ (C21). Note that the data input to the memory (D) is labeled A,22D, meaning that it is address-dependent and control- (C22) dependent. Let us examine how C22 becomes active. Notice that it appears in the label 23C22 by the small triangle at the output of the AND block. As discussed in connection with monostable multivibrators, the triangle means that the output is transitory, or momentary (a pulse), when the output of the AND block goes from low to high. The 23 in the label 23C22 refers to G23, shown at the $\overline{RAS}$ input, so an AND dependency exists with $\overline{RAS}$. Since the two inputs to the AND block are $\overline{CAS}$ and $\overline{W}$, the output of the AND block will go from low to high when one input is low and the other input goes low. Therefore, C22 becomes active when that one input goes low *and* $\overline{RAS}$ is already low. Summarizing, input data can be written when $\overline{CAS} \cdot \overline{W}$ goes from low to high, provided $\overline{RAS}$ is low. To read data, three-state output Q must be enabled. The enabling signal is produced by the latch whose input is labeled 23,21D. (We know it is a latch because of the D.) Since $\overline{RAS}$ is an input to G23, the result of $\overline{RAS}$ *and* $\overline{W}$ is latched when $\overline{CAS}$ (C21) goes low. The output label $\overline{24}EN$ means that the output of the latch is ANDed with the complement of $\overline{CAS}$. Thus, if $\overline{RAS}$ is low and $\overline{W}$ is high, a 1 is latched and EN is active, provided $\overline{CAS}$ remains low.

EXERCISES

Section 16–1 to 16–3

16.1 Briefly distinguish between main and peripheral memory, in terms of applications and characteristics.

16.2 The word size in a certain computer system is 32 bits. How many bytes and how many nibbles constitute one word?

16.3 What is the capacity in Kbits and Kbytes of a memory that can store 2^{14} bits?

16.4 Distinguish between and give an example of a bidirectional bus and a unidirectional bus.

16.5 Define *volatile* memory.

16.6 Identify and briefly define each of the following: RAM, ROM, PROM, EPROM, EEPROM.

16.7 What are the principal differences, in terms of speed and size, between BJT memory and MOSFET memory?

16.8 What technology is used in the construction of dynamic memory? What are its principal advantages? What characteristic creates complications in its use?

16.9 What technology is used in the construction of EPROMs and EEPROMs? What is the principal difference between those types of memories?

16.10 What do the terms SRAM and DRAM refer to?

16.11 With reference to the ROM array in Figure 16.54, determine

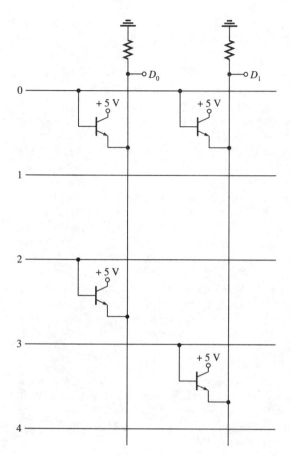

Figure 16.54
Exercise 16–11.

(a) The word size;
(b) The capacity, in words;
(c) The binary contents.

16.12 Using the MOSFET cell shown in Figure 16.5(*e*), draw a diagram of a ROM whose contents are as follows:

Address	D_2	D_1	D_0
0	0	0	0
1	0	0	1
2	0	1	0
3	0	1	1
4	1	0	0
5	1	0	1
6	1	1	0
7	1	1	1

16.13 How many binary address inputs does a ROM have if its capacity is
(a) 32,768 words;
(b) 128K words?

16.14 The total storage capacity of a certain ROM is 1,048,576 bits. If it stores 8-bit words and its X-address decoder has 512 outputs, how many binary inputs does its Y-address decoder have?

16.15 In Figure 16.8, the most significant output of the 3-to-8 decoder is at the right end and the least significant output is at the left end. The most significant output of the 2-to-4 decoder is at the top and the least significant output is at the bottom. List all binary addresses and their contents.

16.16 Using cells like that shown in Figure 16.5(*c*), draw a circuit diagram of a 16 × 1 ROM that uses 2-to-4 row and column decoders and whose contents are as follows:

Address				Contents
A_3	A_2	A_1	A_0	
0	0	0	0	1
0	0	0	1	1
0	0	1	0	0
0	0	1	1	1
0	1	0	0	1
0	1	0	1	0
0	1	1	0	1
0	1	1	1	1
1	0	0	0	0
1	0	0	1	1
1	0	1	0	0
1	0	1	1	1
1	1	0	0	1
1	1	0	1	0
1	1	1	0	0
1	1	1	1	1

Be certain to label each decoder output with the binary combination that selects it.

16.17 List the contents of each memory address in the ROM shown in Figure 16.55.

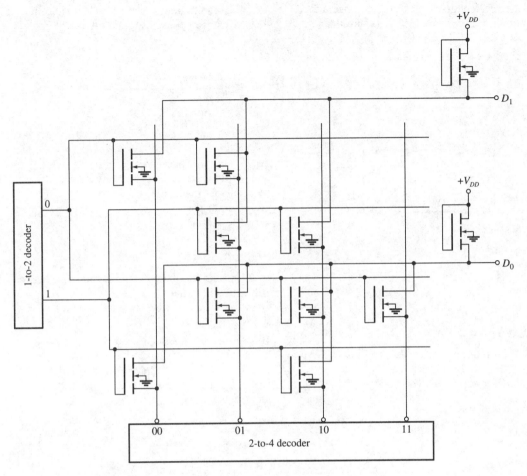

Figure 16.55
Exercise 16–17.

16.18 Draw a circuit diagram of a ROM having the structure shown in Figure 16.9 but containing the following data:

Address			Data	
A_2	A_1	A_0	D_1	D_0
0	0	0	0	1
0	0	1	1	0
0	1	0	0	0
0	1	1	1	1
1	0	0	1	0
1	0	1	0	1
1	1	0	1	1
1	1	1	0	0

16.19 A microprocessor generates a chip-select output to enable a ROM for a read operation. The chip is selected on the leading edge of a clock pulse derived from the system clock. Data must then appear on the data bus on the next trailing edge of the clock. If the access time of the ROM, from chip-select to valid data, is 100 ns, what is the maximum frequency that the clock can have?

16.20 A ROM is constructed using eight 7488 integrated circuits. How many 74187 integrated circuits would be required to construct another ROM having the same capacity?

16.21 The inputs to a 74184 ROM BCD-to-binary decoder are CBA = 011 and ED = 10.

(The letter designations refer to those inside the chip outline in Figure 16.13.) What are the binary outputs Y_5, Y_4, Y_3, Y_2, and Y_1?

16.22 Find the binary outputs Y_1 through Y_8 in Figure 16.56 when
(a) Input $E = 0$; **(b)** Input $E = 1$.

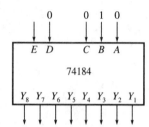

Figure 16.56
Exercise 16–22.

Section 16–4

16.23 With reference to the fusible-link PROM shown in Figure 16.57, determine
(a) Its organization; **(b)** The binary contents of each address.

Figure 16.57
Exercise 16–23.

16.24 Draw a circuit diagram of a fusible-link PROM whose contents are as follows:

	Contents		
Address	D_2	D_1	D_0
0	0	1	0
1	1	0	1
2	1	1	0

16.25 How is charge stored in an EPROM cell? How is it removed? Briefly describe the mechanisms involved.

16.26 How is charge stored in an EEPROM cell? How is it removed?

Section 16–5

16.27 In the bipolar RAM cell shown in Figure 16.18, assume that Q_1 is on and Q_2 is off. Describe what must be done to reverse the states.

16.28 With $R/\overline{W}$ and $\overline{CE}$ both low, the binary input to a 74S189B RAM has decimal value 12. A short time later, $R/\overline{W}$ is made high. What is the decimal value of the data that appear at the output?

16.29 In the NMOS RAM cell in Figure 16.21, Q_1 is off and Q_2 is on. Describe what must be done to reverse the states.

16.30 In the CMOS RAM cell in Figure 16.22, the drains of Q_1 and Q_3 are low. List the states (on or off) of transistors Q_1, Q_2, Q_3, and Q_4.

16.31 Describe what must be done to reverse the states of the transistors in Exercise 16–30.

16.32 Describe the principal features and applications of static ECL RAMs.

Section 16–6

16.33 Explain why data read from the 3-transistor dynamic cell in Figure 16.24(a) is the logical complement of the data stored in it.

16.34 What design consideration limits the number of 1-transistor dynamic cells that can be connected to a single sense amplifier?

16.35 Define the following terms.
(a) Burst-mode refresh
(b) Dynamic memory controller
(c) Pseudodynamic RAM

16.36 Describe how address multiplexing is used in dynamic memory circuits and give an example.

16.37 Name and briefly describe two methods for refreshing a dynamic memory such as the TMS4C1024, using $\overline{\text{CAS}}$ and $\overline{\text{RAS}}$ control signals. What is the principal advantage of each method?

16.38 In the dynamic memory controller shown in Figure 16.28, the 20-bit input address is $A6B93)_{16}$ and the 10-bit counter contents are $3A5)_{10}$. The $R/\overline{C}$ input is low. What is the output, in binary, of the controller when
(a) REFRESH is low;
(b) REFRESH goes high?

16.39 A 256K $\times$ 1 dynamic RAM with address multiplexing is to be refreshed using $\overline{\text{RAS}}$-only burst-mode refreshing. The memory must be refreshed at least every 8 ms and the entire memory must be refreshed in no longer than 200 μs.
(a) How many outputs should a dynamic controller used for this memory circuit have?

(b) At what frequency should the counter in the controller be clocked?
(c) At what frequency should the $\overline{\text{RAS}}$ input to the circuit be pulsed?
(d) At what frequency should the REFRESH input to the controller be driven? What should be the duty cycle of that input?
(e) If the access time for the memory is 250 ns, how many read cycles can occur between refreshes?

16.40 When the counter in a dynamic memory controller used to refresh memories having address multiplexing is clocked at 2 MHz, it takes a total of 512 μs for the count to cycle through all possible states. How many addresses can a dynamic memory used with this controller have?

Section 16–7

16.41 Determine how many 16K $\times$ 4 memory circuits would be required to construct a memory having each given capacity.
(a) 256K $\times$ 8
(b) 128K $\times$ 16
(c) 1M $\times$ 4

16.42 A memory used to store 8-bit words is constructed from eight 256K $\times$ 4 memory circuits.
(a) How many words can the memory store?
(b) How many circuits would be required if the memory were constructed from 512K $\times$ 1 circuits?

16.43 Draw a logic diagram showing how to interconnect 2K $\times$ 8 memory circuits to obtain a 4K $\times$ 8 memory. Each circuit has an active-low chip-select input and common data-in/data-out pins.

16.44 Draw a logic diagram showing how to interconnect 128 $\times$ 4 static RAMs to construct a 256 $\times$ 8 memory. Each circuit has an active-low chip-select input and common data-in/data-out pins.

16.45 A 128K $\times$ 8 memory is to be constructed using 16K $\times$ 8 circuits, each of which has an active-low chip-select input. Draw a logic diagram showing only the connections necessary for the chip-select inputs. It is not necessary to show address or data pins on the individual 16K $\times$ 8 circuits.

Figure 16.58
Exercise 16–47.

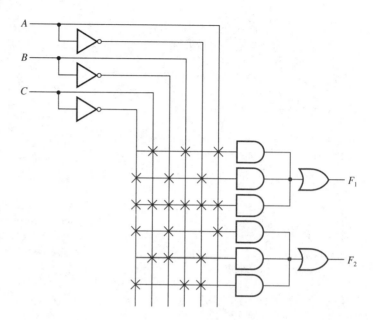

16.46 A 1M × 8 memory is to be constructed using 64K × 8 circuits, each of which has an active-low chip-select input.
(a) How many 64K × 8 circuits are required?
(b) What type of decoder is necessary to select individual 64K × 8 circuits?
(c) What are the address inputs to the decoder?

Section 16–8

16.47 What logic functions are implemented by the programmable logic array shown in Figure 16.58?

16.48 Show how the programmable logic array in Figure 16.33(*b*) would be programmed to implement the functions $F_1 = ABC + \overline{A}\overline{B}C$ and $F_2 = AB + AC + \overline{A}\overline{C}$.

16.49 Show how the FPLA circuit in Figure 16.35 would be programmed to implement $F_1 = A\overline{B}C + AB\overline{C}$, $F_2 = ABC + A\overline{B}\overline{C} + \overline{A}\overline{B}C$, and $F_3 = A + B$.

16.50 Show how the FPLA circuit in Figure 16.35 would be programmed to implement $F_1 = A \odot B \odot \overline{C}$ and $F_2 = \overline{F}_1$.

16.51 Using the simplified connection format of a PLA, show how an 8 × 1 PROM should be programmed to implement logic function F, whose truth table is as follows:

A	B	C	F
0	0	0	1
0	0	1	1
0	1	0	0
0	1	1	1
1	0	0	1
1	0	1	0
1	1	0	1
1	1	1	0

16.52 Show how an 8 × 1 PROM should be programmed to serve as a look-up table for the odd-parity bit of a 3-bit number. Use the simplified connection format of PLA.

Section 16–9

16.53 Compare floppy disks with hard disks, in terms of size, cost, and application.

16.54 What property distinguishes a stored 1 from a stored 0 on a magnetic disk?

16.55 Sketch the sequence of magnetic pulses that represent the bit sequence 00101101 when recorded in an RZ format.

16.56 The bias level to which the magnetic pulses in Figure 16.59 return is the same as the 0-level magnetization. What sequence of bits (from left to right) is represented by this return-to-bias data?

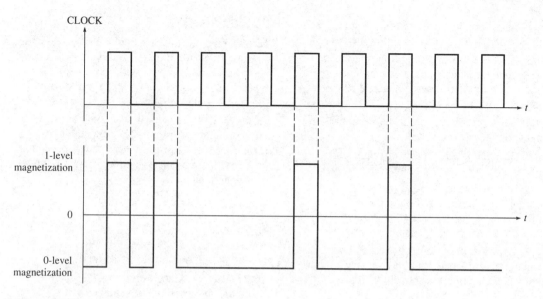

Figure 16.59
Exercise 16–56.

16.57 Sketch the sequence of magnetic pulses that represent the bit sequence 100101110 when recorded in an NRZ format. Identify bit times in your sketch.

16.58 The data in Figure 16.60 was recorded in a biphase format. Make a sketch showing how the same data would be recorded in a Manchester format.

16.59 Define each of the following.
(a) Index hole
(b) Tracks

(c) Sector
(d) Soft-sectored disk
(e) Disk controller

16.60 A disk has tracks numbered 0 through 37 and sectors numbered 0 through 25. Each sector in each track stores 2048 bits. What is the total storage capacity of the disk, in bytes?

Section 16–10

16.61 What is a magnetic bubble? How is it created?

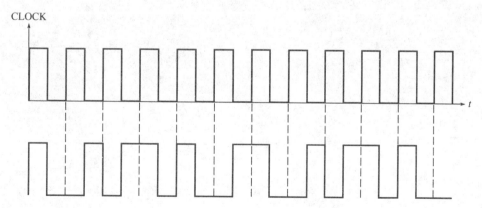

Figure 16.60
Exercise 16–58.

16.62 What property distinguishes a stored 0 from a stored 1 in magnetic bubble memory?

16.63 What are the two shapes used for propagating elements in magnetic bubble memory? From what type of material are the propagating elements constructed?

16.64 What is the purpose of the field coils in a magnetic bubble memory?

16.65 Describe the functions of major and minor loops in a magnetic bubble memory.

16.66 With reference to read and write operations in magnetic bubble memory, what are the roles of the replicator/annihilator? What device creates new bubbles?

16.67 What are the principal features and advantages of magnetic bubble memory? What type of memory is it most likely to replace?

Section 16–11

16.68 What are the principal roles of magnetic tape memory in computer systems? How does its access time compare with those of other types of memory? Briefly describe the format for storing data on tape.

16.69 What type of tape storage is used in microcomputer systems? Name a standard format used for storing data in that type. What property distinguishes a stored 1 from a stored 0 in that standard?

Section 16–12

16.70 What technology is used in the construction of CCD memory? What is a potential well and how is it formed?

16.71 How are data stored in CCD memory? What are the principal advantages and disadvantages of CCD memory?

Chapter 17

INTERFACE CIRCUITS

17–1   DIGITAL-TO-ANALOG CONVERSION

Recall from Chapter 1 that an analog variable is *continuous* in nature, as opposed to discrete. It can have *any* value within whatever its prescribed range of variation may be. Most physical quantities are in fact analog: temperature, pressure, sound and light intensity, position and velocity, and so on. Many digital systems process data representing numerical values of such quantities and must then convert the digital representations back to analog. An example is a motor-speed controller: Digital computations are performed to determine the required speed, and the binary number representing that speed is converted to an analog voltage that drives the motor. Other examples include a digitally controlled voltage source (power supply) and a CRT driver. In the latter example, digital data is converted to analog voltages suitable for controlling the X and/or Y deflection of the electron beam in an oscilloscope.

A digital-to-analog converter (DAC, or D/A converter) produces an output voltage that is proportional to the value of its binary input. As a simple example, a 2-bit DAC might produce 0 V when its input is 00, 1 V when its input is 01, 2 V when its input is 10, and 3 V when its input is 11. The output of a DAC is not truly continuous because it produces only those voltages that correspond to the possible combinations of its binary input. Of course, those are the only possible outputs. The greater the number of binary inputs, the greater the number of output voltages and the more the output resembles a truly analog voltage. This property is called *resolution* and is defined more precisely in a later discussion, along with other DAC properties and specifications.

The R-2R Ladder Converter

The most popular method for converting a digital input to an analog output incorporates a ladder network containing series-parallel combinations of two resistor values: R and $2R$. Figure 17.1 shows an example of an R-$2R$ ladder having a 4-bit digital input.

To understand the operation of the R-$2R$ ladder, let us first determine the output voltage in Figure 17.1 when the input is 1000. We will assume that a logical-1 input is E volts and that logical 0 is 0 V (ground). Figure 17.2(a) shows the circuit with input D_3

Figure 17.1

A 4-bit R-2R ladder network used for digital-to-analog conversion.

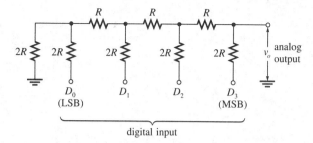

connected to E and all other inputs connected to ground, corresponding to the binary input 1000. We wish to find the total equivalent resistance, R_{eq}, looking to the left from node A. At the left end of the ladder, we see that $2R$ is in parallel with $2R$, so that combination is equivalent to R. That R is in series with another R, giving $2R$. That $2R$ is in parallel with another $2R$, which is equivalent to R once again. Continuing in this manner, we ultimately

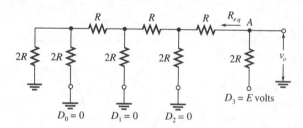

(a) When the input is 1000, D_0, D_1, and D_2 are grounded (0 V) and D_3 is E volts.

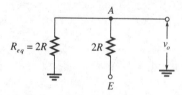

(b) The circuit equivalent to (a) when the network to the left of node A is replaced by its equivalent resistance, R_{eq}.

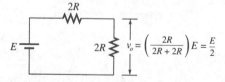

(c) Calculation of v_o using the voltage-divider rule. (Note that v_o in (b) is the voltage across $R_{eq} = 2R$.)

Figure 17.2

Calculating the output of the R-2R ladder when the input is 1000.

find that $R_{eq} = 2R$. In fact, we see that the equivalent resistance looking to the left from every node is $2R$. Figure 17.2(b) shows the circuit when it is redrawn with R_{eq} replacing all of the network to the left of node A. Figure 17.2(c) shows an equivalent way to draw the circuit in (b), and it is now readily apparent that $v_o = E/2$.

Let us now find v_o when the input is 0100. Figure 17.3(a) shows the circuit, with D_2 connected to E and all other inputs grounded. As demonstrated earlier, the equivalent resistance looking to the left from node B is $2R$. The equivalent circuit with R_{eq} replacing the network to the left of B is shown in Figure 17.3(b). We proceed with the analysis by

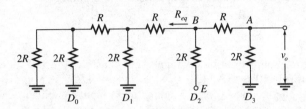

(a) When the input is 0100, D_0, D_1, and D_3 are grounded and D_2 is E volts.

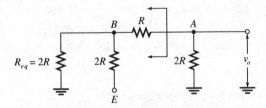

(b) The circuit equivalent to (a) when the network to the left of node B is replaced by its equivalent resistance ($2R$). The Thevenin equivalent circuit to the left of the bracketed arrows is shown in (c).

$$R_{TH} = 2R$$

$$E_{TH} = \frac{E}{2}$$

$$R_{TH} = R + 2R \| 2R = 2R$$

$$E_{TH} = \left(\frac{2R}{2R + 2R} \right) E = \frac{E}{2}$$

(c) The Thevenin equivalent circuit to the left of node A.

$$R_{TH} = 2R$$

$$E_{TH} = \frac{E}{2}$$

$$v_o = \left(\frac{2R}{2R + 2R} \right) \frac{E}{2} = \frac{E}{4}$$

(d) Calculation of v_o using the voltage-divider rule.

Figure 17.3
Calculating the output of the R-2R ladder when the input is 0100.

finding the (Thevenin) equivalent circuit to the left of node A, as indicated by the bracketed arrows. The Thevenin equivalent resistance (with E shorted to ground) is $R_{TH} = R + 2R \parallel 2R = R + R = 2R$. The Thevenin equivalent voltage is $[2R/(2R + 2R)]E = E/2$. The Thevenin equivalent circuit is shown in Figure 17.3(c). Figure 17.3(d) shows the ladder with the Thevenin equivalent circuit replacing everything to the left of node A. It is now apparent that $v_o = E/4$.

By an analysis similar to the foregoing, we find that the output of the ladder is $E/8$ when the input is 0010 and that the output is $E/16$ when the input is 0001. In general, when the D_n input is 1 and all other inputs are 0, the output is

$$v_o = \frac{E}{2^{N-n}} \tag{17.1}$$

where N is the total number of binary inputs. For example, if $E = 5$ V, the output voltage when the input is 0010 is $(5 \text{ V})/(2^{4-1}) = (5 \text{ V})/8 = 0.625$ V.

To find the output voltage corresponding to *any* input combination, we can invoke the principal of superposition and simply add the voltages produced by the inputs where 1s are applied. For example, when the input is 1100, the output is $E/2 + E/4 = 3E/4$. The next example illustrates these computations.

Example 17–1

The logic levels used in a 4-bit R-$2R$ ladder DAC are $1 = +5$ V and $0 = 0$ V.

a. Find the output voltage when the input is 0001 and when it is 1010.
b. Sketch the output when the inputs are driven from a 4-bit binary up counter.

Solution

a. By equation (17.1), the output when the input is 0001 is

$$v_o = \frac{5 \text{ V}}{2^{4-0}} = \frac{5 \text{ V}}{16} = 0.3125 \text{ V}$$

When the input is 1000, the output is $(5 \text{ V})/2 = 2.5$ V, and when the input is 0010, the output is $(5 \text{ V})/8 = 0.625$ V. Therefore, when the input is 1010, the output is $2.5 \text{ V} + 0.625 \text{ V} = 3.125$ V.

b. Figure 17.4 shows a table of the output voltages corresponding to every input combination, calculated using the method illustrated in part (a) of this example. As the counter counts up, the output voltage increases by 0.3125 V at each new count. Thus, the output is the *staircase* waveform shown in the figure. The voltage steps from 0 V to 4.6875 V each time the counter counts from 0000 to 1111.

Drill 17–1

What is the output voltage of an 8-bit DAC when its input is 11010100? Logic levels are $1 = +5$ V and $0 = 0$ V.
Answer: 4.140625 V.

Typical values for R and $2R$ are 10 kΩ and 20 kΩ. For accurate conversion, the output voltage from the R-$2R$ ladder should be connected to a high impedance to prevent loading. Figure 17.5 shows how an operational amplifier can be used for that purpose. The output of the ladder is connected to a unity-gain *voltage follower*, whose input impedance is extremely large and whose output voltage is the same as its input voltage.

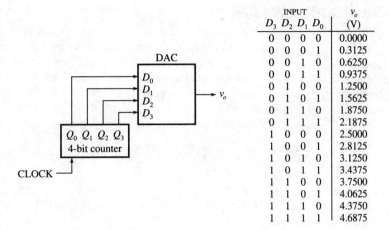

INPUT				v_o
D_3	D_2	D_1	D_0	(V)
0	0	0	0	0.0000
0	0	0	1	0.3125
0	0	1	0	0.6250
0	0	1	1	0.9375
0	1	0	0	1.2500
0	1	0	1	1.5625
0	1	1	0	1.8750
0	1	1	1	2.1875
1	0	0	0	2.5000
1	0	0	1	2.8125
1	0	1	0	3.1250
1	0	1	1	3.4375
1	1	0	0	3.7500
1	1	0	1	4.0625
1	1	1	0	4.3750
1	1	1	1	4.6875

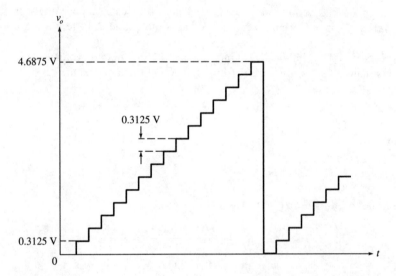

Figure 17.4
Example 17–1.

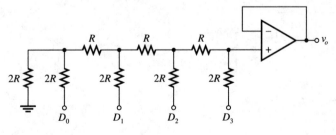

Figure 17.5
Using a unity-gain voltage follower to provide a high impedance for the R-2R ladder.

A Weighted-Resistor DAC

Figure 17.6 illustrates another approach to digital-to-analog conversion. The operational amplifier is used to produce a *weighted sum* of the digital inputs, where the weights are proportional to the weights of the bit positions of the inputs. Recall from Chapter 7 that each input is amplified by a factor equal to the ratio of the feedback resistance divided by the input resistance to which it is connected. Thus, D_3, the most significant bit, is amplified by R_f/R, D_2 by $R_f/2R = 1/2(R_f/R)$, D_1 by $R_f/4R = 1/4(R_f/R)$, and D_o by $R_f/8R = 1/8(R_f/R)$. Since the amplifier sums and inverts, the output is

$$v_o = -\left(D_3 + \frac{1}{2}D_2 + \frac{1}{4}D_1 + \frac{1}{8}D_o\right)\frac{R_f}{R} \tag{17.2}$$

The principal disadvantage of this type of converter is that a different-valued precision resistor must be used for each digital input. In contrast, the R-$2R$ ladder network uses only two values of resistance.

Example 17–2

Design a 4-bit, weighted-resistor DAC whose full-scale output voltage is -10 V. Logic levels are $1 = +5$ V and $0 = 0$ V. What is the output voltage when the input is 1010?

Solution. The full-scale output voltage is the output voltage when the input is maximum: 1111. In that case, from equation (17.2), we require

$$\left(5\text{ V} + \frac{5\text{ V}}{2} + \frac{5\text{ V}}{4} + \frac{5\text{ V}}{8}\right)\frac{R_f}{R} = 10\text{ V}$$

or

$$9.375\,\frac{R_f}{R} = 10$$

Let us choose $R_f = 10$ kΩ. Then

$$R = \frac{9.375(10\text{ k}\Omega)}{10} = 9.375\text{ k}\Omega$$
$$2R = 18.75\text{ k}\Omega$$
$$4R = 37.50\text{ k}\Omega$$
$$8R = 75\text{ k}\Omega$$

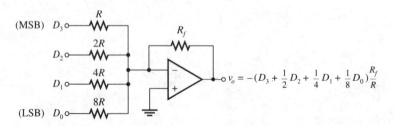

Figure 17.6
A weighted-resistor DAC using an inverting operational amplifier.

When the input is 1010, the output voltage is

$$V_o = -\left(5\ \text{V} + \frac{0\ \text{V}}{2} + \frac{5\ \text{V}}{4} + \frac{0\ \text{V}}{8}\right)\frac{10\ \text{k}\Omega}{9.375\ \text{k}\Omega} = -6.667\ \text{V}$$

Drill
17–2

If $R_f = 10\ \text{k}\Omega$ in Example 17–2, what should be the value of R if it is required that the voltage be -10 V when the input is 1010?
Answer: 6.25 kΩ.

The Switched Current-Source DAC

The D/A converters we have discussed so far can be regarded as switched voltage-source converters: When a binary input goes high, the high voltage is effectively switched into the circuit, where it is summed with other input voltages. Because of the technology used to construct integrated-circuit DACs, currents can be switched in and out of a circuit faster than voltages can. For that reason, most integrated-circuit DACs utilize some form of current switching, where the binary inputs are used to open and close switches that connect and disconnect internally generated currents. The currents are weighted according to the bit positions they represent and are summed in an operational amplifier. Figure 17.7 shows an example. Note that an R-$2R$ ladder is connected to a voltage source identified as E_{REF}. The current that flows in each $2R$ resistor is

$$I_n = \left(\frac{E_{\text{REF}}}{R}\right)\frac{1}{2^{N-n}} \tag{17.3}$$

when $n = 0, 1, \ldots, N - 1$ is the subscript for the current created by input D_n and N is the total number of inputs. Thus, each current is weighted according to the bit position it represents. For example, the current in the $2R$ resistor at the D_1 input in the figure ($n = 1$ and $N = 4$) is $(E_{\text{REF}}/R)(1/2)^8$. The binary inputs control switches that connect

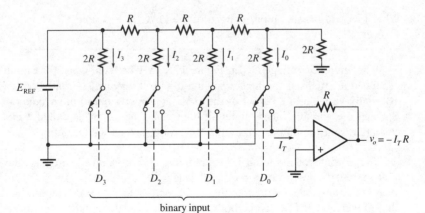

binary input
A high input switches current to the amplifier
input, and a low input switches current to ground.

Figure 17.7
A 4-bit switched current-source DAC.

the currents either to ground or to the input of the amplifier. The amplifier sums all currents whose corresponding binary inputs are high. The amplifier also serves as a current-to-voltage converter. It is connected in an inverting configuration and its output is

$$v_o = -I_T R \tag{17.4}$$

where I_T is the sum of the currents that have been switched to its input. For example, if the input is 1001, then

$$I_T = \frac{E_{REF}}{R}\left(\frac{1}{2}\right) + \frac{E_{REF}}{R}\left(\frac{1}{16}\right) = \frac{E_{REF}}{R}\left(\frac{7}{16}\right)$$

and

$$v_o = -\left(\frac{E_{REF}}{R}\right)\left(\frac{7}{16}\right)R = -\frac{7}{16}E_{REF}$$

Example 17-3

The switched current-source DAC in Figure 17.7 has $R = 10 \text{ k}\Omega$ and $E_{REF} = 10$ V. Find the total current delivered to the amplifier and the output voltage when the binary input is 1010.

Solution. From equation (17.3),

$$I_3 = \left(\frac{10 \text{ V}}{10 \text{ k}\Omega}\right)\frac{1}{2^{4-3}} = 0.5 \text{ mA}$$

and

$$I_1 = \left(\frac{10 \text{ V}}{10 \text{ k}\Omega}\right)\frac{1}{2^{4-1}} = 0.125 \text{ mA}$$

Therefore, $I_T = I_3 + I_1 = 0.5 \text{ mA} = 0.125 \text{ mA} = 0.625 \text{ mA}$. From equation (17.4),

$$v_o = -I_T R = -(0.625 \text{ mA})(10 \text{ k}\Omega) = -6.25 \text{ V}$$

Drill 17-3

What is the full-scale output voltage of the DAC in Example 17.3?
Answer: 9.375 V.

The reference voltage E_{REF} in Figure 17.7 may be fixed—as, for example, when it is generated internally in an integrated circuit—or it may be externally variable. When it is externally variable, the output of the DAC is proportional to the *product* of the variable E_{REF} and the variable binary input. In that case, the circuit is called a *multiplying* D/A converter, and the output represents the product of an analog input (E_{REF}) and a digital input.

In most integrated-circuit DACs utilizing current-source switching, the output is the total current I_T produced in the R-2R ladder. The user may then connect a variety of operational amplifier configurations at the output to perform magnitude scaling and/or phase inversion. If E_{REF} can be both positive and negative and if the binary input always represents a positive number, then the output of the DAC is both positive and negative. In that case, the DAC is called a *two-quadrant multiplier*. A DAC whose output can be both positive and negative is said to be *bipolar,* and one whose output is only positive or only negative is *unipolar.* A four-quadrant multiplier is one in which both inputs can be either negative or positive and in which the output (product) has the correct sign for every

Table 17.1
The offset binary code used to represent positive and negative binary inputs to a D/A converter.

Decimal number	Offset binary code	Analog output
+7	1111	$+(7/8)E_{REF}$
+6	1110	$+(6/8)E_{REF}$
+5	1101	$+(5/8)E_{REF}$
+4	1100	$+(4/8)E_{REF}$
+3	1011	$+(3/8)E_{REF}$
+2	1010	$+(2/8)E_{REF}$
+1	1001	$+(1/8)E_{REF}$
0	1000	0
−1	0111	$-(1/8)E_{REF}$
−2	0110	$-(2/8)E_{REF}$
−3	0101	$-(3/8)E_{REF}$
−4	0100	$-(4/8)E_{REF}$
−5	0011	$-(5/8)E_{REF}$
−6	0010	$-(6/8)E_{REF}$
−7	0001	$-(7/8)E_{REF}$
−8	0000	$-(8/8)E_{REF}$

case. Negative binary inputs can be represented using the *offset binary code,* the 4-bit version of which is shown in Table 17.1. Note that the numbers +7 through −8 are represented by 0000 through 1111, respectively, with $0)_{10}$ represented by 1000. The table also shows the analog outputs that are produced by a multiplying D/A converter. In general, for an N-bit DAC, the offset binary code represents decimal numbers from $+2^{N-1} - 1$ through -2^{N-1} and the analog outputs range from

$$+\left(\frac{2^{N-1}-1}{2^N - 1}\right)E_{REF} \quad \text{through} \quad -E_{REF}$$

in steps of size $(1/2^{N-1})E_{REF}$.

Switched-Capacitor DACs

The newest technology used to construct D/A converters employs weighted capacitors instead of resistors. In this method, charged capacitors form a capacitive voltage divider whose output is proportional to the sum of the binary inputs.

As an aid in understanding the method, let us review the theory of capacitive voltage dividers. Figure 17.8 shows a two-capacitor example. The total equivalent capacitance of the two series-connected capacitors is

$$C_T = \frac{C_1 C_2}{C_1 + C_2} \tag{17.5}$$

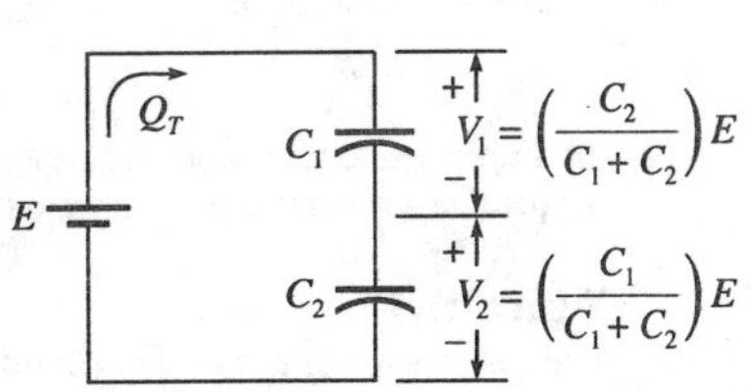

Figure 17.8
The capacitive voltage divider.

Therefore, the total charge delivered to the circuit, which is the same as the charge on both C_1 and C_2, is

$$Q_1 = Q_2 = Q_T = C_T E = \left(\frac{C_1 C_2}{C_1 + C_2}\right) E \tag{17.6}$$

The voltage across C_2 is

$$V_2 = \frac{Q_2}{C_2} = \frac{Q_T}{C_2} = \frac{\left(\dfrac{C_1 C_2}{C_1 + C_2}\right) E}{C_2} = \left(\frac{C_1}{C_1 + C_2}\right) E \tag{17.7}$$

Similarly,

$$V_1 = \left(\frac{C_2}{C_1 + C_2}\right) E \tag{17.8}$$

Figure 17.9(a) shows an example of a 4-bit switched-capacitor DAC. Note that the capacitance values have binary weights. A two-phase clock is used to control switching of the capacitors. When ϕ_1 goes high, all capacitors are switched to ground and

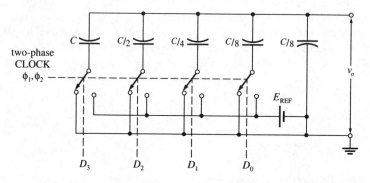

(a) All capacitors are switched to ground by ϕ_1. Those capacitors whose digital inputs are high are switched to E_{REF} by ϕ_2.

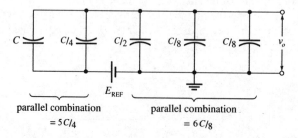

(b) Equivalent circuit when the input is 1010. The capacitors switched to E_{REF} are in parallel as are the ones connected to ground.

Figure 17.9
The switched-capacitor D/A converter.

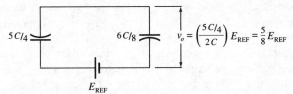

(c) Circuit equivalent to (b). The output is determined by a capacitive voltage divider.

Figure 17.9
(Continued)

discharged. When ϕ_2 goes high, those capacitors where the digital input is high are switched to E_{REF}, whereas those whose inputs are low remain grounded. Figure 17.9(b) shows the equivalent circuit when ϕ_2 is high and the digital input is 1010. We see that the two capacitors whose digital inputs are 1 are in parallel, as are the two capacitors whose digital inputs are 0. The circuit is redrawn in Figure 17.9(c) with the parallel capacitors replaced by their equivalents (sums). The output of the capacitive voltage divider is

$$v_o = \left(\frac{5C/4}{5C/4 + 6C/8}\right)E_{REF} = \left(\frac{5C/4}{2C}\right)E_{REF} = \frac{5}{8}E_{REF} \qquad (17.9)$$

The denominator in (17.9) will always be $2C$, the sum of all the capacitance values in the circuit. From the foregoing analysis, we see that the output of the circuit in the general case is

$$v_o = \left(\frac{C_{eq}}{2C}\right)E_{REF} \qquad (17.10)$$

where C_{eq} is the equivalent (sum) of all the capacitors whose digital inputs are high. Table 17.2 shows the outputs for every possible input combination, and it is apparent that the

Table 17.2
Output voltages produced by a 4-bit switched-capacitor DAC.

Binary input $D_3D_2D_1D_0$	V_0
0000	0
0001	$(1/16)E_{REF}$
0010	$(1/8)E_{REF}$
0011	$(3/16)E_{REF}$
0100	$(1/4)E_{REF}$
0101	$(5/16)E_{REF}$
0110	$(3/8)E_{REF}$
0111	$(7/16)E_{REF}$
1000	$(1/2)E_{REF}$
1001	$(9/16)E_{REF}$
1010	$(5/8)E_{REF}$
1011	$(11/16)E_{REF}$
1100	$(3/4)E_{REF}$
1101	$(13/16)E_{REF}$
1110	$(7/8)E_{REF}$
1111	$(15/16)E_{REF}$

analog output is proportional to the digital input. For the case where the input is 0000, note that the positive terminal of E_{REF} in Figure 17.9 is effectively open-circuited, so the output is 0 V.

Switched-capacitor technology evolved as a means for implementing analog functions in integrated circuits, particularly MOS circuits. It has been used to construct filters, amplifiers, and many other special devices. The principal advantage of the technology is that small capacitors, on the order of a few picofarads, can be constructed in the integrated circuits to perform the function of the much larger capacitors that are normally needed in low-frequency analog circuits. When capacitors are switched at a high enough frequency, they can be effectively "transformed" into other components, including resistors. The transformations are studied from the standpoint of sampled-data theory, which is beyond the scope of this book.

DAC Performance Specifications

The *resolution* of a D/A converter is a measure of the fineness of the increments between output values. Given a fixed output voltage range, say, 0 to 10 V, a DAC that divides that range into 1024 distinct output values has greater resolution than one that divides it into 512 values. Since the output increment is directly dependent on the number of input bits, the resolution is often quoted as simply that total number of bits. The most commonly available integrated-circuit converters have resolutions of 8, 10, 12, or 16 bits. Resolution is also expressed as the reciprocal of the total number of output voltages, often in terms of a percent. For example, the resolution of an 8-bit DAC may be specified as

$$\left(\frac{1}{2^8}\right) \times 100\% = 0.39\%$$

Some DAC specifications are quoted with reference to one or to one-half LSB (least significant bit). In this context, an LSB is simply the increment between successive output voltages. Since an n-bit converter has $2^n - 1$ such increments,

$$LSB = \frac{FSR}{2^n - 1} \qquad \textbf{(17.11)}$$

where FSR is the full-scale range of the output voltage.

When the output of a DAC changes from one value to another, it typically overshoots the new value and may oscillate briefly around that new value before it settles to a constant voltage. The *settling time* of a D/A converter is the total time between the instant that the digital input changes and the time that the output enters a specified error band for the last time, usually $\pm 1/2$ LSB around the final value. Figure 17.10 illustrates the specification. Settling times of typical integrated-circuit converters range from 50 ns to several microseconds. Settling time may depend on the magnitude of the change at the input and is often specified for a prescribed input change.

Linearity error is the maximum deviation of the analog output from the ideal output. Since the output is ideally in direct proportion to the input, the ideal output is a straight line drawn from 0 V. Linearity error may be specified as a percentage of the full-scale range or in terms of an LSB.

Differential linearity error is the difference between the ideal output increment (1 LSB) and the actual increment. For example, each output increment of an 8-bit DAC whose full-scale range is 10 V should be $(10\ V)/(2^8 - 1) = 39.22$ mV. If any one increment between two successive values is, say, 30 mV, then there is a differential

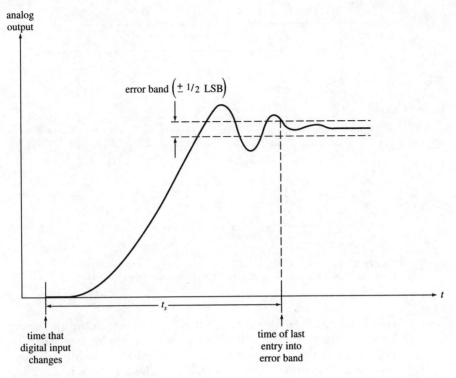

Figure 17.10
The settling time, t_s, of a D/A converter.

linearity error of 9.22 mV. This error is also specified as a percentage of the full-scale range or in terms of an LSB. If the differential linearity error is greater than 1 LSB, it is possible for the output voltage to *decrease* when there is an increase in the value of the digital input or to increase when the input decreases. Such behavior is said to be *non-monotonic*. In a monotonic DAC, the output always increases when the input increases and decreases when the input decreases.

The input of a DAC is said to undergo a *major change* when every input bit changes, as, for example, from 01111111 to 10000000. If the switches in Figure 17.7 open faster than they close or vice versa, the output of the DAC will momentarily go to 0 or to full scale when a major change occurs, thus creating an output *glitch. Glitch area* is the total area of an output voltage glitch in volt-seconds or of an output current glitch in ampere-seconds. Commercial units are often equipped with *deglitchers* to minimize glitch area.

An Integrated-Circuit DAC

The AD7524 CMOS integrated-circuit DAC, manufactured by Analog Devices and available from other manufacturers, is an example of an 8-bit, multiplying D/A converter. Figure 17.11 shows a functional block diagram. Note that the digital input is latched under the control of $\overline{CS}$ and $\overline{WR}$. When both of these control inputs are low, the output of the DAC responds directly to the digital inputs, with no latching occurring. If either control input goes high, the digital input is latched and the analog output remains at the level

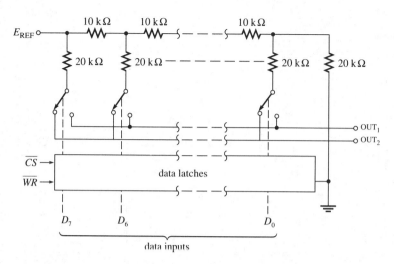

Figure 17.11
Functional block diagram of the AD7524 D/A converter.

corresponding to the latched data, independent of any changes in the digital input. The device is then said to be in a HOLD mode, with the data bus *locked out*. The OUT2 output is normally grounded. Maximum settling time to a $\pm 1/2$ LSB error band for the Texas Instruments version is 100 ns and maximum linearity error is $\pm 0.2\%$ of the full-scale range. The device can be used as a 2- or 4-quadrant multiplier and E_{REF} can vary ± 25 V.

17–2   ANALOG-TO-DIGITAL CONVERSION

Analog data that is to be processed in a digital system appears in the form of a continuously varying voltage. The voltage is often derived from a *transducer* that converts a quantity such as pressure or temperature to a proportional voltage. An analog-to-digital converter (ADC, or A/D converter) is used to convert the analog voltage to a digital form suitable for processing by a digital system.

The Counter-Type A/D Converter

The simplest type of A/D converter employs a binary counter, a voltage comparator, and a D/A converter, as shown in Figure 17.12(*a*). Recall from Chapter 7 that the output of the voltage comparator is high as long as its v^+ input is greater than its v^- input. Notice that the analog input is the v^+ input to the comparator. As long as it is greater than the v^- input, the AND gate is enabled and clock pulses are passed to the counter. The digital output of the counter is converted to an analog voltage by the DAC, and that voltage is the other input to the comparator. Thus, the counter counts up until its output has a value equal to the analog input. At that time, the comparator switches low, inhibiting the clock pulses, and counting ceases. The count it reached is the digital output proportional to the analog input. Control circuitry, not shown, is used to latch the output and reset the counter. The cycle is repeated, with the counter reaching a new count proportional to whatever new value the analog input has acquired. Figure 17.12(*b*) illustrates the output of a 4-bit DAC in an ADC over several counting cycles when the analog input is a slowly

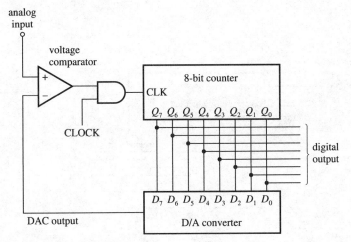

(a) Block diagram of an 8-bit ADC.

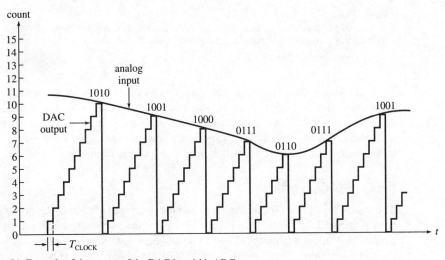

(b) Example of the output of the DAC in a 4-bit ADC.

Figure 17.12
The counter-type A/D converter.

varying voltage. The principal disadvantage of this type of converter is that the conversion time depends on the magnitude of the analog input: the larger the input, the more clock pulses must pass to reach the proper count. An 8-bit converter could require as many as 255 clock pulses to perform a conversion, so the counter-type ADC is considered quite slow in comparison to other types we will study.

Tracking A/D Converter

To reduce conversion times of the counter-type ADC, the up counter can be replaced by an up/down counter, as illustrated in Figure 17.13. In this design, the counter is not reset after each conversion. Instead, it counts up or down from its last count to its new count.

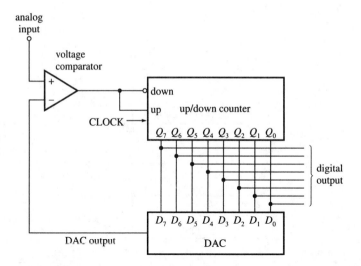

Figure 17.13
The tracking counter-type A/D converter. The counter counts up or down from its last count to reach its next count rather than resetting to 0 between counts.

Thus, the total number of clock pulses required to perform a conversion is proportional to the *change* in the analog input between counts rather than to its magnitude. Since the count more or less keeps up with the changing analog input, this type of ADC is called a *tracking converter*. A disadvantage of the design is that the count may oscillate up and down from a fixed count when the analog input is constant.

Flash A/D Converters

The fastest type of A/D converter is called the *flash* (or simultaneous, or parallel) type. As shown for the 3-bit example in Figure 17.14, a reference voltage is connected to a voltage divider that divides it into 7 ($2^n - 1$) equal-increment levels. Each level is compared to the analog input by a voltage comparator. For any given analog input, one comparator and all those below it will have a high output. All comparator outputs are connected to a priority encoder. Recall from Chapter 13 that a priority encoder produces a binary output corresponding to the input having the highest priority, in this case, the one representing the largest voltage level equal to or less than the analog input. Thus, the binary output represents the voltage that is closest in value to the analog input.

The voltage applied to the v^- input of the uppermost comparator in Figure 17.14 is, by voltage-divider action,

$$\left(\frac{6R + R/2}{6R + R/2 + 3R/2}\right)E_{\text{REF}} = \frac{13R/2}{16R/2}E_{\text{REF}} = \frac{13}{16}E_{\text{REF}} \qquad (17.12)$$

Similarly, the voltage applied to the v^- input of the second comparator is $(11/16)E_{\text{REF}}$, that applied to the third is $(9/16)E_{\text{REF}}$, and so forth. The increment between voltages is seen to be $(2/16)E_{\text{REF}}$, or $(1/8)E_{\text{REF}}$. An n-bit flash comparator has $n - 2$ R-valued resistors, and the increment between voltages is

Figure 17.14
A 3-bit flash A/D converter.

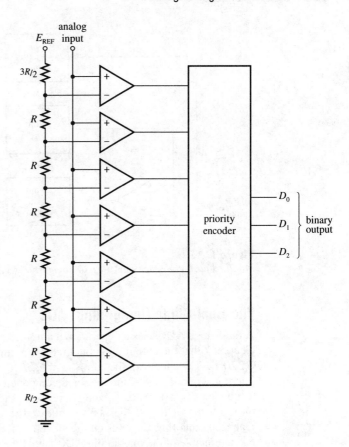

$$\Delta V = \frac{1}{2^n}E_{REF} \qquad (17.13)$$

The voltage levels range from

$$\left(\frac{2^{n+1} - 3}{2^{n+1}}\right)E_{REF} \quad \text{through} \quad \left(\frac{1}{2^{n+1}}\right)E_{REF}$$

The flash converter is fast because the only delays in the conversion are in the comparators and the priority encoder. Under the control of a clock, a new conversion can be performed very soon after one conversion is complete. The principal disadvantage of the flash converter is the need for a large number of voltage comparators ($2^n - 1$). For example, an 8-bit flash ADC requires 255 comparators.

Figure 17.15 shows a block diagram of a modified flash technique that uses 30 comparators instead of 255 to perform an 8-bit A/D conversion. One 4-bit flash converter is used to produce the 4 most significant bits. Those 4 bits are converted back to an analog voltage by a D/A converter and the voltage is subtracted from the analog input. The difference between the analog input and the analog voltage corresponding to the 4 most significant bits is an analog voltage corresponding to the 4 least significant bits. Therefore, that voltage is converted to the 4 least significant bits by another 4-bit flash converter.

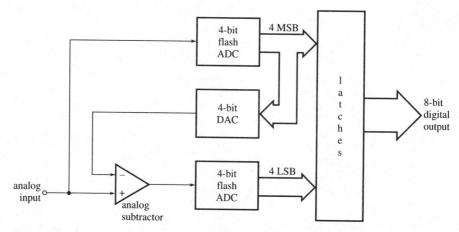

Figure 17.15
Modified flash converter that uses 30 comparators instead of 255 to produce an 8-bit output.

The Dual-Slope (Integrating) ADC

A dual-slope ADC uses an operational amplifier to integrate the analog input. Recall from Chapter 7 that the output of an integrator is a ramp when the input is a fixed level [Figure 7.32(a)]. The slope of the ramp is $\pm E_{in}/R_1C$, where E_{in} is the input voltage that is integrated and R_1 and C are the fixed components of the integrating operational amplifier [(equation (7.44)]. Since R_1 and C are fixed, the slope of the ramp is directly dependent on the value of E_{in}. If the ramp is allowed to continue for a fixed time, the voltage it reaches in that time depends on the slope of the ramp and hence on the value of E_{in}. The basic principle of the integrating ADC is that the voltage reached by the ramp controls the length of time that a binary counter is allowed to count. Thus, a binary number proportional to the value of E_{in} is obtained. In the dual-slope ADC, two integrations are performed, as described next.

Figure 17.16(a) shows a functional block diagram of the dual-slope ADC. Recall that the integrating operational amplifier inverts, so a positive input generates a negative-going ramp and vice versa. A conversion begins with the switch connected to the analog input. Assume that the input is negative, so a positive-going ramp is generated by the integrator. As discussed earlier, the ramp is allowed to continue for a fixed time, and the voltage it reaches in that time is directly dependent on the analog input. The fixed time is controlled by sensing the time when the counter reaches a specific count. At that time, the counter is reset and control circuitry causes the switch to be connected to a reference voltage having a polarity *opposite* to that of the analog input—in this case, a positive voltage. As a consequence, the output of the integrator becomes a negative-going ramp, beginning from the positive value it reached during the first integration. Since the reference voltage is fixed, so is the slope of the negative-going ramp. When the negative-going ramp reaches 0 V, the voltage comparator switches, the clock pulses are inhibited, and the counter ceases to count. The count it contains at that time is proportional to the time required for the negative-going ramp to reach 0 V, which is proportional to the positive voltage reached in the first integration. Thus, the binary count is proportional to the value of the analog input. Figure 17.16(b) shows examples of the ramp waveforms generated by a small analog input and by a large analog input. Note that the slope of the positive-going

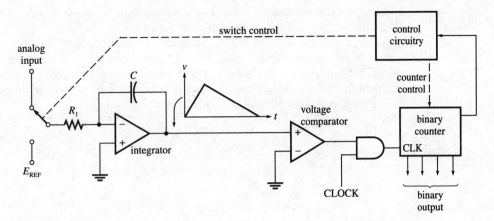

(a) Functional block diagram.

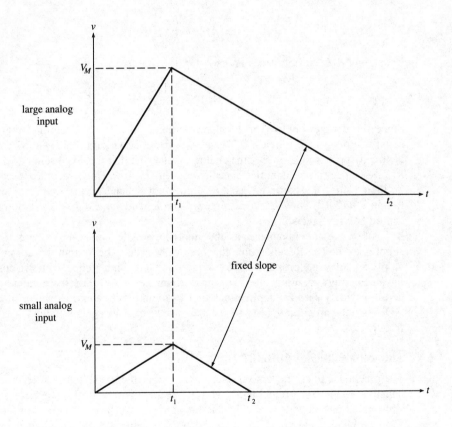

(b) Ramps resulting from large and small (negative) analog inputs.

Figure 17.16
The dual-slope integrating A/D converter.

ramp is variable (depending on E_{in}), and the slope of the negative-going ramp is fixed. The origin of the name *dual-slope* is now apparent.

One advantage of the dual-slope converter is that its accuracy does not depend on the values of the integrator components R_1 and C, nor upon any long-term changes that may occur in them. This fact is demonstrated by examining the equations governing the times required for the two integrations. Since the slope of the positive-going ramp is E_{in}/R_1C, the maximum voltage V_M reached by the ramp in time t_1 is

$$V_M = \frac{|E_{in}|t_1}{R_1C} \tag{17.14}$$

The magnitude of the slope of the negative-going ramp is $|E_{REF}|/R_1C$, so

$$V_M = \frac{|E_{REF}|}{R_1C}(t_2 - t_1) \tag{17.15}$$

Equating (17.14) and (17.15),

$$\frac{|E_{in}|}{R_1C}t_1 = \frac{|E_{REF}|}{R_1C}(t_2 - t_1) \tag{17.16}$$

Cancelling R_1C on both sides and solving for $t_2 - t_1$ gives

$$t_2 - t_1 = \frac{|E_{in}|}{|E_{REF}|}t_1 \tag{17.17}$$

Since the counter contains a count proportional to $t_2 - t_1$ (the time required for the negative-going ramp to reach 0 V) and t_1 is fixed, equation (17.17) shows that the count is directly proportional to E_{in}, the analog input. Note that this expression does not contain R_1 or C, since those quantities cancelled out in (17.16). Thus, accuracy does not depend on their values. Furthermore, accuracy does not depend on the frequency of the clock. Equation (17.17) shows that accuracy does depend on E_{REF}, so the reference voltage should be very precise.

An important advantage of the dual-slope A/D converter is that the integrator suppresses noise. Recall from equation (7.46) that the output of an integrator has amplitude inversely proportional to frequency. Thus, high-frequency noise components in the analog input are attenuated. This property makes it useful for instrumentation systems, and it is widely used for applications such as digital voltmeters. However, the integrating ADC is not particularly fast, so its use is restricted to signals having low to medium frequencies.

The Successive-Approximation ADC

The method called successive approximation is the most popular technique used to construct A/D converters, and, with the exception of flash converters, successive-approximation converters are the fastest of those we have discussed. Figure 17.17(a) shows a block diagram of a 4-bit version. The method is best explained by way of an example. For simplifying purposes, let us assume that the output of the D/A converter ranges from 0 V through 15 V as its binary input ranges from 0000 through 1111, with 0000 producing 0 V, 0001 producing 1 V, and so forth. Suppose the "unknown" analog input is 13 V. On the first clock pulse, the output register is loaded with 1000, which is converted by the DAC to 8 V. The voltage comparator determines that 8 V is less than the

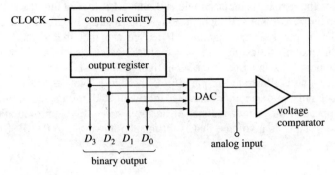

(a) Block diagram of a 4-bit converter.

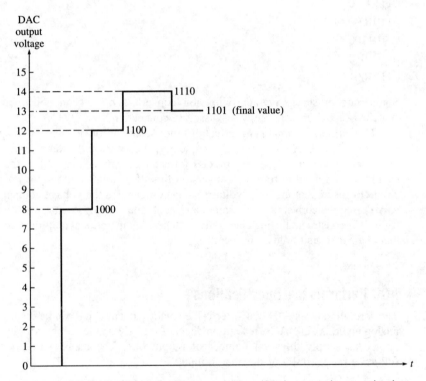

(b) Output of the DAC and contents of the output register, showing successive approximations
     when a 13-V analog input is converted.

Figure 17.17
The successive-approximation A/D converter.

analog input (13 V), so on the next clock pulse, the control circuitry causes the output
register to be loaded with 1100. The output of the DAC is now 12 V, which the
comparator again determines is less than the analog input. Consequently, the register is
loaded with 1110 on the next clock pulse. The output of the DAC is 14 V, which the
comparator now determines is larger than the 13-V analog input. Therefore, the last 1 that

was loaded into the register is replaced with a 0, and a 1 is loaded into the LSB. This time, the output of the DAC is 13 V, which equals the analog input, so the conversion is complete. The output register contains 1101. We see that the method of successive approximation amounts to testing a sequence of trial values, each of which is adjusted to produce a number closer in value to the input than the previous value. This "homing-in" on the correct value is illustrated in part (*b*) of the figure. Note that an *n*-bit conversion requires *n* clock pulses. As another example, following is the sequence of binary numbers that would appear in the output register of an 8-bit successive-approximation converter when the analog input is a voltage that is ultimately converted to 01101001:

10000000

01000000

01100000

01110000

01101000

01101100

01101010

01101001

Some modern successive-approximation converters have been constructed using the switched-capacitor technology discussed in connection with DACs.

The primary component affecting the accuracy of a successive-approximation converter is the D/A converter. Consequently, the reference voltage connected to it and its ladder network must be very precise for accurate conversions. Also, the analog input should remain fixed during the conversion time. Some units employ a *sample-and-hold* circuit to ensure that the input voltage being compared at the voltage comparator does not vary during the conversion. A sample-and-hold is the analog counterpart of a digital data latch. It is constructed using electronic switches, an operational amplifier, and a capacitor that charges to and holds a particular voltage level.

ADC Performance Specifications

The *resolution* of an A/D converter is the smallest change that can be distinguished in the analog input. As in DACs, resolution depends directly on the number of bits, so it is often quoted as simply the total number of output bits. The actual value depends on the full-scale range (FSR) of the analog input:

$$\text{resolution} = \frac{\text{FSR}}{2^n} \tag{17.18}$$

Some A/D converters have 8-4-2-1 BCD outputs rather than straight binary. This is especially true of dual-slope types designed for use in digital instruments. The BCD outputs facilitate driving numerical displays. The resolution of a BCD converter is quoted as the number of (decimal) digits available at the output, where each digit is represented by 4 bits. In this context, the term 1/2 *digit* is used to refer to a single binary output. For example, a 1½-digit output is represented by 5 bits. The 1/2 digit is used as the most significant bit, and its presence doubles the number of decimal values that can be represented by the 4 BCD bits. Some BCD converters employ multiplexers to expand the number of output digits. An example is Texas Instruments' TLC135C 4½-digit A/D converter. Expressed as a voltage, the resolution of a BCD A/D converter is FSR/10^d, where d is the number of output digits.

The time required to convert a single analog input to a digital output is called the *conversion time* of an A/D converter. Conversion time may be quoted as including any other delays, such as access time, associated with acquiring and converting an analog input. In that case, the total number of conversions that can be performed each second is the reciprocal of the conversion time. The reason this specification is important is that it imposes a limit on the rate at which the analog input can be allowed to change. In effect, the A/D converter *samples* the changing analog input when it performs a sequence of conversions. The *Shannon sampling theorem* states that sampled data can be used to faithfully reproduce a time-varying signal provided that the sampling rate is at least *twice* the frequency of the highest-frequency component in the signal. Thus, in order for the sequence of digital outputs to be a valid representation of the analog input, the A/D converter must perform conversions at a rate equal to at least twice the frequency of the highest component of the input.

Example 17-4

What maximum conversion time can an A/D converter have if it is to be used to convert *audio* input signals? (The audio frequency range is considered to be 20 Hz to 20 kHz.)

Solution.   Since the highest frequency in the input may be 20 kHz, conversions should be performed at a rate of at least 40×10^3 conversions/s. The maximum allowable conversion time is therefore equal to

$$\frac{1}{40 \times 10^3} = 25 \ \mu s$$

Drill 17-4

An A/D converter has a total conversion time of 100 μs. What is the highest frequency its analog input should be allowed to contain?
Answer: 5 kHz

One LSB for an A/D converter is defined in the same way it is for a D/A converter:

$$\text{LSB} = \frac{\text{FSR}}{2^{n-1}} \tag{17.19}$$

where FSR is the full-scale range of the analog input. Other A/D converter specifications may be quoted in terms of an LSB or as a percent of FSR.

Integrated-Circuit A/D Converters

A wide variety of A/D converters of all the types we have discussed are available in integrated circuits. Most have additional features, such as latched three-state outputs, that make them compatible with microprocessor systems. Some have *differential* inputs, wherein the voltage converted to a digital output is the difference between two analog inputs. Differential inputs are valuable in reducing the effects of noise, because any noise signal common to both inputs is "differenced out." This property is called *common-mode rejection*. Differential inputs also allow the user to add or subtract a fixed voltage to the analog input, thereby offsetting the values converted. Conventional (single-ended) inputs can be accommodated simply by grounding one of the differential inputs.

The advances in MOS technology that have made high-density memory circuits possible have also made it possible to incorporate many special functions into a single-integrated circuit containing an ADC. Examples include sample-and-hold circuitry and analog multiplexers. With these functions, and versatile control circuitry, complete microprocessor-compatible *data-acquisition* systems have become available in a single-integrated circuit. The microprocessor can be programmed to control the multiplexer so that analog data from many different sources (up to 19 in some versions) can be sampled in a desired sequence. The analog data from an instrumentation system, for example, is sampled, converted, and transmitted directly to the microprocessor for storing in memory or further processing.

17–3   THE SWITCH INTERFACE—DEBOUNCING

Mechanical switches form the interface between human beings and computers or other digital systems. In Chapter 13, we showed an example of a keyboard matrix used to supply alphanumeric data to a computer. Whatever the switch design, it is a potential source of problems due to contact bounce, as discussed in Chapter 6. Recall that contact bounce creates a sense of narrow pulses when a switch is opened or closed (see Figure 6.15). An example of a device whose operation would be adversely affected by contact bounce is a digital counter used to count the number of times a switch is depressed. Eliminating the effects of contact bounce is called *debouncing*.

When data is entered into a computer via a keyboard, a *software debounce* is often used. This type of debouncing is a program that causes the computer to sample the switch terminal (i.e., to input data from it) many times in succession during the interval of time that contact bounce occurs. If the data is sensed to be 1s (or 0s) for a specific number of *consecutive* samples, then it is assumed that contact bounce has ended and the last value sensed is valid.

Hardware debouncing is the use of electronic circuitry to eliminate the effects of contact bounce. There are numerous versions of such circuitry, including those that use monostable multivibrators (one-shots), but the most straightforward is simply an RS latch. Figure 17.18 shows the circuit. When the switch is in position 1, $R = 0$ and $S = 1$, so the latch is set and the output (Q) is 1. When the switch is in position 0, $R = 1$ and $S = 0$, so the latch is reset and the output is 0. When the switch is moved from one position to the other, the latch changes state and bouncing occurs at either the R or the S input. The bouncing does not affect the latch after it has changed state. Recall that an RS latch will remain set, for example, when its R input is 0 and its S input is alternately changed from 1 to 0.

Figure 17.18
Use of an RS latch to debounce a mechanical switch.

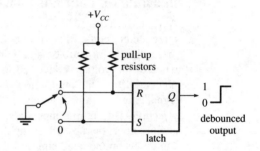

Figure 17.19
An optocoupler consists of an LED packaged with a photodiode or phototransistor.

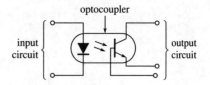

17–4   OPTOCOUPLERS

An optocoupler is a light-activated electronic switch. It consists of a light-emitting diode (LED) packaged with a *photodiode* or *phototransistor,* devices that are activated (turned on) by light energy. See Figure 17.19. A pulse applied to the input side turns on (illuminates) the LED, and the light energy it generates activates the photodiode or phototransistor. The latter devices can be connected to an output circuit to perform switching functions in any of the applications for which conventional diodes and transistors are used. The principal use of optocouplers is to interface circuits where good electrical *isolation* is required. Since there is no physical contact between the input circuit and the output circuit, each can have a separate ground, or reference. Because of this ability to isolate circuits, optocouplers are also called *optoisolators*.

The output circuit in Figure 17.19 contains a phototransistor. In some commercially available units, the base is not an externally accessible terminal, since only the collector and emitter are necessary to serve as switch terminals. In other units, the base terminal is accessible so that the user can optionally connect the output circuit between the base and collector. In this mode, the device serves as a photodiode rather than as a phototransistor. The advantage of the photodiode mode is that switching is much faster than in the phototransistor mode. The advantage of the phototransistor mode is that, as in conventional devices, the transistor has the capability of driving a load.

The *current transfer ratio* (CTR) of an optocoupler is the ratio of output current to input current. Depending on device design and application, it may range from 0.1 or less to several hundred. Optocoupler specifications usually include electrical isolation, expressed as a voltage. This voltage is the input-to-output voltage that the device can withstand without electrical conduction occurring between input and output. A typical value is 2 kV. Other important specifications of an optocoupler relate to switching speed. These may be quoted in terms of rise and fall times. Typical values are 1 μs for an optocoupler using a photodiode and 5 μs for an optocoupler using a phototransistor.

17–5   INTERFACING RELAYS AND SOLENOIDS

A relay is a mechanical switch or set of switches that are opened or closed by a magnetic field generated when an electrical current is passed through a coil. See Figure 17.20. In the context of a relay, the switches are called *contacts*. When no current passes through the coil, it is said to be deenergized, and the contacts are in their "normal" state: normally open or normally closed. When the coil is energized, the contacts switch to the opposite state. Like optocouplers, relays provide electrical isolation between an input circuit (the coil) and output circuit(s), the circuits connected to the contacts. They are used to drive heavy loads. A relatively small voltage applied to the coil circuit opens and closes heavy-duty contacts that can switch high voltages and currents. A very common application of a relay is to start and stop an electrical motor. A solenoid is similar to a

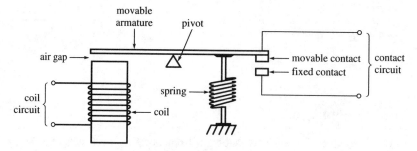

(a) Pictorial representation of a relay. In this diagram, energizing the coil opens the contacts, and the spring force closes them when the coil is deenergized. The contacts are said to be *normally closed* (N.C.). The air gap is shown for clarity, but it disappears when the contacts open.

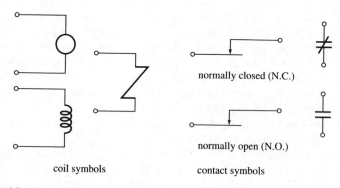

coil symbols   contact symbols

(b) Symbols used for relay coils and contacts.

Figure 17.20
The electromechanical relay.

relay except instead of opening and closing switches when its coil is energized, it opens or closes a mechanical valve. Solenoids are used to control the flow of liquids and gases. Relays and solenoids are both very slow in comparison to electronic switching speeds, so they are used only when the load is to be switched in or out of a circuit for long intervals of time. An example of such a load is the fan motor in an air conditioning system.

Figure 17.21 shows a circuit used to drive a relay coil. One of the problems with driving a relay is that a very large voltage spike appears across the coil terminals when it

Figure 17.21
A relay-driver circuit.

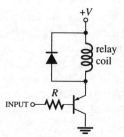

is first deenergized. (This is a consequence of Lenz's law: When the current in an inductance is changed, it induces a voltage across itself to oppose that change; the greater the change in current, the greater the induced voltage.) The voltage spike may be large enough to damage the transistor or to create intolerable noise signals in surrounding circuitry. The purpose of the diode in Figure 17.21 is to suppress the spike by holding the voltage across the coil terminals to the small, forward-biased drop of the diode.

Solid-state relays are available for switching small (light-current) loads. These devices are actually optocouplers, so they have no coils or mechanical contacts. As a result, they do not have contact bounce or generate the noise voltages that electromechanical relays do.

17–6   DIGITAL DATA TRANSMISSION

When digital data is transmitted between computer installations and input/output (I/O) equipment, it is rarely transmitted in the form of low and high voltages (logic levels), except in the case of very short distances. In the most commonly used method of transmission, logic levels are converted to different frequencies (tones), which are transmitted over telephone lines. This is a form of *frequency modulation* similar to that described in Chapter 16 as a data recording format, except different frequencies are used. Figure 17.22 shows an example of a sequence of serial bits transmitted using this method. The lower frequency represents binary 0 and the higher frequency, binary 1.

Data communication systems are classified as belonging to one of the following three categories:

1. Simplex: Data can be transmitted in one direction only.
2. Half-duplex: Data can be transmitted in two directions but not simultaneously.
3. Full-duplex: Data can be transmitted in two directions simultaneously.

Ordinary telephone conversations are full-duplex because both parties can speak and listen simultaneously. In a typical general-purpose, multi-user computer system, users communicate with the system from remote terminals via a telephone line in full-duplex operation. In order to achieve full-duplex communication on a single telephone line—i.e., to transmit data simultaneously in two directions—different frequencies must be used for each direction. Thus, the *originator* of a data transmission must have two distinct frequencies to represent 0s and 1s, and the *answering* transmitter must have two distinct frequencies, different from those of the originator, to represent 0s and 1s. In communications terminology, a 0 is called a *space* and a 1 is called a *mark,* so the

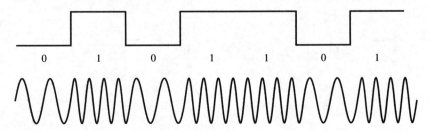

Figure 17.22
An example of digital data transmission using different frequencies to represent 1s and 0s.

frequencies representing 0s and 1s are called the space and mark frequencies. In typical, low-speed communication systems using a telephone line, those frequencies are as follows:

Originating frequencies			*Answering frequencies*
0	(Space)	1070 Hz	2025 Hz
1	(Mark)	1270 Hz	2225 Hz

Note that the originating frequencies are equal to 1170 Hz $\pm$ 100 Hz and the answering frequencies are 2125 Hz $\pm$ 100 Hz. 1170 Hz and 2125 Hz are called the *carrier* frequencies. The transmitting frequencies are generated by *shifting* the frequency of each carrier down 100 Hz for a binary 0 and up 100 Hz for a binary 1. The process is called *frequency-shift keying* (FSK).

Modems

Figure 17.23 shows a block diagram of a two-way communication system that incorporates frequency-shift keying for data transmission. At each end, digital data originates or is received at *data terminal equipment* (DTE), such as a computer, a printer, or a video display terminal (VDT). The purpose of the *modem* at each end is to perform the frequency-shift keying necessary to convert the digital data to tones for transmitting and to convert received tones back to digital data. As mentioned earlier, FSK is a form of frequency *modulation,* and the process of recovering digital data from the tones is called *demodulation.* The term *modem* is coined from modulator-demodulator. A modem is also called a *data set* and is an example of *data communications equipment* (DCE).

The Bell telephone system dominated the development of modems, so many of the standards associated with their use, including the frequencies used in FKS, originated with their equipment. The Bell 103 modem is widely used for low-speed communication.

As discussed in Chapter 12, digital data is most often transmitted in serial form, and a UART (Figure 12.15) may be needed at each end to convert parallel data to serial before modulation or to convert serial data to parallel after demodulation. The rate at which data is transmitted is expressed in bits per second (bps), or *baud*. In a binary system, 1 baud = 1 bps. The Bell 103 modem transmits at speeds up to 300 baud, whereas other FSK models transmit at speeds up to 1800 baud. High-speed modems using different modulation techniques are available to transmit at speeds up to 230,400 baud. Some modems are designed with an *acoustic coupler* as the interface between the modem and the telephone line. See Figure 17.24. When a telephone handset is placed into the coupler, tones representing data to be transmitted are coupled directly into the mouthpiece of the

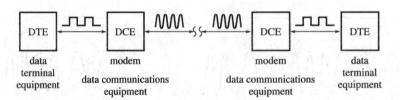

Figure 17.23
A two-way digital communication system using frequency-shift keying.

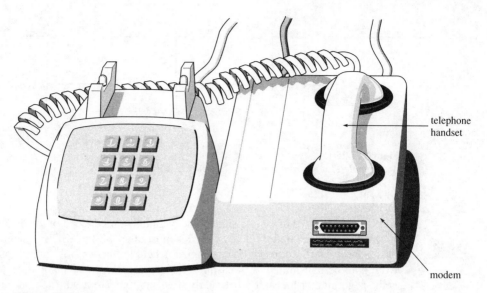

Figure 17.24
A modem equipped with an acoustic coupler for transmitting and receiving FSK tones through a telephone line.

telephone. Tones representing received data are coupled from the earpiece of the telephone into the coupler.

Synchronous and Asynchronous Transmission

In synchronous data transmission, all bits are transmitted at a constant rate under the control of a synchronizing clock. The receiving unit must have some means of acquiring that clock so that serial bits can be recovered at the proper bit times. One way for the receiving unit to acquire the clock is to embed the clock signal in the transmitted data. For example, if a biphase data format is used [see Figure 16.42(a)], then a 0-to-1 or a 1-to-0 transition occurs in the middle of every bit time, and these transitions can be used to generate a clock signal synchronized with the data. Another way to maintain synchronization between transmitting and receiving units is for both to access a common *master clock*. For example, the telephone industry maintains a master clock used for synchronizing transmissions on digital lines leased to its customers. Finally, synchronization can be maintained by simply transmitting the clock signal on a separate line from the data.

The principle difference between synchronous and asynchronous transmission is that the clock in a synchronous system is a stable, *long-term* timing reference that synchronizes both bits and words. For example, in a synchronous system transmitting 8-bit words, it is known that every 8-period interval represents one word. On the other hand, in an asynchronous system, data is transmitted *intermittently,* and although bit times within each word are synchronized to a clock, some means must be used to signify the beginning and end of each word. The most common example of asynchronous transmission occurs when the user of a video display terminal intermittently transmits (or receives) data from a computer. In that application, a *start bit* and one or two *stop bits* are

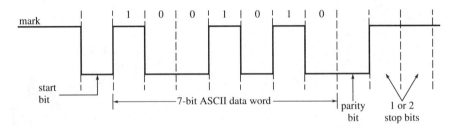

Figure 17.25
An example of data transmitted in an asynchronous system.

added to every word of data to signify the beginning and the end of the data word. Figure 17.25 shows an example. The data word is most often a 7-bit ASCII code group, or 7 bits plus a parity bit. By convention, the signal is held in the *marking* (high) state until a data word is to be transmitted. At that time, the signal goes low. This high-to-low transition is the beginning of the start bit. Usually, two stop bits are used if the bit rate is less than or equal to 300 bps, and one stop bit is used when the rate is higher.

As in a synchronous system, it is still necessary for the receiving unit in an asynchronous system to have a timing reference so that bits within each word can be recovered. Figure 17.26 shows one method that can be used. Two one-shots are connected to operate as a free-running (astable) multivibrator, with the output of one fed back through an OR gate to the input of the other. This circuit is similar to the pulse generator discussed in Section 8–2. The multivibrator is adjusted to oscillate at a frequency slightly lower than the bit rate of the data. V_{REF} is set so that when the start bit goes from high to low, the output of the voltage comparator changes state and triggers a one-shot. This one-shot and the one it triggers when it times out create a delayed pulse, as discussed in Section 8–2. The pulsewidths are set so that a narrow strobing pulse occurs at approximately the middle of the bit time of the start pulse. The output of the AND gate goes high at that time and triggers the free-running multivibrator. All succeeding high-to-low transitions in the data likewise trigger the free-running multivibrator, keeping it more or less in synchronism with the data. If no high-to-low transitions occur, the clock runs at its preset rate. It is apparent that the clock frequency is not perfectly constant, but its pulses occur close enough to mid-bit times to allow recovery of the data.

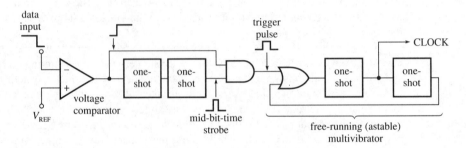

Figure 17.26
Creating a clock signal to synchronize data acquisition in an asynchronous system. The clock is intermittently synchronized by high-to-low transitions in the data.

Interface Standards

To standardize the interface between data terminal equipment and data communications equipment, the Electronics Industry Association (EIA) developed a standard called RS-232C that specifies voltage levels and connector pin assignments for all signals needed at the interface. The standard was developed in the 1960s and is now widely observed for interfacing terminals and modems. According to the standard, the voltage level for a mark (logical 1) shall be between -3 V and -25 V, and that for a space (logical 0) shall be between $+3$ V and $+25$ V. Note the use of negative logic. Typical values are ± 6 V. The maximum recommended cable length between the DTE and the DCE is 50 ft. Table 17.3 lists the 25 standard pin assignments for cable connectors. Not all pins are used in every interface application, but those required in every case are *frame ground, transmitted data, received data,* and *signal ground.*

Although the RS-232C standard is still widely used, it is not suited for modern, high-speed data communications. One of the problems with the standard is that all data and control signals are referenced to a single ground. Data transmission with that characteristic is said to be *unbalanced,* or single-ended, and it is susceptible to noise when the transmission is over a long distance. RS-422 is a new standard that specifies a balanced, or differential, mode of operation to alleviate those problems. Other standards in use include those created by the Consultative Committee for International Telephone and Telegraph (CCITT). The CCITT standard V.24 is basically compatible with RS-232C.

Table 17.3
The RS-232C interface.

Pin number	Mnemonic	Function	Direction
1	FG	Frame ground	
2	TD	Transmitted data	to DCE
3	RD	Received data	to DTE
4	RTS	Request to send	to DCE
5	CTS	Clear to send	to DTE
6	DSR	Data set ready	to DTE
7	SG	Signal ground	
8	DCD	Data carrier detect	to DTE
9		Positive test voltage	to DTE
10		Negative test voltage	to DTE
11	QM	Equalizer mode	to DTE
12	(S)DCE	Secondary data carrier detect	to DTE
13	(S)CTS	Secondary clear to send	to DTE
14	(S)TD	Secondary transmitted data	to DCE
15	TC	Transmitter clock	to DTE
16	(S)RD	Secondary received data	to DTE
17	RC	Receiver clock	to DTE
18	DCR	Divided clock, receiver	to DTE
19	(S)RTS	Secondary request to send	to DCE
20	DTR	Data terminal ready	to DCE
21	SQ	Signal quality detect	to DTE
22	RI	Ring indicator	to DTE
23		Data rate selector	to DCE or to DTE
24	TC	External transmitter clock	to DCE
25		Unassigned	

EXERCISES

Section 17–1

17.1 The logic levels used in an 8-bit R-$2R$ ladder DAC are 1 = 5 V and 0 = 0 V. Find the output voltage for each input:
(a) 00100000
(b) 10100100

17.2 The logic levels used in a 6-bit R-$2R$ ladder DAC are 1 = +5 V and 0 = 0 V. What is the binary input when the analog output is 3.28125 V?

17.3 Design a 5-bit weighted-resistor DAC whose full-scale output voltage is −15 V. Logic levels are 1 = +5 V and 0 = 0 V. What is the output voltage when the input is 01010?

17.4 Design an 8-bit weighted-resistor DAC using operational amplifiers and two *identical* 4-bit DACs of the design shown in Figure 17.6. The output should be +10 V when the input is 11110000. (*Hint:* Sum the outputs of two amplifiers in a third amplifier, using appropriate voltage gains in the summation.) What is the full-scale output of your design?

17.5 In the switched current-source DAC shown in Figure 17.7, R = 10 kΩ and E_{REF} = 20 V. Find the current in each 20-kΩ resistor.

17.6 An 8-bit switched current-source DAC of the design shown in Figure 17.7 has R = 10 kΩ and E_{REF} = 15 V. Find the total current I_T delivered to the amplifier and the output voltage when the input is 01101100.

17.7 An 8-bit switched current-source DAC is operated as a 2-quadrant multiplier. The binary input is positive and the reference voltage can range from −10 V to +10 V. If R = 10 kΩ, what is the total range of the output voltage?

17.8 A 4-bit switched current-source DAC is operated as a 4-quadrant multiplier. The input is offset binary code.
Find the output voltage in each case:
(a) The reference voltage is −10 V and the input is 0011.

(b) The reference voltage is +5 V and the input is 1010.
(c) The reference voltage is +10 V and the input is 0001.

17.9 The binary input to a 4-bit switched-capacitor DAC having C = 2 pF is 0101.
(a) What is the total capacitance connected to E_{REF} when the ϕ_2 clock is high?
(b) What is the total capacitance connected to ground when the ϕ_2 clock is high?
(c) If E_{REF} = 8 V, what is the output voltage?

17.10 Draw a schematic diagram of an 8-bit switched capacitor DAC. Label all capacitor values in terms of capacitance C. What is the total capacitance when all capacitors are in parallel?

17.11 A 12-bit D/A converter has a full-scale range of 15 V. Its maximum differential linearity error is specified to be ±(1/2)LSB.
(a) What is its percent resolution?
(b) What are the minimum and maximum possible values of the increment in its output voltage?

17.12 The LSB of a 10-bit D/A converter is 20 mV.
(a) What is its percent resolution?
(b) What is its full-scale range?
(c) A differential linearity error greater than what percent of FSR could make its output nonmonotonic?

Section 17–2

17.13 The A/D converter in Figure 17.12(*a*) is clocked at 1 MHz. What is the maximum possible time that could be required to perform a conversion?

17.14 The minimum conversion time of a tracking type A/D converter is 400 ns. At what frequency is it clocked?

17.15 A flash-type 5-bit A/D converter has a reference voltage of 10 V.

(a) How many voltage comparators does it have?

(b) What is the increment between the fixed voltages applied to the comparators?

17.16 The largest fixed voltage applied to a comparator in a flash-type A/D converter is 14.824218 V when the reference voltage is 15 V. What is the number of bits in the digital output?

17.17 In an 8-bit, dual-slope A/D converter, $R_1 = 20$ kΩ and $C = 0.001$ μF. An analog input of -0.25 V is integrated for $t_1 = 160$ μs.

(a) What is the maximum voltage reached in the integration?

(b) If the input to the integrator is then switched to a reference voltage of $+5$ V, how long does it take the output to reach 0 V?

(c) If the counter is clocked at 3.125 MHz, what is the digital output after the conversion?

17.18 An 8-bit dual-slope A/D converter integrates analog inputs for $t_1 = 50$ μs. What should be the magnitude of the reference voltage if an input of 25 V is to produce a binary output of 11111111 when the clock frequency is 1 MHz?

17.19 List the sequence of binary numbers that would appear in the output register of a 4-bit successive-approximation A/D converter when the analog input has a value that is ultimately converted to 1011.

17.20 Sketch the output of the DAC in an 8-bit successive-approximation A/D converter when the analog input is a voltage that is ultimately converted to 10101011. Label each step of the DAC output with the decimal number corresponding to the binary value it represents.

17.21 List four types of A/D converters in descending order of speed (fastest converter type first).

17.22 List the principal advantage of each of four types of A/D converters.

17.23 The analog input of an 8-bit A/D converter can range from 0 V to 10 V. Find its resolution in volts and as a percent of full-scale range.

17.24 The resolution of a 12-bit A/D converter is 7 mV. What is its full-scale range?

17.25 The frequency components of the analog input to an A/D converter range from 50 Hz to 10 kHz. What maximum total conversion time should the converter have?

17.26 The analog input to an A/D converter consists of a 500-Hz fundamental waveform and its harmonics. If the converter has a total conversion time of 40 μs, what is the highest-order harmonic that should be in the input?

Section 17–3

17.27 A set of mechanical switches is used to preload a decade counter by switching the direct-set and direct-reset inputs on its flip-flops to appropriate levels. After the counter has been preloaded, an observer presses a pushbutton switch that decrements the counter each time a vehicle passes through a traffic intersection. The total time required for the count to reach $0)_{10}$ is then the total time required for the preloaded number of vehicles to pass through the intersection. The process is then repeated. Which switches, if any, should be debounced? Explain.

17.28 Draw a schematic diagram of a switch debouncer using an RS flip-flop with active-low inputs. Label the 0 and 1 positions of the switch.

Section 17–4

17.29 Name two components of an optocoupler. Optocouplers are used in applications that have what requirement?

17.30 Define the CTR of an optocoupler. In optocouplers that can be operated in one of two modes depending on the output terminals used, which mode provides faster switching times?

Section 17–5

17.31

(a) What is the difference between a relay and a solenoid?

(b) What is a switch in the output circuit of a relay called? Name two types.

(c) Briefly describe the types of applications for which relays are suited.

(d) What is the purpose of a diode connected in parallel with the relay coil in a relay driver circuit?

17.32 The β of the transistor in Figure 17.21 is 100. The relay is energized by applying a 10-V level to the base. If a coil current of 20 mA is required to energize the relay, what should be the value of the series base resistance? Neglect the base-to-emitter voltage drop.

Section 17–6

17.33 Name and define the three categories of data communications.

17.34 Define each of the following.
(a) FSK
(b) DTE
(c) Modem
(d) DCE
(e) Acoustic coupler

17.35 Define the term *baud* as applied to a digital (binary) data communications system. What is the maximum rate at which data can be transmitted by the Bell 103 modem?

17.36 A modem transmits a sequence of marks using an FSK system in which a mark is 2025 Hz. If the data rate is 300 bps, how many cycles represent each mark?

17.37 Distinguish between synchronous and asynchronous data communications.

17.38 How are successive words distinguished in asynchronous data transmission?

17.39 Name the most widely used interface standard for low-speed data communications.

17.40 Distinguish between unbalanced and balanced data communication and give an advantage of the latter.

Appendix A

SPICE

A-1   INTRODUCTION

SPICE—Simulation Program with Integrated Circuit Emphasis—was developed at the University of California, Berkeley, as a computer aid for designing integrated circuits. However, it is readily used to analyze discrete circuits as well and can in fact analyze circuits containing no semiconductor devices at all. In addition to this versatility, SPICE owes its current popularity to the ease with which a circuit model can be constructed and the wide range of output options (analysis types) available to the user.

As a brief note on terminology, let us be aware that SPICE is a computer *program*, stored in computer memory, which we do not normally inspect or alter. As users, we merely supply the program with data, in the form of an *input data file*, which describes the circuit we wish to analyze and the type of output we desire. This input data file is usually supplied to SPICE by way of a keyboard. SPICE then executes a program run, using the data we have supplied, and displays or prints the results on a video terminal or printer. The mechanisms, or *commands*, that must be used to supply the input data to SPICE and to cause it to execute a program run vary widely with the computer system used, so we cannot describe the procedure here. Furthermore, there are numerous versions of SPICE, some designed for use with microcomputers and others, capable of analyzing more complex circuits at greater speed, designed for use on large mainframe type computers. Depending on the version used, minor variations in *syntax* (the format for specifying the input data) may be encountered. The *User's Guide* supplied with most versions should be consulted if any difficulty is experienced with any of the programs in this book. All programs here have been run successfully using SPICE version 2G.6 on a Honeywell DPS 90 computer.

A-2   DESCRIBING A CIRCUIT FOR A SPICE INPUT FILE

The input data file consists of successive lines, which we hereafter refer to as *statements*, each of which serves a specific purpose, such as identifying one component in the circuit. The statements do not have to be numbered and, except for the first and last, can appear in any order.

The Title

The first statement in every input file must be a *title*. Subject only to the number of characters permitted by a particular version, the title can be anything we wish. Examples are:

    AMPLIFIER

and

    EXERCISE 2.25

and

    A DIODE (1N54) TEST

Nodes and Component Descriptions

The first step in preparing the circuit description is to identify and number all the *nodes* in the circuit. It is good practice to draw a schematic diagram with nodes shown by circles containing the node numbers. Node numbers can be any positive integers and one of them must be 0. (The zero node is usually, but not necessarily, the circuit common, or ground.) Node numbers can be assigned in any sequence, such as 0, 1, 2, 3, . . . or 0, 2, 4, 6, ...

Once the node numbers have been assigned, each component in the circuit is identified by a separate statement that specifies the type of component it is and the node numbers between which it is connected. The first letter appearing in the statement identifies the component type. Passive components (resistors, capacitors, and inductors) are identified by the letters R, C, and L. Any other characters can follow the first letter, but each component must have a unique designation. For example, the resistors in a circuit might be identified by R1, R2, RB, and REQUIV.

The node numbers between which the component is connected appear next in the statement, separated by one or more spaces. Except in some special cases, it does not matter which node number appears first. The component value, in ohms, farads, or henries, appears next. Resistors cannot have value 0. Following are some examples:

R25	6	0	100	(Resistor R25 is connected between nodes 6 and 0 and has value 100 Ω.)
R25	0	6	1E2	(Interpreted by SPICE in the same way as the previous notation.)
CIN	3	5	22E−6	(Capacitor CIN is connected between nodes 3 and 5 and has value 22 μF.)
LSHUNT	12	20	0.01	(Inductor LSHUNT is connected between nodes 12 and 20 and has value 0.01H.)

Specifying Numerical Values

Suffixes representing powers of 10 can be appended to value specifications (with no spaces in between). Following are the SPICE suffix designations:

T	(tera: 10^{12})		U	(micro: 10^{-6})
G	(giga: 10^{9})		N	(nano: 10^{-9})
MEG	(mega: 10^{6})		P	(pico: 10^{-12})
K	(kilo: 10^{3})		F	(femto: 10^{-15})
M	(milli: 10^{-3})			

Note that M represents *milli* (10^{-3}) and that MEG is used for 10^6. Following are some examples of equivalent ways of representing values, all of which would be interpreted by SPICE in the same way:

$$0.002 = 2M = 2E{-}3 = 2000U$$
$$5000E{-}12 = 5000P = 5N = .005U = 0.005E{-}6$$
$$0.15MEG = 150K = 150E3 = 150E{+}3 = .15E{+}6$$

Any characters can follow a value specification, and unless the characters are none of the previous powers-of-10 suffixes, SPICE simply ignores them. Characters are often added to designate units. Following are some examples, all of which would be interpreted by SPICE in the same way:

$$100UF = 100E{-}6F = 100U = 100UFARADS$$
$$56N = 56NSEC = 0.056US = 56E{-}9SECONDS$$
$$2.2K = 2200OHMS = 0.0022MEGOHMS = 2.2E3$$
$$0.05MV = 50UVOLTS = 50E{-}6V = 0.05MILLIV$$

DC Voltage Sources

The first letter designating a voltage source, dc or ac, is V. As with passive components, any characters can follow the V, and each voltage source must have a unique designation. Examples are V1, VIN, and VSIGNAL. Following is the format for representing a dc source:

V******* N+ N− <DC> *value*

where ******* are arbitrary characters, N+ is the number of the node to which the positive terminal of the source is connected, and N− is the number of the node to which the negative terminal is connected. The symbol < > enclosing DC means that the specification DC is optional: If a source is not designated DC, SPICE will automatically assume that it is DC. *Value* is the source voltage, in volts. *Value* can be negative, which is equivalent to reversing the N+ and N− node numbers. Following are examples:

VIN 5 0 24VOLTS
VIN 0 5 DC −24

Both of these statements specify a 24-V dc source designated VIN whose positive terminal is connected to node 5 and whose negative terminal is connected to node 0. Both statements would be treated equivalently by SPICE.

Figure A.1 shows some examples of circuits and the statements that could be used to describe them in a SPICE input file. (These examples are not complete input files, because we have not yet discussed *control* statements, used to specify the type of analysis and the output desired.)

DC Current Sources

The format for specifying a dc current source is

I******* N+ N− <DC> *value*

where *value* is the value of the source in amperes and DC is optional. Note the following

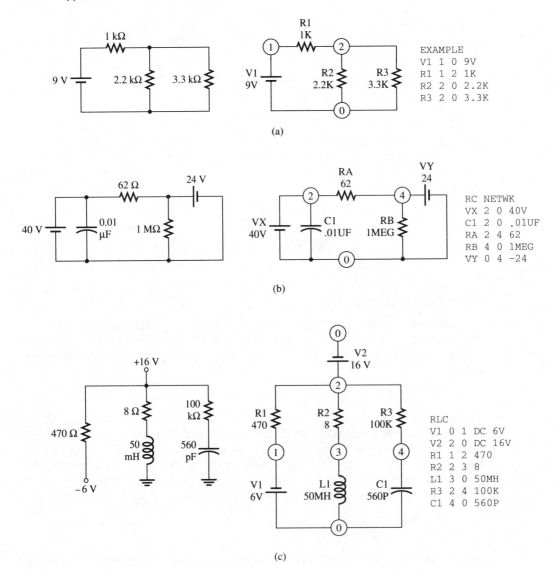

(a)

(b)

(c)

Figure A.1
Examples of circuit descriptions for a SPICE input file.

Figure A.2
Equivalent ways of specifying a dc current source. Note that the "negative" terminal is the one to which current is delivered (N− = 2).

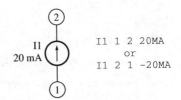

unconventional definition of N+ and N− : *N− is the number of the node to which the source delivers current*. To illustrate, Figure A.2 shows two equivalent ways of specifying a 20-mA dc current source.

A–3   THE .DC AND .PRINT CONTROL STATEMENTS

A control statement is one that specifies the type of analysis to be performed or the type of output desired. Every control statement begins with a period, followed immediately by a group of characters that identifies the type of control statement it is.

The .DC Control Statement

The .DC control statement tells SPICE that a dc analysis is to be performed. This type of analysis is necessary when the user wishes to determine dc voltages and/or currents in a circuit. Although ac sources can be present in the circuit, there must be at least one dc (voltage or current) source present if a dc analysis is to be performed. The format for the .DC control statement is

.DC *name1 start stop incr. <name2 start2 stop2 incr.2>*

where *name1* is the name of one dc voltage or current source in the circuit and *name2* is (optionally) the name of another. The .DC control statement can be used to *step* a source through a sequence of values, a useful feature when plotting characteristic curves. For that use, *start* is the first value of voltage or current in the sequence, *stop* is the last value, and *incr.* is the value of the increment, or step, in the sequence. A second source, *name2,* can also be stepped. If analysis is desired at a *single* dc value, we set *start* and *stop* both equal to that value and arbitrarily set *incr.* equal to 1. In cases where the circuit contains several fixed-value sources, one of them (any one) *must* be specified in the .DC control statement. Following are some examples:

.DC	V1	24	24	1

.DC   V1   24   24   1   V1 is a fixed 24-V dc voltage source.
.DC   IB   5OU 5OU   1   IB is a fixed 50-μA dc current source.
.DC   VCE 0   50   10   VCE is a dc voltage source that is stepped from
                                0 to 50 V in 10-V increments.

When two sources are stepped in a .DC control statement, the first source is stepped over its entire range for each value of the second source. Following is an example:

.DC   VB   0   5   5   VA   0   5   5

In this example, the dc voltage source named VB is set to 0 V while VA is stepped from 0 V to 5 V (in one 5-V step); then VB is set to 5 V while VA is again stepped from 0 V to 5 V. Thus, the sequence of voltages is

VB	VA
0 V	0 V
0 V	5 V
5 V	0 V
5 V	5 V

Identifying Output Voltages and Currents

To tell SPICE the voltages whose values we wish to determine (the *output* voltages we desire), we must identify them in one of the following formats:

$$V(N+, N-) \quad \text{or} \quad V(N+)$$

In the first case, V(N+, N−) refers to the voltage at node N+ with respect to node N−. In the second case, V(N+) is the voltage at node N+ with respect to node 0. Following are examples:

V(5,1)   The voltage at node 5 with respect to node 1.
V(3)   The voltage at node 3 with respect to node 0.

The only way to obtain the value of a current in SPICE is to request the value of the current in a *voltage* source. [In some newer versions of SPICE, the current in a component can be requested directly, as, for example, I(R1), the current in resistor R1.] Thus, a voltage source must be in the circuit at any point where we wish to know the value of the current. The current is identified by I*(Vname),* where *Vname* is the name of the voltage source. *We can insert zero-valued voltage sources (dummy sources) anywhere in a circuit for the purpose of obtaining a current.* These dummy voltage sources serve as ammeters for SPICE and do not in any way affect circuit behavior. Note carefully the following unconventional way that SPICE assigns polarity to the current through a voltage source: Positive current flows *into* the positive terminal (N+) of the voltage source. Figure A.3 shows an example. Here, conventional positive current flows in a clockwise direction, but a SPICE output would show I(V1) equal to −2A. On the other hand, SPICE would show the current in the zero-valued dummy source I(VDUM) to be +2A.

The .PRINT Control Statement

The .PRINT statement tells SPICE to print the values of voltages and/or currents resulting from an analysis. The format is

.PRINT *type out1* <*out2 out3* . . .>

where *type* is the type of analysis and *out1, out2,* . . . identify the output voltages and/or currents whose values we desire. So far, DC is the only analysis type we have discussed. For example, the statement

.PRINT DC V(1,2) V(3) I(VDUM)

tells SPICE to print the values of the voltages V(1, 2) and V(3) and the value of the current I(VDUM) resulting from a dc analysis. If the .DC control statement specifies a stepped

Figure A.3
SPICE treats conventional current flowing out of a positive terminal as negative current and current flowing into a positive terminal as positive current.

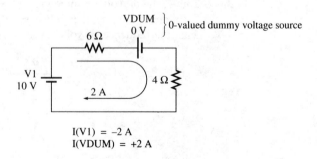

I(V1) = −2 A
I(VDUM) = +2 A

source, the .PRINT statement will print all the output values resulting from all the stepped values. The number of output variables whose values can be requested by a single .PRINT statement may vary with the version of SPICE used (up to eight can be requested in version 2G.6). Any number of .PRINT statements can appear in a SPICE input file.

The .END Statement

The last statement in every SPICE input file must be .END. We have now discussed enough statements to construct a complete SPICE input file, as demonstrated in the next example.

Example A-1

Use SPICE to determine the voltage drop across and the current through every resistor in Figure A.4(a).

Solution.   Figure A.4(b) shows the circuit when redrawn and labeled for analysis by SPICE. Note that two dummy voltage sources are inserted to obtain currents in two branches. The polarities of these sources are such that positive currents will be computed. The current in R1 is the same as the current in V1, and we must simply remember that SPICE will print a negative value for that current.

Figure A.4(c) shows the SPICE input file. Note that it is not necessary to specify a voltage value in the statement defining V1, since that value is given in the .DC statement.

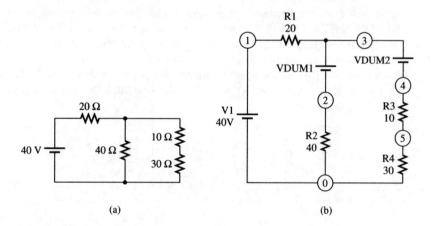

(a)   (b)

```
EXAMPLE A.1
V1 1 0
VDUM1 3 2
VDUM2 3 4
R1 1 3 20
R2 2 0 40
R3 4 5 10
R4 5 0 30
.DC V1 40 40 1
.PRINT DC I(V1) I(VDUM1) I(VDUM2) V(1,3) V(2) V(4,5) V(5)
.END
```

(c)

Figure A.4
Example A-1. (Continued on p. 732.)

```
****    DC TRANSFER CURVES              TEMPERATURE =    27.000 DEG C


***********************************************************************


    V1            I(V1)      I(VDUM1)    I(VDUM2)    V(1,3)      V(2)


  4.000E+01     -1.000E+00   5.000E-01   5.000E-01   2.000E+01   2.000E+01

***********************************************************************


    V1            V(4,5)      V(5)


  4.000E+01     5.000E+00   1.500E+01
```

(d)

Figure A.4
(Continued)

Figure A.4*(d)* shows the results of a program run. The outputs appear under the heading "DC TRANSFER CURVES," which refers to the type of output obtained when the source(s) are stepped. In our case, the heading is irrelevant. Note that the analysis is performed under the (default) assumption that the circuit temperature is 27°C (80.6°F). We will see later that we can specify different temperatures. Referring to the circuit nodes in Figure A.4*(b),* we see that the printed results give the following voltages and currents:

	I	V
R_1	1 A	20 V
R_2	0.5 A	20 V
R_3	0.5 A	5 V
R_4	0.5 A	15 V

A–4   THE .TRAN AND .PLOT CONTROL STATEMENTS

The .TRAN Control Statement

The .TRAN control statement (derived from "transient") is used when we want to obtain values of voltages or currents versus *time* (whether they are transients in the traditional sense or not). We must specify the total time interval over which we wish to obtain the time-varying values and the increment of time between each, using the format:

. TRAN *TSTEP TSTOP* $<$*TSTART*$>$

where *TSTEP* is the time increment and *TSTOP* is the largest value of time at which values will be computed. Unless we optionally specify the start time, *TSTART*, SPICE assumes it to be 0. If we do specify *TSTART*, computations still begin at $t = 0$, but only those in the interval from $t = TSTART$ through $TSTOP$ are provided as output. To illustrate, the statement

       .TRAN 5M 100M

causes SPICE to produce values of the output(s) at the 21 time points 0, 5 ms, 10 ms, . . . , 100 ms. When TRAN is used as the analysis type in a .PRINT statement, 21 values of each output variable specified in the .PRINT statement will be printed. The statement

       .TRAN 2US 200US 150US

will cause SPICE to produce the value(s) of the outputs at 2-μs intervals between 150 μs and 200 μs only. We show an example of a .TRAN analysis (Example A—2) after discussing a few more statement types.

A .TRAN analysis is sometimes difficult to obtain when pulse signals have very small rise and fall times and/or when investigating circuits (such as flip-flops) have regenerative *feedback*. It may be necessary in some cases to experiment with different rise and fall times to ''persuade'' SPICE to reach a solution.

The .PLOT Control Statement

The .PLOT control statement can be used to obtain many different kinds of plots, depending on the analysis type specified. The format is

       .PLOT *type out1* <*out2 out3* . . .>

Where *type* is the analysis type and *out1, out2,* . . . are the outputs whose values are to be plotted. If the analysis type is TRAN, then the output variables are plotted versus time, with time increasing downward along the vertical axis. If more than one output is specified in the .PLOT statement, all of them will be plotted, using different symbols, on the same axes. Although it is possible to specify the scale desired, it is easier to let SPICE automatically determine the scale (using the minimum and maximum values it computes). When more than one output is plotted, SPICE automatically determines and displays all scales needed for all outputs. It also prints the time increments and the values of the points plotted. If more than one output is plotted, the values of the first output specified in the .PLOT statement are the only ones printed. Separate .PLOT statements can be used to obtain separate plots and value printouts if desired.

.DC Plots

When the analysis type is .DC, a .PLOT statement causes SPICE to plot the output(s) specified in that statement versus the values of a stepped source. The stepped source values are printed downward along the vertical axis. When there are two stepped sources, the values of the stepped source appearing first in the .DC statement are printed downward along the vertical axis. These sets of values and the plots are repeated for each value of the stepped source appearing second in the .DC statement.

LIMPTS and the .OPTIONS Control Statement

Normally, SPICE will not print or plot more than 201 values. We can override this limit by specifying a different limit on the number of points (LIMPTS) using the .OPTIONS control statement. The .OPTIONS statement can also be used to change many other operating characteristics and limits that are normally imposed by SPICE, most of them related to the mathematical techniques used in the computations. The majority of these do not concern us. The format for changing the limit on the number of points plotted or printed is

OPTIONS LIMPTS = n

where n is the number of points.

The .NODESET Statement

The dc levels of some circuits are difficult or impossible to predict. In this category are circuits that have substantial feedback or that operate regeneratively, such as flip-flops. The .NODESET control statement can be used to guide SPICE to the solution for the dc voltage at a particular node in those types of circuits. For example, it can be used to specify whether a flip-flop is to be initially set or reset. The format is

.NODESET V(N1)=$value1$ <V(N2)=$value2$ . . .>

where N1, N2, . . . are nodes whose voltages are to be set to $value1, value2, \ldots$.

As an example, suppose the Q output of a flip-flop is at node 3 and the $\overline{Q}$ output is at node 4. If the logic levels are nominally 0 V and 5 V, then the flip-flop can be made to reset when a dc analysis is performed by specifying

.NODESET V(3) = 0 V(4) = 5

Depending on the circuit, the actual node voltages calculated by SPICE may not be exactly those specified in the .NODESET statement, since the statement merely guides SPICE towards the solution. (.NODESET provides a first guess, or approximation, for a node voltage, to help the computations converge.)

A–5   THE SIN AND PULSE SOURCES

If we wish to obtain a printout or plot of an output versus time, as in a .TRAN analysis, at least one source in the circuit should itself be a time-varying voltage or current. In other words, we will not be able to observe and output *waveform* unless an input waveform is defined. It is not sufficient to indicate in a component definition that a particular source is AC instead of DC. A source designated AC is used by SPICE in an .AC analysis, to be discussed presently, and that analysis type does *not* cause SPICE to display time-varying outputs. For a .TRAN analysis, we use a different format to define the time-varying sources. The two sources that are most widely used for that purpose are the SIN (sinusoidal ac) and PULSE sources.

The SIN Source

The format for specifying a sinusoidal voltage source is

V******* $N+$ $N-$ SIN($VO, VP, FREQ, TD, \theta$)

where ******* are arbitrary characters, $N+$ and $N-$ are the node numbers of the positive

Figure A.5
An example of the specifica-tion of a sinusoidal voltage source.

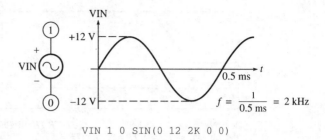

VIN 1 0 SIN(0 12 2K 0 0)

and negative terminals, *VO* is the *offset* (dc, or bias level), *VP* is the peak value in volts, and *FREQ* is the frequency in hertz. *TD* and θ are special parameters related to time delay and damping, both of which are set to 0 to obtain a conventional sinewave. A sinusoidal current source is defined by using I instead of V as the first character. Note that it is not possible to specify a phase shift (other than 180°, by reversing $N+$ and $N-$). Figure A.5 shows an example of a SIN source definition. The zero values for TD and θ can be omitted in the specification, and SPICE will assume they are zero by default.

The PULSE Source

The PULSE source can be used to simulate a dc source that is switched into a circuit at a particular instant of time (a *step* input) or to generate a sequence of square, trapezoidal, or triangular pulses. Figure A.6(a) shows the parameters used to define a voltage pulse or pulse-type waveform: one having time delay *(TD)* that elapses before the voltage begins to rise linearly with time; a *rise-time (TR)* that represents the time required for the voltage to change from *V1* volts to *V2* volts; a *pulsewidth (PW)*; and a *fall-time (TF)*, during which the voltage falls from *V2* volts to *V1* volts. If the pulse is repetitive, a value for the period of the waveform *(PER)* is also specified. If *PER* is not specified, its default value is the value of *TSTOP* in a .TRAN analysis; that is, the pulse is assumed to remain at *V2* volts for the duration of the analysis, simulating a step input. The default value for the rise and fall times is the *TSTEP* time specified in a .TRAN analysis. The figure shows the format for identifying a pulsed voltage source. Current pulses can be obtained by using I instead of V as the first character. Figure A.6(b) shows an example of how the PULSE source is used to simulate a step input caused by switching a 10-V dc source into a circuit at $t = 0$. Figure A.6(c) shows an example of how the PULSE source is used to define a triangular waveform that alternates between ± 5 V with a frequency of 100 Hz. SPICE does not accept a zero pulsewidth, as would be necessary to define an ideal triangular waveform, but *PW* can be made negligibly small. In this example, we set $PW = 10^{-12}$ s = 1 ps, which makes the period 10^{10} times greater than the pulsewidth.

Note that the rise- and fall-times, *TR* and *TF*, in a PULSE specification are *not* de-fined the same way as t_r and t_f; i.e., they are not the times required for the pulse to change between its 10% and 90% values. Example 6–11 shows how to calculate the values that should be used for *TR* and *TF* when the rise- and fall-times, t_r and t_f, are known.

It is not necessary that the "high" level voltage, *V2* in Figure A.6(a), be greater than or more positive than *V1*. For example, we could specify PULSE (10 0 0 1PS 1PS 50US), a pulse that starts (at $t = 0$) with a value of 10 V and then (immediately) switches to 0 V. Note, however, that the pulsewidth specification always refers to *V2*, which is 0 V in this example. Therefore, the pulse in this case would remain at 0 V for 50 μs and then return

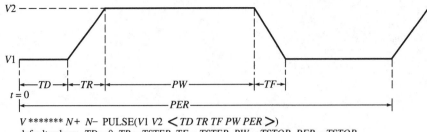

$$V \, ******* \, N+ \ N- \ \text{PULSE}(V1 \ V2 \ \langle TD \ TR \ TF \ PW \ PER \ \rangle)$$
default values: $TD = 0$, $TR = TSTEP$, $TF = TSTEP$, $PW = TSTOP$, $PER = TSTOP$.

(a)

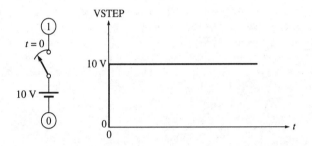

VSTEP 1 0 PULSE(0 10 0 0 0)

(b) Using the PULSE source to simulate a 10-V dc source switched into
a circuit at $t = 0$. Even though TR and TF are set to 0, SPICE assigns
each the default value of $TSTEP$ specified in a . TRAN statement. PW
and PER both default to the value of $TSTOP$.

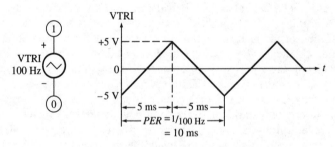

VTRI 1 0 PULSE(–5 5 0 5M 5M 1P 10M)

(c) Using the PULSE source to generate a 100-Hz triangular waveform. Note
that $PW = 1$ ps ≈ 0.

Figure A.6
The PULSE source.

to $+10$ V. Since SPICE always begins an analysis with the (initial condition) voltage
equal to V1, we can make that value either "high" or "low," depending on how V1 and
V2 are specified. Example 7–22 shows a case where it is necessary that V1 be more
positive than V2.

A–6   THE INITIAL TRANSIENT SOLUTION

The next example demonstrates the use of the PULSE source to generate a square wave and contains some important discussion on how SPICE performs a dc analysis in conjunction with every transient analysis.

Example A–2

Use SPICE to obtain a plot of the capacitor voltage versus time in Figure A.7(a). The plot should cover two full periods of the square wave input.

Solution.   Figure A.7(b) shows the circuit when redrawn for analysis by SPICE and the corresponding input data file. The period of the input is T = 1/(2.5 Hz) = 0.4 s, so *TSTOP* in the .TRAN statement is set to 0.8 s to obtain a plot covering two full periods. Note that *PW* in the definition of V1 is set to 0.2 s. The square wave is idealized by setting the rise- and fall-times, *TR* and *TF,* to 0, so the actual value assigned by SPICE to *TR* and *TF* is the value of *TSTEP:* 0.02 s.

Figure A.8 shows the results of a program run. When SPICE performs a .TRAN analysis, it first obtains an "initial transient solution." This solution is obtained from a dc analysis with all time-varying sources set to 0. Thus, the initial solution represents the *quiescent,* or dc operating conditions in the circuit, useful information for determining the bias point(s) in circuits containing transistors. The actual time-varying outputs are computed with the initial voltages and currents as the starting points, so the outputs do not reflect initial transients associated with the charging of capacitors, such as coupling capacitors, in the circuit.

In our example, the initial transient solution gives the dc voltages and currents when V1, the square wave generator, is set to 0. The figure shows that the dc voltages at all nodes are printed, as are the dc currents in all voltage sources and the total dc power

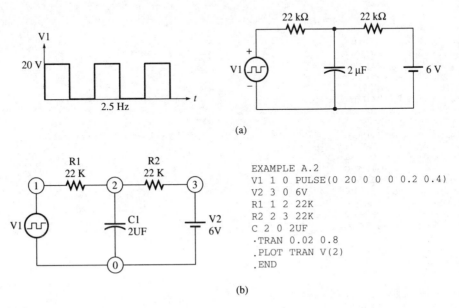

(a)

```
EXAMPLE A.2
V1 1 0 PULSE(0 20 0 0 0 0.2 0.4)
V2 3 0 6V
R1 1 2 22K
R2 2 3 22K
C 2 0 2UF
.TRAN 0.02 0.8
.PLOT TRAN V(2)
.END
```

(b)

Figure A.7
Example A–2.

```
****      INITIAL TRANSIENT SOLUTION          TEMPERATURE =   27.000 DEG C


*****************************************************************************


   NODE    VOLTAGE      NODE    VOLTAGE      NODE    VOLTAGE

  (  1)      .0000     (  2)    3.0000     (  3)    6.0000




      VOLTAGE SOURCE CURRENTS

      NAME        CURRENT


      V1        1.364D-04

      V2       -1.364D-04


      TOTAL POWER DISSIPATION   8.18D-04   WATTS


****      TRANSIENT ANALYSIS              TEMPERATURE =   27.000 DEG C

*****************************************************************************
      TIME      V(2)

                     0.000D+00     5.000D+00     1.000D+01     1.500D+01  2.000D+01
                     - - - - - - - - - - - - - - - - - - - - - - - - - - - - - -
  0.000D+00   3.000D+00 .        *       .       .        .        .
  2.000D-02   6.394D+00 .           .    *   .      .        .        .
  4.000D-02   1.023D+01 .           .        .    *   .        .        .
  6.000D-02   1.191D+01 .           .        .       .*       .        .
  8.000D-02   1.259D+01 .           .        .       . *      .        .
  1.000D-01   1.285D+01 .           .        .       .  *     .        .
  1.200D-01   1.294D+01 .           .        .       .  *     .        .
  1.400D-01   1.298D+01 .           .        .       .  *     .        .
  1.600D-01   1.299D+01 .           .        .       .  *     .        .
  1.800D-01   1.300D+01 .           .        .       .  *     .        .
  2.000D-01   1.300D+01 .           .        .       .  *     .        .
  2.200D-01   1.300D+01 .           .        .       .  *     .        .
  2.400D-01   9.591D+00 .           .        .      *.        .        .
  2.600D-01   5.658D+00 .           .    *   .       .        .        .
  2.800D-01   4.071D+00 .        *     .       .        .        .
  3.000D-01   3.420D+00 .        *    .       .        .        .
  3.200D-01   3.159D+00 .       *     .       .        .        .
  3.400D-01   3.059D+00 .       *     .       .        .        .
  3.600D-01   3.024D+00 .      *      .       .        .        .
  3.800D-01   3.009D+00 .      *      .       .        .        .
  4.000D-01   3.003D+00 .      *      .       .        .        .
  4.200D-01   6.410D+00 .           .   *    .       .        .        .
  4.400D-01   1.034D+01 .           .        .    *   .        .        .
  4.600D-01   1.193D+01 .           .        .       . *      .        .
  4.800D-01   1.258D+01 .           .        .       .  *     .        .
```

Figure A.8
Example A–2. (Continued on p. 739.)

```
5.000D-01   1.284D+01 .            .           .      *   .        .
5.200D-01   1.294D+01 .            .           .      *   .  .     .
5.400D-01   1.298D+01 .            .           .      *   .        .
5.600D-01   1.299D+01 .            .           .      *   .        .
5.800D-01   1.300D+01 .            .           .      *   .        .
6.000D-01   1.300D+01 .            .           .      *   .        .
6.200D-01   1.300D+01 .            .           .      *   .        .
6.400D-01   9.591D+00 .            .           .    *.    .        .
6.600D-01   5.658D+00 .            .      * .         .            .
6.800D-01   4.071D+00 .         * .                  .            .
7.000D-01   3.420D+00 .        * .                   .            .
7.200D-01   3.159D+00 .      * .                     .            .
7.400D-01   3.059D+00 .      * .                     .            .
7.600D-01   3.024D+00 .     * .                      .            .
7.800D-01   3.009D+00 .     * .                      .            .
8.000D-01   3.003D+00 .     * .                      .            .
                      - - - - - - - - - - - - - - - - - - - - - - - -
```

Figure A.8
(Continued)

dissipation in the circuit. (Note that dc current flows *into* V1 when it is set to 0 V.) Since the capacitor is charged to $+3$ Vdc, the time-varying plot shows its voltage to begin at $+3$ V. The actual transient that would occur (beginning at $t = 0$) while the capacitor charged to 3 V does not appear in the output.

A–7   DIODE MODELS

When there is a semiconductor device in a circuit to be analyzed by SPICE, the SPICE input file must contain two new types of statements: one that identifies the device by name and gives its node numbers in the circuit, and another, called a .MODEL statement, that specifies the values of the device *parameters* (saturation current, and so on).

All diode names must begin with D. The format for identifying a diode in a circuit is

D******* *NA NC Nname*

where *NA* and *NC* are the numbers of the nodes to which the anode and cathode are connected, respectively, and *Mname* is the *model name*. The model name associates the diode with a particular .MODEL statement that specifies the parameter values of the diode:

.MODEL *Mname* D <*Pval1=n1 Pval2=n2* . . .>

where D, signifying diode, *must* appear as shown, and *Pval1* = *n1* . . . specify parameter values, to be described shortly. Note that several diodes, having different names, can all be associated with the same .MODEL statement, and other diodes can be associated with a different .MODEL statement. Figure A.9 shows an example: a diode bridge in which diodes D1 and D3 are modeled by MODA and diodes D2 and D4 are modeled by MODB. MODA specifies a diode having saturation current (IS) 0.5 pA and MODB specifies a diode having saturation current 0.1 pA.

Table A.1 lists the diode parameters whose values can be specified in a .MODEL statement and the default values of each. The default values are typical, or average, values

Figure A.9

The parameter values of diodes D1 and D3 are given in the model whose name is MODA, and the parameter values of D2 and D4 are given in the model whose name is MODB.

```
D1  5  6  MODA
D2  6  7  MODB
D3  4  7  MODA
D4  5  4  MODB
.MODEL MODA D IS = 0.5P
.MODEL MODB D IS = 0.1P
```

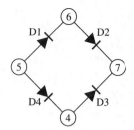

Table A.1

Diode parameters

Parameter	Identification	Units	Default value
Saturation current	IS	A	1×10^{-4}
Ohmic resistance	RS	ohms	0
Emission coefficient	N	—	1
Transit time	TT	s	0
Zero-bias junction capacitance	CJO	F	0
Junction potential	VJ	V	1
Grading coefficient	M	—	0.5
Activation energy	EG	eV	1.11
Saturation current temperature exponent	XTI	—	3
Flicker noise coefficient	KF	—	0
Flicker noise exponent	AF	—	0
Coefficient for forward-bias depletion capacitance equation	FC	—	0.5
Reverse breakdown voltage	BV	V	Infinite
Current at breakdown voltage	IBV	A	1×10^{-3}

and are acceptable for most electronic circuit analysis at our level of study. In practice, some of these diode parameters are very difficult to obtain or measure and are necessary only when a highly accurate model is essential. For example, if we were designing a new diode to have specific low-noise characteristics, we would want to know the noise parameters, KF and AF, very accurately. Example 9–8 illustrates the .DC analysis of a circuit containing diodes.

A–8 BJT MODELS

The format for identifying a bipolar junction transistor in a circuit is

Q******* *NC NB NE   Mname*

Figure A.10
An example of how an NPN and a PNP transistor are identified and modeled.

```
Q1 4 3 2 TYPE1
Q2 0 1 2 TYPE2
.MODEL TYPE1 NPN BF = 150
.MODEL TYPE2 PNP
```

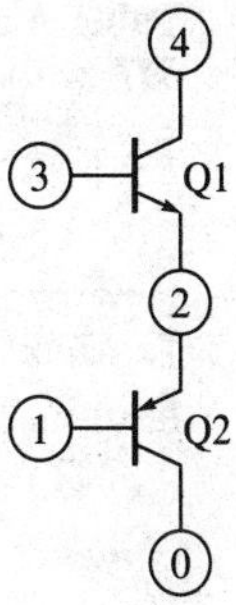

where *NC, NB,* and *NE* are the numbers of the nodes to which the collector, base, and emitter are connected, respectively, and *Mname* is the name of the model that specifies the transistor parameters. The format of the .MODEL statement for a BJT is

.MODEL *Mname type* <*Pval1=n1 Pval2=n2 . . .*>

where *type* is either NPN or PNP. Figure A.10 shows an example of how an NPN and a PNP transistor in a circuit are identified and modeled. The NPN model specifies that the ''ideal maximum forward beta'' (BF) of the transistor is 150, and the PNP model allows that parameter to have its default value of 100.

As shown in Table A.2, p. 742–743, we can specify up to 40 different parameter values for a BJT. The mathematical model used by SPICE to simulate a BJT is very complex but very accurate, provided that all parameter values are accurately known. However, to an even greater extent than in the diode model, many of the parameters are very difficult to measure, estimate, or deduce from other characteristics. Once again, for analysis, it is generally sufficient to allow most of the parameters to have their default values. This approach is further justified by the fact that there is usually a wide variation in parameter values among transistors of the same type. Therefore, it is unrealistic and unnecessary in many practical applications to seek a highly accurate analysis of a circuit containing transistors. On the other hand, there are situations, such as the design and development of a new integrated circuit required to have certain properties, where an accurate determination of parameter values is warranted. Example 9–9 illustrates a dc analysis of a circuit containing a transistor.

A–9   THE .TEMP STATEMENT

As noted earlier, all SPICE computations are performed under the assumption that the temperature of the circuit is 27°C, unless a different temperature is specified. The .TEMP statement is used to request an analysis at one or more different temperatures:

.TEMP *T1* <*T2 T3 . . .*>

where *T1* is the temperature in °C at which an analysis is desired. *T2, T3, . . .* are optional additional temperatures at which SPICE will repeat the analysis, once for each temperature. Temperature is a particularly important parameter in circuits containing semiconductor devices, since their characteristics are temperature-sensitive. However, SPICE will not adjust all device characteristics for temperature unless the parameters in the .MODEL statement that relate to temperature sensitivity are given specific values. A case in point is β in a BJT. The parameter BF is called the *ideal maximum forward beta*.

Table A.2
BJT parameters. (Continued on p. 743.)

Parameter	Identification	Units	Default value
Transport saturation current	IS	A	10^{-16}
Ideal maximum forward beta	BF	—	100
Forward current emission coefficient	NF	—	1
Forward early voltage	VAF	V	Infinity
Corner for forward beta high-current roll-off	IKF	A	Infinity
Base-emitter leakage saturation current	ISE	A	0
Base-emitter leakage emission coefficient	NE	—	1.5
Ideal maximum reverse beta	BR	—	1
Reverse current emission coefficient	NR	—	1
Reverse early voltage	VAR	V	Infinity
Corner for reverse beta high-current roll-off	IKR	A	Infinity
Base-collector leakage saturation current	ISC	A	0
Base-collector leakage emission coefficient	NC	—	2
Zero-bias base resistance	RB	Ω	0
Current where base resistance falls to half of its minimum value	IRB	A	Infinity
Minimum base resistance at high currents	RBM	Ω	RB
Emitter resistance	RE	Ω	0
Collector resistance	RC	Ω	0
Base-emitter zero-bias depletion capacitance	CJE	F	0
Base-emitter built-in potential	VJE	V	0.75
Base-emitter junction exponential factor	MJE	—	0.33
Ideal forward transit time	TF	s	0

The actual value of β used in the computations depends on other factors, including the forward early voltage (VAF) and the temperature. However, no temperature variation in the value of BF will occur unless the parameter XTB (forward and reverse beta temperature exponent) is specified to have a value other than 0 (its default value). When SPICE performs an analysis at a temperature other than 27°C, it prints a list of temperature-adjusted values of device parameters that are temperature-sensitive.

The values of resistors in a circuit are not adjusted for temperature unless first- and/or second-order temperature coefficients of resistance, tc_1 and tc_2, are given values in the statements defining resistors. The format is

R******** *N1 N2 value* TC=tc_1, tc_2

The temperature adjusted value is then computed by

$$R_T = R_{27}[1 + tc_1(T - 27°) + 1 + tc_2(T - 27°)]$$

where R_T is the resistance at temperature T and R_{27} is the resistance at temperature 27°C. The default values of tc_1 and tc_2 are 0.

Table A.2
(Continued)

Parameter	Identification	Units	Default value
Coefficient for bias dependence of TF	XTF	—	0
Voltage describing V_{BC} dependence of TF	VTF	V	Infinity
High-current parameter for effect on TF	ITF	A	0
Excess phase at $f = 1/(2\pi \text{TF})$ hertz	PTF	Degrees	0
Base-collector zero-bias depletion capacitance	CJC	F	0
Base-collector built-in potential	VJC	V	0.75
Base-collector junction exponential factor	MJC	—	0.33
Fraction of base-collector depletion capacitance to internal base node	XCJC	—	1
Ideal reverse transit time	TR	s	0
Zero-bias collector-substrate capacitance	CJS	F	0
Substrate junction built-in potential	VJS	V	0.75
Substrate junction exponential factor	MJS	—	0
Forward and reverse beta temperature exponent	XTB	—	0
Energy gap for temperature effect on IS	EG	eV	1.11
Temperature exponent for effect on IS	XTI	—	3
Flicker-noise coefficient	KF	—	0
Flicker-noise exponent	AF	—	1
Coefficient for forward-bias depletion capacitance formula	FC	—	0.5

A–10   AC SOURCES AND THE .AC CONTROL STATEMENT

The format for identifying an ac voltage source in a circuit is

V******* *N*+ *N*− AC <*mag*> <*phase*>

where ******* is an arbitrary sequence of characters, *N*+ and *N*− are the numbers of the nodes to which the positive and negative terminals are connected, *mag* is the magnitude of the voltage in volts, and *phase* is its phase angle in degrees. All ac sources are assumed to be sinusoidal. If *mag* is omitted, its default value is 1 V, and if *phase* is omitted, its default value is 0. An ac current source is identified by making the first character I instead of V. Note that *mag* may be regarded as either a peak or an rms value, since SPICE output from an ac analysis does not consist of instantaneous (time-varying) values.

The .AC Control Statement

The .AC control statement is used to compute ac voltages and currents in a circuit *versus frequency*. A single frequency or a range of frequencies can be specified. The circuit must contain at least one source designated AC, and all AC sources are assumed to be sinusoidal and to have identical frequencies or to undergo the same frequency variation, if any. The format is

.AC *vartype N fstart fstop*

where *vartype* specifies the way frequency is to be varied in the range from *fstart* through *fstop*. *N* is a number related to the number of frequencies at which computations are to be performed, as discussed shortly. The *vartype* is one of DEC, OCT, or LIN (decade, octave, or linear). If analysis at a single frequency is desired, we can use any *vartype*, set *fstart* equal to *fstop*, and let $N = 1$.

When *vartype* is LIN, the frequencies at which analysis is performed vary linearly from *fstart* through *fstop*. In that case, *N* is the total number of frequencies at which the analysis is performed (counting *fstart* and *fstop*). Thus, the internal between frequencies is

$$\Delta f = \frac{fstop - fstart}{N - 1}$$

For example, the statement

.AC LIN 21 100 1K

tells SPICE to perform an ac analysis at 21 frequencies from 100 Hz through 1 kHz. The frequency interval is $(1000 - 100)/20 = 45$ Hz, so computations are performed at 100 Hz, 145 Hz, 190 Hz, . . . , 1 kHz.

When the *vartype* is DEC or OCT, the analysis is performed at logarithmically spaced intervals and *N* is the total number of frequencies *per decade or per octave*. For example, the statement

.AC DEC 10 100 10K

causes SPICE to analyze the circuit at 10 frequencies in each of the decades 100 Hz to 1 kHz and 1 kHz through 10 kHz. The frequencies are at one-tenth-decade intervals, so each interval is different. The frequencies at one-tenth-decade intervals are 10^x, $10^{x+0.1}$, $10^{x+0.2}$, . . . , where $x = \log_{10}(fstart)$. In general, the frequencies at which SPICE performs an ac analysis using the DEC *vartype* are

$$10^x, \; 10^{x+1/N}, \; 10^{x+2/N}, \; . \; . \; .$$

where $x = \log_{10}(fstart)$. The last frequency in this sequence is not necessarily *fstop*, but SPICE will compute at frequencies in the sequence up through the first frequency that is equal to or greater than *fstop*. The frequencies at which SPICE performs an ac analysis when the *vartype* is OCT are

$$2^x, \; 2^{x+1/N}, \; 2^{x+2/N}, \; . \; . \; .$$

where $x = \log_2(fstart)$.

AC Outputs

AC voltages and currents whose values are desired from an .AC analysis are specified the same way as dc voltages and currents, using V*(N1, N2)*, V*(N1)*, or I*(Vname)* in a .PRINT or .PLOT statement. The values of the magnitudes of these quantities are printed or plotted. In addition, we can request certain other values, as indicated in the following list of voltage characteristics:

VR   real part
VI   imaginary part
VM   magnitude, volts
VP   phase, degrees
VDB $20 \log_{10} |V|$

The same values for ac currents can be obtained by substituting I for V. To illustrate, the statement

```
.PRINT AC V(1) VLP(2,3) II(VDUM) IR(VX)
```

causes SPICE to print the magnitude of the ac voltage between nodes 1 and 0, the phase angle of the ac voltage VL between nodes 2 and 3, the imaginary part of the current in VDUM, and the real part of the current in VX. Note that AC must be listed as the analysis type.

AC Plots

When ac voltages or currents are specified in a .PLOT statement, we obtain a linear, semilog, or log-log plot, depending on the type of voltage or current output requested and the *vartype*. Table A.3 summarizes the types of plots produced for each combination. "Log" in the table means that the scale supplied by SPICE has logarithmically spaced values. Example 7–18(b) illustrates an .AC analysis and an .AC plot.

Small-Signal Analysis and Distortion

When a circuit contains active devices such as transistors, an .AC analysis by SPICE is assumed to be a small-signal analysis. That is, ac variations are assumed to be small enough that the values of device parameters do not change. As in a .TRAN analysis,

Table A.3

Vartype	Output requested	Frequency (vertical) axis	Output (horizontal) axis
LIN	Magnitude	Linear	Log
	Phase	Linear	Linear
	Imaginary part	Linear	Linear
	Real part	Linear	Linear
	dB	Linear	Linear
DEC or OCT	Magnitude	Log	Log
	Phase	Log	Linear
	Imaginary part	Log	Linear
	Real part	Log	Linear
	dB	Log	Linear

SPICE performs an initial, dc analysis to determine quiescent voltages and currents so that the values of those device parameters that are affected by dc levels can be computed. In the ac analysis, the device parameters are assumed to retain those initial values, regardless of the actual magnitudes of the ac variations. *Thus, in an .AC analysis, SPICE does not take into account any distortion, even clipping, that would actually occur if we were to severely overdrive a transistor by specifying very large ac inputs.* For example, if the output swing of an actual transistor circuit were limited to 10 V, this fact would not be "known" to SPICE, and by overdriving the computer-simulated circuit, we could obtain outputs of several hundred volts from an .AC analysis.

A–11   MOS FIELD-EFFECT TRANSISTORS (MOSFETS)

An N-channel or P-channel MOSFET must be identified in a SPICE data file by a name beginning with the letter M, using the format:

M******* *ND NG NS NSS Mname*

where *ND, NG, NS,* and *NSS* are the node numbers of the drain, gate, source, and substrate, respectively, and *Mname* is the model name. The values of certain geometric parameters, such as the length and width of the channel, can be optionally specified with the MOSFET identification, but these do not concern us and we can allow all of them to default.

The MOSFET model is very complex. As with other semiconductor models in SPICE, it involves many parameters whose values are difficult to determine and many that are beyond the scope of our treatment. There are actually three built-in models, referred to as *levels* 1, 2, and 3. LEVEL is one of the parameters that can be specified in a MOSFET .MODEL statement, and its value prescribes the particular model to be used. The default level is 1, which we can assume for all purposes in this book. The format of the MOSFET .MODEL statement is

.MODEL *Mname type* <*Pval1=n1 Pval2=n2 . . .*>

where *type* is NMOS for an N-channel MOSFET and PMOS for a P-channel MOSFET. A MOSFET can be of either the depletion-mode type or of the enhancement-mode type, as discussed shortly.

Since the MOSFET parameters in the three-level SPICE model are so numerous and complex, we do not present a table showing their identifications, units, and default values. (Such information should be available in a user's guide furnished with the SPICE software used.) In any case, there is a significant variation among the several versions of SPICE in the number and type of MOSFET parameters that can be specified. However, there are two fundamental parameters whose values should probably be specified in every MOSFET simulation: β and V_T. In SPICE, β is called the *intrinsic transconductance parameter* and is identified in a .MODEL statement by KP. Its value is always positive. V_T is called the zero-bias threshold voltage and is identified by VTO. Following is an example of a .MODEL statement for an N-channel MOSFET having $\beta = 0.5 \times 10^{-3}$ A/V^2 and $V_T = 2$ V:

.MODEL M1 NMOS KP = 0.5E–3 VTO = 2

The threshold voltage, VTO, is positive or negative, according to the following table:

Mode	Channel	Sign of VTO
Depletion	N	−
Depletion	P	+
Enhancement	N	+
Enhancement	P	−

For example, the value specified for the VTO of an N-channel, enhancement-mode MOSFET should be positive.

Example 9–11 illustrates a SPICE simulation of a circuit containing MOSFETs.

A–12   CONTROLLED (DEPENDENT) SOURCES

A controlled voltage source is one whose output voltage is controlled by (depends on) the value of a voltage or current elsewhere in the circuit. The simplest and most familiar example is a voltage amplifier. It is a voltage-controlled voltage source because its output voltage *depends* on its input voltage: The output voltage equals the input voltage multiplied by the gain. A current-controlled voltage source obeys the relation $v_o = ki$, where i is the controlling current and k is a constant having the units of resistance: $k = v_o/i$ volts per ampere, or ohms. In the context of a current-controlled voltage source, k is called a *transresistance*.

Similarly, a controlled current source produces a current whose value depends on a voltage or a current elsewhere in the circuit. A voltage-controlled current source obeys the relation $i_o = kv$, where v is the controlling voltage and k has the units of conductance: $k = i_o/V$ amperes per volt, or siemens. In the context of a controlled source, k is called a *transconductance*.

The four types of controlled sources, voltage-controlled voltage sources, current-controlled voltage sources, voltage-controlled current sources, and current-controlled current sources, can be modeled in SPICE. Figure A.11 shows the format used to model each type. Note that the controlling voltage in voltage-controlled sources [the voltage between NC+ and NC− in *(a)* and *(b)* of the figure] can be the voltage between any two nodes; it is not necessary that a component be connected between those nodes. Also note that the controlling current in controlled current sources [parts *(c)* and *(d)* of the figure] is

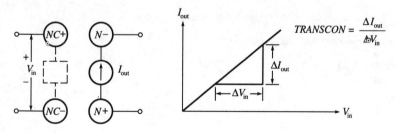

$$G^{*******} N+ \; N- \; NC+ \; NC- \; TRANSCON$$

(a) Voltage-controlled current source.

Figure A.11
Specification of controlled sources in SPICE. (Continued on p. 748.)

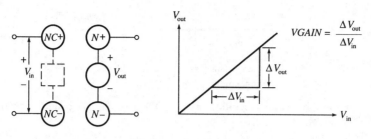

$$E^{*******}\ N+\ N-\ NC+\ NC-\ VGAIN$$

(b) Voltage-controlled voltage source.

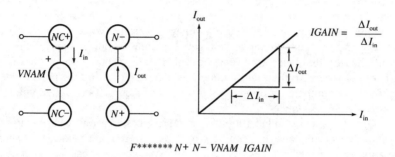

$$F^{*******}\ N+\ N-\ VNAM\ IGAIN$$

(c) Current-controlled current source.

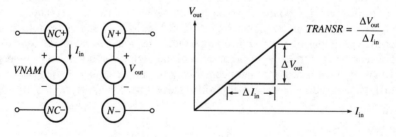

$$H^{*******}\ N+\ N-\ VNAM\ TRANSR$$

(d) Current-controlled voltage source.

Figure A.11
(Continued)

always the current in a voltage source. Thus, it may be necessary to insert a dummy voltage source in a circuit in order to specify a controlling current at a desired point in the circuit. Observe how the "plus" and "minus" nodes are defined in the figure in connection with the polarity assumptions made by SPICE for the currents and current sources. Example 7–21 illustrates the use of a voltage-controlled voltage source in a SPICE program.

A–13   SUBCIRCUITS

A complex electronic circuit often contains several components, or subsections of circuitry, that are identical to each other. An example is a logic circuit containing many identical logic gates. When modeling such circuits in SPICE, it is a tedious and time-consuming task to write numerous sets of identical statements describing identical circuitry. Furthermore, the input data file for such a system may become so long and cumbersome that it is difficult to interpret or modify. To alleviate those kinds of problems, SPICE allows a user to create a *subcircuit:* circuitry that can be defined just once and then, effectively, inserted into a larger system (which we call the *main* circuit) at as many places as desired. The concept is similar to that of a *subroutine* in conventional computer programming. In SPICE, it is possible to define several different subcircuits in one data file, and, in fact, one subcircuit can contain other subcircuits (called *nested* subcircuits).

The first statement of a subcircuit is

.SUBCKT *NAME N1 <N2 N3 . . .>*

where *NAME* is any name chosen to identify the subcircuit and *N1, N2, . . .* , are the numbers of the nodes in the subcircuit that will be joined to other nodes in the main circuit. None of these can be node 0. Components in a subcircuit are defined in exactly the same way they are in any SPICE data file, using successive statements following the .SUBCKT statement. A subcircuit can contain .MODEL statements, but it cannot contain any control statements, such as .DC, .TRAN, .PRINT, or .PLOT. The node numbers in a subcircuit do not have to be different from those in the main program. SPICE treats a subcircuit as a completely separate (isolated) entity, and a node having the same number in a subcircuit as another node in the main circuit is still treated as a different node. The exception is node 0: If node 0 appears in a subcircuit, it is treated as the same node 0 in the main circuit. (In the language of computer science, subcircuit nodes are said to be *local*, except node 0, which is *global*.) The last statement in a subcircuit must be

.ENDS *<NAME>*

If a subcircuit itself contains subcircuits, the *NAME* must be given in the .ENDS statement to specify which subcircuit definition has been ended.

In order to "insert" a subcircuit into the main circuit, we must write a subcircuit *call* statement in the main program. A different call statement is required for each location where the subcircuit is to be inserted. The format of a call statement is

X******* *N1 <N2 N3 . . .> NAME*

where ******* are arbitrary characters that must be different for each call; *N1, N2, . . .* , are the node numbers in the main circuit that are to be joined to the subcircuit nodes specified in the .SUBCKT statement; and *NAME* is the name of the subcircuit to be inserted. The node numbers in the call statement will be joined to the nodes in the .SUBCKT statement in exactly the same order as they both appear. That is, *N1* in the subcircuit will be joined to *N1* in the main circuit, and so forth. Following is an example:

```
.SUBCKT  AND  1  2  3
       —  ⎫
       —  ⎬  statements describing components in the subcircuit
       —  ⎭
.ENDS
```

```
X1  8  4  12  AND
X2  1  4   3  AND
```

In this example, the subcircuit named AND is inserted at two locations in the main program. The first call statement (X1) connects subcircuit nodes 1, 2, and 3 to main circuit nodes 8, 4, and 12, respectively. The X2 call statement connects subcircuit nodes 1, 2, and 3 to main circuit nodes 1, 4, and 3, respectively. (In this case some of the subcircuit and main circuit node numbers are the same.)

Example 11–3 illustrates the use of subcircuits in a SPICE simulation of a bistable latch.

Nested Subcircuits

As already mentioned, a subcircuit is said to be nested when it is wholly contained within another. A nested subcircuit can itself have subcircuits nested within it, which can have

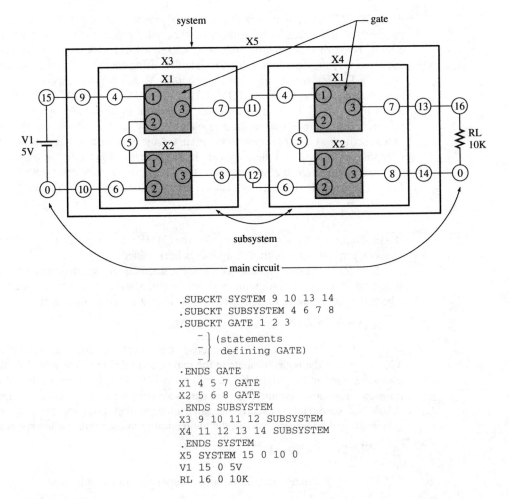

```
.SUBCKT SYSTEM 9 10 13 14
.SUBCKT SUBSYSTEM 4 6 7 8
.SUBCKT GATE 1 2 3
    -  ⎫ (statements
    -  ⎬ defining GATE)
    -  ⎭
.ENDS GATE
X1 4 5 7 GATE
X2 5 6 8 GATE
.ENDS SUBSYSTEM
X3 9 10 11 12 SUBSYSTEM
X4 11 12 13 14 SUBSYSTEM
.ENDS SYSTEM
X5 SYSTEM 15 0 10 0
V1 15 0 5V
RL 16 0 10K
```

Figure A.12
Example of nested subcircuits.

more nested subcircuits, and so forth. When subcircuits are nested, care must be taken to define appropriate nodes at each level of nesting. Figure A.12 shows an example. The innermost nested subcircuit is called GATE and has nodes 1, 2, and 3. It is identified by X1 and X2. Note that two GATEs are nested within another subcircuit, called SUBSYSTEM, which is itself nested in the subcircuit called SYSTEM. Observe that nodes must be defined within SUBSYSTEM and within SYSTEM. To reduce confusion, different sets of node numbers were used for each nesting level, although that is not required. SUBSYSTEM nodes are numbered 4, 5, 6, 7, and 8 and SYSTEM nodes are numbered 9 through 14. The figure shows how the subcircuits are defined in the input data file. Note that each .SUBCKT statement lists only those nodes that are to be connected to other nodes in a higher-level subcircuit. For example, node 5 in SUBSYSTEM does not appear in the .SUBCKT statement for SUBSYSTEM because it is not connected to any nodes within SYSTEM. Similarly, nodes 11 and 12 do not appear in the .SUBCKT statement for SYSTEM.

Appendix B

ANSWERS TO ODD-NUMBERED EXERCISES

CHAPTER 2

2.1 (a) $9)_{10}$ (b) $43)_{10}$ (c) $127)_{10}$
(d) $6.25)_{10}$ (e) $221.75)_{10}$

2.3 (a) 111101 (b) 10100010 (c) 1100110
(d) 110101.100 (e) 1101.10

2.5 (a) 1001 (b) 1000101 (c) 01011
(d) 1000.10 (e) 011011.01

2.7 (a) 11001 (b) 1110101 (c) 1001.1010
(d) 0.00111001

2.9 (a) 11 (b) 1101 (c) 110.1 (d) 1111.01

2.11 (a) $29)_{10}$ (b) $327)_{10}$ (c) $52.3125)_{10}$
(d) $576.5)_{10}$ (e) $4)_{10}$

2.13 (a) $476)_8$ (b) $145)_8$ (c) $102.4)_8$
(d) $1076)_8$

2.15 (a) $47)_8$ (b) $755)_8$ (c) $4535)_8$ (d) $4.04)_8$

2.17 (a) $602)_{10}$ (b) $9)_{10}$ (c) $956.05859375)_{10}$
(d) $4053.875)_{10}$

2.19 (a) $D)_{16}$ (b) $18)_{16}$ (c) $D3C)_{16}$
(d) $822.C)_{16}$ (e) $1480F)_{16}$

2.21

	Incremented Number	Decremented Number
(a)	10100	10010
(b)	100001	011111
(c)	1110000	1101110
(d)	0010101	0010011

2.23

	Incremented Number	Decremented Number
(a)	$1000)_8$	$776)_8$
(b)	$A01)_{16}$	$9FF)_{16}$

(c)	$2100)_{16}$	$20FE)_{16}$
(d)	$1001)_8$	$0777)_8$
(e)	$1001)_{16}$	$0FFF)_{16}$

2.25 (a) $16,777,216)_{10}$ (b) $166,777,215)_{10}$

2.27 (a) 101011 (b) 1100100 (c) 101101101
(d) 1010010.01 (e) 11110001.001 (f) 1.101

2.29 (a) $175)_8$; $7D)_{16}$ (b) $13636)_8$; $179E)_{16}$
(c) $31.4)_8$; $19.8)_{16}$ (d) $1245.44)_8$; $2A5.9)_{16}$

2.31 (a) 110101011 (b) 010000001011.0110
(c) 1010000110010011.1110
(d) 000.101110111

2.33 (a)
$$\begin{array}{r} 38 \\ +\ 82 \\ \hline 120 \\ \hookrightarrow 1 \\ \hline 21 \end{array}$$
(b)
$$\begin{array}{r} 642 \\ +\ 924 \\ \hline 1566 \\ \hookrightarrow 1 \\ \hline 567 \end{array}$$

(c)
$$\begin{array}{r} 1056.0 \\ +\ 9614.7 \\ \hline 10670.7 \\ \longrightarrow 1 \\ \hline 0670.8 \end{array}$$
(d)
$$\begin{array}{r} 812.45 \\ +\ 902.89 \\ \hline 1715.34 \\ \longrightarrow 1 \\ \hline 715.35 \end{array}$$

2.35 (a)
$$\begin{array}{r} 65 \\ +\ 68 \\ \hline \cancel{\times}33 \end{array}$$
(b)
$$\begin{array}{r} 574 \\ +\ 949 \\ \hline \cancel{\times}523 \end{array}$$

(c)
$$\begin{array}{r} 9085.0 \\ +\ 9362.6 \\ \hline \cancel{\times}8447.6 \end{array}$$
(d)
$$\begin{array}{r} 581.075 \\ +\ 904.800 \\ \hline \cancel{\times}485.875 \end{array}$$

2.37 (a)
```
  10110
+ 01100
 100010
   └──→1
  00011
```
(b)
```
   1001100
 + 1100110
  10110010
    └───→1
   0110011
```

(c)
```
  11100.101
+ 11010.101
 110111.010
    └───→1
  10111.011
```
(d)
```
  100000.00
+ 111111.00
 1011111.00
    └───→1
  011111.01
```

2.39 (a)
```
  110011
+ 010100
⊗000111
```
(b)
```
  1001001
+ 1100111
⊗0110000
```

(c)
```
  10111101.1
+ 11111010.0
⊗10110111.1
```
(d)
```
  111111.00
+ 111111.11
⊗111110.11
```

2.41 (a) $9DC)_{16}$ (b) $103.1)_8$ (c) $CE8.B)_{16}$
(d) $352.C)_{16}$

2.43

	1's Complement	2's Complement
(a)	11000111	11001000
(b)	01111100	01111100
(c)	10111011	10111100
(d)	00001100	00001100
(e)	10011011	10011100

2.45

	i	ii	iii
(a)	93	93	93
(b)	223	−32	−33
(c)	228	−27	−28
(d)	142	−113	−114
(e)	26	26	26

2.47

	1's Complement	2's Complement
(a)	00111011	00111011
(b)	01101001	01101001
(c)	10111100	10111101
(d)	10010011	10010100
(e)	11111111	00000000

2.49 684

2.51 110101

2.53 $11342)_5$

2.55 2974

2.57 GO FOR IT

2.59 Overflow has occurred if both numbers have the same sign bit and the sign bit of the sum is different from that of the numbers.

CHAPTER 3

3.1

A	B	C	D	ABCD
0	0	0	0	0
0	0	0	1	0
0	0	1	0	0
0	0	1	1	0
0	1	0	0	0
0	1	0	1	0
0	1	1	0	0
0	1	1	1	0
1	0	0	0	0
1	0	0	1	0
1	0	1	0	0
1	0	1	1	0
1	1	0	0	0
1	1	0	1	0
1	1	1	0	0
1	1	1	1	1

3.3 64

3.5

$OFF = T_1 T_2$

3.7

$\overline{X_1 X_2 Y_1 Y_2}$

X_1	X_2	Y_1	Y_2	$\overline{X_1 X_2 Y_1 Y_2}$
0	0	0	0	1
0	0	0	1	1
0	0	1	0	1
0	0	1	1	1
0	1	0	0	1
0	1	0	1	1
0	1	1	0	1
0	1	1	1	1
1	0	0	0	1
1	0	0	1	1
1	0	1	0	1
1	0	1	1	1
1	1	0	0	1
1	1	0	1	1
1	1	1	0	1
1	1	1	1	0

3.9 (a)

A	B	C	OUT = $(A + B)C$
0	0	0	0
0	0	1	0
0	1	0	0
0	1	1	1
1	0	0	0
1	0	1	1
1	1	0	0
1	1	1	1

(b)

X	Y	OUT = $\overline{\overline{XY}}$
0	0	0
0	1	1
1	0	1
1	1	1

(c)

U	V	W	OUT = $\overline{\overline{U}V} + W$
0	0	0	1
0	0	1	0
0	1	0	0
0	1	1	0
1	0	0	1
1	0	1	0
1	1	0	1

(d)

W	X	Y	Z	OUT = $(W + X)(\overline{\overline{Y} + Z})$
0	0	0	0	0
0	0	0	1	0
0	0	1	0	0
0	0	1	1	0
0	1	0	0	0
0	1	0	1	0
0	1	1	0	1
0	1	1	1	0
1	0	0	0	0
1	0	0	1	0
1	0	1	0	1
1	0	1	1	0
1	1	0	0	0
1	1	0	1	0
1	1	1	0	1
1	1	1	1	0

3.11

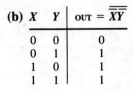

3.13 (a)

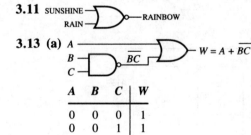

$W = A + \overline{BC}$

A	B	C	W
0	0	0	1
0	0	1	1
0	1	0	1
0	1	1	0
1	0	0	1
1	0	1	1
1	1	0	1
1	1	1	1

(b)

X_1 ▷o— $\overline{X_1}$

$\overline{X_1}X_2$

X_2

X_1

X_3

$\overline{X_1 + X_3}$

$Z = \overline{X_1}X_2(\overline{X_1 + X_3})$

X_1	X_2	X_3	Z
0	0	0	0
0	0	1	0
0	1	0	1
0	1	1	0
1	0	0	0
1	0	1	0
1	1	0	0
1	1	1	0

(c)

A

B ▷o— $\overline{B}$

$\overline{A + \overline{B}}$

X ▷o— $\overline{X}$

Y

$\overline{\overline{X}Y}$

OUT $= \overline{A + \overline{B}} + \overline{\overline{X}Y}$

A	B	X	Y	OUT
0	0	0	0	1
0	0	0	1	0
0	0	1	0	1
0	0	1	1	1
0	1	0	0	1
0	1	0	1	1
0	1	1	0	1
0	1	1	1	1
1	0	0	0	1
1	0	0	1	0
1	0	1	0	1
1	0	1	1	1
1	1	0	0	1
1	1	0	1	0
1	1	1	0	1
1	1	1	1	1

(d)

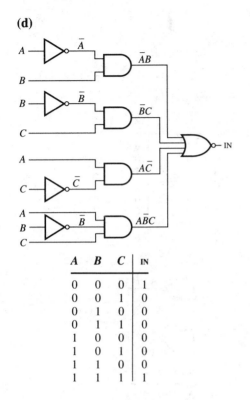

A	B	C	IN
0	0	0	1
0	0	1	0
0	1	0	0
0	1	1	0
1	0	0	0
1	0	1	0
1	1	0	0
1	1	1	1

3.15

A	B	$\bar{A}$	$\bar{B}$	A + B	$\bar{A} + \bar{B}$	$(A + B)(\bar{A} + \bar{B})$
0	0	1	1	0	1	0
0	1	1	0	1	1	1
1	0	0	1	1	1	1
1	1	0	0	1	0	0

A	B	$A \oplus B$
0	0	0
0	1	1
1	0	1
1	1	0

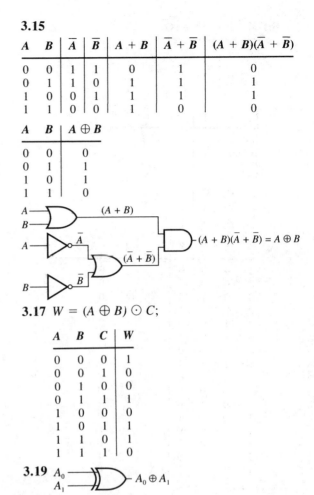

3.17 $W = (A \oplus B) \odot C$;

A	B	C	W
0	0	0	1
0	0	1	0
0	1	0	0
0	1	1	1
1	0	0	0
1	0	1	1
1	1	0	1
1	1	1	0

3.19 $A_0 \oplus A_1$

3.21

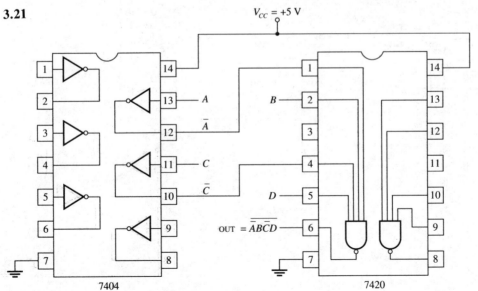

3.23

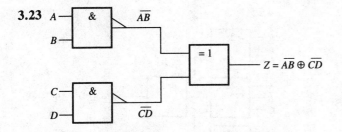

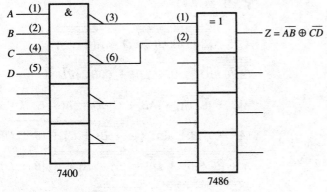

7400

7486

3.25 (cont'd chart 2 at right)

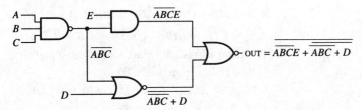

OUT $= \overline{\overline{\overline{ABCE}} + \overline{\overline{ABC}} + D}$

3.25 (*Continued*)

A	B	C	D	E	OUT
0	0	0	0	0	1
0	0	0	0	1	0
0	0	0	1	0	1
0	0	0	1	1	0
0	0	1	0	0	1
0	0	1	0	1	0
0	0	1	1	0	1
0	0	1	1	1	0
0	1	0	0	0	1
0	1	0	0	1	0
0	1	0	1	0	1
0	1	0	1	1	0
0	1	1	0	0	1
0	1	1	0	1	0
0	1	1	1	0	1
0	1	1	1	1	0
1	0	0	0	0	1
1	0	0	0	1	0
1	0	0	1	0	1
1	0	0	1	1	0
1	0	1	0	0	1
1	0	1	0	1	0
1	0	1	1	0	1
1	0	1	1	1	0
1	1	0	0	0	1
1	1	0	0	1	0
1	1	0	1	0	1
1	1	0	1	1	0
1	1	1	0	0	0
1	1	1	0	1	0
1	1	1	1	0	1
1	1	1	1	1	1

3.27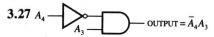

3.29

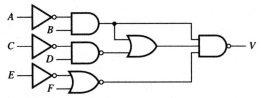

3.31 (a) $A \odot B$ (b) $A \oplus B$ (c) $A \oplus B$
(d) $A \oplus B$ (e) $A \odot B$ (f) $A \odot B$
3.33 The light is never off.

CHAPTER 4

4.1 (a) A (b) $A\overline{B}C$ (c) 0 (d) $T_1\overline{T_2}$
(e) $A\overline{B} + A\overline{C}$
4.3 $A_2\overline{A_1}\overline{A_0} + A_2\overline{A_1}A_0 + A_2A_1\overline{A_0} + A_2A_1A_0 = A_2$
4.5 (a) A (b) A (c) $X + Y + Z$ (d) $\overline{Y} + \overline{X}\overline{Z}$
(e) $X_2 + \overline{X}_1\overline{X}_3$ (f) $\overline{A}_2 + \overline{A}_3$
4.7 $\overline{A \oplus B} = \overline{\overline{A}B + A\overline{B}} = (\overline{\overline{A}B})(\overline{A\overline{B}}) =$
$(\overline{\overline{A}} + \overline{B})(\overline{A} + \overline{\overline{B}}) = (A + \overline{B})(\overline{A} + B) = \overline{A}A +$
$\overline{A}B + AB + B\overline{B} = 0 + \overline{A}B + AB + 0 =$
$\overline{A}B + AB = A \odot B$
$\overline{A \odot B} = \overline{\overline{A}\overline{B} + AB} = (\overline{\overline{A}\overline{B}})(\overline{AB}) = (\overline{\overline{A}} + \overline{\overline{B}}) \cdot$
$(\overline{A} + \overline{B}) = (A + B)(\overline{A} + \overline{B}) = A\overline{A} + A\overline{B} +$
$\overline{A}B + B\overline{B} = 0 + A\overline{B} + \overline{A}B + 0 = A\overline{B} +$
$\overline{A}B = A \oplus B$
4.9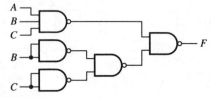

4.11 (a) $A\overline{B} + AB$; $(A + B)(A + \overline{B})$
(b) $\overline{X}_1\overline{X}_2\overline{X}_3 + \overline{X}_1X_2X_3 + X_1\overline{X}_2\overline{X}_3 + X_1\overline{X}_2X_3 +$
$X_1X_2X_3$; $(X_1 + X_2 + \overline{X}_3)(X_1 + \overline{X}_2 + X_3) \cdot$
$(\overline{X}_1 + \overline{X}_2 + X_3)$
4.13 (a) $F = A + B$ (b) $G = \overline{X}\overline{Y} + \overline{Z}$
(c) $H = ABD + \overline{A}\overline{B}D + \overline{B}C$
4.15 (a) $F = A\overline{B} + \overline{A}B + \overline{A}C$; Alternate
solution: $F = A\overline{B} + \overline{A}B + \overline{B}C$ (b) $G = \overline{B}C +$
$C\overline{D} + \overline{A}\overline{B}D$ (c) $T = W + X + Y + \overline{Z}$
(d) $W = B + A\overline{C}$ (e) $B = 1$
4.17 $F = CD + \overline{A}\overline{B}DE + AB\overline{D}E$
4.19 ON $= \overline{M} + T\overline{H} + ST$

4.23

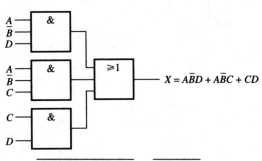
$X = A\overline{B}D + A\overline{B}C + CD$

4.25 $\overline{A[B\overline{(CD)}](\overline{ABC})B\overline{D}} = \overline{A[B\overline{(CD)}]} +$
$\overline{(\overline{ABC})B\overline{D}} = \overline{A[B\overline{(CD)}]} + \overline{(\overline{A}\overline{B}C)B\overline{D}} =$
$A[\overline{B} + \overline{\overline{(CD)}}] + (\overline{A\overline{B}} + \overline{C})B\overline{D} = A(\overline{B} + \overline{C}D) +$
$(A\overline{B} + C)B\overline{D}$; also, $\overline{[(\overline{A} + B) + C] + (\overline{B} + D)} +$
$\overline{A} + [\overline{B} + \overline{(C + D)}] = \overline{[(\overline{A} + B) + C]} \overline{(\overline{B} + D)} +$
$\overline{A[\overline{B} + \overline{(C + D)}]} = \overline{[(\overline{A} + B) + C]} (\overline{\overline{B}D}) +$
$A[\overline{B} + \overline{(C + D)}] = (\overline{A}B + C)(B\overline{D}) +$
$A(\overline{B} + \overline{C}\overline{\overline{D}}) = (A\overline{B} + C)B\overline{D} + A(\overline{B} + \overline{C}D)$;
The expressions are equal, since both equal
$(A\overline{B} + C)B\overline{D} + A(\overline{B} + \overline{C}D)$.
4.27 $W = ABCD + ABC\overline{D} + AB\overline{C}D + \overline{A}BCD$
4.29 $F = \overline{A}B + BC\overline{D} + \overline{B}CD$
4.31 Sum $= \overline{A}\overline{B}C + \overline{A}B\overline{C} + A\overline{B}\overline{C} + ABC$ (no
simplification possible)

CHAPTER 5

5.1 NAND gates:

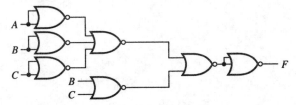

NOR gates:

5.3 $F = AB + BC = B(A + C)$

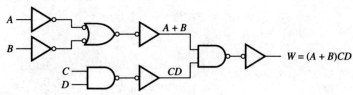

5.5

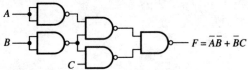

5.7

$F = \overline{\overline{AC} + \overline{AB}}$ 1/2 7451

5.9 $F = B$

5.11 $A_2 + A_4$

5.13 $F = A_8A_4 + A_8A_2 + A_4A_2$

5.15 $X = (A + B + \overline{C} + \overline{D})(E + \overline{F}\overline{G})$

5.17

$F = \overline{A}\overline{B} + \overline{B}C$

5.19

$A \oplus 1 = \overline{A}$ $A \odot 0 = \overline{A}$

5.21

$\overline{(X_1 + X_2)(\overline{X}_1 + \overline{X}_2)}$

CHAPTER 6

6.1 $v_o(\text{low}) = 0.102$ V; $v_o(\text{high}) = 4.992$ V

6.3 $v_o(\text{low}) = 0.235$ V; $v_o(\text{high}) = 3.89$ V

6.5 (a) $v_C(t) = 5(1 - e^{-t/7.5 \times 10^{-8}})$ V
(b) 120.7 ns **(c)** $v_C(t) = 5e^{-t/7.5 \times 10^{-8}}$ V
(d) 68.7 ns

6.7 19.03 ns

6.9 (a)

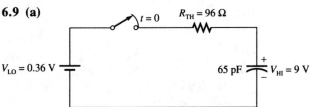

(b) $v_o(t) = 0.36 + 8.64e^{-t/6.24\times 10^{-9}}$ V
(c) 3.88 ns

6.11 (a) 0.8 ms **(b)** 0.4 ms **(c)** 8 V
(d) 9.75 ms

6.13 425 ns

6.15 $t_{pHL} = 4.16$ ns; $t_{pLH} = 6.24$ ns

6.17 (a) 40% **(b)** 6.896%

6.19 (a) 24% **(b)** 8% **(c)** 10 kHz

6.21 No. The low level of a pulse in A is 0.5 V greater than the maximum value interpreted as a low in B.

6.23 625×10^3 PPS

6.25 0.8 μs

6.27 3 V

6.29 16.67%

6.31

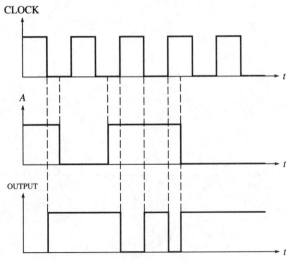

CLOCK

A

OUTPUT

6.33 Charge: $v_C(t) = 12(1 - e^{-t/2\times 10^{-8}})$ V; discharge: $v_C(t) = 7.585e^{-t/2\times 10^{-8}}$ V

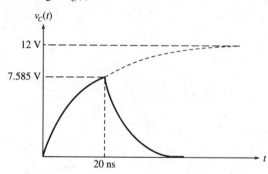

$v_C(t)$

12 V

7.585 V

20 ns

6.35 Let t_1 be the time at which $v_o(t) = 0.9E$ and t_2 be the time at which $v_o(t) = 0.1E$.

$$0.9E = Ee^{-t_1/\tau} \rightarrow 0.9 = e^{-t_1/\tau}$$

$$\ln(0.9) = -\frac{t_1}{\tau} \quad \rightarrow \quad t_1 = -\tau \ln(0.9)$$

$$0.1E = Ee^{-t_2/\tau} \rightarrow 0.1 = e^{-t_2/\tau}$$

$$\ln(0.1) = -\frac{t_2}{\tau} \quad \rightarrow \quad t_2 = -\tau \ln(0.1)$$

$$t_f = t_2 - t_1 = -\tau \ln(0.1) - [-\tau \ln(0.9)]$$
$$= \tau[\ln(0.9) - \ln(0.1)] = \tau \ln\left(\frac{0.9}{0.1}\right)$$
$$= \tau \ln 9 = 2.197\tau$$

6.37 Let T_H = time waveform is high during one cycle and T_L = time waveform is low during one cycle. Then, duty cycle =
$$d = \frac{T_H}{T_H + T_L} = \frac{T_H/T_L}{T_H/T_L + 1} \rightarrow d =$$
$(M/S)/(M/S + 1)$. Solving this equation for M/S gives $M/S = d/(1 - d)$.

6.39 $t_r = t_f = 2.2$ μs

6.41 37.3%

6.43 20%

CHAPTER 7

7.1 (a) 2.5 V **(b)** fundamental: 3.183 V, 4 kHz; f_2: 0 V, 8 kHz; f_3: 1.061 V, 12 kHz

7.3 20.2 MHz

7.5 $V_{avg} = 8$ V; $A_1 = 3.742$ V; $A_2 = 3.027$ V; $A_3 = 2.018$ V

7.7 $v(t) = 0.05 + 0.1 \sin(2\pi \times 10^6 t + 88.2°) + 0.2 \sin(2\pi \times 2 \times 10^6 t + 86.4°) + 0.3 \sin(2\pi \times 3 \times 10^6 t + 84.6°) + \cdots$ V

7.9 4 kHz, 7.07 V

7.11 (a) $19.9 \sin(2\pi \times 100t - 5.71°)$ V
(b) $14.14 \sin(2\pi \times 10^3 t - 45°)$ V
(c) $1.99 \sin(2\pi \times 10^4 t - 84.29°)$ V

7.13 0.04 μF

7.15 (a) 7th harmonic **(b)** 100 Ω

7.17 0.85%

7.19 $V_{avg} = 7.5$ V; $V_{pp} = 0.125$ V

7.21 (a) 15.49 μs **(b)** 0.743 μs

7.23 (a) 3.56 V; 72.72° **(b)** 1.66 V; 63.95°
(c) 24.88 mV; −59.29°

7.25 **(a)** 4% **(b)** 2.55 V **(c)** 200 kHz

7.27 **(a)**

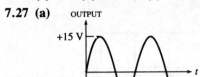

(b)

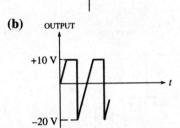

(c)

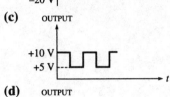

(d)

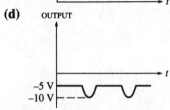

7.29

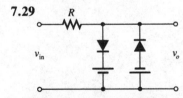

7.31 Each of the following is one of many possible solutions.

(a)

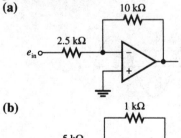

(b)

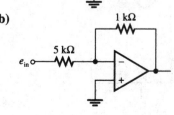

(c)

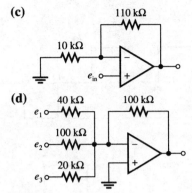

(d)

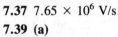

7.33 **(a)** 0.2 **(b)** 0.833 **(c)** 0.0833 **(d)** 0.1052

7.35 4348 Ω

7.37 7.65×10^6 V/s

7.39 **(a)**

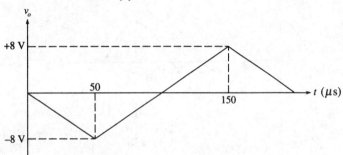

(b) 833 Ω **(c)** 1.6×10^5 V/s

7.41 6.11 V

7.43 (a)

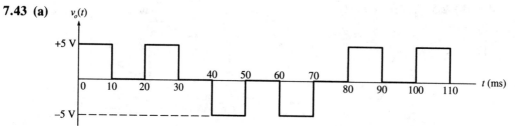

(b)

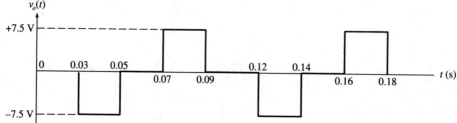

(c)

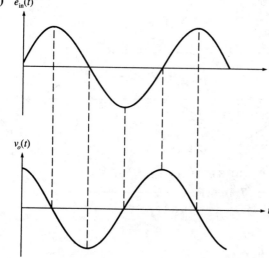

7.45 (a)

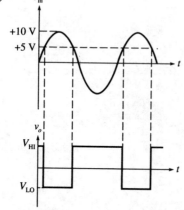

(b)

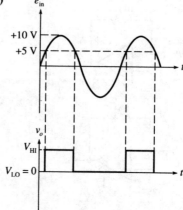

(c)

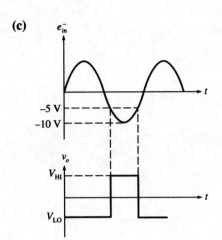

(d)

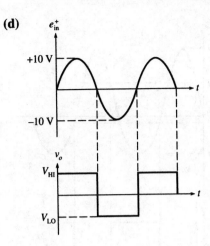

7.47 (a)

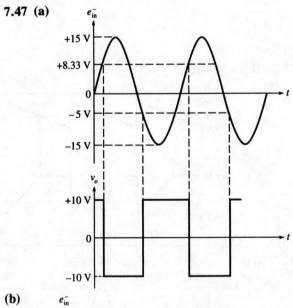

hysteresis = 13.33 V

(b)

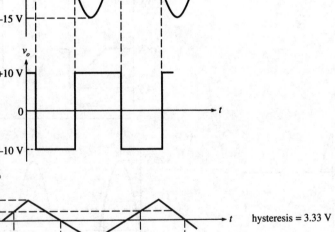

hysteresis = 3.33 V

(c)

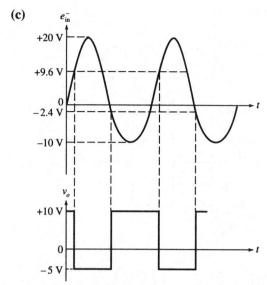

hysteresis = 12 V

7.49 One possible solution:

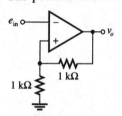

7.51 (a) *(Continued to right.)*

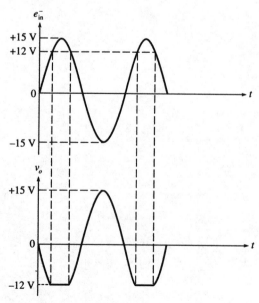

7.51 (b)

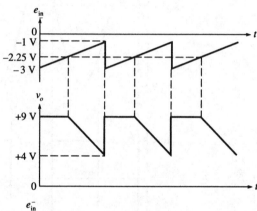

(c)

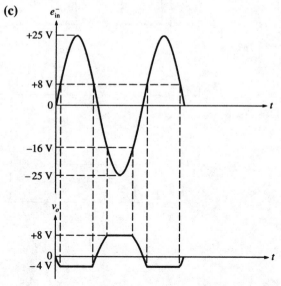

7.53 (a)

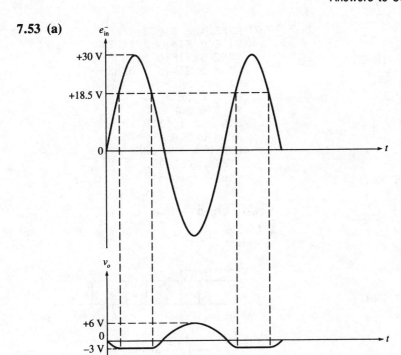

(b)

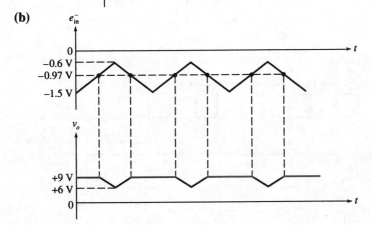

7.55 $1 + 2 \sin(2\pi \times 400t) + 0.1 \sin(2\pi \times 5 \times 10^3 t)$ V

7.57 663.3 ns

7.59 (a) 100 **(b)** 0.25

7.61 EXERCISE 7.61
```
VR 1 0 PULSE(-1 5 0 0 0 1MS 2.5MS)
R 1 0 1K
.TRAN 5US 2.5MS
.OPTIONS LIMPTS = 501
.PRINT TRAN V(1)
.FOUR 400 HZ V(1)
.END
```

7.61 *(Continued)*

	Theoretical	**SPICE**	**Percent error**
V_{avg}	1.4 V	1.412 V	0.857
A_1	3.633 V	3.640 V	0.193
A_2	1.1226 V	1.103 V	−1.75
A_3	0.7482 V	0.7676 V	2.59
A_4	0.9084 V	0.9005 V	−0.870
A_5	0 V	0.02398 V	NA
A_6	0.6054 V	0.6124 V	1.16
A_7	0.321 V	0.3010 V	−6.23
A_8	0.2808 V	0.2996 V	6.70
A_9	0.4038 V	0.3955 V	−2.06

7.63 641.6 Hz; $-65.38°$

7.65 (a) EXERCISE 7.65A
```
VIN 1 0 SIN(0 15 1KHZ)
R 1 2 10
D1 2 3 DIODE
V 3 0 6V
.MODEL DIODE D
.TRAN 25US 1MS
.PLOT TRAN V(2)
.END
```
(b) EXERCISE 7.65B
```
VIN 1 0 SIN(0 15 1KHZ)
R 1 2 10
D1 2 3 DIODE
V 0 3 6V
.MODEL DIODE D
.TRAN 25US 1MS
.PLOT TRAN V(2)
.END
```
(c) EXERCISE 7.65C
```
VIN 1 0 SIN(0 15 1KHZ)
R 1 2 10
D1 3 2 DIODE
V 3 0 6V
.MODEL DIODE D
.TRAN 25US 1MS
.PLOT TRAN V(2)
.END
```

7.67 EXERCISE 7.67
```
V1 1 0 SIN(0 1 1KHZ)
V2 2 0 SIN(0 1 3KHZ)
V3 3 0 SIN(0 1 5KHZ)
R1 1 4 2K
R3 2 4 6K
R5 3 4 10K
RIN 4 0 100MEG
RF 4 5 30K
EOP 0 5 4 0 1MEG
.TRAN 20US 1MS
.PLOT TRAN V(5)
.END
```

CHAPTER 8

8.1

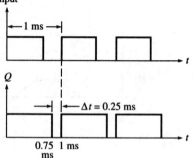

8.3

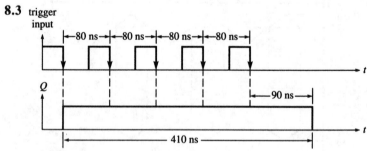

8.5

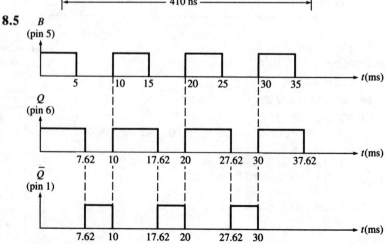

8.7

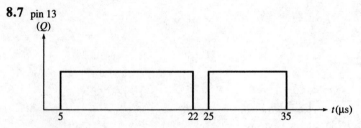

8.9 23,637 Ω

8.11 One possible solution:

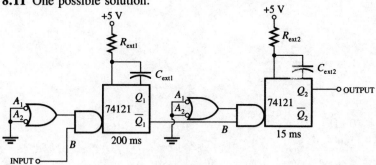

$$C_{ext1} = C_{ext2} = 10 \ \mu F$$
$$R_{ext1} = 28.85 \ k\Omega \quad R_{ext2} = 2164 \ \Omega$$

8.13

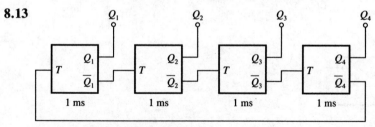

Each monostable (one solution):

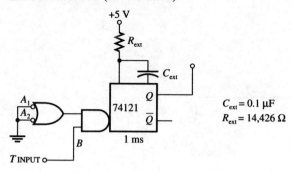

$$C_{ext} = 0.1 \ \mu F$$
$$R_{ext} = 14,426 \ \Omega$$

8.15 One possible solution:

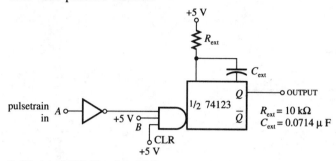

8.17 **(a)** 3607 Hz **(b)** 5 kΩ

8.19 0.02 Hz − 4.1 MHz

8.21 5 MHz

8.23 One possible solution: $C = 10$ μF; $R = 91$ kΩ (see Figure 8.17).

8.25 One possible solution: $R_1 = 10$ kΩ; $R_2 = 3.3$ kΩ; $C = 0.017$ μF (see Figure 8.19).

8.27 One possible solution: $R_2 = 1$ kΩ; $C = 0.1$ μF; $R_1(\text{min}) = 5213$ Ω; $R_1(\text{max}) = 142.27$ kΩ (see Figure 8.19). $d_{\text{min}} = 86.14\%$; $d_{\text{max}} = 99.3\%$

8.29 **(a)** 200 PPS **(b)** 80%

8.31

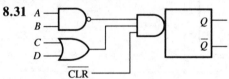

8.33

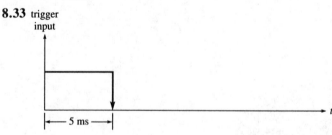

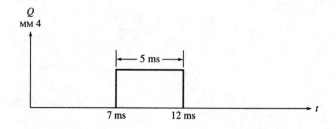

8.35 75%

8.37 −0.52 V

CHAPTER 9

9.1 (a) Nonconducting **(b)** Conducting
(c) Nonconducting **(d)** D_1 nonconducting;
D_2 conducting **(e)** D_1 conducting;
D_2 nonconducting

9.3 (a) Off **(b)** Off **(c)** On **(d)** On **(e)** Off
(f) On

9.5 (a) On **(b)** Off **(c)** On **(d)** On **(e)** Off
(f) On

9.7 (a)

A	B	OUTPUT
−1 V	−1 V	−1 V
−1 V	10 V	−1 V
10 V	−1 V	−1 V
10 V	10 V	10 V

A	B	OUTPUT = A AND B
0	0	0
0	1	0
1	0	0
1	1	1

(b)

A	B	C	OUTPUT
−5 V	−5 V	−5 V	−5 V
−5 V	−5 V	0 V	−5 V
−5 V	0 V	−5 V	−5 V
−5 V	0 V	0 V	−5 V
0 V	−5 V	−5 V	−5 V
0 V	−5 V	0 V	−5 V
0 V	0 V	−5 V	−5 V
0 V	0 V	0 V	0 V

A	B	C	OUTPUT = A AND B AND C
0	0	0	0
0	0	1	0
0	1	0	0
0	1	1	0
1	0	0	0
1	0	1	0
1	1	0	0
1	1	1	1

9.9 (a)

A	B	C	OUTPUT
−5 V	−5 V	−5 V	−5.7 V
−5 V	−5 V	0 V	−0.7 V
−5 V	0 V	−5 V	−0.7 V
−5 V	0 V	0 V	−0.7 V
0 V	−5 V	−5 V	−0.7 V
0 V	−5 V	0 V	−0.7 V
0 V	0 V	−5 V	−0.7 V
0 V	0 V	0 V	−0.7 V

(b)

A	B	OUTPUT
5 V	5 V	4.3 V
5 V	10 V	9.3 V
10 V	5 V	9.3 V
10 V	10 V	9.3 V

9.11 1.5 kΩ

9.13 1 kΩ

9.15 (a) $R_{DC}(min) = 10.75$ kΩ; $R_{DC}(max) = 23$ kΩ **(b)** $r_{ac}(min) = 5$ kΩ; $r_{ac}(max) = 10$ kΩ

9.17 (a) 0.851 V **(b)** 12.5 V

9.19 (a) $V_{GS1} = 0$ V, $V_{GS2} = -10$ V
(b) $V_{GS1} = 10$ V, $V_{GS2} = 0$ V

9.21 (i) (a) NAND **(b)** OR
(ii) (a) NOR **(b)** AND

9.23 $N_L = 0.75$ V; $N_H = 3.75$ V

9.25 $R_B = 143.3$ kΩ; $R_C = 2467$ Ω

9.27 6.185 V

9.29 $V_{HI} = 5.0$ V; $V_{LO} = 0.1622$ V

9.31 $t_r = 599$ ns; $t_f = 135$ ns

CHAPTER 10

10.1 (a) 3.94 V **(b)** 0.1 V

10.3 (a) 6 mA **(b)** 200 μA

10.5 (a) −0.5 mA **(b)** 20 μA **(c)** 20 mA
(d) 3 ns

10.7 (a) 40 **(b)** $N_L = 0.3$ V; $N_H = 0.5$ V
(c) 32.45 mW

10.9 (a) $ABCD$ **(b)** $(\overline{A} + \overline{B})\overline{C}$ **(c)** $\overline{A}\overline{B}\overline{D}$
(d) $A\overline{B}$

10.11 One possible solution:

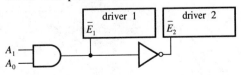

10.13

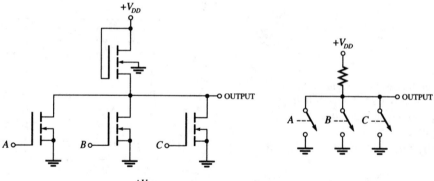

10.15

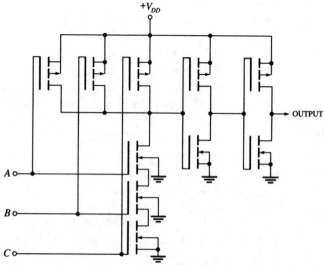

10.17 Coincidence: $A \odot B = AB + \overline{A}\overline{B}$

10.19 $\overline{A}\overline{B}$

10.21

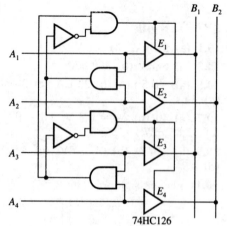

74HC126

10.23 (a) $N_L = 2.95$ V; $N_H = 2.95$ V
(b) $N_L = 1.95$ V; $N_H = 1.95$ V

10.25 (a) 32.5 ns (b) 0.109 mW

10.27

1. The person handling the device and the workplace should be grounded.
2. The device should be kept in conductive foam when out of circuit.
3. All unused pins should be connected to ground or to the supply voltage.
4. Power should be applied before connecting or disconnecting low-impedance signal sources.

10.29 (a)

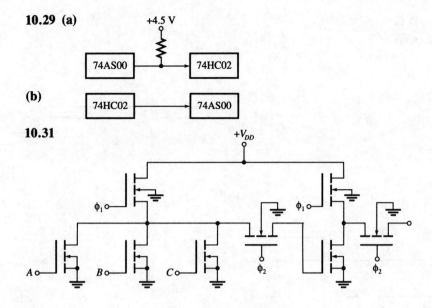

(b)

10.31

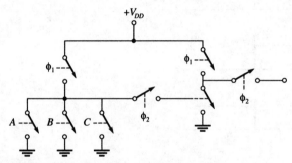

10.33 (a) -1.6 V **(b)** 4.5 mA **(c)** -0.9 V

10.35 Inverter: $\overline{A}$

10.37 42.25 pJ

10.39 (a) $\text{OUT}_1 = \overline{B} + \overline{C}$, $\text{OUT}_2 = 1$

(b) $\text{OUT}_1 = A + B + \overline{C} + \overline{D}$,

$\text{OUT}_2 = \overline{A}\overline{B} + CD$

10.41 $A + B$

10.43 4

10.45

OUTPUT $= \overline{ABC} \cdot \overline{D}$

10.47 $R_1 = 217 \ \Omega$; $R_2 = 245 \ \Omega$

10.49 $V_{IL}(\text{max}) \approx 1$ V; $V_{IH}(\text{min}) \approx 1.6$ V

10.51 55 ns

CHAPTER 11

11.1

R	S	Q	$\bar{Q}$
0	0	0	0
0	1	0	1
1	0	1	0
1	1	No change	

11.3

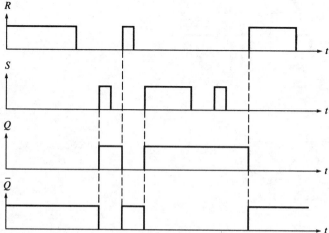

11.5

R	S	Q	$\bar{Q}$
0	0	1	1
0	1	0	1
1	0	1	0
1	1	No change	

11.7

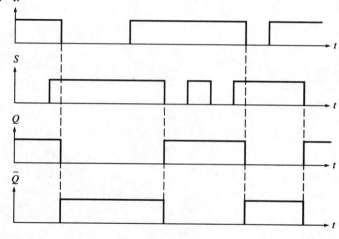

11.9 CLOCK

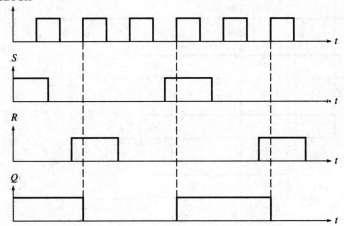

11.11 CLOCK

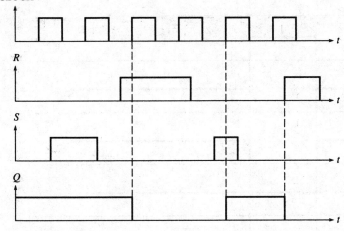

11.13 LATCH ENABLE

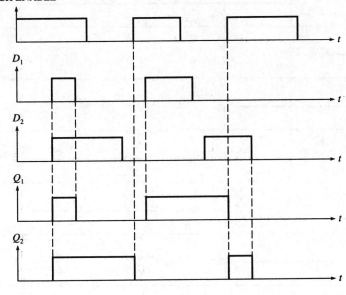

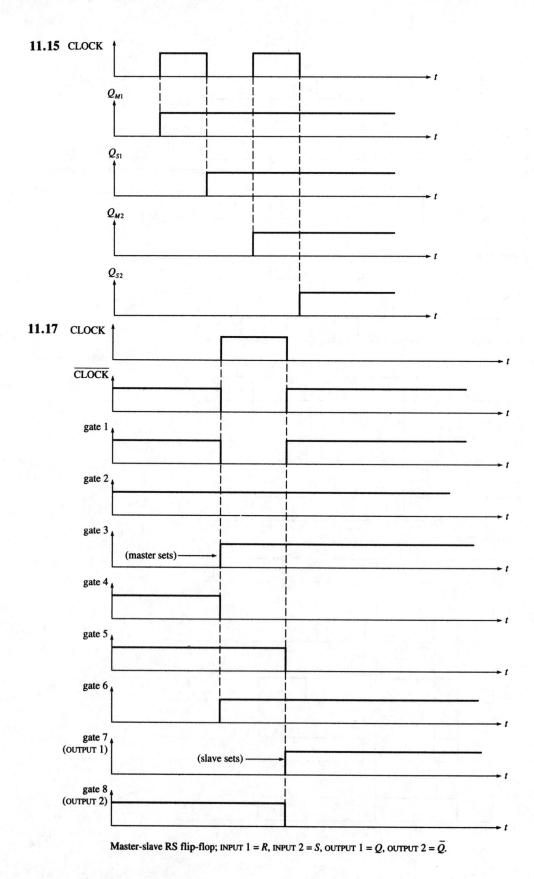

11.15 CLOCK

Q_{M1}

Q_{S1}

Q_{M2}

Q_{S2}

11.17 CLOCK

$\overline{\text{CLOCK}}$

gate 1

gate 2

gate 3

(master sets) ⟶

gate 4

gate 5

gate 6

gate 7
(OUTPUT 1)

(slave sets) ⟶

gate 8
(OUTPUT 2)

Master-slave RS flip-flop; INPUT 1 = R, INPUT 2 = S, OUTPUT 1 = Q, OUTPUT 2 = $\bar{Q}$.

11.19 CLOCK

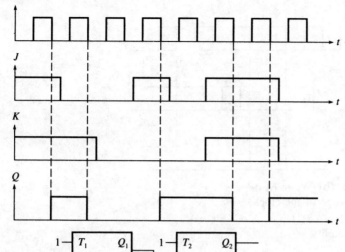

11.21

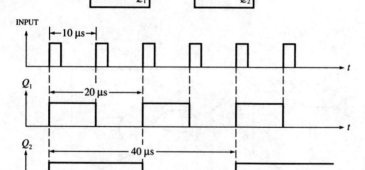

11.23 (a)

Gate	OUTPUT
1	1
2	1
3	0
4	1
5	0
6	0
7	0
8	1

(b)

Gate	OUTPUT
1	1
2	0 or 1
3	1
4	0 or 1
5	0
6	0
7	0
8	1

11.25 (a) 15 ns (b) 4 ns (c) 20

11.27 CLOCK

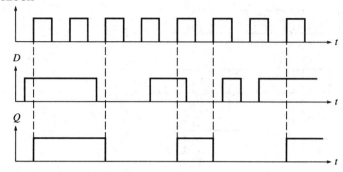

11.29 (a) Four D-type latches **(b)** When M or
N is high **(c)** On the leading edge of the clock,
provided $\overline{G1}$ and $\overline{G2}$ are both low **(d)** No
(e) More than usual drive capability

11.31 NOR-gate latch

IN$_1$	IN$_2$	OUT$_1$	OUT$_2$
0	0	No change	
0	1	1	0
1	0	0	1
1	1	0	0

11.33

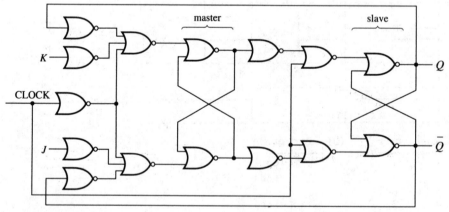

11.35
```
EXERCISE 11.35
.SUBCKT NAND 1 2 4
M1 3 1 0 0 MOS
M2 4 2 3 0 MOS
.MODEL MOS NMOS KP = 0.25E-3 VTO = 2
RD 4 5 10K
VDD 5 0 10V
.ENDS NAND
X1 2 3 5 NAND
X2 1 2 4 NAND
X3 7 5 6 NAND
X4 4 6 7 NAND
VR 1 0 PULSE(0 10 7MS 1US 1US 2MS 6MS)
VCLK 2 0 PULSE(0 10 1MS 1US 1US 4MS 10MS)
VS 3 0 PULSE( 0 10 3MS 1US 1US 2MS)
.NODESET V(7) = 10
.TRAN 0.5MS 15MS
.PLOT TRAN V(6)
.PLOT TRAN V(7)
.END
```

11.37 EXERCISE 11.37
```
.SUBCKT NOR 1 2 3
M1 3 2 0 0 MOSN
M2 3 1 0 0 MOSN
M3 3 1 4 5 MOSP
M4 4 2 5 5 MOSP
.MODEL MOSN NMOS KP = 0.25E-3 VTO = 2
.MODEL MOSP PMOS KP = 0.25E-3 VTO = -2
VDD 5 0 10V
.ENDS NOR
X1 4 2 3 NOR
X2 1 3 4 NOR
VS 1 0
VR 2 0
.NODESET V(3) = 10
.DC VS 0 10 10 VR 0 10 10
.PRINT DC V(3) V(4)
.END
```

The low output levels are very small in comparison to those obtained in NMOS models, and the high output levels are exactly 10 V.

CHAPTER 12

12.1 10011010   (Initial contents)
11001101
11100110
11110011
01111001
00111100
10011110
01001111
10100111

12.3 CLOCK

After four clock pulses, contents = 0110.

12.5 294.912 ms

12.7

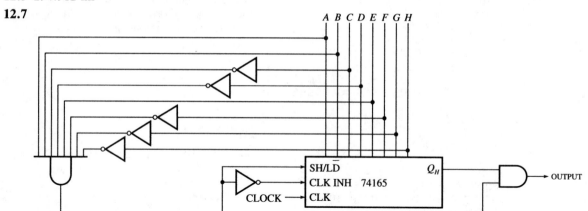

12.9 One possible solution:

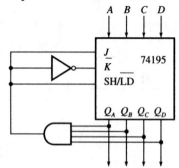

If *ABCD* changes, so does $Q_A Q_B Q_C Q_D$, and that number is loaded into the shift register.

12.11

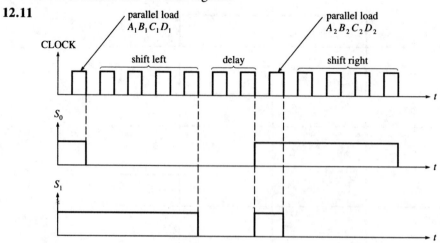

12.13 2.048 ms

12.15

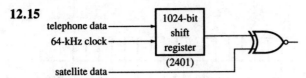

12.17 CLOCK

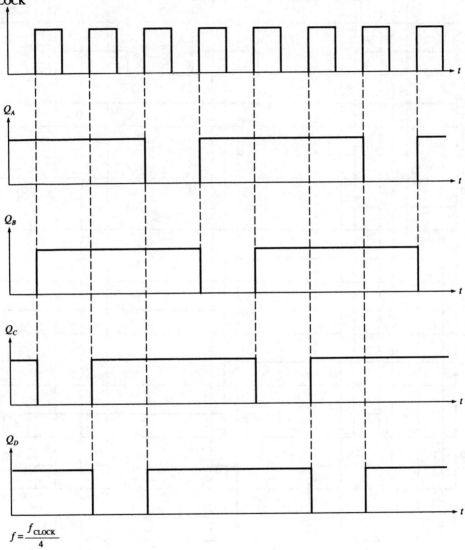

$$f = \frac{f_{CLOCK}}{4}$$

12.19 4

12.21 **(a)** 4-bit, parallel-in, parallel-out; data can also be loaded serially at stage A, and serial output is available at stage D.
(b) Asynchronous **(c)** 2 **(d)** 0s are shifted right into the register at stage A on the leading edges of clock pulses.

12.23

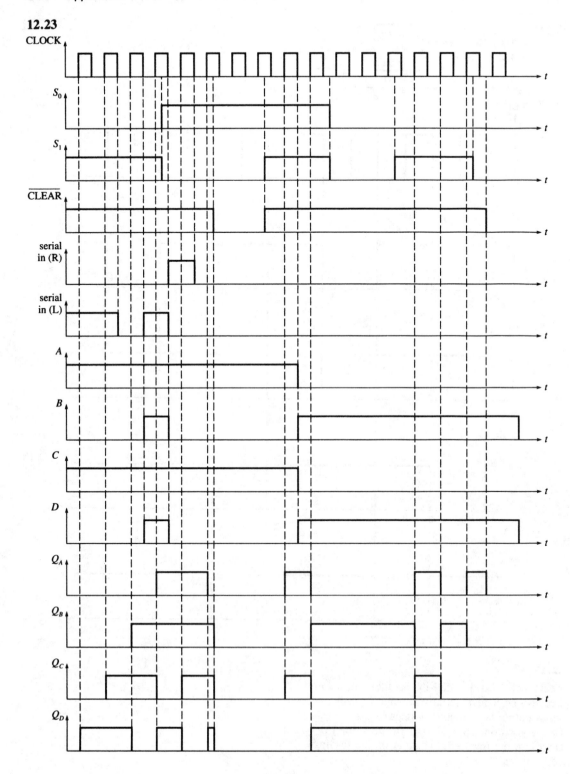

12.25

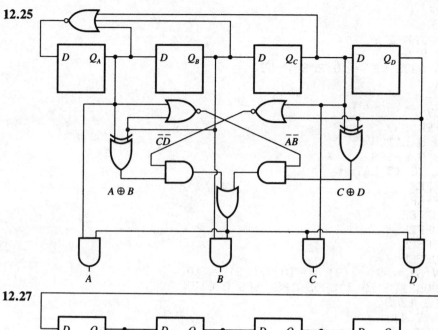

12.27

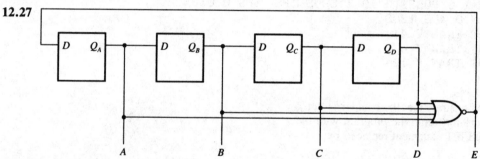

12.29 EXERCISE 12.29
```
.SUBCKT LATCH 7 8 9 12 13
.SUBCKT AND 1 2 3
D1 3 1 DIODE
D2 3 2 DIODE
RD 4 3 20K
VD 4 0 15V
.MODEL DIODE D
.ENDS AND
.SUBCKT NOR 4 5 6
M1 6 4 0 0 MOS
M2 6 5 0 0 MOS
RDD 6 7 10K
VDD 7 0 10V
.MODEL MOS NMOS KP = 0.25E-3 VTO = 2
.ENDS NOR
X1 7 8 10 AND
X2 8 9 11 AND
X3 10 13 12 NOR
X4 12 11 13 NOR
C1 12 0 2PF
```
(Continues on p. 782.)

```
C2 13 0 2PF
.ENDS LATCH
.SUBCKT BUFFER 1 2
M1 2 1 0 0 MOS
.MODEL MOS NMOS KP = 0.25E-3 VTO = 5
RD 2 3 10K
CL 2 0 2PF
VDD 3 0 10V
.ENDS BUFFER
X1 2 1 3 4 5 LATCH
X2 6 1 7 8 9 LATCH
X3 10 1 11 12 13 LATCH
X4 4 6 BUFFER
X5 5 7 BUFFER
X6 8 10 BUFFER
X7 9 11 BUFFER
X8 12 3 BUFFER
X9 13 2 BUFFER
.NODESET V(5) = 10 V(9) = 10 V(13) = 10
VCLK 1 0 PULSE(0 10 1.39US 1PS 1PS 0.01US 2US)
.TRAN 0.4US 12US
.PLOT TRAN V(4)
.PLOT TRAN V(8)
.PLOT TRAN V(12)
.END
```

To demonstrate that the counter is not self-starting, rerun the program with the .NODESET statement replaced by:

.NODESET V(4) = 10 V(9) = 10 V(12) = 10

The counter continually cycles between 101 and 010.

CHAPTER 13

13.1 **(a)** 0010 0101 **(b)** 0110 0100 0000 **(c)** 0001 1000 0110 0011

13.3 **(a)** 20 bits **(b)** 1,048,576 numbers

13.5 **(a)** 0101 1001 **(b)** 1000 0010 **(c)** 1001 1000 **(d)** 1001 0000 0000

13.7 **(a)** 0010 0011 **(b)** 0101 1001 0100 **(c)** 0010 0100 0010

13.9 **(a)** **(i)** 805 **(ii)** 7246 **(b)** **(i)** 039 **(ii)** 8645

13.11 **(a)** 1100 **(b)** 1001 0100 **(c)** 1011 0011 0101

13.13 **(a)** 1000 0101 **(b)** 0110 0110 1010 **(c)** 0110 1010 1011

13.15

	Number	9's Complement
(a)	0010 1100 0011	1101 0011 1100
(b)	0001 0000 0000 0001	1110 1111 1111 1110
(c)	0001 1101 0001	1110 0010 1110
(d)	0100 1011	1011 0100

13.17 3321 and 4221 codes

13.19

Decimal	A	B	C	D	Odd parity P	A ⊙ B ⊙ C ⊙ D
0	0	0	0	0	1	1
1	0	0	0	1	0	0
2	0	0	1	0	0	0
3	0	0	1	1	1	1
4	0	1	0	0	0	0
5	0	1	0	1	1	1
6	0	1	1	0	1	1
7	0	1	1	1	0	0
8	1	0	0	0	0	0
9	1	0	0	1	1	1

The truth table shows that $P = A \odot B \odot C \odot D$.

13.21

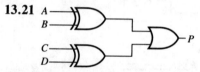

13.23 The 1 in the seventh row and sixth column should be 0.

13.25 The contents of the counter should always be an even number, so the least significant bit of the count should always be 0.

13.27

16	11000	24	10100
17	11001	25	10101
18	11011	26	10111
19	11010	27	10110
20	11110	28	10010
21	11111	29	10011
22	11101	30	10001
23	11100	31	10000

13.29 **(a)** 11000 **(b)** 010100 **(c)** 11101101

13.31 **(a)** 10011 **(b)** 0110101 **(c)** 11101111

13.33 **(a)** 0111 1100 **(b)** 0110 0010 0100 **(c)** 1010 1101 0101 1111

13.35 Route 10

13.37 11000011 10000001 10010011 10010011 01000000 11110010 11111001

13.39 Box 1384

13.41

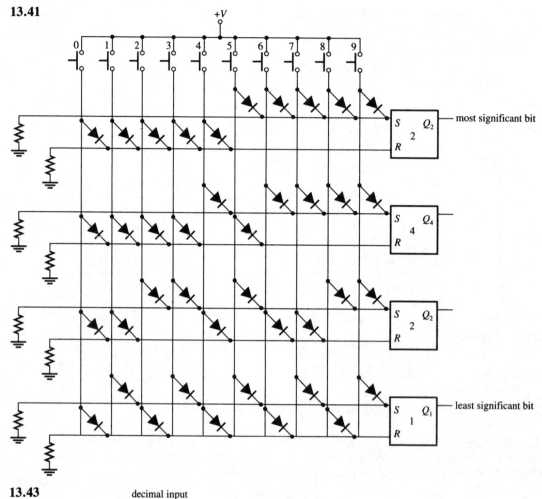

most significant bit

least significant bit

13.43

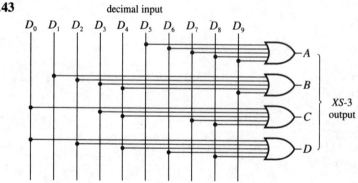

decimal input

XS-3 output

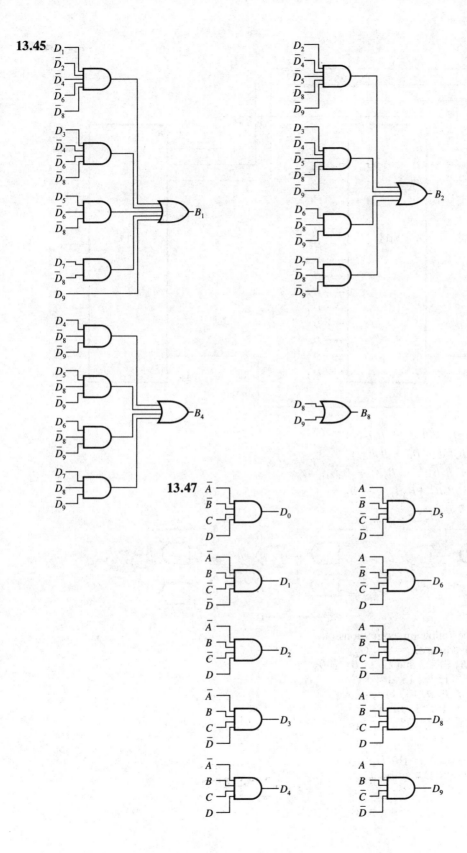

13.49

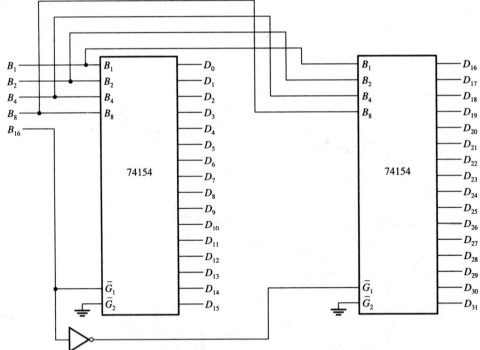

13.51 $b = B_8 + \overline{B}_4 + B_1B_2 + \overline{B}_1\overline{B}_2$;

$c = B_1 + \overline{B}_2 + B_4$; $d = B_8 + B_2\overline{B}_4 + B_4\overline{B}_2B_1 + B_2\overline{B}_1 + \overline{B}_1\overline{B}_4$; $e = B_2\overline{B}_1 + \overline{B}_4\overline{B}_1$;

$f = B_8 + B_4\overline{B}_1 + B_4\overline{B}_2 + \overline{B}_2\overline{B}_1$; $g = B_8 + \overline{B}_4B_2 + B_4\overline{B}_2 + B_2\overline{B}_1$

13.53

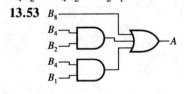
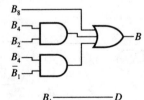
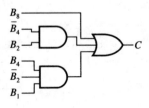

13.55 Plotting the following expressions on Karnaugh maps leads to $G_1 = B_1$, $G_2 = B_1 \oplus B_2$, $G_3 = B_2 \oplus B_3$, and $G_4 = B_3 \oplus B_4$:

$G_1 = \Sigma\ 8, 9, 10, 11, 12, 13, 14, 15$
$G_2 = \Sigma\ 4, 5, 6, 7, 8, 9, 10, 11$
$G_3 = \Sigma\ 2, 3, 4, 5, 10, 11, 12, 13$
$G_4 = \Sigma\ 1, 2, 5, 6, 9, 10, 13, 14$

13.57 10

13.59 **(a)** 4-to-16 decoder **(b)** $\overline{G}_1$ or $\overline{G}_2$ (or both) are high **(c)** Output 6

13.61 One possible solution:

Decimal	4-3-1-1
0	0 0 0 0
1	0 0 0 1
2	0 0 1 1
3	0 1 0 0
4	1 0 0 0
5	0 1 1 1
6	1 0 1 1
7	1 1 0 0
8	1 1 1 0
9	1 1 1 1

13.63

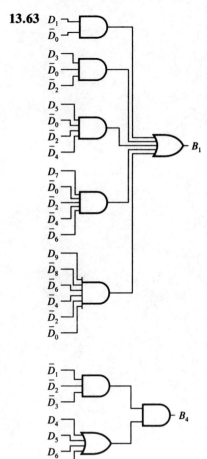

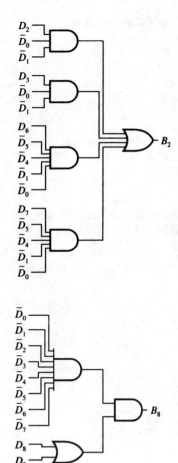

13.65

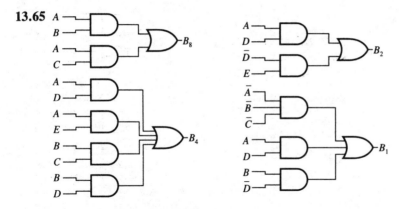

2-out-of-5 input = $ABCDE$ (A=MSB)

13.67
```
EXERCISE 13.67
.SUBCKT XOR 1 2 5
.SUBCKT TG 1 2 3 5
M1 5 3 2 4 MOSP
M2 5 1 2 0 MOSN
VDD 4 0 10V
.ENDS TG
.SUBCKT INV 1 2
M1 2 1 3 3 MOSP
M2 2 1 0 0 MOSN
VDD 3 0 10V
.ENDS INV
.MODEL MOSP PMOS KP = 0.25E-3 VTO = -2
.MODEL MOSN NMOS KP = 0.25E-3 VTO = 2
X1 2 4 INV
X2 1 3 INV
X3 4 1 2 5 TG
X4 2 3 4 5 TG
.ENDS XOR
X1 1 2 4 XOR
X2 2 3 5 XOR
R1 4 0 1MEG
R2 5 0 1MEG
V1 1 0
V2 2 0
V3 3 0
.DC V3 0 10 10 V2 0 10 10
.PRINT DC V(1) V(2) V(3)
.PRINT DC V(1) V(4) V(5)
.END
```

To obtain the second half of the truth
table, change

V1 1 0 to V1 1 0 10

CHAPTER 14

14.1

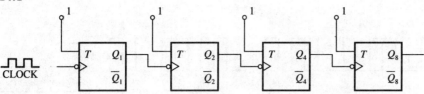

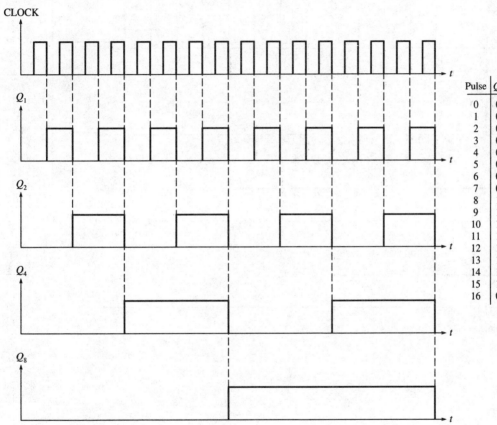

Pulse	Q_8	Q_4	Q_2	Q_1
0	0	0	0	0
1	0	0	0	1
2	0	0	1	0
3	0	0	1	1
4	0	1	0	0
5	0	1	0	1
6	0	1	1	0
7	0	1	1	1
8	1	0	0	0
9	1	0	0	1
10	1	0	1	0
11	1	0	1	1
12	1	1	0	0
13	1	1	0	1
14	1	1	1	0
15	1	1	1	1
16	0	0	0	0

14.3 2.2528 MHz

14.5

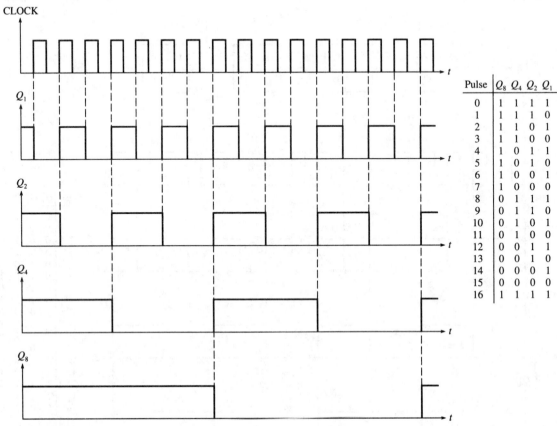

Pulse	Q_8	Q_4	Q_2	Q_1
0	1	1	1	1
1	1	1	1	0
2	1	1	0	1
3	1	1	0	0
4	1	0	1	1
5	1	0	1	0
6	1	0	0	1
7	1	0	0	0
8	0	1	1	1
9	0	1	1	0
10	0	1	0	1
11	0	1	0	0
12	0	0	1	1
13	0	0	1	0
14	0	0	0	1
15	0	0	0	0
16	1	1	1	1

down counter

14.7

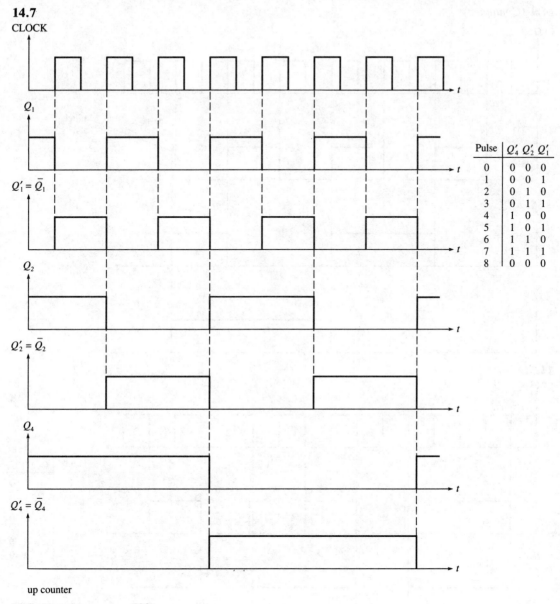

Pulse	Q'_4	Q'_2	Q'_1
0	0	0	0
1	0	0	1
2	0	1	0
3	0	1	1
4	1	0	0
5	1	0	1
6	1	1	0
7	1	1	1
8	0	0	0

up counter

14.9 (Cont'd on page 792)

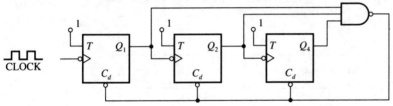

14.9 (*Continued*)

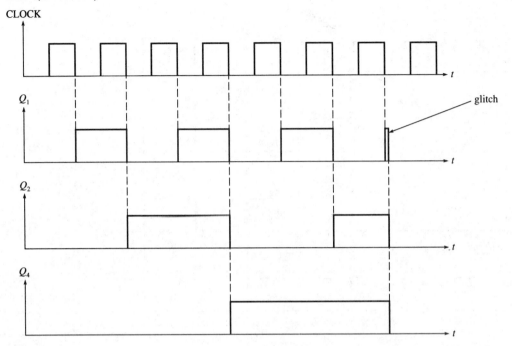

14.11

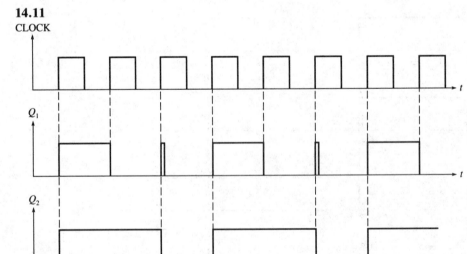

Pulse	Q_4	Q_2	Q_1
0	0	0	0
1	1	1	1
2	1	1	0
3	0	0	0
4	1	1	1
5	1	1	0
6	0	0	0

250 kHz

14.13

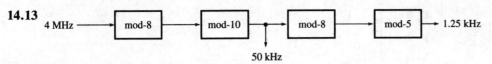

4 MHz → mod-8 → mod-10 → • → mod-8 → mod-5 → 1.25 kHz

50 kHz

14.15

CLOCK

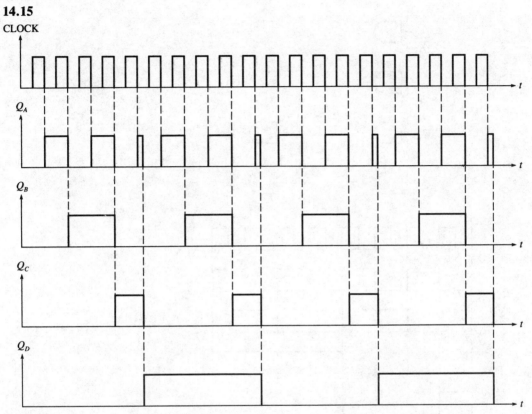

$f = 100$ kHz, $d = 0.5$

14.17 6.25 MHz

14.19 255

14.21

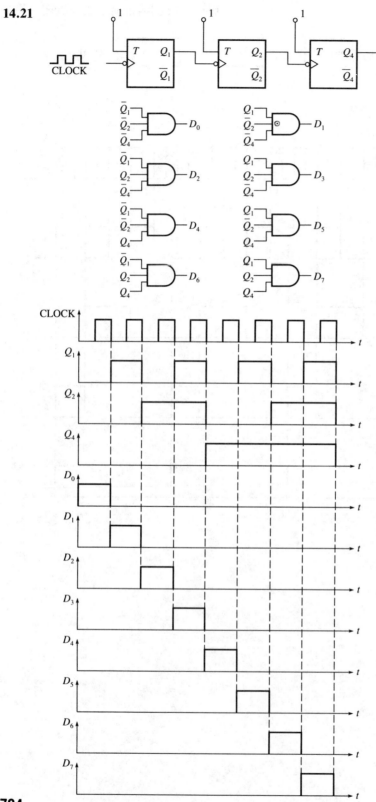

14.23 (a)

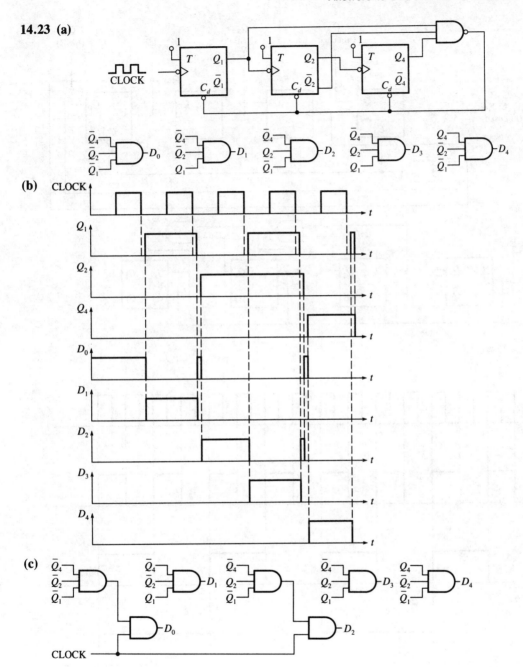

14.25

14.27

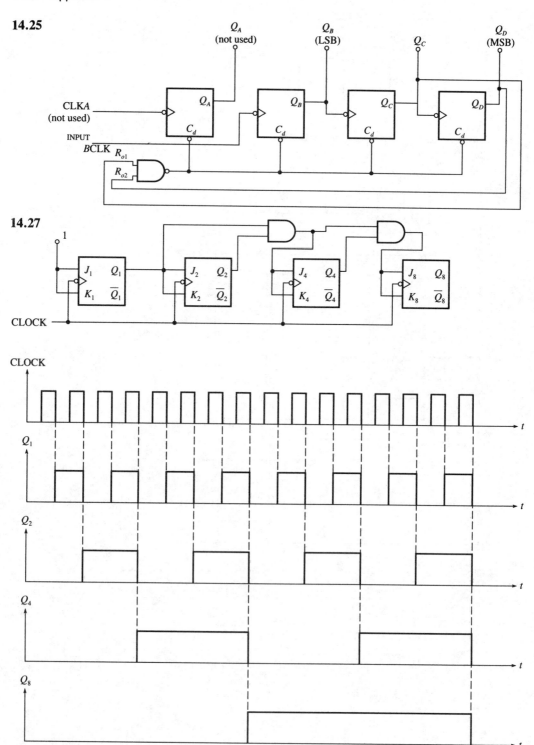

modulus = 16

14.29

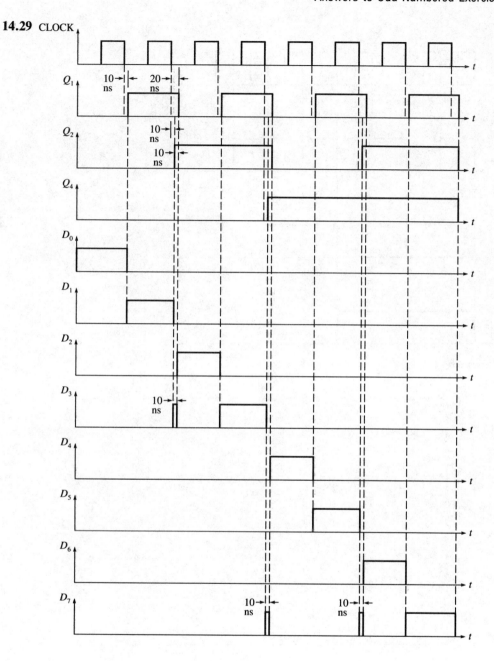

14.31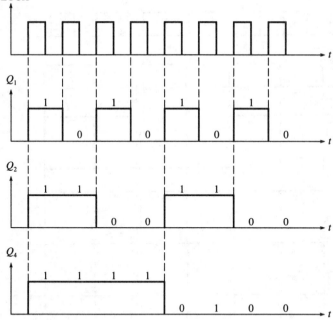

binary down counter

14.33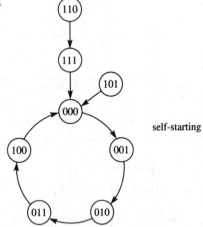

self-starting

14.35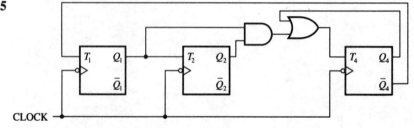

14.37 $\bar{Q}_8\bar{Q}_4\bar{Q}_2Q_1 + Q_8Q_2$ $Q_8 + Q_4 + Q_2 + \bar{Q}_1$ $Q_8Q_1 + Q_4\bar{Q}_2Q_1 + \bar{Q}_8Q_2\bar{Q}_1$ $Q_8Q_1 + Q_4Q_2\bar{Q}_1$

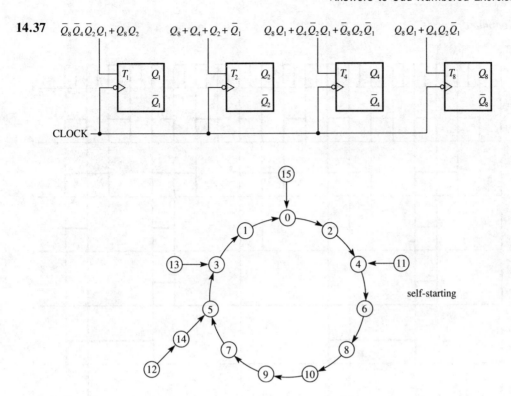

14.39 $J_1 = 1; K_1 = 1; J_2 = Q_1; K_2 = Q_1;$
$J_4 = Q_2Q_1; K_4 = Q_2Q_1$

14.41 (Cont'd on page 800)

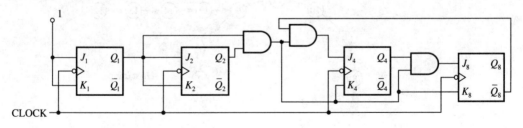

14.41 (*Continued*)

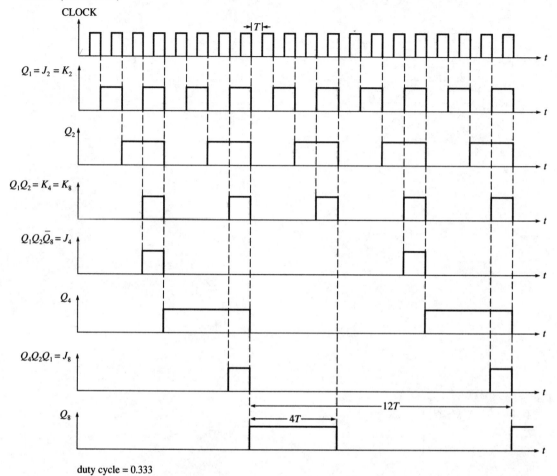

duty cycle = 0.333

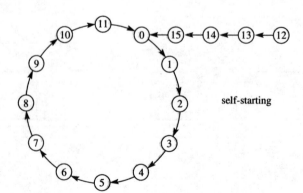

self-starting

14.43

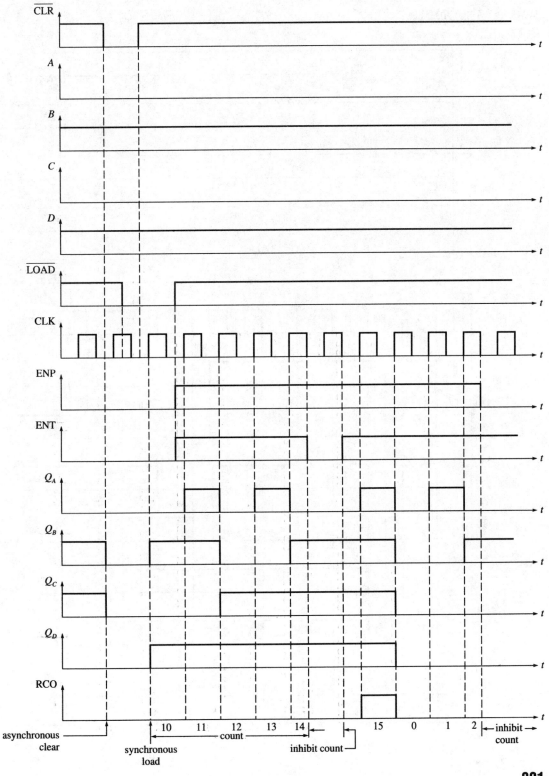

14.45 **(a)** 5 **(b)** 12.5 ns

14.47

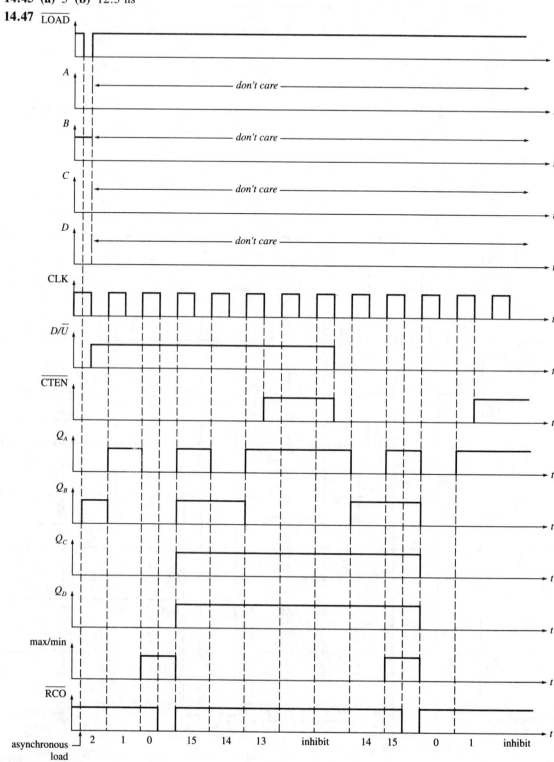

14.49 **(a)** 4-bit asynchronous **(b)** Trailing edge **(c)** No **(d)** Asynchronously **(e)** Pins 6 and 12

14.51 **(a)** CLOCK

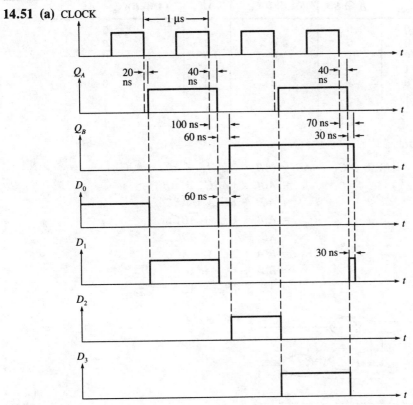

(b) 10 MHz

14.53

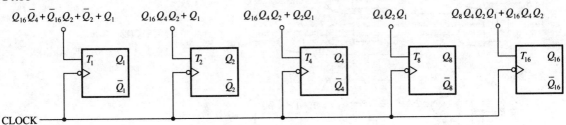

CHAPTER 15

15.1 Using truth tables:

A	B	C_{in}	C_{out}	$A \oplus B$	$(A \oplus B)C_{in}$	AB	$(A \oplus B)C_{in} + AB = C_{out}$
0	0	0	0	0	0	0	0
0	0	1	0	0	0	0	0
0	1	0	0	1	0	0	0
0	1	1	1	1	1	0	1
1	0	0	0	1	0	0	0
1	0	1	1	1	1	0	1
1	1	0	1	0	0	1	1
1	1	1	1	0	0	1	1

Using Boolean algebra:

$$
\begin{aligned}
(A \oplus B)C_{in} + AB &= (A\bar{B} + \bar{A}B)C_{in} + AB \\
&= A\bar{B}C_{in} + \bar{A}BC_{in} + AB \\
&= A(B + \bar{B}C_{in}) + \bar{A}BC_{in} \\
&= A(B + C_{in}) + \bar{A}BC_{in} \\
&= AB + AC_{in} + \bar{A}BC_{in} \\
&= B(A + \bar{A}C_{in}) + AC_{in} \\
&= B(A + C_{in}) + AC_{in} \\
&= AB + BC_{in} + AC_{in} = C_{out}
\end{aligned}
$$

15.3

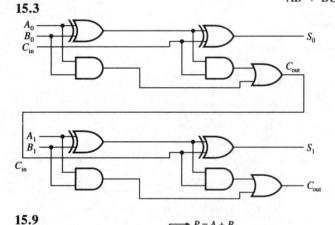

15.5 202

15.7 3.125×10^6 additions per second

15.9

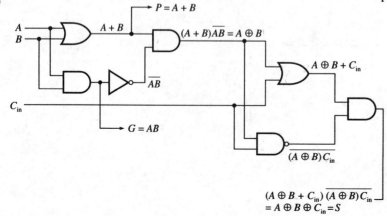

15.11 255

15.13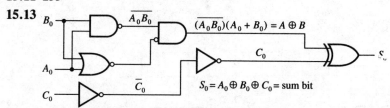

$S_0 = A_0 \oplus B_0 \oplus C_0 =$ sum bit

15.15 (a) $1100 = F_3F_2F_1F_0$ (b) $0100 = F_3F_2F_1F_0$

15.17 A and B are the 1's complements of each other.

15.19 CLOCK

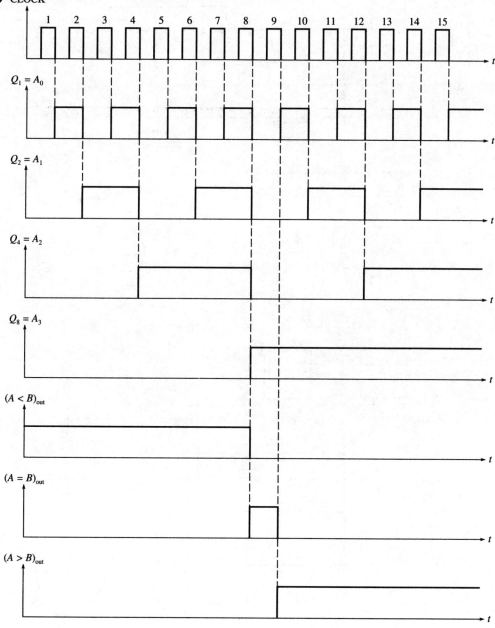

15.21

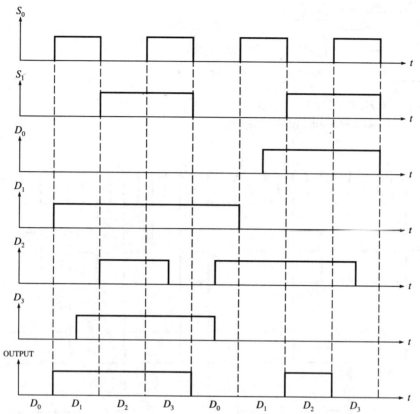

15.23 31.25 kHz

15.25

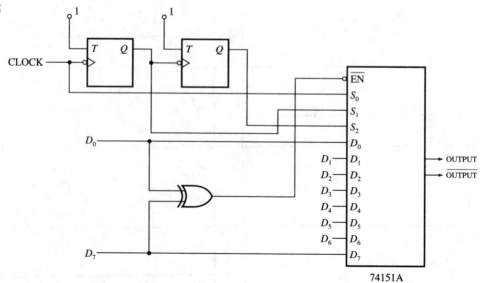

15.27

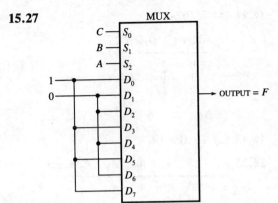

15.29 $Q_0 = B_0$

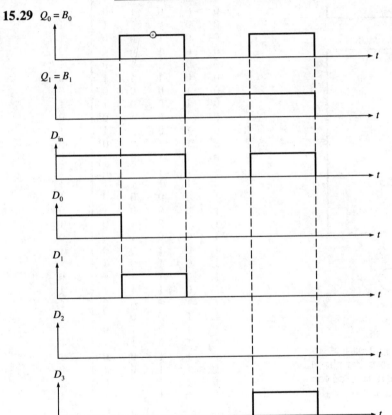

15.31 $S = \overline{A}\,\overline{B}C_{in} + \overline{A}B\overline{C}_{in} + A\overline{B}\,\overline{C}_{in} + ABC_{in}$

$= \overline{A}(\overline{B}C_{in} + B\overline{C}_{in}) + A(\overline{B}\,\overline{C}_{in} + BC_{in})$

Let $Z = \overline{B}C_{in} + B\overline{C}_{in} = B \oplus C_{in}$.

Then $\overline{Z} = \overline{B \oplus C_{in}} = \overline{B}\,\overline{C}_{in} + BC_{in}$

Thus, $S = \overline{A}Z + A\overline{Z} = A \oplus Z =$

$\qquad A \oplus (B \oplus C_{in}) = A \oplus B \oplus C_{in}$.

15.33

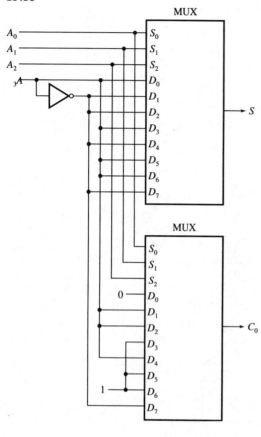

CHAPTER 16

16.1 Main memory is an integral part of a computer system and is very fast. Peripheral memory is slower, add-on memory used for storing large quantities of data.

16.3 16K bits; 2K bytes

16.5 A memory in which contents are lost when power is removed.

16.7 BJT memory is faster but has less storage capacity (per integrated circuit) than MOSFET memory.

16.9 MOSFET technology only. EPROMs are erased by ultraviolet light and EEPROMs are erased electrically.

16.11 (a) 2 (b) 5

(c)

Address	D_1	D_0
0	1	1
1	0	0
2	0	1
3	1	0
4	0	0

16.13 (a) 15 (b) 17

16.15

A_4	A_3	A_2	A_1	A_0	Contents
0	0	0	0	0	1
0	0	0	0	1	1
0	0	0	1	0	1
0	0	0	1	1	1
0	0	1	0	0	1
0	0	1	0	1	0
0	0	1	1	0	1
0	0	1	1	1	0
0	1	0	0	0	0
0	1	0	0	1	1
0	1	0	1	0	0
0	1	0	1	1	1
0	1	1	0	0	0
0	1	1	0	1	0
0	1	1	1	0	1
0	1	1	1	1	1
1	0	0	0	0	1
1	0	0	0	1	0
1	0	0	1	0	1
1	0	0	1	1	0
1	0	1	0	0	1
1	0	1	0	1	1
1	0	1	1	0	0
1	0	1	1	1	1
1	1	0	0	0	1
1	1	0	0	1	1
1	1	0	1	0	1
1	1	0	1	1	0
1	1	1	0	0	1
1	1	1	0	1	0
1	1	1	1	0	1
1	1	1	1	1	1

16.17

Address			Contents	
A_2	A_1	A_0	D_1	D_0
0	0	0	0	1
0	0	1	0	0
0	1	0	1	0
0	1	1	1	0
1	0	0	1	0
1	0	1	0	1
1	1	0	0	0
1	1	1	1	1

16.19 5 MHz

16.21 $Y_5Y_4Y_3Y_2Y_1 = 01101$

16.23 (a) 4×4 (b)

Address	Contents			
	D_3	D_2	D_1	D_0
0	1	1	0	0
1	1	1	0	1
2	0	1	0	0
3	1	0	1	1

16.25 Charge is stored by applying a high drain-to-source voltage, which injects electrons into an electrically isolated floating gate, where they remain trapped. Charge is removed by exposing the floating gate to ultraviolet radiation, which imparts energy to the electrons to allow them to escape.

16.27 Make both address lines high; make the sense-1 line high and the sense-0 line low.

16.29 Make the row-select line and the $\overline{\text{bit line}}$ both high, and make the bit line low.

16.31 Make the row-select line and the $\overline{\text{bit line}}$ both high and the bit line low.

16.33 Transistor Q_2 acts as an inverter: If the capacitance is charged (representing a stored 1),

then Q_2 is on and its drain is low (logical 0) when the cell is read. If the capacitance is discharged (0), Q_2 is off and its drain is high (1).

16.35 (a) Refreshing all cells in a dynamic memory with all other memory operations suspended until the refresh is complete (b) A circuit used to control refresh cycles of dynamic memory (and other functions, such as address multiplexing) (c) Dynamic RAM in which refreshing is controlled by internal circuitry

16.37

1. $\overline{\text{RAS}}$-only refresh: Row addresses are cycled and $\overline{\text{RAS}}$ is pulsed after each new address, while $\overline{\text{CAS}}$ is left inactive (high). Output buffers are inactive, resulting in power conservation.

2. $\overline{\text{CAS}}$-before-$\overline{\text{RAS}}$ refresh: $\overline{\text{CAS}}$ is held low while $\overline{\text{RAS}}$ is pulsed. Rows are refreshed in a sequence of addresses generated internally, so external address cycling is not required.

16.39 (a) 9 (b) 2.56 MHz (c) 2.56 MHz (d) 125 Hz (e) 31,200

16.41 (a) 32 (b) 32 (c) 64

16.43

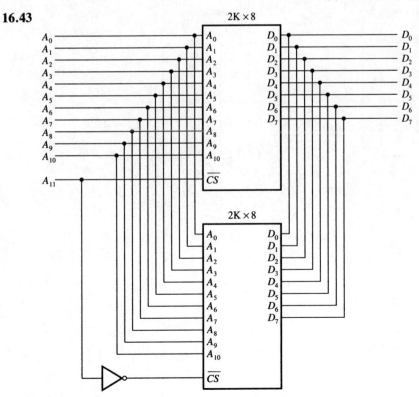

16.45

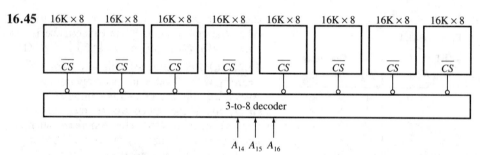

16.47 $F_1 = ABC + \overline{A}\,\overline{B}\,\overline{C}$; $F_2 = A\overline{B}\,\overline{C} + \overline{A}\,\overline{B}C + \overline{A}B\overline{C}$

16.49

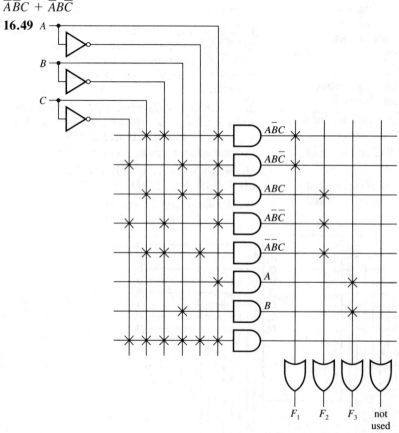

16.51

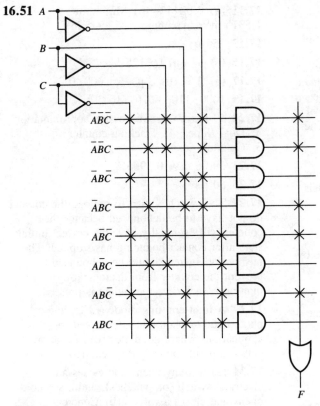

16.53 Floppy disks are smaller and less expensive than hard disks. They are used primarily in small computer systems such as microcomputers.

16.55

16.57

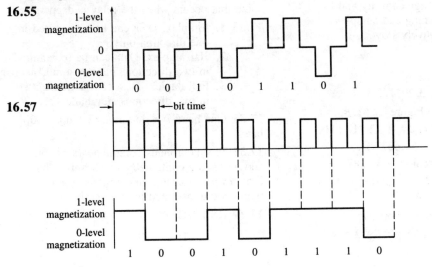

16.59 (a) A hole in a floppy disk used to reference the beginning of stored data **(b)** Concentric circular paths on the surface of a floppy disk, along which data is stored **(c)** A subdivision of storage area on a floppy disk extending radially outward from the center and intersecting all tracks **(d)** A disk whose sector numbers are assigned under the control of a computer before the disk is used **(e)** The electronics and mechanisms used to control reading and writing on a floppy disk

16.61 A small, cylindrically shaped magnetic domain created by subjecting certain materials to a strong magnetic field.

16.63 T-bar and chevron, constructed from permalloy.

16.65 Minor loops are used for data storage by continually circulating magnetic bubbles. Each page of data (all the bits occupying corresponding positions in the minor loops) is transferred to and from the major loop for the purpose of reading and writing.

16.67 Fast in comparison to floppy disk memory and has no moving parts. Non-volatile. Most likely to replace floppy disk memory.

16.69 Cassette tape. Kansas City standard FM. 0 = 4 cycles of 1200 Hz and 1 = 8 cycles of 2400 Hz.

16.71 Serially, in shift registers that are continually clocked. Large capacity and low cost but requires refreshing and complex support circuitry. Relatively slow.

CHAPTER 17

17.1 (a) 0.625 V **(b)** 3.203125 V

17.3 One solution (see Figure 17.6): R_f = 10 kΩ, R = 6.4583 kΩ, $2R$ = 12.917 kΩ, $4R$ = 25.833 kΩ, $8R$ = 51.667 kΩ, $16R$ = 103.33 kΩ; output = -4.8387 V

17.5 I_0 = 0.125 mA; I_1 = 0.250 mA; I_2 = 0.500 mA; I_3 = 1.00 mA

17.7 -9.9609375 V to $+9.9609375$ V

17.9 (a) 1.25 pF **(b)** 2.75 pF **(c)** 2.5 V

17.11 (a) 0.0244% **(b)** Minimum = 1.8315 mV; maximum = 5.4945 mV

17.13 255 μs

17.15 (a) 31 **(b)** 0.3125 V

17.17 (a) 2 V **(b)** 8 μs **(c)** 00011001

17.19 1000, 1100, 1010, 1011

17.21 1. Flash   2. Successive approximation 3. Dual-slope   4. Tracking counter 5. Counter.

17.23 39.06 mV, 0.3906%

17.25 50 μs

17.27 The switches used to preload the counter do not have to be debounced because the counter will eventually have the correct initial state after contact bouncing has stopped. The pushbutton switch should be debounced to prevent the counter from decrementing more than once each time the switch is pressed.

17.29 A light-emitting diode (LED) and a photodiode or phototransistor. Used in applications where good electrical isolation between input and output is required.

17.31 (a) A relay opens and closes an electrical switch (or switches), and a solenoid opens and closes a valve. **(b)** Contact. Normally open and normally closed. **(c)** Applications in which fast switching times are not required, such as starting and stopping an electric motor. **(d)** To suppress the voltage spike that occurs when the relay is deenergized.

17.33 1. Simplex: Data can be transmitted in one direction only.
 2. Half-duplex: Data can be transmitted in two directions but not simultaneously.
 3. Full-duplex: Data can be transmitted in two directions simultaneously.

17.35 One baud is a data rate of 1 bit/s; 300 baud.

17.37 In synchronous communications, bits and words are continually synchronized by a long-term clock. In asynchronous systems, words are transmitted intermittently.

17.39 RS-232C

INDEX